Geophysical Monograph Series

Including

IUGG Volumes
Maurice Ewing Volumes
Mineral Physics Volumes

Geophysical Monograph Series

110 **The Controlled Flood in Grand Canyon** *Robert H. Webb, John C. Schmidt, G. Richard Marzolf, and Richard A. Valdez (Eds.)*

111 **Magnetic Helicity in Space and Laboratory Plasmas** *Michael R. Brown, Richard C. Canfield, and Alexei A. Pevtsov (Eds.)*

112 **Mechanisms of Global Climate Change at Millennial Time Scales** *Peter U. Clark, Robert S. Webb, and Lloyd D. Keigwin (Eds.)*

113 **Faults and Subsurface Fluid Flow in the Shallow Crust** *William C. Haneberg, Peter S. Mozley, J. Casey Moore, and Laurel B. Goodwin (Eds.)*

114 **Inverse Methods in Global Biogeochemical Cycles** *Prasad Kasibhatla, Martin Heimann, Peter Rayner, Natalie Mahowald, Ronald G. Prinn, and Dana E. Hartley (Eds.)*

115 **Atlantic Rifts and Continental Margins** *Webster Mohriak and Manik Talwani (Eds.)*

116 **Remote Sensing of Active Volcanism** *Peter J. Mouginis-Mark, Joy A. Crisp, and Jonathan H. Fink (Eds.)*

117 **Earth's Deep Interior: Mineral Physics and Tomography From the Atomic to the Global Scale** *Shun-ichiro Karato, Alessandro Forte, Robert Liebermann, Guy Masters, and Lars Stixrude (Eds.)*

118 **Magnetospheric Current Systems** *Shin-ichi Ohtani, Ryoichi Fujii, Michael Hesse, and Robert L. Lysak (Eds.)*

119 **Radio Astronomy at Long Wavelengths** *Robert G. Stone, Kurt W. Weiler, Melvyn L. Goldstein, and Jean-Louis Bougeret (Eds.)*

120 **GeoComplexity and the Physics of Earthquakes** *John B. Rundle, Donald L. Turcotte, and William Klein (Eds.)*

121 **The History and Dynamics of Global Plate Motions** *Mark A. Richards, Richard G. Gordon, and Rob D. van der Hilst (Eds.)*

122 **Dynamics of Fluids in Fractured Rock** *Boris Faybishenko, Paul A. Witherspoon, and Sally M. Benson (Eds.)*

123 **Atmospheric Science Across the Stratopause** *David E. Siskind, Stephen D. Eckerman, and Michael E. Summers (Eds.)*

124 **Natural Gas Hydrates: Occurrence, Distribution, and Detection** *Charles K. Paull and Wilam P. Dillon (Eds.)*

125 **Space Weather** *Paul Song, Howard J. Singer, and George L. Siscoe (Eds.)*

126 **The Oceans and Rapid Climate Change: Past, Present, and Future** *Dan Seidov, Bernd J. Haupt, and Mark Maslin (Eds.)*

127 **Gas Transfer at Water Surfaces** *M. A. Donelan, W. M. Drennan, E. S. Saltzman, and R. Wanninkhof (Eds.)*

128 **Hawaiian Volcanoes: Deep Underwater Perspectives** *Eiichi Takahashi, Peter W. Lipman, Michael O. Garcia, Jiro Naka, and Shigeo Aramaki (Eds.)*

129 **Environmental Mechanics: Water, Mass and Energy Transfer in the Biosphere** *Peter A.C. Raats, David Smiles, and Arthur W. Warrick (Eds.)*

130 **Atmospheres in the Solar System: Comparative Aeronomy** *Michael Mendillo, Andrew Nagy, and J. H. Waite (Eds.)*

131 **The Ostracoda: Applications in Quaternary Research** *Jonathan A. Holmes and Allan R. Chivas (Eds.)*

132 **Mountain Building in the Uralides Pangea to the Present** *Dennis Brown, Christopher Juhlin, and Victor Puchkov (Eds.)*

133 **Earth's Low-Latitude Boundary Layer** *Patrick T. Newell and Terry Onsage (Eds.)*

134 **The North Atlantic Oscillation: Climatic Significance and Environmental Impact** *James W. Hurrell, Yochanan Kushnir, Geir Ottersen, and Martin Visbeck (Eds.)*

135 **Prediction in Geomorphology** *Peter R. Wilcock and Richard M. Iverson (Eds.)*

136 **The Central Atlantic Magmatic Province: Insights from Fragments of Pangea** *W. Hames, J. G. McHone, P. Renne, and C. Ruppel (Eds.)*

137 **Earth's Climate and Orbital Eccentricity: The Marine Isotope Stage 11 Question** *André W. Droxler, Richard Z. Poore, and Lloyd H. Burckle (Eds.)*

138 **Inside the Subduction Factory** *John Eiler (Ed.)*

139 **Volcanism and the Earth's Atmosphere** *Alan Robock and Clive Oppenheimer (Eds.)*

140 **Explosive Subaqueous Volcanism** *James D. L. White, John L. Smellie, and David A. Clague (Eds.)*

141 **Solar Variability and Its Effects on Climate** *Judit M. Pap and Peter Fox (Eds.)*

142 **Disturbances in Geospace: The Storm-Substorm Relationship** *A. Surjalal Sharma, Yohsuke Kamide, and Gurbax S. Lakhima (Eds.)*

143 **Mt. Etna: Volcano Laboratory** *Alessandro Bonaccorso, Sonia Calvari, Mauro Coltelli, Ciro Del Negro, and Susanna Falsaperla*

144 **The Subseafloor Biosphere at Mid-Ocean Ridges** *William S. D. Wilcock, Edward F. DeLong, Deborah S. Kelley, John A. Baross, and S. Craig Cary (Eds.)*

145 **Timescales of the Paleomagnetic Field** *James E. T. Channell, Dennis V. Kent, William Lowrie, and Joseph G. Meert (Eds.)*

146 **The Extreme Proterozoic: Geology, Geochemistry, and Climate** *Gregory S. Jenkins, Mark A. S. McMenamin, Christopher P. McKay, and Linda Sohl (Eds.)*

Geophysical Monograph 147

Earth's Climate: The Ocean–Atmosphere Interaction

Chunzai Wang
Shang-Ping Xie
James A. Carton
Editors

American Geophysical Union
Washington, DC

Published under the aegis of the AGU Books Board

Jean-Louis Bougeret, Chair; Gray E. Bebout, Carl T. Friedrichs, James L. Horwitz, Lisa A. Levin, W. Berry Lyons, Kenneth R. Minschwaner, Andy Nyblade, Darrell Strobel, and William R. Young, members.

Library of Congress Cataloging-in-Publication Data

Earth's climate : the ocean-atmosphere interaction : from basin to global scales / Chunzai Wang, Shang-Ping Xie, James A. Carton, editors.
     p. cm. -- (Geophysical monograph ; 147)
  Includes bibliographical references.
  ISBN 0-87590-412-2 (alk. paper)
 1. Ocean-atmosphere interaction. 2. Climatic changes. I. Wang, Chunzai. II. Xie, Shang-Ping. III. Carton, James A. IV. Series.

GC190.2.E37 2004
551.5'246--dc22

                       2004059499

ISBN 87590-412-2
ISSN 0065-8448

Copyright 2004 by the American Geophysical Union
2000 Florida Avenue, N.W.
Washington, DC 20009

Cover: Sunlight Over Northwestern Australia. This image was taken by NASA's Sea WiFS satellite in a visible channel on November 13, 1998, over the Indian Ocean west of northwestern Australia. Clouds on the blue ocean background convey an image of coexistence and interaction of the ocean and atmosphere. The Australian Continent at the bottom-right corner suggests that land is an important component of the climate system and its climate is influenced by ocean–atmosphere interaction in the open ocean (such as during Indian Ocean dipole events). In this image, one also sees rich structures in the ocean, including swells, internal waves, and biological activity, as reflected in ocean color in the coastal waters. The varying shades over the ocean are due to the reflection of direct sunlight.

Figures, tables, and short excerpts may be reprinted in scientific books and journals if the source is properly cited.

Authorization to photocopy items for internal or personal use, or the internal or personal use of specific clients, is granted by the American Geophysical Union for libraries and other users registered with the Copyright Clearance Center (CCC) Transactional Reporting Service, provided that the base fee of $1.50 per copy plus $0.35 per page is paid directly to CCC, 222 Rosewood Dr., Danvers, MA 01923. 1526-758X/04/$01.50+0.35.

This consent does not extend to other kinds of copying, such as copying for creating new collective works or for resale. The reproduction of multiple copies and the use of full articles or the use of extracts, including figures and tables, for commercial purposes requires permission from the American Geophysical Union.

Printed in the United States of America.

CONTENTS

Preface
Chunzai Wang, Shang-Ping Xie, and James A. Carton vii

A Global Survey of Ocean-Atmosphere Interaction and Climate Variability
Chunzai Wang, Shang-Ping Xie, and James A. Carton 1

Theme I: Pacific Climate Variability

Understanding ENSO Physics—A Review
Chunzai Wang and Joel Picaut 21

Westerly Wind Events in the Tropical Pacific and Their Influence on the Coupled Ocean-Atmosphere System: A Review
Matthieu Lengaigne, Jean-Philippe Boulanger, Christophe Menkes, Pascale Delecluse, and Julia Slingo 49

The Control of Meridional Differential Surface Heating Over the Level of ENSO Activity: A Heat-Pump Hypothesis
De-Zheng Sun 71

Broadening the Atmospheric Bridge Paradigm: ENSO Teleconnections to the Tropical West Pacific-Indian Oceans Over the Seasonal Cycle and to the North Pacific in Summer
Michael A. Alexander, Ngar-Cheung Lau, and James D. Scott 85

Predicting Pacific Decadal Variability
Richard Seager, Alicia R. Karspeck, Mark A. Cane, Yochanan Kushnir, Alessandra Giannini,
Alexey Kaplan, Ben Kerman, and Jennifer Velez 105

Theme II: Tropical Atlantic Climate Variability

Tropical Atlantic Variability: Patterns, Mechanisms, and Impacts
Shang-Ping Xie and James A. Carton 121

On the Role of the South Atlantic Atmospheric Circulation in Tropical Atlantic Variability
Marcelo Barreiro, Alessandra Giannini, Ping Chang, and R. Saravanan 143

Toward Understanding Tropical Atlantic Variability Using Coupled Modeling Surgery
Lixin Wu, Qiong Zhang, and Zhengyu Liu 157

Thermodynamic Coupling and Predictability of Tropical Sea Surface Temperature
R. Saravanan and Ping Chang 171

Internal Variability of the Tropical Atlantic Ocean
Markus Jochum, Raghu Murtugudde, Paola Malanotte-Rizzoli, and Antonio J. Busalacchi 181

Theme III: Indian Ocean Climate Variability

Coupled Ocean-Atmosphere Variability in the Tropical Indian Ocean
Toshio Yamagata, Swadhin K. Behera, Jing-Jia Luo, Sebastien Masson, Mark R. Jury,
and Suryachandra A. Rao 189

Role of the Indian Ocean in Regional Climate Variability
H. Annamalai and Raghu Murtugudde..213

Interannual Indian Rainfall Variability and Indian Ocean Sea Surface Temperature Anomalies
Gabriel A. Vecchi and D. E. Harrison...247

Theme IV: Tropical-Extratropical Interaction

Shallow Overturning Circulations of the Tropical-Subtropical Oceans
Friedrich A. Schott, Julian P. McCreary, Jr., and Gregory C. Johnson...261

Seasonal Variation of the Subtropical/Tropical Pathways in the Atlantic Ocean From an Ocean Data Assimilation Experiment
Meyre P. Da Silva and Ping Chang...305

Gyre-Connected Variations Inferred From the Circulation Indices in the Northern Pacific Ocean
Dongxiao Wang, Yun Liu, and Dejun Gu..319

Theme V: Cross-Basin Issues

Observed Associations Among Storm Tracks, Jet Streams and Midlatitude Oceanic Fronts
Hisashi Nakamura, Takeaki Sampe, Youichi Tanimoto, and Akihiko Shimpo..329

The Relationship of Western Boundary Current Heat Transport and Storage to Midlatitude Ocean-Atmosphere Interaction
Kathryn A. Kelly and Shenfu Dong..347

Two Different Regimes of Anomalous Walker Circulation Over the Indian and Pacific Oceans Before and After the Late 1970s
Ryuichi Kawamura, Hiromitsu Aruga, Tomonori Matsuura, and Satoshi Iizuka..365

Tropical Tropospheric Temperature and Precipitation Response to Sea Surface Temperature Forcing
Hui Su, J. David Neelin, and Joyce E. Meyerson 379

Convection, Cloud-Radiative Feedbacks and Thermodynamic Ocean Coupling in Simple Models of the Walker Circulation
Adam H. Sobel, Christopher S. Bretherton, Hezi Gildor, and Matthew E. Peters......................................393

PREFACE

It is more than 30 years since the publication of Jacob Bjerknes' groundbreaking ideas made clear the importance of ocean-atmosphere interaction in the tropics. It is now more than 20 years since the arrival of a massive El Niño in the fall of 1982 set off a cascade of observational and theoretical studies. During the following decades, the climate research community has made exceptional progress in refining our capacity to observe earth's climate and theorize about it, including new satellite-based and *in situ* monitoring systems and coupled ocean-atmosphere predictive numerical models. Of equal importance is the expanding scope of research, which now reaches far beyond the Pacific El Niño and includes climate phenomena in other ocean basins.

In order to cover the now global context of ocean-atmosphere interaction we have organized this monograph around five principal themes, each introduced by one or more broad overview papers. Theme I covers interaction and climate variability in the Pacific sector, with extensive discussion of El Niño-Southern Oscillation, and with the possible causes and consequences of variability on both shorter and longer timescales. Theme II is devoted to interaction in the Atlantic sector. This basin exhibits complex behavior, reflecting its geographic location between two major zones of convection as well as neighboring the tropical Pacific. Theme III reviews the recent, exciting progress in our understanding of climate variability in the Indian sector. Theme IV addresses the interaction between the tropics and the extratropics, which are linked through the presence of shallow meridional overturning cells in the ocean. Finally, Theme V discusses overarching issues of cross-basin interaction.

Indeed, this monograph represents the climate community's first effort to summarize the modern science of ocean-atmosphere interaction and the roles that the interaction plays in climate variability on the basin and global scales. We believe that the material covered here will be of interest to the climate research community as well as members of the broader scientific community who want to learn about the current state of climate research, to students studying climate and related topics, and to those members of the public who find themselves increasingly fascinated by the patterns of climate and climate change now revealed by climate monitoring tools.

This monograph derives from a special session at the American Geophysical Union (AGU) Fall Meeting in December 2002 entitled, "Ocean-Atmosphere Interaction and Climate Variability", which attracted a large audience. The session was remarkable for having brought together many of the scientific leaders in the field, and for providing the first clear overview of this rapidly evolving discipline spanning all three ocean basins—Pacific, Atlantic, and Indian—as well as the interactions among these basins. However, this monograph is not simply a collection of conference papers. Indeed, fewer than half of the papers contained here were drawn from that conference and many others were invited.

We are indebted to a number of people who played a critical role in constructing this monograph. Most importantly, we would like to thank the referees for their time and effort. They are M. Alexander, S.-I. An, M. Barreiro, M. Cai, E. Chang, G. Chepurin, C. Clark, S. Cravatte, H. Dijkstra, A. Fedorov, C. Frankignoul, A. Giannini, B. Giese, B. Goswami, S. Hastenrath, M. Jochum, J. Kinter, B. Kirtman, R. Kleeman, B. Klinger, A. Kumar, N. Larkin, T. Lee, B. Lintner, Z. Liu, C. Meinen, A. Mestas-Nunez, M. Nonaka, Y. Okumura, W. Robinson, R. Saravanan, T. Shinoda, D. Snowden, A. Sobel, H. Su, Y. Tanimoto, A. Timmermann, D. Vimont, R. Wajsowicz, X. Wang, A. Wittenberg, L. Wu, T. Yamagata, and C. Zhang. David B. Enfield served as a guest editor, overseeing the review of the opening overview paper authored by us. Yuko Okumura served as editorial assistant for S.-P. Xie, and carefully proofread the camera-ready text for some papers in the volume. We are also grateful to nine anonymous referees of the original book proposal for useful comments. Finally, we would like to acknowledge Kenneth Minschwaner, the oversight editor for this project and member of AGU's Books Board, and Allan Graubard, our acquisitions editor, along with Maxine Aldred and Pamela Ingate of AGU Books.

Chunzai Wang
Atlantic Oceanographic and Meteorological Laboratory

Shang-Ping Xie
University of Hawaii

James A. Carton
University of Maryland

Earth's Climate: The Ocean-Atmosphere Interaction
Geophysical Monograph Series 147
Copyright 2004 by the American Geophysical Union
10.1029/147GM00

A Global Survey of Ocean–Atmosphere Interaction and Climate Variability

Chunzai Wang

NOAA Atlantic Oceanographic and Meteorological Laboratory, Miami, Florida

Shang-Ping Xie

International Pacific Research Center and Department of Meteorology, University of Hawaii, Honolulu, Hawaii

James A. Carton

Department of Meteorology, University of Maryland, College Park, Maryland

The interaction of the ocean and atmosphere plays an important role in shaping the climate and its variations. This chapter reviews the current state of knowledge of air–sea interaction and climate variations over the global ocean. The largest source of climate variability in the instrumental record is El Niño–Southern Oscillation (ENSO), which extends its reach globally through the ability of the atmosphere to bridge ocean basins. The growth of ENSO owes its existence to a positive ocean–atmosphere feedback mechanism (originally envisioned by J. Bjerknes) that involves the interaction of ocean dynamics, atmospheric convection, and winds in the equatorial Pacific. The Bjerknes feedback and the resultant equatorial zonal mode of climate variability are a common feature to all three tropical oceans despite differences in dimension, geometry and mean climate. In addition to this zonal mode, the tropics also support a meridional mode, whose growth is due to a thermodynamic feedback mechanism involving the interaction of the cross–equatorial gradient of properties such as sea surface temperature and displacements of the seasonal intertropical convergence zone. This meridional mode is observed in the tropical Atlantic, with some evidence of its existence in the Pacific and Indian Oceans. In the extratropics, in contrast, the sources of climate variability are more distributed. Much of climate variability may be explained by the presence of white noise due to synoptic weather disturbances whose impact on climate at longer timescales is due to the integrating effect of the ocean's ability to store and release heat. Still, there is some evidence of a more active role for the mid–latitude ocean in climate variability, especially near major ocean currents/fronts. Finally, various atmospheric and oceanic bridges that link different ocean basins are discussed, along with their implications for paleoclimate changes and the current global warming.

1. INTRODUCTION

The Earth's climate is determined by many complex physical, chemical and biological interactions among the ocean, atmosphere, land and ice/snow, subject to solar and tectonic forcing. The present volume addresses the key physical interactions of the ocean and atmosphere, which affect climate variability on all timescales. The role of the ocean in climate variability results partly from its large capacity to store and distribute heat. While the ocean is 4000 m deep on average, the upper 10 m has the same mass as the entire atmosphere and the upper 4 m has a similar capacity to store heat. The importance of the ocean for climate is evident in the comparison of diurnal and annual ranges of air temperature over land and ocean. For example, air temperature varies more than 50°C in Beijing within a year but only about 10°C at the same latitude off the coast of northern California.

Solar radiation is the ultimate driving force for all motions in the ocean and atmosphere and gives rise to the pronounced and regular diurnal and seasonal cycles throughout much of the world. Even the minute variations in the intensity and distribution of solar radiation due to orbital changes are implicated in the ice age cycles as well as other features of the Earth's paleoclimate. On the other hand, climate does not just display repeating and regular cycles of solar radiation, but also displays variability that is not correlated with solar radiation. The most famous among such climate variations is the atmosphere's Southern Oscillation [*Walker*, 1924] and its oceanic counterpart El Niño.

Historically, the term El Niño refers to extended episodes of anomalous warming of the ocean off the coast of Peru, while the term Southern Oscillation refers to a sea level pressure (SLP) swing between Darwin, Australia and the island of Tahiti in the central tropical Pacific. *Bjerknes* [1969] recognized that El Niño and the Southern Oscillation (ENSO) are in fact just two different aspects of the same phenomenon, and demonstrated a remarkable correlation between Darwin atmospheric pressure and water temperature off Peru, two locations separated by the vast span of the Pacific Ocean. He further realized that ocean–atmosphere interaction is at the heart of the ENSO phenomenon, and described how an initial change in the ocean could affect the atmosphere in such a manner that the altered atmospheric conditions would in turn induce oceanic changes that reinforce the initial change. For example, if at some initial time sea surface temperatures (SSTs) in the equatorial eastern Pacific are anomalously warm, then the east–west gradient in SST will be reduced. The atmosphere will respond by reducing the east–west gradient of SLP, and consequently relaxing the strength of the easterly trade winds. The relaxation of the easterly winds in turn causes an eastward surge of warm water along the equator, positively reinforcing the initial warm SST anomalies. Thus, this positive ocean–atmosphere feedback of Bjerknes amplifies small initial perturbations into large observable amplitudes.

A further step forward was taken as a result of the intense warm episode of the 1982–83 El Niño, which was not recognized until it was well developed, surprising the oceanographic and meteorological communities. Soon after this surprise, the ten–year (1985–94) international TOGA (Tropical Ocean and Global Atmosphere) Program was launched building on earlier efforts such as the Equatorial Pacific Ocean Climate Studies (EPOCS) Program. One of TOGA's goals was "to study the feasibility of modeling the coupled ocean–atmosphere system for the purpose of predicting its variations on timescales of months to years". The 10–year TOGA program greatly advanced the understanding, simulation, and prediction of the coupled system, as summarized in a series of review articles published in the special volume of the *Journal of Geophysical Research (Oceans)* in June 1998.

A great legacy of TOGA has been the Tropical Atmosphere–Ocean (TAO) and TRITON array [*Hayes et al.*, 1991; *McPhaden et al.*, 1998], which provides real–time assessments of the thermal structure, currents, and surface meteorology of the tropical Pacific. Beginning in the 1970s and 1980s with outgoing longwave radiation [*Xie and Arkin*, 1996] and SST [*Reynolds and Smith*, 1994], satellite–based observations have played an important role in air–sea interaction and climate research. Since the 1990s, several new space–borne microwave sensors have allowed all–weather observations of SST, rainfall, surface wind and sea surface height over the global ocean. These instruments collectively provide an unprecedented level of detail over the ocean, which has initiated a wide array of air–sea interaction research as reviewed by *Xie* [2004a]. In addition to modern instruments and data archival, the painstaking compilation of global datasets based on historical ship observations [e.g., *Levitus*, 1982; *Woodruff et al.*, 1987], and global products using dynamical models to assimilate observations [*Kalnay et al.*, 1996; *Gibson et al.*, 1997; *Carton et al.*, 2000] have all aided the rapid progress in describing, understanding and simulating the climate and its variations.

In 1969, the year of publication of Bjerknes' seminal work, Manabe and Bryan published the results from the first coupled ocean–atmosphere general circulation model. While using a simple sector configuration partitioned between land and sea and a coarse numerical resolution, they demonstrated that such a coupled model could produce a climate not far away from observations. Since then, the coupled ocean–atmosphere models have become an important tool for understanding the climate system and predicting its changes, with ever increasing sophistication and realism [*Meehl*, 1992]. Today there are a multitude of climate models developed independently and they sometimes display such diverse behavior that model intercomparison [e.g., *Mechoso et al.*, 1995; *Davey et al.*, 2002] is regularly conducted to determine the causes of their differences.

The years since 1969 have seen a succession of conceptual advances. Notable among these are the discovery of mechanisms for the full cycle of ENSO, the mechanisms behind the strong annual cycle in the tropics [e.g., *Mitchell and Wallace*, 1992; *Xie*, 1994], and the identification of thermodynamic exchanges as an important ocean–atmosphere feedback mechanism. While in the Bjerknes feedback mechanism the role of regulating SST is played by ocean dynamics, recent studies have identified important additional roles for surface heat flux. A positive feedback due to interaction of surface wind, evaporation and SST has been proposed for the northward displacement of the intertropical convergence zone over the Atlantic and eastern Pacific [*Xie and Philander*, 1994] and the meridional gradient mode of tropical Atlantic variability [*Chang et al.*, 1997]. *Philander et al.* [1996] proposed a positive feedback between SST and low–level stratus clouds that shield incoming solar radiation over the cold water region of the southeast tropical Pacific. More recently, *Wang and Enfield* [2003] suggested a positive feedback between SST and downward longwave radiation of deep convective clouds over the Western Hemisphere warm pool.

In the extratropics, the atmosphere is stably stratified and the direct effect of SST changes is limited to a shallow atmospheric boundary layer. This limitation, along with high levels of weather noise, has made it difficult to make a robust identification of mechanisms by which the ocean can induce positive feedback in the coupled system (except near narrow ocean fronts). Indeed, it now appears that much of mid–latitude SST variability can be explained by the null hypothesis of *Hasselmann* [1976], in which the ocean mixed layer integrates white weather noise in time to yield a red spectrum without much feedback to the atmosphere. On the other hand, ocean–atmosphere interaction in the extratropics has been argued to play a role for mid–latitude climate variability, although it may be relatively weak compared to that in the tropics [e.g., *Latif and Barnett*, 1996; *Kushnir et al.*, 2002; *Czaja et al.*, 2003].

This chapter provides an overview of global climate variations, and is a general introduction to the subjects discussed in the papers that follow in this volume. Sections 2 and 3 review our understanding of ocean–atmosphere interaction in the tropics and extratropics, respectively. Section 4 considers interaction among different ocean basins and between the tropics and extratropics. Finally, Section 5 discusses remaining issues and future challenges.

2. TROPICAL CLIMATE VARIABILITY

2.1. Mean State

Easterly trade winds prevail in tropical oceans. This prevalence of the easterlies is the surface wind response to deep atmospheric convection and heavy rainfall taking place in the intertropical convergence zone (ITCZ). The ITCZ is readily recognizable as a bright band of clouds in satellite images. Condensational heating in the ITCZ lowers the local SLP and causes surface wind to converge onto it. The Coriolis force acting on these converging meridional winds induces an easterly component, leading to the southeast and northeast trade winds in the Southern and Northern Hemispheres, respectively (Figure 1).

The southeast trades penetrate slightly north of the equator over the eastern Pacific and Atlantic. These easterlies drive surface water away from the equator because the Coriolis parameter changes sign there, forcing cold deep water to upwell. This upwelled water keeps the eastern equatorial Pacific and Atlantic Oceans cold. Along the equator, the easterly wind stresses acting upon the ocean are to first order balanced by a zonal tilt of the thermocline that shoals in the east (Figure 1). The shoaling of the thermocline in the east brings cool thermocline water to the sea surface, reinforcing the cooling effect of the equatorial upwelling. The deepening of the thermocline in the western Pacific and Atlantic, on the other hand, prevents cold thermocline water from being upwelled to the sea surface, keeping SST warm there. This thermal contrast between the warm western and cold eastern ocean establishes a westward pressure gradient in the lower atmosphere, enhancing the easterly winds on the equator that act to reinforce the sea surface cooling and shoaling thermocline in the eastern ocean. This ocean–atmospheric interaction, in maintaining climatological mean states, is similar to the positive feedback mechanism *Bjerknes* [1969] proposed for ENSO [*Neelin and Dijkstra*, 1995; *Sun and Liu*, 1996].

The tongue of cold SSTs prevents deep atmospheric convection from forming on the equator. This cold tongue displaces atmospheric convection on both sides of the equator in the western Pacific. In the eastern Pacific and Atlantic, by contrast, the atmospheric ITCZ is peculiarly displaced north of the equator. As well as its role in developing the east–west contrast, air–sea interaction also plays a key role in maintaining this climatic asymmetry in the north–south direction. One consequence of the displaced ITCZ is a local reduction of surface wind speed (Figure 1). Winds are weak in the ITCZ

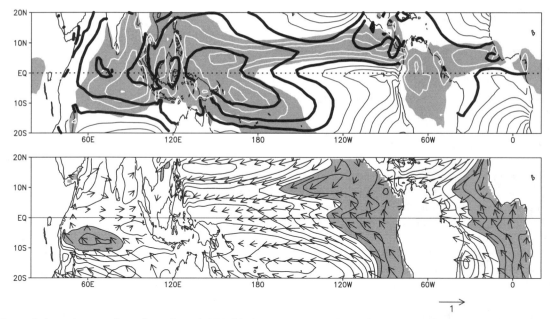

Figure 1. Annual-mean climatology: (upper) SST (black contours at 1°C intervals; contours of SST greater than 27°C thickened) and precipitation (white contours at 2 mm day^{-1}; shade >4 mm day^{-1}), based on the *Reynolds and Smith* [1994] and CMAP [*Xie and Arkin*, 1996] products, respectively; (lower) surface wind stress vectors (N m^{-2}) and the 20°C isothermal depth (contours at 20 m intervals; shade < 100 m), both based on the Simple Ocean Data Assimilation [*Carton et al.*, 2000].

since the Coriolis force, acting upon the southerly cross–equatorial flow, accelerates the easterly trades south and decelerates them north of the equator. In addition to this feedback resulting from the interaction of surface wind, evaporation and SST, the extensive low–level cloud deck that tends to form in the eastern tropical ocean south of the equator is also important in creating the thermal contrast between high SSTs north and low SSTs south of the equator. These positive feedbacks between the ocean and atmosphere help the coupled system to break the equatorial symmetry set by the distribution of annual–mean solar radiation, and to amplify asymmetric perturbations induced by coastline orientation and other aspects of continental geometry. See *Xie* [2004b] for a review of the recent progress in studying this northward–displaced ITCZ.

In the far eastern Pacific and Atlantic where zonal winds are weak, the strong southerly winds displace the upwelling zone south of the equator as indicated by the southward displacement of the meridional minimum of SST. This southerly–induced upwelling helps keep the eastern Pacific and Atlantic cold and, through the Bjerknes feedback mechanism, plays an important role in maintaining the east–west SST gradients and equatorial easterlies. A strong SST front (associated with tropical instability waves) forms north of the equator during boreal summer to fall, with much weaker SST gradients to the south. The southerly winds experience strong modification as they cross this equatorial front, and the associated adjustment has been the focus of a recent observational campaign [*Raymond et al.*, 2004; *Small et al.*, 2004] which further explores the dynamics associated with this process.

2.2. Seasonal Cycle

Solar radiation is dominated by an annual cycle with a spatial structure that is roughly anti-symmetric about the equator. In response, the seasons in the Northern Hemisphere are opposite to those in the Southern Hemisphere. Over the oceans in the Northern Hemisphere, the seasonal maximum (minimum) in SST generally take place in September (March), three months after the summer (winter) solstice, a lag due to the large thermal inertia of the upper ocean.

This local waxing and waning of solar radiation is a reliable predictor of seasons over most of the world except on the equator. While the annual harmonic of solar radiation reaches a minimum near the equator, a pronounced annual cycle in SST is observed in the eastern Pacific [*Horel*, 1982] and Atlantic, both locations where the climatological–mean thermocline is shallow. For example, the annual range of SST near the Galapagos Islands is about 6°C, greater than most of the tropical oceans (Figure 2). *Mitchell and Wallace* [1992] note that there is a westward co-propagation of seasonal SST and zonal wind anomalies along the equator in the eastern

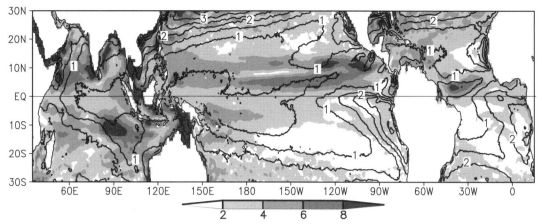

Figure 2. Root-mean-square variance of seasonal variations: SST (contours at 0.5°C intervals) and sea surface height (shade in cm), based on TRMM observations and a combined Topex/POSEIDON and ERS product.

Pacific and suggest that the SST annual cycle results from its interaction with the atmosphere. The northward displacement of the annual ITCZ implies that southerlies prevail in the eastern equatorial Pacific throughout the year, reaching a maximum in September and a minimum in March. A weakening (strengthening) of the cross–equatorial southerlies in response to the anti-symmetric solar forcing acts to decrease (increase) upwelling, vertical mixing, and surface evaporation, all helping to warm (cool) the surface ocean in a cascade of ocean–atmosphere interactions [*Xie*, 1994; *Chang*, 1996; *Nigam and Chao*, 1996].

In the east the annual cycle of equatorial SST is influenced by zonal variations in thermocline depth, but not its temporal variations. In fact, the peak amplitude of the annual cycle of SST is located on or slightly south of the equator while at this latitude the annual harmonic of sea surface height (thermocline depth varies proportionally to sea surface height, with a proportionality factor of roughly 200)—reaches a minimum (Figure 2). The annual harmonic of SST decreases westward along the equator as the thermocline becomes deeper and ocean mixed layer temperature is less sensitive to changes in winds through upwelling, vertical mixing, and/or surface heat flux. Similarly, the annual harmonic of SST reaches a minimum to the north at which latitude the annual harmonic of sea surface height reaches its tropical maximum. Thus, the annual harmonic of thermocline depth variability is largely decoupled from SST, in sharp contrast to their strong coupling on interannual timescales.

The collocation of a variance maximum in sea surface height and a variance minimum in SST along the position of the mean ITCZ is particularly clear over the tropical Atlantic presumably because of weaker interannual variability there than over the Pacific. Unlike the Pacific, there is a marked maximum in sea surface height variance near the equator in the Gulf of Guinea where the thermocline depth shoals by as much as 40 m in July–August relative to March [*Houghton*, 1983]. Thus, thermocline feedback probably plays a more important role in the annual cycle of equatorial SST in the Atlantic than in the Pacific.

Flanked by major continents, the tropical Atlantic is also more strongly influenced by continental monsoons. As an example of this continental influence, *Mitchell and Wallace* [1992] note a strong asymmetry between the seasonal cooling and warming in equatorial Atlantic SST, with the former taking just 3 months and the latter the rest of the year. The abrupt equatorial cooling is initiated by the rapid onset of the West African monsoon and the intense cross–equatorial southerlies [*Okumura and Xie*, 2004].

Among the tropical oceans, the strongest influences of the continental monsoon are found over the tropical North Indian Ocean, where the seasonal winds reverse direction, from southwesterly in summer to northeasterly in winter. This seasonal cycle in wind speed, which peaks twice a year, forces a distinctive semi-annual cycle in SST there. This semi-annual SST cycle is particularly pronounced in the western Arabian Sea and the South China Sea where coastal upwelling induced by the southwest monsoon causes a mid-summer cooling.

Seasonal SST variance reaches a meridional minimum in the Indo–Pacific warm pool region where the thermocline is deep. Equatorial Indian Ocean SST displays a weak annual cycle, which is due to cloud–induced solar forcing [*McPhaden*, 1982]. Twice a year in spring and fall, an intense eastward jet forms in the surface equatorial Indian Ocean. These *Wyrtki* [1973] jets are part of a resonant response of the Indian Ocean to a semi-annual cycle in equatorial zonal winds. *Jensen* [1993] and *Han et al.* [1999] show that the basin mode of the second baroclinic mode is in resonance with semi-annual forcing at the zonal size of the Indian Ocean. *Philander and*

Pacanowski [1986] note a similar resonance at the semi-annual frequency for the tropical Atlantic, where a pronounced semi-annual cycle in thermocline depth is observed in the east.

2.3. Regional Views

This and the next subsections discuss climate variability in tropical oceans from regional and comparative perspectives.

2.3.1. Tropical Pacific. The vast tropical Pacific Ocean hosts ENSO whose influence spans the globe. An extensive literature exists on interactions between the atmosphere and ocean that give rise to ENSO. Several articles review this literature during the past one and a half decades [*Philander*, 1990; *McCreary and Anderson*, 1991; *Battisti and Sarachik*, 1995; *Neelin et al.*, 1998]. The latest review of this subject is provided by *Wang and Picaut* [this volume] who emphasize the progress made after the TOGA era.

The starting point for most of the current literature is a consensus that the Bjerknes positive ocean–atmosphere feedback mechanism, involving the interaction of thermocline adjustment, upwelling, SST, atmospheric convection and winds, is central to the development of the interannual swings of ENSO. There also is a consensus that wind–induced mass exchange between the equatorial and off–equatorial oceans is important in the transition phase, thus possibly determining ENSO's interannual timescales. However, the reader will find differing views regarding the mechanisms for this mass exchange, leading to different negative feedback mechanisms. Among the views being promoted are a wave reflection process at the western boundary (delayed oscillator), a discharge/recharge process due to Sverdrup transport (recharge oscillator), an eastward surge that depends on a wind–forced Kelvin wave from the western Pacific (western Pacific oscillator), and a process that relies on zonal redistribution of heat due to anomalous zonal advection (advective–reflective oscillator). See *Wang and Picaut* [this volume] for a detailed discussion of these ENSO oscillator models.

In addition to its interpretation as a self-sustained mode, ENSO may be viewed as a stable mode whose phase is regulated by stochastic forcing such as that provided by the Madden–Julian Oscillation of the tropical atmosphere (see references listed in *Lengaingne et al.* [this volume] and *Wang and Picaut* [this volume]). This view does not necessarily contradict the view of ENSO as a self-sustained oscillation. After an El Niño reaches its mature phase, negative feedbacks, such as those invoked in aforementioned ENSO oscillators, are still required to terminate its growth. The question of whether ENSO is best viewed as a self-sustained oscillation or as a stable mode triggered by random forcing is not settled yet.

Finally, examination of past records of ENSO frequencies and amplitudes makes clear that there is substantial decadal and interdecadal modulation of ENSO. A rich variety of mechanisms have been proposed, including both mechanisms confined to the tropics and mechanisms that rely on tropical–extratropical exchanges and interactions. *Wang and Picaut* [this volume] discuss both the mechanisms for ENSO and rapidly evolving subject of ENSO low–frequency modulation.

2.3.2. Tropical Atlantic. Research on climate variability in the tropical Atlantic sector has expanded tremendously in recent years due to the recognition of scientifically interesting climate phenomena linked to interactions between the ocean and atmosphere as well as the need for improved prediction capabilities. In comparison with the Pacific Ocean, the Atlantic Ocean circulation has a stronger mean northward component at upper and intermediate depths to compensate for the presence of southward transport at deep levels, i.e., a strong meridional overturning circulation. The climate of the tropical Atlantic is primarily seasonal, but also varies on longer timescales ranging through decadal and beyond. Variability of this climate causes massive disruptions of populations as well as changes to the environment. There is now strong evidence that a significant part of this variability is the result of, or is modified by, local air–sea interaction within the tropical Atlantic sector itself [*Carton et al.*, 1996; *Chang et al.*, 1997].

Like the Pacific, the tropical Atlantic is subject to a Bjerknes-type feedback mechanism [*Zebiak*, 1993]. The resulting Atlantic Niños resemble their Pacific counterparts in that they involve the disappearance of the cold tongue of water along the equator (during boreal summer or fall), a surge of warm tropical water eastward and then southward along the southern coast of Africa, an anomalous reversal of direction of the equatorial trade winds, and shifts of atmospheric convection towards the anomalous warm water in the east. However, they are weaker and more frequent than the Pacific El Niños.

While the Atlantic Niño is most pronounced in the eastern half of the basin, striking variability also occurs in the west. The eastern Nordeste region of Brazil lies at the southern edge of the range of latitudes spanned by the seasonal migration of the ITCZ and receives most of its annual rainfall during March and April. Small shifts in the latitude of convection to the north or south during these months leads to droughts or floods in this sensitive region. Furthermore, it has been known since the 1970s that anomalous disturbances in the latitude of the ITCZ result from anomalous changes in the cross–equatorial SST gradient. Frequently these patterns of anomalous SST resemble a "dipole", a term used to refer to this phenomenon. Furthermore, as reviewed in *Xie and Carton* [this volume], perturbations in the latitudinal position of the ITCZ cause anomalous changes in surface fluxes of heat into the ocean that tend to reinforce the original anomaly. Alternative flux–based feedback mechanisms may also involve low clouds, which influence net surface radiation.

Because of the magnitude of ENSO and its close proximity to the Atlantic basin, the phase of ENSO has a direct impact on Amazonian convection, as well as winds and SSTs in the tropical Atlantic, and most particularly in the northern half of the basin [e.g., *Curtis and Hastenrath*, 1995; *Enfield and Mayer*, 1997]. Several atmospheric bridge mechanisms have been suggested, including the effect of ENSO on temperatures throughout the troposphere, through its effect on the anomalous Walker circulation in the Atlantic sector and through more mid-latitude routes. These issues, as well as the ongoing debate regarding the connection between the tropical Atlantic and the northern mid-latitude basin are reviewed in *Xie and Carton* [this volume].

2.3.3. Tropical Indian Ocean. ENSO exerts a strong influence on the tropical Indian Ocean, causing a basin–wide warming following a Pacific El Niño event. The tropical Indian Ocean has in the past been viewed as uninterestingly non-interactive by part of the atmospheric/climate modeling community, and was often modeled as a slab mixed layer. Recent studies, however, paint a picture of a more dynamic Indian Ocean with variability in thermocline depth and ocean currents that alter the transport of heat. The Bjerknes feedback mechanism operates in this basin as well, but only during boreal summer and fall when the equatorial winds are weakly easterly and winds off the coast of Indonesia favor upwelling. When a strong Indian Ocean dipole develops with anomalous easterlies on the equator, the fall eastward Wyrtki jet often disappears as part of a dynamic response that reinforces the cooling in the east [*Saji et al.*, 1999]. As a result of this disappearance, the Wyrtki jet is much more variable in fall than in spring.

Unlike the Pacific and Atlantic Oceans, the thermocline is flat and deep on the equator in the Indian Ocean, but is shallow in a dome south of the equator (Figure 1). This thermocline dome results from a Sverdrup–type ocean response to basin–wide positive wind curl between the equatorial westerlies and southeast trades to the south. The shallow thermocline and presence of upwelling allow subsurface variability to affect SST in this thermocline dome. Large–amplitude ocean Rossby waves are excited by the curl resulting from zonal wind anomalies associated with ENSO and the Indian Ocean dipole mode. As these waves propagate westward, they induce large SST anomalies in the South Indian Ocean dome, which in turn induce changes in atmospheric convection and winds [*Xie et al.*, 2002]. This coupling of oceanic Rossby waves with the atmosphere is quite strong because the thermocline dome resides within the meridional band encompassed by the annual migrations of the ITCZ. This off–equatorial thermocline dome and the associated maximum in thermocline feedback are unique to the South Indian Ocean. The fact that these Rossby waves take a few months to cross the basin and affect convection and cyclone development in the west may be exploited for useful climate prediction. Elsewhere in this volume, Yamagata et al. and Annamalai and Murtugudde review Indian Ocean variability.

Another important area in which progress is being made is in understanding the Indian Ocean's considerable effect on atmospheric variability. In the Bay of Bengal large SST fluctuations, with peak–to–peak values exceeding 2°C [*Sengupta et al.*, 2001; *Vecchi and Harrison*, 2002], occur in conjunction with the break and active cycle of the Indian summer monsoon. The even greater subseasonal SST anomalies in the summer western Arabian Sea due to the instability of the Somali Current and the Great Whirl induce significant anomalies in atmospheric stability, surface wind speed, and wind curl [*Vecchi et al.*, 2004]. The Findlater wind jet (the southwesterly monsoonal wind off the coast of Somalia and Arabia), for example, is found to slow down as it passes over cold filaments in the ocean, a result of stabilization of the near–surface atmosphere and decoupling of the boundary layer from the faster moving winds aloft. The tropical South Indian Ocean is another region of large subseasonal SST variability in boreal winter and spring when the Indian Ocean ITCZ is roughly collocated with this region of open–ocean upwelling [*Saji et al.*, 2004].

2.4. *A Comparative View*

2.4.1. Equatorial zonal mode. The tropical Pacific and Atlantic share many common features in their climatology, including the northward–displaced ITCZ, the prevailing easterly trades, the associated eastward shoaling of the thermocline, and an eastern cold tongue along the equator in the latter half of the year. Not surprisingly, both oceans feature an equatorial zonal mode of interannual variability that tends to be phase locked to the cold season. As shown in Figure 3, the equator of both oceans stands out as a meridional maximum in interannual SST variance (except in the far western Pacific).

Philander et al. [1984] show that the vanishing Coriolis effect near the equator renders the Bjerknes feedback positive, allowing for unstable growth of coupled ocean–atmospheric disturbances when the presence of a shallow thermocline allows tight coupling between thermocline variations and SST. The Bjerknes feedback becomes negative off the equator because of the changes in the phase of SLP, wind stress and ocean upwelling introduced by the effects of earth rotation.

Yet, despite these common features, the zonal mode in the Atlantic (Atlantic Niño) is considerably weaker in amplitude, occurs more frequently, and has a shorter duration when it does occur than the corresponding mode in the Pacific (El

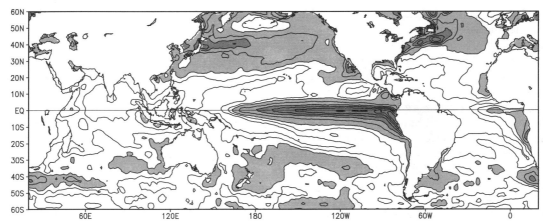

Figure 3. Root-mean-square variance of interannual SST anomalies, based on the Reynolds and Smith [1994] dataset for 1982-2000. Contour intervals are 0.1°C (0.2°C) for values smaller (greater) than 0.6°C. Light (dark) shading denotes values greater than 0.6°C (1.0°C).

Niño). The Atlantic Niño is generally limited to a brief window of June–September when upwelling is normally strong and the thermocline is shallow in the east. The causes of these differences may be found from the results of stability analysis of linear coupled models. These generally give a dispersion relation in which the growth rate of the equatorial mode vanishes at both the long and short wave limits and peaks at a zonal wavelength close to the basin size of the Pacific, much longer than the modest width of the Atlantic Ocean [*Hirst*, 1986]. Drawing much the same conclusion, *Zebiak* [1993] has argued that the Atlantic Niño is in reality the least–damped of a variety of decaying modes, and thus is easy to be excited in response to external forcing.

The annual–mean climate of the equatorial Indian Ocean is very different from that of either the Pacific or Atlantic. In response to the weak westerlies that prevail on annual average, the thermocline is nearly flat along the equator. Because of this deep and flat thermocline and lack of equatorial upwelling (under the mean westerlies), the thermocline feedback mechanism described above is very weak in the equatorial Indian Ocean. This realization of the weakness has led many scientists to model the Indian Ocean as a slab mixed layer that passively responds to remote forcing by ENSO [e.g., *Lau and Nath*, 2000]. Recent studies, as reviewed by Yamagata et al. and Annamalai and Murtugudde in this volume, however, show that during boreal summer and fall, strong anomalies of SST cause east–west SST gradients to occasionally develop and these gradients and the anomalies of convection and zonal winds which may be arranged in such a way as to support positive Bjerknes feedback (akin to that observed in the Pacific and Atlantic).

The eastern Indian Ocean is part of the Indo–Pacific warm pool that hosts a major convection center of the global atmosphere. Because of the intense convection normally occurring here, a unit change in SST induces a large response in atmospheric convection and winds. In fact, the interannual variance of precipitation over the eastern equatorial Indian Ocean is as large as that over the equatorial Pacific despite the much smaller variance of SST [*Saji and Yamagata*, 2003]. In linear coupled models, the growth rate of a coupled mode is dependent on the ocean's response to the atmosphere and the atmosphere's response to the ocean. In the eastern equatorial Pacific and Atlantic, the ocean's response to wind changes, involving thermocline feedback, is very strong because of the prevailing upwelling and shallow thermocline, but the atmosphere's response to SST changes is weak, because the mean SST is low and mean convection is weak. In the equatorial Indian Ocean, in contrast, the atmosphere's response is strong and as a result, a strong growth of an equatorial zonal mode is possible during the seasons when the upwelling favorable winds switch on a thermocline feedback (albeit weak).

Thus, despite large differences in the annual mean winds and thermocline structure, the equatorial mode relying on the Bjerknes feedback mechanism turns out to be a feature common to all the three tropical oceans. The Pacific ENSO displays the largest amplitudes and longest timescales, with significant SST anomalies persisting for a year or more. The Atlantic Niño and the Indian Ocean dipole mode are generally weaker and less regular, with significant SST anomalies limited to a brief window of a few months when the upwelling and thermocline feedback reach their seasonal maximum.

2.4.2. Meridional mode. While thermocline feedback is important in regions of upwelling, Ekman downwelling prevails over the vast off–equatorial and subtropical oceans. In these downwelling regions, subsurface ocean variability is shielded from the sea surface, and other ocean–atmospheric feedback mechanisms involving surface heat flux becomes important. A positive feedback between surface wind, evaporation and SST (WES) favors an anti-symmetric mode that maximizes the cross–equatorial SST gradient [*Chang et al.*, 1997]. Unlike the Bjerknes feedback that favors east–west oriented anomalies, this anti-symmetric mode involves air–sea interaction in the north–south direction, with a growth rate that peaks at zonal wavenumber zero and decreases with increasing wavenumber [*Xie et al.*, 1999]. In the large Pacific basin, the Bjerknes feedback is strong and so ENSO dominates the climate variability. In the smaller Atlantic basin, the Bjerknes feedback weakens sufficiently that it is comparable to the WES feedback. As a result, the two modes co-exist in this basin without one dominating the other.

Very recently, *Chiang and Vimont* [2004] have carried out a careful analysis of tropical Pacific variability and report a meridional mode analogous to that in the tropical Atlantic. Like the Atlantic meridional mode, which is strongly influenced in the subtropics by local atmospheric variability such as that associated with the North Atlantic Oscillation (NAO), the Pacific meridional mode is linked to the North Pacific teleconnection pattern of the atmosphere. This Pacific meridional mode also has a zonal uniform structure. Further confirmation of the presence of this mode in the Pacific comes from some coupled models [*Yukimoto et al.*, 2000; *Okajima et al.*, 2003].

The easterly trades prevail on both sides of the equator/ITCZ in the Pacific and Atlantic, a necessary condition for the WES feedback to be positive. This condition is partially met in the South Indian Ocean where the southeast trades prevail year around. While the annual mean winds are weakly southwesterly in the North Indian Ocean, the northeasterlies prevail during boreal winter and spring, meeting the necessary condition for positive WES feedback. During these seasons, *Kawamura et al.* [2001] show that interannual anomalies of SST and surface winds also sometimes organize themselves into equatorial anti-symmetric patterns indicative of the WES feedback. They suggest that this meridional mode may affect the strength of the subsequent summer monsoon over South Asia, an idea that deserves further attention.

3. EXTRATROPICAL CLIMATE VARIABILITY

3.1. Mid-Latitudes

As we move towards mid-latitudes, the difficulty in identifying ocean–atmosphere interactions increases. The difficulty may be due to more complex meteorology in which local anomalies of SST and surface winds are less strongly linked, to oceanic conditions in which SSTs are cooler and mixed layers are deeper, to the longer timescales of oceanic response to atmospheric conditions, as well as to the momentum constraints of the larger Coriolis term. Still, despite this difficulty, there is considerable evidence of interactions. For example, in their analysis of surface air pressure in the North Pacific sector, *Trenberth and Hurrell* [1994] found a long period of elevated surface pressure in midbasin spanning the decades of the 1950s and 1960s, followed in the mid-1970s by a period of reduced surface pressure. The reduction intensified and caused an eastward shift of the Aleutian Low, bringing warm moist air to the west coast of North America and a southward shift of the mid-latitude storm track. See *Mantua and Hare* [2002] and *Miller et al.* [2003] for recent reviews of Pacific decadal variability and its effects on ocean ecosystems.

In his review of the literature, *Latif* [1998] divides mid-latitude interaction theories into three categories: (1) those occurring through interactions in both mid-latitudes and tropics, (2) those involving changes in the gyre circulation, and (3) those occurring in mid-latitudes and involving changes in the thermohaline circulation. Here we examine each of these.

The possibility of interactions of the first type involving both the tropics and mid-latitudes was proposed by *Gu and Philander* [1997]. In their simple conceptual model water with temperature anomalies of either sign is introduced into the oceanic mixed layer in the extratropical Pacific due to local meteorological conditions. This newly formed water is subducted in the thermocline and then follows Lagrangian pathways called subtropical cells (STCs) equatorward and generally westward [e.g., *McCreary and Lu*, 1994; *Liu and Philander*, 2001; *Schott et al.*, this volume]. Depending on its geographic origin the water may eventually enter the equatorial thermocline, thus altering the stratification along the equator. According to this conceptual model, changes in stratification lead to changes in SST, causing changes in winds that in turn affect the properties of subducted extratropical water. Changes in the stratification of the equatorial Pacific are also surmised to affect the development of ENSO cycles [*Zebiak and Cane*, 1987; *Neelin et al.*, 1994]. Thus, the apparent frequency of El Niño conditions in the 1990s may have resulted from a slow deepening of the thermocline. Some observational evidence to support these theories has been provided by examination of the meridional propagation of temperature anomalies by *Deser et al.* [1996] although the data analysis of *Schneider et al.* [1999] does not find any significant link between the North Pacific and the equator through the subduction of SST anomalies.

Kleeman et al. [1999] proposed an alternative role for STCs in climate, arguing that wind–induced changes in the strength of these shallow overturning cells play a key role in generating equatorial SST anomalies by modulating the amount of cold thermocline water advected into the tropics. This mechanism is supported by a study of *McPhaden and Zhang* [2002] who observed a slowdown of the Pacific STCs from the mid-1970s to the late 1990s associated with a decrease in STC transport. Using an ocean general circulation model (OGCM), *Nonaka et al.* [2002] found that unlike El Niño in which SST anomalies are mostly induced by equatorial winds, off–equatorial winds cause STCs to vary in strength and are thus important for decadal SST variations in the equatorial Pacific. Related studies have examined the pathways of water entering the tropical thermocline in the Atlantic [*Zhang et al.*, 2003; *Schott et al.*, this volume]. Current research on the influence of conditions in the tropical Atlantic Ocean on the mid-latitude atmosphere is reviewed in *Xie and Carton* [this volume].

A flurry of research has been provoked by the joint observational and coupled modeling analysis of *Latif and Barnett* [1996] who found a decadal (~25 year) mode of variability involving the Aleutian Low and the subtropical oceanic gyre. They argue along the lines of the second interaction category. According to their mechanism an intensification of the subtropical gyre will transport more warm water into the central North Pacific, which through interaction with the overlying atmosphere will lead to positive feedback and enhanced SSTs through reduction of meridional temperature gradients and enhancement of net heat flux from the atmosphere. It will also sow the seeds of its own demise by reducing the strength of the wind stress curl, thus ultimately reducing the strength of the subtropical gyre on timescales of a decade or so. The importance of oceanic advection in regulating heat storage, a key aspect of this mechanism, is examined by *Kelly and Dong* [this volume] who show that the relationship between meteorological forcing in the North Pacific and Atlantic Oceans results from changes in advection caused by changes in mid-latitude westerly winds.

Fluctuations in the oceanic thermohaline circulation may also play a role in the third category of coupled interactions. As discussed in Section 4.3, the North Atlantic is the only basin with a northern source of deep water. The rate of formation of this North Atlantic Deep Water appears to be connected to the salinity balance of the North Atlantic and Arctic basins [*Mann and Park*, 1996]. This connection apparently is the source of interdecadal cycles in several coupled models [*Delworth et al.*, 1993; *Lohmann*, 2003] in which changes in the mid-latitude oceanic gyre influence the properties of water entering the Arctic basin and thus the rate of formation of deep water. Fluctuations in the supply of freshwater to the Arctic has been implicated in the Great Salinity anomalies [*Lazier*, 1988] as well as in decadal variations of wintertime SST and ice in the northwest Atlantic [*Deser and Blackmon*, 1993].

Internal variability of the atmosphere generally increases in amplitude toward the poles, and is well organized in space (but highly disorganized in time) in the form of patterns such as the Pacific–North American (PNA) pattern and the NAO. The null hypothesis of no ocean to atmosphere feedback, originally proposed by *Hasselmann* [1976], suggests that the appearance of low frequency signals in the ocean could simply reflect the ability of the ocean to integrate in time the effects of forcing by the variable atmosphere [*Frankignoul*, 1985]. The Hasselmann null hypothesis has recently been expanded upon by *Deser et al.* [2003] to account for the ability of the ocean to sequester thermal anomalies in the summer, only to reveal them again the next winter when surface cooling and enhanced mixing cause the mixed layer to deepen sufficiently. This reemergence mechanism may be generalized by including ocean heat transport changes that vary subsurface temperature throughout the year [e.g., *Schneider and Miller*, 2001; *Tomita et al.*, 2002]. The stochastic model for mid-latitude climate variability has been further developed [*Barsugli and Battisti*, 1998; *Neelin and Weng*, 1999].

High levels of atmospheric internal variability pose a challenge for detecting SST's influence on the atmosphere [*Kushnir et al.*, 2002]. On basin scale, SST and wind speed tend to be negatively correlated in the extratropics, especially in winter, a correlation consistent with the Hasselmann null hypothesis. Traditional climatic datasets, however, do not adequately resolve major ocean currents and fronts that are only a few hundred kilometers wide. Yet it is near these regions that ocean currents are expect to play an important role in SST variability. Indeed, recent high–resolution satellite observations reveal robust patterns of air–sea interaction near major ocean currents like the Kuroshio, the Gulf Stream, and circumpolar currents. Surface wind speeds tend to increase over positive SST anomalies, a positive correlation indicative of ocean–to–atmospheric feedback. *Xie* [2004a] and *Chelton et al.* [2004] discuss these new results from satellite observations and their climatic implications. Oceanic fronts along the extensions of major western boundary currents are often regions of extratropical cyclogenesis. *Nakamura et al.* [this volume] discuss possible interactions of oceanic fronts and atmospheric storm tracks.

3.2. High Latitudes

The leading mode of Northern Hemisphere sea level pressure variability based on an empirical orthogonal function analysis features a seesaw between the polar cap and the mid-latitudes, with a corresponding seesaw in zonal–mean zonal wind between the subtropical and subpolar belts. This northern annular mode, or Arctic Oscillation (AO), is most pronounced in boreal winter and is linked to variability in the upper troposphere and stratosphere [*Thompson and Wallace*, 2000]. In the winter stratosphere, the northern annular mode is nearly zonally symmetric and associated with changes in the strength of the polar vortex. Such an annular mode with strong coupling between the surface and stratosphere in local winter is also observed in the Southern Hemisphere where the zonal symmetry is even stronger than the northern annular mode. The northern annular mode exhibits considerable variation in the zonal direction in the troposphere, with the largest loading over the North Atlantic where it resembles the NAO. This similarity gives rise to alternative interpretations of the same phenomena.

In its positive phase, SLP associated with the northern annular mode drops over the polar cap, and increases over the Azores high in the North Atlantic and the Aleutian low in the North Pacific. More storms reach northern Europe, resulting in wetter conditions; surface air temperature increases over the mid- and high-latitude Eurasia and most of North America and dryer conditions and air temperature decreases over northeastern Canada. SST anomalies generally follow a pattern of negative (positive) values where prevailing winds intensify (weaken). The meteorological conditions associated with the positive phase of the northern annular mode feature positive SST anomalies in the mid-latitude North Pacific as the surface westerly jet weakens and a tripole SST pattern in the North Atlantic (with negative SST anomalies in the subpolar region as the surface westerlies intensify). The lead/lag relationship of this index of meteorological conditions to anomalous SSTs in the North Atlantic suggests the possibility that this meteorological phenomenon is significantly influenced by exchanges with the underlying ocean [*Czaja and Frankignoul*, 2002]. However, as recently reviewed by *Czaja et al.* [2003], the role of the ocean–atmosphere coupling in the northern annular mode may be relatively weak. The annular modes seem to be the internal mode of the atmosphere resulting from the interaction of subpolar jets and storm tracks. They appear as the dominant mode of atmospheric variability in atmospheric models even when forced just with climatological SST. The issue of the importance of oceanic feedback remains one of active debate.

There is some evidence that stratospheric anomalies lead tropospheric changes in time and this lead may be exploited for prediction of subseasonal variability in the troposphere [*Baldwin and Dunkerton*, 2001]. Since anthropogenic climate change due to the observed increase in carbon dioxide and the depletion of ozone is likely to be stronger and more robust in the stratosphere, the annular modes may be an important mechanism by which the first sign of anthropogenic climate change in the stratosphere reaches the earth surface. A recent AGU monograph [*Hurrel et al.*, 2003] is devoted to studies of the NAO/AO.

Finally, we note that in the Southern Ocean, co-varying ocean–atmosphere–cryosphere signals of 4–5 year timescale have been observed propagating slowly eastward, taking 8–10 years to encircle the pole [*White and Peterson*, 1996; *Jacobs and Mitchell*, 1996]. In this Antarctic circumpolar wave, warm (cold) SST anomalies are associated with poleward (equatorward) meridional surface wind anomalies, suggesting that ocean–atmosphere interaction may be at work. Recently, *White et al.* [2002] suggested this high–latitude wave may influence tropical climate.

4. INTER-BASIN INFLUENCES AND INTERACTIONS

Climate in the tropics and extratropics is intimately linked, and from one ocean basin to another. This section discusses the mechanisms for these linkages, including atmospheric bridges within the tropical belt and to the extratropics, and ocean circulations that link the global oceans together. As discussed in Section 3.2, the northern annular mode may provide a mechanism for inter-basin interaction between the North Pacific and North Atlantic, and north branches of the Antarctic circumpolar wave may provide a linkage between the Southern Ocean and tropical oceans.

4.1. Interactions Among Tropical Oceans

It has been noted that SST anomalies among the tropical Pacific, Atlantic, and Indian Oceans are related to one another [e.g., *Hastenrath et al.*, 1987; *Kiladis and Diaz*, 1989; *Tourre and White*, 1995; *Lanzante*, 1996], mainly reflecting the dominant forcing by the Pacific ENSO. The global nature of ENSO is shown in Figure 4, which displays the correlation between the Niño3 SST anomalies during November–December–January (NDJ) and global SST anomalies during the following February–March–April (FMA). These seasons are chosen because ENSO peaks during boreal winter while its influence on other ocean basins normally peaks 1–2 seasons later [e.g., *Alexander et al.*, 2002]. Outside the Pacific, significant warming is found over the tropical North Atlantic and the entire tropical Indian Ocean. Figures 5a and 5b compare SST anomalies in the tropical North Atlantic and Indian Ocean with those in the Niño3 region, showing high correlations among these time series. Figure 4 does not reflect the Indian Ocean dipole mode that sometimes co-occurs with ENSO and usually peaks in September–November.

The influence of ENSO on other tropical oceans is transmitted through the "atmospheric bridge" of atmospheric circulation changes. Based on a correlation analysis of satellite and ship observations, *Klein et al.* [1999] provide a schematic of the Walker and Hadley circulations that accompany El Niño events (Figure 6). The adjustment in these circulations is confirmed by the direct circulation analyses of the NCEP-NCAR reanalysis fields [*Wang*, 2002, 2004]. During the warm phase of ENSO, convective activity in the equatorial western Pacific shifts eastward. This shift in convection leads to an altered Walker circulation, with anomalous ascent over the equatorial central and eastern Pacific, and anomalous descent over the equatorial Atlantic and the equatorial Indo–western Pacific region. Thus, the Hadley circulation strengthens over the eastern Pacific but weakens over the Atlantic and Indo–western Pacific sectors. These anomalous Walker and Hadley circulations result in variations in surface wind speed, humidity, and cloud cover that in turn influence surface heat fluxes and SST over the tropical Indian and Atlantic Oceans. Elsewhere in this volume, Kawamura et al. examine long–term changes in the anomalous Walker circulation linking the Indo–Pacific Oceans; Xie and Carton discuss the tropical Atlantic response to ENSO; and Su et al. study a mysterious lack of correlation between tropical mean temperature and precipitation during ENSO.

The tropical Indian Ocean is likely to affect tropical Pacific variability. On intraseasonal timescales, the atmospheric Madden–Julian Oscillation develops over the Indian Ocean and propagates eastward to the Pacific, which may affect the Pacific El Niño [*Takayabu et al.*, 1999; *Lengaigne et al.*, this volume]. On interannual timescales, *Watanabe and Jin* [2003] suggest that El Niño–induced Indian Ocean warming helps amplify the Philippine Sea anomalous anticyclone that forms due to Pacific Ocean–atmosphere processes [*Wang et al.*, 1999; *Wang et al.*, 2000], a result consistent with atmospheric GCM hindcasts [*Lau and Nath*, 2000]. The Indian Ocean dipole mode is associated with strong cold SST anomalies and suppresses convection in the eastern equatorial Indian Ocean, which may affect the Pacific through the attendant surface wind anomalies, a hypothesis that received some support from *Behera and Yamagata*'s [2003] correlation analysis.

Several studies suggest global propagating waves [e.g., *Barnett*, 1985; *Yasunari*, 1985; *White et al.*, 2003]. In particular, *White et al.* [2003] found that these waves are composed primarily of global zonal wave numbers 1 and 2, traveling eastward. Based on uncoupled and coupled GCM simulations, *Latif and Barnett* [1995] concluded that most of the variability in the tropical Atlantic and Indian Oceans associated with the interannual global wave is forced by Pacific SST anomalies via changed atmospheric circulations, with local air–sea interactions acting as an amplifier of the Pacific–induced signal.

4.2. Atmospheric Bridge to the Extratropics

Changes in tropical convection excite planetary waves that bring about climatic anomalies around the world and thus ENSO's influence extends far beyond the tropics [*Alexander et al.*, this volume]. As a result, SST correlations with ENSO exceed 0.6 in the extratropical North and South Pacific (Figure 4). The North Pacific SST anomaly time series in Figure 5c is best correlated with Niño3 SST anomaly at a three–month lag. However, the nature of the atmospheric response changes with latitude. In contrast to the dominant baroclinic structure

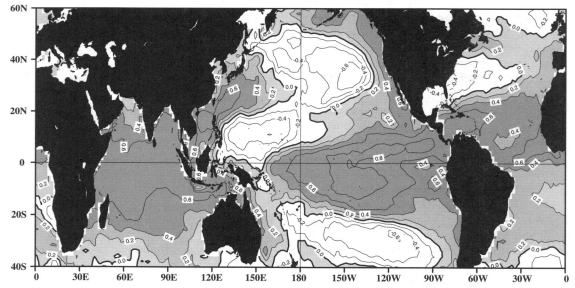

Figure 4. Correlation between the Niño3 (5°S–5°N, 150°W–90°W) SST anomalies during November–December–January (NDJ) and global SST anomalies during the following February–March–April (FMA). The calculation is based on the NCEP SST data from 1950–1999.

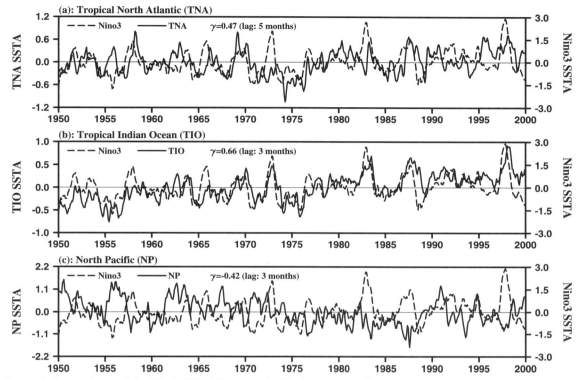

Figure 5. Comparisons of the Niño3 (5°S–5°N, 150°W–90°W) SST anomalies with (a) the SST anomalies in the tropical North Atlantic (5°N–25°N, 55°W–15°W), (b) the SST anomalies in the tropical Indian Ocean (10°S–10°N, 50°E–100°E), and (c) the SST anomalies in the North Pacific (35°N–45°N, 160°E–160°W). All of the time series are three-month running means. The γ represents correlation coefficient.

in the tropics, in the extratropics these waves are barotropic and most pronounced in winter in the Northern Hemisphere [*Trenberth et al.*, 1998]. An example of this atmospheric bridge to the extratropics is the PNA pattern of *Wallace and Gutzler* [1981] and *Horel and Wallace* [1981]. This PNA teleconnection pattern is often associated with ENSO in the tropics, with alternating positive and negative geopotential height anomalies that emanate from the tropical Pacific, pass the North Pacific, curve eastward to northwestern America and then equatorward to reach southeastern United States and the Gulf of Mexico (Figure 7). SLP anomalies are generally of the same sign as geopotential anomalies in the middle and upper troposphere, deepening the Aleutian low during an El Niño. The associated surface wind changes lower SST in the central North Pacific (Figure 4) by increasing surface heat flux from the ocean and Ekman advection [*Alexander et al.*, 2002]. The SLP and wind changes over the tropical North Atlantic contribute to the warming of the tropical North Atlantic and the Western Hemisphere warm pool [*Enfield and Mayer*, 1997; *Wang and Enfield*, 2003]. *Alexander et al.* [this volume] dis-

Figure 6. Schematic diagram of the anomalous Walker and Hadley circulation during an El Niño event [*Klein et al.*, 1999].

cuss ENSO's teleconnection in summer, which is much less studied than that in winter.

The atmospheric bridge may also operate the other way around, from the extratropics to the tropics [*Pierce et al.*, 2000]. On decadal timescales the largest anomalies of SST and ocean heat content occur in mid-latitudes instead of the tropics [*Giese and Carton*, 1999]. *Vimont et al.* [2003] sug-

Figure 7. Schematic diagram of middle and upper tropospheric geopotential height anomalies associated with the Pacific North American (PNA) pattern during boreal winter of the Pacific El Niño year [*Horel and Wallace*, 1981].

gest a seasonal footprinting mechanism in which subtropical SST anomalies in winter persist into spring and summer, inducing a broad–scale atmospheric response in summer with significant wind anomalies near the equator that affect ENSO.

Over the North Atlantic, the Azores high in SLP is an important link between the tropics and mid-latitudes, with its variability affecting the northeast trades on one hand and the mid-latitude westerlies on the other. One possibility is that the Azores high is influenced by both the NAO and the Atlantic ITCZ. Indeed, some observational analyses suggest a pan–Atlantic SST pattern that features zonal bands of anomalies of alternating signs spanning from the South Atlantic to Greenland. *Xie and Carton* [this volume] and *Barreiro et al.* [this volume] discuss the interaction between the tropical and extratropical Atlantic.

4.3. Role of Ocean Circulation

Tropical and extratropical connections can also be through oceanic bridges. One of oceanic bridge mechanisms is through an oceanic wave signal transmitted between the mid-latitude and tropical oceans. Coastal Kelvin waves along the west coast of North America have been proposed as the mechanism for linking tropical and extratropical SST anomalies during ENSO and for longer timescales [e.g., *Enfield and Allan*, 1980; *Clarke and Lebedev*, 1999]. *Jacobs et al.* [1994] suggested that the 1982–83 El Niño could have decadal effects on the northwestern Pacific circulation, through mid–latitude Rossby waves reflected from equatorial Kelvin waves on the American coasts.

Lysne et al. [1997] found a weak decadal signal in their search for another oceanic bridge driven by wave dynamics: anomalous temperature propagated by mid–latitude Rossby waves into the western boundary, then equatorward by coastal Kelvin waves and finally modifying equatorial SST via equatorial Kelvin waves. Recently, observational results of *Hasegawa and Hanawa* [2003] showed anticlockwise propagation of upper–ocean heat content anomalies around the northern tropical Pacific Ocean beginning with an eastward propagation along the equator, northward propagation at the eastern boundary, westward propagation at the subtropics, and finally southward propagation at the western boundary.

As discussed in Section 3.1, shallow overturning ocean circulation may link the tropics and subtropics. Water formed in the subtropics may enter the tropical thermocline within a few years after being subducted. Some of this water returns to the oceanic mixed layer near the equator as the result of entrainment and mixing. Very near the equator quite shallow tropical cells result from the local processes of Ekman divergence and near–equatorial subduction. In Eulerian zonal average these cells appear closed, but in fact involve complex time–dependent pathways.

The thermohaline circulation of the oceans, also known as the meridional overturning circulation, occurs on several vertical levels and at several timescales. Overturning of the lower 2 km of the ocean occurs on long centennial timescales (although the timescales may be much shorter in some regions). This deep and bottom water is formed only at the northern end of the North Atlantic and around Antarctica. Intermediate water masses, which appear at depths of 1 km or so, are formed in several locations. In the Atlantic intermediate water is formed notably by salty water exported from the Mediterranean Sea (Mediterranean water) and also by fresher water formed in the circumpolar current (Antarctic intermediate water).

Changes in the deeper circulation have been implicated in a number of climate anomalies appearing in the geologic record. For example, a shutdown of the deep meridional overturning circulation seems to have occurred at the end of the last ice age due to the release of meltwater from the Laurentide ice sheet some 13–11.5 kyr ago [*Broecker et al.*, 1989; *Ramstorf*, 2000, 2002]. The impact on the climate of high latitudes from this shutdown was prompt, dropping temperatures by more than 5°C. Much more modest freshwater anomalies have appeared at shallow levels in the North Atlantic due to anomalous sea ice export in 1969 and 1994 [*Lazier*, 1988]. A recent study by *Curry et al.* [2003] intriguingly suggests changes in the salinity of surface waters of the Atlantic during the past 30 years (increasing salinity in the tropics and decreases at high latitudes) that one might expect from a slowdown of the meridional overturning circulation.

5. DISCUSSION

For historical reasons the ocean and atmosphere have traditionally been studied separately, assuming that the state of the other fluid is a specified boundary condition. The development of geophysical fluid mechanics that began half a century ago is an attempt to understand the common dynamics shared by the ocean and atmosphere. This development has been a huge success—concepts like quasi-geostrophy, thermal wind, potential vorticity conservation, Kelvin and Rossby waves, and baroclinic instability have helped explain many phenomena in the ocean and atmosphere. The study of climate has triggered a new level of integration of the fields of meteorology and oceanography, demanding that we treat climate phenomenon as a coupled problem. This paradigm of ocean–atmosphere interaction has enabled us to unlock the mysteries of ENSO and other climate phenomena, leading to skillful predictions of ENSO [Cane et al., 1986], its global impacts [Latif et al., 1998], and certain aspects of tropical Atlantic variability [Saravanan and Chang, this volume]. Indeed, the success of TOGA has proved a premise that a better understanding of ocean–atmosphere interaction helps improve model simulation, and along with an adequate observing system leads eventually to useful climate prediction.

While great progress has been made, our understanding of tropical ocean–atmosphere interaction is still incomplete, especially in the Indian and Atlantic sectors. Our understanding of the extratropical ocean–atmosphere system is in an even less satisfactory state, limiting our ability to extend the success of seasonal forecasts into longer time leads. Most of the world's population lives on continents. Past research shows that atmospheric bridges play an important role in influencing continental climate. But such robust atmospheric bridges from the tropics may only apply to limited geographical areas, and are interfered by weather noise and other modes of climate variation. It remains unclear whether land processes have an active role to play in generating or maintaining climatic anomalies on interannual and longer timescales. Webster et al. [1998] gave a comprehensive review of studies of the structure and variability of the Asian–Australian monsoon system and its relationships with ENSO. Another area that is not covered in this volume but of great social and economic importance is the interaction of coastal/marginal seas with the overlying atmosphere. There, ocean bottom topography and land orography are generally important. Recent satellite studies reveal interesting features in coastal and marginal seas that involve interaction with the atmosphere [Xie, 2004a; Chelton et al., 2004; Hu and Liu, 2003].

Anthropogenic climate changes also pose a great challenge for the climate research community as well as for the mankind in general. On the global scale, exchange of carbon dioxide and heat between the atmosphere and ocean has likely slowed down the rate of temperature rise (global warming) so that its effects are just beginning to be observed. On the regional scale, the temperature rise is unevenly distributed, with the greatest rise over continents but some cooling over the North Pacific and North Atlantic Oceans during the second half of the past century.

Human–induced climate change may have some further surprises in store. Some model projections suggest a slow-down of the North Atlantic meridional overturning circulation in a warmer climate due to increased sea surface evaporation, as well as increased rainfall in the mid/high latitudes as a result of a more intense hydrological cycle [Manabe and Stouffer, 1993]. This slow down of the meridional overturning circulation reduces the northward transport of warm surface water and may send the North Atlantic and Europe into a colder climate. Such abrupt climate changes are believed to have happened before as the climate warmed at the end of the last ice age. What causes such a rapid spread of North Atlantic cooling is unclear [Broecker, 2003]. Tropical air–sea interaction may have played an important role in transmitting this regional cooling into a global–scale event [Dong and Sutton, 2002; Xie and Carton, this volume]. Global warming may also affect ENSO variability. However, the relationship between ENSO and global warming is unknown [Wang and Picaut, this volume] and needs further examination.

Coupled ocean–atmosphere general circulation models are an important tool for predicting the evolving climate and projecting the impact of human activity. These models, however, suffer significant biases in their simulation of the current climate and its variations [e.g., Mechoso et al., 1995; Davey et al., 2002]. Such biases are certain to affect the model simulation of the climate response to increased greenhouse gases, casting doubts on the future climate projections produced by these models. These biases generally reflect a poor understanding of physical processes such as clouds, their response to SST changes and treatment in models. We anticipate that the improved understanding of the causes of tropical climate variability as represented by the papers in this volume will help isolate model deficiencies and thus contribute to improved predictive capability.

Acknowledgments. CW was supported by a grant from NOAA Office of Global Programs and by NOAA Environmental Research Laboratories through their base funding of AOML; SPX by grants from NSF, NOAA, NASA, NSFC, and Frontier Research System for Global Change; JAC by grants from the NOAA Office of Global Programs and from NSF. Comments by two anonymous reviewers helped improve the paper. This is IPRC contribution #271 and SOEST contribution #6368.

REFERENCES

Alexander, M. A., et al., The atmospheric bridge: The influence of ENSO teleconnections on air–sea interaction over the global oceans, *J. Clim.*, *15*, 2205–2231, 2002.

Alexander, M. A., N.-C. Lau, and J. D. Scott, Broadening the atmospheric bridge paradigm: ENSO teleconnections to the tropical west Pacific–Indian Oceans over the seasonal cycle and to the North Pacific in summer, this volume.

Annamalai, H., and R. Murtugudde, Role of the Indian Ocean in regional climate variability, this volume.

Baldwin, M. P., and T. J. Dunkerton, Stratospheric harbingers of anomalous weather regimes, *Science*, *294*, 581–584, 2001.

Barnett, T. P., Variations in near global sea level pressure, *J. Atmos. Sci.*, *42*, 478–501, 1985.

Barreiro, M., A. Giannini, P. Chang, and R. Saravanan, On the role of the Southern Hemisphere atmospheric circulation in tropical Atlantic variability, this volume.

Barsugli, J. J., and D. S. Battisti, The basic effects of atmosphere–ocean thermal coupling on midlatitude variability, *J. Atmos. Sci.*, *55*, 477–493, 1998.

Battisti, D. S., and E. S. Sarachik, Understanding and predicting ENSO, *Rev. Geophys., Sup.*, *33*, 1367–1376, 1995.

Behera, S. K., and T. Yamagata, Influence of the Indian Ocean dipole on the Southern Oscillation, *J. Meteorol. Soc. Jpn.*, *81*, 169–177, 2003.

Bjerknes, J., Atmospheric teleconnections from the equatorial Pacific, *Mon. Weather Rev.*, *97*, 163–172, 1969.

Broecker, W. S., Does the trigger for abrupt climate change reside in the ocean or in the atmosphere? *Science*, *300*, 1519–1522, 2003.

Broecker, W. S., et al., Routing of meltwater from the Laurentide ice sheet during the Younger Dryas cold episode, *Nature*, *341*, 318–321, 1989.

Cane, M. A., S. E. Zebiak, and S. C. Dolan, Experimental forecasts of El Niño, *Nature*, *321*, 827–832, 1986.

Carton, J. A., X. Cao, B. Giese, and A. M. Da Silva, Decadal and interannual SST variability in the tropical Atlantic Ocean, *J. Phys. Oceanogr.*, *26*, 1165–1175, 1996.

Carton, J. A., G. Chepurin, X. Cao, and B. Giese, A simple ocean data assimilation analysis of the global upper ocean 1950–95. Part I: Methodology, *J. Phys. Oceanogr.*, *30*, 294–309, 2000.

Chang, P., The role of the dynamic ocean–atmosphere interactions in tropical seasonal cycle, *J. Clim.*, *9*, 2973–2985, 1996.

Chang, P., L. Ji, and H. Li, A decadal climate variation in the tropical Atlantic ocean from thermodynamic air–sea interactions, *Nature*, *385*, 516–518, 1997.

Chelton, D. B., M. G. Schlax, M. H. Freilich, and R. F. Milliff, Satellite radar measurements reveal short–scale features in the wind stress field over the world ocean, *Science*, *303*, 978–983, 2004.

Chiang, J. C. H., and D. J. Vimont, Analogous Pacific and Atlantic meridional modes of tropical atmosphere–ocean variability, *J. Clim.*, submitted, 2004.

Clarke, A. J., and A. Lebedev, Remotely driven decadal and longer changes in the coastal Pacific waters of the Americas, *J. Phys. Oceanogr.*, *29*, 828–835, 1999.

Curry, R., B. Dickson, and I. Yashayaev, A change in the freshwater balance of the Atlantic Ocean over the past four decades, *Nature*, *426*, 826–829, 2003.

Curtis, S., and S. Hastenrath, Forcing of anomalous sea surface temperature evolution in the tropical Atlantic during Pacific warm events, *J. Geophys. Res.*, *100*, 15835–15847, 1995.

Czaja, A., and C. Frankignoul, Observed impact of North Atlantic SST anomalies on the North Atlantic Oscillation, *J. Clim.*, *15*, 606–623, 2002.

Czaja, A., A. W. Robertson, and T. Huck, The role of Atlantic Ocean–atmosphere coupling in affecting North Atlantic oscillation variability, in *The North Atlantic Oscillation: Climatic Significance and Environmental Impact*, edited by J. W. Hurrell, Y. Kushnir, G. Ottersen, and M. Visbeck, pp. 147–172, AGU Geophysical Monograph Series, Washington, D.C., 2003.

Davey, M. K., et al., STOIC: a study of coupled model climatology and variability in tropical ocean regions, *Clim. Dyn.*, *18*, 403–420, 2002.

Delworth, T., S. Manabe, and R. J. Stouffer, Interdecadal variations of the thermohaline circulation in a coupled ocean–atmosphere model, *J. Clim.*, *6*, 1993–2011, 1993.

Deser, C., and M. L. Blackmon, Surface climate variations over the North Atlantic Ocean during winter: 1900–1989, *J. Clim.*, *6*, 1743–1753, 1993.

Deser, C., M. A. Alexander, and M. S. Timlin, Upper–ocean thermal variations in the North Pacific during 1970–1991, *J. Clim.*, *9*, 1841–1855, 1996.

Deser, C., M. A. Alexander, and M. S. Timlin, Understanding the persistence of sea surface temperature anomalies in midlatitudes, *J. Clim.*, *16*, 57–72, 2003.

Dong, B.-W., and R. T. Sutton, Adjustment of the coupled ocean–atmosphere system to a sudden change in the thermohaline circulation, *Geophys. Res. Lett.*, *29*, doi:10.1029/2002GL015229, 2002.

Enfield, D. B., and J. S. Allen, On the structure and dynamics of monthly mean sea level anomalies along the Pacific coast of North and South America, *J. Phys. Oceanogr.*, *10*, 557–578, 1980.

Enfield, D. B., and D. A. Mayer, Tropical Atlantic sea surface temperature variability and its relation to El Niño–Southern Oscillation, *J. Geophys. Res.*, *102*, 929–945, 1997.

Frankignoul, C., Sea surface temperature anomalies, planetary waves, and air–sea feedback in the middle latitudes, *Rev. Geophys.*, *23*, 357–390, 1985.

Gibson, J. K., et al., ERA description, *Re-Analysis (ERA) Project Report Series No. 1*, 72 pp., 1997.

Giese, B. S., and J. A. Carton, Interannual and decadal variability in the tropical and midlatitude Pacific Ocean, *J. Clim.*, *12*, 3402–3418, 1999.

Gu, D., and S. G. H. Philander, Interdecadal climate fluctuations that depend on exchange between the tropics and extratropics, *Science*, *275*, 805–807, 1997.

Han, W., J. P. McCreary, D. L. T. Anderson, and A. J. Mariano, Dynamics of the eastern surface jets in the equatorial Indian Ocean, *J. Phys. Oceanogr.*, *29*, 2191–2209, 1999.

Hasegawa T., and K. Hanawa, Decadal–scale variability of upper ocean heat content in the tropical Pacific, *Geophys. Res. Lett., 30,* doi:10.1029/2002GL016843, 2003.

Hasselmann, K., Stochastic climate models. I: Theory, *Tellus, 28,* 473–485, 1976.

Hastenrath, S., L. C. de Castro, and P. Aceituno, The Southern Oscillation in the tropical Atlantic sector, *Beitr. Phys. Atmos., 60,* 447–463, 1987.

Hayes, S. P., et al., TOGA–TAO: A moored array for real–time measurements in the tropical Pacific Ocean, *Bull. Am. Meteorol. Soc., 72,* 339–347, 1991.

Hirst, A. C., Unstable and damped equatorial modes in simple coupled ocean–atmosphere models, *J. Atmos. Sci., 43,* 606–630, 1986.

Horel, J. D., On the annual cycle of the tropical Pacific atmosphere and Ocean, *Mon. Weather Rev., 110,* 1863–1878, 1982.

Horel, J. D., and J. M. Wallace, Planetary–scale atmospheric phenomena associated with the Southern Oscillation, *Mon. Weather Rev., 109,* 813–829, 1981.

Houghton, R. W., Seasonal variations of the subsurface thermal structure in the Gulf of Guinea, *J. Phys. Oceanogr., 13,* 2070–2081, 1983.

Hu, H., and W. T. Liu, Oceanic thermal and biological responses to Santa Ana winds, *Geophys. Res. Lett., 30*(11), 1596, doi:10.1029/2003GL017159, 2003.

Hurrell, J. W., Y. Kushnir, G. Ottersen, and M. Visbeck (Eds.), *The North Atlantic Oscillation: Climatic Significance and Environmental Impact,* 279 pp., AGU Geophysical Monograph Series, Washington, D.C., 2003.

Jacobs, G. A., and J. L. Mitchell, Ocean circulation variations associated with the Antarctic circumpolar wave, *Geophys. Res. Lett., 23,* 2947–2950, 1996.

Jacobs, G. A., et al., Decade–scale trans-Pacific propagation and warming effects of an El Niño anomaly, *Nature, 370,* 360–363, 1994.

Jensen, T. G., Equatorial variability and resonance in a wind–driven Indian Ocean model, *J. Geophys. Res., 98,* 22533–22552, 1993.

Kalnay, E., et al., The NCEP/NCAR 40–year reanalysis project, *Bull. Am. Meteorol. Soc., 77,* 437–471, 1996.

Kawamura, R., T. Matsumura, and S. Iizuka, Role of equatorially asymmetric sea surface temperature anomalies in the Indian Ocean in the Asian summer monsoon and El Niño–Southern Oscillation coupling, *J. Geophys. Res., 106,* 4681–4693, 2001.

Kawamura, R., H. Aruga, T. Matsuura, and S. Iizuka, Two different regimes of anomalous Walker Circulation over the Indian and Pacific Oceans before and after the late 1970s, this volume.

Kelly, K. A., and S. Dong, The relationship of western boundary current heat transport and storage to mid-latitude ocean–atmosphere interaction, this volume.

Kiladis, G. N., and H. F. Diaz, Global climatic anomalies associated with extremes in the Southern Oscillation, *J. Clim., 2,* 1069–1090, 1989.

Kleeman, R., J. P. McCreary, and B. A. Klinger, A mechanism for generating ENSO decadal variability, *Geophys. Res. Lett., 26,* 1743–1746, 1999.

Klein, S. A., B. J. Soden, and N. C. Lau, Remote sea surface temperature variations during ENSO: Evidence for a tropical atmospheric bridge, *J. Clim., 12,* 917–932, 1999.

Kushnir, Y., et al., Atmospheric GCM response to extratropical SST anomalies: Synthesis and evaluation, *J. Clim., 15,* 2233–2256, 2002.

Lanzante, J. R., Lag relationships involving tropical SSTs, *J. Clim., 9,* 2568–2578, 1996.

Latif, M., Dynamics of interdecadal variability in coupled ocean–atmosphere models, *J. Clim., 11,* 602–624, 1998.

Latif, M., and T. P. Barnett, Interactions of the tropical oceans, *J. Clim., 8,* 952–964, 1995.

Latif, M., and T. P. Barnett, Decadal climate variability over the North Pacific and North Amercia: Dynamics and predictability, *J. Clim., 9,* 2407–2423, 1996.

Latif, M., et al., A review of the predictability and prediction of ENSO, *J. Geophys. Res., 103,* 14375–14393, 1998.

Lau, N.–C., and M. J. Nath, Impact of ENSO on the variability of the Asian–Australian monsoons as simulated in GCM experiments, *J. Clim., 13,* 4287–4309, 2000.

Lazier, J. R. N., Temperature and salinity changes in the deep Labrador Sea, 1962–1986, *Deep–Sea Res., 35,* 1247–1253, 1988.

Lengaigne, M., et al., Westerly wind events in the tropical Pacific and their influence on the coupled ocean–atmosphere system: A review, this volume.

Levitus, S., *Climatological Atlas of the World Ocean,* NOAA Prof. Paper No 13, U.S. Govt. Printing Office, 173 pp, 1982.

Liu, Z., and S. G. H. Philander, Tropical–extratropical oceanic exchange pathways, in *Ocean Circulation and Climate: Observing and Modeling the Global Ocean,* edited by G. Siedler, J. Church, and J. Gould, pp. 247–257, Academic Press, 2001.

Lohmann, G., Atmospheric and oceanic freshwater transport during weak Atlantic overturning circulation, *Tellus A, 55,* 438–449, 2003.

Lysne, J., P. Chang, and B. Giese, Impact of the extratropical Pacific on equatorial variability, *Geophys. Res. Lett., 24,* 2589–2592, 1997.

Manabe, S., and K. Bryan, Climate calculations with a combined ocean–atmosphere model, *J. Atmos. Sci., 26,* 786–789, 1969.

Manabe, S., and R. J. Stouffer, Century–scale effects of increased atmospheric CO_2 on the ocean–atmosphere system, *Nature, 364,* 215–218, 1993.

Mann, M. E., and J. Park, Joint spatiotemporal modes of surface temperature and sea level pressure variability in the northern hemisphere during the last century, *J. Clim., 9,* 2137–2162, 1996.

Mantua, N. J., and S. R. Hare, The Pacific decadal oscillation, *J. Oceanogr., 58,* 35–44, 2002.

McCreary, J. P., and D. L. T. Anderson, An overview of coupled ocean–atmosphere models of El Niño and the Southern Oscillation, *J. Geophys. Res., 96,* 3125–3150, 1991.

McCreary, J. P., and P. Lu, Interaction between the subtropical and equatorial ocean circulations: The subtropical cell, *J. Phys. Oceanogr., 24,* 466–497, 1994.

McPhaden, M. J., Variability in the central equatorial Indian Ocean. Part II: Oceanic heat and turbulent energy balance, *J. Mar. Res., 40*, 403–419, 1982.

McPhaden, M. J., and D. Zhang, Slowdown of the meridional overturning circulation in the upper Pacific Ocean, *Nature, 415*, 603–608, 2002.

McPhaden, M. J., et al., The Tropical Ocean–Global Atmosphere observing system: A decade of progress, *J. Geophys. Res., 103*, 14169–14240, 1998.

Mechoso, C. R., et al., The seasonal cycle over the tropical Pacific in general circulation models, *Mon. Weather Rev., 123*, 2825–2838, 1995.

Meehl, G. A., Global coupled models: Atmosphere, ocean, sea ice, in *Climate System Modeling*, edited by K. E. Trenberth, pp. 555–581, Cambridge University Press, 1992.

Miller, A. J., et al., Potential feedbacks between Pacific Ocean ecosystems and interdecadal climate variations, *Bull. Am. Meteorol. Soc., 84*, 617–633, 2003.

Mitchell, T. P., and J. M. Wallace, The annual cycle in equatorial convection and sea surface temperature, *J. Clim., 5*, 1140–1156, 1992.

Nakamura, H., T. Sampe, Y. Tanimoto, and A. Shimpo, Observed associations among storm tracks, jet streams, and midlatitude oceanic fronts, this volume.

Neelin, J. D., and H. A. Dijkstra, Ocean–atmosphere interaction and the tropical climatology. Part I: The dangers of flux correction, *J. Clim., 8*, 1325–1342, 1995.

Neelin, J. D., and W. J. Weng, Analytical prototypes for ocean–atmosphere interaction at midlatitudes. Part I: Coupled feedbacks as a sea surface temperature dependent stochastic process, *J. Clim., 12*, 697–721, 1999.

Neelin, J. D., M. Latif, and F.–F. Jin, Dynamics of coupled ocean–atmosphere models: The tropical problem, *Annu. Rev. Fluid Mech., 26*, 617–659, 1994.

Neelin, J. D., et al., ENSO theory, *J. Geophys. Res., 103*, 14,262–14,290, 1998.

Nigam, S., and Y. Chao, Evolution dynamics of tropical ocean–atmosphere annual cycle variability, *J. Clim., 9*, 3187–3205, 1996.

Nonaka, M., S.–P. Xie, and J. P. McCreary, Decadal variations in the subtropical cells and equatorial SST, *Geophys. Res. Lett., 29*, 1116, doi:10.1029/2001GL013717, 2002.

Okajima, H., S.–P. Xie, and A. Numaguti, Interhemispheric coherence of tropical climate variability: Effect of climatological ITCZ, *J. Meteorol. Soc. Jpn., 81*, 1371–1386, 2003.

Okumura, Y., and S.–P. Xie, Interaction of the Atlantic equatorial cold tongue and African monsoon, *J. Clim.*, revised, 2004.

Philander, S. G. H., *El Niño, La Niña, and the Southern Oscillation*, Academic Press, London, 289 pp., 1990.

Philander, S. G. H., and R. C. Pacanowski, A model of the seasonal cycle in the tropical Atlantic Ocean, *J. Geophys. Res., 91*, 14192–14206, 1986.

Philander, S. G. H., T. Yamagata, and R. C. Pacanowski, Unstable air–sea interactions in the Tropics, *J. Atmos. Sci., 41*, 604–613, 1984.

Philander, S. G. H., et al., Why the ITCZ is mostly north of the equator, *J. Clim., 9*, 2958–2972, 1996.

Pierce, D. W., T. Barnett, and M. Latif, Connections between the Pacific Ocean tropics and midlatitudes on decadal timescales, *J. Clim., 13*, 1173–1194, 2000.

Raymond, D. J., et al., EPIC2001 and the coupled ocean–atmosphere system of the tropical east Pacific, *Bull. Am. Meteorol. Soc.*, in press, 2004.

Ramstorf, S., The thermohaline ocean circulation—a system with dangerous thresholds? *Clim. Changes, 46*, 247–256, 2000.

Ramstorf, S., Ocean circulation and climate during the past 120,000 years, *Nature, 419*, 207–214, 2002.

Reynolds, R. W., and T. M. Smith, Improved global sea surface temperature analyses using optimal interpolation, *J. Clim., 7*, 929–948, 1994.

Saji, N. H., and T. Yamagata, Possible impacts of Indian Ocean dipole mode events on global climate, *Climate Res., 25*, 151–169, 2003.

Saji, N. H., B. N. Goswami, P. N. Vinayachandran, and T. Yamagata, A dipole mode in the tropical Indian Ocean, *Nature, 401*, 360–363, 1999.

Saji, N. H., et al., Observations of intraseasonal air–sea coupling in the near equatorial Indian Ocean using remote sensed and in-situ data, *J. Clim.*, submitted, 2004.

Saravanan, R., and P. Chang, Thermodynamic coupling and predictability of tropical sea surface temperature, this volume.

Schneider, N., and A. J. Miller, Predicting western North Pacific Ocean climate, *J. Clim., 14*, 3997–4002, 2001.

Schneider, N. S., A. J. Miller, M. A. Alexander, and C. Deser, Subduction of decadal North Pacific temperature anomalies: Observations and dynamics, *J. Phys. Oceanogr., 29*, 1056–1070, 1999.

Schott, F. A., J. P. McCreary, and G. C. Johnson, Shallow overturning circulations of the tropical–subtropical oceans, this volume.

Sengupta, D., B. N. Goswami, and R. Senan, Coherent intraseasonal oscillations of ocean and atmosphere during the Asian summer monsoon, *Geophys. Res. Lett., 28*, 4127–4130, 2001.

Small, R. J., et al., Numerical simulation of boundary layer structure and cross–equatorial flow in the eastern Pacific, *J. Atmos. Sci.*, submitted, 2004.

Su, H., J. D. Neelin, and J. E. Meyerson, Tropical tropospheric temperature and precipitation response to sea surface temperature forcing, this volume.

Sun, D., and Z. Liu, Dynamic ocean–atmosphere coupling: A thermostat for the tropics, *Science, 272*, 1148–1150, 1996.

Takayabu, Y. N., et al., Abrupt termination of the 1997–98 El Niño in response to a Madden–Julian Oscillation, *Nature, 402*, 279–282, 1999.

Thompson, D. W. J., and J. M. Wallace, Annular modes in the extratropical circulation. Part I: Month–to–month variability, *J. Clim., 13*, 1000–1016, 2000.

Tomita, T., S.–P. Xie, and M. Nonaka, Estimates of surface and subsurface forcing for decadal sea surface temperature variability in the mid-latitude North Pacific, *J. Meteorol. Soc. Jpn., 80*, 1289–1300, 2002.

Tourre, Y. M., and W. B. White, ENSO signals in global upper–ocean temperature, *J. Phys. Oceanogr., 25*, 1317–1332, 1995.

Trenberth, K. E., and J. W. Hurrell, Decadal atmosphere–ocean variations in the Pacific, *Clim. Dyn., 9*, 303–319, 1994.

Trenberth, K. E., et al., Progress during TOGA in understanding and modeling global teleconnections associated with tropical sea surface temperature, *J. Geophys. Res., 103*, 14291–14324, 1998.

Vecchi, G. A., and D. E. Harrison, Monsoon breaks and subseasonal sea surface temperature variability in the Bay of Bengal, *J. Clim., 15*, 1485–1493, 2002.

Vecchi, G. A., S.–P. Xie, and A. S. Fischer, Ocean–atmosphere covariability in the western Arabian Sea, *J. Clim., 17*, 1213–1224, 2004.

Vimont, D. J., D. S. Battisti, and A. C. Hirst, The seasonal footprinting mechanism in the CSIRO general circulation models, *J. Clim., 16*, 2653–2667, 2003.

Walker, G. T., Correlation in seasonal variations of weather. IX. A further study of world weather, *Mem. Indian Meteor. Dept., 25*, 275–332, 1924.

Wallace, J. M., and D. S. Gutzler, Teleconnections in the geopotential height field during the Northern Hemisphere winter, *Mon. Weather Rev., 109*, 784–812, 1981.

Wang, B., R. W. Wu, and X. Fu, Pacific–East Asian teleconnection: How does ENSO affect East Asian climate? *J. Clim., 13*, 1517–1536, 2000.

Wang, C., Atmospheric circulation cells associated with the El Niño–Southern Oscillation, *J. Clim., 15*, 399–419, 2002.

Wang, C., ENSO, Atlantic climate variability, and the Walker and Hadley circulations, In *The Hadley Circulation: Past, Present and Future*, edited by H. F. Diaz and R. S. Bradley, Kluwer Academic Publishers, in press, 2004.

Wang, C., and D. B. Enfield, A further study of the tropical Western Hemisphere warm pool, *J. Clim., 16*, 1476–1493, 2003.

Wang, C., and J. Picaut, Understanding ENSO physics—A review, this volume.

Wang, C., R. H. Weisberg, and J. I. Virmani, Western Pacific interannual variability associated with the El Niño–Southern Oscillation, *J. Geophys. Res., 104*, 5131–5149, 1999.

Watanabe, M., and F.–F. Jin, A moist linear baroclinic model: Coupled dynamical–convective response to El Niño, *J. Clim., 16*, 1121–1139, 2003.

Webster, P. J., et al., Monsoon: Processes, predictability, and the prospects for prediction, *J. Geophys. Res., 103*, 14451–14510, 1998.

White, W. B., and R. G. Peterson, An Antarctic circumpolar wave in surface pressure, wind, temperature and sea–ice extent, *Nature, 380*, 699–702, 1996.

White, W. B., S.–C. Chen, R. J. Allan, and R. C. Stone, Positive feedback between the Antarctic circumpolar wave and the global El Niño–Southern Oscillation wave, *J. Geophys. Res., 107(C10)*, 3165, doi:10.1029/2000JC000581, 2002.

White, W. B., Y. M. Tourre, M. Barlow, and M. Dettinger, A delayed action oscillator shared by biennal, interannual, and decadal signals in the Pacific basin, *J. Geophys. Res., 108,* 3070, doi: 10.1029/2002JC001490, 2003.

Woodruff, S. D., R. J. Slutz, R. L. Jenne, and P. M. Steurer, A comprehensive ocean–atmosphere dataset, *Bull. Am. Meteorol. Soc., 68*, 521–527, 1987.

Wyrtki, K., An equatorial jet in the Indian Ocean, *Science, 181*, 262–264, 1973.

Xie, P., and P. A. Arkin, Analyses of global monthly precipitation using gauge observations, satellite estimates, and numerical model predictions, *J. Clim., 9*, 840–858, 1996.

Xie, S.–P., On the genesis of the equatorial annual cycle, *J. Clim., 7*, 2008–2013, 1994.

Xie, S.–P., Satellite observations of cool ocean–atmosphere interaction, *Bull. Am. Meteorol. Soc., 85*, 195–208, 2004a.

Xie, S.–P., The shape of continents, air–sea interaction, and the rising branch of the Hadley circulation, In *The Hadley Circulation: Past, Present and Future*, edited by H. F. Diaz and R. S. Bradley, Kluwer Academic Publishers, in press, 2004b.

Xie, S.–P., and J. A. Carton, Tropical Atlantic variability: Patterns, mechanisms, and impacts, this volume.

Xie, S.–P., and S. G. H. Philander, A coupled ocean–atmosphere model of relevance to the ITCZ in the eastern Pacific, *Tellus, 46A*, 340–350, 1994.

Xie, S.–P., Y. Tanimoto, H. Noguchi, and T. Matsuno, How and why climate variability differs between the tropical Pacific and Atlantic, *Geophys. Res. Lett., 26*, 1609–1612, 1999.

Xie, S.–P., H. Annamalai, F. A. Schott, and J. P. McCreary, Structure and mechanisms of South Indian Ocean climate variability, *J. Clim., 15*, 864–878, 2002.

Yamagata, T., et al., Coupled ocean–atmosphere variability in the tropical Indian Ocean, this volume.

Yasunari, T., Zonally propagating modes of the global east–west circulation associated with the Southern Oscillation, *J. Meteorol. Soc. Jpn., 63*, 1013–1029, 1985.

Yukimoto, S., et al., ENSO–like interdecadal variability in the Pacific Ocean as simulated in a coupled general circulation model, *J. Geophys. Res., 105*, 13945–13963, 2000.

Zebiak, S. E., Air–sea interaction in the equatorial Atlantic region, *J. Clim., 6*, 1567–1586, 1993.

Zebiak, S. E., and M. A. Cane, A model El Niño–Southern Oscillation, *Mon. Weather Rev., 115*, 2262–2278, 1987.

Zhang, D., M. J. McPhaden, and W. E. Johns, Observational evidence for flow between the subtropical and tropical Atlantic: The Atlantic subtropical cells, *J. Phys. Oceanogr., 33*, 1783–1797, 2003.

J. A. Carton, Department of Meteorology, University of Maryland, 3433 Computer and Space Science Bldg., College Park, Maryland 20742. (carton@atmos.umd.edu)

C. Wang, NOAA/AOML, Physical Oceanography Division, 4301 Rickenbacker Causeway, Miami, Florida 33149. (Chunzai.Wang@noaa.gov)

S.–P. Xie, IPRC/SOEST, University of Hawaii at Manoa, 1680 East–West Road, Honolulu, Hawaii 96822. (xie@hawaii.edu)

Understanding ENSO Physics—A Review

Chunzai Wang

NOAA Atlantic Oceanographic and Meteorological Laboratory, Miami, Florida

Joël Picaut

Institut de Recherche pour le Développement, LEGOS, Toulouse, France

Since the TOGA program, and in particular the maintenance of its observing system in the tropical Pacific, significant progress has been made in the understanding of ENSO. ENSO has been viewed as a self-sustained and naturally oscillatory mode or a stable mode triggered by stochastic forcing. Whatever the case, El Niño involves Bjerknes' positive ocean-atmosphere feedback that culminates with warm SST anomalies in the equatorial eastern and central Pacific. After an El Niño reaches its mature phase, negative feedbacks are required to terminate the growth of warm SST anomalies. Four major negative feedbacks have been proposed: wave reflection at the ocean western boundary, a discharge process due to Sverdrup transport, a western Pacific wind-forced Kelvin wave of opposite sign, and anomalous zonal advection. These negative feedbacks may work in varying combinations to terminate El Niño, and reverse it into La Niña.

The seasonal cycle can contribute to the irregularity and phase-locking of ENSO, and the intraseasonal variability can be a source of both ENSO's variability and irregularity. Tropical Pacific decadal-multidecadal variability and warming trends may modulate ENSO. Many mechanisms have been proposed to explain tropical Pacific decadal-multidecadal variability, and they are categorized by their tropical origins and tropical-extratropical connections. Mechanisms of tropical origins include stochastic forcing, interactions between the seasonal and interannual cycles, internal nonlinearity, asymmetry between El Niño and La Niña, and local ocean-atmosphere interaction, while those of tropical-extratropical connections involve oceanic bridges, wave propagation, and atmospheric bridges. Difficulties and uncertainties of studies on low-frequency variability and interpretation of warming trends, global warming, and ENSO are also discussed.

1. INTRODUCTION

At the end of the 19th century, the term El Niño was used to denote the annual occurrence of a warm ocean current that flowed southward along the west coast of Peru and Ecuador around Christmas. The Peruvian geographers noted that in some years the onset of warm conditions was stronger than usual and was accompanied by unusual oceanic and climatic phenomena. Starting with the arrival of foreign-based scientific expeditions off Peru in the early 20th century, the concept of referring to these unusual events as El Niño gradually spread through the world's scientific community. It wasn't until the 1950s/1960s that scientists realized that El Niño is far more than a coastal phenomenon, and that it is associated with basin-scale warming in the tropical Pacific Ocean. Sir Gilbert Walker in the 1920s and 1930s found that notable climate anomalies occur around the world every few years, associated with what he called the Southern Oscillation [*Walker*, 1923, 1924; *Walker and Bliss*, 1932]. The Southern Oscillation is characterized by an interannual seesaw in tropical sea level pressure (SLP) between the Western and Eastern Hemispheres, consisting of a weakening and strengthening of the easterly trade winds over the tropical Pacific. *Bjerknes* [1969] recognized that there is a close connection between El Niño and the Southern Oscillation (ENSO) and they are two different aspects of the same phenomenon.

Bjerknes hypothesized that a positive ocean-atmosphere feedback involving the Walker circulation is a cause of ENSO. An initial positive sea surface temperature (SST) anomaly in the equatorial eastern Pacific reduces the east-west SST gradient and hence the strength of the Walker circulation [*Gill*, 1980; *Lindzen and Nigam*, 1987], resulting in weaker trade winds around the equator. The weaker trade winds in turn drive the ocean circulation changes that further reinforce SST anomaly. This positive ocean-atmosphere feedback leads the equatorial Pacific to a warm state, i.e., the warm phase of ENSO (El Niño). At that time, Bjerknes did not know what causes a turnabout from a warm phase to a cold phase, which has been recently named La Niña [*Philander*, 1990].

After Bjerknes' seminal work, ENSO was not intensively studied until the 1980s. The intense warm episode of the 1982–83 El Niño, which was not recognized until it was well developed, galvanized the scientific community in an effort to understand and predict ENSO. The 1982–83 El Niño onset was not consistent with the prior "buildup" of sea level in the western Pacific by stronger than normal trade winds prior to 1982, presumed to be a necessary precursor of El Niño [*Wyrtki*, 1975]. Also, there was no warming off the west coast of South America in early 1982, considered to be part of the normal sequence of events characterizing the evolution of El Niño [e.g., *Rasmusson and Carpenter*, 1982]. Building on the earlier efforts of the Equatorial Pacific Ocean Climate Studies (EPOCS) program, this motivated a ten-year international Tropical Ocean-Global Atmosphere (TOGA) program (1985–1994) to study ENSO. TOGA built an ocean observing system in the tropical Pacific Ocean, conducted theoretical and diagnostic studies of the ENSO phenomenon, and developed a hierarchy of coupled ocean-atmosphere models to study and predict ENSO. A special volume of the *Journal of Geophysical Research* (volume 103, June 1998) provided a comprehensive review of observations, theory, modeling, and predictability of ENSO during the TOGA decade (also see ENSO reviews of *Enfield* [1989]; *Philander* [1990]; *McCreary and Anderson* [1991]; *Battisti and Sarachik* [1995]). The present paper reviews progress in ENSO understanding, with a major focus on development after the TOGA decade. However, for the sake of continuity it also briefly summarizes the progress made before and during the TOGA decade.

ENSO's low-frequency modulation and the relation of high-frequency influences to ENSO are recent research topics of ENSO. The 1997–98 El Niño was characterized by exceptionally strong high-frequency wind variability during the onset phase. Numerical models, which succeeded in predicting the onset of the 1997–98 El Niño, were unable to forecast its intensity [e.g., *Barnston et al.*, 1999; *Landsea and Knaff*, 2000] until the March 1997 westerly wind burst (WWB) was incorporated. This may suggest the importance of the intraseasonal variability (the WWB and the Madden-Julian Oscillation), and has stimulated scientists to further investigate the roles of high-frequency forcing in ENSO.

ENSO is an irregular oscillation, both in frequency and amplitude. Its recurrence varies usually between two and seven years. Furthermore, its characteristics are modulated on decadal and multidecadal timescales [e.g., *Enfield and Cid-Serrano*, 1991; *Mokhov et al.*, 2000]. In terms of amplitude, there are periods (decades or longer) during which ENSO is more energetic or there are more El Niños than La Niñas (e.g., since the mid-1970s), and vice versa. Such features can be viewed as a nearly regular ENSO oscillation superimposed on natural decadal and multidecadal oscillations and on a warming trend [*Lau and Weng*, 1999; *Cai and Whetton*, 2001a; *Philander and Fedorov*, 2003]. It can also be viewed as a chaotic/irregular ENSO, or stochastic fluctuations, or nonlinear modulation by a changing background state. Decadal-multidecadal variability of ENSO appears to influence the global atmospheric circulation [*Diaz et al.*, 2001], and thus the climate over many parts of the world [*Power et al.*, 1999; *Torrence and Webster*, 1999; *Janicot et al.*, 2001; *Gershunov and Barnett*, 1998]. Such variability appears to alter the ocean productivity of the Pacific Ocean [*Chavez et al.*, 2003] and ENSO predictability [e.g., *Balmaseda et al.*, 1995; *Flugel and Chang*, 1998; *Kirtman and Schopf*, 1998]. Therefore, many studies, especially after the TOGA decade, have focused on ENSO's low-frequency modulation.

The present paper is organized as follows. Section 2 briefly describes observations of ENSO. Section 3 reviews our present understanding of ENSO mechanisms. Section 4 briefly summarizes the effects of high-frequency variability on ENSO. Section 5 reviews a newly and recently developed facet of ENSO, its low-frequency modulation. The paper ends in Section 6 with some discussions and ideas for the future.

2. OBSERVATIONS OF ENSO

2.1. The ENSO Observing System

The backbone of the ENSO observing system (Plate 1) is the TAO (Tropical Atmosphere Ocean) array of about 70 moored buoys [Hayes et al., 1991; McPhaden, 1995]. Most of them are equipped with a 500-m thermistor chain and meteorological sensors. At the equator five to seven moorings are equipped with ADCP (Acoustic Doppler Current Profiler) and current meters [McPhaden, 1995]. Developed during TOGA as a multinational program among France, Japan, South Korea, Taiwan, and United States, this array is now supported by the US with the dedicated R/V Ka'imimoana and by Japan with their TRITON program (hence the official name of TAO/TRITON since January 1, 2000). The ocean observing system is completed by a Voluntary Observing Ship (VOS) program, an island tide-gauge network, and a system of surface drifters. All the data are transmitted in near-real time to the Global Telecommunication System, for research and prediction purposes. A set of meteorological and oceanographic satellites complete all these measurements, with particular emphasis on the TOPEX/Poseidon altimeter that has proved especially useful in observing and analyzing tropical ocean variability [Picaut and Busalacchi, 2001]. Detailed information about the ENSO observing system, such as key variables, sampling requirements and uncertainties, can be found in McPhaden et al. [1998].

The obvious parameters for observing the ENSO coupled phenomenon are surface wind stress and SST (through a combination of satellite and in situ data) but the subsurface temperature has proved to be surprisingly useful in diagnosing ENSO. The basic 2–7 year period of ENSO is set by the thermal inertia of the ocean upper layer. Most of the heat content variability in low-latitude oceans is situated in this layer, and thus is directly reflected in sea-level height. Hence, measurements of the upper ocean thermal field and sea level are also fundamental to ENSO. Upper-layer temperature is mostly controlled by a specific low-latitude dynamics (i.e., equatorial waves), and current measurements are needed, especially near the equator with the vanishing of geostrophy (Coriolis force). Subsurface temperature data can provide estimates of equatorial upwelling and mixing above the thermocline, which are greatly needed. Although less important than in non-equatorial latitudes, surface heat fluxes are also required.

The TOGA observing system was devoted to the large-scale monitoring of the upper tropical oceans, with emphasis on the tropical Pacific. However, there has been considerable controversy regarding the physics that maintain and perturb the western Pacific warm pool, which is believed to be a center of action for ENSO. Hence, a multinational oceanography-meteorology experiment was conceived and carried out in 1992–93, with an intensive observation period (November 1992–February 1993) embedded into a yearlong period of enhanced monitoring. Twelve research vessels, seven research aircrafts, numerous ground-based stations, and additional moored and drifting buoys have collected a unique set of data. The plans for the TOGA Coupled Ocean-Atmosphere Response Experiment (COARE) are listed in Webster and Lukas [1992], and the results are summarized in Godfrey et al. [1998].

2.2. Lessons From the ENSO Observing System and Further Needs

The biggest achievement of TOGA was the installation for the first time of an ocean observing system. It improved the understanding and modeling of ENSO, and proved its prediction capability through the evidence of subsurface memory [Latif et al., 1998]. It provided since 1985 a set of high-quality data, which associated with older and less reliable data showed that El Niño behaves differently over the last decades (Plate 2). The warm SST anomalies associated with El Niño events between 1950 and 1976 first peaked along the South American coast in the boreal spring of the El Niño year and then propagated westward [Rasmusson and Carpenter, 1982]. The El Niño events between 1976 and 1996 seemed to start from the equatorial western and central Pacific, and the coastal warming occurred in the boreal spring subsequent to the El Niño year rather than in the boreal spring of the El Niño year [Wang, 1995a; Wang and An, 2002]. The 1997–98 El Niño started in both the central Pacific and the South American coast during the spring of 1997 and the 2002–03 El Niño started and remained in the equatorial central Pacific. Why El Niños started differently in the last five decades is not understood yet. Being the most likely factor, the role of high- and low-frequency variabilities will be discussed in the following sections.

The duration of TOGA-COARE was not sufficient to understand the link between the intraseasonal westerly winds, such as the Madden-Julian Oscillation (MJO) and Westerly Wind Bursts (WWBs), and El Niño [Lengaigne et al., this volume]. With the discovery of the salinity stratified barrier-layer in the western Pacific warm pool [Lukas and Lindstrom, 1991] and the possibility that it influences the development of El Niño [Maes et al., 2002], there is a strong need for more salinity measurements and in particular sea surface salinity (SSS). The end of TOGA was marked by the progressive replacement of bucket samples on VOS routes by thermosalinograph. Together with satellite missions such as SMOS and Aquarius, these in situ SSS measurements will undeniably improve the ENSO observing system [Lagerloef and Delcroix, 2001].

As compared to the western Pacific warm pool, the eastern tropical Pacific was somewhat neglected during TOGA, and the 5-year experiment EPIC (Eastern Pacific Investigation of Climate processes) was launched in 1999. This experiment was designed to improve the understanding of the intertropical convergence zone (ITCZ), its interaction with the cold water originating from the equatorial upwelling, and the physics of the stratus cloud deck that forms over the cold water off South America [*Cronin et al.*, 2002].

As discussed in Section 5, understanding of the low-frequency variations of ENSO requires an expansion of the present ENSO observing system and its extension toward the western boundary and beyond the tropics. A main goal of the Pacific Basin Extended Climate Study (PBECS) is to provide sufficient additional in situ and satellite observations to constrain data-assimilating models well enough that the processes affecting decadal modulation of ENSO can be studied in detail [*Kessler et al.*, 2001]. This will require a whole set of additional measurements, such as repeated high-resolution expendable and hydrographic sections, several process experiments, and the integration with the Argo program of profiling floats [*Roemmich et al.*, 2001] and the Global Ocean Data Experiment (GODAE) [*Smith et al.*, 2001]. All these efforts are part of the CLIVAR (Climate Variability and Predictability) program.

It will be long before these observing systems and experiments produce sufficient high-quality observations to explain the decadal modulation of ENSO. This strengthens the need for historical and paleoclimate records of ENSO, with coral, tree-ring, tropical ice core, sediment or other proxies [e.g., *Ortlieb*, 2000; *Mann et al.*, 2000; *Markgraf and Diaz*, 2000; *Tudhope et al.*, 2001]. Associated with specific model studies, these records will help understand the evolution of ENSO in the past, present, and future.

3. ENSO MECHANISMS

The theoretical explanations of ENSO can be loosely grouped into two frameworks. First, El Niño is one phase of a self-sustained, unstable, and naturally oscillatory mode of the coupled ocean-atmosphere system. Second, El Niño is a stable (or damped) mode triggered by atmospheric random "noise" forcing. Whatever the case, ENSO involves the positive ocean-atmosphere feedback of *Bjerknes* [1969]. The early idea of *Wyrtki's* [1975] sea level "buildup" in the western Pacific warm pool treats El Niño as an isolated event. Wyrtki suggested that prior to El Niño, the easterly trade winds strengthened, and there was a "buildup" in sea level in the western Pacific warm pool. A "trigger" is a rapid collapse of the easterly trade winds. When this happens, the accumulated warm water in the western Pacific would surge eastward in the form of equatorial downwelling Kelvin waves to initiate an El Niño event. The recent studies have suggested atmospheric stochastic forcing as important "triggers" of El Niño. On the other hand, numerical models with tunable model parameters suggest that ENSO is a self-sustained mode of the coupled ocean-atmosphere system in some parameter regimes. Additionally, many studies have shown that the ocean-atmosphere coupling can produce slow modes that can explain both eastward and westward propagating events.

3.1. ENSO Oscillator Models

Bjerknes [1969] was the first to hypothesize that a positive ocean-atmosphere feedback causes El Niño. Although the starting point is arbitrary, an initial positive SST anomaly in the equatorial eastern Pacific reduces the east-west SST gradient and hence the strength of the Walker circulation, resulting in weaker trade winds along the equator. The weaker trade winds in turn drive the ocean circulation changes that further reinforce SST anomaly. This positive feedback leads the equatorial Pacific to a warm state. For the coupled system to oscillate, a negative feedback is needed to turn the warm state around. Since the 1980s, four major negative feedbacks have been proposed: (1) wave reflection at the western boundary, (2) a discharge process, (3) a western Pacific wind-forced Kelvin wave, and (4) anomalous zonal advection. These negative feedbacks correspond to the delayed oscillator [*Suarez and Schopf*, 1988; *Battisti and Hirst*, 1989], the recharge oscillator [*Jin*, 1997], the western Pacific oscillator [*Weisberg and Wang*, 1997a; *Wang et al.*, 1999b], and the advective-reflective oscillator [*Picaut et al.*, 1997]. Since these models also operate for initial negative SST, they can produce ENSO-like oscillations.

With the different conceptual oscillator models capable of producing ENSO-like oscillations, more than one may operate in nature. Motivated by the existence of different oscillator models, *Wang* [2001a, b] formulated and derived a unified ENSO oscillator from the dynamics and thermodynamics of the coupled ocean-atmosphere system that is similar to the *Zebiak and Cane* [1987] coupled model. The unified oscillator includes the physics of all ENSO oscillator models (Figure 1). As suggested by the unified oscillator, ENSO may be a multi-mechanism phenomenon (see *Picaut et al.* [2002] for observational evidence) and the relative importance of different mechanisms may be time-dependent. Observations show that ENSO displays both eastern and western Pacific interannual anomaly patterns [e.g., *Rasmusson and Carpenter*, 1982; *Weisberg and Wang*, 1997b; *Mayer and Weisberg*, 1998; *Wang et al.*, 1999b; *McPhaden*, 1999; *Wang and Weisberg*, 2000; *Vialard et al.*, 2001]. Thus, the unified oscillator considers both eastern and western Pacific anomaly variations:

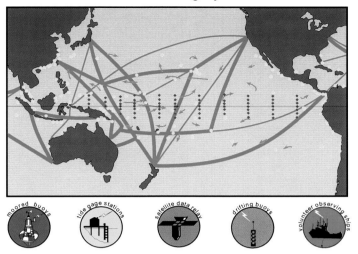

Plate 1. In situ components of the ENSO observing system. The four major elements are the TAO/TRITON array of moored buoys (red diamonds), an island tide-gauge network (yellow circles), surface drifters (arrows), and the volunteer ship program (blue lines). Various satellites are intensively used to complement the in situ network. This ensemble of instruments delivers in near-real time data on surface and subsurface temperature and salinity, wind speed and direction, sea level, and current velocity (Courtesy of Michael J. McPhaden, TAO Project Office).

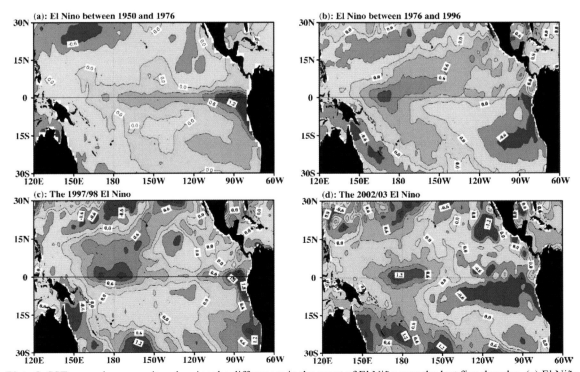

Plate 2. SST anomaly composites showing the differences in the onset of El Niño over the last five decades. (a) El Niño between 1950 and 1976, (b) El Niño between 1977 and 1996, (c) the 1997–98 El Niño, and (d) the 2002–03 El Niño. The composites are calculated by averaging the SST anomalies during March–May of the El Niño year. Since the 2002–03 El Niño started earlier, its composite used the SST anomalies of December 2001 to February 2002.

The Unified Oscillator for ENSO

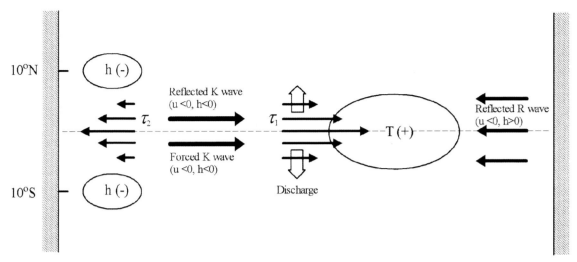

Figure 1. Schematic diagram of the unified oscillator for ENSO. Bjerknes positive ocean-atmosphere feedback leads the equatorial central/eastern Pacific to a warm state (El Niño). Four negative feedbacks, required to turn the warm state around, are (1) reflected Kelvin wave at the ocean western boundary, (2) discharge process due to Sverdrup transport, (3) western Pacific wind-forced Kelvin wave, and (4) reflected Rossby wave at the ocean eastern boundary. These negative feedbacks correspond to the delayed oscillator, the recharge oscillator, the western Pacific oscillator, and the advective-reflective oscillator. The unified oscillator suggests that all of the four negative feedbacks may work together in terminating El Niño warming. The four ENSO oscillators are special cases of the unified oscillator.

$$\frac{dT}{dt} = a\tau_1 - b_1\tau_1(t-\eta) + b_2\tau_2(t-\delta) - b_3\tau_1(t-\mu) - \varepsilon T^3, \quad (1a)$$

$$\frac{dh}{dt} = -c\tau_1(t-\lambda) - R_h h, \quad (1b)$$

$$\frac{d\tau_1}{dt} = dT - R_{\tau 1}\tau_1, \quad (1c)$$

$$\frac{d\tau_2}{dt} = eh - R_{\tau 2}\tau_2, \quad (1d)$$

where T is SST anomaly in the equatorial eastern Pacific, h is thermocline depth anomaly in the off-equatorial western Pacific, and τ_1 and τ_2 are zonal wind stress anomalies in the equatorial central Pacific and in the equatorial western Pacific, respectively. The parameters a, b_1, b_2, b_3, c, d, and e are constants. The parameters η, δ, μ, and λ represent the delay times. The parameters ε, R_h, $R_{\tau 1}$, and $R_{\tau 2}$ are damping coefficients.

The first term on the right-hand side (RHS) of equation (1a) represents the positive feedback in the coupled system. The second term represents the negative feedback due to wave reflection at the western boundary. The third term represents the negative feedback due to the wind-forced wave contribution in the equatorial western Pacific. The fourth term represents the effect of wave reflection at the eastern boundary. The last term is a cubic damping term that does not affect oscillatory behavior, but it limits anomaly growth [*Battisti and Hirst*, 1989; *Wang*, 2001a]. Equation (1b) states that the off-equatorial western Pacific thermocline anomaly is controlled by the wind stress in the equatorial central Pacific, with a damping rate of R_h. Equation (1c) shows that zonal wind stress anomaly in the equatorial central Pacific is related to the eastern Pacific SST anomaly, and equation (1d) states that the zonal wind stress anomaly in the equatorial western Pacific is related to the off-equatorial western Pacific thermocline anomaly. By further simplifications and assumptions, the unified oscillator can reduce to the different ENSO oscillators.

3.1.1. The delayed oscillator. The delayed oscillator (Figure 2) does not consider the coupled role of the western Pacific in ENSO and wave reflection at the eastern boundary. By setting $b_2 = 0$ and $b_3 = 0$ in equation (1a), the western Pacific variables τ_2 and h are decoupled from the coupled system. If we further drop the time derivative of equation (1c), the unified oscillator reduces to:

$$\frac{dT}{dt} = \frac{ad}{R_{\tau 1}}T - \frac{b_1 d}{R_{\tau 1}}T(t-\eta) - \varepsilon T^3. \quad (2)$$

Equation (2) is the delayed oscillator of *Suarez and Schopf* [1988] and *Battisti and Hirst* [1989]. The first term on RHS of equation (2) represents the positive feedback by ocean-atmosphere coupling in the equatorial eastern Pacific, i.e., the Bjerknes feedback. The second term is the delayed nega-

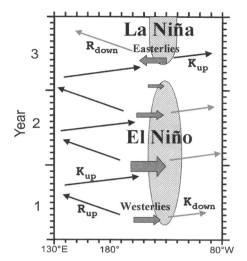

Figure 2. Schematic diagram of the delayed oscillator for ENSO. Positive SST anomalies in the equatorial eastern Pacific cause westerly wind anomalies that drive Kelvin waves eastward and act to increase the positive SST anomalies. The westerly wind anomalies also generate oceanic equatorial Rossby waves, which propagate westward and eventually reflect from the western boundary as equatorial Kelvin waves. Since the thermocline anomalies for the reflected Kelvin waves have an opposite sign to those of the directly forced Kelvin waves, they provide a negative feedback for the coupled system to oscillate.

tive feedback by free equatorial Rossby waves generated in the eastern Pacific coupling region that propagate to and reflect from the western boundary, returning as equatorial Kelvin waves to reverse the anomalies in the eastern Pacific coupling region.

The earliest idea of explaining the oscillatory nature of ENSO was proposed by *McCreary* [1983], based on the reflection of subtropical oceanic Rossby waves at the western boundary. *McCreary* [1983] and *McCreary and Anderson* [1984] explored shallow water ocean dynamics coupled to wind stress patterns that are changed by a discontinuous switch depending on thermocline depth. They showed how oceanic Rossby waves might be involved in generating the interannual oscillations associated with ENSO. In spite of the use of a discontinuous switch in their atmosphere and of reflection of subtropical Rossby waves, the idea of basin adjustment processes has been incorporated by later work. *Suarez and Schopf* [1988] introduced the delayed oscillator model of equation (2) as a candidate mechanism for ENSO. Based on the coupled model of *Zebiak and Cane* [1987], *Battisti and Hirst* [1989] formulated and derived a version of the *Suarez and Schopf* [1988] delayed oscillator model.

Graham and White [1988] presented sparse evidence of off-equatorial Rossby waves and their reflection at the western boundary and then empirically constructed a conceptual oscillator model for ENSO. As shown in *McCreary and Anderson* [1991], the conceptual equations of the *Graham and White* model can be reduced to a single equation that is similar to equation (2) (also see the comments of *Neelin et al.* [1998]).

The work of *McCreary* [1983], *McCreary and Anderson* [1984], and *Graham and White* [1988] emphasized the reflection of off-equatorial Rossby waves at the western boundary whose importance for ENSO has been debated. *Kessler* [1991] and *Battisti* [1989, 1991] argue that the equatorial Kelvin wave results primarily from the reflection of the gravest Rossby wave mode and that off-equator (poleward of ±8°) variations should not be a major factor in ENSO. In contrast, *Graham and White* [1991] contend that coupled model simulations of ENSO are greatly altered if effects poleward of ±8° are neglected. However, all of these studies recognized that wave reflection at the western boundary is important in terminating El Niño. *Li and Clarke* [1994] challenged the validation of the delayed oscillator by noting a low lag correlation between the western Pacific equatorial Kelvin wave amplitude and zonal wind forcing that is inconsistent with the delayed oscillator theory. *Mantua and Battisti* [1994] argued that wave reflection at the western boundary did account for the termination of El Niño and that the low lag correlation is due to irregularity of ENSO. In any case, the reflection efficiency of the western boundary is disrupted by the presence of the throughflow and numerous islands. Its estimation from simple models or observations [e.g., *Clarke*, 1991; *Zang et al.*, 2002; *Boulanger et al.*, 2003] is complicated by the coupled nature of the equatorial western Pacific and not enough observations.

3.1.2. The recharge oscillator. The recharge oscillator (Figure 3) considers variations of eastern Pacific SST and western Pacific thermocline anomalies. As argued by *Jin* [1997], equatorial wave dynamics are important in the adjustment of equatorial ocean, but wave propagations are not explicit in the recharge model. If the time derivatives in equations (1c) and (1d) are dropped and all delay parameters are set to zero (i.e., $\eta = 0$, $\delta = 0$, and $\lambda = 0$) and $b_3 = 0$, the unified oscillator reduces to:

$$\frac{dT}{dt} = \frac{ad - b_1 d}{R_{\tau 1}} T + \frac{b_2 e}{R_{\tau 2}} h - \varepsilon T^3, \quad (3a)$$

$$\frac{dh}{dt} = -\frac{cd}{R_{\tau 1}} T - R_h h. \quad (3b)$$

The mathematical form of equation (3) is the same as the recharge oscillator of *Jin* [1997]. In the Jin's recharge oscillator model, h is the thermocline anomaly in the equatorial western Pacific.

Wyrtki [1975] first suggested a buildup in the western Pacific warm water as a necessary precondition to the development

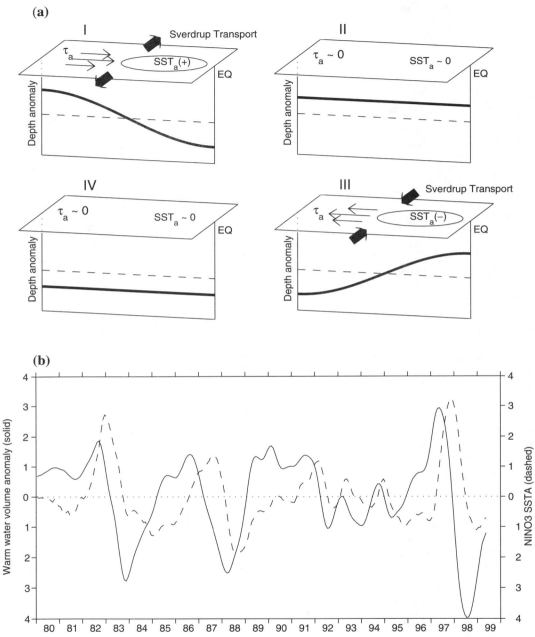

Figure 3. Schematic diagram of the recharge oscillator for ENSO. (a) The four phases of the recharge oscillation: (I) the warm phase, (II) the warm to cold transition phase, (III) the cold phase, and (IV) the cold to warm transition phase. During the warm phase of ENSO, the divergence of Sverdrup transport associated with equatorial central Pacific westerly wind anomalies and equatorial eastern Pacific warm SST anomalies results in the discharge of equatorial heat content. The discharge of equatorial heat content leads to a transition phase in which the entire equatorial Pacific thermocline depth is anomalously shallow due to the discharge of equatorial heat content. This anomalous shallow thermocline at the transition phase allows anomalous cold waters to be pumped into the surface layer by climatological upwelling and then leads to the cold phase. The converse occurs during the cold phase of ENSO. (b) Time series of the Niño3 SST anomalies (dashed; °C) and warm water volume anomalies (solid; 10^{14} m^3) over the entire equatorial tropical Pacific Ocean (5°S–5°N, 120°E–80°W) (courtesy of Christopher S. Meinen).

of El Niño. This concept was later modified by covering the entire tropical Pacific Ocean between 15°S and 15°N [*Wyrtki*, 1985]. Prior to El Niño, upper ocean heat content or warm water volume over the entire tropical Pacific tends to build up (or recharge) gradually, and during El Niño the accumulated warm water is flushed toward (or discharged to) higher latitudes. After the discharge, the eastern tropical Pacific becomes cold (La Niña) with the shallowing of the thermocline and then warm water slowly builds up again (recharge) before the occurrence of the next El Niño (see Figure 3). The recharge and discharge processes have been also examined by *Zebiak* [1989a], *Miller and Cheney* [1990], and *Springer et al.* [1990]. The concept of the recharge and discharge processes is further emphasized by *Jin* [1997]. Based on a coupled system that is similar to the coupled model of *Zebiak and Cane* [1987], *Jin* [1997] formulated and derived the recharge oscillator model.

Many studies have recently attempted to test the validity of the recharge oscillator model by using observational data [e.g., *Meinen and McPhaden*, 2000, 2001; *Hasegawa and Hanawa*, 2003a; *Holland and Mitchum*, 2003; *Sun*, 2003]. These observational studies basically demonstrate the recharge and discharge of the equatorial Pacific warm water during the evolution of ENSO. However, the more appropriate variable in equation (3) may be one that represents the warm water over the entire equatorial Pacific rather than the one only in the equatorial western Pacific. These studies show that the warm water in the entire equatorial Pacific band (for example, 5°S–5°N) highly correlates with the Niño3 SST anomalies, with the former leading the latter by about two seasons (Figure 3b). The correlation between the equatorial western Pacific warm water and the Niño3 SST anomalies is lower but still significant, with the western Pacific warm water leading by five seasons. *Mechoso et al.* [2003] tested the validity of the recharge oscillator model by fitting their coupled GCM output into this model. They suggested that the recharge oscillator could provide a plausible representation of their ENSO simulation. Misfits between the recharge oscillator and the coupled GCM oscillatory mode may be attributed to additional physics that are not included in the recharge oscillator.

There is a debate on the latitudinal bands of the recharge and discharge of warm water. *Wyrtki* [1985] defined the warm water in the tropical Pacific between 15°S and 15°N. *Miller and Cheney* [1990] and *Springer et al.* [1990] showed that, during El Niño, the warm water volume is decreased near the equatorial band (8°S–8°N and 5°S–5°N, respectively), whereas the volume of the tropical Pacific is not affected by ENSO due to water recirculation in the tropical North Pacific. Recently, *Holland and Mitchum* [2003] seem to reconcile this conflict by demonstrating that warm water is indeed lost from the tropical Pacific as a whole over the course of an El Niño event, as suggested by *Wyrtki* [1985]. This loss, however, is relatively small compared to the redistribution within the tropics. *Kug and Kang* [2003] showed that during El Niño meridional transport in the Northern Hemisphere is larger than that into the Southern Hemisphere, and that the asymmetric characteristics are mainly due to a southward shift of the maximum westerly wind anomalies during the mature phase of El Niño [*Harrison and Vecchi*, 1999].

Sun [2003, this volume] presented a "heat pump" hypothesis for ENSO. An increase in the warm pool SST increases the zonal SST contrast that strengthens the easterly trade wind and then helps the ocean to store more heat to the subsurface ocean. Because of the stronger wind and the resulting steeper tilt of the equatorial thermocline, the coupled system is potentially unstable and is poised to release its energy through a stronger warming (i.e., a larger amplitude of El Niño). The occurrence of El Niño pushes the accumulated heat poleward and prevents the further heat buildup in the western Pacific, thereby stabilizing the coupled system. This ENSO "heat pump" hypothesis is conceptually similar to the physics of the recharge oscillator.

3.1.3. The western Pacific oscillator. The western Pacific oscillator (Figure 4) emphasizes the coupled role of the western Pacific anomaly patterns in ENSO. This oscillator model does not necessarily require wave reflections at the western and eastern boundaries. Neglecting the feedbacks due to wave reflections at the western and eastern boundaries in the unified oscillator by setting $b_1 = 0$ and $b_3 = 0$, equations (1a)–(1d) reduce to:

$$\frac{dT}{dt} = a\tau_1 + b_2\tau_2(t-\delta) - \varepsilon T^3, \quad (4a)$$

$$\frac{dh}{dt} = -c\tau_1(t-\lambda) - R_h h, \quad (4b)$$

$$\frac{d\tau_1}{dt} = dT - R_{\tau 1}\tau_1, \quad (4c)$$

$$\frac{d\tau_2}{dt} = eh - R_{\tau 2}\tau_2. \quad (4d)$$

Equation (4) is the western Pacific oscillator of *Weisberg and Wang* [1997a].

Arguing from the vantage point of a *Gill* [1980] atmosphere, condensational heating due to convection in the equatorial central Pacific [*Deser and Wallace*, 1990; *Zebiak*, 1990] induces a pair of off-equatorial cyclones with westerly wind anomalies on the equator. These equatorial westerly wind anomalies act to deepen the thermocline and increase SST in the equatorial eastern Pacific, thereby providing a positive feedback for anomaly growth [represented by the first term of RHS of equation (4a)]. On the other hand, the off-equatorial cyclones raise the thermocline there via Ekman pumping.

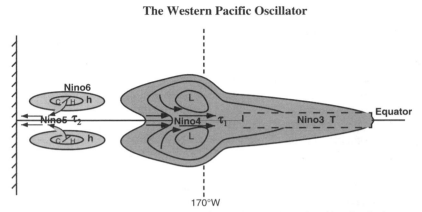

Figure 4. Schematic diagram of the western Pacific oscillator for ENSO. Condensational heating in the central Pacific induces a pair of off-equatorial cyclones with westerly wind anomalies in the Niño4 region. The Niño4 westerly wind anomalies act to deepen the thermocline and increase SST in the Niño3 region. On the other hand, the off-equatorial cyclones raise the thermocline there via Ekman pumping. Thus, a shallow off-equatorial thermocline anomaly expands over the western Pacific leading to a decrease in SST and an increase in SLP in the Niño6 region. During the mature phase of El Niño, the Niño6 anomalous anticyclone initiates equatorial easterly wind anomalies in the Niño5 region. The Niño5 easterly wind anomalies cause upwelling and cooling that proceed eastward as a forced Kelvin wave response providing a negative feedback for the coupled system to oscillate.

Thus, a shallow off-equatorial thermocline anomaly expands over the western Pacific [represented by equation (4b)], leading to a decrease in SST and an increase in sea level pressure in the off-equatorial western Pacific. This results in off-equatorial anomalous anticyclones during the mature phase of El Niño, which initiate equatorial easterly wind anomalies in the western Pacific [*Wang et al.*, 1999b; *Wang*, 2000]. These equatorial easterly wind anomalies cause upwelling and cooling that proceed eastward as a forced ocean response providing a negative feedback [represented by the second term on RHS of equation (4a)]. Equations (4c) and (4d) relate the zonal wind stress anomalies in the equatorial central Pacific to the equatorial eastern Pacific SST anomalies, and the zonal wind stress anomalies in the equatorial western Pacific to the off-equatorial western Pacific thermocline anomalies, respectively. The western Pacific oscillator is also consistent with the onset of El Niño. During the onset and development phases of an El Niño, twin anomalous cyclones in the off-equatorial western Pacific initiate equatorial westerly wind anomalies [e.g., *Wang and Weisberg*, 2000] that produce downwelling Kelvin waves to warm the equatorial central and eastern Pacific.

Earlier studies have shown that the equatorial easterly wind anomalies in the western Pacific can force upwelling Kelvin waves that raise the thermocline in the east [e.g., *Tang and Weisberg*, 1984; *Philander*, 1985]. Recently, *McPhaden and Yu* [1999], *Delcroix et al.* [2000], *Boulanger and Menkes* [2001], *Vialard et al.* [2001], *Picaut et al.* [2002], *Boulanger et al.* [2003], and *Hasegawa and Hanawa* [2003a] have shown that the western Pacific oscillator operates in nature. The western Pacific wind-forced Kelvin waves play an important role in terminating ENSO. For example, *Boulanger and Menkes* [2001] and *Boulanger et al.* [2003] demonstrated that, for the 1997–98 El Niño, about two-thirds of the Kelvin wave amplitude is actually forced by easterly wind in the western Pacific and the other one-third is due to wave reflection at the western boundary. In nature, the equatorial easterly wind anomalies in the western Pacific are observed to become larger and larger (both amplitude and fetch) and move eastward after the mature phase of El Niño. The impact of the easterly wind-forced upwelling Kelvin waves is thus gradually strengthened by the increasing fetch and eastward migration of the easterly wind anomalies [e.g., *Picaut et al.*, 2002].

3.1.4. The advective-reflective oscillator. *Picaut et al.* [1996] found an oceanic convergence zone at the eastern edge of the warm pool, which is advected in phase with the Southern Oscillation Index over thousands of kilometers, eastward during El Niño, westward during La Niña. Based on this finding, the study of *Picaut and Delcroix* [1995] regarding zonal advection and wave reflection, and the fact that westerly (easterly) winds penetrate into the central (western) equatorial Pacific during El Niño (La Niña), *Picaut et al.* [1997] proposed a conceptual advective-reflective oscillator for ENSO (Figure 5). In this concept, they emphasize a positive feedback of zonal currents that advect the western Pacific warm pool toward the east during El Niño. Three negative feedbacks tending to push the warm pool back to its original position and then into the western Pacific are: anomalous zonal current associated with wave reflection at the western boundary, anomalous zonal current associated with wave reflection at the eastern

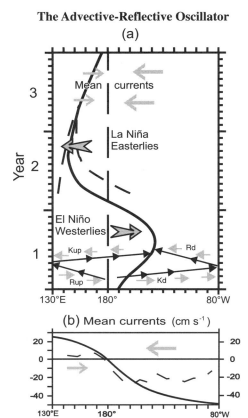

Figure 5. Schematic diagram of the advective-reflective oscillator for ENSO. This oscillator emphasizes a positive feedback of zonal currents that advect the western Pacific warm pool toward the east during El Niño. Three negative feedbacks tending to push the warm pool back to the western Pacific are: anomalous zonal current associated with wave reflection at the western boundary; anomalous zonal current associated with wave reflection at the eastern boundary; and mean zonal current converging at the eastern edge of the warm pool.

boundary, and mean zonal current converging at the eastern edge of the warm pool. During the warm phase of ENSO, equatorial westerly wind anomalies in the central Pacific produce equatorial upwelling Rossby and downwelling Kelvin waves that propagate westward and eastward, respectively. The westward propagating upwelling Rossby waves reflect into upwelling Kelvin waves after they reach the western boundary, whereas the eastward propagating downwelling Kelvin waves reflect into downwelling Rossby waves at the eastern boundary. Since both upwelling Kelvin and downwelling Rossby waves have westward zonal currents in the equatorial band, they tend to push the warm pool back to its original position and then into the western Pacific. These negative feedbacks along with the negative feedback of the mean zonal current make the coupled system to oscillate. Recent observational and modeling support of the advective-reflective oscillator can be found in *Delcroix et al.* [2000], *Clarke et al.*, [2000], *An and Jin* [2001], *Picaut et al.* [2001, 2002], and *Dewitte et al.* [2003].

The advective-reflective oscillator of *Picaut et al.* [1997] can also be represented by a set of simple and heuristic equations. By setting $b_2 = 0$ in equation (1a), the unified oscillator model is reduced to:

$$\frac{dT}{dt} = a\tau_1 - b_1\tau_1(t-\eta) - b_3\tau_1(t-\mu) - \varepsilon T^3, \quad (5a)$$

$$\frac{d\tau_1}{dt} = dT - R_{\tau_1}\tau_1. \quad (5b)$$

In derivation and formulation of the unified oscillator model [*Wang*, 2001a], it is shown that two advection terms of $u\partial \overline{T}/\partial x$ and $\overline{u}\partial T/\partial x$ are included in the first term of $a\tau_1$ in equation (5a) (also see *Battisti and Hirst* [1989]). Thus, the effects of zonal current are included in the term of $a\tau_1$. The effect of anomalous zonal current associated with wave reflection at the western boundary can be explained by the term of $-b_1\tau_1(\tau-\eta)$ in equation (5a) (also see *Clarke et al.* [2000]). The negative feedback of wave reflection at the eastern boundary is represented by the third term of RHS of equation (5a).

3.2. Slow (SST) Modes

Interaction between the tropical Pacific Ocean and atmosphere can produce coupled slow modes. The simple coupled system (with constant mean states) displays a slow westward propagating unstable mode [*Gill*, 1985; *Hirst*, 1986] and a slow eastward propagating unstable mode [*Philander et al.*, 1984; *Yamagata*, 1985; *Hirst*, 1986]. These two modes are further investigated numerically by *Hirst* [1988] and analytically by *Wang and Weisberg* [1996], showing that they can propagate and continuously regenerate on interannual timescales. The delayed oscillator is not relevant to these unstable modes. For example, *Wang and Weisberg* [1994] showed that the evolution of the eastward propagating mode is nearly identical for the closed and open ocean western boundary conditions (the open western boundary does not allow waves to be reflected).

Neelin [1991] introduced a slow SST mode theory by emphasizing physical processes in the oceanic surface layer (not related to wave dynamics). Whether the coupled system favors the SST modes or the ocean-dynamics modes (associated to the delayed oscillator) is determined by the ocean adjustment process. For the ENSO timescale, there are two key adjustments: one associated with the dynamical adjustment of the equatorial ocean, and the other associated with the thermodynamical changes in SST due to air-sea coupling. When the dynamical adjustment of the ocean is fast compared with the changes in SST, the behavior of the coupled ocean-atmosphere system depends critically on the time evolution of the SST, but is less influenced by the ocean-wave dynamics. On the other hand, if the dynamical adjustment of the ocean is slow, the coupled ocean-atmosphere system is dominated by the equatorial wave dynamics that provide the "memory" for an interannual oscillation. *Jin and Neelin* [1993] and *Neelin and Jin* [1993] provided a unified view between the slow SST mode and the ocean-dynamics modes, by arguing that in most of the parameter space the coupled modes will have a mixed nature, i.e., the mixed SST/ocean-dynamics modes. An advantage of the slow modes is that they can explain the propagating property of interannual anomalies whereas the delayed oscillator mode produces a standing oscillation.

A number of physical processes compete in terms of the direction of modes' propagation (eastward and westward). Propagation is first discussed by considering a region with a positive SST anomaly in the equatorial eastern/central Pacific. The wind responses to the west of the region are anomalous westerly, whereas they are easterly to the east [*Gill*, 1980]. This wind distribution drives anomalous zonal currents that advect warm water to the west of the region and cold water to the east. At the same time, the anomalous westerly (easterly) winds to the west (east) induce local anomalous downwelling (upwelling). As a result of both zonal advection and downwelling, the region of positive SST anomaly expands on its western side and the SST anomaly then propagates westward. On the other hand, the anomalous westerly winds induce a deepening of the thermocline in the east, which warms SST in the east through mean upwelling. Additionally, the nonlinearity of the anomalous vertical temperature gradient by the anomalous upwelling can also warm SST in the east [*Jin et al.*, 2003]. Thus, both can make the SST anomaly propagate eastward. Second, propagation of interannual anomalies is considered in the western Pacific Ocean from both dynamical and thermodynamical air-sea coupling [*Wang*, 1995b; *Philander and Fedorov*, 2003]. From the dynamical point of view, a modest disturbance in the form of a brief burst of westerly winds (or the Madden-Julian Oscillation) in the western Pacific will generate currents that transport some of the warm water eastward, thus decreasing the zonal temperature gradient. The resultant weakening of the trade winds will cause more warm water to flow eastward, causing even weaker winds. From the thermodynamical point of view, during the boreal winter and spring, the climatological zonal wind in the equatorial western Pacific varies from a weak westerly at 130°E–150°E to an easterly near the date line, with a direction reversal around 150°E. Superposition of an equatorial westerly anomaly in the above mean zonal wind in the western Pacific will have different effects on SST. In the region of a weak mean easterly between 160E°–170°E, a westerly anomaly implies a reduction in the total wind speed, resulting in an increase in SST due to reduced evaporation. However, in the region of a weak mean westerly at west of 150°E, a westerly wind anomaly increases the total wind speed, inducing the cooling of SST through enhanced evaporation. Therefore, an eastward SST gradient is produced, which in turn reinforces the equatorial westerly wind anomalies [*Lindzen and Nigam*, 1987]. The feedback between the eastward SST gradient and westerly anomalies promotes the eastward propagation of the equatorial westerly anomalies observed during an El Niño event.

3.3. A Stable Mode Triggered by Stochastic Forcing

In the ENSO views of Sections 3.1 and 3.2, the coupled tropical Pacific Ocean-atmosphere system is dynamically unstable. However, as model parameters are changed, the oscillatory and slow modes can become stable (e.g., see model parameter studies of *Battisti and Hirst* [1989]; *Hirst* [1988]; *Neelin and Jin* [1993]; *Wang and Weisberg* [1996]; *Jin* [1997]; *Wang* [2001a]). In this case, a stochastic trigger (forcing term) must be added to an oscillator model to excite an irregular oscillation [e.g., *Graham and White*, 1988; *Jin*, 1997]. ENSO as a stable mode triggered by stochastic forcing (or noise) has been suggested by many authors [e.g., *McWilliams and Gent*, 1978; *Lau*, 1985; *Penland and Sardeshmukh*, 1995; *Blanke et al.*, 1997; *Kleeman and Moore*, 1997; *Eckert and Latif*, 1997; *Moore and Kleeman*, 1999a, b; *Thompson and Battisti*, 2001; *Dijkstra and Burgers*, 2002; *Larkin and Harrison*, 2002; *Kessler*, 2002; *Zavala-Garay et al.*, 2003]. This hypothesis proposes that disturbances, unrelated to internal ENSO dynamics, are the source of stochastic forcing that drives ENSO. It should be pointed out that stochastic forcing might also have a low-frequency spectral tail (as a result of cumulative effect of strong or extended series of random events) that can directly drive ENSO [*Moore and Kleeman*, 1999a]. An attractive feature of this hypothesis is that it offers a natural explanation in terms of noise for the irregular behavior of ENSO variability. Since this view of ENSO requires the presence of atmospheric "noise", it easily explains why each El Niño is distinct and El Niño is so difficult to predict [e.g., *Landsea and Knaff*, 2000; *Fedorov et al.*, 2003].

No matter whether El Niño is a self-sustained mode or a stable mode triggered by stochastic forcing, El Niño matures with warm SST anomalies in the equatorial central and eastern Pacific. After an El Niño reaches its mature phase, negative feedbacks are required to terminate the growth of the mature El Niño anomalies in the central and eastern Pacific. In other words, the negative feedbacks associated with the delayed oscillator, the recharge oscillator, the western Pacific oscillator, and the advective-reflective oscillator may be still valid for demise of an El Niño even if El Niño is regarded as a stable mode triggered by stochastic forcing.

A stable mode can be either oscillatory or non-oscillatory (highly damped). For a non-oscillatory mode, there is no necessary connection between one El Niño event and the next, i.e., El Niño is sporadic, not cyclic [e.g., *Larkin and Harrison*, 2002; *Kessler*, 2002; *Philander and Fedorov*, 2003], and a random disturbance is needed to initiate each new event. For an oscillatory mode, each El Niño is related to the ensuing ENSO phases. *Mantua and Battistti* [1994] discussed three simple ENSO scenarios: (1) periodic ENSO cycle, (2) non-periodic ENSO cycle, and (3) non-periodic, non-cyclic ENSO event. In the latter two cases, the warm SST anomalies in the eastern and central Pacific are initiated by something other than a reflected Kelvin wave issued by the preceding cold event. However, the reflected upwelling Kelvin waves can be responsible for shutting down the growing instability in the equatorial central and eastern Pacific. A sequence of independent warm events can still be consistent with delayed oscillator physics since the termination of individual El Niño can occur as a result of wave reflection at the western boundary.

4. EFFECTS OF HIGH-FREQUENCY VARIABILITY ON ENSO

Variability with frequency higher than the ENSO frequency includes the seasonal cycle and the intraseasonal variability (ISV). Both seasonal cycle and ISV play roles in ENSO.

4.1. Seasonal Cycle

The seasonal cycle can contribute to the irregularity of ENSO and the ENSO phase-locking [e.g., *Jin et al.*, 1994; *Tziperman et al.*, 1995; *Chang et al.*, 1995]. Using numerical models, these studies showed that interannual variability is periodic without seasonal cycle forcing, but as model parameters (related to the seasonal cycle and the ocean-atmosphere coupling) are increased the interannual model solution undergoes a transition from periodic to irregular (or chaotic) through a sequence of rational fractions of the seasonal cycle: ENSO remains phase-locked to the seasonal cycle. In models, the seasonal cycle is influenced through mean background states of atmospheric wind divergence, oceanic upwelling, and so on. *Mantua and Battisti* [1995] hypothesized that interaction between ENSO and the "mobile" mode (a near-annual and westward propagating mode) is the cause for irregular variability in the *Zebiak and Cane* [1987] model simulations. The transition to chaos of a model system can occur in any of three universally recognized scenarios: the period doubling route [*Chang et al*, 1995], the quasi-periodicity route [*Tziperman et al.*, 1995], and the intermittency route [*Wang et al.*, 1999a].

4.2. Intraseasonal Variability (ISV)

The prominent ISV in the western and central Pacific includes westerly wind bursts (WWB) and the Madden-Julian Oscillation (MJO). Although both the WWB and MJO show westerly winds over the western Pacific, they differ temporally and spatially. Based on the region of maximum zonal wind anomalies, *Harrison and Vecchi* [1997] and *Vecchi and Harrison* [2000] identified eight different types of WWB event. On average, WWB has zonal width between 30° and 40° longitude, meridional width between 10° and 15° latitude, and duration between 7 and 10 days. The MJO, a wave-like atmospheric phenomenon, has a time-scale of between 30–90 days and has a much larger structure than the WWB [*Madden and Julian*, 1994; *Slingo et al.*, 1999]. The MJO propagates eastward and the WWB does not necessarily. The WWB tends to develop during active phases of the MJO (also tends to form from paired tropical cyclones and cold surges from mid-latitude), but the exact relationship between the WWB and MJO is not clear. They both have an influence on oceanic variability. However, the quantitative differences between the effects on the ocean by the WWB and MJO have not yet been determined. Therefore, we herein collectively review their roles in the ocean and ENSO, while *Lengaigne et al.* [this volume] provide a detailed review of the WWBs and their influence on the tropical Pacific Ocean-atmosphere system.

The ISV, associated with WWB and MJO, has a local effect on the western Pacific and a remote effect on the eastern Pacific. The local effect includes a change in mixed layer depth, surface jets, and an oceanic cooling in the western Pacific that can be explained by varying both shortwave radiation and latent heat flux. Convective activity associated with ISV increases atmospheric cloudiness that reduces shortwave radiation and then cools the western Pacific Ocean [e.g., *Weller and Anderson*, 1996]. During the boreal winter and spring, the climatological zonal wind in the equatorial western Pacific west of 150°E is a weak westerly. Thus, a westerly wind anomaly (associated with ISV) increases the total wind speed, inducing the cooling of SST through enhanced evaporation.

The remote effect of ISV on the eastern Pacific is via downwelling Kelvin waves generated by westerly wind anomalies in the western Pacific [e.g., *Kessler et al.*, 1995; *Hendon et al.*, 1998; *Zhang*, 2001; *Zhang and Gottschalck*, 2002; *Kutsuwada and McPhaden,* 2002; *Cravatte et al.,* 2003]. The generated downwelling Kelvin waves propagate eastward along the thermocline to the central and eastern Pacific. The resulting rises in sea level along the South American coast are observed to occur approximately 6–7 weeks following the WWB events that generate them. Along the equator the Kelvin waves are also accompanied by anomalous surface currents that induce an eastward displacement of the eastern edge of the western Pacific warm pool [*Matsuura and Iizuka,* 2000; *Picaut et al.,* 2002; *Lengaigne et al.,* 2002]. The combined effects of zonal advection and thermocline depression increase the SST in the central and eastern Pacific and thus decrease the zonal SST gradient and weaken the trade winds [*Lindzen and Nigam,* 1987]. The weakening of the trade winds will cause more warm water to flow eastward, causing even weaker trade winds. This positive feedback can result in the onset of an El Niño event, as hypothesized by *Bjerknes* [1969]. As an example, both observations and numerical models have shown that the westerly wind anomalies in the western Pacific during the boreal winter and spring of 1996–97 played an important role in the onset of the 1997–98 El Niño [e.g., *McPhaden,* 1999; *McPhaden and Yu,* 1999; *Wang and Weisberg,* 2000; *van Oldenborgh,* 2000; *Boulanger et al.,* 2001; *Bergman et al.,* 2001; *Picaut et al.,* 2002]. *Cravatte et al.* [2003] recently noticed an oscillation in the surface winds over the warm pool with an approximate 120-day period. This oscillation, of unknown origin, generates equatorial Kelvin waves as strong as those excited by the MJO. The two sets of downwelling Kelvin waves induced by the MJO and the 120-day wind oscillation seem to be stronger during the onset of El Niño and may interfere in its development. Note that easterly winds associated with the MJO generate upwelling Kelvin waves that can participate in the demise of El Niño [*Takayabu et al.,* 1999].

As discussed in Section 3.3, the ISV has been treated as noise or disturbances that can drive or sustain ENSO. Considering the ENSO oscillator models discussed in Section 3.1, noise terms can be incorporated through equations that control the variations of atmospheric winds [e.g., *Graham and White,* 1988; *Jin,* 1997]. If the ISV is acting on a self-sustained oscillatory system, then it is a source of the irregularity of ENSO. On the other hand, if the ISV is acting on a stable system, then it is the source of both its variability and irregularity. In addition to the stability of the coupled system, the temporal and spatial structures of noise may also determine the impact of noise on ENSO [e.g., *Bergman et al.,* 2001; *Fedorov,* 2002]. For example, strong MJO activity was also evident during the boreal winter of 1989–90 and the early stage of development was similar to that of 1996–97. However, the development of El Niño was aborted in May 1990. The MJO was relatively quiescent during the boreal winter of 1981–82. A strong El Niño developed during 1982, but not as rapidly as it did during 1997. All of these suggest a complex relationship between MJO and ENSO.

A more theoretical approach to El Niño as a stable mode driven by stochastic forcing has been provided in the framework of generalized stability theory [*Farrell and Ioannou,* 1996a, b]. It is argued that the coupled tropical Pacific Ocean-atmosphere system is non-normal (i.e., its low-frequency eigenvectors are non-orthogonal) [e.g., *Moore and Kleeman,* 1996; *Moore and Kleeman,* 1999a, b]. In a non-normal ENSO system, a perturbation can still experience transient growth (grow to a finite amplitude and then decay) even if the system is asymptotically stable. This is because the low-frequency eigenvectors can interfere constructively with one another since they are non-orthogonal. The spatial structure of stochastic forcing is important if it is to increase variability on seasonal-to-interannual timescales. The stochastic optimals are the spatial patterns that stochastic forcing must have in order to produce large response in a coupled model, and the optimal perturbations represent the fastest growing perturbations that can exist in the coupled system [*Moore et al.,* 2003]. Using an intermediate coupled ocean-atmosphere model, *Moore and Kleeman* [1999a, b] showed that when the coupled model is subjected to stochastic noise forcing that projects on the stochastic optimals, perturbations with initial structures that are similar to the optimal perturbations are excited and subsequently grow rapidly. The stochastic component of the NCEP/NCAR reanalysis has been shown to possess such optimal structures [*Zavala-Garay et al.,* 2003]. The idea is also supported by *Penland and Sardeshmukh* [1995] and *Penland* [1996] who analyzed the optimal perturbations of observed SST in the tropical Pacific and Indian Oceans.

Some studies have argued that ISV does not play a critical role in ENSO [e.g., *Zebiak,* 1989b; *Slingo et al.,* 1999; *Syu and Neelin,* 2000; *Kessler and Kleeman,* 2000]. *Zebiak* [1989b] showed that, in his intermediate model, the atmospheric ISV does not seem to affect ENSO. *Syu and Neelin* [2000] demonstrated that a noisier signal with shorter timescales does not appear to have an obvious relation to the ENSO cycle in their model. *Kessler and Kleeman* [2000] concluded that the MJO can interact constructively with the onset of El Niño to amplify a developing warm event, however, the MJO on its own does not appear to be the cause of El Niño. *Slingo et al.* [1999] could not find an interannual relationship or linkage between the MJO and El Niño, based on the MJO index defined from global winds and convection. When the MJO index was based on local signals of the MJO in the Pacific, *Zhang and Gottschack* [2002] found a relation between Kelvin wave ISV forcing and SST anomalies in the eastern equatorial Pacific during El Niño, at least for the 1980–99 period. Obviously, the MJO-ENSO relationship depends on indices that measure variations of the MJO.

5. LOW-FREQUENCY VARIABILITY OF ENSO

In this section, the observational evidence of decadal-multidecadal variability and warming trends in both the tropical and mid-latitude Pacific are first discussed. The mechanisms proposed for tropical Pacific decadal-multidecadal variability are summarized, and the interpretation of tropical Pacific warming trends, global warming, and ENSO are reviewed. The difficulties and uncertainties of the studies of low-frequency variability are finally discussed. *Seager et al.*, in this volume, discuss predictability of Pacific decadal variability.

5.1. Observational Evidence of Decadal-Multidecadal Variability and Warming Trends in the Tropical and Mid-latitude Pacific

Decadal and multidecadal variability in the North Pacific has been analyzed for more than a decade [e.g., *Nitta and Yamada*, 1989; *Trenberth*, 1990; *Minobe*, 2000]. Although this low-frequency variability is relatively well documented, it is still unclear if one major mode or several co-equal decadal modes of variability affect this region. For example, there is some evidence of four decadal ocean-atmosphere statistical modes that occupy a thick layer of the North Pacific Ocean [*Luo and Yamagata*, 2002]. The most studied signal has been called the PDO for Pacific (inter) Decadal Oscillation [*Mantua et al.*, 1997] or NPO for North Pacific decadal-multidecadal Oscillation [*Gershunov and Barnett*, 1998]. Both correspond to the leading EOF of SST North of 20°N. The PDO appears as a recurring pattern of ocean-atmosphere variability centered over the mid-latitudes of the North Pacific. Cold PDO regimes prevailed in 1880–1924, and in 1947–1976, while warm regimes prevailed in 1925–1946 and from 1977 to the mid-1990s. Despite sparse data coverage, there is evidence of decadal variability in the mid-latitudes and subtropics of the Southern Pacific [*Garreaud and Battisti*, 1999; *Linsley et al.*, 2000; *Chang et al.*, 2001; *Mantua and Hare*, 2002]. In particular, the position of the South Pacific convergence zone is subject to an interdecadal oscillation, in addition to an ENSO oscillation. Both oscillations have similar amplitudes, but they appear independent [*Folland et al.*, 2002]. The interdecadal variability in the South Pacific can be regarded as the quasi-symmetric manifestation of the PDO.

The tropics and in particular the tropical Pacific are marked by several decadal-multidecadal coupled modes [*Goswami and Thomas*, 2000; *White et al.*, 2003]. Examination of the PDO over the entire Pacific basin reveals that its spatial signature in SST, SLP, and wind stress is somewhat similar to the "horse shoe" signature of ENSO [*Mantua et al.*, 1997; *Zhang et al.*, 1997; *Garreaud and Battisti*, 1999; *Mestas-Nuñez and Enfield*, 2001; *Salinger et al.*, 2001]. It is marked by an equator-straddling SST anomaly in the eastern tropical Pacific less confined than those of ENSO, and by a relatively greater SST anomaly of opposite sign in the North Pacific (Plate 3). However, the PDO may also be distinct from the ENSO-like Pacific-wide decadal oscillation, as they appear dominated by 50 years and 20–30 years oscillations, respectively [*Minobe*, 2000; *Liu et al.*, 2002]. On the other hand, *Tourre et al.* [2001] found two distinct decadal (9–12 years) and interdecadal (12–25 years) signals in the Pacific basin. Excepting this last study, the spectral peaks found on decadal and interdecadal timescales are not significantly different from red noise, and it is proper to refer to variability rather than oscillation for these timescales. In any case, the links between these tropical and/or Pacific-wide decadal-multidecadal variability and ENSO may be crucial, either through the modulation of the basic ENSO oscillation in the tropical Pacific or through their teleconnections [*Gershunov and Barnett*, 1998; *Alexander et al.*, this volume].

Decadal-multidecadal variability is difficult to comprehend from temporally limited data, and several authors have thus focused on the recent 1976 global climate shift [*Guilderson and Schrag*, 1998; *Zhang et al.*, 1998; *Karspec and Cane*, 2002; *Giese et al.*, 2002]. Its signature in the tropical Pacific is particularly important with a rapid increase of SST over the span of a year. This warming is associated with an increase in the amplitude and period of ENSO, and an eastward displacement along the equator of the maximum anomalies of SST gradient, westerly wind, and thermocline slope [*Wang and An*, 2002]. The origin of this warming and climate shift is still unclear. *Zhang et al.* [1998] suggest that subducted warm-water issued from the North Pacific perturbed the tropical thermocline (a hypothesis refuted by *Guilderson and Schrag* [1998]), while *Giese et al.* [2002] consider also a subsurface bridge but originating from the subtropical South Pacific. Note that other SST shifts in the last century may have occurred around 1924–25, 1941–42 and 1957–58 [*Chao et al.*, 2000], most probably as phase transitions of several decadal-multidecadal oscillations [*Minobe*, 2000]. The dominance of the 1976 shift may be related to the acceleration of the 20th century warming trend observed in the tropical Pacific. *Knutson and Manabe* [1998] noted that this warming trend in a broad triangular region of the eastern tropical and subtropical Pacific increases from 0.41°C (100 yr)$^{-1}$ since 1900 to 2.9°C (100 years)$^{-1}$ since 1971. Like the Pacific-wide decadal mode, this warming trend has an El Niño-like structure.

5.2. Mechanisms of Tropical Pacific Decadal-Multidecadal Variability

As discussed above, both the tropical and mid-latitude Pacific show decadal-multidecadal variability. *Latif* [1998],

36 UNDERSTANDING ENSO

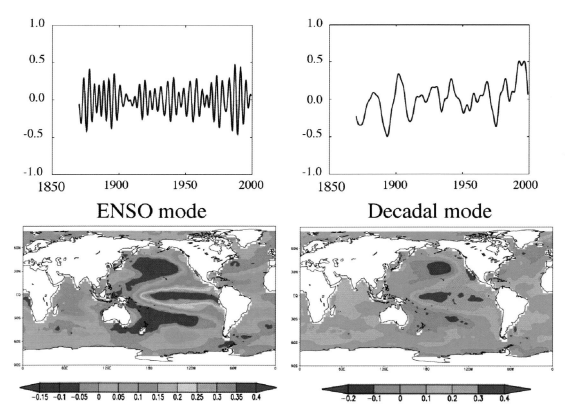

Plate 3. The two leading modes of Niño4 SST anomalies extracted through singular spectrum analysis. The upper panels represent the reconstruction of the Niño4 SST anomaly index using the two leading modes (20% of variance for ENSO mode and 25% for decadal mode); units are in [K]. The lower panels represent the regression patterns of SST anomalies using the upper two time-series as indices; units are [K]/standard deviation (courtesy of Mojib Latif).

Miller and Schneider [2000], *Minobe* [2000], and *Mantua and Hare* [2002] review mechanisms of North Pacific decadal-multidecadal variability. This subsection reviews the mechanisms of tropical Pacific decadal-multidecadal variability that can be divided into two categories: (1) tropical origins and (2) tropical-extratropical connections.

5.2.1. Tropical origins. Tropical Pacific decadal-multidecadal variability can be generated in the tropics only, without involving extratropical processes. Stochastic atmospheric forcing can lead to decadal-multidecadal variability in the tropical Pacific [e.g., *Kirtman and Schopf*, 1998; *Latif et al.*, 1998; *Burgers*, 1999; *Thompson and Battisti*, 2001]. Using a simple model, *Wang et al.* [1999a] showed that tropical Pacific decadal-multidecadal variability might result from the nonlinear interactions between the seasonal and interannual cycles. *Timmermann et al.* [2003] hypothesized that ENSO variations can grow until they reach the maximum intensity of El Niño, then a quick reset takes place and small ENSO variations grow again. Due to the asymmetry of El Niño and La Niña, the decadal amplitude modulation of ENSO is translated into decadal background changes. Such nonlinear behaviors were found in coupled model simulations by *Timmermann* [2003] and *Rodgers et al.* [2004].

Linear dynamics and local ocean-atmosphere interaction can be at the origin of decadal variability in the tropical Pacific Ocean. Tropical local wind may force the decadal variability in the tropical Pacific Ocean [*Schneider et al.*, 1999a; *Karspeck and Cane*, 2002]. The decadal changes in the background wind, before and after the 1976 climate shift, qualitatively reproduce the observed changes in ENSO properties noted above [*Wang and An*, 2002]. The origin of the changes in the winds is unclear, with a mid-latitude SST influence suggested by *Pierce et al.* [2000] and tropical ocean-atmosphere coupling proposed by *Liu et al.* [2002]. Yet, these last authors suggest that the decadal variability in the tropical Pacific can be enhanced by extratropical oceanic teleconnection. Using a coupled GCM, *Schneider* [2000] suggests an interesting decadal mode effective within the tropical Pacific, in which advection of salinity compensated temperature along isopycnals (termed spiciness anomalies) sets the decadal timescale.

Several authors have considered the inclusion of higher vertical and horizontal modes in the oceanic part of the ENSO delayed action oscillator to tentatively explain the decadal tropical variability, through wider ocean-atmosphere coupling, longer time in Rossby wave propagation and a reflected slow equatorial coupled wave. Using coupled models, *Knutson and Manabe* [1998], *Yukimoto et al.* [2000], and *Jin et al.* [2001] note westward phase propagations of decadal upper ocean temperature or thermocline depth around 9–12°N, 20°N and 15–25°N, respectively. Similar decadal propagating signals appear in observations [*White et al.*, 2003] and in a model forced over the 1958–97 period [*Capotondi and Alexander*, 2001]. Similarly, the ENSO recharge oscillator has been amended to include off equatorial Rossby waves and eastern subtropical wind variability [*Jin*, 2001; *Wang et al.*, 2003]. This amendment is supported by the observation and simulation of decadal variability in upper-heat content in the tropical Pacific [*Hasegawa and Hanawa*, 2003b; *Alory and Delcroix*, 2002].

5.2.2. Tropical-extratropical connections. *Gu and Philander* [1997] consider an oceanic bridge that subducts and advects, in about 10 years, mid-latitude surface waters of anomalous temperature all the way to the Equatorial Undercurrent (EUC) via shallow subtropical cells (STCs) [*Schott et al.*, this volume]. The anomalous waters are subsequently brought to the surface by equatorial upwelling and finally moved poleward by Ekman divergence [*Johnson*, 2001]. The circuit can be closed through this poleward surface oceanic bridge. It can also be closed through the upwelling-induced changes in eastern equatorial SST that influence the tropical and extratropical winds, which in turn affect the initial mid-latitude surface water anomalies. There is some evidence that anomalous surface water masses may subduct from the Northern and Southern Hemispheres toward the equator, using observations [*Deser et al.*, 1996; *Zhang et al.*, 1998; *Johnson and McPhaden*, 1999] and using models [*McCreary and Lu*, 1994; *Liu*, 1994; *Rothstein et al.*, 1998; *Harper*, 2000; *Solomon et al.*, 2003]. However, the detailed data analysis of *Schneider et al.* [1999a] does not find any significant decadal link between the North Pacific and the equator through anomalous subduction. Moreover, the temperature advected by the EUC is subject to strong seasonal and interannual variations that probably blur any remaining decadal signal [*Izumo et al.*, 2002]. Finally, model studies [*Schneider et al.*, 1999b; *Hazeleger et al.*, 2001] indicate that decadal variability in the tropics is largely independent of the arrival of water anomalies subducted from the mid-latitudes. While *Gu and Philander* [1997] suggested anomalous temperature transported by STCs as a mechanism of tropical decadal variability, *Kleeman et al.* [1999] proposed that changes in STC strength vary the amount of cold water transported into the equatorial thermocline. This mechanism is supported by the observation of a slowdown of STCs from the PDO related regime shift of mid-1970s to the late 1990s together with a decrease in equatorial upwelling [*McPhaden and Zhang*, 2002]. The results of an OGCM forced by observed winds are also consistent with the mechanism of STC strength [*Nonaka et al.*, 2002]. However, the two-year lag, found by these last authors between STC-induced and wind-forced equatorial SSTs, suggests that STC strength is not dominant in generating tropical decadal oscillations, acting more to amplify than to initiate them.

A number of the previous studies have focused on adiabatic oceanic processes. For example, in *Gu and Philander* [1997] the oceanic bridge is assured adiabatically through the subduction and advection of temperature anomalies. On the other hand, the surface forcing in the subduction region of the central North Pacific seems predominately diabatic [*Schneider et al.*, 1999a]. The theoretical study of *Boccaletti et al.* [2004] indicates that basin-wide diabatic processes may control the tropical thermocline on decadal timescales, without involving explicit connection between the tropics and mid-latitudes. The relative importance of diabatic and adiabatic processes may depend on timescales. In a Pacific study using expendable bathythermograph data and an oceanic model over 1970–88, *Auad et al.* [1998] found that the 10-year variability is strongly influenced by adiabatic heat changes, while 20-year variability is more influenced by diabatic processes, such as direct thermal forcing by the atmosphere.

Another mechanism involves a wave signal exchanged between mid-latitude and the tropics. *Jacobs et al.* [1994] suggested that the 1982–83 El Niño could have decadal effects on the northwestern Pacific circulation, through mid-latitude Rossby waves reflected from equatorial Kelvin waves impinging on the American coasts. *Lysne et al.* [1997] found a weak decadal signal in their search for an oceanic bridge driven by wave dynamics: anomalous temperature is propagated by mid-latitude Rossby waves to the western boundary, then equatorward by coastal Kelvin waves, and finally modifies equatorial SST via equatorial Kelvin waves.

Since the PDO is one of the most important oscillations of decadal-multidecadal timescales on earth, it is an obvious candidate for forcing the tropical decadal variability through an atmospheric bridge. Moreover, on decadal timescale the largest SST anomalies and ocean heat content occur at the mid-latitude not in the tropics [*Giese and Carton*, 1999]. The decadal change in the northern atmosphere is wide enough to alter the wind stress over the equatorial Pacific, hence the mean state of the equatorial thermocline and upwelling, and ultimately ENSO activity [*Barnett et al.*, 1999; *Pierce et al.*, 2000; *Wang and An*, 2002]. The atmospheric bridge between the decadal variability of the North Pacific and ENSO may well be the imprint of a common internal variability in the atmosphere [*Pierce*, 2002]. In fact, *Wang and Weisberg* [1998] found that the out-of-phase SST decadal signal in the mid-latitudes and the tropics (Plate 3) is the result of a tropical-extratropical oscillation that involves feedbacks from the atmospheric Hadley and Walker circulations. As noted above, the PDO has a notable signature in the Southern Hemisphere, and the studies of *Luo and Yamagata* [2001] and *Luo et al.* [2003] illustrate the role of the South Pacific in sustaining tropical decadal variability through anomalous cyclonic or anticyclonic atmospheric circulation. The recent study of *Yu and Boer* [2004] on heat content anomalies forced in the western South Pacific by surface heat forcing and in the western North Pacific by ocean dynamics, underlines the combined role of the two hemispheres in generating Pacific decadal variability.

5.3. Interpretation of Tropical Pacific Warming Trends, Global Warming, and ENSO

Despite the observational difficulties in separating decadal variability from long-term trends, there is an undeniable acceleration of the warming trend over the last 50 years in the eastern tropical and subtropical Pacific [e.g., *Knutson and Manabe*, 1998] but its origins remain uncertain. Using SST and several atmospheric parameters, *Curtis and Hastenrath* [1999] find long-term trends in the tropical Pacific compatible with the radiative but not with the wind forcing. They also note the resemblance of these trends to El Niño patterns. *Liu and Huang* [2000] attribute the SST warming trend to the weakening trade wind, which reduces the vertical and zonal advection of cold water in the eastern and western equatorial Pacific, respectively. *Cane et al.* [1997] argue that the eastern equatorial Pacific has instead cooled since 1900, under increasing trade winds (difficulties in building historical surface fields and in particular correcting wind products are briefly discussed in the next sub-section). In fact, *Lau and Weng* [1999] found a secondary cooling trend centered near the Niño3 region, superimposed on a general warming trend.

Using global coupled ocean-atmosphere models, *Knutson and Manabe* [1998] did not find that the observed warming trends are quantitatively consistent in magnitude and duration with internal climate variability alone. They conclude that part of the warming trend may be attributed to sustained thermal forcing, such as greenhouse warming. In the same way, statistics on the Southern Oscillation behavior by *Trenberth and Hoar* [1996, 1997] open the possibility, challenged by *Harrison and Larkin* [1997] and *Rajagopalan et al.* [1997], that greenhouse gas is involved in the tendency for more frequent El Niño since the late 1970s. *Meehl and Washington* [1986] was one of the first model studies that looked at the changes within the tropics under an increase of atmospheric CO_2. Most coupled models in the late 1990s [e.g., *Meehl and Washington*, 1996; *Knutson and Manabe*, 1998; *Timmermann et al.*, 1999] suggested that the eastern equatorial Pacific would warm more rapidly than the west. The SST gradient along the equator slackened together with the easterlies, and this results in an El Niño-like pattern of changes. Another school of studies suggests that the CO_2 warming response should be La Niña-like, with an increase of the equatorial SST gradient [*Cane et al.*, 1997; *Seager and Murtugudde*, 1997] and maximum warming in mid-latitudes. There are a number of arguments for an El Niño-like pattern in response to global warming. Under this scenario the cloud-shielding thermostat over the warm pool [*Ramanathan and Collins*, 1991; *Meehl and Washington*, 1996] or the evaporative surface cooling [*Knutson and Manabe*, 1995] will make the warming less efficient in the west than the east, and the SST equatorial gradient will decrease. The warm pool can also expand toward the east, thus increasing the overlying atmospheric convection and westerly winds [*Yu and Boer*, 2002]. In the absence of ocean dynamics, the atmospheric response to global warming over the equatorial Pacific is a decrease of easterlies [*Vavrus and Liu*, 2002]. On the contrary, the La Niña-like pattern is due to equatorial upwelling that reduces the surface warming in the east, leading to an increased SST gradient along the equator and thus stronger easterlies. A coupled model forced by historical (1880–1990) and future greenhouse gas concentrations results in a warming trend, which initially is larger in the extra-tropics and has a La Niña-like pattern in the tropics [*Cai and Whetton*, 2000]. The pattern becomes El Niño-like after the 1960s and remains in this state during the 21st century. In this simulation, the shift (analogous to the observed 1976 shift) is explained by the delayed arrival of warm waters from the initial extra-tropical warming by STCs in the equatorial thermocline [*Cai and Whetton*, 2001b]. As noted by *Liu* [1998], any global warming study of the equatorial SST gradient must consider both the relatively rapid air-sea interaction process and the long-term thermocline process.

The plausible tendency for an El Niño-like pattern under global warming does not mean that the tropical Pacific will stay in a permanent El Niño. Superimposed on the new mean warm state, ENSO oscillations may continue but with probable changes in its behavior (possibly due to the change of the mean state). Using a low-resolution coupled model, *Knutson and Manabe* [1997] found a slight decrease in ENSO amplitude, no significant change in ENSO frequency and more pronounced multidecadal modulation of ENSO, in response to doubling or quadrupling of CO_2. With a finer model resolution, *Timmermann et al.* [1999] found more frequent El Niños and stronger La Niñas. *Collins* [2000a] had to quadruple the concentration of greenhouse gas in order to see ENSO changes. More frequent El Niños and La Niñas occur with 20% larger amplitude for both. Increases in meridional temperature gradients on either side of the equator and in the vertical gradient of temperature in the thermocline are respectively responsible for the increases of ENSO frequency and amplitude. The recent coupled model study of *Hu et al.* [2001] also results in an El Niño-like mean pattern but with stronger La Niñas and weaker El Niños. These incoherent results, which underline the complexity of coupled model behavior under greenhouse warming, are discussed briefly in the following.

5.4. Difficulties and Uncertainties of Studies of Low-Frequency Variability

A number of studies of low-frequency modulation of ENSO rely on the analyses of historical surface data (mostly SST, SLP, surface winds) that have been interpolated in time and space in a drastic way. It is recognized that global data coverage is adequate after 1950, if not after 1980 with the satellite era. Several research groups have built global products on a monthly basis and on a latitude-longitude grid that varies from 5° down to 1°. Most have extended the 1950 limit back to 100 years where volunteer observing ships were very rare, particularly in their journey through the equatorial and southern Pacific. For example, *Kaplan et al.* [1998] pointed out the discrepancies that arise from using several interpolated fields in the search of warming or cooling trends in the eastern tropical Pacific since 1900 (from −0.3°C/100 years to +0.3°C/100 years). A disputable assumption for building these products is the stability of their structural relations over the last one and a half century, i.e., stationarity. While these research groups are very aware of the errors associated with these fields, other (internet) users may forget to consider such errors in their analyses. Other articles on the low-frequency modulation of ENSO have been written solely on the Southern Oscillation Index that is extended to 1866. However, it appears difficult to imagine that decadal or multidecadal oscillations have not displaced the centers of action of the Southern Oscillation from Tahiti and/or Darwin.

Similarly, several articles discussing decadal variations of ENSO during the last millennium rely on either a single set of paleoclimate proxies over an extended period of time [e.g., *Linsley et al.*, 2000] or several sets of proxies in the same location over separate periods of time [e.g., *Cobb et al.*, 2003]. Paleoclimate proxies are typically diverse, involving tropical corals, mid-latitude tree rings and tropical or polar/subpolar ice cores. The labor, time and cost involved in data collection and processing render multi-proxy analyses of ENSO decadal variability uncommon as yet [e.g., *Evans et al.*, 2001]. Multi-proxy reconstruction of ENSO time series suffers from technical and stability problems but also from assumptions about climate influence. For example, ice cores from tropical ice caps in Peru were originally thought to reflect ENSO variability, but recently they have been shown to be more affected by the large-scale atmospheric variability over Amazonian and the western tropical Atlantic than over the eastern tropical Pacific [*Thompson et al.*, 2000]. In any case, the potential of paleoclimate indicators is tremendous for understanding the decadal variability and long-term trend of ENSO and thus for separating the natural contributions from the anthropogenic contribution [e.g., *Cole*, 2001].

Statistical tools are sometimes not adequate in extracting and explaining decadal and multidecadal ENSO signals, because of the shortness and uncertainty of the data and products or the difficulty in separating signals that have similar patterns (e.g., ENSO and ENSO-like), or the skewed nature of the signals. As noted by *Liu et al.* [2002], these tools are important for the diagnosis of decadal variability, but they may not be able to identify the true physical modes of variability (see also the series of papers pursuant to the article "A cautionary note on the interpretation of EOFs" by *Dommenget and Latif* [2002]).

Other difficulties in the search for decadal mechanisms arise from the use of observations. Oceanic bridges between the mid-latitudes and the equator are so far impossible to prove given the reduced number of hydrographical or CTDs observations in these regions over the last four decades. Delayed-type decadal oscillators cannot be truly established due to the complexity of extracting propagating signals from sparse data near the Swiss-cheese western boundary. As a consequence, many of the previous studies have used simplified or sophisticated models. Simplified models, such as those used by McCreary and collaborators (see *Schott et al.* [this volume]), have the great advantage of pinpointing mechanisms. Sophisticated ocean models suffer less from insufficient physics, but they are still unable to reproduce subsurface equatorial countercurrents and thus complete STC patterns. In any case, the use of variable forcing instead of seasonal forcing in models that simulate STCs may result in an open circuit rather than a closed circuit [*Fukumori et al.*, 2004], hence making STC-induced decadal variability difficult or impossible to prove. Wind-forced models with or without data assimilation are also subject to spurious decadal variability and long-term trends due to unrealistic flux corrections, changes in the density and quality of data or measurement methods.

Coupled models have their own flaws. Many of them suffer from climate drifts, which are often corrected in dubious ways, and hardly reproduce realistic ENSO and annual cycles [*AchuaRao and Sperber*, 2002; *Davey et al.*, 2002; *Latif et al.*, 2001]. Most of the simulated ENSOs are too close to a biennial cycle, or have weak amplitudes, or cannot reproduce the well-known horseshoe pattern of SST anomalies over the tropical Pacific (Plate 3). Hence, their ability to reproduce realistic decadal variability or warming trends in the tropical Pacific must be questioned. As a result, the projections of coupled models into the 21st century with and without CO_2 are as yet hard to believe. As an example, using version 2 of the Hadley Centre coupled model, *Collins* [2000a] found that the amplitude and frequency of ENSO increase with a quadrupling of CO_2. In version 3, *Collins* [2000b] attributed the lack of significant modification of ENSO behavior to subtle non-linear changes in the physical parametrization schemes, rather than the main differences between the two versions of the model (horizontal resolution and flux adjustments). A promising way to infer ENSO response to global warming is through international multi-model intercomparison projects such as CMIP [*AchutaRao and Sperber*, 2002], which use a quantitative probabilistic approach that takes into account model errors.

6. DISCUSSIONS AND THE FUTURE

As detailed above, significant advances have been made in understanding ENSO, especially during and after the TOGA decade. However, there are many issues or topics that that are still under debate and/or need to be further investigated. This section discusses some issues and questions that should probably be resolved and addressed in the near future.

In addition to the different origins and development of El Niño events as discussed in Section 2, El Niño events show some property changes during the last five decades. The changes include an increase of ENSO period and amplitude, changes in the propagation of interannual anomalies, and a zonal shift of maximum equatorial westerly wind anomalies [e.g., *Wang and An*, 2002; *Wittenberg*, 2004]. In spite of these differences, the equatorial westerly wind anomalies in the western Pacific preceding the warm SST anomalies in the eastern Pacific by 3–4 months is a common characteristic for all El Niño events. Some studies have attempted to understand some changes of El Niño characteristics or properties [e.g., *Fedorov and Philander*, 2001; *Wang and An*, 2002]. However, our understanding of El Niño property changes is poor. In particular, why El Niño events originate differently in the last five decades is not known yet.

The issue of ENSO as a self-sustained oscillation mode or a stable mode triggered by random forcing is not settled. *Philander and Fedorov* [2003] showed that the stability and period of ENSO depend on the mean states of the tropical Pacific Ocean-atmosphere system. It is possible that ENSO is a self-sustained mode during some periods, a stable mode during others, or a mode that is intermediate or mixed between the former and the latter. The predictability of ENSO is more limited if ENSO is a stable mode triggered by stochastic forcing than if ENSO is a self-sustained mode, because then its irregularity depends on random disturbances.

Since 1988, four concepts have been proposed for the oscillatory and self-sustained nature of ENSO. They also represent the negative feedbacks of a growing ENSO stable mode triggered by stochastic forcing, and are unified in a single concept. More data and model diagnoses are needed to test these concepts or to discover others. Several authors have extended three concepts of ENSO to explain the decadal variability in the tropical Pacific. Off-equatorial Rossby waves are at the root of the modified delayed action oscillator of *White et al.* [2003] and the recharge oscillator of *Jin* [2001] for decadal variability. *Yu and Boer* [2004] note the resemblance of the ENSO western Pacific oscillator of *Weisberg and Wang* [1997a] with their findings on decadal variability and heat content anomalies in the western North and South Pacific. A unified concept of decadal oscillator in the tropical Pacific has yet to be determined.

Since the seasonal cycle can greatly affect interannual variations, a critical feature for coupled ocean-atmosphere models is the ability to simulate both seasonal and interannual variations. The El Niño simulation intercomparison project (ENSIP) showed that almost all twenty-four coupled ocean-atmosphere models still have problems in simulating the SST climatology [*Latif et al.*, 2001]. No model is able to simulate realistically all aspects of the interannual SST variability. Therefore, coupled model development and improvement are still major issues.

Interactions among intraseasonal variability (e.g., WWBs and MJO), interannual ENSO variability, and decadal-multidecadal variability have recently received considerable attention. The unexpected intensity of the 1997–98 El Niño underlines the role of intraseasonal variability in the triggering and growth of El Niño. On the other hand, the two major El Niño events over the last two decades of the 20th century may be the signature of decadal variability or global warming. Research on these topics is still in an early stage, and it will likely need additions to the ENSO observing system. Intraseasonal studies will demand finer resolution (in time and space) and improved sampling aloft (in the atmosphere) over the tropical Indo-Pacific region, while decadal ENSO variability will demand wider (in latitude) and deeper (in the ocean) extensions together with better and more comprehensive paleoclimate proxies.

The relationship between ENSO and global warming is largely unknown. We are not even sure if greenhouse warming will result in an El Niño-like or La Niña-like pattern in the tropical Pacific. To understand the relationship between anthropogenic and natural climate variability, global coupled ocean-atmosphere models must be greatly improved and simulate both ENSO and the response to greenhouse warming.

Although it is an important topic of the CLIVAR-GOALS (Global Ocean Atmosphere Land System) program, ENSO is no longer the focus of major international process studies. Keeping in mind that one of the major successes of TOGA was the synergy among observationalists, theoreticians and modelers, every effort should be made to keep this synergy alive.

Despite great progress since the beginning of TOGA, understanding the mechanisms of ENSO is far from completed. In view of the rate of advancement in modeling and assimilation over the same time, it is expected that realistic simulation of ENSO will be feasible by the end of CLIVAR. This goal is linked for obvious socio-economic reasons to the improvement of ENSO prediction. However, one should never forget that understanding a fascinating puzzle of nature, like ENSO, must remain a key driver of future research.

Acknowledgments. CW was supported by grants from NOAA Office of Global Programs and from NSFC (Grant: 40176003) and by NOAA Environmental Research Laboratories through their base funding of AOML. JP was supported by IRD (Institut de Recherche pour le Développement) and PNEDC (Programme National d'Etude de la Dynamique du Climat). Comments by D. B. Enfield, C. Maes, M. J. McPhaden, J. Zavala-Garay, and most especially the two anonymous reviewers helped improve the manuscript. S. K. Lee assisted in plotting Figure 1.

REFERENCES

AchutaRao, K., and K. R. Sperber, Simulation of the El Niño Southern Oscillation: results from the coupled model intercomparison project, *Clim. Dyn., 19,* 191–209, 2002.

Alexander, M. A., N.-C. Lau, and J. D. Scott, Broadening the atmospheric bridge paradigm: ENSO teleconnections to the tropical west Pacific-Indian Oceans over the seasonal cycle and to the North Pacific in summer, this volume.

Alory, G., and T. Delcroix, Interannual sea level changes and associated mass transport in the tropical Pacific from TOPEX/Poseidon data and linear model result (1964–1999), *J. Geophys. Res., 107,* C10, 3153, doi:10.1029/2001JC001067, 2002.

An, S.-I., and F.-F. Jin, Collective role of thermocline and zonal advective feedbacks in the ENSO mode, *J. Clim., 14,* 3421–3432, 2001.

Auad, G., A. J. Miller, and W. B. White, Simulation of heat storages and associated heat budget in the Pacific Ocean 2. Interdecadal timescale, *J. Geophys. Res., 103,* 27,621–27,635, 1998.

Balmaseda, M. A., M. K. Davey, and D. L. T. Anderson, Decadal and seasonal dependence of ENSO prediction skill, *J. Clim., 8,* 2705–2715, 1995.

Barnett, T. P., et al., Interdecadal interactions between the tropics and midlatitudes in the Pacific basin, *Geophys. Res. Lett., 26,* 615–618, 1999.

Barnston, A. G., Y. He, and M. H. Glantz, Predictive skill of statistical and dynamical climate models in SST forecasts during the 1997–98 El Niño episode and the 1998 La Niña onset, *Bull. Am. Meteorol. Soc., 80,* 217–244, 1999.

Battisti, D. S., On the role of off-equatorial oceanic Rossby waves during ENSO, *J. Phys. Oceanogr., 19,* 551–559, 1989.

Battisti, D. S., Reply, *J. Phys. Oceanogr., 21,* 461–465, 1991.

Battisti, D. S., and A. C. Hirst, Interannual variability in the tropical atmosphere-ocean model: influence of the basic state, ocean geometry and nonlinearity, *J. Atmos. Sci., 45,* 1687–1712, 1989.

Battisti, D. S., and E. S. Sarachik, Understanding and predicting ENSO, *Rev. Geophys., Sup., 33,* 1367–1376, 1995.

Bergman, J. W., H. Hendon, and K. M. Weickmann, Intraseasonal air-sea interactions at the onset of El Niño, *J. Clim., 14,* 1702–1719, 2001.

Bjerknes, J., Atmospheric teleconnections from the equatorial Pacific, *Mon. Weather Rev., 97,* 163–172, 1969.

Blanke, B., J. D. Neelin, and D. Gutzler, Estimating the effect of stochastic wind stress forcing on ENSO irregularity, *J. Clim., 10*, 1473–1486, 1997.

Boccaletti, G., R. C. Pacanowski, and S. G. H. Philander, A diabatic mechanism for decadal variability in the tropics, *J. Clim.*, submitted, 2004.

Boulanger, J.-P., and C. Menkes, The TRIDENT Pacific model. Part II: The thermodynamical model and the role of long equatorial wave reflection during the 1993–1998 TOPEX/POSEIDON period, *Clim. Dyn., 17*, 175–186, 2001.

Boulanger, J.-P., S. Cravatte, and C. Menkes, Reflected and locally wind-forced interannual equatorial Kelvin waves in the western Pacific Ocean, *J. Geophys. Res., 108 (C10)*, 3311, doi:10.1029/2002JC001760, 2003.

Burgers, G., The El Niño stochastic oscillator, *Clim. Dyn., 15*, 521–531, 1999.

Cai, W., and P. H. Whetton, Evidence for a time-varying pattern of greenhouse warming in the Pacific Ocean, *Geophys. Res. Lett., 27*, 2577–2580, 2000.

Cai, W., and P. H. Whetton, Modes of SST variability and the fluctuation of global mean temperature, *Clim. Dyn., 17*, 889–901, 2001a.

Cai, W., and P. H. Whetton, A time varying greenhouse warming pattern and the tropical-extratropical circulation linkage in the Pacific Ocean, *J. Clim., 14*, 3337–3355, 2001b.

Cane, M. A., et al., Twentieth-century sea surface temperature trends, *Science, 275*, 957–960, 1997.

Capotondi, A., and M. A. Alexander, Rossby waves in the tropical North Pacific and their role in decadal thermocline variability, *J. Phys. Oceanogr., 31*, 3496–3515, 2001.

Chang, P., L. Ji, B. Wang, and T. Li, Interactions between the seasonal cycle and El Niño-Southern Oscillation in an intermediate coupled ocean-atmosphere model, *J. Atmos. Sci., 52*, 2353–2372, 1995.

Chang, P., B. S. Giese, L. Ji, and H. F. Seidel, Decadal change in the south tropical Pacific in a global assimilation analysis, *Geophys. Res. Lett., 28*, 3461–3464, 2001.

Chao, Y., M. Ghil, and J. C. McWilliams, Pacific interdecadal variability in this century's sea surface temperatures, *Geophys. Res. Lett., 27*, 2261–2264, 2000.

Chavez, F. P., J. Ryan, S. E. Lluch-Cota, and M. Niquen, From Anchovies to sardines and back: multidecadal change in the Pacific Ocean, *Science, 299*, 217–221, 2003.

Clarke, A. J., On the reflection and transmisson of low frequency energy at the irregular western Pacific Ocean boundary, *J. Geophys. Res., 96 (Suppl.)*, 3289–3305, 1991.

Clarke, A. J., J. Wang, and S. Van Gorder, A simple warm-pool displacement ENSO model, *J. Phys. Oceanogr., 30*, 1679–1691, 2000.

Cobb, K. M., C. D. Charles, H. Cheng, and R. L. Edwards, El Niño-Southern Oscillation and tropical Pacific climate during the last millennium, *Nature, 424*, 271–276, 2003.

Cole, J., A slow dance for El Niño, *Science, 291*, 1496–1497, 2001.

Collins, M., The El Niño-Southern Oscillation in the second Hadley Centre coupled model and its response to greenhouse warming, *J. Clim., 13*, 1299–1312, 2000a.

Collins, M., Understanding uncertainties in the response of ENSO to greenhouse warming, *Geophys. Res. Lett., 27*, 3509–3512, 2000b.

Cravatte, S., J. Picaut, and G. Eldin, Second and first baroclinic Kelvin modes in the equatorial Pacific at intraseasonal timescales, *J. Geophys. Res., 108(C8)*, 3266, doi:10.1029/2002JC001511, 2003.

Cronin, M. F., et al., Enhanced oceanic and atmospheric monitoring underway in eastern Pacific, *Eos, 83*, 210–211, 2002.

Curtis, S., and S. Hastenrath, Long-term trends and forcing mechanisms of circulation and climate in the equatorial Pacific, *J. Clim., 12*, 1134–1144, 1999.

Davey, M. K., et al., STOIC: a study of coupled model climatology and variability in tropical ocean regions, *Clim. Dyn., 18*, 403–420, 2002.

Delcroix, T., et al., Equatorial waves and warm pool displacements during the 1992–1998 El Niño-Southern Oscillation events: observations and modeling, *J. Geophys. Res, 105*, 26045–26062, 2000.

Deser, C., and J. M. Wallace, Large-scale atmospheric circulation features of warm and cold episodes in the tropical Pacific, *J. Clim., 3*, 1254–1281, 1990.

Deser, C., M. Alexander, and M. Timlin, Upper-ocean thermal variations in the North Pacific during 1970–1991, *J. Clim., 9*, 1840–1855, 1996.

Dewitte, B., et al., Tropical Pacific baroclinic mode contribution and associated long waves for the 1994–999 period from an assimilation experiment with altimetric data, *J. Geophys. Res., 108*, doi:10.1029/2002JC001362, 2003.

Diaz, H. F., M. P. Hoerling, and J. K. Eischeid, ENSO variability, teleconnections and climate change, *Int. J. Climatol., 21*, 1845–1862, 2001.

Dijkstra, H. A., and G. Burgers, Fluid dynamics of El Niño variability, *Annu. Rev. Fluid Mech., 34*, 531–558, 2002.

Dommenget, D., and M. Latif, A cautionary note on the interpretation of EOFs, *J. Clim., 15*, 216–225, 2002.

Eckert, C., and M. Latif, Predictability of a stochastically forced hybrid coupled model of El Niño, *J. Clim., 10*, 1488–1504, 1997.

Enfield, D. B, El Niño, past and present, *Rev. Geophys., 27*, 159–187, 1989.

Enfield, D. B., and L. Cid-Serrano, Low-frequency changes in El Niño/Southern Oscillation, *J. Clim., 4*, 1137–1146, 1991.

Evans, M. N., et al., Support for tropically-driven Pacific decadal variability based on paleoproxy evidence, *Geophys. Res. Lett., 28*, 3689–3692, 2001.

Farrell, B. F., and P. J. Ioannou, Generalized stability theory. Part I: Autonomous Operators, *J. Atmos. Sci., 53*, 2025–2040, 1996a.

Farrell, B. F., and P. J. Ioannou, Generalized stability theory. Part II: Nonautonomous Operators, *J. Atmos. Sci., 53*, 2041–2053, 1996b.

Fedorov, A. V., The response of the coupled tropical ocean-atmosphere to westerly wind bursts, *Q. J. R. Meteorol. Soc., 128*, 1–23, 2002.

Fedorov, A. V., and S. G. Philander, A stability analysis of tropical ocean-atmosphere interactions: Bridging measurements and theory for El Niño, *J. Clim., 14*, 3086–3101, 2001.

Fedorov, A. V., et al., How predictable is El Niño? *Bull. Am. Meteorol. Soc., 84*, 911–919, 2003.

Flügel, M., and P. Chang, Does the predictability of ENSO depend on the seasonal cycle? *J. Atmos. Sci., 55,* 3230–3243, 1998.

Folland, C. K., J. A. Renwick, M. Salinger, and A. B. Mullan, Relative influence of the interdecadal Pacific oscillation and ENSO on the South Pacific convergence zone, *Geophys. Res. Lett., 29,* 13, doi:10.1029/2001GL014201, 2002.

Fukumori, I., T. Lee, B. Cheng, and D. Menemenlis, The origin, pathway, and destination of Niño3 water estimated by a simulated passive tracer and its adjoint, *J. Phys. Oceanogr., 34,* 582–604, 2004.

Garreaud, R. D., and D. S. Battisti, Interannual (ENSO) and interdecadal (ENSO-like) variability in the Southern tropospheric circulation, *J. Clim., 12,* 2113–2123, 1999.

Gershunov, A., and T. Barnett, Interdecadal modulation of ENSO teleconnection, *Bull. Am. Meteorol. Soc., 79,* 2715–2725, 1998.

Giese, B. S., and J. A. Carton, Interannual and decadal variability in the tropical and midlatitude Pacific Ocean, *J. Clim., 12,* 3402–3418, 1999.

Giese, B. S., S. C. Urizar, and N. S. Fuckar, Southern hemisphere origins of the 1976 climate shift, *Geophys. Res. Lett., 29,* 10.1029/2001GL013268, 2002.

Gill, A. E., Some simple solutions for heat-induced tropical circulation, *Q. J. R. Meteorol. Soc., 106,* 447–462, 1980.

Gill, A. E., Elements of coupled ocean-atmosphere models for the tropics, in *Coupled Ocean-Atmosphere Models,* edited by J. C. Nihoul, pp. 303–327, Vol. 40, Elsevier Oceanography Series, Elsevier, 1985.

Godfrey, J. S., et al., Coupled Ocean-Atmosphere Response Experiment (COARE): an interim report, *J. Geophys. Res., 103,* 14,395–14,450, 1998.

Goswami, B. N., and M. A. Thomas, Coupled ocean-atmosphere inter-decadal modes in the tropics, *J. Meteorol. Soc. Jpn., 78,* 765–775, 2000.

Graham, N. E., and W. B. White, The El Niño cycle: A natural oscillator of the Pacific Ocean- atmosphere system, *Science, 24,* 1293–1302, 1988.

Graham, N. E., and W. B. White, Comments on "On the role of off-equatorial oceanic Rossby waves during ENSO", *J. Phys. Oceanogr., 21,* 453–460, 1991.

Gu, D., and S. G. H. Philander, Interdecadal climate fluctuations that depend on exchange between the tropics and extratropics, *Science, 275,* 805–807, 1997.

Guilderson, T. P., and D. P. Schrag, Abrupt shift in subsurface temperatures in the tropical Pacific associated with changes in El Niño, *Science, 281,* 240–243, 1998.

Harper, S., Thermocline ventilation and pathways of tropical-subtropical water mass exchange, *Tellus, 52A,* 330–345, 2000.

Harrison, D. E., and N. K. Larkin, Darwin sea level pressure, 1876–1996: Evidence for climate change? *Geophys. Res. Lett., 24,* 1779–1782, 1997.

Harrison, D. E., and G. A. Vecchi, Westerly wind events in the tropical Pacific, 1986–95, *J. Clim., 10,* 3131–3156, 1997.

Harrison, D. E., and G. A. Vecchi, On the termination of El Niño, *Geophys. Res. Lett., 26,* 1593–1596, 1999.

Hasegawa T., and K. Hanawa, Heat content variability related to ENSO events in the Pacific, *J. Phys. Oceanogr., 33,* 407–421, 2003a.

Hasegawa T., and K. Hanawa, Decadal-scale variability of upper ocean heat content in the tropical Pacific, *Geophys. Res. Lett., 6,* doi:10.1029/2002GL016843, 2003b.

Hayes, S. P., et al., TOGA-TAO: A moored array for real-time measurements in the tropical Pacific Ocean, *Bull. Am. Meteorol. Soc., 72,* 339–347, 1991.

Hazeleger, W., et al., Decadal upper ocean temperature variability in the tropical Pacific, *J. Geophys. Res., 106,* 8971–8988, 2001.

Hendon, H. H., B. Liebmann, and J. D. Glick, Oceanic Kelvin waves and the Madden-Julian Oscillation, *J. Atmos. Sci., 55,* 88–101, 1998.

Hirst, A. C., Unstable and damped equatorial modes in simple coupled ocean-atmosphere models, *J. Atmos. Sci., 43,* 606–630, 1986.

Hirst, A. C., Slow instabilities in tropical ocean basin-global atmosphere models, *J. Atmos. Sci., 45,* 830–852, 1988.

Holland, C. L., and G. T. Mitchum, Interannual volume variability in the tropical Pacific, *J. Geophys. Res., 108 (C11),* 3369, doi:10.1029/2003JC001835, 2003.

Hu, Z.-Z., et al., Impact of global warming on the interannual and interdecadal climate modes in a coupled GCM, *Clim. Dyn., 17,* 361–374, 2001.

Izumo, T., J. Picaut, and B. Blanke, Tropical pathways, equatorial undercurrent variability and the 1998 La Niña, *Geophys. Res. Lett., 29,* doi:10.1029/2002GL015073, 2002.

Jacobs, G. A., et al., Decade-scale trans-Pacific propagation and warming effects of an El Niño anomaly, *Nature, 370,* 360–363, 1994.

Janicot, S., S. Trzaska, and I. Poccard, Summer Sahel-ENSO teleconnection and decadal time scale SST variations, *Clim. Dyn., 18,* 303–320, 2001.

Jin, F.-F., An equatorial ocean recharge paradigm for ENSO. Part I: Conceptual model, *J. Atmos. Sci., 54,* 811–829, 1997.

Jin, F.-F., Low-frequency modes of tropical ocean dynamics, *J. Clim., 14,* 3874–3881, 2001.

Jin, F.-F., and J. D. Neelin, Modes of interannual tropical ocean-atmosphere interaction—a unified view. Part I: Numerical Results, *J. Atmos. Sci., 50,* 3477–3503, 1993.

Jin, F.-F., J. D. Neelin, and M. Ghil, El Niño on the Devil's Staircase: annual subharmonic steps to chaos, *Science, 264,* 70–72, 1994.

Jin, F.-F., M. Kimoto, and X. Wang, A model of decadal ocean-atmosphere interaction in the North Pacific basin, *Geophys. Res. Lett., 28,* 1531–1534, 2001.

Jin, F.-F., S.-I. An, A. Timmermann, and J. Zhao, Strong El Niño events and nonlinear dynamical heating, *Geophys. Res. Lett., 30(3),* 1120, doi:10.1029/2002GL016356, 2003.

Johnson, G. C., The Pacific Ocean subtropical cell surface limb, *Geophys. Res. Lett., 9,* 1771–1774, 2001.

Johnson, G. C., and M. J. McPhaden, Interior pycnocline flow from the subtropical to the equatorial Pacific Ocean, *J. Phys. Oceanogr., 29,* 3073–3089, 1999.

Kaplan, A., et al., Analyses of global sea surface temperature 1856–1991, *J. Geophys. Res., 103,* 18,567–18,589, 1998.

Karspeck, A. R., and M. A. Cane, Tropical Pacific 1976–77 climate shift in a linear, wind-driven model, *J. Phys. Oceanogr., 32,* 2350–2360, 2002.

Kessler, W. S., Can reflected extra-equatorial Rossby waves drive ENSO? *J. Phys. Oceanogr.*, *21*, 444–452, 1991.

Kessler, W. S., Is ENSO a cycle or a series of events? *Geophys. Res. Lett.*, *29(23)*, 2125, doi:10.1029/2002GL015924, 2002.

Kessler, W. S., and R. Kleeman, Rectification of the Madden-Julian Oscillation into the ENSO cycle, *J. Clim.*, *13*, 3560–3575, 2000.

Kessler, W. S., M. J. McPhaden, and K. M. Weickmann, Forcing of intraseasonal Kelvin waves in the equatorial Pacific, *J. Geophys. Res.*, *100*, 10,613–10,631, 1995.

Kessler, W. S., R. E. Davis, K. Takeuchi, and R. Lukas, The Pacific Basin Extended Climate Study, in *Observing the Ocean of the 21st Century*, edited by C. J. Koblinsky and N. R. Smith, pp. 473–484, Aust. Bur. Meteor. Melbourne, 2001.

Kirtman B. P., and P. S. Schopf, Decadal variability in ENSO predictability and prediction, *J. Clim.*, *11*, 2804–2822, 1998.

Kleeman, R., and A. M. Moore, A theory for the limitation of ENSO predictability due to stochastic atmospheric transients, *J. Atmos. Sci.*, *54*, 753–767, 1997.

Kleeman, R., J. P. McCreary, and B. A. Klinger, A mechanism for generating ENSO decadal variability, *Geophys. Res. Lett.*, *26*, 1743–1746, 1999.

Knutson, T. R., and S. Manabe, Time-mean response over the tropical Pacific to increased CO_2 in a coupled ocean-atmosphere model, *J. Clim.*, *8*, 2181–2199, 1995.

Knutson, T. R., and S. Manabe, Simulated ENSO in a global coupled ocean-atmosphere model: multidecadal amplitude modulation and CO_2 sensitivity, *Bull. Am. Meteorol. Soc.*, *10*, 138–161, 1997.

Knutson, T. R., and S. Manabe, Model assessment of decadal variability and trends in the tropical Pacific Ocean, *J. Clim.*, *11*, 2273–2294, 1998.

Kug, J.-S., and I.-S. Kang, Symmetric and antisymmetric mass exchanges between the equatorial and off-equatorial Pacific associated with ENSO, *J. Geophys. Res.*, *108(C8)*, 3284, doi:10.1029/2002JC001671, 2003.

Kutsuwada, K., and M. J. McPhaden, Intraseasonal variations in the upper equatorial Pacific Ocean prior to and during the 1997–98 El Niño, *J. Phys. Oceanogr.*, *32*, 1133–1149, 2002.

Lagerloef, G. S., and T. Delcroix, Sea surface salinity: a regional case study for the tropical Pacific, in *Observing the Ocean of the 21st Century*, edited by C. J. Koblinsky and N. R. Smith, pp. 231–246, Aust. Bur. Meteor. Melbourne, 2001.

Landsea, C. W., and J. A. Knaff, How much skill was there in forecasting the very strong 1997–98 El Niño, *Bull. Am. Meteorol. Soc.*, *81*, 2107–2119, 2000.

Larkin, N. K., and D. E. Harrison, ENSO warm (El Niño) and cold (La Niña) event life cycles: Ocean surface anomaly patterns, their symmetries, asymmetries, and implications, *J. Clim.*, *15*, 1118–1140, 2002.

Latif, M., Dynamics of interdecadal variability in coupled ocean-atmosphere models, *J. Clim.*, *11*, 602–624, 1998.

Latif, M., et al., A review of the predictability and prediction of ENSO, *J. Geophys. Res.*, *103*, 14375–14393, 1998.

Latif, M., et al., ENSIP: the El Niño simulation intercomparison project, *Clim. Dyn.*, *18*, 255–276, 2001.

Lau, K. M., Elements of a stochastic-dynamical theory of long-term variability of the El Niño-Southern Oscillation, *J. Atmos. Sci.*, *42*, 1552–1558, 1985.

Lau, K.-M., and H. Weng, Interannual, decadal-interdecadal, and global warming signals in sea surface temperature during 1955–97, *J. Clim.*, *12*, 1257–1267, 1999.

Lengaine, M., et al., Ocean response to the March 1997 westerly wind event, *J. Geophys. Res.*, *107(C12)*, 8015, doi:10.1029/2001JC000841, 2002.

Lengaigne, M., et al., Westerly wind events in the tropical Pacific and their influence on the coupled ocean-atmosphere system: A review, this volume.

Li, B., and A. J. Clarke, An examination of some ENSO mechanisms using interannual sea level at the eastern and western equatorial boundaries and the zonally averaged equatorial wind, *J. Phys. Oceanogr.*, *24*, 681–690, 1994.

Lindzen, R. S., and S. Nigam, On the role of sea surface temperature gradients in forcing low-level winds and convergence in the Tropics, *J. Atmos. Sci.*, *44*, 2418–2436, 1987.

Linsley, B. K., G. M. Wellington, and D. P. Schrag, Decadal sea surface temperature variability in the subtropical South Pacific from 1726 to 1997 A.D., *Science*, *290*, 1145–1148, 2000.

Liu, Z., A simple model of the mass exchange between the subtropical and tropical ocean, *J. Phys. Oceanogr.*, *24*, 1153–1165, 1994.

Liu, Z., The role of ocean in the response of tropical climatology to global warming: the west-east SST contrast, *J. Clim.*, *11*, 864–875, 1998.

Liu, Z., and B. Huang, Cause of tropical Pacific warming trend, *Geophys. Res. Lett.*, *27*, 1935–1938, 2000.

Liu, Z., L. Wu, R. Gallimore, and R. Jacobs, Search for the origins of Pacific decadal climate variability, *Geophys. Res. Lett.*, *29*, doi:10.1029/2001GL013735, 2002.

Lukas, R., and E. Lindstrom, The mixed layer of the western equatorial Pacific Ocean, *J. Geophys. Res.*, *96*, 3343–3357, 1991.

Luo, J.-J., and T. Yamagata, Long-term El Niño-Southern Oscillation (ENSO) like variation with special emphasis on the South Pacific, *J. Geophys. Res.*, *106*, 22,211–22,227, 2001.

Luo, J.-J., and T. Yamagata, Four decadal ocean-atmosphere modes in the North Pacific revealed by various analysis method, *J. Oceanogr.*, *58*, 861–876, 2002.

Luo, J.-J., et al., South Pacific origin of the decadal ENSO-like variation as simulated by a coupled GCM, *Geophys. Res. Lett.*, *30*, 24, 2250, doi:10.1029/2003GL018649, 2003.

Lysne, J., P. Chang, and B. Giese, Impact of the extratropical Pacific on equatorial variability, *Geophys. Res. Lett.*, *24*, 2589–2592, 1997.

Madden, R. A., and P. R. Julian, Observations of the 40–50-day tropical oscillation—A review, *Mon. Weather Rev.*, *122*, 814–837, 1994.

Maes, C., J. Picaut, and S. Belamari, Barrier layer and onset of El Niño in a Pacific coupled model, *Geophys. Res. Lett.*, *29*, 24, 2206, doi:10.1029/2002GL016029, 2002.

Mann, M. E., R. S. Bradley, and M. K. Hughes, Long-term variability in the El Niño/Southern Oscillation and associated teleconnections, in *El Niño and the Southern Oscillation, multiscale variability and regional impacts,* edited by H. Diaz and V. Markgraf, pp. 357–412, Cambridge University Press, 2000.

Mantua, N. J., and D. S. Battisti, Evidence of the delayed oscillator mechanism for ENSO: the "observed" oceanic Kelvin mode in the far western Pacific, *J. Phys. Oceanogr., 24,* 691–699, 1994.

Mantua, N. J., and D. S. Battisti, Aperiodic variability in the Zebiak-Cane coupled ocean-atmosphere model: Air-sea interactions in the western equatorial Pacific, *J. Clim., 8,* 2897–2927, 1995.

Mantua, N. J., and S. R. Hare, The Pacific Decadal Oscillation, *J. Oceanogr., 58,* 35–44, 2002.

Mantua, N. J., et al., A Pacific interdecadal climate oscillation with impacts on salmon production, *Bull. Am. Meteorol. Soc., 78,* 1069–1079, 1997.

Markgraf, V., and H. Diaz, The past ENSO record: a synthesis, in *El Niño and the Southern Oscillation, multiscale variability and regional impacts,* edited by H. Diaz and V. Markgraf, pp. 465–488, Cambridge University Press, 2000.

Matsuura, T., and S. Iizuka, Zonal migration of the Pacific warm pool tongue during El Niño events, *J. Phys. Oceanogr., 30,* 1582, 2000.

Mayer, D. A., and R. H. Weisberg, El Niño-Southern Oscillation-related ocean-atmosphere coupling in the western equatorial Pacific, *J. Geophys. Res., 103,* 18635–18648, 1998.

McCreary, J. P., A model of tropical ocean-atmosphere interaction, *Mon. Weather Rev., 111,* 370–387, 1983.

McCreary, J. P., and D. L. T. Anderson, A simple model of El Niño and the Southern Oscillation, *Mon. Weather Rev., 112,* 934–946, 1984.

McCreary, J. P., and D. L. T. Anderson, An overview of coupled ocean-atmosphere models of El Niño and the Southern Oscillation, *J. Geophys. Res., 96,* 3125–3150, 1991.

McCreary, J. P., and P. Lu, Interaction between the subtropical and equatorial ocean circulations: the subtropical cell, *J. Phys. Oceanogr., 24,* 466–497, 1994.

McPhaden, M. J., The Tropical Atmosphere Ocean Array is completed, *Bull. Am. Meteorol. Soc., 76,* 739–741, 1995.

McPhaden, M. J., Genesis and evolution of the 1997–1998 El Niño, *Science, 283,* 950–954, 1999.

McPhaden, M. J., and X. Yu, Equatorial waves and the 1997–98 El Niño, *Geophys. Res. Lett, 26,* 2961–2964, 1999.

McPhaden, M. J., and D. Zhang, Slowdown of the meridional overturning circulation in the upper Pacific Ocean, *Nature, 415,* 603–608, 2002.

McPhaden, M. J., et al., The Tropical Ocean-Global Atmosphere observing system: A decade of progress, *J. Geophys. Res., 103,* 14169–14240, 1998.

McWilliams, J., and P. Gent, A coupled air-sea model for the tropical Pacific, *J. Atmos. Sci., 35,* 962–989, 1978.

Mechoso, C. R., J. D. Neelin, and J.-Y. Yu, Testing simple model of ENSO, *J. Atmos. Sci., 60,* 305–318, 2003.

Meehl, G. A., and W. M. Washington, Tropical response to increased CO_2 in a GCM with a simple mixed layer ocean: similarities to an observed Pacific warm event, *Mon. Weather Rev., 114,* 667–674, 1986.

Meehl, G. A., and W. M. Washington, El Niño-like climate change in a model with increased atmospheric CO_2 concentrations, *Nature, 382,* 56–60, 1996.

Meinen, C. S., and M. J. McPhaden, Observations of warm water volume changes in the equatorial Pacific and their relationship to El Niño and La Niña, *J. Clim., 13,* 3551–3559, 2000.

Meinen, C. S., and M. J. McPhaden, Interannual variability in warm water volume transport in the equatorial Pacific during 1993–99, *J. Phys. Oceanogr., 31,* 1324–1345, 2001.

Mestas-Nuñez A. M., and D. B. Enfield, Eastern equatorial Pacific SST variability: ENSO and non-ENSO components and their climatic associations, *J. Clim., 14,* 391–402, 2001.

Miller, A. J., and N. Schneider, Interdecadal climate regime dynamics in the North Pacific Ocean: theories, observations and ecosystem impacts, *Prog. Oceanogr., 47,* 355–379, 2000.

Miller, G. T., and R. Cheney, Large-scale meridional transport in the tropical Pacific Ocean during the 1986–87 El Niño from Geosat, *J. Geophys. Res., 95,* 17905–17919, 1990.

Minobe, S., Interannual to interdecadal changes in the Bering Sea and concurrent 1998/99 changes over the North Pacific, *Prog. Oceanogr., 55,* 45–64, 2000.

Mokhov, I. I., A. V. Eliseev, and D. K. Khvorost'yanov, Evolution of the characteristics of interannual climate variability associated with the El Niño and La Niña phenomenon, *Izv. Atmos. Ocean. Phys., 36,* 681–690, 2000.

Moore, A. M., and R. Kleeman, The dynamics of error growth and predictability in a coupled model of ENSO, *Q. J. R. Meteorol. Soc., 122,* 1405–1446, 1996.

Moore, A. M., and R. Kleeman, Stochastic forcing of ENSO by intraseasonal oscillations, *J. Clim., 12,* 1199–1220, 1999a.

Moore, A. M., and R. Kleeman, The nonnormal nature of El Niño and intraseasonal variability, *J. Clim., 12,* 2965–2982, 1999b.

Moore, A. M., et al., The role of air-sea interaction in controlling the optimal perturbations of low-frequency tropical coupled ocean-atmosphere modes, *J. Clim., 16,* 951–968, 2003.

Neelin, J. D., The slow sea surface temperature mode and the fast-wave limit: analytic theory for tropical interannual oscillations and experiments in a hybrid coupled models, *J. Atmos. Sci., 48,* 584–606, 1991.

Neelin, J. D., and F.-F. Jin, Modes of interannual tropical ocean-atmosphere interaction—a unified view. Part II: Analytical results in the weak-coupling limit, *J. Atmos. Sci., 50,* 3504–3522, 1993.

Neelin, J. D., et al., ENSO theory, *J. Geophys. Res., 103,* 14,262–14,290, 1998.

Nitta, T., and S. Yamada, Recent warming of tropical sea surface temperature and its relationship to the northern hemisphere circulation, *J. Meteorol. Soc. Jpn., 67,* 375–383, 1989.

Nonaka, M., S.-P. Xie, and J. P. McCreary, Decadal variations in the subtropical cells and equatorial SST, *Geophys. Res. Lett., 29,* doi:10.1029/2001GL013717, 2002.

Ortlieb, L., The documented historical record of El Niño events in Peru: an update of the Quinn record (sixteen through nineteen centuries), in *El Niño and the Southern Oscillation, multiscale variability and regional impacts,* edited by H. Diaz and V. Markgraf, pp. 207–295, Cambridge University Press, 2000.

Penland, C., A stochastic model of IndoPacific sea surface temperature anomalies, *Physica D, 98,* 534–558, 1996.

Penland, C., and P. D. Sardeshmukh, The optimal growth of tropical sea surface temperature anomalies, *J. Clim., 8,* 1999–2024, 1995.

Philander, S. G., El Niño and La Niña, *J. Atmos. Sci., 42,* 2652–2662, 1985.

Philander, S. G., *El Niño, La Niña, and the Southern Oscillation,* Academic Press, London, 289 pp, 1990.

Philander, S. G., and A. Fedorov, Is El Niño sporadic or cyclic? *Ann. Rev. Earth Planet. Sci., 31,* 579–594, 2003.

Philander, S. G. H., T. Yamagata, and R. C. Pacanowski, Unstable air-sea interactions in the tropics, *J. Atmos. Sci., 41,* 604–613, 1984.

Picaut, J., and T. Delcroix, Equatorial wave sequence associated with warm pool displacement during the 1986–1989 El Niño and La Niña, *J. Geophys. Res., 100,* 18,398–18,408, 1995.

Picaut, J., and A. J. Busalacchi, Tropical ocean variability, *in Satellite Altimetry and Earth Sciences*, Academic Press, edited by L.-L. Fu and A. Cazenave, pp. 217–236, 2001.

Picaut, J., F. Masia, and Y. du Penhoat, An advective-reflective conceptual model for the oscillatory nature of the ENSO, *Science, 277,* 663–666, 1997.

Picaut, J., et al., Mechanism of the zonal displacements of the Pacific warm pool: Implications for ENSO, *Science, 274,* 1486–1489, 1996.

Picaut, J., et al., The oceanic zone of convergence on the eastern edge of the Pacific warm pool: a synthesis of results and implications for ENSO and biogeochemical phenomena, *J. Geophys. Res., 106,* 2363–2386, 2001.

Picaut, J., et al., Mechanisms of the 1997–98 El Niño-La Niña as inferred from space-based observations, *J. Geophys. Res., 107,* doi:10.1029/2001JC000850, 2002.

Pierce, D. W., The role of sea surface temperatures in interactions between ENSO and the North Pacific Oscillation, *J. Clim., 15,* 1295–1308, 2002.

Pierce, D. W., T. Barnett, and M. Latif, Connections between the Pacific Ocean tropics and midlatitudes on decadal timescales, *J. Clim., 13,* 1173–1194, 2000.

Power, S., et al., Inter-decadal modulation of the impact of ENSO on Australia, *Clim. Dyn., 15,* 319–324, 1999.

Rajagopalan, B., U. Lall, and M. A. Cane, Anomalous ENSO occurrences: An alternate view, *J. Clim., 10,* 2351–2357, 1997.

Ramanathan, V., and W. Collins, Thermodynamic regulation of ocean warming by cirrus clouds deduced from observations of the 1987 El Niño, *Nature, 351,* 27–32, 1991.

Rasmusson, E. M., and T. H. Carpenter, Variations in tropical sea surface temperature and surface wind fields associated with the Southern Oscillation/El Niño, *Mon. Weather Rev., 110,* 354–384, 1982.

Rodgers, K. B., P. Friederichs, and M. Latif, Tropical Pacific decadal variability and its relation to decadal modulation of ENSO, *J. Clim.,* in press, 2004.

Roemmich, D., et al., Argo: the global array of profiling floats, in *Observing the Ocean of the 21st Century*, edited by C. J. Koblinsky and N. R. Smith, pp. 248–257, Aust. Bur. Meteor. Melbourne, 2001.

Rothstein, L. M., R.-H. Zhang, A. J. Busalacchi, and D. Chen, A numerical simulation of the mean water pathways in the subtropical and tropical Pacific Ocean, *J. Phys. Oceanogr., 28,* 322–342, 1998.

Salinger, M. J., J. A. Renwick, and A. B. Mullan, Interdecadal Pacific oscillation and South Pacific climate, *Int. J. Climatol., 21,* 1705–1721, 2001.

Schneider, N., A decadal spiciness mode in the tropics, *Geophys. Res. Lett., 27,* 257–260, 2000.

Schneider, N., A. J. Miller, M. A. Alexander, and C. Deser, Subduction of decadal North Pacific temperature anomalies: Observations and dynamics, *J. Phys. Oceanogr., 29,* 1056–1070, 1999a.

Schneider, N., et al., Pacific thermocline bridge revisited, *Geophys. Res. Lett., 26,* 1329–1332, 1999b.

Schott, F. A., J. P. McCreary, and G. C. Johnson, Shallow overturning circulations of the tropical-subtropical oceans, this volume.

Seager, R., and R. Murtugudde, Ocean dynamics, thermocline adjustment, and regulation of tropical SST, *J. Clim., 10,* 521–534, 1997.

Seager, R., et al., Predicting Pacific decadal variability, this volume.

Slingo, J. M., D. P. Rowell, K. R. Sperber, and F. Nortley, On the predictability of the interannual behavior of the Madden-Julian Oscillation and its relationship with El Niño, *Q. J. R. Meteorol. Soc, 125,* 583–560, 1999.

Smith, N. R, et al., The upper ocean thermal network, in *Observing the Ocean of the 21st Century*, edited by C. J. Koblinsky and N. R. Smith, pp. 259–283, Aust. Bur. Meteor. Melbourne, 2001.

Solomon, A., J. P. McCreary, R. Kleeman, and B. A. Klinger, Interannual and decadal variability in an intermediate coupled model of the Pacific region, *J. Clim., 16,* 383–405, 2003.

Springer, S. R., M. J. McPhaden, and A. J. Busalacchi, Oceanic heat content variability in the tropical Pacific during the 1982/1983 El Niño, *J. Geophys. Res., 95,* 22089–22101, 1990.

Suarez, M. J., and P. S. Schopf, A delayed action oscillator for ENSO, *J. Atmos. Sci., 45,* 3283–3287, 1988.

Sun, D.-Z., A possible effect of an increase in the warm-pool SST on the magnitude of El Niño warming, *J. Clim., 16,* 185–205, 2003.

Sun, D.-Z., The control of meridional differential surface heating over the level of ENSO activity: A heat pump hypothesis, this volume.

Syu, H.-H., and J. D. Neelin, ENSO in a hybrid coupled model. Part II: Prediction with piggyback data assimilation, *Clim. Dyn., 16,* 35–48, 2000.

Takayabu, Y. N., et al., Abrupt termination of the 1997–98 El Niño in response to a Madden-Julian-Oscillation, *Nature, 402,* 279–282, 1999.

Tang, T. Y., and R. H. Weisberg, On the equatorial Pacific response to the 1982/1983 El Niño-Southern Oscillation event, *J. Mar. Res., 42,* 809–829, 1984.

Thompson, C. J., and D. S. Battisti, A linear stochastic dynamical model of ENSO. Part II: Analysis, *J. Clim., 14,* 445–466, 2001.

Thompson, L. G., K. A. Henderson, E. Mosley-Thompson, and P.-N. Lin, The tropical ice record of ENSO, in *El Niño and the Southern Oscillation, multiscale variability and global and regional impacts*, edited by H. F. Diaz and V. Markgraf, pp. 325–356, Cambridge Univ. Press, 2000.

Timmermann, A., Decadal ENSO amplitude modulation: a nonlinear paradigm, *Glob. Plan. Change, 37*, 135–156, 2003.

Timmermann, A., F.-F. Jin, and J. Abshagen, A nonlinear theory for El Niño bursting, *J. Atmos. Sci., 60*, 165, 2003.

Timmermann, A., et al., Increased El Niño frequency in a climate model forced by future greenhouse warming, *Nature, 398*, 694–696, 1999.

Torrence, C., and P. J. Webster, Interdecadal changes in the ENSO-Monsoon system, *J. Clim., 12*, 2679–2690, 1999.

Tourre, Y., et al., Patterns of coherent decadal and interdecadal climate signals in the Pacific basin during the 20th century, *Geophys. Res. Lett., 28*, 2069–2072, 2001.

Trenberth, K. E., Recent observed interdecadal climate changes in the Northern Hemisphere, *Bull. Am. Meteorol. Soc., 71*, 988–993, 1990.

Trenberth, K. E., and T. J. Hoar, The 1990–95 El Niño-Southern Oscillation event: longest on record, *Geophys. Res. Lett., 23*, 57–60, 1996.

Trenberth, K. E., and T. J. Hoar, El Niño and climate change, *Geophys. Res. Lett., 24*, 3057–3060, 1997.

Tudhope, A. W., et al., Variability in the El Niño-Southern Oscillation through a glacial-interglacial cycle, *Science, 291*, 1511–1517, 2001.

Tziperman, E., M. Cane, and S. E. Zebiak, Irregularity and locking to the seasonal cycle in the ENSO-prediction model as explained by the quasi-periodicity route to chaos, *J. Atmos. Sci., 52*, 293–306, 1995.

van Oldenborgh, G. J., What caused the onset of the 1997–1998 El Niño? *Mon. Weather Rev., 128*, 2601–2607, 2000.

Vavrus, S., and Z. Liu, Toward understanding the response of the tropical atmosphere-ocean system to increased CO_2 using equilibrium asynchronous coupling, *Clim. Dyn., 19*, 355–369, 2002.

Vecchi, G. A., and D. E. Harrison, Tropical Pacific sea surface temperature anomalies, El Niño, and equatorial westerly wind events, *J. Clim., 13*, 1814–1830, 2000.

Vialard, J., et al., A model study of oceanic mechanisms affecting equatorial Pacific sea surface temperature during the 1997–98 El Niño, *J. Phys. Oceanogr., 31*, 1649–1675, 2001.

Walker, G. T., Correlation in seasonal variations of weather VIII: A preliminary study of world weather, *Mem. Indian Meteor. Dept., 24*, 75–131, 1923.

Walker, G. T., Correlation in seasonal variations of weather IX: A further study of world weather, *Mem. Indian Meteor. Dept., 25*, 275–332, 1924.

Walker, G. T., and E. Bliss, World Weather V, *Mem. Roy. Meteor. Soc., 4*, 53–84, 1932.

Wang, B., Interdecadal changes in El Niño onset in the last four decades, *J. Clim., 8*, 267–285, 1995a.

Wang, B., Transition from a cold to a warm state of the El Niño-Southern Oscillation cycle, *Meteorol. Atmos. Phys., 56*, 17–32, 1995b.

Wang, B., and S.-I. An, A mechanism for decadal changes of ENSO behavior: roles of background wind changes, *Clim. Dyn., 18*, 475–486, 2002.

Wang, C., On the atmospheric responses to tropical Pacific heating during the mature phase of El Niño, *J. Atmos. Sci., 57*, 3767–3781, 2000.

Wang, C., A unified oscillator model for the El Niño-Southern Oscillation, *J. Clim., 14*, 98–115, 2001a.

Wang, C., On the ENSO mechanisms, *Adv. Atmos. Sci., 18*, 674–691, 2001b.

Wang, C., and R. H. Weisberg, On the 'slow mode' mechanism in ENSO-related coupled ocean-atmosphere models, *J. Clim., 7*, 1657–1667, 1994.

Wang, C., and R. H. Weisberg, Stability of equatorial modes in a simplified coupled ocean-atmosphere model, *J. Clim., 9*, 3132–3148, 1996.

Wang, C., and R. H. Weisberg, Climate variability of the coupled tropical-extratropical ocean-atmosphere system, *Geophys. Res. Lett., 25*, 3979–3982, 1998.

Wang, C., and R. H. Weisberg, The 1997–98 El Niño evolution relative to previous El Niño events, *J. Clim., 13*, 488–501, 2000.

Wang, C., R. H. Weisberg, and H. Yang, Effects of the wind speed-evaporation-SST feedback on the El Niño-Southern Oscillation, *J. Atmos. Sci., 56*, 1391–1403, 1999a.

Wang, C., R. H. Weisberg, and J. I. Virmani, Western Pacific interannual variability associated with the El Niño-Southern Oscillation, *J. Geophys. Res., 104*, 5131–5149, 1999b.

Wang, X., F.-F. Jin, and Y. Wang, A tropical ocean recharge mechanism for climate variability. Part II: a unified theory for decadal and ENSO modes, *J. Clim., 16*, 3599–3616, 2003.

Webster, P. J. and R. Lukas, TOGA-COARE: The Coupled Ocean-Atmosphere Response Experiment, *Bull. Am. Meteorol. Soc., 73*, 1377–1416, 1992.

Weisberg, R. H., and C. Wang, A western Pacific oscillator paradigm for the El Niño-Southern Oscillation, *Geophys. Res. Lett., 24*, 779–782, 1997a.

Weisberg, R. H., and C. Wang, Slow variability in the equatorial west-central Pacific in relation to ENSO, *J. Clim., 10*, 1998–2017, 1997b.

Weller, R. A., and S. P. Anderson, Surface meteorology and air-sea fluxes in the western equatorial Pacific warm pool during TOGA COARE, *J. Clim., 9*, 1959–1990, 1996.

White, W. B., Y. M. Tourre, M. Barlow, and M. Dettinger, A delayed action oscillator shared by biennal, interannual, and decadal signals in the Pacific basin, *J. Geophys. Res., 108*, 3070, doi:10.1029/2002JC001490, 2003.

Wittenberg, A. T., On extended wind stress analyses for ENSO, *J. Clim.*, in press, 2004.

Wyrtki, K., El Niño—The dynamic response of the equatorial Pacific Ocean to atmospheric forcing, *J. Phys. Oceanogr., 5*, 572–584, 1975.

Wyrtki, K., Water displacements in the Pacific and genesis of El Niño cycles, *J. Geophys. Res., 90*, 7129–7132, 1985.

Yamagata, T., Stability of a simple air-sea coupled model in the tropics, In *Coupled Ocean-Atmosphere Models*, edited by J. C. J. Nihoul, pp. 637–657, Vol. 40, Elsevier Oceanography Series, Elsevier, 1985.

Yu, B., and G. J. Boer, The roles of radiation and dynamical processes in the El Niño-like response to global warming, *Clim. Dyn., 19*, 539–553, 2002.

Yu, B., and G. J. Boer, The role of the western Pacific in decadal variability, *Geophys. Res. Lett., 31*, L02204, doi:10.1029/2003GL018471, 2004.

Yukimoto, S., et al., ENSO-like interdecadal variability in the Pacific Ocean as simulated in a coupled general circulation model, *J. Geophys. Res., 105*, 13,945–13,963, 2000.

Zang, X., L.-L. Fu, and C. Wunsch, Observed reflectivity of the western boundary of the equatorial Pacific Ocean, *J. Geophys. Res., 107(C10),* 3150, doi:10.1029/2000JC000719, 2002.

Zavala-Garay, J., A. M. Moore, C. L. Perez, and R. Kleeman, The response of a coupled mode of ENSO to observed estimates of stochastic forcing, *J. Clim., 16*, 2827–2842, 2003.

Zebiak, S. E., Oceanic heat content variability and El Niño cycles, *J. Phys. Oceanogr., 19*, 475–486, 1989a.

Zebiak, S. E., On the 30–60 day oscillation and the prediction of El Niño, *J. Clim., 2*, 1381–1387, 1989b.

Zebiak, S. E., Diagnostic studies of Pacific surface winds, *J. Clim., 3*, 1016–1031, 1990.

Zebiak, S. E., and M. A. Cane, A model El Niño-Southern Oscillation, *Mon. Weather Rev., 115*, 2262–2278, 1987.

Zhang, C., Intraseasonal perturbations in sea surface temperatures of the equatorial eastern Pacific and their association with the Madden-Julian Oscillation, *J. Clim., 14*, 1309–1322, 2001.

Zhang, C., and J. Gottschalck, SST anomalies of ENSO and the Madden-Julian Oscillation in the equatorial Pacific, *J. Clim., 15*, 2429–2445, 2002.

Zhang, R.-H., L. M. Rothstein, and A. J. Busalacchi, Origin of upper-ocean warming and El Niño change on decadal scales in the tropical Pacific Ocean, *Nature, 391*, 879–882, 1998.

Zhang, Y., J. M. Wallace, and D. S. Battisti, ENSO-like interdecadal variability: 1900–93, *J. Clim., 10*, 1004–1020, 1997.

J. Picaut, LEGOS, 18 Av. Edouard Belin, 31401 Toulouse cedex 09, France. (Joel.Picaut@cnes.fr)

C. Wang, NOAA/AOML, Physical Oceanography Division, 4301 Rickenbacker Causeway, Miami, Florida 33149. (Chunzai.Wang@noaa.gov)

Westerly Wind Events in the Tropical Pacific and Their Influence on the Coupled Ocean-Atmosphere System: A Review

Matthieu Lengaigne[1], Jean-Philippe Boulanger, Christophe Menkes, and Pascale Delecluse

Laboratoire d'Océanographie Dynamique et de Climatologie, Université Pierre et Marie Curie, Paris, France

Julia Slingo

Center for Global Atmospheric Modelling, University of Reading, Reading, United Kingdom

Observational and modeling aspects about Westerly Wind Events (WWEs) and their influence on the tropical Pacific ocean-atmosphere system are reviewed. WWEs are a large part of the intraseasonal zonal wind activity over the warm pool. They have typical amplitudes of 7 m s^{-1}, zonal width of 20° longitude and duration of about 8 days. Their root causes are often a combination of various factors including the Madden-Julian Oscillation, cold surges from mid-latitudes, tropical cyclones and other mesoscale phenomena. The relationship between WWEs and the ENSO cycle is complex, involving among others the equatorial characteristics of the WWEs, the oceanic background state and the internal atmospheric variability. Both observational and modeling studies demonstrate that WWEs tend to cool the far western Pacific, shift the warm pool eastward and warm the central-eastern Pacific through the generation of Kelvin waves. They are therefore important processes for the central and eastern Pacific warming during the onset and development phase of El Niño. The strong atmospheric feedbacks that are likely to be generated by the ocean response to WWEs even suggest that a single WWE is capable of establishing the conditions under which El Niño can occur. The important role played by WWEs in the evolution and amplitude of recent El Niño events may therefore strongly limit the predictability of El Niño.

1. INTRODUCTION

The tropical Pacific coupled ocean-atmosphere system exhibits large-scale variability on both the seasonal and interannual timescales. The strongest climate variation on interannual timescale is the El Niño/Southern Oscillation (ENSO) phenomenon [*Cane*, 1983]. The air-sea interactions that bring about El Niño equatorial Pacific sea surface temperature (SST) changes are the focus of much interest [see *Wang and Picaut*, this volume]. At present, there are two main para-

---

[1]Now at Center for Global Modelling, University of Reading, Reading, United Kingdom.

Earth's Climate: The Ocean-Atmosphere Interaction
Geophysical Monograph Series 147
Copyright 2004 by the American Geophysical Union
10.1029/147GM03

digms of the evolution of the coupled ENSO system. The first paradigm considers ENSO as a cyclic quasi-periodic self-sustained phenomenon [see *Neelin et al.*, 1998 for a review], which implies El Niño predictability up to one year or so in advance [*Latif et al.*, 1998]. The second views ENSO as a damped phenomenon excited by sub-seasonal atmospheric forcing, in which each El Niño is a relatively independent event [*Kessler*, 2002]. During the past two decades, high-frequency wind variability has indeed been suggested to play a leading role in the evolution and amplitude of the observed warm ENSO events [*Lau and Chan*, 1988; *Kleeman and Moore*, 1997; *Fedorov*, 2002]. Support for this hypothesis has increased following the intense 1997–1998 El Niño, whose growing phase was particularly rapid. This event was characterized by exceptionally strong high-frequency wind variability during the 1996–1997 winter and spring, leading *McPhaden and Yu* [1999] to speculate that this high-frequency wind activity was responsible for the timing and amplitude of the event. To improve our understanding and prediction of El Niño, it is thus necessary to better understand to what extent each of these paradigms controls the variability of ENSO.

At the high frequency end of the sub-seasonal wind band, pulses of westerly wind anomalies, referred to as westerly wind events (WWEs) or westerly wind bursts (WWBs), are prominent features of the western Pacific warm pool. Such episodes are marked by temporally well-defined shifts in wind direction, with speeds exceeding the magnitude of the prevailing trade winds [*Luther and Harrison*, 1984; *Harrison and Vecchi*, 1997]. WWEs are usually associated with convective features [*Fasullo and Webster*, 2000], although they are not the surface expression of any single mode of atmospheric variability and have been observed with a variety of atmospheric circulation phenomena [*Yu and Rienecker*, 1998].

WWEs are of primary interest because they can force substantial local and remote oceanic responses in a region of the world known to be very sensitive to ocean-atmosphere interactions [*Webster and Lukas*, 1992]. However, even if on some occasions WWEs have been potentially implicated in the onset and development of El Niño events (e.g., 1982–83, 1997–98), on other occasions, the bursts appear to have little or no effect (e.g., 1967, 1989–90). Such a variation in the response of the tropical Pacific to WWEs illustrates the complex relationship between high frequency wind variability and the ENSO phenomenon, involving among others the equatorial characteristics (amplitude, duration and fetch) of the WWEs, the oceanic background state (warm pool location, salinity structure, strength of the equatorial SST gradient, depth of the thermocline) and the internal atmospheric variability.

In this paper, the discussion concentrates on WWEs and their impact on the ocean-atmosphere coupled system. This review deals with observational studies, as well as modeling works. Section 2 briefly summarizes our knowledge on the WWEs horizontal and vertical structures, the way they can be classified and their root causes. Section 3 focuses on the remote and local oceanic impact of WWEs, with special attention devoted to the SST response that is likely to modify the atmospheric circulation. The role of high-frequency wind variability on the ENSO phenomenon is discussed in Section 4. This section focuses on the role of high-frequency wind variability as a source of stochastic forcing on the coupled system, as well as the specific impact of WWEs and the MJO on the coupled system trough complex scale interactions. Finally, Section 5 offers a summary and conclusion.

2. MAIN CHARACTERISTICS

The existence of anomalous, short-lived, but strong westerlies in the western equatorial Pacific has been known to meteorologists since the middle of the century [*Palmer*, 1952]. However, detailed studies of the spatial and temporal characteristics of WWEs have for a long time been hindered by a lack of high-quality data and a limited number of observations. First analyses focused on case-studies of individual WWEs, such as those observed during the Winter Monsoon Experiment [*Chang et al.*, 1979], at the onset of the 1982/83 El Niño [*Chu and Fredericks*, 1990] or of the 1986/87 El Niño [*Nitta*, 1989]. Some others, based on short-term observations [*Nitta and Motoki*, 1987; *Kindle and Phoebus*, 1995] or even long-term data records [*Keen*, 1982; *Chu et al.*, 1991; *Verbickas*, 1998], revealed the highly variable characteristics of WWEs. The first more systematic study has been carried out by *Harrison and Giese* [1991], who used a 30-year long island network of near-dateline surface wind observations (1950–1980) to characterize meridional and temporal scales of WWEs in that area. They suggested that near-dateline WWEs could be classified into four relatively distinct types according to distance from the equator of the strongest WWEs anomalies. However, the limited spatial coverage of the island dataset was inadequate to provide a full characterization of the events.

Based on the examination of 10 years of 10-meter winds from the European Center for Medium-Range Weather Forecasts (ECMWF) analyses, *Haarten* [1996] proposed a subjective classification based on large-scale aspects of the circulation associated with periods of WWEs. According to her classification, nine typical patterns can represent the near-surface flow during 90% of the synoptic westerly wind variability. A single cyclone or a series of cyclones and several different types of cross-equatorial flow are the major components of the patterns. Using the same dataset, *Harrison and Vecchi* [1997] (hereafter HV97) extended *Harrison and Giese*'s [1991] previous analysis to the entire Tropical Pacific. They

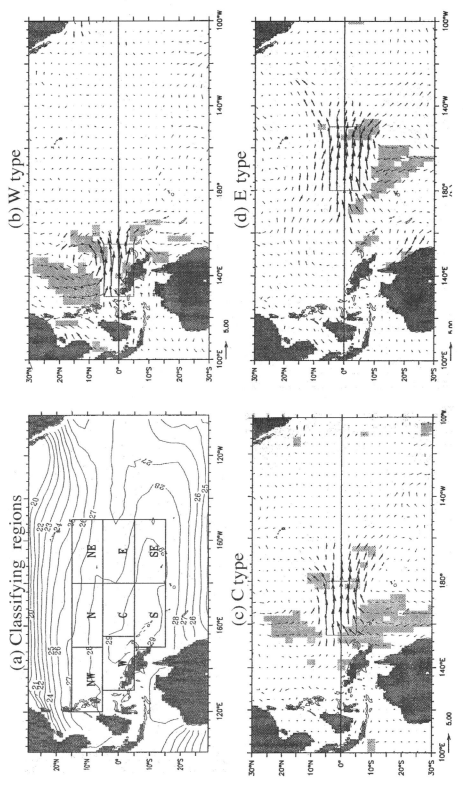

Figure 1. Panel (a) shows the arrangement of the classifying regions for the eight WWE types of HV97. Composite 10-m wind anomalies for the three centered WWE types of HV97: (b) type W, (c) type C, (d) type E. The classifying region is indicated by the thin-lined box. The scale vector is 5 m s^{-1}. Zonal wind anomalies significant at 99% are indicated by bold vectors; meridional wind anomalies significant at 99% are indicateed by shaded background. [from *Harrison and Vecchi*, 1997]

identified eight different types of events depending on the region where the maximum zonal wind anomalies are reached (Figure 1). Using this classifying scheme, HV97 generated composites of surface wind anomaly for each of the WWE types, and found that the wind anomalies of each type were compact and almost stationary during the lifetime of the event. They therefore modeled the WWE zonal wind anomalies using gaussian structures in space and time with e-folding scales fitted to each observed class. Typical zonal scales are between 10° and 30° longitude, meridional scales between 10° and 15° latitude, duration between 5 and 10 days and amplitude of about 6–7 m s^{-1}. HV97 also suggested WWEs to be highly variable on seasonal and interannual time-scales. Moderate to strong WWEs occurrence frequencies are indeed suggested to be influenced by the seasonal cycle: for example, type C events tend to occur preferentially in boreal winter, coinciding with the wet season of the Australian monsoon. Moreover, WWEs occur less often during cold ENSO conditions and are displaced eastward during warm ENSO conditions. Additionally, the vertical structure of these WWEs has been observed to vary significantly, and their coincidence with enhanced clouds and precipitation, whilst evident, can vary in phase, magnitude and duration [*Gutzler et al.*, 1994]. A recent analysis by *Fasullo and Webster* [2000] of 44 WWEs that occurred from 1979 to 1995 in the Intensive Flux Array (151°E–158°E, 5°S–2°N) in the western Pacific enabled a description of their vertical structure. Based on earlier findings by *Kiladis et al.* [1994], this study distinguishes brief (5–25 days) and sustained (30–90 days) WWEs. While brief WWEs have a deep tropospheric structure, sustained WWEs are more baroclinic, with westerlies below and strong easterlies aloft of ~300 hPa. Due to their larger size and duration, *Fasullo and Webster* [2000] suggest that sustained WWEs may be more effective in generating a large-scale coherent oceanic response.

If the basic spatial and temporal characteristics of the WWEs in the Tropical Pacific have been relatively well documented and classified, attempts to explain the existence of WWEs are more problematic. The root cause of convective blowups and WWEs over Pacific warm waters was initially attributed to propagation of cold surges from the northern hemisphere into the tropics [*Chang and Lau*, 1980; *Arkin and Webster*, 1985; *Love*, 1985a; *Chu*, 1988]. In fact, pressure surges especially from eastern Asia are frequently initiating near equatorial surface pressure gradients, surface convergence, enhanced deep convection over the maritime continent and the western Pacific and therefore subsequent WWEs. Modeling and diagnostic studies of extratropical-tropical interactions suggest that such interactions do exist and that they may be an important cause of WWEs [*Lim and Chang*, 1981; *Kiladis et al.*, 1994; *Slingo*, 1998].

However, other bursts appear to be independent of higher latitude forcing [*Compo et al.*, 1999]. Twin or individual tropical cyclones have also been identified as a major atmospheric condition leading to WWEs [*Keen*, 1982; *Love*, 1985b; *Nitta*, 1989; *Lander*, 1990]. The majority (>60%) of the WWEs that are west of the dateline are in fact related to tropical cyclone activity [*Vecchi*, 2001], and the strongest, longer lasting, and more horizontally extensive WWEs are often those related to nearby twin tropical cyclones [*Harrison and Giese*, 1991; *Haarten*, 1996]. It has been suggested that these cyclones pairs could result from vorticity intrusions into the tropical eastern Pacific associated with upper level mid-latitude Rossby wave activity [*Kiladis and Wheeler*, 1995; *Numaguti*, 1995].

In the western and central Pacific, subseasonal wind and convective variability is also an integral part of the Madden-Julian Oscillation [MJO; *Madden and Julian*, 1994], which is the dominant mode of oscillation of the tropical troposphere at intraseasonal timescales. The MJO is characterized by spectral peak in the 30–90 day range in global tropical 850 hPa and 150 hPa zonal wind divergence. This tropics-wide phenomenon is associated with large-scale (wave number 1 and 2) variations in wind and atmospheric convection, which propagate eastward with a characteristics timescale of 40–50 days. As WWEs are associated with atmospheric convection, it has been suggested that the reduction and enhancement of convection associated with MJO over the western Pacific warm waters modulate the intensity and frequency of WWEs [*Lau et al.*, 1989; *Gutzler*, 1991; *Sui and Lau*, 1992; *Godfrey et al.*, 1998] but the exact connection between WWEs and the MJO is still unclear. WWEs commonly occur during the active phase of the MJO but are not characteristic of this oscillation [*Vecchi*, 2001], and none of the equatorial WWEs in HV97 classification shows eastward translation as would be expected from a surface expression of the MJO envelope. However, there are considerable time and space scales that play a role in the MJO. For instance, this oscillation is composed of convective synoptic systems of smaller time and space scales embedded within its eastward moving envelop [*Nakazawa*, 1988] and WWEs could be one surface manifestation of these smaller scale processes. To cope with this issue, *Fasullo and Webster* [2000] examined the relationship between the phase of the MJO and WWEs in the TOGA-COARE IFA (Intensive Flux Array) region, centered south of the equator. Their results support the finding that brief WWEs (6–30 days) occur frequently, independently of the MJO, while sustained WWEs are strongly associated with the MJO active phase. This result is consistent with the baroclinic vertical structure of sustained WWEs, typical of a convectively coupled phenomenon such as the MJO. This suggests, in agreement with *Vecchi* [2001], that the strongest southern hemisphere WWEs might be related to the MJO.

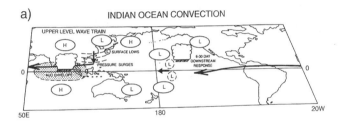
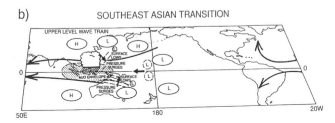
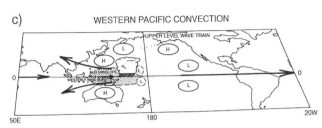

Figure 2. Schematic diagram of time and space scale interactions in the transition of a maximum of convective activity (stippled area) from (a) the eastern Indian ocean, (b) the Maritime Continent and (c) into the western equatorial Pacific. The H and L in ellipses indicate upper tropospheric circulation anomalies depicting major features of intraseasonal timescales. The L in dashed circles near the equator denotes westward moving equatorial Rossby waves. L in solid circles indicates surface low-pressure areas. Thick solid arrows represent major features of anomalous upper tropospheric winds. Thin dashed lines near Australia and eastern Asia denote pressure surges. [from *Meehl et al.*, 1996]

The WWEs characteristics therefore result from complex interactions between MJO, tropical mid-latitude interaction, equatorial waves, tropical cyclones and convective 'blowups' in the western Pacific [*Kiladis and Wheeler*, 1995; *Yu and Rienecker*, 1998]. The schematic diagram on Figure 2 illustrates these interactions during northern winter [*Meehl et al.*, 1996]. As the MJO envelope develops over the Indian Ocean, an extratropical wave train is excited over the northern hemisphere. This propagates eastward and re-enters the upper troposphere of the eastern Pacific in the region of upper level westerlies, the so-called Pacific waveguide [*Webster and Holton*, 1982]. This upper tropospheric incursion can excite equatorial Rossby waves, which propagate westward, interacting with the convectively active envelope of the MJO, as it moves over the West Pacific. At the same time, the perturbation to the extratropics produced by enhanced convection over the Indian Ocean, can excite a cold surge down to the South China sea which further amplifies the western Pacific convective activity and therefore the characteristics of the associated WWEs. The linkages described in Figure 2 have also been described in GCM experiments by *Slingo* [1998]. It should however be kept in mind that the real system contains considerable noise and other possible outcomes for WWEs formation mechanisms.

3. THE FORCED OCEAN RESPONSE

The ocean response to WWEs has been analyzed in many studies using observations [*McPhaden et al.*, 1992; *Delcroix et al.*, 1993; *Feng et al.*, 1998] or ocean models [*Giese and Harrison*, 1991; *Kindle and Phoebus*, 1995; *Richardson et al.*, 1999]. In fact, WWEs are of major interest as they can force substantial oceanic thermodynamical and dynamical response, both locally and remotely. First, they force eastward propagating downwelling equatorial Kelvin waves, which increase the sea surface height, deepen the thermocline and induce eastward surface current anomalies along their paths. Under certain circumstances, downwelling Kelvin waves can also favor the development of positive intense SST anomalies mainly in the central and eastern Pacific. Locally, WWEs affect the structure of the western Pacific warm-pool because they significantly alter surface heat, fresh and momentum fluxes. The local oceanic response to WWEs involves a surface cooling and freshening, a deepening of the mixed layer, and a tightening of the thermocline, along with an eastward surface and westward subsurface current jet. In this section, each aspect of the ocean response to WWEs is detailed.

3.1. Remote Kelvin Wave Response

Linear theory predicts that WWEs will drive current and thermocline depth changes across the tropical Pacific through the excitation of Kelvin and Rossby waves packet [*Matsuno*, 1966; *McCreary*, 1976]. The predicted eastward propagating Kelvin wave will lead to a thermocline deepening and anomalous eastward zonal currents, which amplitude depends on the wind patch properties. Observed evidence for first baroclinic mode Kelvin waves has been first documented through *in situ* observations [*Ripa and Hayes*, 1981; *Knox and Halpern*, 1982; *Eriksen et al.*, 1983; *Hayes and Halpern*, 1984; *Lukas et al.*, 1984; *Enfield*, 1987]. Then, the unprecedented spatiotemporal sea level coverage provided by satellites such as GEOSAT and TOPEX/POSEIDON allowed to describe the propagation, the reflection and the role of the equatorial waves in the sea-level variability [*Miller et al.*, 1988; *Delcroix et* al., 1994; *Boulanger and Menkes*, 1995; *Boulanger et al.*, 2003] as well as to identify their contribution to equatorial zonal currents [*Picaut and Delcroix*, 1995; *Boulanger and Menkes*, 1999; *Delcroix et al.*, 2000]. Observational studies also allowed to estimate the phase speed of the Kelvin wave to be around 2.82 +/- 0.96 m s^{-1} [*Delcroix et al.*, 1991] suggesting the first baroclinic mode to travel 10–30% faster than predicted by the linear theory. Numerical studies have suggested such a large phase speed to be either due to a Doppler shifting by the equatorial undercurrent [*McPhaden et al.*, 1986] or to non-linear auto advection [*Ripa*, 1982; *Fedorov and Melville*, 2000]. Several studies also noted evidence for a second baroclinic mode excited by the observed wind forcing [*Busalacchi and Cane*, 1985; *Cravatte et al.*, 2003] and its significant contribution in the oceanic variability has been underlined in numerical studies [*Giese and Harrison*, 1990; *Kindle and Phoebus*, 1995].

Then, the development of the TOGA-TAO mooring array allowed to elucidate the vertical structure of the Kelvin baroclinic modes [*McPhaden and Taft*, 1988]. *Johnson and McPhaden* [1993a] extracted the zonal velocity and temperature signals of intraseasonal Kelvin waves in central and eastern Pacific (Figure 3). The observed Kelvin wave signal displays a thermocline downwelling and an eastward surge of zonal current. However, this signal shows significant departures from the first-mode theoretical vertical structures due to interactions of the waves with the mean flow. Indeed, zonal velocity shows a maximum at the surface and a second maximum in and below the core of the EUC, while the observed temperature amplitudes are more than twice as large as expected for a first vertical mode Kelvin wave in an otherwise motionless ocean. Linear waves interacting with a zonally invariant zonal mean flow explain some of these features. However, *Johnson and McPhaden* [1993b] suggested that the increased temperature amplitudes induced by the wave in the eastern Pacific are actually largely produced by the effects of mean vertical advection of temperature anomalies associated to the wave induced thermocline displacement. They similarly attributed the second maximum in zonal velocity vertical profile to be mainly due to the vertical advection of the wave-induced signal.

At surface, observations show a significantly nonzero temperature signal associated to the propagating Kelvin wave (Figure 3). This effect of Kelvin waves on SST in the central and eastern Pacific has been noticed and quantified in several observational studies [*Kessler et al.*, 1995; *Zhang*, 2001]. Using a composite technique for the 1986–98 period, a recent observational study [*Vecchi and Harrison*, 2000] examined statistical relationships between three particular WWE types (cases W, C and E such as defined in VH97 classification; Fig. 1) and SST anomaly variability. Their results demonstrate that equatorial WWEs substantially alter the SST field over the entire equatorial Pacific (Plate 1). They show that, on average, when the tropical Pacific has near-normal eastern equatorial Pacific SST (Plate 1a), all equatorial WWE types are followed by a substantial equatorial waveguide warming (up to 1°C) in the central and eastern Pacific 80 days after the WWE (Plate 1c). Moreover, when the eastern equatorial Pacific is initially warmer than usual (Plate 1b), warm conditions are maintained in presence of WWEs (Plate 1d), whereas they tend to disappear in absence of such WWEs. They also underlined that more than 50% of large amplitude WWEs are followed by Niño3 SST warming in excess of 0.5°C. Different mechanisms have been suggested to be responsible for these equatorial SST variations induced by the intraseasonal atmospheric variability. *Johnson and McPhaden* [1993b] and *Kessler and McPhaden* [1995] have pointed out the importance of zonal advection in the SST variations east of the dateline and *Zhang* [2001] related these intraseasonal SST variations to Kelvin wave induced vertical processes. However, *Zhang* [2001] noted that, whereas thermocline

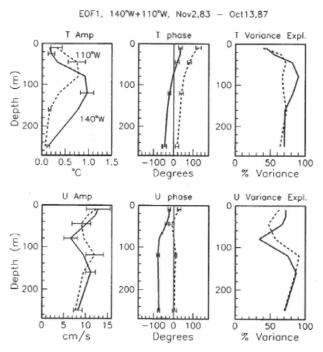

Figure 3. The first frequency domain empirical orthogonal function analysis of zonal velocity and temperature at 140°W (solid line) and 110°W (dashed line) over the 4-year study period (1983–87). The analysis includes 13 frequencies (period from 59 to 125 days) and accounts for a large fraction of the intraseasonal variance, effectively extracting the coherent Kelvin wave signal. Error bars represent one standard error. [from *Johnson and McPhaden*, 1993a]

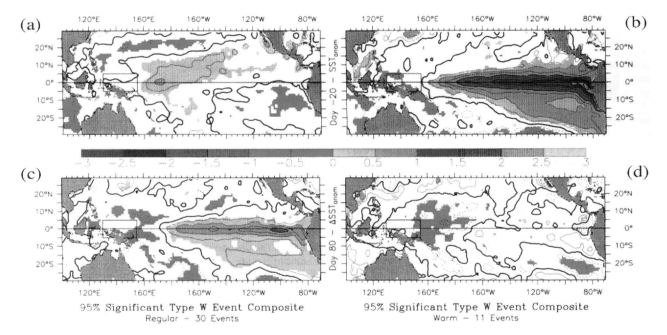

Plate 1. Composite associated with a W-type WWE when $-0.75°C <$Niño3$<+0.75°C$: (a) SST anomaly 20 days before the WWE (SSTA(-20)) and (c) difference between SSTA(+80) and SSTA(-20). Composite associated with a W-type WWE when Niño3$>+0.75°C$: (b) SSTA(-20) and (d) difference between SSTA(+80) and SSTA(-20). Values significantly different from zero are highlighted by color shading. Units (°C). The contour interval is 0.25°C. The shading interval is 0.5°C. The classifying region is indicated by the purple box in each plate. [from *Vecchi and Harrison*, 2000]

depth variations associated with Kelvin waves propagate eastward, SST on intraseasonal timescales tends to be more zonally stationary. *McPhaden* [2002] explained this contrast in behaviour by arguing that the combination of Kelvin wave induced advection in the central Pacific plus vertical advection and entrainment in the eastern Pacific can plausibly produce SST variations that are nearly in phase across a wide range of longitudes.

Modeling studies of the oceanic response to WWEs confirmed that WWEs can indeed induce large oceanic changes by modifying the zonal current and temperature structure in the central and eastern Pacific [*Harrison and Giese*, 1988; *Giese and Harrison*, 1990, 1991; *Kindle and Phoebus*, 1995]. Although all types of WWEs excite a western Pacific response, the results of *Giese and Harrison* [1991] indicate that equatorial WWEs (W, C, and E WWEs types in HV97 classification) are the most efficient in exciting a strong eastern Pacific ocean response. A description of the WWE impact in a model study using an ocean GCM can be illustrated from *Lengaigne et al.* [2002] who analyzed the oceanic response to a strong WWE in March 1997, suggested to have played an important role in the onset of the 1997–98 El Niño [*Boulanger and Menkes*, 1999]. As shown in Figure 4, this WWE is associated with an eastward propagating signal in zonal current (Figure 4c; dashed line) at a speed of about 2.8 m s^{-1} that crosses the basin in about two months. In agreement with other primitive equation ocean models [*Harrison and Giese*, 1988; *Giese and Harrison*, 1990, 1991; *Boulanger et al.*, 2001; *Vecchi*, 2001], the remote modeled response to WWEs involves SST changes of about 0.5°C along the Kelvin wave path (Figure 4a), in qualitative agreement but substantially smaller than the observed one (0.5°C vs. 1°C).

Several numerical studies confirmed that zonal advection by the eastward propagating current pulses of the background temperature gradient in the central Pacific [e.g., *Schopf and Harrison*, 1983; *Giese and Harrison*, 1990; *Lengaigne et al.*, 2002] and reduction of the entrainment cooling through the downwelling associated with the Kelvin waves in the eastern Pacific [*Vialard et al.*, 2001; *Belamari et al.*, 2003] was the main mechanisms responsible for the raise of the equatorial SST. In addition, another unexpected contribution to this warming has been proposed by *Harrison and Giese* [1988]: they suggested that anomalous meridional advection of heat through the interaction of the Kelvin pulse with the background tropical instability waves (TIWs) can lead to enhanced wave guide warming. They suggested that the increased meridional shear of zonal current on both side of the equator due to the Kelvin pulse enhances the TIWs activity [*Weidman et al.*, 1999]. As TIWs act to transport heat equatorward, an increased instability wave field will therefore cause anomalous warming. Recently, *Vecchi* [2001] tempered these results by showing that the TIW/WWE interactions contributes to warm the central Pacific (155°W) but also to cool the central-eastern Pacific (140°W, 125°W). In fact, in the observations, WWEs (and more generally subseasonal winds) are suggested to strongly control the TIWs phase [*Allen et al.*, 1995; *Benestad et al.*, 2001] but their impact on the amplitude of the TIWs activity is however more controversial. Contrary to the numerical results of *Harrison and Giese* [1988], the observational analysis of *Qiao and Weisberg* [1995] suggested that a downwelling Kelvin pulse could result in a reduction of TIWs activity by temporally halting the SEC whose shear may be a source for TIWs. Overall, these results suggest that the response of TIWs activity (and thus their heat induced action) to high fre-

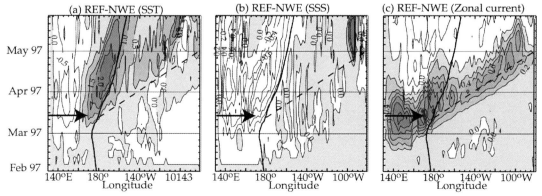

Figure 4. Time-longitude evolution averaged over 1°N–1°S of (a) SST, (b) SSS and (c) surface zonal current differences between the reference experiment (REF) that is forced with the observed ERS1-2 wind stresses and the perturbed experiment (NEW) which has been forced with modified ERS wind in which the March 1997 WWE has been eliminated. The contour interval is 0.5°C for temperature, 0.2 psu for salinity and 0.2 m s^{-1} for currents. The thick line represents the eastern of the warm-pool simulated in the REF experiment. The dashed line represents the path of the Kelvin wave. The arrow indicates the March 1997 WWE. [from *Lengaigne et al.*, 2002]

quency wind variability strongly depends on the oceanic background conditions.

Such a strong influence of the ocean background state (e.g., the magnitude of the zonal equatorial temperature gradient and the mean thermocline depth) on the characteristics of Kelvin waves has been noticed in several modeling studies [*Schopf and Harrison*, 1983; *Harrison and Schopf*, 1984]. In particular, *Benestad et al.* [2002] have found that Kelvin waves excited by identical intraseasonal wind variations are strongly damped during La Niña compared to El Niño, such that little wave energy reaches the eastern coast when the mean conditions are cold. They attributed this effect to an enhanced viscous dissipation in La Niña conditions, primarily due to an increase of the vertical current shear but also to the interaction of the wave with the enhanced TIW field observed during La Niña. This strong influence of the ENSO cycle on the ocean response to WWEs could therefore be a significant source of non-linearity that could partially explain the diversity of the tropical Pacific response to WWEs observed during the past two decades.

3.2. Local Response

WWEs have also been identified as playing a leading role in the momentum and heat balances of the warm pool [*Meyers et al.*, 1986; *Feng et al.*, 1998]. During WWE conditions, observations show that the thermocline is depressed (by 10 to 40m) due to locally wind-forced downwelling signal and the SST over the western Pacific warm-pool is characterized by a local cooling [*McPhaden et al.*, 1988, 1992; *Delcroix et al.*, 1993; *Eldin et al.*, 1994; *Cronin and McPhaden*, 1997; *Vecchi and Harrison*, 2000; see Plate 1b–d]. The amplitude of this local cooling under individual or series of WWEs ranges from 0.2° to 1°C.

Many mechanisms have been suggested to explain such a local SST cooling under westerly wind conditions in the western Pacific. During the TOGA-COARE Intensive Observation Period (IOP; November 1992–February 1993), microstructure measurements designed to yield estimates of the turbulent heat flux were made in concert with measurements of various components of the surface heat flux. It allowed observing the SST evolution during a series of WWEs more thoroughly than ever before [*Smyth et al.*, 1996a; *Cronin and McPhaden*, 1997; *Feng et al.*, 1998; *Richards and Inall*, 2000]. Numerous studies in the western Pacific indicated that one-dimensional processes, mostly latent heat flux and shortwave radiation, are of greatest importance in generating intraseasonal SST variations [*Anderson et al.*, 1996; *Cronin and McPhaden*, 1997; *Shinoda and Hendon*, 1998, *Zhang and McPhaden*, 2000]. *Smyth et al.* [1996a] indeed observed that turbulent entrainment at the base at the base of the mixed layer caused cooling but also heating due to the reversal of the vertical temperature gradient during rain events, minimizing the effect of turbulent entrainment. However, *Feng et al.* [2000] also underlined the strong episodic role of 3-D heat advection oceanic processes in modulating SST variability during their 3 months study and even argue that heat advection may be as important as net air-sea fluxes during WWEs period [see also *Ralph et al.*, 1997]. It is worth noting that all these observational studies are conducted at one location over the warm pool and therefore do not allow to study the spatial variability of the results. Numerical experiments allowed such an investigation. Hence, by using a regional ocean GCM over the COARE domain, *Dourado and Caniaux* [2000] confirmed that all the terms of heat budget contribute to the temperature variations in the oceanic mixed layer during WWEs episodes and underlined the strong spatial inhomogeneity of the results mainly because of advective processes. In conclusion, if local cooling is a robust feature of the far-western Pacific under WWE conditions and air-sea heat fluxes have a crucial role in controlling this SST decrease, other processes can contribute to this cooling, depending on both the WWE temporal and spatial characteristics and the warm pool structure at the time of the WWE.

Compared to temperature, the salinity evolution under WWE conditions is even more complex. *McPhaden et al.* [1992] noticed a salinity drop of 1 psu in November 1989, mainly attributed to meridional advection. In contrast, *Delcroix et al.* [1993] only observed weak SSS variations during February 1991. During the TOGA-COARE IOP, *Feng et al.* [1998] even observed a salinity increase during the December 1992 WWE. In order to explain the ocean freshwater balance, 1-D processes are not adequate. Horizontal advection terms, as well as freshwater fluxes, have been both identified to be important to the long term fresh water balance near the equator [*Cronin and McPhaden*, 1998; *Feng*, 2000] but there is, for the moment, too few observational studies on the salinity variability under WWEs conditions to draw definitive conclusions on this topics.

The dynamical response to a westerly wind burst over the warm-pool is mainly observed above the thermocline. The first hydrographic measurements under strong westerly winds were made by *Hisard et al.* [1970]. They reported a vertical profile consisting of three currents: an eastward surface current in the direction of the winds, a preexisting eastward EUC, and a subsurface westward jet (SSWJ) between the two eastward currents. Subsequent observational studies provided more information on the temporal evolution of this vertical structure of alternated jets under WWEs conditions [*Lindstrom et al.*, 1987; *McPhaden et al.*, 1988, 1990, 1992; *Kuroda and McPhaden*, 1993; *Delcroix et al.*, 1992, 1993; *Eldin et al.*, 1994; *Weisberg and Hayes*, 1995; *Smith et al.*, 1996b]. For

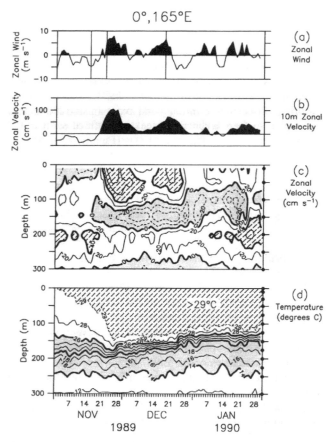

Figure 5. Daily average of (a) zonal wind, (b) 10-m zonal currents, (c) 10- to 300-m zonal currents and (d) 0- to 300-m temperatures from the current meter mooring at 0°, 165°E, for November 1989 to January 1990. The contour interval in Figure 5c is 20 cm s^{-1} with westward flow shaded and eastward flow >40 cm s^{-1} hatched. The contour interval in Figure 5d is 2°C, except for the 29°C isotherm which is indicated by a dashed contour. [from *McPhaden et al.*, 1992]

instance, observations by *McPhaden et al.* [1992] obtained from the TOGA-TAO Array mooring present simultaneous measurements of zonal surface wind, zonal velocity, and temperature during a period that encompasses several WWEs (Figure 5). As the first WWE began in late November 1989 (Figure 5a), an eastward jet developed with a 1.0 m s^{-1} maximum amplitude (Figure 5b). Meanwhile, a SSWJ developed in the upper thermocline (Figure 5c) as a local response to the WWE. This SSWJ developed 8–10 days after the WWE peak and persisted after the WWE end. The observational analysis of *Cronin et al.* [2000] mainly attributed these currents to be resulting from the interplay between wind forcing and compensating pressure gradients. Below the surface layer, these pressure gradients tend to accelerate the upper-thermocline flow in a direction opposite to the local winds. Using a primitive equation model, *Zhang and Rothstein* [1998] and

Richarson et al. [1999] elucidated the principal features and the mechanism behind the SSWJ, in agreement with *Cronin et al.*'s [2000] observational results.

The surface jet, the so-called "Yoshida jet" [*Yoshida*, 1959], which is trapped in the first 100 meters and is confined around the equator (3°N–3°S), is also a common feature associated with WWEs. It has long been recognized that strong westerly wind forcing on the equator leads to the rapid development of these eastward surface jets, nearly in phase with the wind [*McPhaden et al.*, 1988; *Ralph et al.*, 1997; *Cronin et al.*, 2000]. In addition, salinity can play a decisive role in intensifying and prolonging these jets over the warm pool through the existence of a zonal salinity gradient accelerating the surface layers eastward [*Roemmich et al.*, 1994] and through the formation of barrier layers trapping the zonal momentum into the upper layer [*Cronin and McPhaden*, 2002]. Barrier layers are indeed climatological features of the western equatorial Pacific [*Sprintall and Tomczak*, 1992] even if there was not any barrier layer present during the TOGA-COARE IOP. Figure 6 illustrates mechanisms of barrier layer formation and growth. Under WWE conditions, the presence of the barrier layer prevents upward mixing of cold thermocline water although the

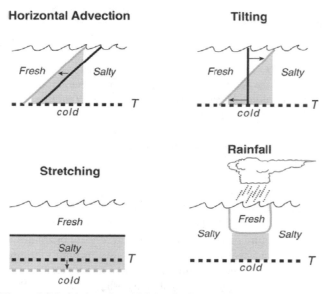

Figure 6. Mechanisms by which barrier layer can form and grow, i.e. (a) uniform horizontal advection in the direction from which the water flows, (b) tilting of a vertical salinity front into the horizontal by a vertically sheared horizontal flow, (c) vertical streching by a nonuniform vertical advection and (d) rain-formed fresh surface lense. The initial halocline and thermocline are indicated in by a black solid line and a black dashed line. The resulting halocline and thermocline are indicated by a grey solid line and grey dashed line, respectively. Stippling indicates the resulting barrier layer. [from *Cronin and McPhaden*, 2002]

very energetic turbulent mixing produced by the wind tends to gradually destroy it in the region of the forcing. However even if WWEs occurrence leads locally to a barrier layer erosion, they can sometimes effectively contribute at forming thick barrier layers to the east of the forcing region. If local surface processes (e.g., rain) are suggested to be the main formation mechanism for salinity barrier over the far western Pacific, subduction and horizontal advection processes are the dominant processes responsible for the formation of barrier layer at the Eastern Edge of the Warm Pool (EEWP; Figure 6). In this case, the vertically sheared horizontal flow generated by the WWE advects the horizontal salinity gradient within the isothermal surface layers. Near-vertical salinity contours tend to tilt into the horizontal and therefore generate a shallow halocline above the top of the thermocline. This barrier layer, by trapping the zonal momentum and heat input into the ocean upper layer, acts to intensify the local ocean response to WWEs, and particularly the surface eastward jet at the EEWP [*Cronin and McPhaden*, 2002].

Some numerical studies allowed confirming and completing these observational studies focusing on the local dynamical response of the ocean surface to WWEs. If the Yoshida jets are strongly wind-driven, modelling studies suggests that two other processes contribute are suggested to contribute to intensify and/or prolong these eastward surface currents: the existence of a barrier layer [*Vialard and Delecluse*, 1998] and non-linear processes [*Harrison et al.*, 2000; *Boulanger et al.*, 2001]. An example of such processes is provided by the modelling study of *Lengaigne et al.* [2002] who detailed the mechanisms responsible for the intensity and structure of the surface Yoshida jet generated by the March 1997 WWE, during the onset of the 1997–98 El Niño. As displayed in Figure 4, this jet is divided in two dynamically different regions. A first strong intensification is located west of 160°E which is essentially a direct response to the westerly wind forcing and a second intensification up to 0.8 m s^{-1} occurs along the EEWP. This zonal current signal in the eastern portion of the Pacific warm pool is concomitant with a strong SST (Figure 4a) and SSS signal (Figure 4b). The eastern part of the jet is not a direct local response to the March WWE since it is in fact developing east of the WWE maximum. This eastward moving jet is driven by a non-linear interaction between the wind forcing and the thermohaline front that results in a strong negative zonal gradient of surface currents at this location, allowing the momentum zonal advection terms to develop. This non-linear acceleration strongly contributes to the intensity and persistence of the zonal current front at the EEWP.

Similar intense surface jets during ENSO periods have already been studied in previous modeling works. In fact, *Harrison and Craig* [1993] reported the existence of such strong surface jets during November–December 1982 and in early 1983. Through the use of a primitive equation ocean model, they underlined the important contribution of zonal momentum advection in the jet formation and their impact on the temperature changes. Similarly, *Harrison et al.* [2000] pointed out the strong contribution of zonal momentum as well as zonal pressure gradient in the formation and duration of the intense jets (up to 1.2 m s^{-1}) that were observed during November1991–January 1992. These long-lasting eastward jets developing in the eastern part of the Pacific warm pool are likely to be important features of the El Niño onset and development phase as they contribute to the eastward zonal advection of warm waters, which is the dominant mechanism in determining the zonal migration of the EEWP [*Picaut et al.*, 1996].

4. THE RELATION BETWEEN WWES AND THE ENSO PHENOMENON

WWEs are therefore responsible for significant SST modifications to which the atmosphere is likely to respond. Over the western Pacific, WWEs are associated with a cooling of the far western Pacific and a displacement of the eastern edge of the warm pool. Moreover, the downwelling Kelvin waves excited by the WWE promote a substantial surface warming in the central and eastern Pacific. Air-sea interactions involving WWEs, oceanic Kelvin waves and western Pacific warm pool have therefore been suggested to be instrumental to the fast growth of El Niño events and the modulation of ENSO characteristics in the observations [*Kessler et al.*, 1995; *McPhaden*, 1999; *Boulanger and Menkes*, 1999; *Bergman et al.*, 2001] as in models [*Moore and Kleeman*, 1999; *Kessler and Kleeman*, 2000; *Fedorov*, 2002; *Lengaigne et al.*, 2003, 2004]. This section will therefore highlight the impact of WWEs on the ENSO phenomenon in both observations and models.

4.1. Observational Results

The substantial surface warming often induced by WWEs [*Vecchi and Harrison*, 2000] raises the question of the WWEs abilities to trigger and/or amplify El Niño events. *McPhaden et al.* [1988] discussed the role of a western Pacific WWE during May 1986, at the onset of the 1986/1987 El Niño. This event generated an eastward propagating current pulse that was associated with a 1°C warming at 0°, 110°W in mid-June 1986. According to *McPhaden et al.* [1988], the warming pulse following the WWE had little effect on the evolution of the subsequent 1986/87 El Niño event. However, *Harrison and Giese* [1989] offer another interpretation of the same event. They agree that the June 1986 warming at 110°W resulting from the WWE wind-forced first baroclinic mode Kelvin pulse is short-lived, but they argue that a second warming occurring in mid-July 1986 resulted from the WWE wind-forced second baroclinic mode. According to them, this event induced a warming larger than 2°C at the onset of the 1986/1987 El Niño. However, local wind forcing in the eastern Pacific could also have contributed to such a warming, making the actual impact of the first two baroclinic wind-forced Kelvin waves difficult to quantify accurately. Then, the frequent observations of enhanced tropical convection associated with WWEs activity observed prior and/or during the 1982–83, the 1991–92 and 1997–98 El Niño events [*Harrison*, 1984; *Lau and Chan*, 1988; *Kessler et al.*, 1995; *McPhaden*, 1999; *Bergman et al.*, 2001; *Kutsuwada and McPhaden*, 2002] supported the hypothesis that WWEs could be key ingredient for both El Niño timing and amplitude. Confirmation of the potential for WWEs to influence the onset and development stage of El Niño has been provided, indirectly, by the predictions of the 1997–1998 El Niño. Most of the forecasts failed to capture the onset of the El Niño prior to the March 1997 WWE, and it was not until after June that the forecasts managed to represent the intensity of the event at a time when the El Niño was already very strong [*Barnston et al.*, 1999]. In particular, no model predicted the extremely steep rise of the central and eastern Pacific SSTs that took place in March–June 1997. These shortcomings may be attributed to the fact that most of these models do not simulate well, if at all, WWEs and their associated atmospheric causes (e.g., MJO, tropical cyclones; see Section 2).

Using surface winds from islands near the dateline, *Luther et al.* [1983], *Gutzler* [1991] and *Harrison and Giese* [1991] also noted significant tendency for near dateline WWEs to be associated preferentially with warm ENSO periods. These results have been further confirmed and detailed through the analysis of *Harrison and Vecchi* [1997]. Based on their WWEs classification, they found that some WWEs type were significantly correlated with the Southern Oscillation Index (normalized sea level pressure difference at Darwin minus Tahiti), adding to the suggestion of a relationship between equatorial WWE types and El Niño.

As MJO is one of the major mechanisms in generating WWEs, the connection between MJO activity and ENSO could also be indicative for the possible links between WWEs and ENSO. Significant correlations can hardly be found between ENSO indices and conventional MJO indices based on global winds and convection [*Slingo et al.*, 1999; *Hendon et al.*, 1999]. From a global index based on the modulation of the upper-tropospheric equatorial zonal mean of the zonal wind, *Slingo et al.* [1999] could not find significant correlations between the interannual variation of the MJO and ENSO signals. However, these empirical relationships strongly depend on the MJO indices used. Global indices measure the global activity of the MJO well, but do not necessarily depict local effects of the MJO in the Pacific ocean. Several studies have shown that MJO envelope and surface zonal winds shift eastward during El Niño events over the tropical Pacific, following the EEWP displacement [*Gutzler*, 1991; *Hendon et al.*, 1999; *Kessler*, 2001]. Also MJO effects on ENSO, if any, must take place through air-sea interactions in the Pacific, such as the generation of WWEs over the warm pool warm waters. In fact, based on an MJO index derived from fields that directly represent physical air-sea interactions in the equatorial Pacific, *Zhang and Gottschalck* [2002] suggested that stronger Kelvin wave forcing in the western Pacific associated with intraseasonal signal precedes greater SST anomalies in the eastern Pacific by 6–12 months for the 1980–1999 period (Figure 7). This result suggests that the surface signature of the MJO can accelerate the growth and amplify the strength of an ENSO warm event. Although this work indicated a statistical relationship between intraseasonal and interannual timescales, it should be noticed that strong MJO and WWEs activities are also regularly seen during non-El Niño years (e.g., 1989/1990), suggesting complex WWEs/ENSO relationship.

4.2. Modeling Results

WWEs can be considered as part of a high-frequency transient atmospheric forcing. *Hasselmann* (1976) first suggested that stochastic forcing could act to modulate the low-frequency component of the climate system. Then, simple models confirm that addition of noise modify a regular modelled ENSO structure into an irregular ENSO behaviour that is con-

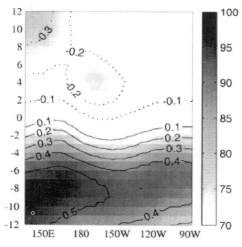

Figure 7. Coefficients (contours) and confidence levels (shading) of the lag correlation between the interannual anomaly in seasonal variance of Kelvin wave forcing by the MJO ($\Delta\sigma_{K(MJO)}$) and interannual SST anomaly in Niño3 region (ΔSST) for the time period from 1980 to 1999. Dashed contours are for negative correlation coefficients. Positive lags indicate ΔSST leading $\Delta\sigma_{K(MJO)}$. Maximum correlation is located at 152.5°E and lag –9. [from *Zhang and Gottschalck*, 2002]

siderably richer in spatial and temporal structures [*Penland and Sardeshmukh*, 1995; *Blanke et al.*, 1997; *Kestin et al.*, 1998; *Roulston and Neelin*, 2000]. Using an intermediate coupled model, *Moore and Kleeman* [1999] explore this point farther by calculating the stochastic optimals of the coupled system growth, i.e. the spatial structure of stochastic noise forcing that produces the largest response in the variability of the coupled system on seasonal-interannual timescales. The SST and wind stress patterns of a typical optimal perturbation of their coupled system is shown in Figure 8. This structure bears a remarkable resemblance with the observed characteristics of WWEs and MJO structures (duration, longitudinal extent, associated cyclone activity; see Section 2). This suggests that MJO/WWE type structures are the part of the stochastic forcing that is the most important for influencing ENSO variability. Based on their findings, they concurred with the view that ENSO may be explained, at least partially, as a stochastically forced phenomenon, the source of noise in the tropics being mainly MJOs and WWEs. They further underlined that the ability of these noise-induced perturbations to impact the coupled system depends on various factors, including the phase of the seasonal cycle, the presence of non-linearities in the system, the past history of the stochastic noise forcing and the stability of the coupled ocean-atmosphere system. However, their results do not allow concluding whether ENSO is a self-sustained oscillating phenomenon partially perturbed by stochastic atmospheric noise or whether it is a naturally damped phenomenon excited by westerly wind events or other specific atmospheric high-frequency wind variability.

If these studies indicate a strong influence of high frequency wind forcing on the low frequency SST evolution, the rectifying mechanisms still need to be explain. Several attempts have tried to elucidate these interactions [*Harrison and Schopf*, 1984; *Kessler et al.*, 1995; *Kessler and Kleeman*, 2000]. One example of such suggested non-linear mechanism has been proposed by *Kessler et al.* [1995] using a simple model. The key dynamics of this process is that the atmosphere responds rapidly to the state of the ocean, but the ocean response is lagged because it is an integral of the wind forcing. This process could result in a slow progression of warm waters and westerly winds to the east as observed during the 1991–92 El Niño. *Kessler and Kleeman* [2000] suggested other processes by which a forcing at intraseasonal frequencies can force a net nonzero low-frequency ocean signal. By forcing an OGCM with idealized oscillating wind stresses over the western Pacific, they demonstrated that a rectified low-frequency anomaly in SST develops. This rectification resulted, among other things, from stronger zonal wind speeds, and therefore higher evaporation. This flattening of the zonal SST gradient results in an hindcast El Niño about 30% stronger, suggesting that intraseasonal forcing can interact constructively with the ENSO cycle.

The effect of a single WWE (i.e. a short-term positive zonal wind stress anomaly) on the coupled system in models has first been explored from a state of rest and served to kick ENSO-like oscillations in simple coupled ocean-atmosphere models [*Zebiak and Cane*, 1987; *Battisti*, 1988]. The coupled response to a WWE over the western Pacific has then been assessed with intermediate coupled models using more realistic mean states [*Chen et al.*, 1999; *Fedorov*, 2002; *Boulanger et al.*, 2004]. For example, *Perigaud and Cassou* [2000] demonstrated that WWEs have an impact on the coupled system in El Niño forecasting. Adding westerly wind anomalies in boreal winter 1981–82 and 1996–97 greatly improved the prediction of the 1982–83 and 1997–98 El Niño (Figure 9). Moreover, as suggested by *Boulanger et al.* [2004], WWEs may favor an increase in the warming growth rate leading to a shift in the peak of El Niño occurring earlier than in the absence of WWEs and therefore interfere with the regular phase-locking of El Niño. Overall, all these results obtained with intermediate coupled models strongly suggest that WWEs can significantly impact the coupled system and amplify/trigger El Niño events. However, complex coupled processes that occur over the warm-

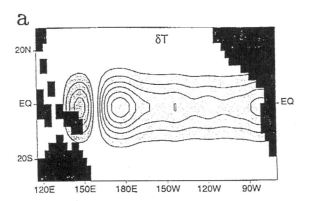

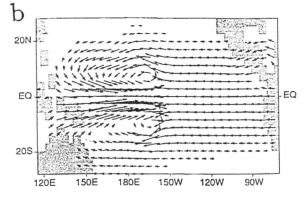

Figure 8. (a) SST and (b) surface wind stress structures associated with a typical optimal perturbation of the coupled model over the tropical Pacific. [from *Moore and Kleeman*, 1999]

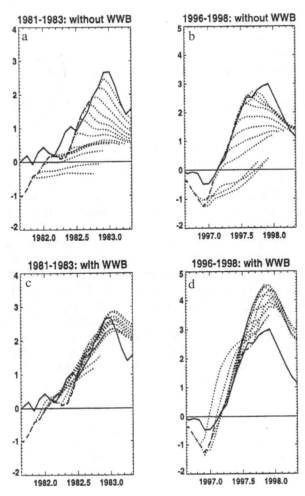

Figure 9. Time series of SST Niño3 indices derived from observations (solid), model initial conditions (dashed) or predictions (dotted) over various two-year long periods for (a) the 1981–83 and (b) the 1996–98 cases. (c) and (d) correspond to forecasts where a westerly wind anomaly is introduced as described in the text. [from *Perigaud and Cassou*, 2000]

pool following a WWE have been suggested to be important features of the El Niño onset and growth rate [*Picaut et al.*, 1996; *McPhaden et al.*, 1999]. Recently, *Lengaigne et al.* [2003] using a atmospheric general circulation model suggested that the eastward shift of warm waters induced by the March 1997 WWE was responsible for the eastward migration of the main convective area and the persistence of a strong WWE activity over the western Pacific in the following months, that is likely to be responsible for the rapid growth and the exceptional intensity of the 1997–98 El Niño event [*Van Oldenbourgh*, 2000]. These coupled interactions that strongly constrain the El Niño characteristics are not represented by the simple physics of intermediate coupled models and a more accurate representation of these coupled processes requires the use of coupled general circulation models.

A first attempt to study the response of a coupled ocean-atmosphere general circulation model to a westerly wind event was made by *Latif et al.* [1988]. To that end, they introduced a westerly wind stress anomaly over the western Pacific for a month and compared the results to their control run without wind stress perturbation (Figure 10). They demonstrated that the WWE promoted an eastward shift of the warm-pool persisting for about 12 months due to the feedback of the atmosphere (Figure 10a–b). This eastward shift was accompanied by westerly surface winds in the western Pacific (Figure 10c) and by a displacement of the convective rainfall zone (Figure 10d). An interesting result was the occurrence of spontaneous WWEs in the western Pacific, induced by the introduction of the initial disturbance. This result suggest that the coupled system is capable of rectifying short-term wind anomalies into low-frequency changes as the atmospheric response to the SST anomalies act to reinforce the initial oceanic response. Nevertheless, their SST and zonal wind stress anomalies did not propagate and there was not any warming in the eastern Pacific (Figure 10b). They associated this weak eastern Pacific response to their model biases (coarse atmospheric resolution, weak atmospheric response to SST anomalies). Moreover, the flux correction method they used to avoid climate drift within the coupled model could act to damp the SST anomalies.

A recent study using a fully coupled model with a reasonable mean climate and seasonal cycle [*Lengaigne et al.*, 2004] further pursued this investigation. In that study, two ten-member ensemble experiments were performed to study the dynamical response to a strong WWE. In the reference ensemble, no wind perturbation were applied, whereas a strong westerly wind event anomaly were introduced in boreal winter over the western Pacific in the perturbed ensemble. As shown in Figure 11, 4 members develop into strong El Niño warming in the perturbed ensemble (up to 3°C anomaly in the Niño3 region at the end of the year; gray curves in Figure 11). This result suggests that a strong WWE is able to create the conditions under which a strong El Niño event can occur. First, it generates a strong downwelling Kelvin wave that induces a positive SST anomaly in the central-eastern Pacific and a weakening of the trade winds, as suggested by *Lengaigne et al.* [2003]. Secondly, the inserted WWE also initiates an eastward displacement of the warm-pool that promotes the occurrence of subsequent WWEs in the following months. These events reinforce the initial warming through the generation of additional Kelvin waves and generate intense surface jets at the eastern edge of the warm-pool, similar to those already described in forced ocean models [*Harrison and Craig*, 1993; *Harrison et al.*, 2000; *Lengaigne et al.*, 2002]. However, the use of an ensemble strategy in this study reveals substantial dif-

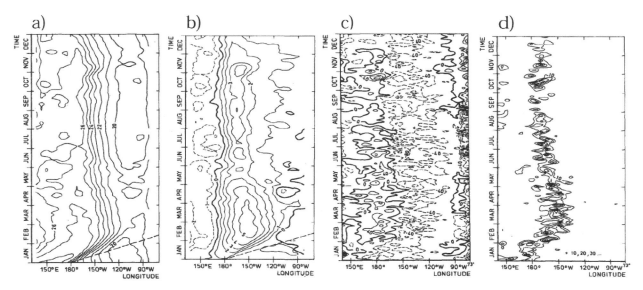

Figure 10. Time-longitude evolution along the equator of (a) SST, (b) SSTA, (c) zonal wind stress and (d) convective precipitation. Contour interval is 1°C for (a) and (b), 40 mPa for (c) and 10 mm day^{-1} for (d). [from *Latif et al.*, 1988]

ferences in the coupled ocean-atmosphere response to a WWE (Figure 11). Over the ten members, 4 other members only display a moderate warming (gray dashed curves) and 2 remain in neutral conditions (black curves). This diversity between the members appears to be due to the internal atmospheric variability during and following the inserted WWE (intensity of the trade winds, intensity of the WWEs activity following the insertion of the strong WWE...). This study therefore suggests that the occurrence of an intense WWE in the early year therefore favors a stronger subsequent WWE activity by shifting warm waters eastwards in the following months, but does not seem to totally constrain it. In addition, another recent study with the ECMWF coupled model [*Vitart et al.*, 2003] also suggested that addition of the observed wind stress anomalies in May–June 1997 produces significantly better forecasts for the 1997–98 El Niño, whereas *Maes et al.* [2002] underlined the importance of the barrier layer in coupled mode to amplify ocean-atmosphere response to WWEs during the onset of El Niño events.

However, if WWEs strongly contributed to the onset and development of El Niño events on some occasions (e.g., 1982–83, 1997–98), WWEs failed to generate El Niño on other occasions (e.g., in 1967 and in 1989–90). In fact, as suggested by *Lengaigne et al.* [2004] study, the response of the coupled system to a strong WWE is complex and strongly depends on the internal atmospheric variability. In addition, this response also depends on the mean background conditions in the tropical Pacific. Using an intermediate coupled model, *Perigaud and Cassou* [2000] suggested that the impact of WWEs depends on the ocean heat content at the time of the WWE: adding a WWE when the ocean heat content is low does not modify the coupled system evolution, whereas the high ocean heat content in winter 1981–82 and 1996–97 allowed WWEs to onset the corresponding El Niño events. A recent study [*Fedorov*, 2002] using a simple coupled tropical ocean-atmosphere model showed that adding a WWE in the coupled system impacts the magnitude and other characteristics of ENSO by modifying the energetics of the ocean-atmosphere interactions. They suggest that this impact depends crucially on two factors: the background state of the system as described by the mean depth of the thermocline and the intensity of the mean winds, and the timing of the event with respect to the ENSO phase. The sensitivity to this timing is illustrated in Figure 12. The

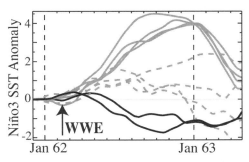

Figure 11. Time series of Niño3 SST anomalies for the members of the perturbed ensemble where a WWE is introduced, compared to the reference ensemble mean. The grey lines correspond to the strong warming cases, the grey dashed ones to moderate warming cases and the black ones to neutral cases. The arrow denotes the artificially introduced WWE. [from *Lengaigne et al.*, 2004]

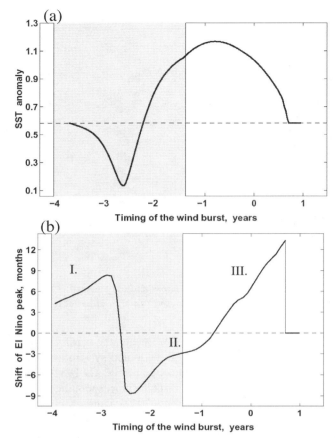

Figure 12. (a) The impact, on the magnitude of El Niño, of WWEs that happen at different times of one cycle of a pre-existing Southern Oscillation. the vertical axis displays the strength of the warming of the eastern Pacific ocean (in °C); while the horizontal axis shows the timing of the bursts within this cycle. Time zero corresponds to the instance when El Niño would have its peak in the absence of the wind bursts. Negative/positive times correspond to the times before/after such peak. The magnitude of El Niño is shown by the shaded line. Shading indicates times when the burst occurs during the cold phase of the cycle, i.e. La Niña. (b) The impact, on the timing of El Niño, of WWEs that happen at different times of a pre-existing southern oscillation. The vertical axis displays the shift of the peak of El Niño in time, i.e. its delay (positive values) or its advancement in time (negative values); while the horizontal axis is as in (a). In the absence of WWEs, El Niño would have peaked at time 0. Shading indicates times when the WWE occurs during the cold phase of the cycle, i.e. La Niña. [from *Fedorov*, 2002]

results in Figure 12a indicate that WWEs can significantly amplify the strength of El Niño if it occurs approximately 6 to 10 months in advance of the upcoming El Niño (as in 1997–98 and 1982–83). In contrast, the WWE that happens during the cold phase of the cycle have either negligible or negative impact of the upcoming El Niño, by decreasing the available potential energy of the system. Another important effect of WWEs is to shift the peak of El Niño in time (Figure 12b). Depending on the ENSO phase, a WWE can either delay the following El Niño (region I), accelerate the onset of the warming (region II) or generate a second peak (region III). Even if the model used is too crude to describe the real system in its full complexity, this study therefore suggests that WWEs impact strongly depend on the ocean state at the time the WWE occurs.

5. CONCLUSION

In this paper, observational and modeling aspects about Westerly Wind Events (WWEs) and their influence on the tropical Pacific ocean-atmosphere system are reviewed. WWEs are a large part of the intraseasonal zonal wind activity over the warm pool. They have a significant seasonality, typical amplitudes of 7 m s^{-1}, zonal widths of 20° longitude and durations of about 8 days. Their root causes are often suggested to be a combination of various factors including the Madden-Julian Oscillation, cold surges from extra-tropical regions, tropical cyclones and other mesoscale phenomena. However, the exact relationships between these atmospheric phenomena and WWEs remain often unclear. Future work is therefore needed to better understand, for instance, the relationship between the MJO and WWEs, to further document the role of mid-latitude pressure surges in forcing tropical convection, to examine the role of atmospheric equatorial Rossby wave disturbances (associated with the intrusion of upper level mid-latitude Rossby wave activity into the tropical eastern Pacific) that can play a role in the subsequent western Pacific activity.

The ocean response to WWEs is complex, involving among other processes the equatorial characteristics of the WWEs, the oceanic background state and the internal atmospheric variability. Both observational and modeling studies however reveal systematic ocean evolution under WWEs conditions. As a remote response, they force eastward propagating downwelling equatorial Kelvin waves that favor the development of positive intense SST anomalies in the central and eastern Pacific through zonal advection of the SST gradient in the central Pacific and/or thermocline deepening in the eastern Pacific. Locally, WWEs affect the structure of the western Pacific warm-pool because they significantly alter surface heat, fresh and momentum fluxes. The local oceanic response to WWEs shows a surface cooling over the far-western Pacific and the generation of a westward subsurface and an eastward surface jets. Under certain circumstances, the eastward jet advects warm waters to the central Pacific.

As both observational and modeling studies demonstrate that equatorial WWEs are associated with the appearance and maintenance of warm SST conditions, it has therefore been assumed that WWEs might influence ENSO. Observational studies suggest a link between the WWEs activity and El Niño occurrences but the lack of long-term reliable data records prevents to draw definitive conclusions on this topics. However, modelling studies suggest that WWEs can significantly contribute to the irregularity of ENSO as atmospheric feedbacks are suggested to amplify the initial ocean response to a WWE, allowing a strong WWE to establish the conditions under which El Niño can occur. In fact, WWEs act to shift warm waters eastward promoting the occurrence of further WWE activity, which can generate other downwelling Kelvin waves and therefore further amplify the initial warming. It is therefore likely that WWE activity during the development phase of an El Niño event strongly influence its evolution and amplitude but the impact of WWEs on the coupled system has been shown to crucially depend on the oceanic background state and the internal atmospheric variability. However, whether ENSO is a self-sustained oscillating phenomenon perturbed by stochastic atmospheric noise or whether it is a damped phenomenon excited by WWEs is still an open question and no consensus has thus far been reached on this subject. Future work is therefore needed to better understand the influence of the mean state on the coupled response to WWEs and to precise to what extend high frequency wind variability influences ENSO.

The strong sensitivity of the coupled system response to WWEs raises the question of the predictability of the tropical Pacific evolution. Our inability to predict WWE occurrences and intensities are a strong limitation to El Niño predictability. Indeed, the leading role played by high frequency wind in determining the evolution and intensity of the 1997–98 El Niño suggests that the exceptionally large amplitude of this El Niño, as well as the 1982–83's, could not have been anticipated far in advance because it depended on the occurrence of successive westerly wind bursts [*McPhaden and Yu*, 1999; *Lengaigne et al.*, 2004]. This led *Fedorov et al.* [2003] to argue that the use of probabilistic forecast is necessary to cope with such high frequency wind variability. In fact, high frequency wind variability is currently unpredictable, even at short lead time. However, whether such a high-frequency wind activity can be considered as purely stochastic is an open question as it is likely to be, at least partially, influenced by large-scale conditions. It is evident that the predictability of synoptic features such as individual WWEs is very limited and very complex as not only the WWE occurrences should be predicted, but also their equatorial characteristics (fetch, duration, intensity). However, even if individual WWEs remain unpredictable beyond a few days, one can certainly improve the probabilistic predictability of the seasonal WWEs activity as its characteristics are partially controlled by large-scale atmospheric and oceanic variability. Preliminary results by *Yu et al.* [2003] suggest for instance that the WWEs activity over the western Pacific could be partially controlled by the equatorial sea level pressure gradient in the far western Pacific, which is correlated with the location of the warm pool. This means that the characteristics of the western Pacific warm pool could modulate the WWEs activity. Improved understanding and prediction of El Niño SST changes likely depends on a better understanding and prediction of the entire range of oceanic and atmospheric variability, which brings about WWEs.

Acknowledgments. Alexey Fedorov and an anonymous reviewer are gratefully acknowledged for helpful comments on this manuscript. This work was supported by the Programme National d'Etude du Climat (PNEDC).

REFERENCES

Allen, M. R., S. P. Lawrence, M. J. Murray, C. T. Mutlow, T. N. Stockdale, D. T. Llewellyn-Jones, and D. L. Anderson, Control of Tropical Instability Waves in the Pacific, *Geophys. Res. Lett., 22*, 2581–2584, 1995.

Anderson, S. P., R. A. Weller, and R. B. Lukas, Surface buoyancy forcing and the mixed layer of the western equatorial Pacific warm pool: observations and 1-D model results, *J. Clim., 9*, 3056–3085, 1996.

Arkin, P., and P. J. Webster, Annual and interannual variability of tropical-extratropical interactions : An empirical study, *Mon. Weather Rev., 113*, 1510–1523, 1985.

Barnston, A. G., M. H. Glantz, and Y. He, Predictive skill of statistical and dynamical climate models in SST forecasts during the 1997/98 El Niño episode and the 1998 La Niña onset, *Bull. Am. Meteor. Soc., 80*, 217–243, 1999.

Battisti, D. S., Dynamics and thermodynamics of a warming event in a coupled tropical atmosphere-ocean model, *J. Atmos. Sci., 45*, 2889–2919, 1988.

Belamari, S., J.-L. Redelsperger, and M. Pontaud, Dynamic Role of a westerly wind burst in triggering an equatorial Pacific warm event, *J. Clim., 16*, 1869–1890, 2003 .

Benestad, R. E., R. T. Sutton, M. R. Allen, and D. L. T. Anderson, The influence of subseasonal wind variability on tropical instability waves in the Pacific, *Geophys. Res. Lett., 28*, 2041–2044, 2001.

Benestad, R. E., R. T. Sutton, and D. L. T. Anderson, The effect of El Niño on intraseasonal Kelvin waves, *Q. J. R. Meteorol. Soc., 128*, 1277–1291, 2002.

Bergman, J. W., H. H. Hendon, and K. M. Weickmann, Intraseasonal Air-Sea Interactions at the onset of El Niño, *J. Clim., 14*, 1702–1719, 2001.

Blanke, B., J. D. Neelin, and D. Gutzler, Estimating the effect of stochastic wind stress forcing on ENSO irregularity, *J. Clim., 10*, 1473–1486, 1997.

Boulanger, J.-P., and C. Menkes, Propagation and reflection of long equatorial waves in the Pacific ocean during the 1992–1993 El Niño, *J. Geophys. Res., 100*, 25087–25099, 1995.

Boulanger, J.-P., and C. Menkes, Long equatorial wave reflection in the Pacific ocean during the 1992/1998 TOPEX/POSEIDON period, *Clim. Dyn.*, *15*, 205–225, 1999.

Boulanger, J.-P., E. Stretta, J.-P. Duvel, P. Delecluse, C. Menkes, M. Lengaigne, G. Madec, M. Imbard, and S. Masson, Westerly wind bursts and ENSO: Role of the March 1997 wind event in the 1997–1998 El Niño onset, *Geophys. Res. Lett.*, *28*, 1603–1606, 2001.

Boulanger, J.-P., S. Cravatte, and C. Menkes, Reflected and locally wind-forced interannual equatorial Kelvin waves in the western Pacific Ocean, *J. Geophys. Res.*, *108*, 3311, doi:10.1029/2002JC001760, 2003.

Boulanger, J.-P., C. Menkes, and M. Lengaigne, Role of high-frequency wind variability and other potential mechanisms in the onset, growth and termination phases of the 1997–1998 El Niño, *Clim. Dyn.*, in press, 2004.

Busalacchi, A. J., and M. A. Cane, Hindcasts of sea level variations during the 1982–83 El Niño, *J. Phys. Oceanogr.*, *15*, 213–221, 1985.

Cane, M. A., Oceanographic events during El Niño, *Science*, *222*, 1189–1194, 1983.

Chang, C. P., J. E. Erickson, and K. M. Lau, Northeasterly cold surges and near-equatorial disturbances over the Winter MONEX area during December 1974, I, Synoptic aspects, *Mon. Weather Rev.*, *107*, 812–829, 1979.

Chang, C. P., and K. M. Lau, Northeasterly cold surges and near-equatorial disturbances over the Winter MONEX area during December 1974, II, Planetary aspects, *Mon. Weather Rev.*, *108*, 298–312, 1980.

Chen, D., M. A. Cane, and S. E. Zebiak, The impact of NSCAT winds on predicting the 1997/98 El Niño: A case study with the Lamont-Doherty Earth Observatory model, *J. Geophys. Res.*, *104*, 11321–11327, 1999.

Chu, P.-S., Extratropical forcing and the burst of equatorial westerlies in the western Pacific : A synoptic study, *J. Meteorol. Soc. Jpn.*, *66*, 549–564, 1988.

Chu, P.-S., and J. Frederick ,Westerly wind bursts and surface heat fluxes in the equatorial western Pacific in May 1982, *J. Meteorol. Soc. Jpn.*, *68*, 523–536, 1990.

Chu, P.-S., J. Frederick, and A. J. Nash, Exploratory analysis of surface winds in the western equatorial Pacific and El Niño, *J. Clim.*, *4*, 1088–1101, 1991.

Compo, G. P., G. N. Kiladis, and P. J. Webster , The horizontal and vertical structure of east asian summer monsoon pressure surges. *Q. J. R. Meteorol. Soc.*, *125*, 29–54, 1999.

Cravatte, S., J. Picaut, and G. Eldin, Second and first baroclinic modes in the equatorial Pacific at intraseasonal timescales, *J. Geophys. Res.*, *108*, 3226, doi:10.1029/2002JC001511, 2003.

Cronin, M. F., and M. J. Mc Phaden, The upper ocean heat balance in the western equatorial warm pool durin September–December 1992, *J. Geophys. Res.*, *102*, 8533–8553, 1997.

Cronin, M. F., and M. J. McPhaden , Upper ocean salinity balance in the western equatorial Pacific, *J. Geophys. Res.*, *103*, 27567–27587, 1998.

Cronin, M. F., M. J. McPhaden, and R. H. Weisberg , Wind-forced reversing jets in the western equatorial Pacific, *J. Phys. Oceanogr.*, *30*, 657–676, 2000.

Cronin, M. F., and M. J. McPhaden, Barrier layer formation during westerly wind bursts, *J. Geophys. Res.*, *107*, 8020, doi:10.1029/2001JCC001171, 2002 .

Delcroix, T., J. Picaut, and G. Eldin , Equatorial Kelvin and Rossby waves evidenced in the Pacific Ocean through Geosat Sea Level and surface current anomalies, *J. Geophys. Res.*, *96*, 3249–3262, 1991.

Delcroix, T., G. Heldin, M.-H. Radenac, J. Toole, and E. Firing, Variation of the western equatorial Pacific Ocean, 1986–1988, *J. Geophys. Res.*, *97*, 5423–5446, 1992.

Delcroix, T., and G. Heldin, Effects of Westerly Wind Bursts Upon the Western Equatorial Pacific Ocean, February–April 1991, *J. Geophys. Res.*, *98*, 16379–16385, 1993.

Delcroix T., J.-P. Boulanger, F. Masia, and C. Menkes, GEOSAT–derived sea level and surface-current anomalies in the equatorial Pacific, during the 1986–1989 El Niño and La Niña, *J. Geophys. Res.*, *99*, 25093–25107, 1994.

Delcroix T., B. Dewitte, Y. Dupenhoat, F. Masia, and J. Picaut, Equatorial waves and warm pool displacements during the 1992–98 ENSO events: observations and modelling, *J. Geophys. Res.*, *105*, 26045–26062, 2000.

Dourado, M., and G. Caniaux, Surface heat budget in an oceanic simulation using data from Tropical Ocean-Global Atmosphere Coupled Ocean-Atmosphere Response Experiment, *J. Geophys. Res.*, *106*, 16623–16640, 2001.

Eldin, G., T. Delcroix, C. Henin, K. Richards, Y. Du Penhoat, J. Picaut, and P. Tual, Large-scale current and thermohaline structures along 156°E during the COARE intensive observation period, *Geophys. Res. Lett.*, *21*, 2681–2684, 1994.

Enfield, D. B., The intraseasonal oscillation in eastern Pacific sea levels: How is it forced? *J. Phys. Oceanogr.*, *17*, 1860–1876, 1987.

Eriksen, C. C., M. B. Blumenthal, S. P. Hayes, and P. Ripa, Wind generated equatorial Kelvin waves observed across the Pacific Ocean, *J. Phys. Oceanogr.*, *13*, 1622–1640, 1983.

Fasullo, J., and P. J. Webster, Atmospheric and surface variations during westerly wind bursts in the tropical Pacific, *Q. J. R. Meteorol. Soc.*, *126*, 899–924, 2000.

Fedorov, A. V. and W. K. Melville, Kelvin fronts on the equatorial thermocline. *J.Phys.Oceanogr.*, *30*, 1692–1705, 2000.

Fedorov, A. V., The response of the coupled tropical ocean-atmosphere to westerly wind bursts. *Q. J. R.. Meteorol. Soc.*, *128*, 1–23, 2002.

Fedorov, A. V., S. L. Harper, S. G. Philander, B. Winter, and A. Wittenberg, How predictable is El Niño? *Bull. Am. Meteorol. Soc.*, *84*, 911–919, 2003.

Feng M., P. Hacker, and R. Lukas, Upper ocean heat and salt balances in response to a westerly wind burst in the western equatorial Pacific during TOGA COARE, *J. Geophys. Res.*, *103*, 10289–10311, 1998.

Feng, M., R. Lukas, P. Hacker, R. A. Weller, and S. P. Anderson, Upper-Ocean heat and salt balances in the western equatorial Pacific in response to the intraseasonal oscillation during TOGA-COARE. *J. Clim.*, *13*, 2409–2427, 2000.

Giese, B. S., and D. E. Harrison, Aspects of the Kelvin wave response to episodic forcing, *J. Geophys. Res*, *95*, 7289–7312, 1990.

Giese, B. S., and D. E. Harrison, Eastern Equatorial Pacific Response to Three Composite Westerly Wind Types, *J. Geophys. Res.*, *96*, 3239–3248, 1991.

Godfrey, J. S., R. A. Houze, R. H. Johnson, R. Lukas, J. L. Redelsperger, A. Sumi, and R. Weller, Coupled Ocean-Atmosphere Response Experiment (COARE): An interim report, *J. Geophys. Res.*, *103*, 14395–14450, 1998.

Gutzler, D. S., Interannual fluctuation of intraseasonal variance of near-equatorial zonal winds, *J. Geophys. Res.*, *96*, 3172–3195, 1991.

Gutzler, D. S., G. N. Kiladis, G. A. Meehl, K. M. Weickmann, and M. Wheeler, The global climate of December 1992–February 1993. Part II : Large scale variabililty across the tropical western Pacific during TOGA COARE. *J. Clim.*, *7*, 1606–1622, 1994.

Haarten, L. M., Synoptic settings of westerly wind bursts, *J. Geophys. Res.*, *101*, 16997–17019, 1996.

Harrison, D. E., On the appearance of sustained equatorial westerlies during the 1982 Pacific warm event, *Science*, *225*, 1099–1102, 1984.

Harrison, D. E., and P. S. Schopf, Kelvin wave-induced advection and the onset of SST warming in El Niño events, *Mon. Weather Rev.*, *112*, 923–933, 1984.

Harrison, D. E., and B. S. Giese, Remote westerly wind forcing of the eastern equatorial Pacific: Some model results, *Geophys. Res. Lett.*, *15*, 804–807, 1988.

Harrison D. E., and B. S. Giese, Comment on "The response of the Equatorial Pacific Ocean to a Westerly Wind Burst in May 1986", *J. Geophys. Res.*, *94*, 5024–5026, 1989.

Harrison, D. E., and B. S. Giese, Episodes of surface westerly winds as observed from islands in the western tropical Pacific. *J. Geophys. Res.*, *96*, 3221–3237, 1991.

Harrison, D. E., and A. P. Craig, Ocean Model Studies of Upper-Ocean Variability at 0°, 160°W during the 1982–1983 ENSO : Local and Remotely Forced Response, *J. Phys. Oceanogr*, *23*, 426–451, 1993.

Harrison, D. E., and G. A. Vecchi, Westerly wind events in the tropical Pacific, 1986–1995, *J. Clim.*, *10*, 3131–3156, 1997.

Harrison, D. E., G. A. Vecchi, and R. H. Weisberg, Eastward surface jets in the central equatorial Pacific, November 1991–March 1992, *J. Mar. Res.*, *58*, 735–754, 2000.

Hasselmann, K., Stochastic climate models. Part I : Theory. *Tellus*, *28*, 289–305, 1976.

Hayes, S., and D. Halpern, Correlation of current and sea level in the eastern equatorial Pacific. *J. Phys. Oceanogr.*, *14*, 811–824, 1984.

Hendon, H. H., C. Zhang, and J. D. Glick, Interannual variability of the Madden-Julian Oscillation during austral summer, *J. Clim.*, *12*, 2358–2550, 1999.

Hisard, P., J. Merle, and B. Voituriez, The equatorial undercurrent observed at 170°E in March and April 1967, *J. Mar. Res.*, *28*, 281–303, 1970.

Johnson, E. S., and M. J. McPhaden, Structure of intraseasonal Kelvin waves in the equatorial Pacific ocean, *J. Phys. Oceanogr.*, *23*, 608–625, 1993a.

Johnson, E. S., and M. J. McPhaden, Effects of a three dimensional mean flow on intraseasonal Kelvin waves in the equatorial Pacific ocean, *J. Geophys. Res.*, *98*, 10185–10194, 1993b.

Keen, R. A., The role of cross-equatorial tropical cyclone pairs in the Southern Oscillation, *Mon. Weather Rev.*, *110*, 1405–1416, 1982.

Kessler, W. S., M. J. McPhaden, and K. M. Weickmann, Forcing of intraseasonal Kelvin waves in the equatorial Pacific. *J. Geophys. Res.*, *100*, 10613–10631, 1995.

Kessler, W. S., and M. J. McPhaden, Oceanic Equatorial Waves and the 1991–93 El Niño, *J. Clim.*, *8*, 1757–1776, 1995.

Kessler, W. S., and R. Kleeman, Rectification of the Madden-Julian Oscillation into the ENSO cycle, *J. Clim.*, *13*, 3560–3575, 2000.

Kessler, W. S., EOF representations of the Madden-Julian Oscillation and its connection with ENSO, *J. Clim.*, *14*, 3055–3061, 2001.

Kessler, W. S., Is ENSO a cycle or a series of events? *Geophys. Res. Lett.*, *29*, 2125, doi:10.1029/ 2002GL015924, 2002.

Kestin, T. S., D. J. Karoly, J.-I. Yano, and N. A. Rayner, Time-frequency variability of ENSO and stochastic simulations, *J. Clim.*, *11*, 2258–2272, 1998.

Kiladis, G. N., G. A. Meehl, and K. M. Weickmann, Large-scale circulation associated with westerly wind bursts and deep convection over the western equatorial Pacific, *J. Geophys. Res.*, *99*, 18,527–18,544, 1994.

Kiladis, G. N., and K. M. Wheeler, Horizontal and vertical structure of observed tropospheric equatorial Rossby waves, *J. Geophys. Res.*, *100*, 22981–22997, 1995.

Kindle, J. C., and P. A. Phoebus, The ocean response to operational westerly wind bursts during the 1991–1992 El Niño, *J. Geophys. Res.*, *100*, 4893–4920, 1995.

Kleeman, R., and A. M. Moore, A theory for the limitation of ENSO predictability due to stochastic atmospheric transients, *J. Atmos. Sci.*, *54*, 753–767, 1997.

Knox, R., and D. Halpern, Long range Kelvin wave propagation of transport variations in Pacific ocean equatorial currents, *J. Mar. Res.*, *40*, 329–339, 1982.

Kuroda, Y., and M. J. McPhaden, Variability in the western equatorial Pacific Ocean during JAPACS cruises in 1989 and 1990, *J. Geophys. Res.*, *98*, 4747–4759, 1993.

Lander, M. A., Evolution of the cloud pattern during the formation of tropical cyclone twins symetrical with respect to the equator, *Mon. Weather Rev.*, *118*, 1194–1202, 1990.

Latif, M., J. Biercamp, and H. von Storch, The response of a coupled Ocean-Atmosphere General Circulation Model to Wind Bursts, *J. Atmos. Sci.*, *45*, 964–976, 1988.

Latif, M., D. Anderson, T. Barnett, M. A. Cane, R. Kleeman, A. Leetmaa, J. J. O'Brien, A. Rosati, E. Schneider, TOGA review paper:"Predictability and prediction", *J. Geophys. Res.*, *103*, 14375–14393, 1998.

Lau, K. M., and P. H. Chan, Intraseasonal and interannual variations of tropical convection: A possible link between 40–50 Day Oscillation and ENSO? *J. Atmos. Sci.*, *45*, 506–521, 1988.

Lau, K. M., L. Peng, C. H. Sui, and T. Nakazawa, Dynamics of super cloud clusters, westerly wind bursts, 30–60 day oscillation and ENSO : A unified view, *J. Meteorol. Soc. Jpn.*, *67*, 205–219, 1989.

Lengaigne, M., J.-P. Boulanger, C. Menkes, S. Masson, G. Madec, and P. Delecluse, Ocean Response to the March 1997 Westerly Wind Event, *J. Geophys. Res.*, *107*, 8015, doi:10.1029/2001JC000841, 2002.

Lengaigne, M., J.-P. Boulanger, C. Menkes, G. Madec, P. Delecluse, E. Guilyardi, and J. Slingo: The March 1997 Westerly Wind Event and the onset of the 1997/98 El Niño: Understanding the atmospheric response, *J. Clim.*, *16*, 3330–3343, 2003.

Lengaigne, M., E. Guilyardi, J.-P. Boulanger, C. Menkes, P. Inness, J. Cole, P. Delecluse, and J. Slingo, Triggering of El Niño by Westerly Wind Events in a Coupled General Circulation Model, *Clim. Dyn.*, accepted, 2004.

Lim, H., and C. P. Chang, A theory for mid-latitude forcing of tropical motion during the winter monsoon, *J. Atmos. Sci.*, *38*, 2377–2392, 1981.

Lindstrom, E., R. Lukas, R. Fine, E. Firing, J. S. Godfrey, G. Meyers, and M. Tsuchiya, The western equatorial Pacific Ocean circulation study, *Nature*, *330*, 533–537, 1987.

Love, G., Cross-equatorial influence of winter hemisphere sub-tropical cold surges, *Mon. Weather Rev.*, *113*, 1487–1498, 1985a.

Love, G., Cross-equatorial interactions during tropical cyclogenesis, *Mon. Weather Rev.*, *113*, 1499–1509, 1985b.

Lukas, R., S. P. Hayes, and K. Wyrtki, Equatorial sea level response during the 1982–1983 El Niño, *J. Geophys. Res.*, *89*, 10425–10430, 1984.

Luther, D. S., Harrison, B. S., and R. A. Knox, Zonal winds in the central equatorial Pacific and El Niño. *Science*, *222*, 327–330, 1983.

Luther, D. S., and D. E. Harrison, Observing Long-Period Fluctuations of Surface Winds in the Tropical Pacific : Initial results from Island Data. *Mon. Weather Rev.*, *112*, 285–302, 1984.

Madden, R. A., and P. R. Julian, Observations of the 40–50-day tropical oscillation: A review, *Mon. Weather Rev.*, *122*, 814–837, 1994.

Maes C., J. Picaut, and S. Belamari, Salinity barrier layer and onset of El Niño in a Pacific coupled model, *Geophys. Res. Lett.*, *29*, 2206, doi : 10.1029/2002GL016029, 2002.

Matsuno, T., Quasi-geostrophic motions in the equatorial area. *J. Meteorol. Soc. Jpn.*, *44*, 25–42, 1966.

McCreary, J. P., Eastern tropical ocean response to changing wind systems : With application to El Niño. *J. Phys. Oceanogr.*, *6*, 632–645, 1976

McPhaden, M. J., J. P. Proehl, and L. M. Rothstein, The interaction of equatorial Kelvin waves with realistically sheared zonal currents. *J. Phys. Oceanogr.*, *16*, 1499–1515, 1986.

McPhaden, M. J., H. P. Freitag, S. P. Hayes, B. A. Taft, Z. Chen, and K. Wyrtki, The response of the Equatorial Pacific Ocean to a Westerly Wind Burst in May 1986, *J. Geophys. Res.*, *93*, 10589–10603, 1988.

McPhaden, M. J., and B. A. Taft, Dynamics of seasonal and intraseasonal variability in the eastern equatorial Pacific, *J. Phys. Oceanogr.*, *18*, 1713–1732, 1988.

McPhaden, M. J., S. P. Hayes, L. J. Mangum, and J. M. Toole, Variability in the western equatorial Pacific Ocean during the 1986–1987 El Niño/Southern Oscillation event, *J. Phys. Oceanogr.*, *20*, 190–208, 1990.

McPhaden, M. J., F. Bahr, Y. du Penhoat, E. Firing, S. P. Hayes, P. P. Niiler, P. L. Richardson, and J. M. Toole, The response of the western equatorial Pacific Ocean to Westerly Wind Bursts During November 1989 to January 1990, *J. Geophys. Res.*, *97*, 14289–14303, 1992.

McPhaden, M. J., Genesis and evolution of the 1997–98 El Niño, *Science*, *283*, 950–954, 1999.

McPhaden, M. J., and X. Yu, Equatorial waves and the 1997–98 El Niño, *Geophys. Res. Lett.*, *26*, 2961–2964, 1999.

McPhaden, M. J., Mixed layer temperature balance on intraseasonal time scales in the equatorial Pacific Ocean, *J. Clim.*, *15*, 2632–2647, 2002.

Meyers, G., J. R. Donguy, and R. K. Reed, Evaporative cooling of the western equatorial Pacific Ocean by anomalous winds, *Nature*, *323*, 523–526, 1986.

Meehl G. A., G. N. Kiladis, K. M. Weickmann, M. Wheeler, D. S. Gutzler, and G. P. Compo, Modulation of equatorial subseasonal convective episodes by tropical-extratropical interaction in the Indian and Pacific Ocean regions, *J. Geophys. Res.*, *101*, 15,033–15,049, 1996.

Miller, L., R. E. Cheney, and B. C. Douglas, Geosat altimeter observations of Kelvin waves and the 1986–1987 El Niño, *Science*, *239*, 52–54, 1988.

Moore, A. M., and R. Kleeman, Stochastic forcing of ENSO by Intraseasonal Oscillation, *J. Clim.*, *12*, 1199–1220, 1999.

Nakazawa, T., Intraseasonal oscillations during the TOGA-COARE IOP, *J. Meteorol. Soc. Jpn.*, *73*, 305–319, 1995.

Neelin, J. D., D. S. Battisti, A. C. Hirst, F. F. Jin, Y. Wakata, T. Yamagata, and S. E. Zebiak, ENSO theory, *J. Geophys. Res.*, *103*, 14261–14290, 1998.

Nitta, T., and T. Motoki, Abrupt enhancement of convective activity and low-level westerly bursts during the onset phase of the 1986–1987 El Niño, *J. Meteorol. Soc. Jpn.*, *65*, 497–506, 1987.

Nitta, T., Development of twin cyclone and twin cyclone and westerly bursts during the initial phase of the 1986–1987 El Niño, *J. Meteorol. Soc. Jpn.*, *67*, 677–681, 1989.

Numaguti, A., characterictics of 4-to-20-day period disturbances observed in the equatorial Pacific during TOGA COARE IOP, *J. Meteorol. Soc. Jpn.*, *73*, 353–377, 1995.

Palmer, C., Tropical Meteorology, *Q. J. R. Meteorol. Soc.*, *78*, 128–154, 1952.

Penland, C., and P. D. Sardeshmukh, The optimal growth of tropical sea surface temperature anomalies, *J. Clim.*, *8*, 1999–2024, 1995.

Perigaud, C. M., and C. Cassou, Importance of oceanic decadal trends and westerly wind bursts for forcasting El Niño, *Geophys. Res. Lett.*, *27*, 389–392, 2000.

Picaut, J., and T. Delcroix, Equatorial wave sequence associated with warm pool displacements during the 1986–1989 El Niño-La Niña, *J. Geophys. Res.*, *100*, 18393–18408, 1995.

Picaut, J., M. Ioualanen, C. Menkes, T. Delcroix, and M. J. McPhaden, Mechanism of the zonal displacements of the Pacific warm pool : Implications for ENSO, *Science*, *274*, 1486–1489, 1996.

Qiao, L., and R. H. Weisberg, Tropical instability wave kinematics: Observations from the Tropical Instability Wave Experiment (TIWE), *J. Geophys. Res.*, *100*, 8677–8693, 1995.

Ralph, E. A., K. Bi, P.P. Niiler, and Y. du Penhoat, A lagrangian description of the western Equatorial Pacific response to the wind burst of December 1992: Heat advection of the warm pool, *J. Clim.*, *10*, 1706–1721, 1997.

Richards, K. J., and M. E. Inall, The upper ocean heat content of the western equatorial Pacific: Processes controlling its change during TOGA-COARE, *J. Geophys. Res.*, *105*, 19575–19590, 2000.

Richardson, R. A., I. Ginis, and L. M. Rothstein, A numerical investigation of the local ocean response to westerly wind burst forcing in the western equatorial Pacific, *J. Phys. Oceanogr.*, *29*, 1334–1352, 1999.

Ripa, P., Nonlinear wave-wave interactions in a one-layer reduced-gravity model on the equatorial β-plan, *J. Phys. Oceanogr.*, *12*, 97–111, 1982.

Ripa, P., and S. P. Hayes, Evidence for equatorial trapped waves at the Galapagos Islands, *J. Geophys. Res.*, *86*, 6509–6516, 1981.

Roemmich, D., M. Morris, W. R. Young, and J. R. Donguy, Fresh equatorial jets, *J. Phys. Oceanogr.*, *24*, 540–558, 1994.

Roulston, M., and J. D. Neelin, The response of an ENSO model to climate noise, weather noise and intraseasonal forcing, *Geophys. Res. Lett.*, *27*, 3723–3726, 2000.

Schopf, P. S., and D. E. Harrison, On equatorial Kelvin waves and El Niño, I, Influence of initial states on wave-induced current and warming, *J. Phys. Oceanogr.*, *13*, 936–948, 1983.

Shinoda, T., and H. H. Hendon, Mixed layer modeling of intraseasonal variability in the tropical western Pacific and Indian Oceans, *J. Clim.*, *11*, 2668–2685, 1998.

Slingo, J. M., Extratropical forcing of tropical convection in a northern winter simulation with the UGAMP GCM, *Q. J. R.. Meteorol. Soc.*, *124*, 27–51, 1998.

Slingo, J. M., D. P. Rowell, K. R. Sperber, and F. Nortley, On the predictability of the interannual behaviour of the Madden-Julian Oscillation and its relationship with El Niño, *Q. J. R.. Meteorol. Soc.*, *125*, 583–609, 1999.

Smyth, W. D., D. Hebert, and J. N. Moum, Local ocean response to a multiphase westerly windburst, 2, Thermal and freshwater responses, *J. Geophys. Res.*, *101*, 22513–22533, 1996a.

Smyth, W. D., D. Hebert, and J. N. Moum, Local ocean response to a multiphase westerly windburst, 1, Dynamic response, *J. Geophys. Res.*, *101*, 22495–22512, 1996b.

Sprintall, J., and M. Tomczak, Evidence of the barrier layer in the surface layer of the tropics, *J. Geophys. Res.*, *97*, 7305–7316, 1992.

Sui, C. H., and K. M. Lau, Multiscale phenomena in the tropical atmosphere over the western Pacific, *Mon. Wea. Rev.*, *120*, 407–430, 1992.

Van Oldenborgh, G. J., What caused the Onset of the 1997–1998 El Niño? *J. Phys. Oceanogr.*, *128*, 2601–2607, 2000.

Vecchi, G. A., and D. E. Harrison, Tropical Pacific Sea Surface Temperature Anomalies, El Niño and Equatorial Westerly Wind Events, *J. Clim.*, *13*, 1814–1832, 2000.

Vecchi, G. A., Sub-seasonal wind variability and El Niño. Ph. D. Diss., Univ. Of Washington, 2001.

Verbickas, S., Westerly wind bursts in the Tropical Pacific, *Weather*, *53*, 282–284, 1998.

Vialard, J., and P. Delecluse, An OGCM study for the TOGA decade. Part II: Barrier-layer formation and variability, *J. Phys. Oceanogr.*, *28*, 1089–1106, 1998.

Vialard, J., C. Menkes, J.-P. Boulanger, P. Delecluse, E. Guilyardi, M. J. McPhaden, and G. Madec, Oceanic mechanisms driving the SST during the 1997–1998 El Niño, *J. Phys. Oceanogr.*, *31*, 1649–1675, 2001.

Vitart, F., M. A. Balmaseda, L. Ferranti, and D. Anderson, Westerly Wind Events and the 1997/98 El Niño Event in the ECMWF Seasonal Forcasting System: A Case Study, *J. Clim.*, *16*, 3153–3170, 2003.

Wang, C., and J. Picaut, Understanding ENSO Physics—A review, this volume.

Webster, P. J., and J. R. Holton, Cross-equatorial response to middle-latitude forcing in a zonally varying basic state, *J. Atmos. Sci.*, *39*, 722–733, 1982.

Webster, P. J., and R. Lukas, TOGA COARE : The Coupled Ocean-Atmosphere Response Experiment, *Bull. Am. Meteorol. Soc.*, *73*, 1377–1416, 1992.

Weidman, P. D., D. L. Mickler, B. Dayyani, and G. H. Born, Analysis of Legeckis eddies in the near-equatorial Pacific, *J. Geophys. Res.*, *104*, 7865–7887, 1999.

Weisberg, R. H., and S. P. Hayes, Upper ocean variability on the equator in the west-central Pacific at 170°W, *J. Geophys. Res.*, *100*, 20485–20498, 1995.

Yoshida, K., A theory of the Cromwell Current and equatorial upwelling, *J. Oceanogr. Soc. Jpn.*, *15*, 154–170, 1959.

Yu, L., and M. M. Rienecker, Evidence of an extratropical atmospheric influence during the onset of the 1997–8 El Niño, *Geophys. Res. Lett.*, *25*, 3537–3540, 1998.

Yu, L., R. A. Weller, and W. T. Liu, Case analysis of a role of ENSO in regulating the generation of westerly wind bursts in the Western Equatorial Pacific, *J. Geophys. Res.*, *108*, 3128, doi:10.1029/2002JC001498, 2003.

Zhang, K. Q., and L. M. Rothstein, Modelling the oceanic response to westerly wind bursts in the western equatorial Pacific, *J. Phys. Oceanogr.*, *28*, 2227–2249, 1998.

Zhang, C., and M. J. McPhaden, Intraseasonal surface cooling in the equatorial western Pacific, *J. Clim.*, *13*, 2261–2276, 2000.

Zhang, C., Intraseasonal perturbations in sea surface temperatures of the equatorial eastern Pacific and their association with the Madden-Julian Oscillation, *J. Clim.*, *14*, 1309–1322, 2001.

Zhang, C., and J. Gottschalck, SST anomalies of ENSO and the Madden-Julian Oscillation in the equatorial Pacific, *J. Clim.*, *15*, 2429–2445, 2002.

Zebiak, S. E., and M. A. Cane, A model of El Niño-Southern Oscillation, *Mon. Wea. Rev.*, *115*, 2262–2278, 1987.

J.-P. Boulanger, P. Delecluse, and C. Menkes, LODYC, Université Pierre et Marie Curie, Case 100, 4 Place Jussieu, 75252 Paris, Cedex 05, France. (boulanger@lodyc.jussieu.fr; pna@lodyc.jussieu.fr; menkes@lodyc.jussieu.fr)

M. Lengaigne and J. Slingo, CGAM, Department of Meteorology, University of Reading, Early Gate, Reading, Berkshire RG6 6BB, U.K. (lengaign@met.reading.ac.uk; j.m.slingo@reading.ac.uk)

The Control of Meridional Differential Surface Heating Over the Level of ENSO Activity: A Heat-Pump Hypothesis

De-Zheng Sun

NOAA-CIRES/Climate Diagnostics Center, Boulder, Colorado

Numerical experiments with a coupled model have been carried out to test the heat-pump hypothesis for ENSO. The hypothesis states that the level of ENSO activity is controlled by the meridional differential surface heating over the Pacific: either an enhanced surface heating over the equatorial region or an enhanced cooling over the subtropical/extratropical ocean may result in a regime with stronger ENSO events. Moreover, ENSO may be a mechanism that regulates the long-term stability of the coupled equatorial ocean-atmosphere system. The results from the numerical experiments are shown to be consistent with this hypothesis. A stronger tropical heating or a stronger subtropical/extratropical cooling tends to increase the contrast between the SST in the tropical western Pacific warm-pool and the temperature of the equatorial thermocline water and thereby destabilize the coupled equatorial ocean-atmosphere system. In response, a regime with stronger ENSO events sets in. The stronger ENSO events transport more heat downward and poleward, cooling the warm-pool SST and warming the equatorial thermocline water. In the presence of ENSO, the difference between the time-mean warm-pool SST and the time-mean temperature of the equatorial thermocline water is found to be insensitive to changes in the external forcing.

1. INTRODUCTION

The level of the ENSO activity has varied over the past [*Tudhope et al.*, 2001; *Cobb et al.*, 2003], suggesting it may vary in the future. What controls the level of ENSO activity, however, has not been well understood [*Wang and Picaut*, this volume]. For example, there is no consensus on the causes of the strengthening of the ENSO activity during the last 20 years relative to the previous two decades. At least three different explanations for this change in the level of ENSO activity have been put forward. *Wang and An* [2001] have suggested that the strengthening of ENSO activity during the last 20 years is due to a change in the background winds. *Jin et al.* [2003] attributes this strengthening of ENSO activity to non-linear dynamic heating. *Sun* [2003] suggests that the strengthening of ENSO may be a consequence of an increase in the equatorial radiative heating or equivalently an increase in the tropical maximum SST. The understanding of the changes in the level of ENSO activity on the longer time-scales is not better. For example, there is also no agreement on the causes of the suppression of ENSO activity during the mid Holocene [*Rondel et al.*, 1999; *Sandweiss et al.*, 1996]. *Sun* [2000] suggests that the warmer equatorial thermocline water could be a cause—the reduced difference between the tropical maximum SST and the temperature of the equatorial thermocline water makes the coupled system more stable. *Liu et al.* [2000] did find warmer equatorial thermocline water in their model simulation of the mid-Holocene, but suggest a role of the Asian summer monsoon. *Clement et al.* [2000] proposed another mechanism: orbitally driven changes in the seasonal cycle of solar radiation in the tropics. Because of the differences in the models used, whether these proposed mechanisms are really independent of each other or overlapping is not known.

The inadequacy in our understanding of what controls the level of ENSO activity is also reflected in the lack of consistency in the predictions by coupled GCMs. Some early models suggest that the level of ENSO activity may reduce in

Earth's Climate: The Ocean-Atmosphere Interaction
Geophysical Monograph Series 147
Copyright 2004 by the American Geophysical Union
10.1029/147GM04

response to global warming [*Meehl et al.*, 1993; *Tett*, 1995; *Knutson et al.*, 1997]. The more recent experiments by *Timmerman et al.* [1999] suggest the opposite effect: the level of ENSO activity will increase in response to global warming. *Collins* [2001] even found different responses of ENSO to global warming in two different versions of the same model. The credibility of the predictions of the response to global warming by these coupled GCMs is also undermined by the lack of confidence in their simulation of cloud feedbacks [*Cess et al.*, 1989; *Sun et al.*, 2003].

The rather scattered results concerning the causes of the variability in the level of ENSO highlight a need for a better delineation of the fundamental forces controlling the level of ENSO activity. The purpose of this article is to highlight recent progress in understanding the influence of surface heating over the level of ENSO activity, specifically, the heat-pump hypothesis [*Sun*, 2003; *Sun et al.*, 2004]. The hypothesis states that ENSO amplitude is proportional to the meridional differential surface heating over the Pacific: either an enhanced surface heating over the equatorial region or an enhanced cooling over the subtropical/extratropical ocean may result in a regime with stronger El Niño events. Moreover, El Niño may be a mechanism that regulates the stability of the time-mean state of the equatorial Pacific. This hypothesis, if proved to be true, could become a step stone to better understand why the level of ENSO activity has varied over the past and how it may change in the future as the CO_2 concentration in the atmosphere increases. The heat-pump hypothesis deals with the question of what controls the amplitude of ENSO. Therefore, it complements existing theories of ENSO as these theories mainly provide an explanation for the phase transition of ENSO.

The paper is organized as follows. The observational background for the conception of the heat-pump hypothesis is introduced in Section 2. The numerical experiments that have been carried out to test this hypothesis are highlighted in Section 3. The implications of the heat-pump hypothesis are discussed in Section 4.

2. OBSERVATIONAL BACKGROUND

The distribution of the net surface heat flux over the Pacific is characterized by heating in the equatorial region and cooling in the higher latitudes (Figure 1a). This implies that the ocean has to remove heat away from the equatorial Pacific to the higher latitudes. Extending previous calculations by *Wyrtki* [1985], *Meinen and McPhaden* [2000, 2001], *Sun and Trenberth* [1998], and *Sun* [2000], *Sun* [2003] calculated the poleward heat transport in the tropical Pacific over the last 20 years using a single ocean data set. The calculation showed that the required poleward heat removal is achieved episodically and those episodes correspond with the occurrence of El Niño events (see Figure 5 in *Sun* [2003]). A basin-wide view of the anomalous heat transfer during El Niño is shown in Figure1b. Shown is the divergent component of the heat transport vertically averaged over the upper ocean. As there is anomalous eastward heat transport to feed the El Niño warming, the poleward heat transport is enhanced across the equatorial Pacific. The center of the divergence in the transport occurs slightly west to the dateline while the surface heating is peaked in the far eastern Pacific. This indicates that the heat also has to be first transported westward, consistent with the observation that heat is transported to the subsurface ocean of the western Pacific during the quiescent periods—the La Niña events [*Wyrtki*, 1985; *Sun*, 2000; *Sun*, 2003]. Therefore, both phases of ENSO are fundamentally involved in the planetary heat transport in the tropical Pacific. The fact that El Niño corresponds to an elevated poleward heat transport has been noted earlier [*Wyrtki*, 1985; *Meinen and McPhaden*, 2000, 2001; *Sun* 2001]. *Jin* [1997a, b] has suggested that this elevated poleward heat transport provides the negative feedback that terminates the anomalous surface warming. Therefore, the extended calculation by *Sun* [2003] may be viewed as evidence for the relevance of Jin's theory for the phase transition of ENSO. The key point in *Sun* [2003], however, is the connection between ENSO and the surface heating over the tropical Pacific and the implied long-term heat balance of the tropical Pacific.

Sun [2003] has also noted in his heat budget analysis that the two strongest El Niño events, the 1982–83 El Niño event and the 1997–98 El Niño event, are also accompanied by the strongest poleward heat transport. The peak value of the heat transport out of the equatorial Pacific (5°S–5°N) associated with the 1997–98 El Niño event almost doubles the mean peak value associated with the 4 weaker El Niño events (1986–87, 1991–92, 1993, and 1994–95). Figure 2a, b contrasts the heat transport during the two strongest El Niño events with a more moderate one, the 1991–92 El Niño event. As there is more heat transport from the equatorial western Pacific to the equatorial eastern Pacific, the poleward heat transport is enhanced across the bulk of the tropical Pacific.

If all the El Niño events in the record were as strong as the 1997–98 El Niño event, we would have seen a stronger poleward heat transport in the time-mean. Conversely, if the time mean poleward heat transport is forced to increase for some reason, the mean level of ENSO activity may have to increase to maintain the time-mean heat balance. This consideration led to the conception that the meridional differential surface heating may act as a fundamental factor influencing the level of ENSO activity.

Admittedly, the above consideration only points to a possibility, not a necessity. For example, to achieve a higher poleward heat transport in the mean, the system could increase

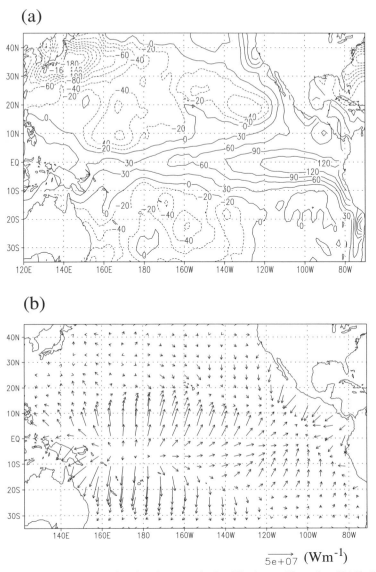

Figure 1. (a) Distribution of annual mean surface heating over the Pacific Ocean (from Sun 2003). (b) A basin-wide view of the anomalous heat transport during El Niño. Shown are the average differences in the divergent component of the ocean transport in the upper ocean between El Niño and non-El Niño periods. The divergent component of the ocean transport in the upper ocean is obtained by solving a Laplace equation using the heat divergence data (D_o) calculated by Sun (2003) (see Equation (1) in that paper). Realistic topography is used for the lateral boundary conditions. The meridional domain is from 35°S to 45°N. The model outside the analysis model domain is fixed to the climatology over the 1980–98 period. The definition of El Niño and La Niña periods follows Trenberth (1997).

the poleward transport during the relative quiescent periods—the La Niña events. This scenario, however, is unlikely, because heat has to be pumped to the subsurface ocean for a substantial poleward ocean heat transport. Note that the tropical Pacific is stably stratified in the vertical except the shallow surface mixed layer. It needs Ekman pumping to do the work to pump heat down to the subsurface ocean. The pattern of transport during El Niño shown in Figure 2 also indicates that heat has to be transported from the eastern Pacific to the western Pacific. Pumping heat down to the subsurface ocean of the western Pacific appears to occur primarily during La Niña events [*Sun* 2001, 2003]. To put these arguments and inferences from observations on a firmer ground, experiments with a coupled model have been carried out.

(a)

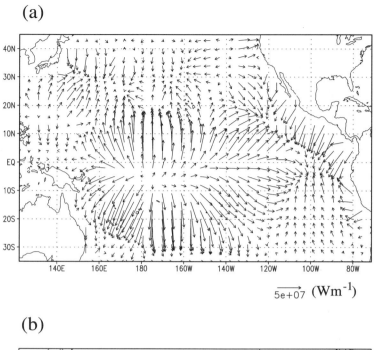

(b)

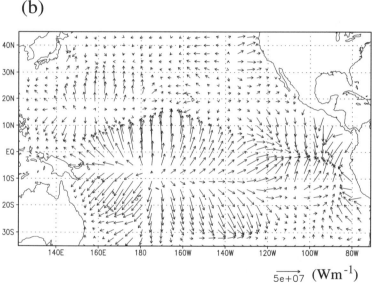

Figure 2. A basin-wide view of the anomalous heat transport during the two exceptionally strong El Niño events: the 1997–98 (a) and the 1982–83 (b) events. Shown are the divergent component of the mean upper ocean heat transport during these two events relative to that during the 1991–92 El Niño event (the 97–98 event minus the 91–92 event and the 82–83 event minus the 91–92 event respectively). The 1991–92 El Niño event has a similar life cycle to the two strongest El Niño events.

3. NUMERICAL RESULTS

3.1. The Model

The model is described in *Sun* [2003] and *Sun et al.* [2004]. The atmospheric model is an empirical one. The surface heat flux is parameterized in the same way as previous theoretical studies of ENSO: it is proportional to the difference between a prescribed radiative convective equilibrium SST (SST_p) and the actual SST predicted by the coupled model,

$$F_s(\lambda,\phi) = C_p \rho c H_m (SST_p(\phi) - SST(\lambda,\varphi)) \quad (1)$$

where F_s is the net surface heat flux into the ocean, λ is the longitude, ϕ is the latitude, C_p is the specific heat, ρ is the density, c is the restoring coefficient, and H_m is the depth of the mixed layer (50 m). The SST_p in the equation is prescribed empirically such that the model ocean is heated in the equatorial region and cooled in the higher latitudes (see Equation (5) in *Sun* [2003] for the exact form of the prescribed SST_p). The treatment of the coupling of winds is also in line with what has been done previously: the equatorial zonal wind stress is proportional to the equatorial zonal SST contrast,

$$\tau^x(\lambda,\phi) = \tau^x_{ref}(\lambda,\phi) - \mu(\phi)(\Delta T - \Delta T_{ref}) \quad (2)$$

where τ^x is the zonal wind stress, τ^x_{ref} is the zonal wind stress that is used to spin up the ocean model to obtain a reference state. ΔT_{ref} is a measure of the equatorial zonal SST contrast of the reference state. It is defined as the area averaged SST difference between (5°S–5°N, 130°E–180°E) and (5°S–5°N, 230°E–280°E). ΔT is the same measure of the actual equatorial zonal SST contrast predicted by the model. μ measures the coupling strength and has a prescribed meridional profile.

The ocean component of the coupled model is the NCAR Pacific Basin model—the model of *Gent and Cane* [1989]. This ocean model, though less used so far for ENSO studies than the phenomenally successful model of *Zebiak and Cane* [1987], explicitly calculates the heat budget of the entire upper ocean. The ocean component of the model of *Zebiak and Cane* [1987], in contrast, has the mean temperature structure of the subsurface ocean fixed. For our present purpose, which is to examine how the coupled ENSO system responds to an increase in the radiative heating, it is crucially important to explicitly calculate the heat budget of the entire upper ocean. The NCAR Pacific basin model also features a fine spatial resolution in the equatorial waveguide (about 0.25°) and therefore ensures accurate simulation of the equatorial waves.

The model simulates the major observed characteristics of ENSO and the mean climate [*Sun*, 2003]. As in many other models, ENSO in this model is more regular than in the real world. With instantaneous coupling, the model does have some internal variability on multi-decadal time-scales apparently because of the presence of noise in the SST field and the resulting noisy winds. In the experiments we report in the present article, weekly mean SST is used to compute the winds in the coupling and this largely suppresses the internal variability of ENSO on multi-decadal time-scales. The lack of decadal variability in the amplitude of ENSO in the model helps to identify the effect of an external forcing, such as an increase in the tropical heating or subtropical cooling, on the amplitude of ENSO. The goal here is limited to isolating the mechanisms for further tests using observations or more sophisticated models.

3.2. Tropical Heating Experiments

A typical response of the Niño3 SST variability to an increase in the tropical heating is shown in Figure 3. Figure 3a shows the Niño3 SST time series from a control run and a perturbed run. The increase in the tropical heating is introduced through an increase in the SST_p in that region. The exact form of the increase in the SST_p in the perturbed run is shown in Figure 3b. (The heating in this particular case is confined to the equatorial region (5°S–5°N). Extending to a broader region (10°S–10°N) has essentially the same results). Figure 3c shows the corresponding time series of the transport. The numerical results apparently support the hypothesis: ENSO becomes more energetic—the amplitude has become considerably larger—and the poleward heat transport becomes more episodical. The ENSO events are stronger and have a longer duration.

To understand the response of ENSO to the tropical heating, we have conducted experiments in which the equatorial coupling is turned off—setting the coupling strength parameter in Equation (2) to zero—so that the tendency created by the imposed surface heating can be isolated. Figure 4a, b shows the response of the upper ocean temperature to an increase in the tropical heating from such experiments. There is a considerable increase in the warm-pool SST. but there is little change in the temperature of the equatorial thermocline water. The effect of the imposed surface heating is confined to the surface mixed layer. As we will see later, this confinement is due to the absence of the coupling between the atmosphere and the ocean. The increase in the SST in the eastern equatorial Pacific is less than in the western Pacific. This is because the thermocline in the eastern Pacific is shallower and consequently the upwelling has more influence over the SST in that region. As the temperature of the source water for the equatorial upwelling—the temperature of the equatorial thermocline water—remains unaffected by the imposed surface heating, the upwelling reduces the sensitivity of the SST in the eastern Pacific to the increase in the surface heating. The resulting increase in the zonal SST contrast, measured by in Equation (2), is about 0.50°C. It should be emphasized that such an increase in the zonal SST contrast is fundamentally linked to the increase in the contrast between the warm-pool SST and the temperature of the equatorial thermocline water that feeds the equatorial upwelling. The difference between the western Pacific warm-pool SST and the characteristic temperature of the equatorial thermocline water is a fundamental parameter in determining the stability of the coupled equatorial ocean-atmosphere [*Sun* 2000; *Sun*, 1997; *Jin*, 1996; *Sun*, 1996]. We use the mean SST over the region (5°S–5°N, 120°E–160°E) Tw to measure the warm-pool SST and the core temperature of the equatorial undercurrent Tc [*Sun et al.*, 2004] to represent the characteristic temperature of the equatorial thermocline water. The perturbation from the enhanced tropical heating to the value of Tw–Tc without equatorial coupling is about 1.0°C (Figure 4a). Thus, the imposed tropical heating tends to reduce the stability of the coupled equatorial ocean-atmosphere system.

In the presence of coupling, the perturbation to the zonal SST contrast by the increase in the difference between Tw

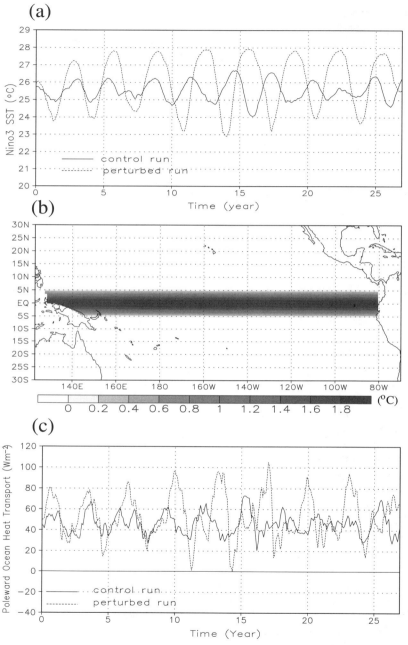

Figure 3. (a) Response of ENSO in the coupled model to an increase in the tropical heating. Shown are time series of Niño3 SST from a control run (solid line) and a perturbed run (dashed line). (b) The differences in the radiative convective equilibrium SST between the perturbed run and the control run (the perturbed run minus the control run). (c) Response in the poleward heat transport out of the equatorial region (5°S–5°N) to an increase in the tropical heating. The solid line is for the control run; the dashed line is for the perturbed run.

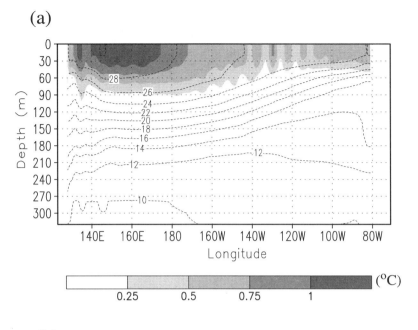

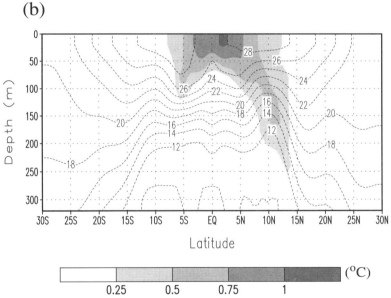

Figure 4. The equilibrium response in the upper ocean temperature to an increase in the tropical heating in the absence of ENSO. Shown are the differences between a control run and a perturbed run in which there is no equatorial ocean-atmosphere coupling. The perturbed run is subject to the same tropical heating as the coupled perturbed run (Figure 3b). Both the control run and the perturbed run are 27 years long. Shown are the mean differences over the last 3 years. The dashed lines are the mean isentropes of the control run. (a) A zonal section for the equatorial Pacific (averaged over 5°S–5°N). (b) A meridional section for the central Pacific (averaged over 160°E–210°E).

and Tc is expected to be amplified by the Bjerknes feedback loop: stronger zonal SST contrast results in stronger winds and stronger upwelling which in turn enhances the zonal SST contrast. Moreover, because this perturbation depends on the equatorial upwelling, it has more effect during the La Niña phase; namely, this perturbation will result in stronger La Niña. Indeed, the zonal SST contrast during the cold phase of the coupled run with the enhanced tropical heating is much larger than the control run. Measured by ΔT in Equation (2), the zonal SST contrast has increased by 2.0°C during the cold phase (Figure 5a).

Because of the effect of the enhanced upwelling and the enhanced zonal advection during the cold phase, the eastern equatorial Pacific is colder in the perturbed run than in the control run despite the increase in the surface heating (Figure 5a). This regulatory effect has been noted before [*Clement et al.*, 1996; *Sun and Liu*, 1996]. The SST in the far western Pacific does increase significantly during the cold phase. The stronger La Niña in the perturbed run results in a higher upper ocean heat content in the western Pacific apparently because of a stronger equatorial zonal wind and a stronger Ekman pumping in the off-equatorial region. This higher heat content further leads to stronger El Niño. *Sun* [2003] noted in the observations that stronger El Niño tends to be preceded by a higher heat content in the western Pacific. The stronger El Niño then transports the accumulated heat in the western Pacific subsurface ocean eastward and largely reverses the large increase in the zonal SST contrast during the cold-phase (Figure 5b). During and immediately flowing this zonal redistribution of heat, more heat is also transported poleward (Figure 3c).

Figure 6a, b further shows the time-mean upper ocean temperature differences between the coupled perturbed run and the coupled control run. By comparing Figure 6 with Figure 4, one sees the effect of the equatorial ocean-atmosphere coupling on the response of the equatorial upper ocean temperature to the enhanced surface heating. In the uncoupled case, the effect of heating is confined to the mixed layer. There is a significant increase in the warm-pool SST. In the coupled case, heat is transported downward all the way to the thermocline. The temperature of the thermocline water is increased considerably. The core temperature of the equatorial undercurrent Tc is increased by 0.80°C. At the same time, the increase in Tw is reduced by 0.20°C. The change in the value of Tw-Tc in the coupled case is thus negligibly small. Therefore allowing ocean-atmosphere to couple—allowing the presence of ENSO—reduces the sensitivity of the difference between Tw and Tc in the time-mean state to the increase in the surface heating, offering evidence for a stabilizing role of ENSO in maintaining the mean climate. Comparing Figure 6b with Figure 4b, it further reveals that heat is not only transported to a deeper depth, but also to higher latitudes in the coupled case than in the uncoupled case.

3.3. Subtropical Cooling Experiments

A typical response of ENSO to an increase in the subtropical cooling is shown in Figure 7. There is a considerable delay in the response of the ENSO amplitude (about 15 years), but eventually a regime with stronger ENSO develops. In the regime with strong ENSO, the poleward heat transport is also more episodical. The duration of El Niño events appears to become longer also.

To understand the response of the amplitude of ENSO to subtropical surface cooling, we have also conducted runs in which the coupling between the surface winds and the zonal SST gradients is turned off so that the tendency created by the subtropical surface cooling can be isolated. Figure 8 shows the effect of the subtropical surface cooling on the equatorial upper ocean temperature. Shown are differences in the upper ocean temperature between a control run and a perturbed run. The perturbed run is subject to the cooling shown in Figure 7b. The temperature of the equatorial thermocline water is considerably colder (The core temperature of the equatorial undercurrent Tc is about 0.75°C colder). The cooling of the equatorial thermocline is through the subtropical cell or the "ocean tunnel" [*Sun et al.*, 2004]. The

(a)

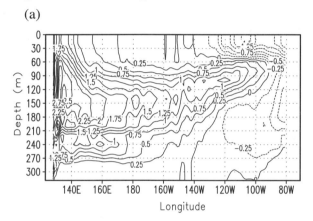

(b)

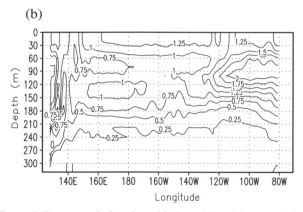

Figure 5. Response during the cold phase (a) and the warm phase (b) in the equatorial upper ocean temperature (5°S–5°N) to an increase in the tropical heating when the equatorial ocean atmosphere is allowed to produce ENSO. The definition of the cold and warm phase of the ENSO in the model is the same as in Sun (2003). Shown are the differences of the phase-averaged temperature between the control run and the perturbed run whose Niño3 SST time series are shown in Figure 3a. The last 4 cycles of ENSO in the time series are used for the calculation.

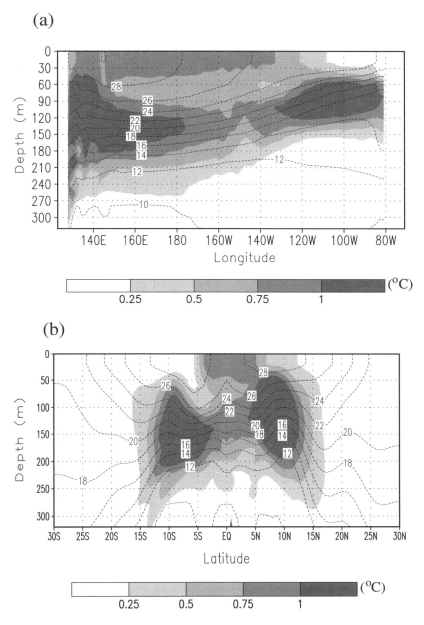

Figure 6. The response in the time-mean upper ocean temperature to an increase in the tropical heating in the presence of ENSO. Shown are the time-mean differences between the control run and the perturbed run whose Niño3 SST time series are shown in Figure 3a. The entire run (27 years) is used for computing the time-mean. The dashed lines are the mean isentropes of the control run. (a) A zonal section for the equatorial Pacific (averaged over 5°S–5°N). (b) A meridional section for the central Pacific (averaged over 160°E–210°E).

cooling is largely confined to the thermocline water. The change in the warm-pool SST Tw is small (about −0.13°C). Therefore, the subtropical cooling increases the difference between Tw and Tc and has the same effect on the stability of the coupled equatorial ocean atmosphere as the tropical heating. The subsequent upwelling of the colder thermocline water perturbs the zonal SST contrast and triggers stronger coupled instability—ENSO. Figure 9a, b shows respectively the equatorial upper ocean temperature differences during the cold phase and the warm phase. Again, the zonal SST contrast is much enhanced during the cold phase— is increased by about 2.1°C. The cold phase ΔT in the per-

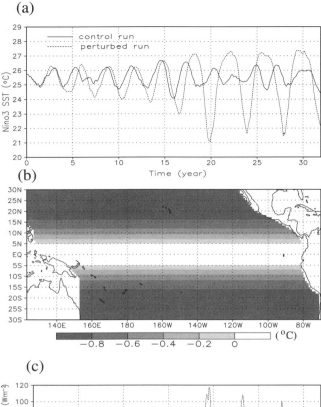
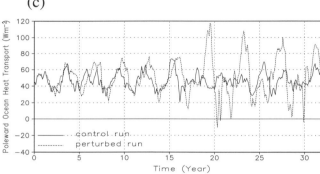

temperature to the enhanced subtropical cooling. In the uncoupled case, the effect of cooling to the equatorial upper ocean is confined to the thermocline. There is little change in the warm-pool SST. In the coupled case, however, the cooling effect is commuted upward all the way to the surface. Consequently, there is a significant decrease in the warm-pool SST. (The value of Tw is lowered by about 0.67°C.) The cooling to the temperature of the equatorial thermocline water is at the same time reduced (the cooling to Tc is reduced from about 0.75°C to 0.45°C). Therefore, the equatorial ocean-atmosphere coupling or the presence of ENSO

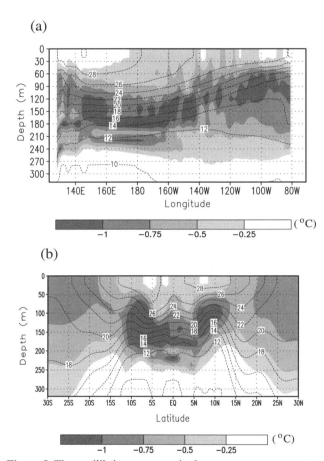

Figure 7. (a) Response of ENSO in the coupled model to an increase in the subtropical cooling. Shown are time series of Niño3 SST from a control run (solid line) and a perturbed run (dashed line). (b) The differences in the radiative convective equilibrium SST between the perturbed run and the control run (the perturbed run minus the control run). (c) Response in the poleward heat transport out of the equatorial region (5°S–5°N). The solid line is for the control run; the dashed line is for the perturbed run.

turbed run is also accompanied with greater upper ocean heat content in the western Pacific, which is responsible for stronger El Niño events.

Figure 10 further shows the time-mean upper ocean temperature differences between the coupled perturbed run and the coupled control run. By comparing Figure 10 with Figure 8, one sees the effect of equatorial ocean-atmosphere coupling—the presence of ENSO—on the response of the upper ocean

Figure 8. The equilibrium response in the upper ocean temperature to an increase in the subtropical cooling in the absence of ENSO. Shown are the differences between a control run and a perturbed run in which there is no equatorial ocean-atmosphere coupling. The perturbed run is subject to the same subtropical cooling as the coupled perturbed run (Figure 7b). Both the control run and the perturbed run are 27 years long. Show are the mean differences over the last 3 years. The dashed lines are the mean isentopes of the control run. (a) A zonal section for the equatorial Pacific (averaged over 5°S–5°N). (b) A meridional section for the central Pacific (averaged over 160°E–210°E).

reduces the sensitivity of the difference between Tw and Tc to an external forcing. Therefore in response to either an increase in the equatorial surface heating or an increase in the subtropical surface cooling, the onset of the coupled instability—ENSO—plays as a negative feedback mechanism, preventing increases in the value of Tw-Tc in the time-mean state. Comparing Figure 10b with Figure 8b, it further reveals that the subsurface cooling in the off-equatorial region is also significant reduced by the coupling.

4. DISCUSSION

We have presented numerical evidence supporting the heat-pump hypothesis. Due to limited space, we only reported two cases. We have done more experiments and found that the results do not qualitatively depend on the details of the heating profile used. ENSO become stronger so long as the heating or cooling increases the contrast between the warm-pool SST and the temperature of the equatorial thermocline water.

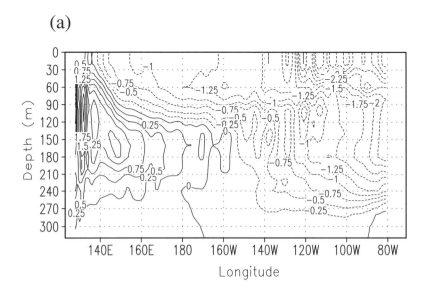

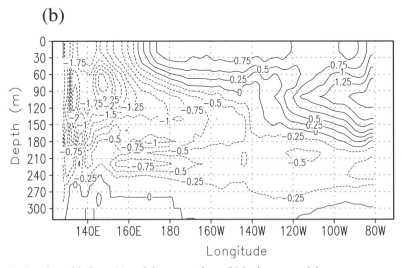

Figure 9. Response during the cold phase (a) and the warm phase (b) in the equatorial upper ocean temperature (5°S–5°N) to an increase in the subtropical cooling when the equatorial ocean-atmosphere is allowed to couple to produce ENSO. Shown are the differences of the phase-averaged temperature between the control run and the perturbed run whose Niño3 SST time series are shown in Figure 7a. The last 4 cycles of ENSO in time series are used for the calculation.

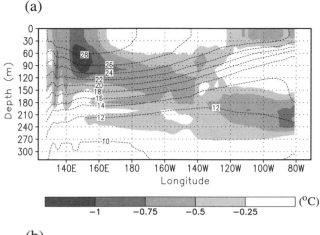

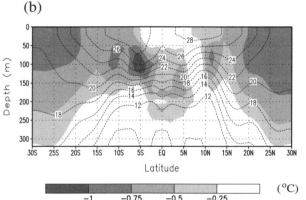

Figure 10. The response in the time-mean upper ocean temperature to an increase in the subtropical cooling in the presence of ENSO. Shown are the time-mean differences between the control run and the perturbed run whose Niño3 SST time series are shown in Figure 7a. The last 20 years of the 32 year long run are used to compute the time mean. The dashed lines are the mean isentropes of the control run. (a) A zonal section for the equatorial Pacific (averaged over 5°S–5°N). (b) A meridional section for the central Pacific (averaged over 160°E–210°E).

Conversely, we have also found that a decrease in the equatorial heating or a decrease in the subtropical cooling reduces the amplitude of ENSO.

This article is motivated to further delineate the role of surface heating in controlling the level of ENSO activity. It is also motivated to highlight some potential inaccuracies in some popular notions about ENSO. For example, ENSO has been largely regarded as an adiabatic phenomenon—it results from an adiabatic redistribution of warm water in the ocean (see review by *Neelin et al., 1998*). In light of the present results, this notion about an adiabatic ENSO only has ground in a model with a prescribed mean climate. The present results suggest that ENSO is in fact fundamentally diabatic: it is a coupled instability in response to the destabilizing effect of the meridional differential surface heating.

The present results also challenges describing ENSO as an oscillator about an independent mean climate. We see evidence from the numerical experiments for a regulatory role of ENSO in determining the long-term stability of the coupled equatorial ocean-atmosphere, specifically, the difference between the warm-pool SST and the temperature of the water feeding the equatorial undercurrent (the value of Tw−Tc).

The present results have significant implications for the response of ENSO to global warming. Since the heating in the higher latitudes may have the opposite effect on the level of ENSO activity from the effect of a local heating over the equatorial ocean, the response of the level of ENSO activity to global warming could be complicated, but should depend strongly on the effect of global warming on the meridional differential surface heating over the Pacific ocean. In the same vein, in understanding why different coupled GCMs give different predictions of the response of ENSO activity to global warming, one may need to pay attention to the change in the meridional differential surface heating over the Pacific due to global warming in the models. Because the surface heating distribution is greatly affected by clouds that are a major uncertainty in the GCMs [*Cess et al., 1989*], the response in the meridional differential surface heating to global warming may be significantly different in the GCMs, causing different response of ENSO to global warming in different models.

It has to be mentioned that the coupled model used for the numerical experiments presented in this article is still very idealized. The parameterization of the surface heating and the wind-SST coupling particularly need improvement. Therefore, the results presented in this article are only suggestive at present. Further experiments are needed to further substantiate them. In particular, future experiments need to take into account the feedback from the Hadley circulation in the atmosphere as its strength also depends strongly on the meridional differential heating. The focus of the present analysis is also limited—it is on the changes in the amplitude of ENSO. More analysis is also needed to understand the changes in the period of ENSO in response to changes in the meridional differential heating. In both the tropical heating and subtropical cooling experiments, we find that the period of ENSO increases.

The ultimate test of the heat-pump hypothesis has to come from observations. In this connection, it may be worth noting the recent data from *Cobb et al.* [2003]. Their coral records appear to suggest that ENSO during the little ice age was stronger than in the medieval warm period. They have also noted in the same record that the equatorial time-mean climate has little change from periods with strong ENSO activity to periods with weak ENSO activity.

Acknowledgments. This research was supported by the NOAA Office for Global Programs and by the NSF Climate Dynamics Program (ATM-9912434 and ATM-0331760). The author thanks Dr. Tao Zhang and Mr. Andres Roubicek for the help with the calculations and the figures. The computing support from NCAR SCD is also gratefully acknowledged. The writing of this article has also benefited from reviewers' comments and the editor's suggestions.

REFERENCES

Cess, R. D., et al., Intercomparison and interpretation of climate feedback processes in 19 atmospheric general circulation models, *J. Geophys. Res.*, *95*, 16601–16615, 1990.

Clement, A., R. Seager, and M. A. Cane, Suppression of El Niño during the mid-Holocence by changes in the earth's orbit, *Paleoceanography*, *15*, 731–737, 2000.

Clement, A., R. Seager, M. A. Cane, S. E. Zebiak, An ocean dynamical thermostat, *J. Clim.*, **9**, 2190–2196, 1996.

Cobb, K. M, C. D. Charles, H. Cheng, and R. L. Edwards, Coral records of the El Niño-Southern Oscillation and tropical Pacific climate over the last millennium, *Nature*, *424*, 271–276, 2003.

Collins, M., Understanding uncertainties in the response of ENSO to greenhouse warming, *Geophys. Res. Lett.*, *27*, 3509–3512, 2001.

Gent, P. R., and M. A. Cane, A reduced gravity, primitive equation model of the upper equatorial ocean, *Comp. Phys.*, *81*, 444–480, 1989.

Ji, M., A. Leetmaa, and J. Derber, An ocean analysis system for seasonal to interannual climate studies, *Mon. Weather Rev.*, *123*, 460–481, 1995.

Jin, F.F., S.-I. An, A.Timmermann and J. Zhao, Strong El Niño events and nonlinear dynamical heating, *Geophys. Res. Lett.*, *30*, 1120–1123, 2003.

Jin, F.-F., An equatorial ocean recharge paradigm for ENSO. Part I: Conceptual model, *J. Atmos. Sci.*, *54*, 811–829, 1997a.

Jin, F.-F., An equatorial ocean recharge paradigm for ENSO. Part II: A stripped-down coupled model, *J. Atmos. Sci.*, *54*, 830–847, 1997b.

Kalnay, E., et al., The NCEP/NCAR 40-year reanalysis project, *Bull. Am. Meteorol. Soc.*, *77*, 437–471, 1996.

Knutson,T.R., S. Manabe, and D. Gu, Simulated ENSO in a global coupled ocean-atmosphere model: Multidecadal amplitude modulation and CO_2 sensitivity, *J. Clim.*, *10*, 131–161, 1997.

Latif, M., et al., ENSIP: The El Niño simulation intercomparison project, *Clim. Dyn.*, *18*, 255–276, 2002.

Liu Z. Y., J. Kutzbach, and L. X. Wu, Modeling climate shift of El Niño variability in the Holocene, *Geophys. Res. Lett.*, *27*, 2265–2268, 2000.

Meehl, G. A., P. R., Brantstator, and W. M. Washington, Tropical Pacific interannual variability and CO_2 climate change, *J. Clim.*, *6*, 42–63, 1993.

Meinen, C, M. J. McPhaden, Observations of warm water volume changes in the equatorial Pacific and their relationship to El Niño and La Niña, *J. Clim.*, *13*, 3551–3559, 2000.

Meinen, C, and M. J. McPhaden, Interannual variability in warm water volume transports in the equatorial Pacific during 1993–99, *J. Phys. Oceanogr.*, *31*, 1324–1345, 2000.

Rodbell, D. T., et al., An similar to 15,000-year record of El Nino-driven alluviation in southwestern Ecuador, *Science*, *283*, 516–520, 1999.

Sandweiss, D. H., J. B. Richardson, E. J. Reitz, H. B. Rollins, K. A. Maasch, Geoarchaeological evidence from Peru for a 5000 years BP onset of El Nino, *Science*, *273*, 1531–1533, 1996.

Sun, D.-Z., The heat sources and sinks of the 1986–87 El Niño, *J. Clim.*, *13*, 3533–3550, 2000.

Sun, D.-Z., A possible effect of an increase in the warm-pool SST on the magnitude of El Nino warming, *J. Clim.*, *16*, 185–205, 2003.

Sun, D.-Z., J. Fasullo, T. Zhang, and A. Roubicek, On the radiative and dynamical feedbacks over the equatorial Pacific cold-tongue, *J. Clim.*, *16*, 2425–2432, 2003.

Sun, D.-Z., and Z. Liu, Dynamic ocean-atmosphere coupling: a thermostat for the tropics, *Science*, *272*, 1148–1150, 1996

Sun, D.-Z., and K. E. Trenberth, Coordinated heat removal from the equatorial Pacific during the 1986–87 El Niño, *Geophys. Res. Lett.*, *25*, 2659–2662, 1998.

Sun, D.-Z., T. Zhang, and S.-I. Shin, The effect of subtropical cooling on the amplitude of ENSO: a numerical study, *J. Clim.*, in press, 2004.

Tett, S., Simulation of El Nino-Southern Oscillation-like variability in a global AOGCM and its response to CO_2 increase, *J. Clim.*, *8*, 1473–1502, 1995.

Timmermann, A., et al., Increased El Niño frequency in a climate model forced by future greenhouse warming, *Nature*, *398*, 694–697, 1999.

Trenberth, K. E., The definition of El Nino, *Bull. Am. Meteorol. Soc.*, *78*, 2771–2777, 1997.

Tudhope, A. W., et al., Variability in the El Niño-Southern Oscillation Through a Glacial-Interglacial Cycle, *Science*, *291*, 1511–1517, 2001.

Wang, B., and S.-I. An, A mechanism for decadal changes of ENSO behavior: Roles of background wind changes, *Clim. Dyn.*, *18*, 475–486, 2002.

Wang, C., and J. Picaut, Understanding ENSO physics—a review, this volume.

Wyrtki, K., Water displacements in the Pacific and the genesis of El Niño cycle, *J. Geophys. Res.*, *90*, 7129–7132, 1985.

Zebiak, S .E., and M. A. Cane, A model El Niño-Southern Oscillation, *Mon. Weather Rev.*, *115*, 2262–2278, 1987.

D.-Z. Sun, NOAA-CIRES/Climate Diagnostics Center, R/CDC1, 325 Broadway, Boulder, Colorado 80305. (Dezheng.Sun@noaa.gov)

Broadening the Atmospheric Bridge Paradigm: ENSO Teleconnections to the Tropical West Pacific-Indian Oceans Over the Seasonal Cycle and to the North Pacific in Summer

Michael A. Alexander

NOAA-CIRES, Climate Diagnostics Center, Boulder, Colorado

Ngar-Cheung Lau

NOAA/Geophysical Fluid Dynamics Laboratory, Princeton, New Jersey

James D. Scott

NOAA-CIRES, Climate Diagnostics Center, Boulder, Colorado

During El Niño–Southern Oscillation (ENSO) events, atmospheric teleconnections associated with sea surface temperature (SST) anomalies in the equatorial Pacific can influence the ocean thousands of kilometers away. We use several data sets to delineate this "atmospheric bridge" between ocean basins, focusing on two emerging research areas: 1) the evolution of atmosphere–ocean interactions in the tropical Indian–Western Pacific Oceans over the full ENSO cycle and 2) the formation of large amplitude SST anomalies in North Pacific in the summer before ENSO peaks. In ENSO composites [where events peak near the end of Yr(0)], an east–west SST dipole develops in the Indian Ocean during the summer–fall of Yr(0), followed by basin-wide warming through spring of Yr(1). The SST anomalies over most of the tropical west Pacific also reverse sign, from negative in summer of Yr(0) to positive in the following summer. Local air–sea interactions influence the evolution of these ENSO-induced SST anomalies and related sea level pressure (SLP) and precipitation anomalies. Over the western North Pacific, the southward displacement of the jet stream and storm track in the summer of Yr(0) changes the solar radiation and latent heat flux at the surface, which results in anomalous cooling (and deepening) of the oceanic mixed layer at ~40°N. The potential impact of both the tropical and North Pacific SST anomalies on the broader climate is discussed.

1. INTRODUCTION

While the essential atmospheric and oceanic processes responsible for El Niño and the Southern Oscillation (ENSO) are contained within the tropical Pacific, ENSO impacts the global climate system. Some of the ENSO signal is communicated to remote locations via coastally trapped ocean waves that propagate poleward along the west coast of North and South America [*Enfield and Allen*, 1980; *Pares-Sierra and O'Brien*, 1989], but most of the teleconnections are through the atmosphere [*Alexander* 1992; *Lau and Nath*, 1996; *Alexander et al.*, 2002; *Lau and Nath* 2003]. The global atmospheric response to SST anomalies in the equatorial Pacific includes changes in the wind, temperature, moisture and cloud cover, which then alter the fluxes of heat, momentum and fresh water into the ocean. Through this "atmospheric bridge", changes in the central and eastern equatorial Pacific Ocean are communicated to the global oceans via atmospheric teleconnections associated with ENSO.

In general, ENSO begins in boreal spring, peaks in late fall/early winter and decays in the following spring. In addition, the extratropical atmospheric circulation anomalies associated with ENSO are strongest in northern winter. As a result, most studies of the atmospheric bridge have focused on boreal winter through the following spring. The roughly one season lag in the SST response to the atmospheric forcing is due to the large thermal inertia of the ocean. However, significant bridge-related changes in the climate system also occur in other seasons, including both the previous and following summers.

The evolution of the atmospheric bridge is illustrated in Plate 1 by the difference in sea surface temperature (SST) between composites of El Niño and La Niña events for five two-month periods during the ENSO cycle: July–August [JA(0)], October–November [ON(0)], January–February [JF(1)], April–May [AM(1)] and July–August [JA(1)], where 0 denotes the year ENSO peaks and 1 the following year. Composites are constructed from ten El Niño (warm) events: 1957, 1965, 1969, 1972, 1976, 1982, 1987, 1991, 1997, and 2002; and ten La Niña (cold) events: 1950, 1954, 1955, 1964, 1970, 1973, 1975, 1988, 1998, and 1999 during 1950–2003.

If we first focus on boreal fall and winter (Plate 1b–c), the ENSO-related signal is strong with positive SSTs (>2.0°C) over the eastern half of the tropical Pacific. Beyond the ENSO region, a reduction in the strength of the trade winds and the amount of cloud cover contribute to abnormally warm water in the tropical Atlantic by AM(1) [e.g., *Covey and Hastenrath*, 1978; *Alexander and Scott*, 2002; *Wu et al.*, this volume] and over most of the Indian Ocean from ON(0) to AM(1) [e.g., *Cadet*, 1985; *Nicholson*, 1997; *Klein et al.*, 1999]. In the North Pacific, strong cyclonic flow around an anomalously deep Aleutian low during El Niño events cools the central North Pacific and warms the water along the west coast of North America [e.g., *Alexander*, 1992; *Luksch*, 1992; *Lau and Nath*, 1996; *Alexander et al.*, 2002; our Plates 1 and 2 b–d].

El Niño events, however, are already well established by JA(0), when warm water covers the equatorial Pacific (Plate 1a) and positive sea level pressure (SLP) anomalies are located over the eastern hemisphere and negative anomalies over the western hemisphere, characteristic of the negative phase of the Southern Oscillation [e.g. *Wang and Picaut*, this volume; our Plate 2a]. Unlike JF(1), SST anomalies are negative to the south of Indonesia in JA(0), but like the subsequent winter, they are positive over the western part of the tropical Indian Ocean, leading to an east–west dipole across the basin. The negative anomalies near Indonesia are part of an inter-hemispheric "horseshoe" pattern that extends from the North Pacific to the south–central Pacific. Large amplitude negative SST anomalies have already begun to form in the western North Pacific by JA(0); indeed, one of the largest bridge-related signals occurs during late summer/early fall along ~40°N (Plate 1a and 1b). The composite El Niño minus La Niña SSTs normalized by the monthly standard deviation, shown in Figure 1, indicates the SST anomaly in the western North Pacific region reaches a minimum in September(0) of approximately –1.5 times the standard deviation (a non-normalized value of ~–1.5°C).

By JA(1) ENSO has all but disappeared with negative SST anomalies along the equator in the eastern Pacific (Plate 1e). Nevertheless, many of the SST anomalies created by the atmospheric bridge, including those in the Indian Ocean, South China Sea and eastern North Pacific, peak in late winter or spring and then persist into early summer, albeit at a smaller amplitude (Plate 1 and Figure 1).

Plate 1 and Figure 1 indicate that the bridge-related SST anomalies vary greatly depending on the region and the phase of the ENSO cycle. Some of these features are fairly well understood, such as the atmospheric teleconnections to the North Pacific Ocean during boreal winter [as reviewed by *Alexander et al.*, 2002] and the bridge to the tropical Atlantic [*Wu et al.*, this volume, and references therein]. In this article, we examine two emerging research foci of the atmospheric bridge phenomena. First, we review recent literature concerning the evolution of the atmosphere–ocean system in the tropical Indian and western Pacific Oceans over the course of the ENSO cycle. Second, we perform new analyses of processes that cause large-amplitude SST anomalies in the western North Pacific during the summer of Yr(0). In both regions, air–sea interactions in seasons other than boreal winter have the potential to feedback on the broader climate system.

2. INDIAN AND SUBTROPICAL WESTERN PACIFIC OCEANS

To delineate the role of the atmosphere in communicating ENSO's effect on the Indo–Pacific sector, the composite procedure adopted in Plate 1 has been applied to selected mete-

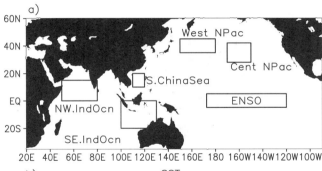

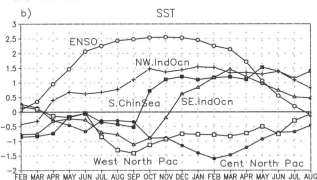

Figure 1. (a) Indo–Pacific regions where the ENSO SST signal is strong. (b) El Niño–La Niña composite SST from Feb(0) to Aug(1) normalized by the SST standard deviation in each calendar month for the regions shown in (a). The regions are located in the central equatorial Pacific (ENSO; 5°S–5°N, 172°E–120°W), Western North Pacific (WNP; 35°N–45°N, 150°E–180°), Central North Pacific (28°N–42°N, 170°W–150°W), northwest Indian Ocean (0°–15°N, 50°E–80°E), southeast Indian Ocean (0°–20°S, 100°E–130°E), and the South China Sea (10°N–20°N, 110°E–120°E). While the magnitude of the ENSO anomalies is much larger in winter than in summer, the weak summer variability results in nearly uniform normalized ENSO SST anomalies from August(0) to January(1).

orological fields. The results for various stages of the ENSO life cycle are displayed in Figure 2 for surface wind vectors and Plate 2 for SLP (contours) and precipitation (shading). The patterns in Figure 2 are based on NCEP reanalysis data for the 10 warm and 10 cold events. The SLP and precipitation composites in Plate 2 are computed with NCEP reanalysis and Climate Prediction Center (CPC) Merged Analyses of Precipitation [CMAP, *Xie and Arkin*, 1997] data, respectively, for the ENSO events occurring in 1982–2002. We use the CMAP data as they are considered more reliable than the precipitation estimates from NCEP reanalysis. Similar ENSO-related anomalies were found using precipitation and SLP for the composite of the 10 warm and 10 cold ENSO events from reanalysis (not shown).

The tropical atmospheric response to ENSO, the first element in the atmospheric bridge, is well established even during the very early stages of the ENSO cycle: surface convergence and precipitation are enhanced across the tropical Pacific from the South American coast to 160°E between 0°–10°N in JA(0) (Figure 2a and Plate 2a). These anomalies increase in magnitude and meridional extent through JF(1), then weaken by AM(1) and dissipate by JA(1). Changes in diabatic heating associated with the precipitation anomalies over the equatorial Pacific drive atmospheric circulation changes connecting the ENSO region to the tropical portion (25°N–25°S) of the Indian and West Pacific Oceans, as discussed in the following subsections.

2.1. Indian Ocean

The most notable development in the SLP pattern during the JA(0)–JF(1) period is the emergence of a high anomaly center (denoted as S1 in Plate 2a–c) off the northwestern Australian coast. Stationary wave modeling [e.g., *Wang et al.* 2000, 2003; *Lau et al.*, 2004a] indicates that this feature is a Rossby wave response to the reduced latent heating over the equatorial western Pacific and Indonesia (mainly due to the reduction in precipitation seen in Plate 2a–c), where the subsiding branch of the anomalous Walker Circulation resides during El Niño events. In the northern summer and early autumn, the climatological flow is directed northwestward over much of the Indian Ocean (IO) basin south of the Equator, and eastward just north of the Equator [e.g., Figure 1 in *Schott and McCreary*, 2001]. Hence the anomalous circulation in the vicinity of S1 from June to October (Figure 2a–b) is associated with above-normal wind speeds (not shown) over the waters off the Sumatra/Java coasts, which enhance heat loss from the ocean as well as coastal upwelling [e.g. *Murtugudde et al.*, 2000; *Iizuka and Matsura*, 2000; *Li et al.*, 2002; *Lau and Nath*, 2004]. These processes contribute to the occurrence of cold SST anomalies forming off the southern coasts of Java and Sumatra, and the northwestern coast of Australia (Plate 1a–b and Figure 1). Conversely, the below-normal wind speeds on the western flank of S1 as well as over the central equatorial IO during austral spring are accompanied by decreases in the oceanic heat loss and SST warming. This east–west SST contrast peaks in boreal fall, as indicated by the difference between the composite SST anomalies in the northwest and southeast IO regions (Plate 1b and Figure 1).

The overall SST, wind circulation, SLP and precipitation patterns in Plate 1b, Figure 2b and Plate 2b bear a strong resemblance to those associated with a prominent "dipole mode" of atmosphere–ocean variability in the IO basin, as described by *Saji et al.* [1999] and *Webster et al.* [1999]. The appearance of this characteristic mode in the ENSO-based composites presented here suggests that ENSO events could play a considerable role in the zonally asymmetric SST anomaly pattern occurring in the IO during boreal summer and

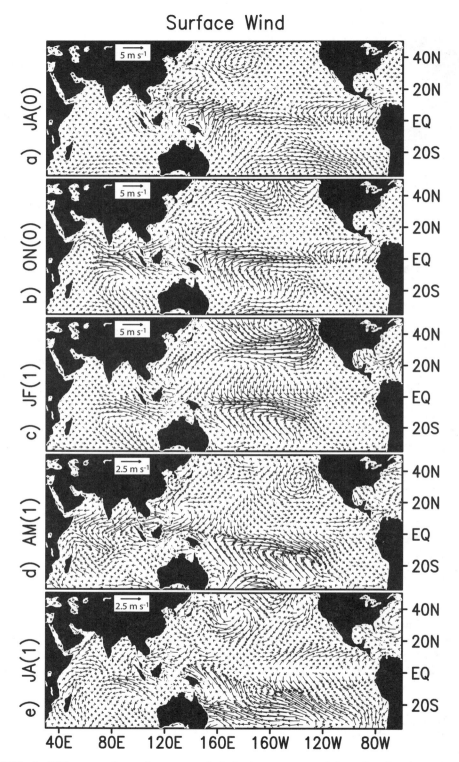

Figure 2. El Niño–La Niña composite surface vector winds (scale vector upper left corner of each panel), for (a) JA(0), (b) ON(0), (c) JF(1), (d) AM(1), and (e) JA(1). Values are from NCEP/NCAR reanalysis data for the period 1950–2003. Note that scale vector in (a)–(c) differs from that in (d)–(e).

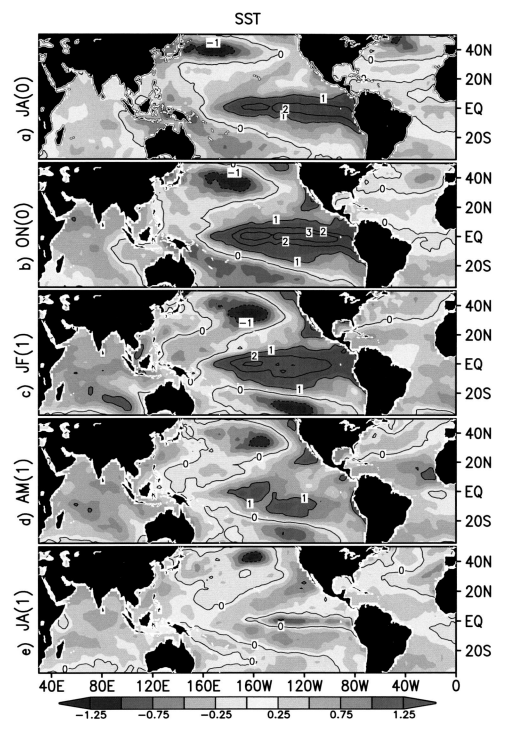

Plate 1. Anomalous SST is constructed by subtracting the composite of 10 El Niño events from the composite of 10 La Niña events in the 1950–2003 period, for (a) July–August [JA(0)], (b) October–November [ON(0)], (c) January–February [JF(1)], (d) April–May [AM(1)] and (e) July–August [JA(1)], where 0 indicates the ENSO year and 1 the following year. The shading (contour) interval is 0.25 (1.0) °C. The values are obtained from the National Center for Environmental Prediction (NCEP) reanalysis [*Kalnay et al.*, 1996; *Klister et al.*, 2001].

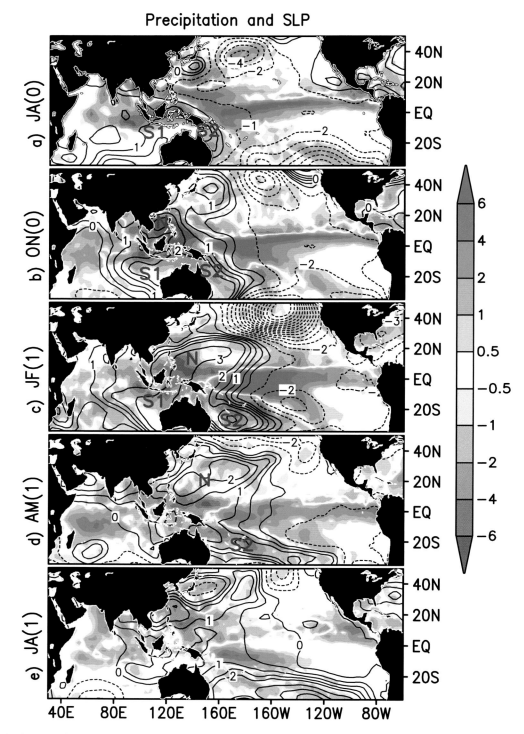

Plate 2. El Niño–La Niña composite sea level pressure (contours, base interval: 1.0 mb; with additional contours for +0.5 and +1.5 mb) and precipitation (shading, see scale at side), based on ENSO events in the 1982–2003 period, for (a) JA(0), (b) ON(0), (c) JF(1), (d) AM(1), (e) JA(1). The subtropical high pressure anomaly centers are identified by the labels N, S_1 and S_2. Results for the pressure and precipitation fields are obtained using NCEP/NCAR reanalysis and CMAP data, respectively.

autumn. The influence of ENSO on atmosphere–ocean variability in the IO sector have been emphasized in the observational analysis of *Allan et al.* [2001] and *Hendon* [2003], and the modeling studies of *Lau and Nath* [2004] and *Shinoda et al.* [2004]. There exists, however, empirical and model evidence for the occurrence of anomalous events with an east–west SST contrast when ENSO is absent in the tropical Pacific [*Saji and Yamagata*, 2003; *Lau and Nath*, 2004; *Yamagata et al.*, this volume].

With the approach of the summer monsoon season over northern Australia, the climatological flow above the waters south of Sumatra/Java switches from easterly to westerly by November and remains westerly through February [e.g., *Shinoda et al.*, 2004]. The southeasterly wind anomaly occurring at this time and location (Figure 2b–c) is therefore directed against the time mean circulation. The resulting decrease of the local wind speed reduces both oceanic heat loss and upwelling, reversing the cold SST anomaly in the eastern IO south of the Equator [*Li et al.*, 2003; *Lau and Nath*, 2004]. The northwesterly wind anomalies situated to the west of S1, suppress the wind speed over the south central IO through JF(1), and thus enhance the warm SST anomaly in that region. The contributions of surface latent heat fluxes to SST changes in various parts of the IO during ENSO have previously been emphasized in the observational analyses of *Yu and Rienecker* [1999] and modeling studies of *Behera et al.* [2000] and *Venzke et al.* [2000]. Rossby waves generated by anomalous winds in the southwestern IO may also lead to basin-wide warming by JF(0) [*Chambers et al.*, 1999].

During ON(0), below normal precipitation prevails over the cold SST anomaly in the eastern tropical IO, whereas wet conditions occur over the warm anomaly off the eastern African coast (Plates 1b and 2b). The dryness over the eastern IO is accompanied by reduced cloud cover and enhanced shortwave heating of the ocean surface, resulting in the dissipation of the cold SST anomaly in that region [*Klein et al.*, 1999; *Li et al.*, 2002; *Shinoda et al.*, 2004].

In the boreal winter and spring of Yr(1), a broad positive precipitation anomaly extends across the IO south of the Equator (Plate 2c–d), where above-normal SST prevails (Plate 1c–d). During the AM(1) period (Figure 2d), both the orientation of the anomalous cross-equatorial flow from the northern IO towards this rainbelt, and the development of cyclonic circulation in the surface wind field over 0°–30°S, 60°E–90°E, situated to the west of the precipitation maximum, suggest that these atmospheric circulation features are the response to the underlying SST anomaly pattern and the associated condensational heating aloft [e.g., see discussions in *Hoskins and Karoly*, 1981; *Xie et al.*, 2002]. Hence the SST changes in the IO basin, which were partially driven by the atmospheric bridge mechanism in the previous seasons, influence the atmospheric circulation in AM(1).

2.2. Subtropical Western Pacific

The atmospheric pattern over the South China and Philippine Seas during the summer of Yr(0) is characterized by cyclonic wind anomalies (westward winds near 20°N and eastward winds near 10°N in Figure 2a) and more intense rainfall (Plate 2a). These features are indicative of a more eastward extension of the summer monsoon trough to the subtropical western North Pacific, and are probably associated with the Rossby wave response to the latent heat release accompanying the enhanced precipitation near the dateline during El Niño events (Plate 2a). The attendant increase in surface wind speed is conducive to below normal SST along the Chinese coast and the Philippine Sea.

High pressure anomalies, denoted by S2 and N, appear over the northeastern Australian coast and the South China Sea in JA(0) and ON(0), respectively (Plates 2a and 2b), these features migrate eastward in the following months (Plates 2c–2d). Results from mechanistic models [*Wang et al.*, 2003; *Lau et al.*, 2004a] indicate that, in analogy with the forcing of S1, the high centers in the western Pacific are also Rossby-wave responses to the below-normal condensational heating in the Indonesian sector. Over the subtropical northwestern Pacific, the time mean circulation in the northern autumn and winter seasons is dominated by the northeasterly monsoon off the eastern Asian seaboard [e.g., *Lau and Nath*, 2000]. The southwesterly anomalous flow on the northwestern flank of the anticyclonic center N impedes the strength of the climatological monsoon (Figures 2b and 2c), thereby warming the waters in the South China and East China Seas in ON(0) and JF(1) [Plate 1, also see *Wang et al.*, 2000; *Lau et al.*, 2004a]. On the other hand, the northwesterly anomalous circulation located southeast of N in ON(0) is coincident with increased wind speeds and SST cooling; as discussed in greater detail by *Wang et al.* [2000]. An analogous set of local relationships between the mean circulation and the anomalies in the wind and SST fields is discernible among the features associated with S2. The superposition of the anomalous counterclockwise circulation on the climatological southeasterly flow in that region results in below normal wind speeds and warm SSTs off the southeast Australian coast.

The anomalous precipitation pattern over the northwestern subtropical Pacific in ON(0)–JF(1) (Plate 2b–2c) is characterized by dryness in the vicinity of N. Due to weakening of the dry winter monsoon over East Asia [*Wang et al.*, 2000], above-normal precipitation amounts are seen over the East China Sea.

In AM(1), the anomalous anticyclonic circulation around the high pressure anomaly N remains clearly evident (Figure 2d and Plate 2d), with the center of N extending slightly farther to the northeast relative to its locations in the previous autumn and winter. The observational and model evidence presented by *Wang and Zhang* [2002], *Wang et al.* [2003] and *Lau et al.* [2004a] imply that local air–sea thermodynamic feedbacks play a crucial role in the sustenance and eastward migration of N. Anomalous high pressure and dry conditions are still evident in JA(1) over much of the subtropical western North Pacific from the Chinese coast to ~170°E (Plate 2e).

2.3. Discussion

Comparison of the composite patterns over both the Indian and the subtropical western Pacific Oceans for JA(0) and JA(1) reveals interesting differences between anomalies occurring during northern summer in the two consecutive years. Of particular note is the transition to the east of the Phillipines from cyclonic circulation and enhanced rainfall in Yr(0) to anticyclonic circulation and reduced rainfall in Yr(1). The switch in polarity (from cold to warm) of the SST anomaly in the eastern portion of the IO basin is related to a local reduction in precipitation in Yr(0) and enhanced precipitation in Yr(1). The tendency for the anomalies in the above locations to change their polarities from one year to the next is one facet of the Tropospheric Biennial Oscillation (TBO) of Asian–Australian monsoon system [see the review by *Meehl*, 1997]. The results and discussions presented in this section suggest that the seasonal dependence of local air–sea coupling as well as responses of the atmosphere–ocean system in the Indo–Pacific to remote ENSO forcing is an important factor for understanding the origin of some of the phenomena associated with the TBO.

Tropical SST anomalies associated with the atmospheric bridge influence the large-scale atmospheric circulation well after ENSO has peaked. The persistence of tropical SST anomalies outside the equatorial east Pacific contributes to the delayed atmospheric response to ENSO: e.g. the zonal mean 200 mb height anomalies between 30°N–30°S are three times stronger in the summer of Yr(1) than the summer of Yr(0), despite stronger SST anomalies in the Niño region in Yr(0) [*Kumar and Hoerling*, 2003]. ENSO-induced SST anomalies in the Indo–Western Pacific sector also exert a strong influence on the Asian monsoon system, where the accompanying redistribution of condensational heat sources and sinks leads to marked changes in the strength and position of the Asian jet stream [*Lau et al.*, 2004b]. The effects of these atmospheric perturbations are extended eastward across the extratropical North Pacific through dynamical interactions between the quasi-stationary flow and synoptic-scale transient eddies, such as those described in Section 3.1. *Lau et al.* [2004b] further noted that this chain of processes contributes to summertime anomalies in the zonally averaged circulation in midlatitudes, as well as the regional climate over North America (e.g., occurrence of droughts and prolonged heat waves).

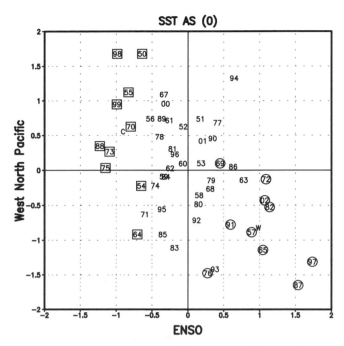

Figure 3. Scatter diagram of the SST anomalies (°C) in AS(0) relative to the 1950–2003 mean, in the ENSO and WNP regions (domains shown in Figure 1a). The values of the ten individual El Niño (La Niña) events are enclosed in circles (squares) and the composite average is denoted by a W (C).

3. NORTH PACIFIC IN SUMMER OF YR(0)

3.1. SST

In overviews of the SST and atmospheric surface changes that typically occur during ENSO periods, *Harrison and Larkin* [1998], *Wang* [2002], and *Park and Leovy* [2004] found that cold SST anomalies are centered along ~40°N in the western half of the North Pacific in summer and fall of Yr(0). These negative summertime SST anomalies are also readily apparent from the leading mode of SST variability based on rotated empirical orthogonal functions (EOFs) of Pacific SST variability in all calendar months [*Barlow et al.*, 2001]. We examine the association between SST anomalies in the central–eastern equatorial Pacific and the western North Pacific (the ENSO and WNP regions shown in Figure 1a) via a scatter diagram of the departures of SSTs in

Aug–Sep in the years 1950–2002 from their long-term mean over this period in Figure 3. The relationship between SSTs in the ENSO and WNP regions is quite strong: *i)* the correlation between the two regions over all 52 summers is –0.49 (significant at the 99% level), *ii)* the composite difference between the warm and cold phases of ENSO is about –1.45°C in the northwest Pacific (significant at the 99% level), and *iii)* the SST anomalies in the northwest region are negative in 9 of 10 El Niño events and positive in 8 of 10 La Niña events. While the negative SST anomalies in the northwest Pacific region are somewhat larger in amplitude and exhibit less scatter then the positive anomalies, it is unclear whether this represents a non-linear difference in the remote oceanic response to the warm and cold phases of ENSO, or is merely due to the limited number of samples. The atmospheric and oceanic processes that contribute to the formation of North Pacific SST anomalies during the summer of ENSO events are examined in the following subsections.

3.2. Atmospheric Circulation

The atmospheric response to ENSO is not confined to the tropics, and teleconnections to the Pacific North American sector occur during much of the ENSO cycle. The dynamical link between the tropics and extratropics [as reviewed by *Trenberth et al.*, 1998] are complex involving the excitation of Rossby waves by tropical convection, the propagation of these waves to midlatitudes and their subsequent interaction with asymmetries in the zonal mean flow and with midlatitude storm tracks.

While most studies of the atmospheric processes linking the tropics and extratropics were based on conditions during boreal winter, the few investigations that considered other seasons found that there could be strong teleconnections in summer as well. The initial analyses of ENSO-related teleconnections in summer were motivated by the severe drought in the central United States during the summer of 1988 [*Trenberth et al.*, 1988; *Mo et al.*, 1991; *Palmer and Brankovic*, 1989; *Trenberth and Branstator*, 1992]. Several factors may enable ENSO to influence the extratropical circulation in summer despite the broad latitudinal extent of the mean easterlies in the tropical troposphere that act as a barrier to Rossby wave propagation. These include SST and precipitation (heating) anomalies in the subtropics (e.g. see Plates 1a and 2a), Rossby waves created by local divergent circulations, and longitudinal and/or height variations in the mean zonal wind that allow Rossby waves to propagate to the midlatitudes [*Nitta* 1986, 1987; *Lau and Peng*, 1992; *Chen and Yen*, 1993; *Trenberth and Branstator*, 1992; *Grimm and Silva Dias*, 1995; *Newman and Sardeshmukh*, 1998]. Heating anomalies associated with the Asian monsoon [*Lau and Weng*, 2002] and ENSO induced changes in extratropical transients [*Kok and Opsteegh*, 1985] may also contribute to the height anomalies over the North Pacific in summer.

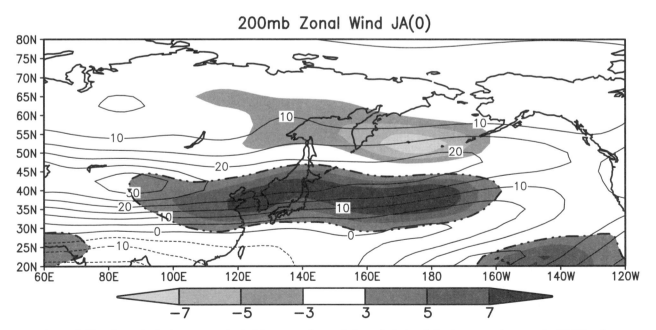

Figure 4. The climatological mean zonal wind (contours, interval: 5 m s^{-1}) during July–August and composite El Niño–La Niña zonal wind (shading, scale at bottom; regions >3 m s^{-1} enclosed by a dot-dash line) during JA(0) from NCEP reanalysis for ENSO events between 1950–2003.

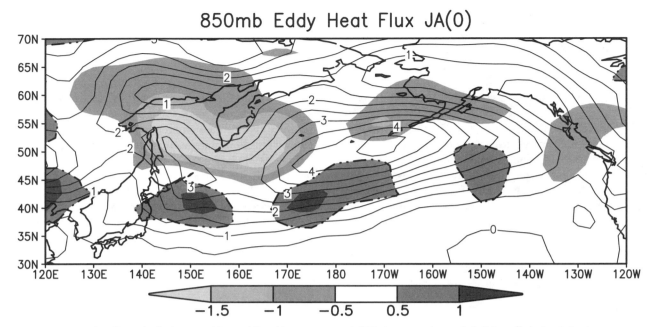

Figure 5. The climatological mean eddy meridional heat transport ($\overline{v'T'}$) (contours, interval: 0.5°Cm s^{-1}) during July–August and El Niño–La Niña composite $\overline{v'T'}$ (shading, scale at bottom; regions >0.5°Cm s^{-1} enclosed by a dot-dash line) at 850 mb during JA(0). The prime indicates the band pass (2–8 day) filtered anomaly and an overbar denotes the time average over JA. The values were obtained from NCEP reanalysis for ENSO events between 1950–2003.

Here we examine the circulation and storm track anomalies over the North Pacific in summer of Yr(0). During JA(0), positive (negative) zonal wind anomalies form to the south (north) of the climatological mean jet axis at ~40°N over eastern Asia and the western half of the North Pacific (Figure 4). The anomalies are approximately 25–50% of the mean winds, representing a substantial southward shift in the Asian–Pacific jet. (The mean jet is about 1/2 as strong and ~15° farther north in summer than in midwinter). The El Niño–La Niña composite difference in the Pacific storm track, as measured by the band pass (2–8 day) filtered meridional heat transport at 850 mb ($\overline{v'T'}$), is shown in Figure 5. Like the seasonal mean changes in the jet stream, the anomalous eddy heat transport also indicates a southward shift in the storm track over much of the North Pacific.

Following previous studies of the relationship between the low frequency circulation anomalies and synoptic eddies in winter [e.g. *Lau and Holopainen*, 1984; *Lau*, 1988], we compare the anomalous geopotential height (z) and eddy induced height tendency (z_t) in the upper troposphere in summer, presented here as the difference between El Niño and La Niña composites of both quantities at 200 mb averaged over JA(0). The synoptic forcing is derived from the contribution of the band pass filtered eddy vorticity flux divergence ($\nabla \cdot \overline{\vec{v}'\zeta'}$) to the vorticity tendency at the 200 mb level, and then converting the vorticity tendency to height tendency by assuming geostrophy, e.g.

$$z_t = \frac{f}{g}\nabla^{-2}\left(\nabla \cdot \overline{\vec{v}'\zeta'}\right), \quad (1)$$

where f the Coriolis parameter, g the acceleration due to gravity, $\vec{v}$ the horizontal wind vector, ζ the vorticity. The changes in the jet (Figure 4) are consistent with geostrophic flow around a negative z anomaly that extends from central Asia to the central Pacific along 45°N and positive anomaly over eastern Siberia (contours in Figure 6); this pair of anomalies resembles the Pacific–Japan pattern [*Nitta*, 1987] and the North Pacific teleconnection pattern [*Barnston and Livezey*, 1987]. The synoptic forcing (shading) is large (z_t >5 m d^{-1}) and nearly collocated with the height anomalies, suggesting that the synoptic eddies strongly contribute to the seasonal circulation anomalies. The anomalies, however, are displaced slightly to the west of the z_t anomalies. The time scale of the eddy forcing, z/z_t, is ~10 days, indicating that horizontal eddy forcing in the upper troposphere is rapid enough to generate the observed summer height anomalies. Results from previous studies [e.g. *Ting and Lau*, 1993] indicate that other processes, such as eddy heat flux forcing, diffusion, etc., somewhat compensate the height changes initiated by z_t.

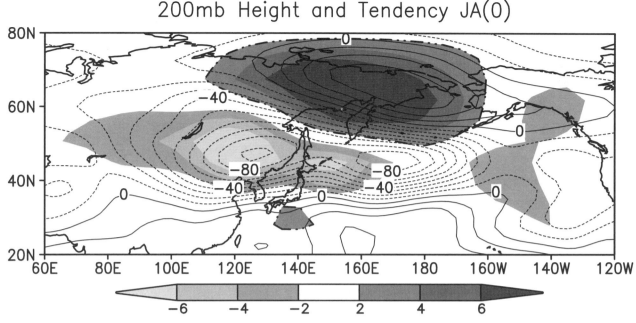

Figure 6. The El Niño–La Niña composite geopotential height (contours, interval: 10 m) and the eddy induced height tendency (shading, scale at bottom; regions >2m d^{-1} enclosed by a dot-dash line) at 200 mb during JA(0). The height tendency, is given by equation (1). The values were obtained from NCEP reanalysis for ENSO events between 1950–2003.

In addition to interacting with the large-scale circulation patterns, synoptic eddies also influence clouds. Low clouds associated with traveling cyclones are prevalent over the North Pacific during summer, where the cloud fraction exceeds 60% north of ~40°N and rapidly decreases towards the subtropics [*Weaver and Ramanathan*, 1996; *Norris and Leovy*, 1994]. The leading interannual pattern of stratus cloud variability in summer is located where the gradient in the mean cloud amount is strongest, i.e. the central and west Pacific between about 30°N–45°N, and is associated with meridional displacements in the storm track and the underlying SST [*Norris*, 2000]. Recently, *Park and Leovy* [2004] found that much of the interannual fluctuations in the storm track, cloud amount and SST over the North Pacific in summer are associated with ENSO variability. The changes in the storm track and cloud amount (not shown) are consistent with enhanced precipitation and reduced SLP in the vicinity of 40°N, 170°W during El Niño relative to La Niña events (Plate 2a). The passage of more and/or stronger synoptic disturbances cools the ocean by increasing cloudiness, which reduces the solar radiation reaching the surface.

3.3. Surface Fluxes

The ENSO-driven atmospheric circulation changes over the North Pacific influence the SST directly via the net heat flux and indirectly via momentum and fresh water fluxes that subsequently affect ocean currents and turbulent mixing. Here we consider two key factors that influence SST anomalies on interannual time scales: the net surface heat flux (Q_{net}) and the Ekman heat transport in flux form:

$$Q_{ek} = \frac{c}{f}\left[-\tau_y \frac{\partial}{\partial x} SST + \tau_x \frac{\partial}{\partial y} SST\right], \quad (2)$$

where c is the specific heat of seawater, and τ_x, τ_y are the zonal and meridional components of the surface wind stress.

The net surface heat flux depends on four components: the short wave (Q_{sw}) and long wave (Q_{lw}) radiation, and the sensible (Q_{sh}) and latent (Q_{lh}) heat flux. The Q_{sh} and Q_{lh} values used here are from NCEP reanalysis. However, the Q_{sw} values from reanalysis have been shown to have large errors [e.g. *Scott and Alexander*, 1999], so we use the Q_{sw} and Q_{lw} estimates derived from the International Satellite Cloud Climatology Project [ISCCP; *Zhang et al.*, 1995; *Rossow and Zhang*, 1995; *Zhang and Rossow*, 2004]. The flux composites presented in Figures 7 and 8 are based on ENSO events from 1983–2000, the period when radiative fluxes derived from ISSCP data are available. The El Niño–La Niña composites of Q_{net} and Q_{ek} for JA(0) are displayed in Figure 7, when the SST decreases rapidly in the western North Pacific (WNP) region (Plate 1 and Figure 1).

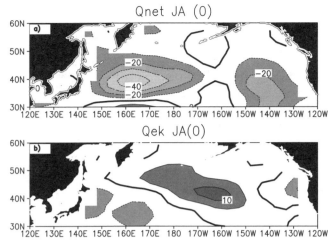

Figure 7. Composite El Niño–La Niña (a) net heat flux into the ocean (Q_{net}) and b) Ekman heat transport in flux form (Q_{ek}) in JA(0) during ENSO events from 1983–2000. In (a) the contour interval is 10 W m^{-2} and shading transitions occur at –10, –20 and –40 W m^{-2}, in (b) the shading and contour intervals are both 5 W m^{-2}. The radiative fluxes used in Q_{net} are derived from satellite data [*Zhang and Rossow*; 2003], while the sensible and latent fluxes used to compute Q_{net} are derived from NCEP reanalysis. Reanalysis is also used to compute Q_{ek} in equation (2).

The WNP region is strongly cooled by Q_{net}, as indicated by negative values (< –40 W m^{-2}) between approximately 150°E–180°, 35°N–42°N. The Q_{ek} anomalies, while much smaller, also cool the ocean between 30°N–45°N from Japan to the dateline (Figure 7b). This anomalous cooling is located about 5° further south in the ENSO compo-site based on all events between 1950–2003 (not shown), which is consistent with the anomalous westerly winds between approximately 25°N–38°N (Figure 2a) that enhance the southward transport of cold water in the western North Pacific.

The ENSO composite anomalies for each of the four flux components are presented in Figure 8. Unlike winter, when Q_{sw} anomalies are negligible, the ENSO-related shortwave radiation anomalies during JA(0) have the largest magnitude of the four components. Q_{sw} cools the ocean in the vicinity of 35°N, 170°E, consistent with the ENSO-induced increase in low-level clouds [*Park and Leovy*, 2004] and precipitation over the northwest Pacific (Plate 2a). The Q_{lw} and Q_{sw} radiation anomalies tend to be of opposite sign but the former are of modest amplitude. The anomalous Q_{lh} and to a lesser degree Q_{sh} also cool the WNP region, although they are located slightly west of the main Q_{sw} anomaly center.

3.4. Mixed Layer Depth

The surface layer of the ocean is generally well mixed, and as a result, the temperature and salinity is nearly uniform over this layer. In winter, the climatological MLD over the North Pacific ranges from approximately 100 m along the west coast of the United States to 250 m to the east of Japan [e.g. *White*, 1995; *Monterey and Levitus*, 1997], and the mixed layer depth (MLD) anomalies associated with El Niño events exceed 15 m in the central North Pacific [*Alexander et al.*, 2002]. As the wind stirring and negative buoyancy forcing (surface cooling) decrease from winter to spring, the climatological mixed layer shoals rapidly and is on the order of 15–20 m during the summer over the North Pacific Ocean [*Monterey and Levitus*, 1997; our Figure 9]. During El Niño events, the southward shift in the storm track, and the associated changes in surface winds and heat flux, act to deepen (shoal) the mixed layer from 35°N–46°N (48°N–55°N) across much of the Pacific in JA(0) (Figure 9). While the amplitude of the ENSO-related mixed layer depth anomalies is smaller in summer compared to winter, the MLD is approximately 10–30% greater during El Niño than La Niña events during late summer/early fall in the WNP region and 30–50% in a sub-region centered slightly to the east (Figures 9 and 10), comparable to or even larger than the fractional change in MLD during winter.

3.5. SST Tendency

The vertical distribution of the temperature anomalies in the WNP region as a function of the ENSO cycle is shown in Figure 10a, based on output from the NCEP Ocean assimilation system [*Derber and Rosati*, 1989; *Ji et al.*, 1995] during 1982–2003. Clearly, the ENSO-induced temperature anomalies in the WNP region are confined to the mixed layer through the first year of the ENSO cycle. The mixed layer temperature is influenced by the surface heat flux, Ekman and geostrophic transports, penetrating solar radiation, and entrainment of water through the base of the mixed layer [e.g. *Frankignoul*, 1985]. For well-mixed surface layers the SST tendency is determined by the fluxes into the mixed layer integrated over the MLD. As a result, the ENSO-induced SST response to the same forcing is much greater in AS(0) compared to JF(1), since the mean MLD in the northwest Pacific is approximately an order of magnitude smaller during summer compared to winter.

The ENSO-related ocean temperature anomalies are likely driven by local processes, since they develop rapidly and are confined to the thin surface layer; advection by geostrophic currents is a relatively slow process and operates over depths that are much greater than the MLD. We examine the extent to which the North Pacific SST anomalies are driven by surface fluxes and Ekman transport by comparing the inferred SST tendency, given by

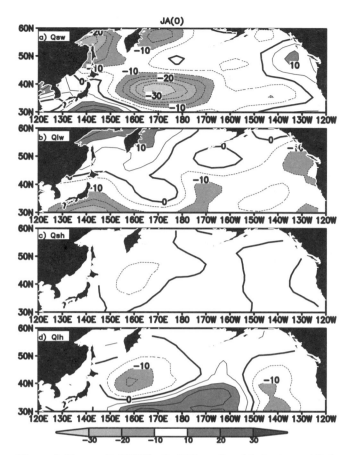

Figure 8. Composite El Niño–La Niña surface (a) shortwave (Q_{sw}), (b) longwave (Q_{lw}) (c) sensible (Q_{sh}) and (d) latent (Q_{lh}), fluxes in JA(0) during ENSO events from 1983–2000. The contour interval is 5 Wm^{-2}. Shading interval indicated by scale at bottom.

$$Q_t = [Q_{net} + Q_{ek}]/\rho c MLD, \qquad (3)$$

where ρ is the density of seawater, to the actual SST tendency (SST_t). Q_t is computed using three data sets: Q_{lh} and Q_{sh} are from NCEP reanalysis [*Kistler et al.*, 2001], Q_{sw} and Q_{lw} are derived from ISCCP data [*Zhang and Rossow*, 2003] and the MLD is based on ocean temperature profiles [*White*, 1995]. The composite ENSO anomalies of Q_t (contours) and SST_t (shaded) are shown over the North Pacific during JA(0) in Plate 3. Both Q_t and SST_t indicate rapid cooling in the western North Pacific (32°N–45°N, 150°E–180°), with smaller areas of anomalous warming and cooling in the central and eastern parts of the basin, respectively. In the WNP, the deeper MLD during El Niño compared to La Niña events (Figures 9 and 10) acts to amplify the anomalous SST tendency in the latter, since the forcing is integrated over a thinner layer. During JA(0) the anomalous Q_t and SST_t are –2.4°C and –2.2°C, respectively, when averaged over the northwest Pacific region. The agreement between the observed forcing and the SST response is surprisingly good, given the errors inherent in observations (especially considering that three independent data sources were used to compute Q_t) and that several forcing terms were neglected in equation (3).

3.6. Discussion

The atmospheric teleconnections associated with ENSO appear to strongly influence North Pacific SST anomalies in the summer of Yr(0), but do these anomalies have a broader impact on climate variability? In other words, does the summertime bridge influence the basin-wide SST characteristics and to what extent does the ENSO-generated SST changes feedback on the atmosphere?

Zhang et al. [1998] examined the seasonal persistence of North Pacific SST anomalies based on EOF and singular value decomposition (SVD) analyses. They found the leading pattern in summer was similar to that in winter, but with the largest signal located along 40°N between 160°E–180°, approximately 30° west of its wintertime position. In addition, the summer and winter patterns were highly correlated, which led them to conclude that the SST anomalies persisted from one season to the next, possibly due to SST-stratus cloud interactions [*Norris et al.*, 1998]. However, the leading EOF of North Pacific SSTs in summer and winter [Figures 4 and 6 in *Zhang et al.*, 1998] closely resembles the SST anomalies induced by the atmospheric bridge (Plate 1), including the westward displacement of the anomaly center in summer relative to winter. The rapid decorrelation time of summertime SST anomalies [*Deser et al.*, 2003] and the strong relationship between tropical and North Pacific SST anomalies in all seasons suggests that the connection between summer and winter SSTs in the North Pacific is not primarily due to local persistence, but rather to atmospheric forcing associated with ENSO, a conclusion reached by *Newman et al.* [2003] as well.

Once SST anomalies form in the North Pacific, they can influence the atmosphere both locally and perhaps remotely as well. The basic local atmospheric response to extratropical SST anomalies, is such that the near-surface air temperature and underlying ocean will adjust to each other, reducing the ocean-to-atmosphere damping of surface air temperature anomalies by Q_{sh}, Q_{lh}, and Q_{lw}, which all depend on the air–sea temperature difference [e.g., *Barsugli and Battisti*, 1998; *Bladé*, 1999]. The effect of this "reduced thermal damping" is to enhance air temperature variance at interannual and longer time scales in coupled atmosphere–ocean models relative to atmospheric GCM simulations in which climatological SSTs are specified as boundary conditions.

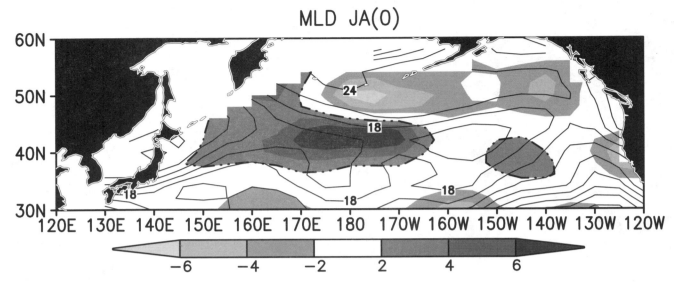

Figure 9. The climatological mean mixed layer depth (contours: interval 3 m) during Jul–Aug and composite El Niño–La Niña mixed layer depth (shading: scale at bottom; values >2 m are enclosed by a dot–dash line) during JA(0). The mean and ENSO MLD values, derived from ocean temperature observations as described by *White* [1995] for the period 1955–2001, have been spatially smoothed using a 9-point filter.

An additional atmosphere–ocean interaction process in the North Pacific during summer involves positive feedbacks between stratus clouds and SSTs [e.g., *Klein and Hartmann*, 1993; and *Norris and Leovy*, 1994]. An increase in low clouds reduces Q_{sw}, thereby cooling the SST, while colder SSTs enhance the atmosphere's static stability, which increases the strength of the surface inversion trapping the moisture that forms stratus clouds. In addition to the direct generation of clouds by atmospheric processes associated with storms, enhanced cloudiness over the North Pacific in summer may be due in part to the advection of warm air over cold SSTs in the regions of southerly flow between storms. Thus, the initial ENSO-driven increase in clouds and decrease in SSTs over the northwest Pacific in summer may be enhanced by positive stratus cloud–SST feedbacks.

Another local sea–air feedback involves the influence of static stability in the atmospheric boundary layer on vertical mixing of momentum. As the SST–air temperature difference increases, the static stability decreases, which enhances the vertical mixing of strong upper-level winds down to the surface. Recent satellite data indicates, that while this process is strongest in winter, it also operates in the western North Pacific during summer [*Nonaka and Xie*, 2003]. Thus, surface wind speeds tend to be higher (lower) above warm (cold) SSTs, which would tend to reduce the wind speed in the WNP during El Niño events. The negative ENSO-related SST anomalies also enhance (reduce) the strong mean meridional SST gradient from 140°E–180° to the south (north) of ~40°N (not shown). In regions where ocean temperature gradients are strong, SST-induced changes in static stability can influence the divergence and curl of the near-surface winds [*Chelton et al.*, 2001], which in turn can drive circulation changes beyond the boundary layer in both the atmosphere and ocean.

Several recent studies of extratropical atmosphere–ocean interaction in summer have investigated the relationships between, rainfall over Asia and/or North America, the large-scale atmospheric circulation and Pacific SST anomalies. Of the two leading patterns of precipitation and circulation variability, one involves a wave train that extends from eastern Asia to North America that is primarily associated with North Pacific SST anomalies, while the other resembles the ENSO-signal with zonally elongated height anomalies over the western North Pacific and SST anomalies of opposite sign in the tropical and North Pacific Ocean [*Ting and Wang*, 1997; *Lau and Weng*, 2002; *Lau et al.*, 2003]. Both patterns affect the North Pacific SST gradients, which in turn, may influence the near surface baroclinicity and thus the strength/position of the storm track and jet stream [*Tanimoto et al.*, 2003].

4. CONCLUSIONS

While the atmospheric circulation anomalies associated with ENSO are strongest in boreal winter, significant SST anomalies develop outside the equatorial Pacific in summer

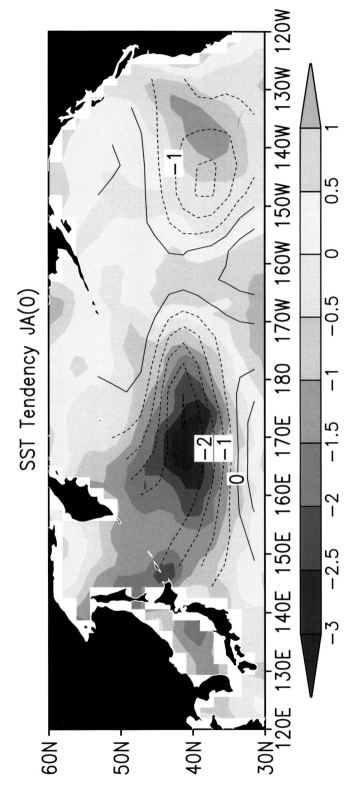

Plate 3. Composite El Niño–La Niña observed SST tendency and the SST tendency inferred from the net heat flux and Ekman transport (Q_t) during JA(0) for ENSO events between 1983–2000. SST tendency is shaded and Q_t is contoured, both with an interval of 0.5°C over the 2 month period. The SST tendency is based on the difference in temperature between AS(0) and JA(0). Both fields have been spatially smoothed using a 9-point filter.

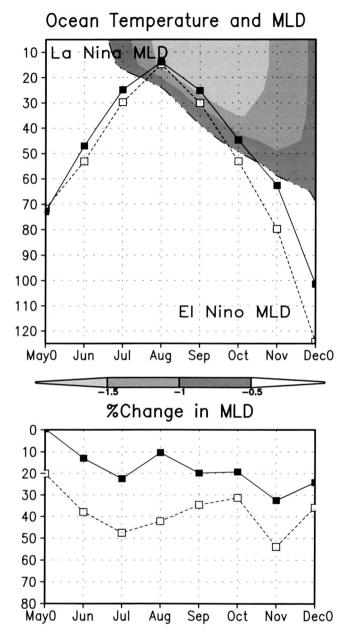

Figure 10. (a) Composite El Niño–La Niña ocean temperature (shading scale at bottom; values <–0.5°C are enclosed by a dot-dash line) from May(0) to Dec(0) and the composite MLD (m) during El Niño and La Niña events in the western North Pacific region (see Figure 1). (b) The percent change in the composite MLD during El Niño relative to La Niña events in the Western North Pacific region (black squares) and the north eastern portion of that region (170°E–180°, 39°N–43°N; open squares). The temperatures are from the NCEP Ocean Analyses [*Derber and Rosati*, 1989; *Ji et al.*, 1995] and the MLD from [*White*, 1995] for the period 1980–2001, the period when the ocean analyses are available.

soon after ENSO events begin, while others persist into the following summer, well after ENSO has dissipated. Here we have focused on two aspects of the remote atmosphere and ocean processes during ENSO: air–sea interaction in the tropical Indian and West Pacific Oceans over the seasonal cycle and the atmospheric bridge to the North Pacific in the summer of the ENSO year.

Anomalous cold (warm) water forms on the eastern (western) side of the tropical Indian Ocean, suggesting that ENSO contributes to the Indian Ocean SST dipole, the leading pattern of variability in boreal summer and fall. The eastern side of the basin warms rapidly in late fall and early winter resulting in positive SST anomalies across the entire Indian Ocean by the following spring. The SST anomalies in the tropical west Pacific also reverse sign, from negative in summer of Yr(0) to positive in the summer of Yr(1). The evolution of SST, SLP and precipitation anomalies shown here, and additional observational and modeling studies by *Wang et al.* [2003], *Lau and Nath* [2004] and *Lau et al.* [2004a], suggest that local air–sea interactions play an important role in the progression of ENSO-related anomalies over the tropical Indian and west Pacific Oceans.

During the summer of Yr(0), the atmospheric response to ENSO includes a southward shift in the Pacific storm track and jet stream, where the changes in synoptic eddy activity appear to strongly influence the large-scale circulation. An increase in cloudiness along 40°N, to the west of the dateline, accompanies the bridge-induced circulation anomalies. While the atmospheric surface circulation anomalies are much weaker than in winter, SST anomalies can develop rapidly in the western North Pacific during summer as the solar radiation and latent heat flux anomalies are large and this surface forcing is integrated over the relatively shallow mixed layer. ENSO likely influences the large-scale upper-ocean variability in the North Pacific, as the bridge-related SST anomaly pattern in AS(0) is very similar to the leading EOF of SST poleward of 20°N in summer.

Acknowledgments. We thank Shiling Peng and Jeff Whitaker for their guidance in computing the eddy induced height tendency and Mary Jo Nath for her assistance in preparing several of the figures. We also thank Ileana Bladé, Clara Deser, Dan Vimont and an anonymous reviewer for their suggestions that improved the manuscript. This work was supported in part by a grant from NOAA's CLIVAR–Pacific program.

REFERENCES

Alexander, M. A., Midlatitude atmosphere–ocean interaction during El Niño. Part I: the North Pacific Ocean, *J. Clim.*, 5, 944–958, 1992.

Alexander, M. A., I. Bladé, M. Newman, J. R. Lanzante, N.-C. Lau, and J. D. Scott, The atmospheric bridge: the influence of ENSO teleconnections on air–sea interaction over the global oceans, *J. Clim.*, *15*, 2205–2231, 2002.

Alexander, M. A., and J. D. Scott, The influence of ENSO on air–sea interaction in the Atlantic, *Geophys. Res. Lett.*, *29*, 5.1–5.4, 2002.

Allan, R. J., D. Chambers, W. Drosdowsky, H. Hendon, M. Latif, N. Nicholls, I. Smith, R. Stone, and Y. Yourre, Is there an Indian Ocean dipole, and is it independent of the El Niño–Southern Oscillation?, *CLIVAR Exch*, *6* (3), 18–22, 2001.

Barlow, M., S. Nigam, and E. H. Berbery, ENSO, Pacific decadal variability, and U.S. summertime precipitation, drought, and stream flow, *J. Clim.*, *14*, 2105–2128, 2001.

Barnston, A. G., and R. E. Livezey, Classification, seasonality and persistence of low-frequency atmospheric circulation patterns, *Mon. Weather Rev.*, *115*, 1083–1126, 1987.

Barsugli, J. J., and D. S. Battisti, The basic effects of atmosphere–ocean thermal coupling on midlatitude variability, *J. Atmos. Sci.*, *55* (4), 477–493, 1998.

Behera, S. K., P. S. Salvekar, and T. Yamagata, Simulation of interannual SST variability in the tropical Indian Ocean, *J. Clim.*, *13*, 3487–3499, 2000.

Bladé, I., The influence of midlatitude ocean–atmosphere coupling on the low-frequency variability of a GCM. Part II: Interannual variability induced by tropical SST forcing, *J. Clim.*, *12*, 21–45, 1999.

Cadet, D. L., The Southern Oscillation over the Indian Ocean, *Int. J. Climatol.*, *5*, 189–212, 1985.

Chambers, D. P., B. D. Tarpley, and R. H. Stewart, Anomalous warming in the Indian Ocean coincident with El Niño, *J. Geophys. Res.*, *104*, 3035–3047, 1999.

Chelton, D. B., and Coauthors, Observations of coupling between surface wind stress and sea surface temperature in the eastern tropical Pacific, *J. Clim.*, *14*, 1479–1498, 2001.

Chen, T.-C., and M.-C. Yen, Interannual variation of summertime stationary eddies, *J. Clim.*, *6*, 2263–2277, 1993.

Covey, D. L., and S. Hastenrath, The Pacific El Niño phenomenon in the Atlantic Sector, *Mon. Weather Rev.*, *106*, 1280–1287, 1978.

Derber, J. D., and A. Rosati, A global oceanic data assimilation system, *J. Phys. Oceangr.*, *19*, 1333–1347, 1989.

Deser, C., M. A. Alexander, and M. S. Timlin, Understanding the persistence of sea surface temperature anomalies in midlatitudes, *J. Clim.*, *16* (1), 57–72, 2003.

Enfield, D. B., and J. S. Allen, On the structure and dynamics of monthly mean sea level anomalies along the Pacific coast of North and South America, *J. Phys. Oceanogr.*, *10*, 557–588, 1980.

Frankignoul, C., Sea surface temperature anomalies, planetary waves, and air–sea feedback in the middle latitudes, *Rev. Geophys.*, *23*, 357–390, 1985.

Grimm, A. M., and P. L. Silva Dias, Analysis of tropical–extratropical interactions with influence functions of a barotropic model, *J. Atmos. Sci.*, *52*, 3538–3555, 1995.

Harrison, D. E., and N. K. Larkin, El Niño–Southern Oscillation sea surface temperature and wind anomalies, *Rev. Geophys.*, *36*, 353–399., 1998.

Hendon, H. H., Indonesian rainfall variability: Impacts of ENSO and local air–sea interaction, *J. Clim.*, *16*, 1775–1790, 2003.

Hoskins, B. J., and D. J. Karoly, The steady linear response of a spherical atmosphere to thermal and orographic forcing, *J. Atmos. Sci.*, *38*, 1179–1196, 1981.

Iizuka, S., and T. Matsura, The Indian Ocean SST dipole simulated in a coupled general circulation model, *Geophys. Res. Lett.*, *27*, 3369–3372, 2000.

Ji, M., A. Leetmaa, and J. Derber, An ocean analyses system for seasonal to interannual climate studies, *Mon. Weather Rev.*, *123*, 460–480, 1995.

Kistler, R., E. Kalnay, W. Collins, S. Saha, G. White, J. Woollen, M. Chelliah, W. Ebisuzaki, M. Kanamitsu, V. Kousky, H. van den Dool, R. Jenne, and M. Fiorino, The NCEP–NCAR 50–Year Reanalysis: Monthly means CD–ROM and documentation, *Bull. Am. Meteorol. Soc.*, *82*, 247–268, 2001.

Klein, S. A., B. J. Soden, and N.-C. Lau, Remote sea surface variations during ENSO: evidence for a tropical atmospheric bridge, *J. Clim.*, *12*, 917–932, 1999.

Klein, S. A., and D. L. Hartmann, The seasonal cycle of low stratiform clouds, *J. Clim.*, *6*, 1587–1606, 1993.

Kok, C. J., and J. D. Opsteegh, Possible causes of anomalies in seasonal mean circulation patterns during the 1982–83 El Niño event, *J. Atmos. Sci.*, *42*, 677–694, 1985.

Kumar, A., and M. P. Hoerling, The nature and causes for the delayed atmospheric response to El Niño, *J. Clim.*, *16*, 1391–1403, 2003.

Lau, K. M., K. M. Kim, I.-S. Kang and J. Y. Lee, The North Pacific as a regulator of summertime climate over Eurasia and North America, *J. Clim.*, accepted, 2003.

Lau, K.-M., and L. Peng, Dynamics of atmospheric teleconnections during the northern summer, *J. Clim.*, *5*, 140–158, 1992.

Lau, K.-M., and H. Weng, Recurrent teleconnection patterns linking summertime precipitation variability over east Asia and North America, *J. Meteorol. Soc. Japan*, *80*, 1129–1147, 2002.

Lau, N.-C., Variability of the observed midlatitude storm tracks in relation to low-frequency changes in the circulation pattern, *J. Atmos. Sci.*, *45*, 2718–2743, 1988.

Lau, N.-C., and E. O. Holopainen, Transient eddy forcing of the time mean flow as identified by geopotential tendencies, *J. Atmos. Sci.*, *41*, 313–328, 1984.

Lau, N.-C., and M. J. Nath, The role of the 'atmospheric bridge' in linking tropical Pacific ENSO events to extratropical SST anomalies, *J. Clim.*, *9* (9), 2036–2057, 1996.

Lau, N.-C., and M. J. Nath, Impact of ENSO on the variability of the Asian–Australian Monsoons as simulated in GCM experiments, *J. Clim.*, *13* (24), 4287–4309, 2000.

Lau, N.-C., and M. J. Nath, Atmosphere–ocean variations in the Indo–Pacific sector during ENSO episodes, *J. Clim.*, *16*, 3–20, 2003.

Lau, N.-C., and M. J. Nath, Coupled GCM simulation of atmosphere–ocean variability associated with zonally asymmetric SST changes in the tropical Indian Ocean, *J. Clim.*, *17*, 245–265, 2004.

Lau, N.-C., M. J. Nath, and H. Wang, Simulations by a GFDL GCM of ENSO-related variability of the coupled atmosphere–ocean system in the East Asian Monsoon region., in *East Asian Monsoon*, in press, World Scientific Publishing Company, 2004a.

Lau, N.-C., A. Leetmaa, M. J. Nath, and H. Wang, Influences of ENSO-induced Indo–Western Pacific SST anomalies on extratropical atmospheric variability during the boreal summer, *J. Clim.*, submitted, 2004b.

Li, T., Y. Zhang, E. Liu, and D. Wang, Relative role of dynamic and thermodynamic processes in the development of the Indian dipole: An OGCM diagnosis. *Geophys. Res. Lett.*, 29, 2110, doi:10.1029/2002GL015789.

Li, T., B. Wang, C.-P. Chang, and Y. Zhang, A theory for the Indian Ocean dipole/zonal mode, *J. Atmos. Sci.*, 60, 219–235, 2003.

Luksch, U., and H. v. Storch, Modeling the low-frequency sea surface temperature variability in the North Pacific, *J. Clim.*, 5, 893–906, 1992.

Meehl, G. A., The South Asian Monsoon and the tropospheric biennial oscillation, *J. Clim.*, 10, 1921–1943, 1997.

Mo, K.-C., J. R. Zimmerman, E. Kalnay, and M. Kanamitsu, A GCM study of the 1988 United Stated drought, *Mon. Weather Rev.*, 119, 1512–1532, 1991.

Monterey, G. I., and S. Levitus, *Climatological Cycle of Mixed Layer Depth in the World Ocean*, 5 pp., 87 figs. pp., U.S. Gov. Printing Office, NOAA NESDIS, Wash., D.C, 1997.

Murtugudde, R., J. P. McCreary, and A. J. Busalacchi, Oceanic processes associated with anomalous events in the Indian Ocean with relevance to 1997–1998, *J. Geophys. Res.*, 105, 3295–3306, 2000.

Newman, M., and P. D. Sardeshmukh, The impact of the annual cycle on the North Pacific/North American response to remote low frequency forcing, *J. Atmos. Sci.*, 55, 1336–1353, 1998.

Newman, M., G. P. Compo, and M. A. Alexander, ENSO-forced variability of the Pacific Decadal Oscillation, *J. Clim.*, 16, 3853–3857, 2003.

Nicholson, S. E., An analysis of the ENSO signal in the tropical Atlantic and western Indian Oceans, *Int. J. Climatol.*, 17, 345–375, 1997.

Nitta, T., Convective activities in the tropical western Pacific and their impact on the Northern Hemisphere summer circulation, *J. Meteor. Soc. Japan*, 65, 373–390, 1987.

Nitta, T., Long term variations of cloud amount in the western Pacific region, *J. Meteor. Soc. Japan*, 64, 373–390, 1986.

Nonaka, M., and S.-P. Xie, Covariations of sea surface temperature and wind over the Kuroshio and its extension: Evidence for ocean-to-atmosphere feedback, *J. Clim.*, 16, 1404–1413, 2003.

Norris, J. R., Interannual and interdecadal variability in the storm track, cloudiness, and sea surface temperature over the summertime North Pacific, *J. Clim.*, 13 (2), 422–430, 2000.

Norris, J. R., and C. Leovy, Interannual variability in stratiform cloudiness and sea surface temperature, *J. Clim.*, 7, 1915–1925, 1994.

Norris, J. R., Y. Zhang, and J. M. Wallace, Role of clouds in summertime atmosphere–ocean interactions over the North Pacific, *J. Clim.*, 11, 2482–2490, 1998.

Palmer, T. N., and C. Brankovic, The 1988 United States drought linked to anomalous sea–surface temperature, *Nature*, 338, 54–57, 1989.

Pares-Sierra, A., and J. J. O'Brien, The seasonal and interannual variability of the California Current System: a numerical model, *J. Geophys. Res.*, 94, 3159–3180, 1989.

Park, S., and C.B. Leovy, Marine cloud variations associated with ENSO, *J. Clim.*, submitted, 2004.

Rossow, W.B., and Y.-C. Zhang, Calculation of surface and top-of-atmosphere radiative fluxes from physical quantities based on ISCCP: Part II. Validation and first results, *J. Geophys. Res.*, 100, 1167–1197, 1995.

Saji, N. H., B. N. Goswami, P. N. Vinayachandran, and T. Yamagata, A dipole mode in the tropical Indian Ocean, *Nature*, 401, 360–363, 1999.

Saji, N. H., and T. Yamagata, Structure of SST and surface wind variability during Indian Ocean dipole mode events: COADS observations, *J. Clim.*, 16, 2735–2751, 2003.

Schott, F. A., and J. P. McCreary, The monsoon circulation of the Indian Ocean, *Prog. Oceanogr.*, 51, 1–123, 2001.

Scott, J. D., and M. A. Alexander, Net shortwave fluxes over the ocean., *J. Phys. Oceanogr.*, 29, 3167–3174, 1999.

Shinoda, T., M. A. Alexander, and H. H. Hendon, Remote response of the Indian Ocean to interannual SST variations in the tropical Pacific, *J. Clim.*, 17, 362–372, 2004.

Tanimoto, Y., H. Nakamura, T. Kagimoto, and S. Yamane, An active role of extratropical sea surface temperature anomalies in determining anamolous turbulent heat flux, *J. Geophys. Res.*, 108(C10), doi:10.1029/2002JC001750, 2003.

Ting, M. F. and N.-C. Lau, A Diagnostic and Modeling Study of the Monthly Mean Wintertime Anomalies Appearing in a 100-Year GCM Experiment, *J. Atmos. Sci*, 50, 2845–2867, 1993.

Ting, M., and H. Wang, Summertime U.S. precipitation variability and its relation to Pacific sea surface temperature, *J. Clim.*, 10, 1853–1873, 1997.

Trenberth, K. E., and G. W. Branstator, Issues in establishing causes of the 1988 drought over North America, *J. Clim.*, 5, 159–172, 1992.

Trenberth, K. E., G. W. Branstator, and P. A. Arkin, Origins of the 1988 North American drought, *Science*, 242, 1640–1645, 1988.

Trenberth, K. E., G. W. Branstator, D. Karoly, A. Kumar, N-C. Lau, and C. Ropelewski, Progress during TOGA in understanding and modeling global teleconnections associated with tropical sea surface temperatures., *J. Geophys. Res.*, 103, 14,291–14324, 1998.

Venzke, S., M. Latif, and A. Villwock, The coupled GCM ECHO-2. Part II Indian Ocean response to ENSO, *J. Clim.*, 13, 1384–1405, 2000.

Wang, B., R. Wu, and T. Li, Atmosphere–warm ocean interaction and its impacts on the Asian–Australian monsoon variation, *J. Clim.*, 16, 1195–1211, 2003.

Wang, B., and Q. Zhang, Pacific–East Asian teleconnection. Part II: How the Philippine Sea anomalous anticyclone is established during El Niño development, *J. Clim.*, 15, 3252–3265, 2002.

Wang B., R. Wu, X. Fu, Pacific–East Asian teleconnection: How does ENSO affect East Asian climate?, *J. Clim.*, 13, 1517–1536, 2000.

Wang, C., Atmospheric circulation cells associated with the El Niño–Southern Oscillation, *J. Clim.*, *15*, 399–419, 2002.

Wang, C., and J. Picaut, Understanding ENSO physics–A review, this volume.

Weaver, C. P., and V. Ramanathan, The link between summertime cloud radiative forcing and extratropical cyclones in the North Pacific, *J. Clim.*, 9 (9), 2093–2109, 1996.

Webster, P. J., A. M. Moore, J. P. Loschnig, and R. R. Leben, Coupled ocean–atmosphere dynamics in the Indian Ocean during 1997–98, *Nature*, *401*, 356–360, 1999.

White, W. B., Design of a global observing system for gyre-scale upper ocean temperature variability, *Prog. Oceanogr.*, *36*, 169–217, 1995.

Wu, L., Q. Zhang, and Z. Liu, Toward understanding tropical Atlantic variability using coupled model surgery, this volume.

Xie, P., and P. A. Arkin, Global precipitation: a 17-year monthly analysis based on gauge observations, satellite estimates, and numerical model outputs., *Bull. Am. Meteorol. Soc.*, *78*, 2539–2558, 1997.

Xie, S.-P., H. Annamalai, F. A. Schott, and J. P. McCreary, Structure and mechanisms of South Indian Ocean climate variability, *J. Clim.*, *15*, 864–878, 2002.

Yamagata, T., S. K. Behera, J., J. Luo, S. Masson, M. R. Jury, and S. A. Rao, Coupled ocean–atmosphere variability in the tropical Indian Ocean, this volume.

Yu, L., and M. M. Rienecker, Mechanisms for the Indian Ocean warming during the 1997–1998 El Niño, *Geophys. Res. Lett.*, *26*, 735–738, 1999.

Zhang, Y., J. R. Norris, and J. M. Wallace, Seasonality of large-scale atmosphere–ocean interaction over the North Pacific, *J. Clim.*, *11*, 2473–2481, 1998.

Zhang, Y.-C., and W. B. Rossow, New ISCCP global radiative flux data products, http://isccp.giss.nasa.gov/projects/flux.html, 2004.

Zhang, Y.-C., W. B. Rossow, and A. A. Lacis, Calculation of sur-face and top of atmosphere radiative fluxes from physical quantities based on ISCCP data sets. I. Method and sensitivity to input data uncertainties, *J. Geophys. Res.*, *100*, 1149–1165, 1995.

M. A. Alexander and J. D. Scott, National Oceanic and Atmospheric Administration–Cooperative Institute for Research in Environmental Sciences, Climate Diagnostics Center, R/CDC1, 325 Broadway, Boulder, Colorado 80305. (Michael.Alexander@noaa.gov; James.D.Scott@noaa.gov).

N.-C. Lau, National Oceanic and Atmospheric Administration, Geophysical Fluid Dynamics Laboratory, Princeton University, P.O. Box 308, Princeton, New Jersey 08542. (Gabriel.Lau@noaa.gov)

Predicting Pacific Decadal Variability

Richard Seager, Alicia R. Karspeck, Mark A. Cane, Yochanan Kushnir

Lamont-Doherty Earth Observatory of Columbia University

Alessandra Giannini

International Research Institute for Climate Prediction

Alexey Kaplan, Ben Kerman, and Jennifer Velez

Lamont-Doherty Earth Observatory of Columbia University

The case is advanced that decadal variability of climate in the Pacific sector is driven by tropical atmosphere-ocean interactions and communicated to the extratropics. It is shown that tropical decadal variations in the last century *could* arise as a consequence of the regional subset of physics contained within an intermediate model of the El Niño-Southern Oscillation. These decadal changes in ENSO and tropical mean climate are more predictable than chance years in advance but even in these idealized experiments forecast skill is probably too small to be useful. Nonetheless, forecasts of the next two decades indicate that, according to this model, the 1998 El Niño marked the end of the post 1976 tropical Pacific warm period.

Observations and atmosphere general circulation models are interpreted to suggest that decadal variations of the atmosphere circulation over the North Pacific between the 1960s and the 1980s are explained by a mix of tropical forcing and internal atmospheric variability. This places a limit on their predictability. The ocean response to extratropical atmosphere variability consists of a local response that is instantaneous and a delayed response of the subtropical and subpolar gyres that is predictable a few years in advance.

It is shown that the wintertime internal variability of the Aleutian Low can weakly impact the ENSO system but its impact on decadal predictability is barely discernible.

1. INTRODUCTION

For four years prior to fall 2002 the mid-latitudes of both the Northern and Southern Hemisphere experienced substantially less rain than usual. In the United States and across Southern Europe into Central Asia wells ran dry, crops failed and forests caught fire. The causes of this dry period have been linked to variations of the tropical atmosphere-ocean system in the Indo-Pacific region [*Hoerling and Kumar,* 2003]. After the enormous El Niño of winter 1997/98 the equatorial Pacific remained cooler than usual until early 2002 when a weak El Niño developed. It could be that this cold period marks the end of the most celebrated decadal variation in the Pacific sector: the warm shift in 1976 [*Zhang et al.,* 1997].

After 1976 the tropical Pacific Ocean has been warmer than in the preceding decades while the central and western North Pacific Ocean have been colder and atmospheric pressure has been lower over the mid-latitude North and South

Pacific Oceans. In the early 1940s the climate of the Pacific went through a shift in the opposite direction. These characteristics of Pacific Decadal Variability (PDV) have been described by, among others, Graham [1994], Trenberth and Hurrell [1994], Zhang et al. [1997], Mantua et al. [1997] and Garreaud and Battisti [1999].

Decadal variations of the Pacific climate have important consequences for climate over land analogous, but not identical, to the impacts of the El Niño-Southern Oscillation (ENSO) on interannual timescales. For example Mantua et al. [1997] show that when the tropical Pacific is warm (e.g., after 1976) winters are warm across most of North America but cold in the southeastern United States. Winters are dry across mid-latitude North America but are wet in the southwestern United States and Mexico. The persistent anomalies in atmosphere and ocean exert an impact on energy usage, power generation, agriculture, water resources, North Pacific fish stocks and marine ecosystems [*Mantua et al.*, 1997; *Miller and Schneider*, 2000]. Decadal variations of ENSO have also been associated with decadal variations in Australian climate [*Power et al.*, 1999] and the strength of the Indian monsoon [*Krishnamurthy and Goswami*, 2000; *Kumar et al.*, 1999]. Predictions of the state of the Pacific climate on timescales of years to a decade or more could have significant human benefits.

Most work has been motivated by the idea that the adjustment time for the tropical Pacific Ocean is on the order of years and explains interannual variability and that, analogously, PDV must be associated with a different, decadal timescale, ocean process. This led to explanations that PDV originated in the mid-latitude ocean-atmosphere system [*Latif and Barnett*, 1994, 1996]. It was then postulated that changes in the mid-latitude ocean were communicated through the ocean to the tropics, introducing a delay of several years and coupling together mid-latitude and tropical variability [*Gu and Philander*, 1997]. This has been shown, quite conclusively, not to work because the subsurface temperature signal becomes too weak [*Schneider et al.*, 1999]. In contrast, it has been shown that decadal variations of the tropical Pacific Ocean can be accounted for by tropical and subtropical wind forcing alone [*Schneider et al.*, 1999; *Karspeck and Cane*, 2002; *McPhaden and Zhang*, 2002; *Nonaka et al.*, 2002; *Schott et al.* this volume]. Many proposed mechanisms are reviewed in this volume by Wang and Picaut.

To go with the idea that PDV can originate in the tropics, there is ample evidence that extratropical climate variability in the Pacific can be explained in terms of tropical forcing. Trenberth and Hurrell [1994] made this case on the basis of observational analysis while Alexander et al. [2002] used coupled climate models to demonstrate that much of the North Pacific SST variability on both interannual and decadal timescales can be explained as a remote response to tropical forcing.

Here we will continue the argument that PDV originates in the tropics. Section 2 will examine whether decadal changes of the tropical Pacific atmosphere and ocean are predictable. Section 3 will examine whether the extratropical atmospheric response to decadal variations of tropical SST can be simulated. Section 4 will examine whether the response of the extratropical oceans to wind stress variations forced from the tropics can be predicted some years in advance.

This leaves one interesting stone unturned. Pierce et al. [2000], Vimont et al. [2001] and Vimont et al. [2003b] have argued that variability of the North Pacific atmosphere circulation can cause trade wind variability that changes subtropical SSTs and impacts ENSO. Consequently, in Section 5, we will examine the impact on coupled tropical Pacific climate variability of that part of trade wind variability that is associated with the internal, unforced, variability of the extratropical atmosphere.

2. TROPICALLY GENERATED PACIFIC DECADAL VARIABILITY AND ITS PREDICTABILITY

Until proven otherwise, a valid hypothesis for the origin of PDV is that it originates in the tangle of coupled atmosphere-ocean processes within the tropical Pacific that also give rise to interannual ENSO variability. This could arise in two ways. First, the longer timescale modes may arise deterministically (albeit chaotically) from nonlinear interactions among components of interannual variability or via very low frequency modes in the ocean dynamics [*Jin*, 2001]. Second, the application of noise to a system that can only oscillate on interannual timescales will generate variability on decadal timescales. The second method is by definition not predictable on decadal timescales while the first may be if the slow evolution of the ocean state can be predicted.

2.1. Decadal Variability in the Zebiak-Cane Model

Karspeck et al. [2004] (KSC hereafter) considered the first possibility and examined decadal variations within the Zebiak-Cane (ZC) model and their predictability. The ZC model is a geophysical model of the tropical Pacific Ocean and the atmosphere above that is used for studies of ENSO and ENSO prediction (see *Zebiak and Cane* [1987], *Cane et al.* [1986] for a complete model description).

KSC demonstrated that the model is capable of creating realistic decadal variability by searching 150,000 years of simulated unforced natural variability for periods that matched observations. Figure 1 (top) shows one of dozens of 30 year model segments that resemble the observed NINO3 record

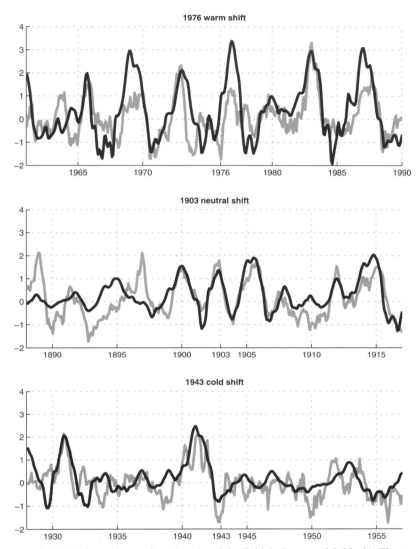

Figure 1. Time series of NINO3 from observations (gray) and the Zebiak-Cane model (black). The model segments were chosen from the long run to match the observed interannual variability and the decadal shifts. The first has a warm shift across 1976 with the 15 years after being warmer than the 15 years before by 0.38°C (observed) and 0.41°C (model), the third has a cold shift across 1943 of −0.32°C (model) and −0.36°C and, for comparison, the second has no shift at all. Taken from KSC.

(the SST anomaly averaged over 5°S–5°N and 90°W–150°W) for the 1961 to 1991 period containing the 1976 warm shift, the unquestioned star of tropical decadal variability. This example has a correlation coefficient with the observed record of 0.59 (using unfiltered monthly data) and has a post 1976 warming (relative to the 15 years before) of 0.41°C compared to the observed 0.38°C. The other two examples shown have equally high correlation to observations and are the 1942 'cold shift' (−0.36°C shift in the model compared to the observed −0.32°C) and, for comparison, a 'neutral shift' centered on 1903 (neither model nor observations had a noticeable shift). Each of these decadal variations can be mimicked by the model. The lesson is that the regional subset of tropical climate physics contained within the ZC model may be sufficient to generate the decadal variations that have occurred in the last 150 years.

2.2. Predictability of Decadal Variability in the Zebiak-Cane Model

KSC identified twenty-four 30 year segments (hereafter called model analogs) for each of the three observed segments (i.e., those centered on 1976, 1942 and 1903). The twenty-four chosen were those with the highest correlation coefficient with the observed record (all ≥0.5) and with an appropriate size shift in the average temperature between the last and first 15 years of the record (0.3°C for warm shift, –0.3°C for the cold shift and absolute value ≤0.1°C for the neutral shift).

To assess the predictability of these events an ensemble of 100 forecasts was run for each of the model analogs. Each forecast was initialized 5 years prior to the shift and integrated for the subsequent 20 years. The initial condition was the exact state of the model analog at the beginning of the forecast plus a random perturbation in the SST anomaly field. The SST perturbation at each grid point equals a random number sampled from a uniform distribution with zero mean and a standard deviation equal to that at the same place of the 150,000 year model integration. Hence perturbations are uncorrelated in space and between ensemble members.

The criteria for evaluating the forecasts are shown in Figure 2. A forecast of the warm shift analogs would be 'correct' if the later 15 years were warmer than the earlier 15 years by more than 0.21°C, 'weakly correct' if they were between 0.06°C and 0.2°C warmer and 'wrong' if they were less than 0.06°C warmer. These numbers were taken by dividing the statistical distribution of shifts in the mean NINO3 value between concurrent 15 year segments in the 150,000 year run into quintiles with equal numbers of shifts in each. The 'wrong' category covers three quintiles. The success of forecasts of the cold and neutral shifts are defined analogously and also illustrated in Figure 2.

Table 1 shows the percentage of correct, weakly correct and wrong forecasts for the warm, cold and neutral shifts. The model has definite skill at predicting warm shifts: less than 20% of the forecasts are wrong while almost 60% are correct. The model seems to have less skill at predicting cold or neutral shifts but still two thirds of the forecasts shift in at least the right sense.

2.3. Comparing Decadal Predictability in the Zebiak-Cane Model With That From Statistical Forecasting Strategies

We know that the irregularity in the ZC model arises not from noise but from its internal, deterministic dynamics [*Tziperman et al.,* 1995]. Do these dynamics provide any predictability beyond what is expected by chance?

Two statistical prediction schemes were used as strawman null hypotheses. One uses the distribution of 20 year means of NINO3 in the model and randomly grabs the twenty years after a forecast start from this distribution. Forecast skill in this scheme rests on a statistical tendency to shift away from extreme states toward the model mean. The second scheme uses a noise-forced, seasonal, second order autoregressive [AR(2)] model (which allows an oscillation) fitted to the ZC

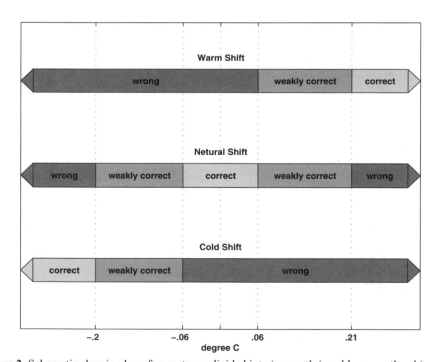

Figure 2. Schematic showing how forecasts are divided into 'correct', 'weakly correct' and 'wrong'.

Table 1. Performance of the dynamical model and two naive forecasting strategies presented as a function of the sense of the shift (warm, neutral or cold). Results for the dynamical model are based on 100 member ensembles for each of the 72 analog series (24 each of warm, neutral and cold shifts). Ensembles of size 500,000 were used for the naive forecasts. Taken from KSC.

	ZC Dynamical Forecasts			Naive Reference Forecasts					
				ZC-Long distribution			AR(2)		
	correct	weak	wrong	correct	weak	wrong	correct	weak	wrong
warm shift	59%	23%	18%	42%	26%	32%	47%	16%	37%
neutral shift	21%	45%	34%	18%	39%	43%	14%	30%	56%
cold shift	41%	23%	36%	30%	22%	48%	34%	14%	52%

model. The first ten years were identical to the model analogs and ensemble forecasts were then performed using different sequences of noise forcing for the 20 years after the forecast initialization.

As shown in Table 1 the ZC model outperforms the two statistical forecasting schemes for all shifts by a modest amount. To assess by how much the ZC model skill exceeds what would be expected by chance we used the Ranked Probability Score (RPS, *Wilks* [1995]), which accounts for how far the forecast is from what actually happened. For each forecast system we compute the fraction correct (f_c), weakly correct (f_{wc}) and wrong (f_w) with $f_c + f_{wc} + f_w = 1$. Then the RPS is given by:

$$RPS = (f_c - 1)^2 + (f_c + f_{wc} - 1)^2.$$

If the forecasts are all correct then RPS = 0 and if they are all wrong then RPS = 2. In between the RPS will get less (i.e. the forecast is better) even when f_c remains the same if f_{wc} increases, thus measuring that the forecasts became closer to the observed state.

The statistical distributions of the RPSs from 5000 100-member ensemble forecasts with each statistical schemes for the warm shift, cold shift and neutral shift are shown in Figure 3. The ZC model forecasts are more skillful than could be accounted for by chance, albeit by a modest amount. The excess predictability of the ZC model must arise from deterministic large scale and coherent evolution of the coupled system.

Instead the statistical model predictive skill, when it arises, comes from knowing at forecast initialization time that conditions have been unusual and that the subsequent 20 years are, statistically speaking, likely to be more akin to climatology (KSC). With only 150 years of observed ENSO variations it is impossible to know the true statistical distribution of decadal shifts and a statistical model based on only that data would probably be a poor tool for decadal prediction, notably worse than the ZC model.

2.4. Forecasting the Future

Extending the work of KSC, we performed a 1000 member ensemble of 30-year ZC model forecasts initialized in December 2002 using the operational data assimilation method outlined in *Chen et al.* [2000]. The initial state of each ensemble member differed by the addition of a random (uncorrelated in space) perturbation of the SST and the sea level height fields with standard deviation of 3°C and 3 cm respectively. Of most interest was the difference in NINO3 for the 15 year period after 1998 minus the 15 year period. Using the same division of shifts into quintiles as before, none of the 1000 forecasts went warm or weakly warm, 2.2% showed no shift, 56.8% went weakly cold and 41% went cold. The model, at least, is convinced that the 1997/8 El Niño marked the end of the post-1976 warm period.

3. PREDICTING DECADAL VARIATIONS OF EXTRATROPICAL ATMOSPHERE CIRCULATION FROM KNOWN SSTS

If decadal variations of tropical Pacific climate are predictable to a modest degree years in advance does this allow prediction of decadal climate variations outside of the tropical Pacific? On interannual timescales the movement of regions of deep convection that occurs within the ENSO cycle forces changes in the tropospheric stationary and transient eddies that create climate anomalies worldwide [*Horel and Wallace*, 1981; *Ropelewski and Halpert*, 1987, 1989; *Sardeshmukh and Hoskins*, 1988; *Held et al.*, 1989; *Hoerling and Ting*, 1994]. It is not so well established that decadal variations of tropical SSTs have an analogous impact.

3.1. Observed Decadal Variations of Atmospheric Circulation

In Figure 4 we show the differences in SST and 500 mb geopotential height during November to March (the season

110 PREDICTING PACIFIC DECADAL VARIABILITY

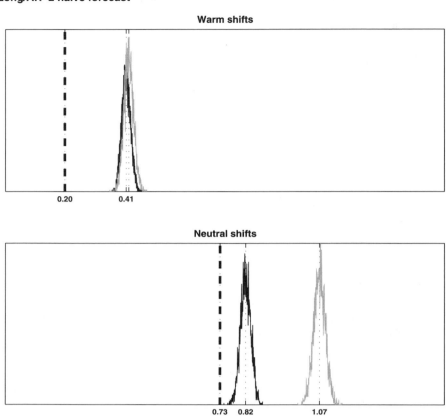

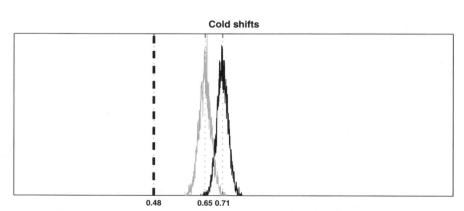

Figure 3. Ranked probability score for the warm shift (top panel), neutral shift (middle panel) and cold shift (bottom panel) forecasts by the ZC model, the AR model (gray) and the naive strategy (black). The RPS scores for the two statistical methods are shown as a distribution of 5000 scores. The ZC scores are shown as a single vertical line. In each case the ZC forecasts are unambiguously more skillful (i.e. RPS closer to 0) than the statistical forecasts. Taken from KSC.

when ENSO-related SST anomalies peak and the one in which the influence of the tropics on the Northern Hemisphere extratropical atmosphere circulation is most marked) for the decade 1977/1978 to 1986/1987 minus the decade 1966/1967 to 1975/1976.

After 1977 the Aleutian Low was anomalously deep (and slightly further south). This was a typical equivalent barotropic signal as evidenced by anomalously low sea level pressure (SLP) below and shifted to the east (not shown). There was anomalous high geopotential height over North America centered in the northwest. Associated with this circulation shift there was cold water stretching from the coast of Japan to the central North Pacific while the tropics were warm, especially in the east.

This pattern of decadal variability of atmosphere circulation could arise from either internal variability or boundary forcing by variations in SST. A useful first step is to compare the decadal variations with the interannual ones. Figure 5 shows the 500 mb height anomaly for the November through March season regressed onto the NINO3 index, using data from the NCEP-NCAR Reanalysis for 1959 to 1999 thus providing an estimate of the ENSO-forced 500mb variability. It is quite similar to the pattern of decadal variability. Both have low geopotential over the North Pacific (although the decadal low is 20° west of the interannual low) and both have high geopotential over North America and in the tropics. The proportionality between the tropical and mid-latitude height anomalies is similar for both patterns which is strong circumstantial evidence for tropical forcing of each.

3.2. Causes of Decadal Variations of the Atmospheric Circulation

To further examine whether the observed decadal variations of the extratropical atmosphere are caused by internal variability or are boundary forced we performed an ensemble

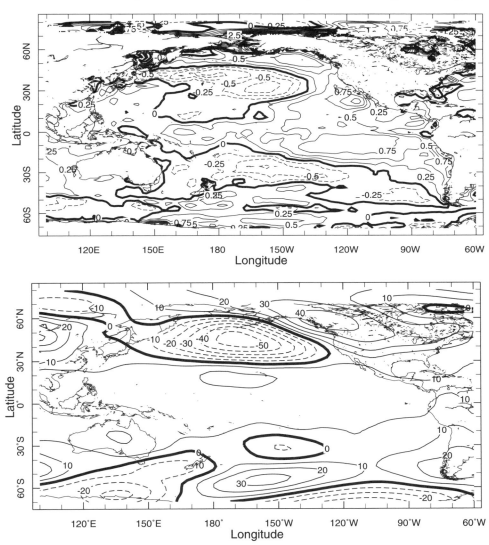

Figure 4. The difference in SST (top panel, contours in Kelvin) and 500mb height (lower panel, contours in meters) for the November to March season for the average of winters from 1977/78 to 1986/87 minus the average of winters from 1966/67 to 1975/76. The data are from the NCEP Reanalysis.

of 16 simulations with the National Center for Atmospheric Research Community Climate Model version 3 (NCAR CCM3, *Kiehl et al.* [1998]). Each ensemble member had different initial conditions but each used the same history of observed SSTs as a lower boundary condition from 1959 to 2000 [*Rayner et al., 2003*].

For a sufficiently large ensemble, taking the ensemble mean removes the internal variability leaving behind the boundary-forced component. This was confirmed in that two subsequent 16 member ensembles contained temporal and spatial patterns of variability in the ensemble mean that were nearly identical to those shown here. The pattern of modeled interannual variability is shown in Figure 6a and is very similar to that observed (Figure 5) both in terms of the spatial location of principal features and in their amplitude. This confirms that the observed pattern is boundary-forced and that the model has some skill at reproducing this signal. On the other hand, the model decadal difference of 500mb height over the tropical and North Pacific (Figure 6b) is only about one half of that observed (Figure 4b) even though the spatial patterns are very similar. Three explanations for why the modeled shift is weak come to mind.

First, the observed decadal difference could be the sum of a small SST-forced component and a much larger component due to internal atmosphere variability. Individual members of the ensemble do capture the pattern and magnitude of the extratropical decadal shift (Figure 7 shows an example). However, in these, as in every ensemble member, the amplitude of the associated tropical shift remains weak. This suggests that the extratropical decadal shift in these ensemble members is the sum of large internal variability and a small SST-forced shift.

Second, the weak tropical 500mb height shift in the model may itself cause the weak extratropical response along the lines proposed by Seager et al. [2003b]. The weak tropical shift is caused, primarily, by the model's failure to capture the increase in surface to 500mb thickness temperature that occured in 1977 and persisted through 1984 (Figure 8a) and, secondly, by the model's failure to simulate the high surface pressure in the seven years following 1976 (Figure 8b)[1]. Both model failures were reproduced in a similarly-forced 24 member ensemble using the ECHAM4.5 atmosphere model conducted by the International Research Institute for Climate Prediction.

The third possibility for why the decadal shift in the modeled height anomalies over the North Pacific is smaller than observed involves the post 1976 warming of the Indian Ocean. Deser et al. [2004] have shown that the post 1976 period had less cloud cover, and therefore presumably less precipitation, in the North Indian Ocean where the SST was warmer. This is the same relationship as occurs on interannual timescales [*Klein et al., 1999*]. In the model, however, specifying the warm SST anomalies in the Indian Ocean caused increased precipitation after 1976 (not shown). Barsugli and Sardesmukh [2002] have shown that increased atmospheric heating in this region should cause higher 500 mb heights over the North

---

[1]The observed increase in surface pressure at this time appears real as it is found in the Reanalysis, the recent Hadley Centre SLP analysis [Basnett and Parker, 1977] and the analysis (following the procedure of Kaplan et al. [2000]) of the most recent release, in 2001, of the COADS surface pressure data from ships and buoys [Woodruff et al., 1998] (also shown in Figure 8b)

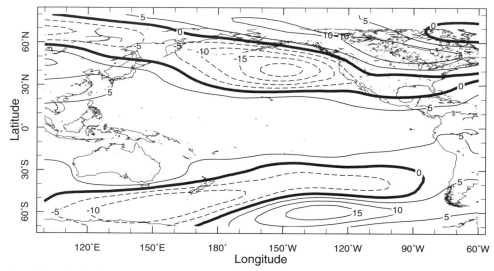

Figure 5. The 500mb height for the November to March season regressed onto the NINO3 index. Contours are meters and the data are from the NCEP Reanalysis.

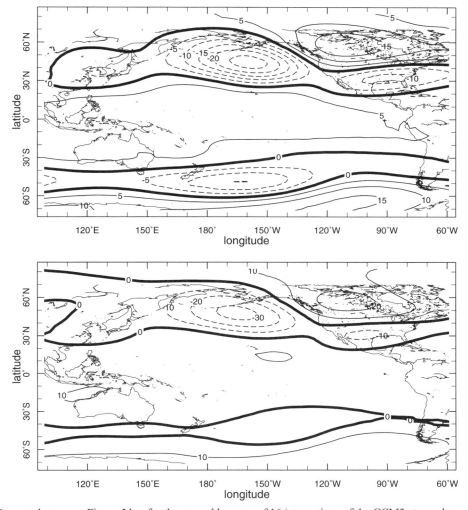

Figure 6. Top panel, same as Figure 5 but for the ensemble mean of 16 integrations of the CCM3 atmosphere model using observed SST in the surface boundary conditions and, lower panel, same as Figure 4b but for the model ensemble mean.

Pacific, opposing the impact of the increased heating over the central tropical Pacific Ocean.

This idea gains some support from a comparison of two 5 member ensembles conducted with CCM3 at the National Center for Atmospheric Research (not shown). One ensemble was forced with observed SSTs everywhere and, akin to our ensemble, had increased precipitation over the Indian Ocean after 1976. The other was forced with observed SSTs in the tropical Pacific Ocean and computed the SST anomalies with a mixed layer ocean elsewhere. This one had no increase in precipitation over the Indian Ocean after 1976 and, consistent with the reasoning above, had a larger drop in 500 mb height over the North Pacific. However, a larger ensemble is needed to prove that imposing Indian Ocean SST anomalies can lead to a mis-estimate of the extratropical response to tropical forcing.

4. PREDICTING THE RESPONSE OF THE NORTH PACIFIC OCEAN TO DECADALLY VARYING WINDS

Tropically-forced wind anomalies over the North Pacific will generate an ocean response. For example, if the tropics force a deeper Aleutian Low, stronger westerlies on the southern side of the Low drive an anomalous southward Ekman flow that cools the SST [*Miller et al.,* 1994; *Seager et al.,* 2001b]. To the east, along the North American coast, southerly wind anomalies warm the SST by warm, moist advection and in the western North Pacific cold, dry advection cools the SST [*Cayan,* 1992a, 1992b]. If the anomalous winds over the North Pacific can be predicted so can these local and instantaneous components of the SST response.

Non-local ocean dynamics also come into play [*Seager et al.,* 2001b; *Schneider and Latif,* 2004]. The shift south of the

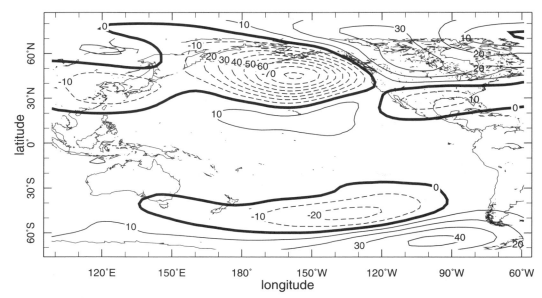

Figure 7. The decadal shift (for the same time period as in Figure 4) in 500mb height for a single CCM3 ensemble member. Units are meters. This member was chosen for its reasonable match in the extratropics to observations in amplitude and pattern and indicates that a mix of internal and boundary-forced variability could create decadal shifts in the extratropics of the observed size.

Aleutian Low after 1976 caused the latitude separating the subtropical and subpolar gyres, marked by the Kuroshio-Oyashio Extension (KOE), to also shift south. Since this separates warm subtropical waters from cool subpolar waters to the north, a potent cold SST anomaly developed in the region it had evacuated as the latitudinal distribution of ocean heat transport altered. This was well captured in the ocean hindcasts of Seager et al. [2001b].

The KOE SST anomalies developed a few years after those in the central Pacific, consistent with the time for westward propagation of oceanic Rossby waves from the region of wind stress forcing. Thus SST variability in the North Pacific involves a response to wind variations that is instantaneous and local in the central Pacific and is delayed and remote in the KOE region (cf. *Nakamura et al.* [1997]).

The delayed response of the gyre circulation has been exploited by Schneider and Miller [2001] to attempt hindcasts of SST anomalies east of Japan. They used a simple time dependent model of thermocline depth in which anomalies at any longitude and time are related to the anomalous wind stress forcing to the east by integrating back in time along Rossby wave characteristics. Hindcasts were performed by integrating the model forward with observed wind stress forcing to the time of the beginning of the hindcast and then continuing with zero wind stress anomaly. As the now unforced Rossby waves continued to propagate west they created thermocline and, by implication, SST anomalies. Schneider and Miller [2001] validated the hindcast SST anomalies against observations to show that the hindcasts have modest skill out to a few years ahead.

5. THE IMPACT OF INTERNAL VARIABILITY OF THE NORTH PACIFIC ATMOSPHERE ON THE TROPICS AND TROPICAL DECADAL VARIABILITY

Vimont et al. [2001, 2003b] have proposed that internal variability of the atmosphere over the North Pacific is associated with variations of the northeast trade wind strength during winter and forces subtropical SST anomalies. They argue that the SST anomalies persist into spring and summer and are damped by anomalous surface fluxes forcing an atmosphere response that includes zonal wind anomalies on the Equator that excite the coupled ocean-atmosphere dynamics familiar in ENSO.

It is difficult to use the observational record to separate between the patterns of internal and boundary-forced variability in the atmosphere. Hence we use the same 16 member SST-forced atmosphere model ensemble as before. Removing the ensemble mean—the boundary-forced component—leaves behind 16 40-year long records of variability generated by internal atmosphere processes alone. We concatenated these records and performed a singular value decomposition (SVD) analysis on the fields of surface wind speed and stress over the subtropical and tropical North Pacific (0°–30°N, to match the tropics-only extent of the ZC model) for the December through February mid winter period. The patterns are shown in Figure 9 and represent a strengthening and weakening of the northeast trades. These are associated with variations of the Aleutian Low (not shown).

Stronger (weaker) trade winds will tend to cool (warm) the subtropical SSTs via an SST tendency due to increased (decreased) latent heat flux (Q_{LH}) as approximated by $\rho_a C_E L \overline{\Delta q} U'/\rho_w c_p H$ where ρ_a and ρ_w are the surface air and water densities, $\overline{\Delta q}$ is the mean air-sea humidity difference, H is the ocean mixed layer depth, U' is the surface wind speed anomaly and other terms have their usual meaning. To first order wind-forced Q_{LH} variations will be balanced by a compensating change in Q_{LH} due to the SST change (*Seager et al.* [2000b]). In the ZC model that we will use this latter term is represented by $-\alpha T'_s$, where T'_s is the SST anomaly and α is an inverse timescale (*Zebiak and Cane* [1987]).

If U' is varying in mid-winter, T'_s will develop into spring and then decrease as U' goes to zero and the SST anomaly is damped provoking an atmospheric response. First we computed a Q_{LH} anomaly equal to $\rho_a C_E L \overline{\Delta q} U'$ using U' derived from the SVD analysis (Figure 9) and climatological mean winter values of $\overline{\Delta q}$. We then imposed this flux anomaly as an initial condition in the Zebiak-Cane model and examined the response and the subsequent evolution of the coupled model over the next several months.

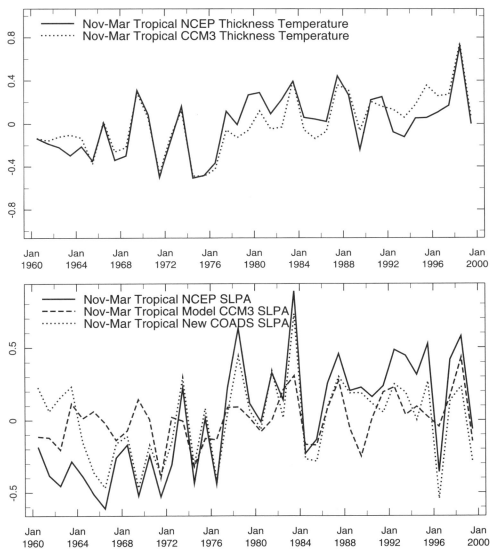

Figure 8. The time history of seasonal means of the tropical mean (20°N to 20°S) anomalies of the thickness temperature (K) between the surface and 500mb from the NCEP Reanalysis and the CCM3 model ensemble mean (top panel) and the SLP (mb) from NCEP Reanalysis, the CCM3 model ensemble mean and a new analysis of COADS ship data (lower panel).

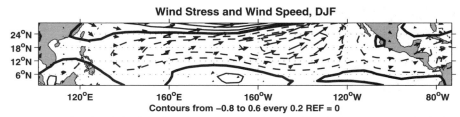

Figure 9. The pattern associated with the first singular vector of an analysis of wind stress (arrows) and wind speed (contours) within an SST-forced ensemble of atmosphere GCM integrations. Before the analysis was performed the ensemble mean, representing the SST anomaly-forced component of variability, was removed. The pattern therefore represents the dominant pattern in the model of internal atmosphere variability over the subtropical North Pacific. It accounts for 28% of the total variance in the fields. The maximum anomalous wind speeds are about 1ms^{-1} and the maximum stress anomalies, corresponding to the longest arrows, are of magnitude about 0.02 Nm^{-2}.

As shown in Figure 10, the Q_{LH} anomaly induced a subtropical warming. Heating north of the Equator forces ascent with the vortex stretching being balanced by meridional advection of planetary vorticity, requiring southerly flow, and the westerly component arising from conservation of angular momentum [*Gill*, 1980]. Zebiak [1982] shows that, when the

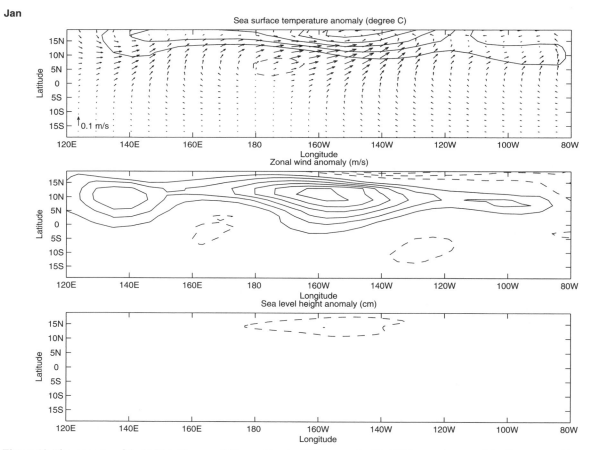

Figure 10. The response of the Zebiak-Cane model to a subtropical SST anomaly. The SST anomaly was originally generated by an imposed wind speed anomaly that begins in November and lasts through March. The figures are for January. The top panel shows the initial wind response of the model and the SST anomaly (°C) which has already begun to evolve according to the model physics. The middle panel shows the associated zonal wind speed response (ms^{-1}) in the model and the lower panel shows the sea level height response (cm). The arrow length of the wind vector is indicated in (a), the contour intervals are 0.03K for the SST anomaly, 0.02ms^{-1} for the zonal wind speed anomaly, 0.3cm for the sea level height anomaly and the zero contours are supressed.

heating north of the Equator is localized in longitude, the southwesterly anomaly extends onto the Equator. The initial ocean response has higher sea level height (SLH), or deeper thermocline, in the central Pacific. As the ocean response evolves (Figure 11), the equatorial southwesterlies force a downwelling oceanic Kelvin wave that propagates east raising SLH and depressing the thermocline in the eastern equatorial Pacific. This immediately causes SST warming in the east and a few months later a classic El Niño pattern develops (Figure 12), albeit with small amplitude, ~0.3°C. Is this impact potent enough to influence ENSO evolution and decadal predictability?

To examine this we add a Q_{LH} anomaly proportional to U' to the SST equation of the Zebiak-Cane model:

$$\frac{\partial T'_s}{\partial t} + dynamics = -\alpha T'_s + bU'f(t).$$

The term 'dynamics' is standing in for the three dimensional advection processes within the model mixed layer. The coefficient b is, for simplicity, taken to be a constant. The function $f(t)$ accounts for the time dependence of the wind speed forcing. It is zero between April and October and then increases to a maximum absolute value in January before declining. It has the same sign throughout each winter season to represent low frequency variability of the trade winds but its value varies randomly from year to year according to a white noise process. Forcing is only imposed in the Northern Hemisphere.

A long run of the ZC model with the subtropical wind forcing imposed was generated and searched for analogs of the observed decadal variability. Hindcasts of these were conducted as before except that, rather than imposing SST perturbations at the start of the forecast, each forecast was continued with a different sequence of subtropical wind speed forcing. Forecast skill was assessed as before and is shown in Table 2 along with the performance of the two statistical forecasting strategies, both of which were regenerated using the data from the long, subtropical forced, model run.

The percentage in each forecast category is very similar to that in the forecasts without noise forcing (Table 1) and the ZC model skill remains modestly higher than that of the statistical schemes. These experiments demonstrate that the skill of the ZC

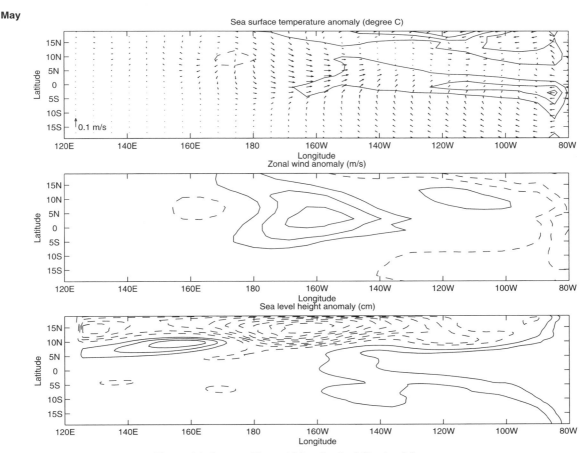

Figure 11. Same as Figure 10 but for the following May.

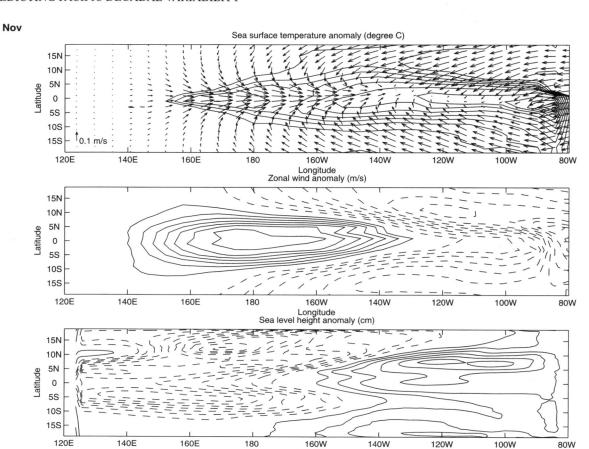

Figure 12. Same as Figure 10 but for the following November

model, such as it is, is retained when realistic noise is continually applied during the forecast. That is, the continual noise is a no more disruptive process than the initial condition perturbations considered in Section 2 and does not alter the mechanisms in the model that generate decadal variability.

6. CONCLUSIONS

We have investigated the causes of decadal variability of Pacific climate and its potential predictability. The conclusions are as follows:

· Decadal variations resembling observations in the tropical Pacific over the last 150 years or so can be generated by the regional subset of climate physics contained within a familiar intermediate model of the tropical Pacific alone (KSC). This does not prove that this physics is the cause of tropical Pacific decadal variability in the real

Table 2. Same as Table 1, with a subtropical wind forcing imposed on the dynamical model.

	ZC Dynamical Forecasts			Naive Reference Forecasts					
				ZC-Long distribution			AR(2)		
	correct	weak	wrong	correct	weak	wrong	correct	weak	wrong
warm shift	54%	27%	19%	43%	24%	33%	48%	17%	35%
neutral shift	23%	42%	35%	20%	37%	43%	14%	30%	56%
cold shift	40%	24%	36%	31%	21%	48%	33%	16%	51%

world but it does make a case that it is not immediately necessary to invoke processes in other regions of the world.

· The model analogs of observed decadal variations, such as the 1976 warm shift, are predictable years in advance. The skill of the geophysical model significantly exceeds that of statistical schemes but is too modest to hold out much hope for useful decadal forecasts (KSC). For what it is worth the model predicts that the 1998 El Niño ended the post-1976 tropical Pacific warm period.

· Decadal variations of tropical Pacific climate drive decadal variations of extratropical climate that could be predicted if the changes in tropical SST were known. Unpredictable internal atmospheric processes can cause equivalent size decadal variations of extratropical climate over the North Pacific. Furthermore, atmosphere models forced by observed SSTs everywhere poorly simulate the tropical response to decadal variations of SST.

· Part of the response of the extratropical North Pacific ocean to decadal variations of the winds involves a delayed adjustment of the subtropical and subpolar gyres. This component causes SST anomalies in the Kuroshio-Oyashio Extension region and is predictable on the timescale it takes for Rossby waves to propagate west to the Asian coast, that is, years [*Seager et al., 2001b*; *Schneider and Miller, 2001*].

· Internal wintertime variability of the atmosphere over the North Pacific is capable of generating a coupled dynamical response that is ENSO-like but it is weak and does not significantly perturb the decadal predictability of the ZC model.

In summary, a case can be made that Pacific Decadal Variability arises through coupled interactions in the tropics and is communicated to the extratropics. Aspects of this variability are predictable years in advance but the skill is so low that the prospects for useful operational prediction are poor.

Acknowledgments. This work was supported by National Oceanic and Atmospheric Administration grants UCSIO CU 02165401SCF and NA16GP2024 and by National Science Foundation grant ATM-9986072. We thank David Battisti and Clara Deser for valuable discussions, Dake Chen for guidance in hindcasting and forecasting and Ed Sarachik for first suggesting this work. This is Lamont Doherty Earth Observatory Contribution Number 6578.

REFERENCES

Alexander, M. A., I. Blade, M. Newman, J. R. Lanzante, N.-C. Lau, and J. D. Scott, The atmosphere bridge: The influence of ENSO teleconnections on air-sea interaction over the global ocean, *J. Clim.*, *15*, 2205–2231, 2002.

Barsugli, J. J., and P. D. Sardeshmukh, Global atmospheric sensitivity to tropcial SST anomalies throughout the Indo-Pacific basin, *J. Clim.*, *15*, 3427–3442, 2002.

Basnett, T. A., and D. E. Parker, Development of the global mean sea level pressure data set GMSLP2., *Tech. Rep. 79*, Hadley Center for Climate Research, 1997.

Cane, M. A., S. E. Zebiak, and S. C. Dolan, Experimental forecasts of El Niño, *Nature*, *321*, 827–832, 1986.

Cayan, D., Latent and sensible heat flux anomalies over the northern oceans: The connection to monthly atmospheric circulation, *J. Clim.*, *5*, 354–369, 1992a.

Cayan, D., Latent and sensible heat flux anomalies over the northern oceans: Driving the sea surface temperature, *J. Phys. Oceanogr.*, *22*, 859–881, 1992b.

Chen, D., M. A. Cane, S. E. Zebiak, R. Canizares, and A. Kaplan, Bias correction of an ocean-atmosphere coupled model, *Geophys. Res. Lett.*, *27*, 2585–2588, 2000.

Deser, C., A. S. Phillips, and J. W. Hurrell, Pacific interdecadal climate variability: Linkages between the tropics and the North Pacific during boreal winter since 1900, *J. Clim.*, in press, 2004.

Garreaud, R. D., and D. S. Battisti, Interannual (ENSO) and interdecadal (ENSO-like) variability in the southern hemisphere tropospheric circulation., *J. Clim.*, *12*, 2113–2123, 1999.

Gill, A. E., Some simple solutions for heat induced tropical circulation, *Q. J. R. Meteorol. Soc.*, *106*, 447–462, 1980.

Graham, N., Decadal-scale climate variability in the tropical and North Pacific during the 1970s and 1980s: observations and model results, *Clim. Dyn.*, *10*, 135–162, 1994.

Gu, D., and S. G. H. Philander, Interdecadal climate fluctuations that depend on exchanges between the tropics and extratropics, *Science*, *275*, 805–807, 1997.

Held, I., S. W. Lyons, and S. Nigam, Transients and the extratropical response to El Niño, *J. Atmos. Sci.*, *46*, 163–176, 1989.

Hoerling, M. P., and A. Kumar, The perfect ocean for drought, *Science*, *299*, 691–694, 2003.

Hoerling, M. P., and M. Ting, Organization of extratropical transients during El Niño, *J. Clim.*, *7*, 745–766, 1994.

Horel, J. D., and J. M. Wallace, Planetary scale atmospheric phenomena associated with the Southern Oscillation, *Mon. Weather Rev.*, *109*, 813–829, 1981.

Jin, F.-F., Low-frequency modes of tropical ocean dynamics, *J. Clim.*, *14*, 3874–3881, 2001.

Kaplan, A., Y. Kushnir, and M. A. Cane, Reduced space optimal interpolation of historical marine sea level pressure: 1854-1992, *J. Clim.*, *13*, 2987–3002, 2000.

Karspeck, A., and M. A. Cane, Tropical Pacific 1976/77 climate shift in a linear wind-driven model, *J. Phys. Oceanogr.*, *32*, 2350–2360, 2002.

Karspeck, A., R. Seager, and M. A. Cane, Predictability of tropical Pacific decadal variability in an intermediate model, *J. Clim.*, in press, 2004.

Kiehl, J. T., J. J. Hack, G. B. Bonan, B. A. Bovile, D. L. Williamson, and P. J. Rasch, The National Center for Atmospheric Research Community Climate Model: CCM3, *J. Clim.*, *11*, 1131–1149, 1998.

Klein, S. A., B. J. Soden, and N. Lau, Remote sea surface temperature variations during ENSO: Evidence for a tropical atmospheric bridge, *J. Clim.*, *12*, 917–932, 1999.

KrishnaKumar, K., B. Rajagopalan, and M. A. Cane, On the weakening relationship between the Indian monsoon and ENSO, *Science*, *284*, 2156–2159, 1999.

Krishnamurthy, V., and B. N. Goswami, Indian monsoon-ENSO relationship on interdecadal timescale, *J. Clim.*, *13*, 579–595, 2000.

Latif, M., and T. P. Barnett, Causes of decadal climate variability over the North Pacific/North American sector, *Science*, *266*, 634–637, 1994.

Latif, M., and T. P. Barnett, Decadal climate variability over the North Pacific and North America: Dynamics and predictability, *J. Clim.*, *9*, 2407–2423, 1996.

Mantua, N. J., S. R. Hare, Y. Zhang, J. M. Wallace, and R. C. Francis, A Pacific interdecadal climate oscillation with impacts on salmon production, *Bull. Am. Meteorol. Soc.*, *78*, 1069–1079, 1997.

McPhaden, M. J., and D. Zhang, Slowdown of the meridional overturning circulation in the upper Pacific Ocean, *Nature*, *415*, 603–608, 2002.

Miller, A. J., and N. Schneider, Interdecadal climate regime dynamics in the North Pacific Ocean: Theories, observations and ecosystem impacts, *Prog. Oceanogr.*, *47*, 355–379, 2000.

Miller, A. J., D. R. Cayan, T. P. Barnett, N. E. Graham, and J. M. Oberhuber, Interdecadal variability of the Pacific Ocean: Model response to observed heat flux and wind stress anomalies, *Clim. Dyn.*, *9*, 287–302, 1994.

Nakamura, H., G. Lin, and T. Yamagata, Decadal climate variability in the North Pacific during the recent decades, *Bull. Am. Meteorol. Soc.*, *78*, 2215–2225, 1997.

Nonaka, M., S.-P. Xie, and J. P. McCreary, Decadal variations in the subtropcial cells and equatorial Pacific SST, *Geophys. Res. Lett.*, *29*, doi: 10.1029/2001GL013676, 2002.

Pierce, D. W., T. P. Barnett, and M. Latif, Connections between the Pacific Ocean tropics and midlatitudes on decadal timescales, *J. Clim.*, *13*, 1173–1194, 2000.

Power, S., F. Tseitkin, V. Mehta, B. Lavery, S. Trock, and N. Holbrook, Decadal climate variability in Australia during the twentieth century, *Int. J. Climatol.*, *19*, 169–184, 1999.

Rayner, N., D. Parker, E. Horton, C. Folland, L. Alexander, D. Rowell, E. Kent, and A. Kaplan, Global analyses of sea surface temperature, sea ice, and night marine air temperature since the late nineteenth century, *J. Geophys. Res.*, *108*, doi: 10.1029/2002JD002670, 2003.

Ropelewski, C. F., and M. S. Halpert, Global and regional scale precipitation patterns associated with the El Niño/Southern Oscillation, *Mon. Weather Rev.*, *114*, 2352–2362, 1987.

Ropelewski, C. F., and M. S. Halpert, Precipitation patterns associated with the high index phase of the Southern Oscillation, *J. Clim.*, *2*, 268–284, 1989.

Sardeshmukh, P. D., and B. J. Hoskins, The generation of global rotational flow by steady idealized tropical divergence, *J. Atmos. Sci.*, *45*, 1228–1251, 1988.

Schneider, N., and A. J. Miller, Predicting western North Pacific Ocean climate, *J. Clim.*, *14*, 3997–4002, 2001.

Schneider, N., S. Venzke, A. J. Miller, D. W. Pierce, T. O. Barnett, C. Deser, and M. Latif, Pacific thermocline bridge revisted, *Geophys. Res. Lett.*, *26*, 1329–1332, 1999.

Schott, F. A., J. P. McCreary, and G. C. Johnson, Shallow overturning circulations of the tropical-subtropical oceans, this volume.

Seager, R., Y. Kushnir, M. Visbeck, N. Naik, J. Miller, G. Krahmann, and H. Cullen, Causes of Atlantic Ocean climate variability between 1958 and 1998, *J. Clim.*, *13*, 2845–2862, 2000.

Seager, R., Y. Kushnir, N. Naik, M. A. Cane, and J. Miller, Wind-driven shifts in the latitude of the Kuroshio-Oyashio extension and generation of SST anomalies on decadal timescales, *J. Clim.*, *14*, 4249–4265, 2001.

Seager, R., N. Harnik, Y. Kushnir, W. Robinson, and J. Miller, Mechanisms of hemispherically symmetric climate variability, *J. Clim.*, *16*, 2960–2978, 2003.

Trenberth, K., and J. W. Hurrell, Decadal atmosphere-ocean variations in the Pacific, *Clim. Dyn.*, *9*, 303–319, 1994.

Tziperman, E., M. A. Cane, and S. E. Zebiak, Irregularity and locking to the seasonal cycle in an enso prediction model as explained by the quasi-periodicity route to chaos, *J. Atmos. Sci.*, *52*, 293–306, 1995.

Vimont, D., D. S. Battisti, and A. C. Hirst, Footprinting: a seasonal link between the mid-latitudes and tropics, *Geophys. Res. Letters*, *28*, 3923–3926, 2001.

Vimont, D., J. M. Wallace, and D. S. Battisti, The seasonal footprinting mechanism in the Pacific; implications for ENSO, *J. Clim.*, *16*, 2668–2675, 2003.

Wang, C., and J. Picaut, Understanding ENSO physics—A review, this volume.

Wilks, D. S., *Statistical methods in the atmospheric sciences*, 467 pp, Academic Press, San Diego, 1995.

Woodruff, S., R. Slutz, R. Jenne, and P. Steurer, COADS Release 2: Data and metadata enhancements for improvements of marine surface flux fields, *Phys. Chem. Earth*, *2*, 517–527, 1998.

Zebiak, S. E., A simple atmospheric model of relevance to El Niño, *J. Atmos. Sci.*, *39*, 2017–2027, 1982.

Zebiak, S. E., and M. A. Cane, A model El Niño-Southern Oscillation, *Mon. Weather Rev.*, *115*, 2262–2278, 1987.

Zhang, Y., J. M. Wallace, and D. S. Battisti, ENSO-like decade-to-century scale variability: 1900-93, *J. Clim.*, *10*, 1004–1020, 1997.

A. Giannini, International Research Institute for Climate Prediction, Palisades, New York 10964-8000. (alesall@iri.columbia.edu)

R. Seager, A. R. Karspeck, M. A. Cane, Y. Kushnir, A. Kaplan, B. Kerman and J. Velez, Lamont-Doherty Earth Observatory of Columbia University, Palisades, New York 10964-8000. (rich@maatkare.ldeo.columbia.edu)

Tropical Atlantic Variability: Patterns, Mechanisms, and Impacts

Shang-Ping Xie

International Pacific Research Center and Department of Meteorology, University of Hawaii, Honolulu, Hawaii

James A. Carton

Department of Meteorology, University of Maryland, College Park, Maryland

This chapter reviews the progress made in the past decade in understanding tropical Atlantic climate variability. In addition to an equatorially anti-symmetric seasonal cycle forced directly by the seasonal march of the sun, Atlantic sea surface temperature (SST) displays a pronounced annual cycle on the equator that results from continental monsoon forcing and air–sea interaction. This cycle interacts with and regulates the meridional excursions of the Atlantic intertropical convergence zone (ITCZ). On interannual timescales, there is an equatorial mode of variability that is similar to El Niño/Southern Oscillation (ENSO) in the Pacific. This Atlantic Niño is most pronounced in boreal summer coinciding with the seasonal development of the equatorial cold tongue. In boreal winter, both ENSO and the North Atlantic Oscillation exert a strong influence on the northeast trades and SST over the northern tropical Atlantic. In boreal spring when the equatorial Atlantic is uniformly warm, anomalies of cross-equatorial SST gradient and the ITCZ are closely coupled, resulting in anomalous rainfall over northeastern Brazil. There is evidence for a positive air–sea feedback through wind-induced surface evaporation that organizes off-equatorial SST anomalies to maximize their cross-equatorial gradient. The resultant anomalous shift of the ITCZ may affect the North Atlantic Oscillation, helping to organize ocean-atmospheric anomalies into a pan-Atlantic pattern.

1. INTRODUCTION

The Atlantic Ocean is flanked by two large tropical continents, which host major centers of atmospheric convection. As early as 320 years ago, *Halley* [1686] recognized the important influence of these continents on climate in the Atlantic sector and suggested that the intense surface heating over North Africa drives the southerly winds in the Gulf of Guinea. It was not until 1970s, however, that the influence of the tropical Atlantic Ocean on continental climate variability began to come to light. The studies that followed showed that interannual variability in rainfall over the semi-arid regions of South America and Africa is associated with well-organized, repeating patterns of sea surface temperature (SST) and trade wind anomalies over the tropical Atlantic. Furthermore, these patterns of ocean and atmospheric anomalies are so arranged that their interaction gives rise to positive feedback acting to amplify each other. Rapid progress has been achieved in the past decade in understanding these air–sea interaction mechanisms and modeling the resulting variability in climate over the tropical Atlantic and beyond. This chapter reviews the progress in describing the patterns and understanding the mechanisms for tropical Atlantic variability (TAV).

We begin with a brief overview of the seasonal cycle, which dominates tropical Atlantic variability. This is followed by a survey in Section 3 of interannual variability in the equatorial Atlantic, which is akin to the El Niño and Southern Oscillation (ENSO) phenomenon in the Pacific. Section 4 concerns off-equatorial SST variability regarding which opposing views exist. We review recent efforts to understand air–sea interaction from the oceanic, atmospheric, and coupled points of view. The tropical Atlantic is not isolated, but is influenced by, and may influence climate variability in other regions, in particular ENSO and the North Atlantic Oscillation (NAO). The NAO is of central importance for climate variability in the extratropical North Atlantic and Europe. Extensive literature on NAO research exists, recently summarized in an American Geophysical Union monograph by *Hurrell et al.* [2003]. Sections 5 and 6 discuss how the NAO and ENSO influence TAV, respectively. Section 7 is a summary and includes discussion of the challenges ahead.

2. SEASONAL CYCLE

SST in the eastern equatorial Atlantic is dominated by the annual cycle. Temperatures reach their maximum in boreal spring when the equatorial winds are weakest and the thermocline is deepest in the east. During this season the sun is directly overhead, providing maximum incident solar radiation. The band of high SSTs exceeding 27°C occupies an equatorial region extending from 8°S to 5°N (Figure 1a). As the year progresses the trade winds along the equator intensify. The resulting zonal pressure gradient in the ocean and associated uplifting thermocline leads to seasonal cooling of SSTs in the eastern equatorial Atlantic. The SSTs reach their minimum along the eastern coast of Africa in July as a result of intensified coastal upwelling (Figure 2a), and then in the southeastern Gulf of Guinea a month later. In July and August, a distinct cold tongue forms across the basin, centered slightly south of the equator (Figure 1b).

The northeast and southeast trade wind systems meet at the narrow, roughly zonally oriented intertropical convergence zone (ITCZ). The time–mean latitude of the ITCZ and the collocated rain band, often called the thermal equator or climatic axis of symmetry, is displaced 5–10 degrees north of the geographical equator over the Atlantic [*Hastenrath*, 1991; *Mitchell and Wallace*, 1992; references therein], despite the fact that solar radiation at the top of the atmosphere is nearly symmetric about the equator on annual mean (see *Xie* [2004] for the latest review of research on this climatic asymmetry over

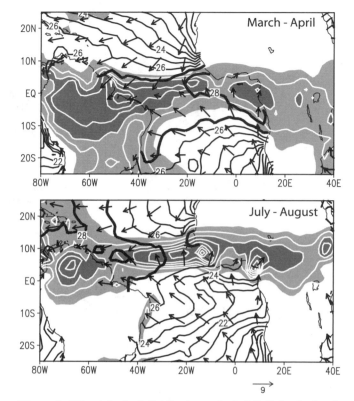

Figure 1. Climatological distributions of rainfall (light shade >2 mm/day; dark shade >6 mm/day), SST (contours in °C) and surface wind velocity (vectors in m/s) for March-April (upper) and July-August (lower panel), based on the Climate Prediction Center Merged Analysis of Precipitation (CMAP; *Xie and Arkin* 1996) and Comprehensive Ocean-Atmospheric Data Set (COADS; *Woodruff et al.* 1987).

the Atlantic and eastern Pacific). The ITCZ is also associated with the latitude of minimum seasonal variance of SST and the latitude of maximum vertical displacement of the thermocline [*Houghton*, 1991].

The ITCZ and its associated band of continental convection display large seasonal excursions over the Atlantic sector. Over the continents, the rain band largely follows the seasonal march of the sun, reaching its northernmost (southernmost) position in July–September (December–February) [*Mitchell and Wallace*, 1992; *Biasutti et al.*, 2003]. Since dry soil has a negligible heat capacity, the apparent lag in the meridional excursion of the continental rain band behind the sun may result from other heat reservoirs such as soil moisture and oceanic influences. For example, northeastern Brazil is in its wet season at the spring equinox (March) but is kept dry at the fall equinox (September) as strong northward SST gradients prevent the oceanic ITCZ from moving south of the equator [*Fu et al.*, 2001].

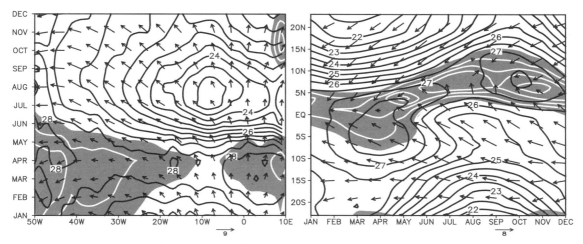

Figure 2. Left: longitude-time sections of COADS SST (black contours in °C) and surface wind velocity (vectors in m/s) at 1°S, and CMAP rainfall in 1.25°S-1.25°N. Right: time-latitude section of SST, surface wind velocity, and rainfall, averaged in 30-25°W. Rainfall are in white contours at 2.5 mm/day intervals with shade >5 and 2.5 mm/day in the left and right panels, respectively.

Over the ocean, patterns of SST and the position of the ITCZ are tightly coupled, with major rainfall confined to a band of high SSTs above 27°C. In March–April, the rain band is located nearly on the equator onto which the trades converge from both hemispheres. SST is uniformly warm in the equatorial zone of 10°S–5°N, making March–April the time when the Atlantic ITCZ is very sensitive to even small changes in interhemispheric SST gradient (Section 4). As the equatorial cold tongue develops in June and persists through September, the ITCZ is kept north of the equator following the northward movement of the high-SST band while a second, smaller, convective zone develops south of the equator west of 30°W [*Grodsky and Carton*, 2003]. The oceanic ITCZ reaches its northernmost position in September, lagging its northward movement over the continents because of the large heat capacity of the ocean mixed layer. In July–August, rainfall in the ITCZ is considerably stronger than in March–April, despite a 1°C drop in SSTs beneath the ITCZ (Figure 2). This strengthening of ITCZ convection may be due to the abundance of strong westward propagating easterly wave disturbances that help trigger convection over the ocean. These disturbances originate from the African rain band [e.g., *Thorncroft et al.*, 2003] and grow in the tropical Atlantic, some into tropical storms and hurricanes that devastate the Caribbean and southern United States [*Gray and Landsea*, 1992]. In addition to the 3–9 day African easterly waves the tropics also support a nearly stationary pattern of summer winds and precipitation with periods of two weeks that appears to result from land–atmosphere interaction [*Janicot and Sultan*, 2001; *Grodsky and Carton*, 2001].

Along the equator SST varies with a strong annual cycle despite the primarily semiannual nature of solar heating (this is also true of the eastern Pacific). At 10°W, equator, SST reaches 28°C in March–April and drops to below 23°C in July–August. This seasonal warming and cooling is highly asymmetric, with the latter taking only three months and the former taking seven months. From the oceanic point of view the rapid cooling is attributed to the sudden onset of the West African monsoon and the rapid intensification of the southerly winds in May–June in the Gulf of Guinea. These southerly winds cause upwelling slightly south and downwelling slightly north of the equator, and this upwelling cools the equatorial ocean [*Philander and Pacanowski*, 1981]. They also have strong convergence/divergence, decelerating over the cold tongue and then accelerating again over the warmer water a few degrees north of the equator. Equatorial zonal wind variations also play an important role in the equatorial SST annual cycle by inducing upwelling (through zonal and meridional divergence) and tilting the thermocline depth on the equator. From April to August, the thermocline shoals more than 60 m in the equatorial Gulf of Guinea [*Houghton*, 1983; *Philander and Pacanowski*, 1986]. Changes in zonal wind strength also affect wind-induced evaporation.

From the coupled point of view, *Xie* [1994] shows that the northward displacement of the climatological ITCZ is the ultimate cause of the annual cycle in equatorial SST in both the Pacific and Atlantic by maintaining southerly cross-equatorial winds that intensify in boreal summer/fall and relax in boreal spring [see also *Giese and Carton*, 1994]. *Mitchell and Wallace*'s [1992] observational analysis sug-

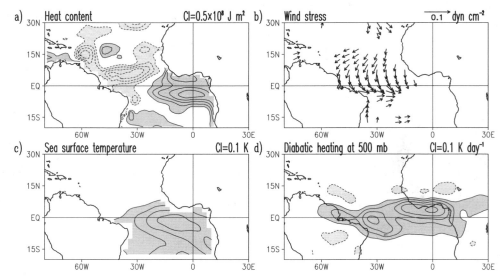

Figure 3. Anomalies associated with the Atlantic Niño principal component from a five-variable rotated principal component analysis: (a) observed heat content (10^8 J m^{-2}), (b) vector wind stress (dyn cm^{-2}), (c) SST (°C), and (d) diabatic heating at 500 mb (°C day^{-1}). Dark (light) shading denotes positive (negative) anomalies, with the zero contours omitted. Contour intervals (CI) are shown at the upper right corner. From *Ruiz-Barradas et al.* [2000].

gests that the annual cycle in the equatorial Pacific is initiated in the east by the seasonal monsoonal winds and propagates westward as the result of air–sea interactions. In contrast to the equatorial Pacific where air–sea interaction is the leading mechanism for the annual cycle, the narrow width of the tropical Atlantic and the presence of strong continental convective zones mean that continental monsoons play a much more important role. Atmospheric general circulation model (GCM) experiments show that the seasonal variations in the cross-equatorial winds in the Gulf of Guinea are mostly due to the continental monsoon [*Li and Philander*, 1997]. The annual cycle in equatorial zonal wind is driven both by the continental monsoon and by the interaction with equatorial SST, mechanisms that are important in the eastern and western half of the basin, respectively [*Okumura and Xie*, 2004]. In an experiment that removes the seasonal development of the equatorial cold tongue, anomalous easterlies still appear in May and June in the eastern equatorial Atlantic as a result of the increased cross-equatorial advection of zonal momentum and a redistribution of monsoonal rainfall.

3. EQUATORIAL VARIABILITY

Superimposed on these primarily annual variations of SST are anomalies during the boreal summer months (JJA) that frequently exceed 1°C during the peak month. The warm anomalies are generally maximum in the zone of the boreal summer cold tongue between 6°S and 2°N, and between 20°W and 5°E (corresponding cool anomalies are less geographically oriented and less limited to the boreal summer). During some years, but not all, the warm anomalies appear along the southwestern coast of Africa as well. The period of the warm events appears to be approximately 30 months with 13 such warm events having occurred in the 40-year period since 1961 ('63, '66, '68, '73, '74, '81, '84, '87, '88, '93, '96, '97, '99). This past summer of 2003 provides just the most recent example.

The first well-documented event in 1963 received attention partly because of its magnitude, and partly because it coincided with the EQUALANT observational program [*Katz et al.*, 1977; *Merle*, 1980]. The coincidence of warming sea surface temperatures, a relaxation of the trade winds and shifts in convection during that summer caused *Merle* [1980] and *Hisard* [1980] to dub this phenomenon the 'Atlantic Niño'. Further observational results by *Servain et al.* [1982] made clear the connection between changes in the trade winds and changes in SST.

The 1984 event occurred during another observational program called SEQUAL/FOCAL (summarized in the 1984 SEQUAL/FOCAL issue of *Geophysical Research Letters*) and just after the massive 1982–83 Niño. The extensive array of subsurface observations showed that the warming of the mixed layer occurred in conjunction with an anomalous deepening of the oceanic thermocline in the eastern basin [*Philander*, 1986], which resulted from an eastward shift of anomalous heat within the equatorial waveguide [*Carton and Huang*, 1994].

Associated with the warming SSTs are changes in the overlying atmosphere. The equatorial trade winds relax west of 20°W while further eastward the meridional winds associated with the North African summer monsoon also weaken [*Horel et al.*, 1986; *Zebiak*, 1993]. Figure 3 shows the anomaly pattern during Atlantic Niños based on a recent joint ocean–atmospheric analysis [*Ruiz-Barradas et al.*, 2000]. Corresponding increases in diabatic heating in the mid-troposphere occur along with a southward shift of tropical convection [*Wagner and da Silva*, 1994; *Carton et al.*, 1996]. In particular, rainfall tends to increase on the Guinea coast during an Atlantic Niño [*Hirst and Hastenrath*, 1983]. These anomalous shifts in tropical convection and equatorial winds are well captured by a number of atmospheric GCMs that are forced by Atlantic Niño SST anomalies [*Chang et al.*, 2000; *Sutton et al.*, 2000; *Okumura and Xie*, 2004], confirming that they result from air–sea interaction much like their El Niño counterparts in the Pacific [*Zebiak* 1993]. This success is not shared by all the GCMs, however, which led *Vizy and Cook* (2002) to explore the use of a regional atmospheric model to simulate the atmospheric response to Atlantic Niño SST anomalies.

The periodicity of the Atlantic Niño seems to vary considerably. The decade beginning in 1974 had few warm events relative to the surrounding decades of the 1960s and 1980s. The reasons for these changes are still poorly understood. Key parameters such as the heat content of the tropical thermocline have only recently come to be measured regularly, while theoretical attention seems to be focusing on changes in the rates of subduction within the tropical thermocline.

As in the case of the eastern Pacific, eastward surges of warm water have important consequences along the southeastern boundary. During 'normal' austral winters the intensification of the North African monsoon as well as the tilting of the equatorial thermocline induces upwelling of cool nutrient-rich water along the coast. The result is a highly productive commercial fishery [*Crawford et al.*, 1990; *Boyd et al., 1992*]. Relaxation of the equatorial trade winds and the meridional winds of the North African monsoon causes a southward surge of warm, saline tropical water at least as far south as Namibia, raising sea level at Walvis Bay (23°S) by an observable 5 cm [*Brundrit*, 1995]. During the years of the Benguela Niño, the length of the upwelling season may be reduced by a factor of two [*Hagen et al.*, 2001].

4. OFF-EQUATORIAL VARIABILITY

4.1. Empirical Studies

Early interest in the relationship between tropical rainfall and SST anomalies was motivated by observational studies of rainfall fluctuations in Northeast Brazil [*Markham and McLain*, 1977; *Hastenrath and Heller*, 1977; *Moura and Shukla*, 1981]. The Nordeste, a part of Brazil dependent on agriculture, has a strongly seasonal cycle in which much of the annual rainfall occurs in the months of March through May when the ITCZ is at its southernmost position. The great drought of 1958 forced 10 million people to emigrate from the Nordeste [*Namias*, 1972]. By matching wet and dry years in the Nordeste with patterns of SST, these studies found that drought associated with an anomalous northward shift of the ITCZ occurred in conjunction with an anomalous northward gradient of SST, an association often referred to as the Atlantic dipole. A second intense example of this circumstance occurred in 1993 [*Rao et al.*, 1995] (this dipole pattern maximizes cross-equatorial gradient but does not necessarily imply strong correlations between its centers of action). Somewhat weaker relationships have also been identified between the northward gradient of SST and rainfall anomalies in West Africa [*Folland et al.*, 1986; *Lough*, 1986; *Hastenrath*, 1990; *Lamb and Pepper*, 1992]. Nordeste rainfall is also influenced by Pacific El Niño, as in 1958. Section 6 discusses such El Niño effects in more detail.

Many observational studies that followed can be roughly divided into those limited to examining oceanic variables and those looking for covariability between the atmosphere and ocean. Early principal component analyses of SST variability that followed [*Weare*, 1977; *Servain*, 1991] seemed to confirm the presence of a pattern of variability in SST that was geographically stationary, with decadal time-scales. However, it was pointed out by *Houghton and Tourre* [1992] and confirmed by *Mehta* [1998] that when the assumption of spatial orthogonality of the principal components is dropped the Northern and Southern Hemispheres appear to act independently. However in observational studies in which both atmospheric and oceanic variables were included such as *Nobre and Shukla* [1996], *Chang et al.* [1997] and *Ruiz-Barradas et al.* [2000], the results again indicated the presence of a stable pattern of variability across the equator.

The pattern identified by *Ruiz-Barradas et al.* [2000] and presented in Figure 4 appears in five variables, anomalous wind stress components, diabatic heating, SST and thermocline heat content. The pattern is most pronounced in spring when it is the primary principal component. The SST pattern is most pronounced in the Northern Hemisphere and is accompanied by meridional wind anomalies along the equator heading down the pressure gradient and thus into the warmer hemisphere. Away from the equator the pattern of anomalous wind stress corresponds to an increase in surface winds in the cool hemisphere and a decrease in the warm hemisphere. A dipole pattern of diabatic heating is in its positive phase, reflecting enhanced convection, in the warm hemisphere, also

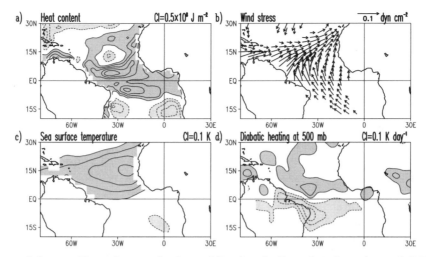

Figure 4. Same as Figure 3 except for the meridional mode. From *Ruiz-Barradas et al.* [2000].

associated with anomalous deepening of the mixed layer. Heat content anomalies seem to follow the thermocline across the equator leading to hemispheric symmetry [see *Ruiz-Barradas et al.*, 2000, their Figure 7]. The nodal line of SST anomalies is displaced north of the equator, roughly coinciding with the mean ITCZ. There are considerable anomalies of wind and thermocline depth on the equator associated with this meridional mode. *Servain et al.* (1999) found significant correlation between the meridional and equatorial modes in certain frequency bands.

4.2. Ocean Response

While observational studies disagree on how to characterize TAV in terms of empirical modes, they agree on the following points: i) the meridional position of the Atlantic ITCZ is sensitive to the anomalous cross-equatorial SST gradient (CESG), especially in February–April when the ITCZ is at its southernmost position and the climatological CESG is weak; ii) a meridional dipole configuration of SST anomalies, although it rarely occurs, maximizes the anomalous CESG; iii) off-equatorial SST anomalies are associated with changes in the strength of the easterly trades on either side of the equator/ITCZ.

The third point is addressed by *Carton et al.* [1996] who present several experiments to examine the relative importance of the mechanical effects of wind stress, surface heating, and internal dynamics in controlling the model CESG. They find that wind-induced changes in surface turbulence heat flux are the dominant mechanism for off-equatorial SST variability. When the effect of wind variability on surface latent heat flux is artificially suppressed, the model CESG variability is substantially reduced (Figure 5). By contrast, when *Carton et al.* remove interannual variability in wind stress but not in wind effect on latent heat flux, the model reproduces CESG variability despite a marked reduction in variability in ocean dynamic fields such as the thermocline depth. These results show that surface heat flux is the leading order process there, in contrast to the equatorial region where ocean dynamics are important [e.g. *Carton and Huang* 1994]. Subsequent calculations using ocean mixed layer models [*Xie and Tanimoto*, 1998; *Czaja et al.*, 2002; *Kushnir et al.*, 2002a] and a different ocean GCM [*Seager et al.* 2001] confirm the major role of wind-induced evaporation in off-equatorial SST variability, a result consistent with the ocean mixed layer heat budget analysis based on observations [*Wagner*, 1996].

The situation in the South Atlantic is less clear. *Hakkinen and Mo* [2002] suggest that ocean circulation changes may be important for southern tropical Atlantic SST variability. But lack of observations may also be responsible for the apparent weakening of the trade wind–SST relation there. *Tanimoto and Xie* [2002] point out that south of 10°S, anomalies of surface wind velocity and sea level pressure based on historical ship observations are often not even in geostrophic balance indicating that the data coverage may be insufficient to draw meaningful conclusions there. *Wu et al.* [this volume] offer further evidence for the effect of anomalous vertical heat advection on southern tropical Atlantic SST variability.

All the above model studies are based on coarse-resolution simulations that do not resolve mesoscale ocean eddies. *Jochum et al.* [this volume] suggest that these eddies due to hydrodynamic instabilities of equatorial currents be a significant source of interannual variability in SST, especially in boreal spring when the ITCZ is sensitive to SST anomalies.

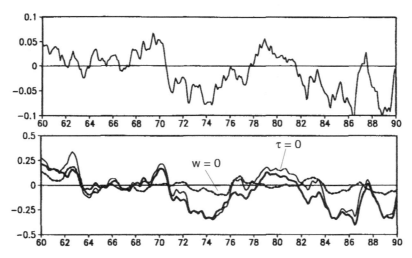

Figure 5. Time series of 12-month smoothed interhemispheric differences in surface wind stress (upper panel in 10^{-1} N m^{-2}) and SST (lower in °C). In the lower panel, results from the full simulation are shown as a thick solid line; from runs removing wind variability in latent heat and momentum fluxes are in dashed and thin solid lines, respectively. From Carton et al. [1996].

4.3. Air–Sea Feedback

Chang et al. [1997] combined the wind-induced evaporation mechanism (point iii) with the direct CESG–atmospheric pressure gradient mechanism of driving cross-equatorial winds (point i) to provide an air–sea interaction scenario for Atlantic CESG variability. They hypothesize that the trade wind anomalies such as those in Figure 4 are forced by SST anomalies with a strong CESG, an assumption supported to various degrees by Moura and Shukla [1981] and subsequent atmospheric GCM studies (the atmospheric response to off-equatorial SST anomalies is a complex issue by its own and will be discussed in detail in Subsection 4.5). A positive anomalous CESG sets up an anomalous southward pressure gradient in the atmospheric boundary layer [e.g., Lindzen and Nigam, 1987], inducing southerly cross-equatorial winds that decelerate the easterly trades north of the equator because of the Coriolis effect. Chang et al. [1997] suggest that these weakened trades north of the equator reduce surface evaporation, thereby acting to strengthen the initial CESG. South of the equator, the southeasterly trades accelerate, increasing surface evaporative cooling and the northward CESG. This positive thermodynamic wind-evaporation-SST (WES) feedback was originally proposed to explain the northward displacement of the climatological ITCZ over the eastern Pacific and Atlantic [Xie and Philander, 1994].

Besides WES, there seem to be additional feedback mechanisms that act between SST and clouds. In a composite analysis of historical ship observations based on a CESG index, a quadrupole banded structure emerges from the cloudiness field [Tanimoto and Xie, 2002]. Near the equator, a dipole of cloudiness anomaly appears in association with the shift of convective clouds in the ITCZ, acting as a negative feedback onto SST as more clouds form over the warmer side of the SST dipole. In the subtropics, more low-level clouds form over negative SST anomalies, reducing net radiation into the oceanic mixed layer causing more cooling and more clouds, etc. South of the equator, the positive feedback resulting from the negative SST–low cloud correlation results in roughly a 10% increase in cloud cover or a 20 W m^{-2} reduction in incoming solar radiation at the surface for each 1°C increase in SST. This SST–low cloud feedback mechanism is significantly weaker in the northern tropical Atlantic.

While the mechanisms underlying air–sea interaction have emerged only recently, their potential for maintaining CESG anomalies was recognized much earlier. For example, Hastenrath and Greischar [1993] state "the SST pattern—itself affected by the surface wind field—...is conducive to a steeper meridional pressure gradient, which in turn favors a stronger southerly wind component." Our expectation is that improved understanding of the physical mechanisms underlying air–sea interaction will lead to improved physically based numerical models, which may then have benefit for prediction systems.

4.4. Free Mode Analysis

Linear stability analysis of the coupled equatorial ocean–atmosphere system has yielded useful insights into the

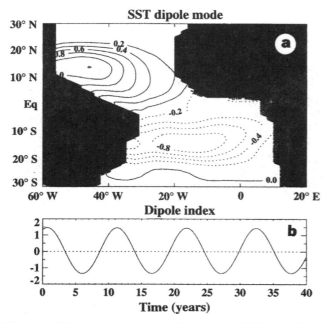

Figure 6. SST regression (upper panel) against a CESG index (lower panel) in *Chang et al.*'s [1997] intermediate coupled model.

dynamics of ENSO [*Neelin et al.*, 1998]. By including the effect of wind-induced evaporation on SST, *Zhou and Carton* [1998] and *Xie et al.* [1999] extend such stability analyses with a more sophisticated surface heat flux formulation including latent heat loss. Two types of coupled modes emerge from the latter extended analysis: one arising from the *Bjerknes* [1969] feedback involving interaction of the thermocline depth, upwelling, SST and zonal winds along the equator; and one due to the thermodynamic WES feedback involving air-sea interaction in the meridional direction. The zonal and meridional modes differ not only in spatial structure—the former with maximum amplitude at the equator while the latter off the equator—but also in the growth rate dependence on zonal wavenumber. The zonal mode favors a zonal wavelength about the size of the Pacific, the meridional mode grows fastest at zonal wavenumber zero, a property consistent with the fact that off-equatorial anomalies of Atlantic SST and wind are nearly zonally uniform in phase. In *Xie et al.*'s [1999] calculations, the growth rate of the equatorial mode at the size of the Atlantic basin is comparable to that of the zonally uniform meridional mode, being 0.6 and 0.8 year^{-1}, respectively, in the absence of SST damping. (The equatorial mode's growth rate is 1.5 year^{-1} at the Pacific basin size.) The wind-induced evaporation anomalies are about 10–20 Wm^{-2} in amplitude in the deep tropics and thus the WES feedback is only weakly positive. In fact, the net feedback from the latent heat flux may even be negative when the dependence of SST on surface evaporation is considered.

Chang et al. [1997] were the first to demonstrate the role of WES in CESG variability in a coupled model. Their atmospheric model is empirically constructed based on a singular value decomposition analysis of surface momentum/heat fluxes and SST over the tropical Atlantic. The coupling of this atmospheric model with either an intermediate ocean model of *Zebiak and Cane* [1987] or an ocean GCM yields oscillations on decadal timescales, in which SST anomalies are organized to maximize CESG with opposite polarities on either side of the equator (Figure 6). Chang et al. show that the growth rate of this meridional mode is highly sensitive to the coupling of surface heat flux with SST but not so to the coupling with momentum flux, a result consistent with *Carton et al.*'s [1996] ocean GCM experiments. In the Chang et al. model the CESG feedback is limited by cross-equatorial advection by the northward flowing North Brazil Current, which helps switch phase of the coupled oscillation, thus setting the timescale of reversal.

The SST advection by surface Ekman flow acts to dampen the growth caused by the positive WES feedback, an effect evident in *Xie*'s [1999] energy equation analysis and in a recent coupled model study of *Kushnir et al.* [2002a]. *Seager et al.*'s [2001] ocean GCM calculations also show that the advection by the mean ocean currents acts as a damping on CESG variability.

One important feature of the observed SST "dipole" pattern is the nodal line's displacement north of the equator along the mean ITCZ. *Okajima et al.* [2003] explore the importance of this asymmetry with an atmospheric GCM coupled to a Zebiak–Cane-type ocean model modified to allow the thermocline depth to vary in space but not in time. This modification suppresses the thermocline feedback and virtually eliminates the ENSO mode in the system, allowing a close look into air-sea interaction in the meridional direction. In a run of this hybrid coupled GCM with a perfectly symmetric land–sea distribution, SST variability in the tropics organizes itself into a distinct dipole pattern with its nodal line on the equator (Figure 7a). The corresponding SST–wind speed relationship is consistent with positive WES feedback. When Okajima et al. perturb the shape of continents to force the mean position of the ITCZ into the Northern Hemisphere, the line of minimum SST variance shifts northward as well (Figure 7b). Furthermore, SST variability becomes much less coherent across the mean ITCZ/SST nodal line with the SST correlation decreasing from 0.7 when the mean ITCZ is symmetric about the equator to 0.2 when the mean ITCZ is displaced a realistic distance off the equator.

The WES feedback owes its positive sign to the sign change in the Coriolis parameter across the equator. *Okajima et al.* [2003] suggest that the departure of the climatic equator from the geographical equator weakens the WES feedback and

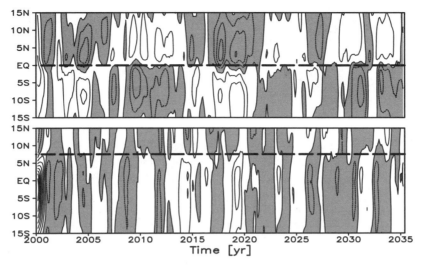

Figure 7. Time-latitude sections of zonal-mean SST anomaly (contours at 0.3°C interval, with negative values shaded) in a coupled model. The thick dashed line indicates the latitude of the climatological ITCZ, which is symmetric about and displaced north of the equator in the upper and lower panels, respectively. From *Okajima et al.* [2003].

reduces the coherence between variability north and south of the ITCZ. This impact of the shift of the mean ITCZ may explain the lack of significant interhemispheric coherence in TAV [*Houghton and Tourre*, 1992; *Enfield et al.*, 1999; *Mehta*, 1998] and why CESG variability and associated atmospheric anomalies are strongest in March–April, the time when the climatological ITCZ is nearly symmetric about the equator.

4.5. Atmospheric Response

In the extratropics, a negative correlation between anomalies of SST and wind speed is now recognized as evidence for atmospheric forcing of the ocean mixed layer rather than the other way around [*Frankignoul*, 1985; *Kushnir et al.*, 2002b]. In the tropical Atlantic, however, the shift in dynamics toward a direct atmospheric response to SST suggests that the observed negative SST–wind speed correlation may support the hypothesis of a positive WES feedback that organizes SST and wind anomalies into a dipole pattern that maximizes CESG, as discussed in the previous two subsections.

To prove this hypothesis of the existence of a coupled meridional mode, it still needs to be shown that the atmosphere responds to CESG anomalies (rather than causes them) and that the sign of the response is such that it leads to positive WES feedback. The basic physics of the atmospheric response was originally examined by *Moura and Shukla* [1981] who explored the response of an atmospheric GCM to an imposed CESG. They report a decrease in sea level pressure over the hemisphere with positive SST and an anomalous shift of the ITCZ in the direction of the imposed CESG. Recent studies conduct an ensemble of multi-decade hindcasts forced by observed SST and use a signal-to-noise maximizing EOF technique to extract SST-forced signals [*Venzke et al.*, 1999; *Chang et al.*, 2000; *Sutton et al.*, 2000; *Terray and Cassou*, 2002]. Some other studies impose time-invariant SST anomalies and integrate models for a long period of time to increase the sample size and thereby reduce the noise due to atmospheric internal variability [*Dommenget and Latif*, 2000; *Okumura et al.*, 2001; *Sutton et al.*, 2001; *Terray and Cassou*, 2002]. The difficulty of the current atmospheric GCMs in reproducing the WES feedback is discussed in *Wang and Carton* [2003].

Nearly all the GCMs agree in their response in the deep tropics within 10° latitude [*Chang et al.*, 2000; *Sutton et al.*, 2000; *Okumura et al.*, 2001; *Terray and Cassou*, 2002]. In response to an SST dipole and the associated changes in CESG, these models generate cross-equatorial winds directed from the colder to the warmer hemisphere. The resultant low-level convergence causes the Atlantic ITCZ to move toward the warmer side of the SST dipole. The change in precipitation is not limited to the ocean but extends considerably inland over South America, probably as a result of the westward propagation of baroclinic Rossby waves forced over the ocean. The presence of substantial diabatic heating over the Amazon basin indicates a possible role for interactions with land surface processes.

In the deep tropics, the atmospheric response is baroclinic with wind anomalies being out of phase in the lower and upper troposphere. The easterly trade winds tend to weaken on the equatorward side of positive SST anomalies and strengthen on the equatorward side of negative SST anomalies. This trade wind response supports the WES feedback as envisioned by

Chang et al. [1997]. While Chang et al. [2000] find the significant response confined to 10°S–10°N in their model, other models show a broader response in latitude. Based on a single multi-decadal integration, results of Robertson et al. [2000] and Watanabe and Kimoto [1999] hint at a response of the NAO to tropical Atlantic SST anomalies. Model studies with large ensemble members/long integrations seem to support this extratropical response [Okumura et al., 2001; Sutton et al., 2001; Terray and Cassou, 2002]. In response to a positive CESG, a barotropic low develops in the mid-latitude North Atlantic centered around 45°N, in addition to a baroclinic response in the deep tropics [Okumura et al., 2001]. This barotropic response is strongest in boreal winter and spring [Venzke et al., 1999; Okumura et al., 2001; Sutton et al., 2001; Terray and Cassou, 2002] and allows the relaxed trades and hence the positive WES feedback to cover the entire northern tropical Atlantic.

Intermediate baroclinic models of the atmosphere are very useful in studying ENSO over the tropical Pacific but they are much less successful in reproducing the surface wind response to tropical Atlantic SST variations, especially in the subtropics [Chiang et al., 2001; Chung et al., 2002]. One likely explanation is because the barotropic response in the subtropical/midlatitude Atlantic modulates the trade winds. The exact mechanism for this barotropic response needs further study. One example is barotropic Rossby wave excitation by upper-tropospheric convergence/divergence associated with the anomalous shift of the ITCZ. Much as in the extratropical response to ENSO, the North Atlantic storm track varies in such a way as to reinforce barotropic stationary eddies [Watanabe and Kimoto, 1999; Okumura et al., 2001]. Another important factor may be diabatic heating over the surrounding continents that alters the zonal pressure gradient and preferentially affects the zonal component of the near-equatorial winds.

A related line of research is the use of atmospheric GCMs to investigate the relationship between the NAO and a pattern of anomalous SST in the form of a tripole (the tropical extension of which represents the CESG), which are the dominant modes of the atmosphere and ocean over the North Atlantic, respectively. Several studies show that the observed NAO time series can be reproduced, albeit at reduced amplitudes, in atmospheric GCMs forced by observed SST [Venzke et al., 1999; Rodwell et al., 1999; Mehta et al., 2000; Peng et al., 2002; Lin and Derome, 2003], a result that Bretherton and Battisti [2000] suggest is consistent with the null hypothesis of atmospheric stochastic forcing of the ocean. It is also likely, as GCM experiments of Sutton et al. [2001] and Terray and Cassou [2002] show, that the tropical part of the SST tripole is what forces the NAO-like response in the extratropics, perhaps by shifting the ITCZ. Indeed, Watanabe and Kimoto [1999] and Okumura et al. [2001] show that the extratropical part of the SST tripole can be reproduced in an ocean mixed layer coupled with an atmospheric GCM that is forced by a tropical SST dipole. To the extent that this tropical forcing scenario holds and that the tropical Atlantic anomalies result from local air–sea interaction, a certain degree of predictability may be achieved for the NAO–tripole pair. In ensemble atmospheric GCM experiments, percentage of SLP variance due to imposed SST variability generally decreases poleward, being 30–60% in the subtropics and 10–30% in the mid-latitudes [Kushnir et al., 2002b; Rodwell, 2003].

Figure 8 shows the regressions of observed SST, surface wind velocity and net surface heat flux upon a northern tropical Atlantic SST index for March–May [see also Kushnir et al., 2002a]. Large anomalies of trade winds take place during January–March, preceding the large SST anomalies during March–May. Czaja et al. [2002] attribute these trade wind anomalies in January–March exclusively to external forcing but the atmospheric GCM studies mentioned above suggest that these wind anomalies may partly result from the mid-latitude barotropic response to a northward shift of the ITCZ. A quantitative estimate of the importance of tropical SST forcing is difficult to make from observations. Czaja et al. [2002] note that wind anomalies in the deep tropics are likely a response to the northern tropical Atlantic SST anomalies, which appear beginning in January and persist into boreal summer. Despite an SST regression pattern confined north of the equator, the wind regression extends well into the Southern Hemisphere and is consistent with the phase relationship expected from WES feedback. Such a cross-equatorial response is seen in atmospheric GCM results when SST anomalies are imposed only on one side of the equator [Sutton et al., 2001].

5. INTERACTION WITH THE EXTRATROPICS

In the tropics SST variations are important in shaping the spatial pattern of variations in convection and hence other atmospheric fields. In the extratropics, on the other hand, the atmospheric dynamics organize low-frequency variability into large-scale patterns, even without feedback from the ocean. The NAO is such a preferred pattern that dominates the month-to-month atmospheric variability over the North Atlantic [Hurrell et al., 2003]. Atmospheric GCM simulations with climatological SST as the surface boundary condition confirm that the NAO is a dominant mode of atmospheric internal variability but that in the absence of SST variability its spectrum is likely white in time. The observed NAO shows enhanced power at decadal timescales, which may result from air–sea interaction within the extratropical North Atlantic [Marshall et al., 2001] or from teleconnections excited by

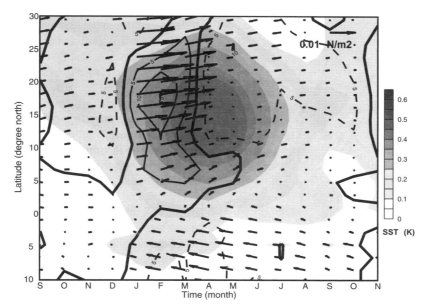

Figure 8. Regression map of surface wind stress (arrows, scale in the top-right corner); net surface heat flux (contoured every 5 W m^{-2}, positive into the ocean, dashed when negative, zero contour thickened); and SST (shaded, in K) onto the NTA SST index time series in MAM. From *Czaja et al.* [2002].

anomalous shifts in the Atlantic ITCZ [*Okumura et al.*, 2001; *Sutton et al.*, 2001; *Terry and Cassou*, 2002], or both.

5.1. A Pan-Atlantic Pattern

The NAO is correlated with the SST tripole over the North Atlantic in boreal winter/spring, with the tropical/subtropical lobe of the latter centered at 10–20°N. *Rajagopalan et al.* [1998] find that the NAO is also correlated with both southern tropical Atlantic SST and CESG variability at decadal timescales. In fact, *Xie and Tanimoto*'s [1998] composite analysis based on a CESG index reveals a pan-Atlantic pattern, with bands of SST anomalies of alternating signs that span from the South Atlantic to Greenland (Figure 9). This so-called pan-Atlantic decadal oscillation pattern features anomalous wind/sea level pressure (SLP) patterns similar to the NAO and a SST tripole over the North Atlantic [*Tanimoto and Xie*, 1999], and is captured in a joint analysis of SST and SLP over the whole Atlantic basin using a frequency domain method [*Tourre et al.*, 1999].

There is some evidence that the above pan-Atlantic pattern favors decadal timescales. *Enfield et al.* [1999] show that SST variability in the northern and southern tropical Atlantic displays marginally significant coherence with anti-symmetric phase in the 8–12 year band in the boreal winter–spring. *Chu* (1984) reports a spectral peak in the frequency band of 12.7–14.9 years in northeast Brazil rainfall. *Mehta* [1998] notes a similar decadal peak in this regional

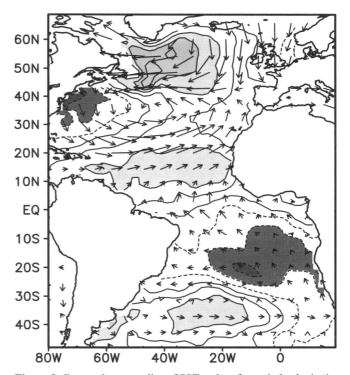

Figure 9. Composite anomalies of SST and surface wind velocity in boreal winter based on a pan-Atlantic decadal oscillation index of *Tanimoto and Xie* [2002]. Global Sea Ice and SST (GISST) dataset [*Parker et al.*, 1994] and National Centers for Environmental Prediction (NCEP) Reanalysis [*Kalnay et al.*, 1996] are used for SST and surface wind, respectively.

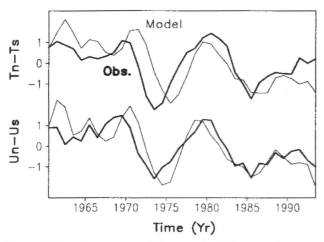

Figure 10. Interhemispheric differences in SST and surface zonal wind velocity in COADS observations (thick) and simulated by a coupled model (thin). All the time series are normalized by their respective standard deviations. From *Xie and Tanimoto* [1998].

rainfall variability as well as in Atlantic CESG. Such a decadal (12–13 years) peak is found in an 825-year long sediment core in the Cariaco Basin north of the Venezuelan coast, which *Black et al.* [1999] interpret as resulting from variability in the northern tropical Atlantic trades. *Tourre et al.*'s [1999] joint SST–SLP pattern is associated with a spectral peak centered at a period of 11.4 years. When the empirical orthogonal function analysis is performed separately for SST and SLP and over the separate northern and southern tropical Atlantic domains, the four independently obtained leading principal components are highly correlated at decadal timescales [*Tanimoto and Xie*, 2002]. Instrumental records, however, are too short to test further this hypothesis for a pan-Atlantic decadal oscillation.

5.2. Subtropical High as the Bridge

This statistical relationship between the CESG and NAO suggests an interaction between them. In one direction, as atmospheric GCM studies suggest, CESG variability affects the NAO through its effect on the Atlantic ITCZ and upper-tropospheric divergence (see Subsection 4.5). *Xie and Tanimoto* [1998], *Chang et al.* [2000], *Czaja et al.* [2002], and *Kushnir et al.* [2002a] investigate this interaction from the other direction by asking how the NAO affects the CESG? The subtropical/mid-latitude center of action of the NAO in SLP is located at 40°N, driving changes in the strength of the northeasterly trades to the south and hence affecting northern tropical Atlantic SST. Assuming that the direct NAO influence on SST is strongest in the subtropics, *Xie and Tanimoto* [1998] apply an external forcing that is random in time and confined poleward of 20° latitude and show that the WES feedback organizes the tropical response into a coherent dipole structure that favors low-frequency (interannual) variability [see also *Kushnir et al.*, 2002a]. In particular, when the observed winds are used as the subtropical forcing, the coupled model reproduces the observed CESG evolution quite well despite the fact that the tropics are free of external forcing (Figure 10). Thus, air–sea feedback may act to transfer the impact of subtropical anomalies like those associated with the NAO into the deep tropics.

Chang et al. [2001] derive an atmospheric noise field empirically based on a 145-year atmospheric GCM run that is forced by the monthly SST climatology. The dominant noise pattern resembles the NAO and features wind anomalies with large amplitudes in the subtropics that decay rapidly toward the deep tropics (Figure 11, left). They force a hybrid coupled ocean GCM with this noise field with a white spectrum in time. In the absence of air–sea feedback, SST anomalies are confined to the subtropics with little response near the equator. When moderate coupling is allowed in the model, regions of maximum SST variability shift toward the equator to 10–20o in latitude, accompanied by large CESG variations reddened at interannual and lower frequencies. Associated with the change in SST spatial structure, the trade winds on both sides of the equator show large variations in the deep tropics, with stronger cross-equatorial coherence than in the uncoupled run. This result supports the notion that NAO' influence on CESG is rather indirect and requires the bridging effect of air–sea interaction to reach the deep tropics.

Namias [1972] reports a correlation between the North Atlantic subtropical high and northeast Brazil rainfall in boreal winter and spring. He notes "the northeast trades are often regulated by the Atlantic subtropical anticyclone" and "these variations alter the intensity of convergence into the ITCZ and perhaps change its position". *Czaja et al.* [2002] construct a simple model for northern tropical Atlantic SST and show that subtropical SLP variations—the forcing in the model—explain most of the observed SST variability. This seems to suggest that air–sea interaction within the tropical Atlantic is secondary, but as *Xie and Tanimoto* [1998] and *Chang et al.* [2001] show, the influence radius of subtropical forcing may be a function of local air–sea feedback; without the feedback, its effect may well be confined and not extend into the deep tropics to affect CESG and ITCZ.

The NAO and the SST tripole emerge as the leading mode from joint ocean–atmospheric analyses of coupled GCM simulations [*Grötzner et al.*, 1998; *Delworth and Mehta*, 1998]. However, different studies disagree on how far the NAO influence can penetrate toward the south. *Delworth and Mehta* [1998] report that it is limited to north of the equator but a

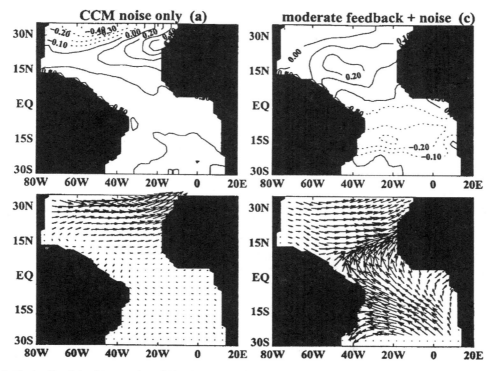

Figure 11. The leading joint SVD modes of SST (upper panels), surface wind stress (lower), and heat flux in coupled model runs forced by atmospheric noise, without (left panels) and with (right) feedbacks onto the atmosphere. From *Chang et al.* [2001], with panels for heat flux omitted.

pan-Atlantic pattern emerges from *Watanabe et al.*'s [1999] simulation, with an SST dipole in the tropics.

The interaction of the TAV with the extratropics is much less well studied in the Southern than in the Northern Hemisphere. Based on a singular value decomposition analysis, *Venegas et al.* [1997] show that there is a meridional SST dipole pattern in the South Atlantic and that it is associated with variations in subtropical SLP, a co-variation pattern similar to that over the North Atlantic. They note that the SST–SLP pattern is most pronounced in the southern summer, a result that they suggest is indicative of "possible links with major climatic oscillations observed in the Northern Hemisphere". CESG/ITCZ variability and air–sea interaction in the tropical Atlantic may well be the mechanism for such interhemispheric links. In fact, the South Atlantic SST dipole of *Venegas et al.* [1997] is part of the pan-Atlantic pattern (Figure 9) as documented by *Tanimoto and Xie* [1999; 2002]. *Barreiro et al* [this volume] investigate further the influence of South Atlantic extratropical variability on TAV.

5.3. Oceanic Pathways

So far, we have examined the link between TAV and the extratropics via the atmosphere. There are also oceanic pathways that link the subtropical with the equatorial Atlantic, via so-called subtropical cells [*Schott et al.*, this volume] . These pathways carry water subducted in the subtropics during winter into the equatorial upwelling zones. In the Atlantic, these subtropical cells are highly asymmetric about the equator because of the deep meridional overturning circulation [MOC; *Jochum and Malanotte-Rizzoli*, 2001] and are sensitive to changes in wind stress [*Inui et al.*, 2002].

Changes in the deepwater formation in the high-latitude North Atlantic can induce changes in cross-equatorial ocean heat transport, which *Yang* [1999] suggests give rise to a dipole SST pattern in the tropical Atlantic. Using a coupled GCM, *Dong and Sutton* [2002] show that this MOC-induced SST dipole amplifies in the tropical Atlantic by interacting with the atmosphere. In particular, in response to a sudden weakening of the MOC, a SST dipole develops in year 4–6, with a strong cooling over the northern and a weak warming over the southern tropical Atlantic. SLP increases over the region of sea surface cooling and decreases over warming. The resulting anomalous CESG causes the Atlantic ITCZ to shift southward, triggering further changes over the tropical Pacific in their model. *Dong and Sutton* [2002] suggest that the effect of changes in the high-latitude North Atlantic and the MOC can be felt quickly through the globe via such atmospheric feedback in the tropical Atlantic, and that this process occurs in years instead of the hundreds of years one would expect if only ocean processes were involved.

Analysis of paleo-proxies shows a strong correlation between the position of the Atlantic ITCZ and Greenland climate conditions [*Peterson et al.*, 2000]. *Chiang et al.* [2003] suggest that the pan-Atlantic pattern discussed earlier in this section is a useful model, with the interaction and feedback between CESG and the position and strength of the ITCZ as a possible mechanism for this link between the high-latitude and tropical North Atlantic. Using an atmospheric GCM coupled with a slab ocean mixed layer, Chiang et al. show that continental ice sheets present during the last glacier maximum could trigger tropical air–sea interaction, by altering atmospheric stationary wave patterns, giving rise to pan Atlantic scale changes with a large anomalous CESG that is coupled with the oceanic ITCZ. Chiang et al. conclude that their model prefers the meridional mode in the tropics in response to various surface forcing terms during the last glacier maximum. Since their ocean model is one-dimensional and does not include any dynamics, they suggest that the atmospheric response to high-latitude changes in sea and land ice is an alternative means of triggering changes in the tropical air–sea system, besides the MOC mechanism of *Yang* [1999] and *Dong and Sutton* [2002]. In both the studies of *Dong and Sutton* [2002] and *Chiang et al.* [2003] the WES feedback seems to be a key to communicating the high-latitude changes to the deep tropics, leading to changes in CESG and ITCZ.

6. ENSO INFLUENCE

It has been known for some time that a basin-wide warming takes place in the tropical Atlantic a few months after the El Niño in the Pacific peaks in December–January. During and immediately following an El Niño event, precipitation generally decreases over the equatorial Atlantic. This section reviews studies of ENSO influence in the tropical Atlantic.

The Atlantic response to ENSO shows strong seasonality because both ENSO and its influence on the Pacific North American (PNA) teleconnection are seasonally phase-locked. The Atlantic response to La Nina is similar in spatial pattern to that to El Niño, albeit with anomalies reversing signs. For this reason, the following discussion describes the response to El Niño.

6.1. SST

In December and January when El Niño peaks in the Pacific, SLP in the subtropical and mid-latitude North Atlantic drops while increasing in the equatorial Atlantic [*Covey and Hastenrath*, 1978; *Aceituno*, 1988; *Giannini et al.*, 2000; *Mestas-Nunes and Enfield*, 2001; *Alexander and Scott*, 2002]. The resultant anomalous pressure gradient drives anomalous southwesterlies over the tropical North Atlantic north of 10°N. These anomalous winds are particularly strong in the western half of the basin, acting to weaken the prevailing northeast trades on the background and hence surface latent and sensible heat flux [*Aceituno*, 1988; *Curtis and Hastenrath*, 1995; *Lanzante*, 1996; *Enfield and Mayer*, 1997; *Klein et al.*, 1999]. This reduced heat release from the ocean gives rise to a delayed warming of the ocean mixed layer that peaks in April–June in a zonal band between 20°N and the latitude of the climatological ITCZ (Figure 12). The decrease in surface evaporation prior to this tropical North Atlantic warming is captured in *Klein et al.*'s [1999] calculations based on ship observations. In addition, *Klein et al.* [1999] report a modest reduction in cloud cover south of 20°N that further contributes to the ocean warming.

Based on atmospheric GCM simulations, *Saravanan and Chang* [2000] suggest that in addition to wind-induced evaporation variations, changes in air–sea difference in surface temperature and humidity are also important for the sea surface warming in the tropical North Atlantic in boreal spring following an El Niño event. *Chikamoto* [2002] confirm this air–sea temperature/humidity difference effect by performing a heat flux analysis based on historical ship reports. They show that much of the decrease in turbulent heat flux over the tropical northwestern Atlantic is due to an increase in air–sea temperature difference. Normally, SST anomalies are slightly higher than surface air temperature anomalies, but in this region and in the boreal winter–spring following an El Niño, anomalies of air temperature are larger and lead those of SST (Figure 13), thereby suppressing surface heat release from the ocean.

ENSO-induced tropical North Atlantic warming induces further air–sea interaction within the tropical Atlantic. In April–June when this warming is at its maximum, significant southeasterly wind anomalies form in a region between the latitude of the mean ITCZ and 10–15°S, apparently in response to the decrease in SLP over the band of positive SST anomalies to the north (Figure 12). These anomalous southeasterlies are in the general direction of the climatological background winds and induce negative SST anomalies south of the equator, through the dependence of evaporation on wind speed. This cooling increases the northward SST gradient and hence the anomalous southeasterly cross-equatorial winds, implying a positive WES feedback discussed earlier. *Enfield and Mayer* [1997] discuss this tendency for the tropical Atlantic to develop a cross-equatorial SST gradient in the boreal summer following an El Niño [see also *Chiang et al.*, 2002]. The SST correlation with ENSO is generally weaker in the South than in the North Atlantic. The abovementioned cooling south

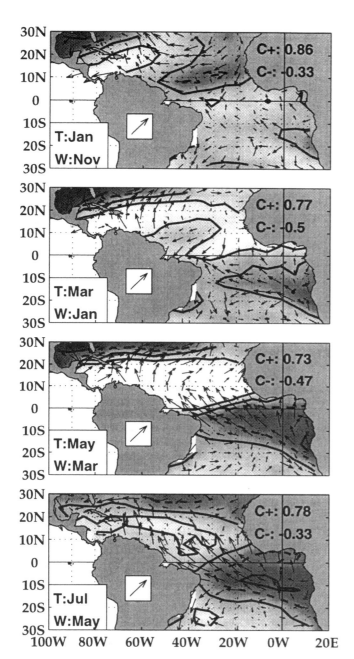

Figure 12. Sequence of development of SST (T) and surface wind (W) correlations in the tropical Atlantic in response to ENSO. The zero contours are dashed, and negative values are dark shaded. Contour intervals are 0.25 (95% significance = 0.25). From *Enfield and Mayer* [1997].

of the equator shows a correlation just above 0.25 (vs. well above 0.5 in the tropical North Atlantic). A positive correlation of around 0.5 develops in the subtropical South Atlantic southwest of the above cooling region [*Enfield and Mayer*, 1997; *Klein et al.*, 1999].

Thus, local air–sea interaction is apparently involved in the Atlantic response to Pacific variability, but the feedback is not strong enough to sustain the cross-equatorial SST gradient anomalies through the boreal summer. In contrast to this ENSO influence on cross-equatorial SST variability, the correlation between the Atlantic Niño and ENSO is generally insignificant [*Zebiak*, 1993] despite the presence of significant southeasterly wind anomalies near the equator (Figure 12; see also *Latif and Barnett* [1995]) that by themselves tend to induce a cold event in the equatorial Atlantic by increasing equatorial upwelling and shoaling the thermocline in the east. (The 1984 Atlantic Niño is one exception taking place following the major El Niño in 1982–83.) Between the anomalous southwesterlies north of 10°N and southeasterlies near the equator, there is a significant band of negative wind curl during January–March that excites downwelling Rossby waves. The opposing effects of these Rossby waves and anomalous equatorial easterlies may be responsible for the lack of correlation between equatorial Pacific and Atlantic SST.

6.2. Precipitation

During an El Niño event, atmospheric convection intensifies in the central and eastern equatorial Pacific. The increased convective heating warms the Pacific troposphere. These positive tropospheric temperature anomalies created in the Pacific are rapidly spread along the equatorial belt via equatorial wave adjustment and occupy the global tropics [*Yulaeva and Wallace*, 1994; *Chiang and Sobel*, 2002; *Su et al.*, this volume]. Outside the tropical Pacific, this tropical tropospheric warming is associated by the anomalous descending motion as part of the anomalous Walker circulation associated with ENSO. This anomalous subsidence, along with the increased static stability associated with the tropospheric warming over the global tropical belt, suppresses atmospheric convection and reduces precipitation over the equatorial Atlantic. During March–May as the tropical North Atlantic warms up, the Atlantic ITCZ shows a tendency to shift anomalously northward, with a dipole in the precipitation anomaly field. This precipitation dipole is not limited to the oceanic sector but extends into the South American continent as well, with a large decrease in rainfall over the Brazil's Nordeste region and a modest increase over the continent north of the equator. By comparing two atmospheric hindcasts with SST forcing prescribed over the global tropics and the tropical Atlantic, respectively, *Saravanan and Chang* [2000] show that the rainfall reduction over the equatorial Atlantic is the direct response to ENSO (via anomalous downdraft and tropospheric warming) while the northward shift of the ITCZ is an indirect response forced by Atlantic SST anomalies (notably the tropical North Atlantic warming and the attendant cross-equatorial SST gradient).

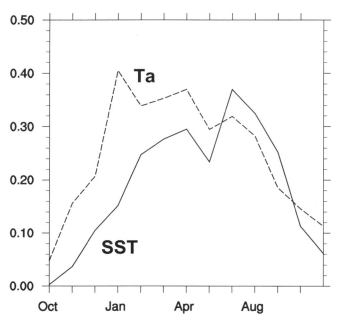

Figure 13. Composite anomalies of SST and surface air temperature averaged in the Caribbean Sea (80°W-60°W, 10°N-20°N), associated with an El Niño in the Pacific. GISST and NCEP reanalysis for 1948-99 are used. Courtesy of Y. Chikamoto and Y. Tanimoto.

Chiang et al. [2002] confirm this sequence of rainfall response based on observational analysis.

March–May rainfall over the Caribbean Sea increases following El Niño, while rainfall response over the land surrounding the Caribbean is also affected by orography [*Giannini et al.*, 2000; *Taylor et al.*, 2002].

6.3. Teleconnection Mechanism

Besides the adjustment through an anomalous Walker circulation, the PNA teleconnection is an additional mechanism by which ENSO affects the North Atlantic in boreal winter and early spring [*Nobre and Shukla*, 1996; *Klein et al.*, 1999; *Lau and Nath*, 2001]. In particular, the center of action over Florida associated with the barotropic PNA pattern contributes to the lowering of SLP there, and to the weakening of the northeasterly trades and warming of the ocean mixed layer in the tropical North Atlantic. Based on an ensemble hindcast with an atmospheric GCM that is forced by tropical Pacific SST variations and coupled with a slab ocean mixed layer model, *Lau and Nath* [2001] show that the influence of ENSO is not limited to the tropical Atlantic but is also significant in the extratropical North Atlantic, a result that further supports the idea that the PNA mechanism plays an important role in the Atlantic response to ENSO [*Alexander et al.*, 2002].

Whereas to first order ENSO is symmetric about the equator in the Pacific, the response of the Atlantic is quite equatorially asymmetric, with the strongest anomalies of SST and rainfall observed in the northern tropics. The PNA mechanism accounts partly for this asymmetry in the Atlantic response. Recently, *Chiang and Sobel* [2002] suggest that the tropical tropospheric warming associated with ENSO—an effect of the anomalous Walker circulation mechanism—is effectively communicated to the ocean surface through vertical mixing by moist convection. Thus, the resultant ocean mixed layer warming is confined to the convective regions such as the northward-displaced oceanic ITCZ (see Figure 13).

6.4. Coupled Model Studies

Using a coupled GCM, *Huang et al.* [2002] carry out an experiment in which only the tropical Atlantic between 30°S–30°N is coupled with the atmosphere and the observed SST history is prescribed elsewhere. Huang et al. report that the prescribed ENSO exerts a strong influence on the tropical North Atlantic and explains up to 50% of the variance in their model. Consistent with *Enfield and Mayer* [1997], the model tropical North Atlantic warming subsequently induces southerly cross-equatorial winds, which interact further with the ocean, leading to SST anomalies south of the equator. In *Huang et al.*'s [2002] model ENSO's influence is weak on and south of the equator, where most of SST variability is due to air–sea interaction local to the tropical Atlantic.

In an independent study with a different coupled GCM, *Wu and Liu* [2002] confirm the importance of local air–sea interaction and in particular the WES feedback in TAV. In a so-called partially coupled experiment in which the active feedback onto the atmosphere is removed over the northern tropical Atlantic, SST variability is reduced more than half compared with a control. This leads Wu and Liu to suggest that the tropical North Atlantic is not just passively responding to external forcing such as ENSO and NAO but positive feedback arising from air–sea interaction is necessary to produce the right level of variability there [see also *Wu et al.*, this volume].

7. SUMMARY AND DISCUSSION

The seasonal cycle is by far the largest source of climate variability in the tropical Atlantic. The seasonal cycle consists of a north–south anti-symmetric annual component that is forced directly by the seasonal march of the sun, and a north–south symmetric component with a maximum on or slightly south of the equator. This air–sea interaction component is triggered by the onset of the West African monsoon causing rapid equatorial cooling in May and June, which is further amplified by air–sea interaction along the equator through a mechanism similar to that *Bjerknes* [1969] envisioned for ENSO.

On interannual and longer timescales, no single mode seems to dominate. Instead, several mechanisms are responsible for tropical Atlantic variability. On the equator, both observational and modeling studies indicate that there is a Bjerknes-type air–sea coupled mode arising from the interaction of the equatorial zonal SST gradient, ITCZ convection, zonal wind, and thermocline depth. The resulting positive feedback here in the Atlantic is weaker than the corresponding feedback in the Pacific probably because of the smaller zonal width of the Atlantic basin. Thus, the anomalous warming on the equator, which occurs every few years, is modest in amplitude and lasts only for a few months in boreal summer. This warming is generally associated with an increase in rainfall along the coasts of Guinea and Angola.

In addition to an equatorial mode, observational and modeling studies generally support the notion that interannual variability in the cross-equatorial SST gradient and the position and strength of the oceanic ITCZ are coupled and that this coupling results to some degree from their mutual interaction. The interaction involves positive WES feedback between anomalous trades, wind-induced changes in surface evaporation, and SST anomalies. The coupled ITCZ/CESG variability affects rainfall over the surrounding continents, in particular over northeastern Brazil and to a lesser extent over the Sahel.

In addition to local interactions the tropical Atlantic is also subject to strong external forcing. ENSO warming in the equatorial Pacific reduces the northeasterly trades and gives rise to a delayed warming in the northern tropical Atlantic through the PNA teleconnection and subsidence associated with an anomalous Walker circulation. The NAO also modulates the strength of the northeast trades and hence SST in the subtropical North Atlantic. Such external forcing of the northeast trades explains a large percentage of observed SST variability in the northern tropical Atlantic, which subsequently triggers the ITCZ/CESG interaction in the deep tropics and induces changes on and across the equator. All these TAV mechanisms are highly seasonal: ENSO and NAO forcing is strongest in boreal winter; the ITCZ/CESG interaction in March–May when the equator is uniformly warm; and the equatorial mode is most pronounced in the boreal summer coinciding with the season of the cold tongue and the shallow thermocline in the east.

While the ITCZ/CESG interaction almost certainly exists, many uncertainties remain. It is unclear, for example, how far the ITCZ/CESG interaction extends toward the poles. There is observational evidence for a positive SST–low cloud feedback in the subtropics, indicative of an atmospheric reaction in the planetary boundary layer to SST anomalies. Some modeling studies suggest that the CESG-induced shift of the ITCZ and the resultant shift of upper-tropospheric divergence force a barotropic response that modulates the strength of the North Atlantic subtropical high and the northeast trades. This suggests that ITCZ/CESG in the deep tropics might interact with the subtropical Atlantic, a mechanism that may give rise in turn to the observed pan-Atlantic pattern of anomalies of SST, SLP, and surface wind. Interestingly, this pan-Atlantic pattern has been used to explain a link between the tropical and high-latitude North Atlantic observed in paleoclimate records.

There is a paradox in TAV research. While theoretical studies indicate that the WES feedback favors a dipole mode antisymmetric about the mean position of the ITCZ, observed SST variability is not significantly correlated across this latitude. This paradox may be reconciled by considering the departure of the position of the climatological ITCZ from the geographical equator, which acts to reduce interhemispheric coherence of WES-induced variability. Other mechanisms for reducing this correlation include interference with other modes of variability like the Atlantic Niño, and disruption by external forcing that is generally not projected optimally onto the meridional mode.

Ocean–atmospheric interaction and feedback, when they exist, offer hope for useful predictability, a subject of *Saravanan and Chang*'s chapter in this volume. Indeed, *Hastenrath and Greischar* [1993] and *Folland et al.* [2001] show that northeast Brazil rainfall in boreal spring is quite predictable in their empirical models using SST in the tropical Pacific and Atlantic as predictors. Recent predictability studies using dynamical models support this conclusion and show improved hindcast skills if they are initialized with SSTs in the eastern equatorial Pacific and tropical Atlantic [*Chang et al.*, 1998; *Penland and Matrosova*, 1998; *Chang et al.*, 2003]. Initial SST anomalies in these regions allow inclusion of ENSO teleconnection and the ITCZ/CESG interaction within the tropical Atlantic, respectively. This result is consistent with diagnostic/modeling studies showing the importance of both ENSO forcing and local air–sea feedback.

A further prerequisite for successful dynamic prediction is the use of an air–sea–land coupled model that is unbiased. Unfortunately, strong biases persist in nearly all current climate models in the tropical Atlantic sector. Chief among these biases are the failure to keep the mean ITCZ north of the equator and to maintain the equatorial cold tongue. In most models, the ITCZ moves back and forth across the equator following the sun, and stays far too long south of the equator. The modeled zonal SST gradient on the equator is opposite to observations, with higher SSTs in the Gulf of Guinea than east of South America. Peculiarly, this reversal of SST gradient occurs despite prevailing easterly winds on the equator in some models [*Davey et al.*, 2002]. The seasonal northward-displacement of the ITCZ and corresponding development of the equatorial cold tongue are features of the seasonal climate that are necessary to the development, structure, and timing of the interannual/decadal TAV, as has been discussed in this review. It is thus a high priority to reduce and remove these biases in climate models.

The tropical Atlantic is a small ocean basin flanked by major continents that host major convection centers of the global atmosphere. Continents exert a strong influence on the annual-mean state and seasonal cycle of the tropical Atlantic, as exemplified by the northward displacement of the climatological ITCZ and the annual cycle in equatorial SST and zonal wind. We also know that interannual variability of the tropical Atlantic exerts a significant effect on the rainfall over both South America and Africa. Unclear is what role the continents play in TAV. Questions that remain to be explored include whether variability on continents provides any feedback to the TAV and to what extent internal variability of the continental monsoon can affect TAV. An accurate representation of the interaction of ocean, atmosphere, and land is imperative for a realistic simulation of the mean state and variability of the tropical Atlantic Ocean.

Acknowledgments. SPX is supported by the NOAA CLIVAR Atlantic Program and Frontier Research System for Global Change. JAC is supported by the NOAA CLIVAR Atlantic Program. We thank Y. Okumura and J. Hafner for graphics, Y. Tanimoto and Y. Chikamoto for providing Figures 9 and 13. This is IPRC contribution #267 and SOEST contribution #6360.

REFERENCES

Aceituno, P., On the functioning of the Southern Oscillation in the South American sector: Part I: Surface climate, *Mon. Weather Rev.*, 97, 505–524, 1988.

Alexander, M. A., and J. D. Scott, The influence of ENSO on air–sea interaction in the Atlantic, *Geophys. Res. Lett.*, 29(14), 1701, doi:10.1029/2001GL014347, 2002.

Alexander, M. A., I. Bladé, M. Newman, J. R. Lanzante, N.-C. Lau, and J. D. Scott, The atmospheric bridge: The influence of ENSO teleconnections on air–sea interaction over the global oceans, *J. Clim.*, 15, 2205–2231, 2002.

Barreiro, M., A. Giannini, P. Chang, and R. Saravanan, On the role of the Southern Hemisphere atmospheric circulation in tropical Atlantic variability, this volume.

Biasutti, M., D. S. Battisti, and E. S. Sarachik, The annual cycle over the tropical Atlantic South America, and Africa, *J. Clim.*, 16, 2491–2508, 2003.

Bjerknes, J., Atmospheric teleconnections from the equatorial Pacific, *Mon. Weather Rev.*, 97, 163–172, 1969.

Black, D. E., L. C. Peterson, J. T. Overpeck, A. Kaplan, M. N. Evans, and M. Kashgarian, Eight centuries of North Atlantic Ocean atmosphere variability, *Science*, 286, 1709–1713, 1999.

Boyd, A. J., J. Taunton-Clark, and G. P. J. Oberholster, Spatial features of the near-surface and midwater circulation patterns off western and southern South Africa and their role in the life histories of various commercially fished species, *S. Afr. J. Mar. Sci.*, 12, 189–206, 1992.

Bretherton, C. S., and D. S. Battisti, An interpretation of the results from atmospheric general circulation models forced by the time history of the observed sea surface temperature distribution, *Geophys. Res. Lett.*, 27, 767–770, 2000.

Brundrit, G. B., Trends of southern African sea level: Statistical analysis and interpretation, *S. Afr. J. Mar. Sci.*, 16, 9–17, 1995.

Carton, J. A., and B. Huang, Warm events in the tropical Atlantic, *J. Phys. Oceanogr.*, 24, 888–903, 1994.

Carton, J. A., X. Cao, B. S. Giese, and A. M. da Silva, Decadal and interannual SST variability in the tropical Atlantic Ocean, *J. Phys. Oceanogr.*, 26, 1165–1175, 1996.

Chang, P., L. Ji, and H. Li, A decadal climate variation in the tropical Atlantic ocean from thermodynamic air–sea interactions, *Nature*, 385, 516–518, 1997.

Chang, P., L. Ji, H. Li, C. Penland, and L. Matrosova, Prediction of tropical Atlantic sea surface temperature, *Geophys. Res. Lett.*, 25, 1193–1196, 1998.

Chang, P., R. Saravanan, L. Ji, and G. C. Hegerl, The effects of local sea surface temperatures on atmospheric circulation over the tropical Atlantic sector, *J. Clim.*, 13, 2195–2216, 2000.

Chang, P., L. Ji, and R. Saravanan, A hybrid coupled model study of tropical Atlantic variability, *J. Clim.*, 14, 361–390, 2001.

Chang, P., R. Saravanan, and L. Ji, Tropical Atlantic seasonal predictability: The roles of El Niño remote influence and thermodynamic air–sea feedback, *Geophys. Res. Lett.*, 30(10), 1501, doi:10.1029/2002GL016119, 2003.

Chiang, J. C. H., and A. H. Sobel, Tropical tropospheric temperature variations caused by ENSO and their influence on the remote tropical climate, *J. Clim.*, 15, 2616–2631, 2002.

Chiang, J. C. H., S. E. Zebiak, and M. A. Cane, Relative roles of elevated heating and sea surface temperature gradients in driving anomalous surface winds over tropical oceans, *J. Atmos. Sci.*, 58, 1371–1394, 2001.

Chiang, J. C. H., Y. Kushnir, and A. Giannini, Deconstructing Atlantic ITCZ variability: influence of the local cross-equatorial SST gradient, and remote forcing from the eastern equatorial Pacific, *J. Geophys. Res.*, 107(D1), 4004, doi:10.1029/2000JD000307, 2002.

Chiang, J. C. H., M. Biasutti, and D. S. Battisti, Sensitivity of the Atlantic ITCZ to conditions during Last Glacial Maximum, *Paleoceanograophy*, 18(4), 1094, doi:10.1029/2003PA000916, 2003

Chikamoto, Y., *Tropical Atlantic Ocean–Atmospheric Response to Tropical Pacific SST Variations*, Master's Thesis, pp. 90, Hokkaido University, Sapporo, Japan, 2002.

Chu, P.-S., Time and space variability of rainfall and surface circulation in the northeast Brazil–tropical Atlantic sector, *J. Meteorol. Soc. Jpn.*, 26, 363–369, 1984.

Chung, C., S. Nigam, and J. A. Carton, SST-forced surface wind variability in the tropical Atlantic: An empirical model, *J. Geophys. Res.*, *107*(D15), 4244, doi:10.1029/2001JD000324, 2002.

Covey, D. L., and S. Hastenrath, The Pacific El Niño phenomenon and the Atlantic circulation, *Mon. Weather Rev.*, *106*, 1280–1287, 1978.

Crawford, R. J. M., W. R. Siegfried, L. V. Shannon, C. A. Villacastin-Herroro, and L. G. Underhill, Environmental influences on marine biota off southern Africa, *S. Afr. J. Sci.*, *86*, 330–339, 1990.

Curtis, S., and S. Hastenrath, Forcing of anomalous sea surface temperature evolution in the tropical Atlantic during Pacific warm events. *J. Geophys. Res.*, *100*, 15,835–15,847, 1995.

Czaja, A., P. van der Vaart, and J. Marshall, A diagnostic study of the role of remote forcing in tropical Atlantic variability, *J. Clim.*, *15*, 3280–3290, 2002.

Davey, M. K., et al., STOIC: a study of coupled model climatology and variability in tropical ocean regions, *Clim. Dyn.*, *18*, 403–420, 2002.

Delworth, T. L., and V. M. Mehta, Simulated interannual to decadal variability in the tropical and sub-tropical North Atlantic, *Geophys. Res. Lett.* *25*, 2825–2828, 1998.

Dommenget, D., and M. Latif, Interannual to decadal variability in the tropical Atlantic, *J. Clim.*, *13*, 777–792, 2000.

Dong, B.-W., and R. T. Sutton, Adjustment of the coupled ocean–atmosphere system to a sudden change in the thermohaline circulation, *Geophys. Res. Lett.*, *29*(15), 1728 doi:10.1029/2002GL015229, 2002.

Enfield, D. B., and D. A. Mayer, Tropical Atlantic SST variability and its relation to El Niño–Southern Oscillation, *J. Geophys. Res.*, *102*, 929–945, 1997.

Enfield, D. B., and A. M. Mestas-Nuez, Multiscale variabilities in global sea surface temperatures and their relationships with tropospheric climate patterns, *J. Clim.*, *12*, 2719–2733, 1999.

Enfield, D. B., A. M. Mestas-Nunez, D. A. Mayer, and L. Cid-Serrano, How ubiquitous is the dipole relationship in tropical Atlantic sea surface temperatures?, *J. Geophys. Res.* *104*, 7841–7848, 1999.

Folland, C. K., T. N. Palmer, and D. E. Parker, Sahel rainfall and world-wide sea temperatures, *Nature*, *320*, 602–607, 1986.

Folland, C. K., A. W. Colman, D. P. Powell, and M. K. Davey, Predictability of northeast Brazil rainfall and real-time forecast skill, 1987–98. *J. Clim.*, *14*, 1937–1958, 2001.

Frankignoul, C., Sea surface temperature anomalies, planetary waves, and air–sea feedback in the middle latitudes, *Rev. Geophys.*, *23*, 357–390, 1985.

Fu, R., R. E. Dickinson, M. Chen, and H. Wang, How do tropical sea surface temperatures influence the seasonal distribution of precipitation in the equatorial Amazon?, *J. Clim.*, *14*, 4003–4026, 2001.

Giannini, A., Y. Kushnir, and M. A. Cane, Interannual variability of Caribbean rainfall, ENSO, and the Atlantic Ocean, *J. Clim.*, *13*, 297–311, 2000.

Giese, B. S., and J. A. Carton, The seasonal cycle in a coupled ocean–atmosphere model, *J. Clim.*, *7*, 1208–1217, 1994.

Gray, W. M., and C. W. Landsea, African rainfall as a precursor of hurricane-related destruction on the U.S. east coast, *Bull. Am. Meteorol. Soc.*, *75*, 1352–1364, 1992.

Grodsky, S. A., and J. A. Carton, Coupled land/atmosphere interactions in the West African monsoon, *Geophys. Res. Letts.*, *28*, 1503–1506, 2001.

Grodsky S. A., and J. A. Carton, Intertropical Convergence Zone in the South Atlantic and the equatorial cold tongue, *J. Clim.*, *16*, 723–733, 2003.

Grötzner, A., M. Latif, and T. P. Barnett, A decadal climate cycle in the North Atlantic Ocean as simulated by the ECHO coupled GCM, *J. Clim.*, *11*, 831–847, 1998.

Hagen, E., R. Feistel, J. J. Agenbag, and T. Ohde, Seasonal and interannual changes in intense Benguela upwelling (1982–1999), *Oceanol. Acta*, 24, 557–567, 2001.

Hakkinen, S., and K.C. Mo, The low-frequency variability of the tropical Atlantic Ocean, *J. Clim.*, *15*, 237–250, 2002.

Halley, E., A historical account of the trade winds, and monsoons, observable in the seas between and near the Tropicks, with an attempt to assign the phisical cause of the said winds, *Philos. Trans. R. Soc. London*, *16*, 153–168, 1686.

Hastenrath, S., Interannual variability and annual cycle: Mechanisms of circulation and climate in the tropical Atlantic sector, *Mon. Weather Rev.*, *112*, 1097–1107, 1984.

Hastenrath, S., Decadal-scale changes of the circulation in the tropical Atlantic sector associated with Sahel drought, *Int. J. Climatol.*, *10*, 459–472, 1990.

Hastenrath, S., *Climate dynamics of the tropics*, 488 pp., Kluwer Academic, Boston, 1991.

Hastenrath, S., and L. Greischar, Further work on the prediction of northeast Brazil rainfall anomalies, *J. Clim.*, *6*, 743–758, 1993.

Hastenrath, S., and L. Heller, Dynamics of climate hazards in Northeast Brazil, *Q. J. R. Meteorol. Soc.*, *103*, 77–92, 1977.

Hirst, A., and S. Hastenrath, Atmosphere–ocean mechanisms of climate anomalies in the Angola–tropical Atlantic sector, *J. Phys. Oceanogr.*, *13*, 1146–1157, 1983.

Hisard, P., Observation de response du type "El Niño" dans l'Atlantique tropical oriental—Golfe de Guinee, *Oceanol. Acta*, *3*, 69–78, 1980.

Horel, J. D., V. E. Kousky, and M. T. Kagano, Atmospheric conditions in the Atlantic sector during 1983 and 1984, *Nature*, *310*, 248–251, 1986.

Houghton, R. W., Seasonal variations of the subsurface thermal structure in the Gulf of Guinea, *J. Phys. Oceanogr.*, *13*, 2070–2081, 1983.

Houghton, R. W., The relationship of sea surface temperature to thermocline depth at annual and interannual time scales in the tropical Atlantic Ocean, *J. Geophys. Res.*, *96*, 15173–15185, 1991.

Houghton, R. W., and Y. M. Tourre, Characteristics of low-frequency sea surface temperature fluctuations in the tropical Atlantic, *J. Clim.*, *5*, 765–771, 1992.

Huang, B., P. S. Schopf, and Z. Pan, The ENSO effect on the tropical Atlantic variability: A regionally coupled model study, *Geophys. Res. Lett.*, *29*(21), 2039, doi:10.1029/2002GL 014872, 2002.

Hurrell, J. W., Y. Kushnir, G. Ottersen, and M. Visbeck (Eds.), *The North Atlantic Oscillation: Climatic Significance and Environmental Impact*, 279 pp., American Geophysical Union, Washington, D.C., 2003.

Inui, T., A. Lazar, P. Malanotte-Rizzoli, and A. Busalacchi, Wind Stress Effects on Subsurface Pathways from the Subtropical to Tropical Atlantic, *J. Phys. Oceanogr.*, *32*, 2257–2276, 2002.

Janicot S., and B. Sultan, Intra-seasonal modulation of convection in the West African monsoon, *Geophys. Res. Letts.*, *28*, 523–526, 2001.

Jochum, M., and P. Malanotte-Rizzoli, Influence of the meridional overturning circulation on tropical–subtropical pathways, *J. Phys. Oceanogr.*, *31*, 1313–1323, 2001.

Jochum, M., R. Murtugudde, P. Malanotte-Rizzoli, and A. J. Busalacchi, Intrinsic variability of the tropical Atlantic Ocean, this volume.

Kalnay, E., et al., The NCEP/NCAR 40-year reanalysis project, *Bull. Am. Meteorol. Soc.*, *77*, 437–471, 1996.

Klein, S. A., B. J. Soden, and N.-C. Lau, Remote sea surface temperature variations during ENSO: Evidence for a tropical atmospheric bridge, *J. Clim.*, *12*, 917–932, 1999.

Kushnir, Y, R. Seager, J. Miller, and J. C. H. Chiang, A simple coupled model of tropical Atlantic decadal climate variability, *Geophys. Res. Lett.* *29*(23), 2133, doi:10.1029/2002GL015874, 2002a.

Kushnir, Y, W. A. Robinson, I. Bladé, N. M. J. Hall, S. Peng, and R. Sutton, Atmospheric GCM response to extratropical SST anomalies: Synthesis and evaluation, *J. Clim.*, *15*, 2233–2256, 2002b.

Lamb, P. J., and R. A. Peppler, Further case studies of tropical Atlantic surface atmospheric and oceanic patterns associated with sub-Saharan drought, *J. Clim.*, *5*, 476–488, 1992.

Lanzante, J. R., Lag relationships involving tropical sea surface temperatures, *J. Clim.*, *9*, 2568–2578, 1996.

Latif, M., and T. P. Barnett, Interactions of the tropical oceans. *J. Clim.*, *8*, 952–964, 1995.

Lau, N.-C., and M. J. Nath, Impact of ENSO on SST variability in the North Pacific and North Atlantic: Seasonal dependence and role of extratropical sea–air coupling, *J. Clim.*, *14*, 2846–2866, 2001.

Li, T., and S. G. H. Philander, On the seasonal cycle of the equatorial Atlantic Ocean, *J. Clim.*, *10*, 813–817, 1997.

Lin, H., and J. Derome, The atmospheric response to North Atlantic SST anomalies in seasonal prediction experiments, *Tellus*, *55A*, 193–207, 2003.

Lindzen, R. S., and S. Nigam, On the role of sea surface temperature gradients in forcing low-level winds and convergence in the tropics, *J. Atmos. Sci.*, *44*, 2418–2436, 1987.

Lough, J. M., Tropical Atlantic sea surface temperature and rainfall variations in subsaharan Africa, *Mon. Weather Rev.*, *114*, 561–570, 1986.

Luksch, U., Simulation of North Atlantic low-frequency SST variability, *J. Clim.*, *9*, 2083–2092, 1996.

Marengo, J. A., and S. Hastenrath, Case studies of extreme climatic events in the Amazon basin, *J. Clim.*, *6*, 617–627, 1993.

Markham, C. G., and D. R. McLain, Sea surface temperature related to rain in Ceara, Northeastern Brazil, *Nature*, *265*, 320–323, 1977.

Marshall, J., et al., North Atlantic climate variability: Phenomena, impacts and mechanisms, *Int. J. Climatol.*, *21*, 1863–1898, 2001.

Mehta, V. M. Variability of the tropical ocean surface temperatures at decadal–multidecadal time scales. Part I: The Atlantic Ocean, *J. Clim.*, 11, 2351–2375 1998.

Mehta, V. M., M. J. Suarez, J. Manganello, and T. L. Delworth, Oceanic influence on the North Atlantic Oscillation and associated Northern Hemisphere climate variations: 1959–1993, *Geophys. Res. Lett.*, *27*, 121–124, 2000.

Merle, J., Annual and interannual variability of temperature in the eastern equatorial Atlantic Ocean – hypothesis of an Atlantic El Niño, *Oceanol. Acta*, *3*, 209–220, 1980.

Mestas-Nunez, A. M., and D. B. Enfield, Eastern equatorial Pacific SST variability: ENSO and non-ENSO components and their climatic associations, *J. Clim.*, *14*, 391–402, 2001.

Mitchell, T. P., and J. M. Wallace, The annual cycle in equatorial convection and sea surface temperature, *J. Clim.*, *5*, 1140–1156, 1992.

Moura, A., and J. Shukla, On the dynamics of droughts in northeast Brazil: Observations, theory and numerical experiments with a general circulation model, *J. Atmos. Sci.*, *38*, 2653–2675, 1981.

Namias, J., Influence of northern hemisphere general circulation on drought in Northeast Brazil, *Tellus*, *24*, 336–42, 1972.

Neelin, J. D., D. S. Battisti, A. C. Hirst, F.-F. Jin, Y. Wakata, T. Yamagata, and S. E. Zebiak, ENSO theory, *J. Geophys. Res.*, *103*, 14,261–14,290, 1998.

Nobre, P., and J. Shukla, Variations of sea surface temperature, wind stress, and rainfall over the tropical Atlantic and South America, *J. Clim.*, *9*, 2464–2479, 1996.

Okajima, H., S.-P. Xie, and A. Numaguti, Interhemispheric coherence of tropical climate variability: Effect of climatological ITCZ, *J. Meteorol. Soc. Jpn.*, *81*, 1371–1386, 2003.

Okumura, Y. and S.-P. Xie, Interaction of the Atlantic equatorial cold tongue and African monsoon, *J. Clim.*, *17*, in press, 2004.

Okumura, Y., S.-P. Xie, A. Numaguti, and Y. Tanimoto, Tropical Atlantic air–sea interaction and its influence on the NAO, *Geophys. Res. Lett.*, *28*, 1507–1510, 2001.

Parker, D. E., P. D. Jones, A. Bevan, and C. K. Folland, Interdecadal changes of surface temperature since the late 19th century, *J. Geophys. Res.*, *99*, 14373–14399, 1994.

Peng, S., W. A. Robinson, and S. Li, North Atlantic SST forcing of the NAO and relationships with intrinsic hemispheric variability. *Geophys. Res. Lett.*, *29*(8), 1276, doi:10.1029/2001 GL014043, 2002.

Penland, C., and L. Matrosova, Prediction of tropical Atlantic sea surface temperatures using linear inverse modeling, *J. Clim.*, *11*, 483–496, 1998.

Peterson, L. C., G. H. Huag, K. A. Hughen, and U. Rohl, Rapid changes in the hydrologic cycle of the tropical Atlantic during the last glacial, *Science*, *290*, 1947–1951, 2000.

Philander, S. G. H., Unusual conditions in the tropical Atlantic Ocean in 1984, *Nature*, *222*, 236–238, 1986.

Philander, S. G. H., and R. C. Pacanowski, The oceanic response to cross-equatorial winds (with application to coastal upwelling in low latitudes), *Tellus*, *33*, 201–210, 1981.

Philander, S. G. H., and R. C. Pacanowski, A model of the seasonal cycle in the tropical Atlantic Ocean, *J. Geophys. Res.*, *91*, 14192–14206, 1986.

Rajagopalan, B., Y. Kushnir, and Y. Tourre, Observed decadal midlatitude and tropical Atlantic climate variability, *Geophys. Res. Lett.*, *25*, 3967–3970, 1998.

Rao, V. B., K. Hada, and D. L. Herdies, On the severe drought of 1993 in Northeast Brazil, *Int. J. Climatol.*, *15*, 697–704, 1995.

Robertson, A. W., C. R. Mechoso, and Y.-J. Kim, The influence of Atlantic sea surface temperature anomalies on the North Atlantic Oscillation, *J. Clim.*, *13*, 122–138, 2000.

Rodwell, M. J., On the predictability of North Atlantic climate, in *The North Atlantic Oscillation: Climatic Significance and Environmental Impact*, edited by J. W. Hurrell et al., pp. 173–192, American Geophysical Union, Washington, DC, 2003.

Rodwell, M. J., D. P. Rowell, and C. K. Folland, Oceanic forcing of the wintertime North Atlantic Oscillation and European climate, *Nature*, *398*, 320–323, 1999.

Ruiz-Barradas, A., J. A. Carton, and S. Nigam, Structure of interannual-to-decadal climate variability in the tropical Atlantic sector, *J. Clim.*, *13*, 3285–3297, 2000.

Ruiz-Barradas, A., J. A. Carton, and S. Nigam, Role of the atmosphere in climate variability of the tropical Atlantic, *J. Clim.*, *16*, 2052–2065, 2003.

Saravanan, R., and P. Chang, Interaction between tropical Atlantic variability and El Niño–Southern Oscillation, *J. Clim.*, *13*, 2177–2194, 2000.

Saravanan, R., and P. Chang, Thermodynamic coupling and predictability of tropical sea surface temperature, this volume.

Schott, F. A., J. P. McCreary, and G. C. Johnson, Shallow overturning circulations of the tropical–subtropical oceans, this volume.

Seager, R., Y. Kushnir, P. Chang, N. Naik, J. Miller, and W. Hazeleger, Looking for the role of the ocean in tropical Atlantic decadal climate variability, *J. Clim.*, *14*, 638–655, 2001.

Servain, J., Simple climate indices for the tropical Atlantic Ocean and some applications, *J. Geophys. Res.*, *96*, 15,137–15,146, 1991.

Servain, J., J. Picaut, and J. Merle, Evidence of remote forcing in the equatorial Atlantic Ocean, *J. Phys. Oceanogr.*, *12*, 457–463, 1982.

Servain, J., I. Wainer, J. P. McCreary, and A. Dessier, Relationship between the equatorial and meridional modes of climatic variability in the tropical Atlantic, *Geophys. Res. Lett.*, *26*, 485–488, 1999.

Su, H., J. D. Neelin, and J. E. Meyerson, Tropical tropospheric temperature and precipitation response to sea surface temperature forcing, this volume.

Sutton, R. T., S. P. Jewson, and D. P. Rowell, The elements of climate variability in the tropical Atlantic region, *J. Clim.*, *13*, 3261–3284, 2000.

Sutton, R. T., W. A. Norton, and S. P. Jewson, The North Atlantic Oscillation—What role for the ocean?, *Atmospheric Science Letters*, *1*(2), doi:10.1006/asle.2000.0018, 2001.

Tanimoto, Y., and S.-P. Xie, Ocean–atmosphere variability over the pan-Atlantic basin, *J. Meteorol. Soc. Jpn.*, *77*, 31–46, 1999.

Tanimoto, Y., and S.-P. Xie, Inter-hemispheric decadal variations in SST, surface wind, heat flux and cloud cover over the Atlantic Ocean, *J. Meteorol. Soc. Jpn.*, *80*, 1199–1219, 2002.

Taylor, M. A., D. B. Enfield, and A. A. Chen, Influence of the tropical Atlantic versus the tropical Pacific on Caribbean rainfall, *J. Geophys. Res.*, *107*(C9), 3127 doi:10.1029/2001JC 001097, 2002.

Terray, L., and C. Cassou, Tropical Atlantic sea surface temperature forcing of quasi-decadal climate variability over the North Atlantic–European region, *J. Clim.*, *15*, 3170–3187, 2002.

Thorncroft, C. D., et al., The JET 2000 Project: Aircraft observations of the African easterly jet and African easterly waves, *Bull. Am. Meteorol. Soc.*, *86*, 337–351, 2003.

Tourre, Y. M., B. Rajagopalan, and Y. Kushnir, Dominant patterns of climate variability in the Atlantic Ocean during the last 136 years, *J. Clim.*, *12*, 2285–2299, 1999.

Venegas, S. A., L. A. Mysak, and D. N. Straub, Atmosphere–ocean coupled variability in the South Atlantic, *J. Clim.*, *10*, 2904–2920, 1997.

Venzke, S., M. R. Allen, R. T. Sutton, and D. P. Rowell, The atmospheric response over the North Atlantic to decadal changes in sea surface temperature, *J. Clim.*, *12*, 2562–2584, 1999.

Vizy, E. K., and K. H. Cook, Development and application of a mesoscale climate model for the tropics: Influence of sea surface temperature anomalies on the West African monsoon. *J. Geophys. Res.*, *107*(D3), 4023, doi:10.1029/2001JD000686, 2002.

Wagner, R. G., Mechanisms controlling variability of the interhemispheric sea surface temperature gradient in the tropical Atlantic, *J. Clim.*, *9*, 2010–2019, 1996.

Wagner, R. G., and A. da Silva, Surface conditions associated with anomalous rainfall in the Guinea coastal region, *Int. J. Climatol.*, *14*, 179–200, 1994.

Wang, J., and J.A. Carton, Modeling climate variability in the tropical Atlantic atmosphere, *J. Clim.*, *16*, 3858–3876, 2003.

Watanabe, M., and M. Kimoto, Tropical–extratropical connection in the Atlantic atmosphere–ocean variability, *Geophys. Res. Lett.*, *26*, 2247–2250, 1999.

Watanabe, M., M. Kimoto, T. Nitta, and M. Kachi, A comparison of decadal climate oscillations in the North Atlantic detected in observations and a coupled GCM, *J. Clim.*, *12*, 2920–2940, 1999.

Weare, B. C., Empirical orthogonal analysis of Atlantic Ocean surface temperatures, *Q. J. R. Meteorol. Soc.*, *103*, 467–478, 1977.

Wu, L., and Z. Liu, Is tropical Atlantic variability driven by the North Atlantic Oscillation? *Geophys. Res. Lett.* *29*(13), 1653, doi:10.1029/2002GL014939, 2002.

Wu, L., Q. Zhang, and Z. Liu, Toward understanding the tropical Atlantic variability using coupled modeling surgery, this volume.

Woodruff, S. D., R. J Slutz, R. L. Jenne, and P. M. Steurer, A comprehensive ocean–atmosphere dataset, *Bull. Am. Meteorol. Soc.*, *68*, 521–527, 1987.

Xie, P., and P. A. Arkin, Analyses of global monthly precipitation using gauge observations, satellite estimates, and numerical model predictions, *J. Clim.*, *9*, 840–858, 1996.

Xie, S.-P., On the genesis of the equatorial annual cycle. *J. Clim.*, *7*, 2008–2013, 1994.

Xie, S.-P., A dynamic ocean–atmosphere model of the tropical Atlantic decadal variability, *J. Clim.*, *12*, 64–70, 1999.

Xie, S.-P., The shape of continents, air–sea interaction, and the rising branch of the Hadley circulation, in *The Hadley Circulation: Past, Present and Future*, edited by H. F. Diaz and R. S. Bradley, Kluwer Academic Publishers, Dordrecht, in press, 2004.

Xie, S.-P., and S.G.H. Philander, A coupled ocean–atmosphere model of relevance to the ITCZ in the eastern Pacific, *Tellus*, *46A*, 340–350, 1994.

Xie, S.-P., and Y. Tanimoto, A pan-Atlantic decadal climate oscillation, *Geophys. Res. Lett.*, *25*, 2185–2188, 1998.

Xie, S.-P., Y. Tanimoto, H. Noguchi, and T. Matsuno, How and why climate variability differs between the tropical Pacific and Atlantic, *Geophys. Res. Lett.*, *26*, 1609–1612, 1999.

Yang, J.Y., A linkage between decadal climate variations in the Labrador Sea and the tropical Atlantic Ocean, *Geophys. Res. Lett.*, *26*, 1023–1026, 1999.

Yulaeva, E., and J. M. Wallace, The signature of ENSO in global temperature and precipitation fields derived from the microwave sounding unit, *J. Clim.*, *7*, 1719–1736, 1994.

Zebiak, S. E., air–sea interaction in the equatorial Atlantic region, *J. Clim.*, *6*, 1567–1586, 1993.

Zebiak, S. E., and M. A. Cane, A model El Niño–Southern Oscillation, *Mon. Weather Rev.*, *115*, 2262–2278, 1987.

Zhou, Z., and J. A. Carton, Latent heat flux and interannual variability of the coupled atmosphere–ocean system, *J. Atmos. Sci.*, *55*, 494–501, 1998.

James A. Carton, Department of Meteorology, University of Maryland, College Park, Maryland 20742. (carton@atmos.umd.edu)

Shang-Ping Xie, International Pacific Research Center, SOEST, University of Hawaii, Honolulu, Hawaii 96822. (xie@hawaii.edu)

On the Role of the South Atlantic Atmospheric Circulation in Tropical Atlantic Variability

Marcelo Barreiro,[1,2] Alessandra Giannini,[3] Ping Chang,[1] and R. Saravanan[4]

One dominant manifestation of tropical Atlantic variability (TAV) takes place in March–April–May in the form of a strong inter-hemispheric sea surface temperature gradient coupled to a cross-equatorial near surface atmospheric flow. The variability of this circulation pattern affects the position of the intertropical convergence zone and the regional climate in the surrounding areas. In this study, we investigated the effect of the South Atlantic atmospheric variability on this phenomenon. We found that southern summer atmospheric variability (and to a lesser extent winter variability) can play a pre-conditioning role in the onset of inter-hemispheric anomalies in the deep tropics during the following austral fall. It does so by inducing a sea surface temperature anomaly in the southern tropics that initiates local thermodynamic air–sea feedbacks. This remote influence of the Southern Hemisphere on TAV is contrasted with the remote influence of El Niño-Southern Oscillation (ENSO) and the North Atlantic Oscillation (NAO) during austral summer. The results suggest that to fully understand TAV and its predictability it is necessary to consider not only the remote influences from ENSO and NAO, but also the influence from the South Atlantic atmospheric circulation.

1. INTRODUCTION

The coupled variability of an anomalous inter-hemispheric sea-surface temperature (SST) gradient and a cross-equatorial atmospheric circulation that is associated with a displacement of the intertropical convergence zone (ITCZ) during March-April-May (MAM) has long been recognized as one of the most important components of tropical Atlantic variability (TAV) [e.g., *Hastenrath*, 1985]. This pattern of variability, which we will refer to as the "gradient mode", is known to have a significant impact on rainfall patterns over Northeastern Brazil [e.g., *Hastenrath and Greischar*, 1993]. Understanding and predicting the onset of this phenomenon is one of the central foci of TAV research.

The physical mechanism behind the gradient mode was first outlined by *Hastenrath and Greischar* [1993]: an anomalous cross-equatorial SST gradient generates surface winds from the cooler to the warmer hemisphere through the hydrostatic effect of SST on sea level pressure [*Lindzen and Nigam*, 1987]. The anomalous cross-equatorial winds influence convection by changing the position of moisture convergence. The observed seasonality of this coupling is argued to be a result of the spatially uniform warm climatological SST conditions in austral fall that make the Atlantic ITCZ highly sensitive to small perturbations in the meridional direction [*Chiang et al.*, 2002]. Since the South Atlantic influence is the focus of this work, throughout this paper the seasons refer to those of the Southern Hemisphere (SH), unless explicitly noted.

The genesis and evolution of the gradient mode is shown in Figure 1. This figure is constructed by lag-regressing observed 1000 hPa winds, surface downward heat flux and SST (see section 2) onto an index characterizing the cross-equatorial SST

---

[1]Department of Oceanography, Texas A&M University, College Station, Texas.
[2]Now at Program in Atmospheric and Oceanic Sciences, Princeton University, Princeton, New Jersey.
[3]International Research Institute for Climate Prediction, Columbia University, Palisades, New York.
[4]National Center for Atmospheric Research, Boulder, Colorado.

Earth's Climate: The Ocean-Atmosphere Interaction
Geophysical Monograph Series 147
Copyright 2004 by the American Geophysical Union
10.1029/147GM08

144 ROLE OF SOUTH ATLANTIC VARIABILITY IN TAV

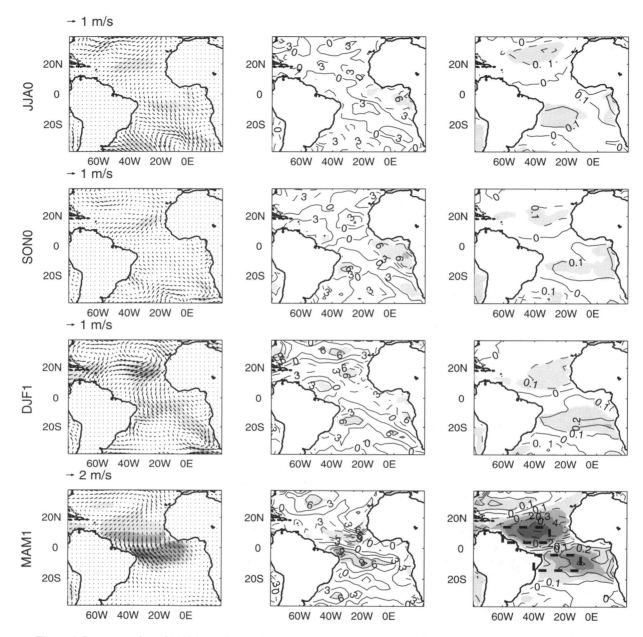

Figure 1. Lag-regression of 1000 hPa winds (left panels), surface downward heat flux (middle panels) and SST (right panels) anomaly onto the GI for seasons JJA0, SON0, DJF1 and MAM1. The GI index is constructed in MAM1. The regions used to construct the index are shown as boxes in the lower right panel. Arrows indicate the scale of 1000 hPa winds. The contour interval for heat flux is 3 W m^{-2}, and for SST is 0.1 K. Shading indicates the percentage of variance explained in intervals of 10%, 30%, 50%, and 70%. Explained variance of 10% is slightly above the 95% significance level. For surface winds the shading indicates explained variance in wind speed.

gradient during MAM. The gradient index used here, GI, is constructed as the average of the SST anomaly over 4°–14°S, 40°–10°W minus the average of the SST anomaly over 4°–14°N, 60°–30°W during MAM (hereafter MAM1). The seasons preceding MAM1 are denoted as JJA0, SON0 and DJF1 for June–July–August, September–October–November of the previous year, and December–January–February of the same year. The evolution shows that large cross-equatorial gradients are preceded by SST anomalies on both sides of the equator in the previous seasons. The maps suggest that the initial subtropical SST anomaly is generated through wind-induced changes in the surface heat fluxes during JJA0, SON0 and DJF1 [*Wagner*, 1996]. The magnitude of the tropical SST anomaly in the SH tends to be larger than that in the Northern Hemisphere (NH) up to DJF1. Also, in the SH the SST anomaly does not grow as much as in the NH from DJF1 to MAM1. This suggests that in the NH atmospheric variability during boreal winter plays the most important role in producing the northern SST anomaly that forms the gradient, while in the SH atmospheric variability during several seasons plays a role. In MAM1 surface winds are concentrated in the deep tropics, flowing from the cooler to the warmer hemisphere and inducing changes in the surface heat fluxes that strengthen the already existing cross-equatorial SST gradient. This in turn strengthens the winds, closing the loop of a positive feedback. This feedback mechanism was formally put forth by *Chang et al.* [1997] to explain the gradient mode, and is also known as Wind-Evaporation-SST (WES) feedback [*Xie and Philander*, 1994]. Evidence of local air–sea feedbacks in the tropical Atlantic comes from observational [*Chiang et al.*, 2002] and modeling studies [*Carton et al.*, 1996; *Chang et al.*, 1997, 2000; *Xie*, 1999]. The existence of this feedback may be fundamental for extending seasonal prediction beyond persistence.

From Figure 1, it is clear that identifying and understanding the sources of the SST anomalies associated with the gradient mode are important aspects of TAV research. Among the known sources of SST anomalies in the tropical North Atlantic are ENSO [e.g., *Curtis and Hastenrath*, 1995; *Enfield and Mayer*, 1997; *Saravanan and Chang*, 2000], and the NAO [e.g., *Marshall et al.*, 2001]. The impact of ENSO on TAV is believed to be as follows: During warm ENSO events diabatic heating anomalies in the tropical Pacific cause the northeasterly trades in the tropical Atlantic to weaken during summer, which coincides with the mature phase of ENSO. This reduces the evaporational cooling, and generates a positive SST anomaly in the northern tropics. *Chiang et al.* [2002] and *Giannini et al.* [2001] show that ENSO generally tends to aid the development of an SST gradient in the following fall. However, the anomalous gradient can also form in the absence of ENSO [*Chiang et al.*, 2002]. This latter result underlies the importance of local feedbacks. In fact, a recent study by *Giannini et al.* [2004] suggests that the local feedbacks can interfere with the remote influence of ENSO.

The NAO is the dominant mode of internal atmospheric variability during boreal winter. The oceanic response to the NAO forcing consists of a tripole pattern in SST anomalies, extending from the high latitudes of the North Atlantic to the northern tropics with decreasing amplitude [*Visbeck et al.*, 1998]. Observational and modeling studies show that during a strong NAO year northeasterly trade winds strengthen and generate a negative SST anomaly in the tropical North Atlantic via changes in latent heat flux. The resulting SST anomaly has its largest amplitude near the African coast north of 10°N [*Chang et al.*, 2001; *Czaja et al.*, 2002]. Studies show that NAO plays an important role in initiating local air–sea feedbacks in the deep tropical Atlantic region [*Halliwell*, 1997; *Xie and Tanimoto*, 1998; *Chang et al.*, 2001]. However, *Wu and Liu* [2002] found that although NAO can enhance variability in the tropical North Atlantic, its occurrence is not a requirement for the development of the gradient mode.

Until now, the origin of SST anomalies in the southern tropical Atlantic has been largely ignored, probably due to the lack of reliable observational data. The origin of the southern SST anomaly may be tied to the northern SST anomaly through the WES feedback. Several authors have argued, however, that the two lobes of the gradient in MAM1 seen in Figure 1 are largely independent and that a dipole does not exist as a physical mode, but is an artifact of construction [e.g., *Enfield et al.* 1999; *Dommenget and Latif*, 2000; *Czaja et al.*, 2002]. This, however, does not exclude the existence of the WES mechanism which may act to strengthen an already existing cross-equatorial gradient.

In a recent observational study, *Sterl and Hazeleger* [2003] found that SST anomalies in the South Atlantic are generated mainly through atmosphere-induced latent heat fluxes and wind-induced mixed layer deepening, and damped by latent heat fluxes. Their results and those of *Venegas et al.* [1997] agree on that the leading mode of observed coupled variability in the basin consists of a weakened/strengthened subtropical anticyclone forcing the ocean, independent of ENSO. *Venegas et al.* [1997] further found that this coupling is strongest during summer, and argued this is due to a link between the SH atmosphere and climate fluctuations in the NH, such as the NAO. On the other hand, the South Atlantic anticyclone is most energetic during winter [*Satyamurty et al.*, 1998], when it reaches its westernmost and northernmost position [*Hastenrath*, 1985]. Therefore, variability of the subtropical high during this season is likely to have an influence on TAV.

In this work, we take a further look at the atmospheric variability in the South Atlantic and explore its influence on TAV. We show that TAV is influenced not only by ENSO and the NAO, but also by atmospheric variability in the South Atlantic during austral winter and summer. Moreover, our results suggest that local air–sea feedbacks play an important role in maintaining the tropical SST anomaly and strengthening the cross-equatorial gradient.

The paper is organized as follows. In the next section we study the evolution of the SST anomaly generated by the leading mode of winter atmospheric variability in the South Atlantic. A mechanism through which this atmospheric pattern can affect the development of the gradient mode in the following fall is proposed and tested using a model. Section 3 investigates the observed evolution of the SST anomaly generated by the leading mode of atmospheric variability in SH summer, and its influence on TAV. In section 4 we compare the relative influence of ENSO, NAO and the South Atlantic atmospheric circulation on the development of the gradient mode. The last section summarizes the main results.

2. SOUTH ATLANTIC WINTER ATMOSPHERIC VARIABILITY AS EXTERNAL FORCING OF TAV

Figure 1 suggests that atmospheric variability in SH winter and summer plays a role in forcing the SST anomaly in the southern tropics that later becomes a part of the cross-equatorial gradient. This SST anomaly is created in the subtropics and afterward strengthens in the equatorial region, particularly from DJF1 to MAM1. In this paper we define winter to go from June to September (JJAS), and summer from November to February (NDJF). This section considers the winter, while section 3 is devoted to the influence of the summer season.

We use the reconstructed SST data set of *Smith et al.* [1996]. We also use 1000 hPa winds, sea level pressure and surface heat fluxes from the NCEP/NCAR Reanalysis [*Kalnay et al.*, 1996]. All data sets have a spatial resolution of approximately 2.8° (T42) and span from January 1950 to December 1994. It should be noted that the quality of the data especially in the SH is more reliable after the introduction of satellites in the 1980s. To diagnose the atmospheric response to oceanic conditions during MAM we also consider the Xie-Arkin precipitation data set [*Xie and Arkin*, 1997], which is available since 1979 on a T42 grid. The regression plots in Figure 4 below were constructed for the common period, 1979 to 1994. Significance levels throughout this study are calculated according to a two-sided Student's t-test, assuming no year-to-year correlation.

2.1. Observed Evolution

We calculated the Empirical Orthogonal Functions (EOFs) of (detrended) 1000 hPa wind speed during JJAS within 60°W–20°E, 40°–0°S, the region that contains the seasonal mean position of the anticyclone. The leading EOF (hereafter winter-EOF) explains 31% of the total variance and is well separated from the second one. A simultaneous regression of sea level pressure shows that the winter-EOF represents a weakening/strengthening of the subtropical high in the South Atlantic with weak correlations outside this basin except in the Southern Indian Ocean (Figure 2a). This is consistent with *Sterl and Hazeleger* [2003]. The principal component (PC) associated with the winter-EOF (hereafter P1JJAS) presents mainly interannual time scales (Figure 2b). This PC has no significant correlations (above the 95% level) with ENSO or with equatorial Atlantic SST during winter. In this study we use Niño3.4 (the average of SST anomalies in 120°–170°W, 5°S–5°N) to characterize ENSO, and ATL3 (the average of SST anomalies in 20°W–0°E, 3°S–3°N) to characterize equatorial Atlantic SST [*Zebiak*, 1993]. The P1JJAS index has no significant correlation with the GI in the next austral fall.

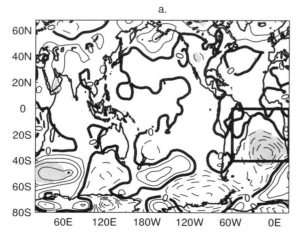
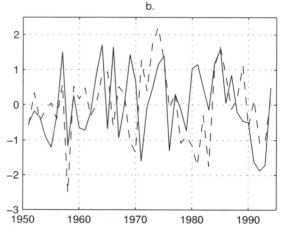

Figure 2. (a) Regression of sea level pressure onto the PC time series (P1JJAS) of winter-EOF during JJAS season. The box marks the region in which the EOF analysis of 1000 hPa wind speed was performed. Contour interval is 20 Pa, and shading indicates significance at the 95% level. (b) Index P1JJAS (solid line) and GI time series (dashed line). P1JJAS is shifted 1-year so that it can be compared with GI index.

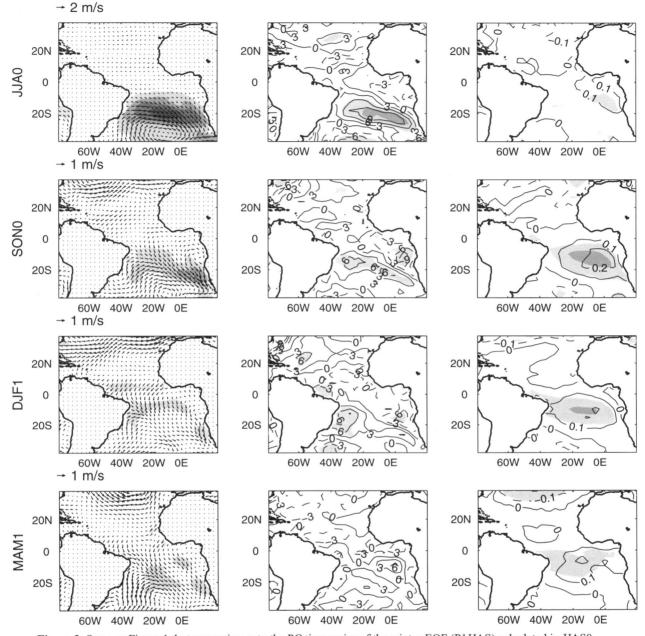

Figure 3. Same as Figure 1, but regressing onto the PC time series of the winter-EOF (P1JJAS) calculated in JJAS0.

However, it is significantly correlated at $r = 0.4$ with the southern tropical index used to construct GI. Thus, it appears that the winter atmospheric variability can influence southern tropical SST and thus TAV three seasons in advance. To investigate this issue, we regress 1000 hPa winds, surface heat flux and SST on P1JJAS for the seasons JJA0, SON0, DJF1 and MAM1 ("0" indicates the year in which the EOF is calculated). This is shown in Figure 3. The maps show, that during winter, cyclonic wind anomalies force the ocean through changes in the heat fluxes in the subtropical South Atlantic. A decomposition into the different heat flux components indicates that latent heat flux changes dominate. The ocean responds with a 2- to 3-month lag in SST change. The subtropical SST anomaly tends to persist into the next season (DJF1) with a noticeable northwestward shift. During this season the SST gradient created by the existence of the southern SST anomaly induces cross-equatorial northerly winds which tend to cool the northern deep tropics and warm the southern tropics through changes in the latent heat flux. This suggests the possibility of a WES feedback. Changes in the solar radiation tend to oppose the creation of SST anomalies in the northern tropics, but help maintaining the SST anomaly in the southern tropics. As a result, the SST anomaly south of the equator persists into MAM1, when it is accompanied by cross-equatorial winds. Subtropical wind anomalies are also present south of the maximum SST anomaly. Since Figure 3 is constructed by regressing onto an index characterizing the atmosphere 3 seasons before, these subtropical winds can only be interpreted as a response to the SST anomaly unless they are a statistical artifact and occur by chance. During MAM1 the tropical atmosphere is so sensitive to ocean conditions that even the small SST gradient tends to shift the ITCZ toward the SH (Figure 4a).

The persistence of the SST anomaly between 5°S–20°S from the coast of Brazil to 20°W may also be attributed to a relatively deep mixed layer in the region. The annual mean mixed layer depth is larger than 40m, while in the western tropical North Atlantic the mixed layer depth is less than 30m (see Figure 1 of *Saravanan and Chang* [this volume]).

2.2. Simulated Evolution Using an Atmospheric General Circulation Model Coupled to a Slab Ocean

In the previous section we found that the evolution of the SST anomaly in the subtropical South Atlantic is mainly governed by surface heat flux anomalies. This result is consistent with *Sterl and Hazeleger* [2003], although other processes like mixed layer deepening or Ekman pumping may also play a role. In this section we use an atmospheric general circulation model coupled to a slab ocean to show that the main characteristics of the observed evolution of Figure 3 can indeed be ascribed to thermodynamic ocean–atmosphere coupling. We use the Community Climate Model 3.6 (CCM3) developed at the National Center for Atmospheric Research coupled to a slab ocean with an annual mean mixed layer depth. The model has been shown to represent TAV realistically [*Saravanan and Chang*, 1999], and is described by *Saravanan and Chang* [this volume]. We use a 100 year control run in which CCM3 is coupled everywhere to the slab ocean (referred as experiment MIXL in *Saravanan and Chang* [this volume]). This setup excludes any dynamical ocean–atmosphere interaction, and thus ENSO and the Atlantic zonal mode [*Zebiak*, 1993] are not present. To assure the correct simulation of the annual cycle of SST the slab ocean uses a Q-flux correction that accounts for the missing climatological ocean dynamics in the model. For a detailed model description we refer the readers to the above referenced article.

We applied the same analysis of Figures 2 and 3 to the model output. The leading EOF of 1000 hPa wind speed in the South Atlantic during JJAS explains 32% of the total variance. The anomalous pattern of sea level pressure is similar to the observed one (not shown). It represents a weakening/strengthening of the South Atlantic anticyclone, and is correlated with pressure anomalies in the South Indian Ocean. It is also significantly correlated with pressure anomalies over the Pacific Ocean at about 160°E, 20°N that is not found in observations. This may be related to the absence of ENSO in the model.

Figure 5 shows the simulated evolution of 1000 hPa winds, surface heat flux and SST by regressing these fields onto the PC of the leading winter EOF. Clearly, the simulated evolution is similar to the observed one. During winter cyclonic winds associated with a weakened subtropical high generate a positive downward heat flux anomaly in the southern subtropics,

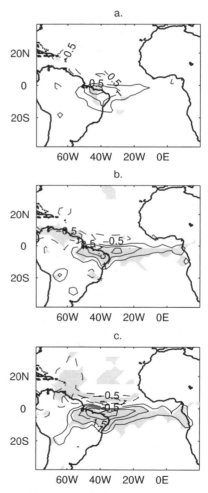

Figure 4. Regression of rainfall anomaly during MAM onto (a) P1JJAS, (b) P1NDJF, and (c) GI indices. The regression is performed during the period of 1979 to 1994. Shading indicates statistical significance at the 95% level. Contour interval is 0.5 mm day^{-1}.

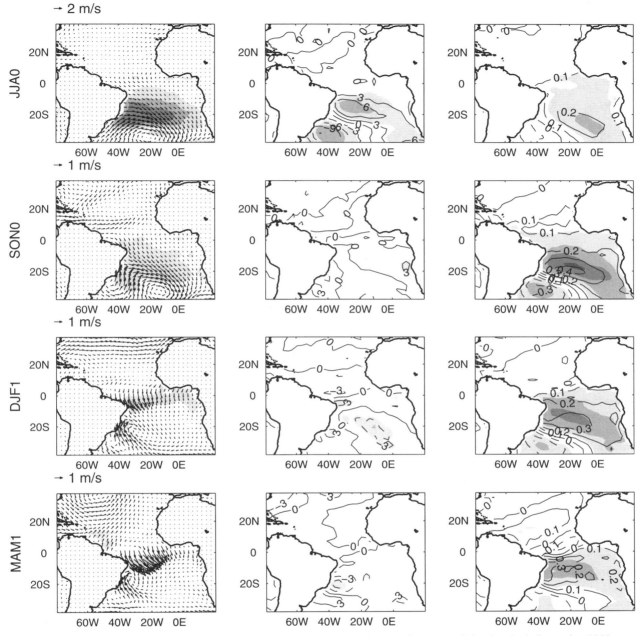

Figure 5. Same as Figure 3, but regressing onto the PC time series of the leading EOF of simulated wind-speed at 1000 hPa during JJAS0.

producing the largest response 2 or 3 months later. In DJF1 the SST anomaly is damped in the subtropics, but maintained and even strengthened in the western deep tropics, presumably through a positive feedback involving surface winds, heat flux and SST. This results in a shift of the position of maximum SST anomaly toward the equator. The cross-equatorial southward surface flow induces a negative heat flux in the northern deep tropics that tends to cool the region. As in observations, changes in latent heat flux are opposed by the negative feedback provided by solar radiation changes due to changes in cloud cover as the ITCZ shifts toward warm waters. This result is also consistent with the observational work of *Tanimoto and Xie* [2002]. The cross-equatorial SST gradient and accompanied surface winds are strongest in MAM1. Note that simulations do not show significant subtropical wind anomalies during MAM1, as observations do (see Figure 3).

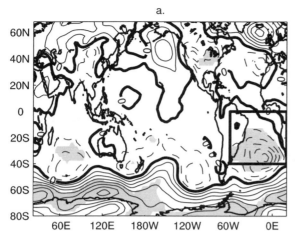 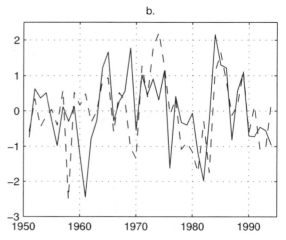

Figure 6. Same as Figure 2, but for the summer-EOF.

The fact that we are able to reproduce the main characteristics of the observed evolution supports the idea that thermodynamic interaction dominates ocean–atmosphere coupling, in general agreement with *Sterl and Hazeleger* [2003]. Here, however, we find evidence for local air–sea feedbacks in the deep tropics. Results also support that winter atmospheric variability can influence the development of the gradient mode.

It is worth pointing out that there are some important differences between simulated and observed fields. First, the simulated wind anomaly in winter is shifted about 10° west compared to observations, hence the maximum in SST anomaly is also shifted to the west. Second, the SST variance explained by the atmospheric forcing is larger in the simulation because SST anomalies in the model ocean only occur through surface heat fluxes, while in the real ocean other dynamical processes are present. The exaggerated SST response in the model may also be attributed to the use of an annual mean mixed layer depth which underestimates its value during winter (see Figure 1 of *Saravanan and Chang* [this volume]). This effect, together with the absence of ocean processes that tend to damp the SST anomaly in the equatorial region [*Chang et al.*, 2001], overestimates the importance of the local feedback and exaggerates the simulated cross-equatorial SST gradient and winds in the fall.

3. SOUTH ATLANTIC SUMMER ATMOSPHERIC VARIABILITY AS EXTERNAL FORCING OF TAV

We next consider the summer season. The leading EOF of 1000 hPa wind speed during NDJF in the South Atlantic explains 28% of the total variance and is well separated from the second one. We will refer to it as the summer-EOF and to its PC as P1NDJF. The regression of sea level pressure shows a weakened/strengthened anticyclone in the South Atlantic (Figure 6a). The regression map does not show any significant correlation with NH anomalies, except maybe over North America. On the other hand, the variability of the subtropical high seems to be part of a global pattern resembling the southern annular mode [*Thompson and Wallace*, 2000]. It also presents a wavenumber 4 structure with maximum amplitude at about 45°S. Recently, *Fauchereau et al.* [2003] reported a similar wavenumber 4 pattern of sea level pressure variability in the SH. They found that during summer the surface winds associated with this structure force the South Atlantic and South Indian Ocean simultaneously through changes in the latent heat flux. Their analysis indicates that the structure is not significantly correlated with the annular mode. Here, we calculated an index for the annular mode as the first principal component of sea level pressure south of 20°S. This time series is correlated with P1NDJF at $r = 0.33$ (just above the 95% significance level).

The summer-EOF shows longer time scales than the winter-EOF (Figure 6b), and P1NDJF is not correlated to ENSO or the NAO. It is, however, correlated with GI at $r = 0.46$, indicating that this mode influences TAV. Figure 7 shows the regression of 1000 hPa winds, surface heat flux and SST onto P1NDJF for DJF1 and MAM1. As in the winter case, there is an eddy-like circulation over the subtropical Atlantic which induces changes in the surface heat fluxes, primarily latent heat. During summer, however, the ocean responds faster and the maximum SST anomaly occurs within the season. The initial SST anomaly is created south of 10°S, but in MAM1 it has reached the equatorial region. This is a consequence of the response of surface winds to the southern SST anomaly, and the interplay between heat flux, SST and winds—a characteristic of the WES feedback: The cross-equatorial gradient forces northerly winds that tend to cool the SST in the north-

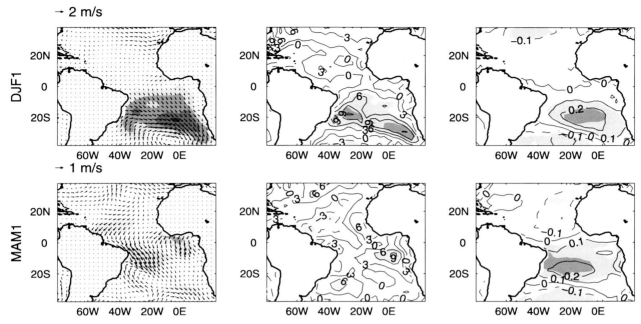

Figure 7. Lag-regression of 1000 hPa winds (left panels), surface downward heat flux (middle panels) and SST (right panels) anomaly onto the PC time series of summer-EOF (P1NDJF) for seasons DJF1 and MAM1. The P1NDJF index is constructed in NDJF1. Arrows indicate the scale of 1000 hPa winds. The contour interval for heat flux is 3 W m^{-2}, and for SST is 0.1 K. Shading indicates the percentage of variance explained in intervals of 10%, 30%, 50%, and 70%. Explained variance of 10% is slightly above the 95% significance level. For surface winds the shading indicates explained variance in wind speed.

ern tropics through changes in latent heat. At the same time weakened southerly trades tend to induce positive heat flux anomalies in the southern deep tropics that maintain the southern SST anomaly and strengthen the SST gradient. On the other hand, in the southern subtropics the SST anomaly appears to be damped. It is interesting to note that *Czaja et al.* [2002] find evidence of WES feedback in the deep northern tropics when the tropical North Atlantic is forced by anomalous winds during boreal winter.

Thus, the SST anomalies created in summer follow a similar evolution as the anomalies created during winter. In the summer case, however, the SST anomaly in MAM1 is larger because it was created just one season before. This generates a larger atmospheric response, as can be seen in the regression of rainfall onto P1NDJF from 1979 to 1994 (Figure 4b). Clearly, rainfall anomalies are larger than in the winter case (Figure 4a), and are of similar magnitude as those created by the cross-equatorial gradient of Figure 1 (Figure 4c).

Note that the SST anomaly in the southern lobe of the dipole in MAM1 of Figure 1 can be explained to first order by considering the SST anomaly generated by summer atmospheric variability. The SST anomaly generated by winter atmospheric variability plays a secondary role, and is most important between 0 and 10°S.

One interesting feature is that the summer PC, P1NDJF, is marginally correlated at the 95% level with the winter PC, P1JJAS. Since the atmosphere does not have a long memory, this suggests that the SST anomaly created by winter atmospheric variability forces an atmospheric response in the following summer, and the summer-EOF has a small contribution of this signal. Modeling studies have suggested the existence of an SST-forced response in this region during summer [e.g., *Barreiro et al.*, 2002]. We repeated the calculation of the summer-EOF after linearly removing the contribution of the winter-EOF to the 1000 hPa wind speed anomaly. The new leading EOF correlates at $r = 0.95$ with P1NDJF. Also, the evolution of SST, surface winds and heat flux does not change significantly from that of Figure 7. Thus, in this work we consider the summer-EOF to mainly represent internal atmospheric variability.

4. RELATIONSHIP BETWEEN THE CROSS-EQUATORIAL GRADIENT AND ATMOSPHERIC VARIABILITY IN BOTH HEMISPHERES

According to the results of the previous sections, to first order the SST anomaly in the tropical North Atlantic is independent of the SST in the tropical South Atlantic. Neverthe-

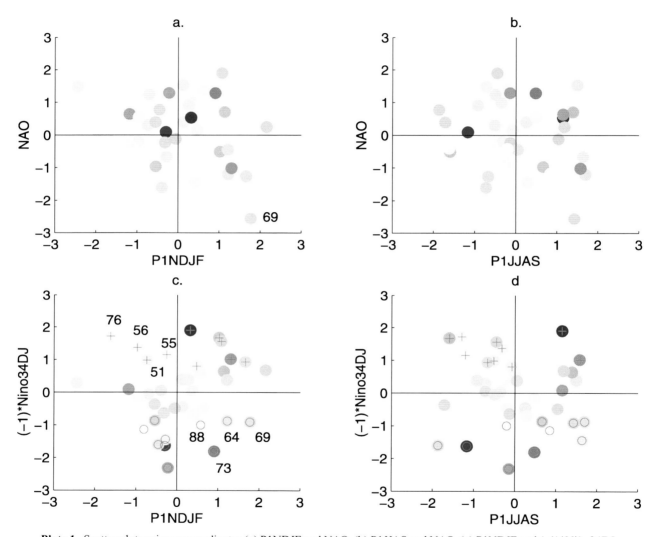

Plate 1. Scatter plots using as coordinates (a) P1NDJF and NAO, (b) P1JJAS and NAO, (c) P1NDJF and (−1)*Niño34DJ, and (d) P1JJAS and (−1)*Niño34DJ. In all plots the color of the markers indicates the intensity and sign of the GI: blue is negative, red is positive. In (c) and (d) "o" indicate El Niño years, and "+" indicate La Niña years. Specific ENSO events discussed in the text are identified by their years. Indices are normalized.

less, an SST anomaly generated in the southern tropics tends to induce an SST anomaly of opposite sign in the northern tropics through changes in the surface heat fluxes, suggesting the existence of the WES feedback.

To further look into this issue we classify the cross-equatorial gradient index as a function of indices of southern and northern atmospheric variability. The largest gradients are expected when the atmospheric anomalies on both sides of the equator are such that they tend to generate SST anomalies of opposite sign. If hemispheres are independent, the existence of an SST anomaly in one hemisphere would not affect the development of the SST anomaly in the other hemisphere. We use GI to characterize the cross-equatorial gradient in MAM. Atmospheric variability in the NH is characterized by the influences of ENSO and the NAO. To characterize ENSO we consider the Niño3.4 index during December–January (Niño34DJ), which is correlated with GI at $r = -0.5$. To characterize the NAO we consider the time series calculated as the difference of normalized sea level pressure between Lisbon, Portugal, and Reykjavik, Iceland, from December to March. The time series was obtained from the web site http://www.cgd.ucar.edu/~hurrell/nao.stat.winter.html, and was detrended by removing the least squares linear fit prior to the analysis. This NAO index is not significantly correlated with GI. We use P1JJAS and P1NDJF to characterize winter and summer atmospheric variability in the South Atlantic, respectively. These indices are not significantly correlated to Niño34DJ or the NAO index.

We investigate the relationship among the atmospheric indices and the sign and intensity of the gradient in MAM by constructing scatter plots (Plate 1). In these plots the color of the marker indicates the sign and intensity of GI, while different pairs of indices are used as coordinate axes. Since the atmospheric indices are not correlated the distribution of points tends to be circular. The scatter plots show how different combinations of atmospheric indices determine the sign and magnitude of GI.

We first consider the summer-EOF (P1NDJF) and the NAO (Plate 1a). The plot clearly shows that the sign of the gradient is mainly given by the sign of the summer-EOF. Most positive cross-equatorial gradients correspond to a situation when the summer-EOF is positive, that is, when the South Atlantic atmospheric variability induces a positive SST anomaly in the southern subtropics. The NAO index does not have an appreciable influence on GI, consistent with the non-significant correlation between these two indices. This is in part due to our choice of GI. The NAO tends to influence SST north of 10°N [e.g., *Ruiz-Barradas et al.*, 2000, 2003; *Czaja et al.*, 2002], and the induced SST anomaly has maximum amplitude near the African coast. The GI was constructed using SST anomalies in the western tropical Atlantic. This is the region of maximum seasonal precipitation, where SST anomalies can more easily induce a response, and away from the region of maximum NAO influence. Note that the intensity of the GI is not clearly related to P1NDJF. For positive GI there is a tendency for large gradients to occur when P1NDJF is large (close to 1), but that is not true for negative GI. An analogous scatter plot using NAO and the winter-EOF does not show a clear relationship between the indices and GI (Plate 1b). There is a tendency, however, for positive gradients to occur when both NAO and P1JJAS are positive (upper right quadrant), that is, when the NAO (SH winter) cools (warms) the northern (southern) tropics.

We now turn to the relationship between GI, ENSO and the SH indices. Plate 1c shows the scatter plot using Niño34DJ and P1NDJF (summer-EOF) indices. Note that in the plot we changed the sign of Niño3.4, so that El Niño events correspond to negative values. We define ENSO events as those when Niño34DJ is larger than 0.75K. This classification leads to the following El Niño events: 1958, 1964, 1966, 1969, 1970, 1973, 1983, 1987, 1988 and 1992, and the following La Niña events: 1951, 1955, 1956, 1965, 1971, 1972, 1974, 1976, 1985 and 1989 (years refer to the January month). In Plates 1c,d El Niño events are marked with a "o", while La Niña events are marked with a "+". El Niño (La Niña) events tend to warm (cool) the tropical North Atlantic. Thus, El Niño (La Niña) events are expected to induce negative (positive) GI values during MAM. From Plate 1c it is clear that the cross-equatorial gradient is strongest when $(-1)*$Niño34DJ and the summer-EOF (P1NDJF) indices are both of the same sign, that is, GI is largest when atmospheric anomalies on both sides of the equator induce SST anomalies of opposite sign (upper right and lower left quadrants). The largest positive values of GI occur when both P1NDJF and $(-1)*$Niño34DJ are large. On the other hand, the largest negative values of GI occur for small negative values of P1NDJF, but for large (negative) values of $(-1)*$Niño34DJ. This suggests that the amplitude of the large negative cross-equatorial gradients is set up mainly by the remote signal from El Niño events.

When the southern atmosphere in summer generates an SST anomaly of the same sign as the SST anomaly to be created by ENSO in the northern tropics (lower right and upper left quadrants in Plate 1c), the resulting gradient in MAM is usually weak. There are 8 ENSO events in this category: El Niño years 1964, 1969, 1973 and 1988, and La Niña years 1951, 1955, 1956 and 1976 (see Plate 1c). Seven of these years have GI values close to neutral or of opposite sign as those expected from ENSO forcing acting alone. Moreover, during most of these years the sign of the cross-equatorial gradient is given by the sign of P1NDJF. To investigate if the pre-existing southern SST anomaly influenced the northern tropics, we look at the sign of the northern index used to construct GI. We found that during El Niño of 1988 and La Niña of 1955 and 1976 the sign is as expected from ENSO forcing. However, during the ENSO events of 1951, 1956, 1964 and 1973 the northern tropics have (weak) SST anomalies of the opposite sign from the expected ENSO signal. This may be explained by the NAO forcing acting against the ENSO forcing in the northern tropics. However, only 1973 is a strong NAO year (> 0.5 standard deviation) that acts against the ENSO forcing. Thus, this suggests that during these years the hemispheres were not independent, but that local feedbacks initiated by the southern tropical SST influenced the northern tropics and worked against the remote ENSO forcing.

These results agree with those of *Giannini et al.* [2004] and *Barreiro et al.* [2004], who suggest that the WES feedback and the remote influence of ENSO can interfere constructively or destructively. Constructive interference occurs when both processes tend to create a cross-equatorial gradient of the same sign. Destructive interference occurs when the WES feedback and the remote ENSO signal act against each other. Our findings suggest that during 1951, 1956, 1964 and 1973 the WES feedback initiated by the SH summer atmospheric variability worked against the remote ENSO forcing, changing the gradient to that expected from ENSO acting alone. During other years the local feedback can be overpowered by ENSO, but may still reduce the SST anomaly in the northern tropics.

Year 1969 is the only case in which a relatively large negative gradient occurs when the South Atlantic summer atmospheric circulation induces a positive SST anomaly in the southern tropics. During this year the remote forcing from El Niño superposed with the weakest NAO on record (see Plate 1a). Thus, the northern tropics developed a large positive SST anomaly that overpowered the local feedbacks and changed the sign of the GI.

These findings show that the summer-EOF pre-conditions the development of the gradient mode in MAM by initiating local air–sea feedbacks, which can interfere constructively or destructively with the remote ENSO signal. Consequently, although to first order the hemispheres are independent, they tend to be connected during the time when the local WES feedback is important. Plate 1d shows the case using Niño34DJ and P1JJAS. As before, ENSO is a good predictor of GI, and the strongest cross-equatorial gradients tend to occur when the indices have the same sign. The pre-conditioning of the gradient from the winter-EOF is, however, less clear than that from the summer-EOF probably because the induced summer SST anomaly is weaker. Thus, the results from this section show that Niño34DJ and P1NDJF are the best predictors for the GI.

5. SUMMARY

We presented an exploratory study of the influence of the South Atlantic atmospheric circulation on TAV. We first calculated the leading patterns of austral winter and summer atmospheric variability in the South Atlantic. They consist of a weakening/strengthening of the anticyclone forcing the ocean below, consistent with the literature [*Venegas et al.*, 1997; *Sterl and Hazeleger*, 2003]. Using regression analysis, we next studied the evolution of the SST anomalies generated, and their possible role in the development of the gradient mode. The relative importance of South Atlantic atmospheric variability in TAV, and specifically in determining the sign of the cross-equatorial SST gradient during fall, is addressed by comparing its role versus that of ENSO and the NAO. The main results are the following (seasons refer to those of the SH):

• Winter atmospheric variability in the South Atlantic induces an SST anomaly that tends to persist until the next fall. Observational and modeling studies suggest that ocean and atmosphere interact mainly through surface heat fluxes. This is consistent with the literature, although other processes like wind-induced mixed layer deepening may also play a role in generating the initial SST anomaly [*Sterl and Hazeleger*, 2003]. The following scenario is proposed: In winter the weakened southeasterly trade winds remove less heat from the ocean, generating a warm SST anomaly. In the following seasons the SST anomaly is damped in the subtropics, but not in the western deep tropics. The existing SST gradient induces southward cross-equatorial winds during summer, which tend to cool the SST in the northern deep tropics and maintain the SST anomaly in the south through the WES feedback mechanism. The weak SST gradient that persists until MAM favors a southward shift of the ITCZ. Although the influence of winter atmospheric variability on TAV is weaker than that of summer, it may be important for enhancing seasonal prediction. More research on this subject is clearly needed. But, if this conjecture turns out to be true then knowledge of the atmospheric conditions during austral winter would allow prediction of the SST anomaly, and thus of the location of the ITCZ, more than two seasons in advance.

• Summer atmospheric variability in the South Atlantic also forces the ocean below through changes in the latent heat fluxes, in agreement with *Fauchereau et al.* [2003]. In this season the ocean responds faster than in winter, perhaps due to a shallower mixed layer depth. The SST anomaly is created south of 10°S, but moves northwestward by means of the WES feedback mechanism. As a result, the SST anomaly induces a strong atmospheric response in the austral fall, shifting the ITCZ toward anomalously warm waters.

• Summer atmospheric variability (and to a lesser extent winter variability) in the South Atlantic pre-conditions the development of strong cross-equatorial gradients during the next fall. It does so by generating an SST anomaly in the southern tropics during summer, which initiates the WES feedback in the deep tropics. This mechanism will in turn tend to strengthen the cross-equatorial gradient. When the ENSO remote influence creates an SST anomaly of opposite sign as that already present in the southern tropics the resulting gradient in MAM is strongest. When the ENSO forced signal is of the same sign the outcome will depend on the relative strengths of ENSO and the WES feedback. An interesting result, also found in *Giannini et al.* [2004], is that the constructive/destructive interference seems to depend only on the southern tropical SST in DJF, and not on the cross-equatorial SST gradient on that season. A possibility is that ENSO and NAO forcings are so strong that they can wipe out any SST anomaly that was present in the northern tropics during DJF. Thus, to first order, the cross-equatorial gradient will only be pre-conditioned by the southern tropics. Further study is needed to address this issue.

Results show that atmospheric circulation in the South Atlantic is capable of initiating local air–sea feedbacks in the tropics, and affecting the development of the gradient mode, as previously found for the NH atmosphere [e.g., *Xie and Tanimoto*, 1998]. These findings suggest that to fully understand TAV it is necessary to consider not only the remote influence of ENSO and the NAO, but also of the southern atmospheric circulation during winter and summer. These remote phenomena, together with the local feedbacks, control the evolution of SST in the tropical Atlantic. The existence of so many players involved in TAV makes prediction of tropical Atlantic climate very challenging, and calls for sustained observations in the region.

Acknowledgments: The authors would like to thank two anonymous reviewers for their constructive criticism that greatly improved the original manuscript. M.B. was supported by NASA Headquarters under the Earth System Science Fellowship Grant NGT5-30417. This study is supported by the NOAA and NSF grants: NA16GP1572 and ATM-99007625. P.C. also acknowledges the support from the National Natural Sciences Foundation of China (NSFC) through Grant 40128003.

REFERENCES

Barreiro, M., P. Chang, and R. Saravanan, Variability of the South Atlantic Convergence Zone simulated by an atmospheric general circulation model, *J. Clim.*, 15, 745–763, 2002.

Barreiro, M., P. Chang, L. Ji, R. Saravanan, and A. Giannini, Dynamical elements of predicting boreal spring tropical Atlantic sea–surface temperatures, *Dyn. Atmos. Oceans*, submitted, 2004.

Carton, J. A., X. Cao, B. Giese, and A. M. Da Silva, Decadal and interannual SST variability in the tropical Atlantic Ocean, *J. Phys. Oceanogr.*, 26, 1165–1175, 1996.

Chang, P., L. Ji, and H. Li, A decadal climate variation in the tropical Atlantic ocean from thermodynamic air–sea interactions, *Nature*, 385, 516–518, 1997.

Chang, P., R. Saravanan, L. Ji, and G. C. Hegerl, The effect of local sea surface temperatures on atmospheric circulation over the tropical Atlantic sector, *J. Clim.*, 13, 2195–2216, 2000.

Chang, P., L. Ji, and R. Saravanan, A hybrid coupled model study of tropical Atlantic variability, *J. Clim.*, 14, 361–390, 2001.

Chiang, J. C. H., Y. Kushnir, and A. Giannini, Deconstructing Atlantic ITCZ variability: Influence of the local cross-equatorial SST gradient, and remote forcing from the eastern equatorial Pacific, *J. Geophys. Res.*, 107(D1), 1–19, 2002.

Curtis, S., and S. Hastenrath, Forcing of anomalous sea surface temperature evolution in the tropical Atlantic during Pacific warm events, *J. Geophys. Res.*, 100, 15,835–15,847, 1995.

Czaja, A., P. van der Vaart, and J. Marshall, A diagnostic study of the role of remote forcing in tropical Atlantic variability, *J. Clim.*, 15, 3280–3290, 2002.

Dommenget, D., and M. Latif, Interannual to decadal variability in the tropical Atlantic, *J. Clim.*, 13, 777–792, 2000.

Enfield, D. B., and D. A. Mayer, Tropical Atlantic sea surface temperature variability and its relation to El Niño-Southern Oscillation, *J. Geophys. Res.*, 102, 929–945, 1997.

Enfield, D. B., A. M. Mestas-Nunez, D. A. Mayer, and L. Cid-Serrano, How ubiquitous is the dipole relationship in tropical Atlantic sea surface temperatures?, *J. Geophys. Res.*, 104, 7841–7848, 1999.

Fauchereau, N., S. Trzaska, Y. Richard, P. Roucou, and P. Chamberlin, Sea surface temperature co-variability in the Southern Atlantic and Indian Oceans and its connection with the atmospheric circulation in the southern hemisphere, *Int. J. Climatol.*, 23, 663–677, 2003.

Giannini, A., J. C. H. Chiang, M. Cane, Y. Kushnir, and R. Seager, The ENSO teleconnection to the tropical Atlantic Ocean: contributions of the remote and local SSTs to rainfall variability in the tropical Americas, *J. Clim.*, 14, 4530–4544, 2001.

Giannini, A., R. Saravanan, and P. Chang, The preconditioning role of tropical Atlantic variability in the development of the ENSO teleconnection: Implications for the prediction of Nordeste rainfall, *Clim. Dyn.*, submitted, 2004.

Halliwell Jr., G. R. H., Decadal and multidecadal North Atlantic SST anomalies driven by standing and propagating basin-scale atmospheric anomalies, *J. Clim.*, 10, 2405–2411, 1997.

Hastenrath, S., *Climate and circulation of the tropics*, 455 pp., D. Reidel, 1985.

Hastenrath, S., and L. Greischar, Circulation mechanisms related to northeast Brazil rainfall anomalies, *J. Geophys. Res.*, 98, 5093–5102, 1993.

Kalnay, E., et al., The NCEP/NCAR 40-year reanalysis project, *Bull. Am. Meteorol. Soc.*, 77, 437–471, 1996.

Lindzen, R. S., and S. Nigam, On the role of sea surface temperature gradients in forcing low level winds and convergence in the tropics, *J. Atmos. Sci.*, 44, 2418–2436, 1987.

Marshall, J., Y. Kushnir, D. Battisti, P. Chang, A. Czaja, R. Dickson, J. Hurrel, M. McCartney, R. Saravanan, M. Visbeck, North Atlantic climate variability: Phenomena, impacts and mechanisms, *Int. J. Climatol.*, 21, 1863–1898, 2001.

Ruiz-Barradas, A., J. A. Carton, and S. Nigam, Structure of interannual-to-decadal climate variability in the tropical Atlantic sector, *J. Clim.*, 13, 3285–3297, 2000.

Ruiz-Barradas, A., J. A. Carton, and S. Nigam, Role of the atmosphere in climate variability of the tropical Atlantic, *J. Clim.*, 12, 2052–2065, 2003.

Saravanan R., and P. Chang, Oceanic mixed layer feedback and tropical Atlantic variability, *Geophys. Res. Lett.*, 26, 3629–3633, 1999.

Saravanan R., and P. Chang, Interaction between tropical Atlantic variability and El Niño-Southern Oscillation, *J. Clim.*, 13, 2177–2194, 2000.

Saravanan, R., and P. Chang, Thermodynamic coupling and predictability of tropical sea surface temperature, this volume.

Satyamurty, P., C. A. Nobre, and P. L. Silva Dias, South America, in *Meteorology of the Southern Hemisphere*, D. J. Karoly and D. G. Vincent, Eds., Meteorol. Monogr., No. 49, pp. 119–139, Am. Meteorol. Soc., 1998.

Smith, T. M., R. W. Reynolds, R. E. Livezey, and D. C. Stokes, Reconstruction of historical sea surface temperatures using empirical orthogonal functions, *J. Clim.*, 9, 1403–1420, 1996.

Sterl, A., and W. Hazeleger, Coupled variability and air–sea interaction in the south Atlantic Ocean, *Clim. Dyn.*, 21, 559–571, 2003.

Tanimoto, Y., and S.-P. Xie, Inter-hemispheric decadal variations in SST, surface winds and cloud cover over the Atlantic Ocean, *J. Meteorol. Soc. Jpn.*, 80, 1199–1219, 2002.

Thompson, D. W. J., and J. M. Wallace, Annular modes in the extratropical circulation. Part I: Month-to-month variability, *J. Clim.*, 13, 1000–1016, 2000.

Venegas, S. A., L. A. Mysak, and D. N. Straub, Atmosphere–ocean coupled variability in the south Atlantic, *J. Clim.*, 10, 2904–2920, 1997.

Visbeck, M., H. Cullen, G. Krahmann, and N. Naik, An ocean model's response to North Atlantic Oscillation-like wind forcing, *Geophys. Res. Lett.*, 25, 4521–4524, 1998.

Wagner R. G., Mechanisms controlling variability of the interhemispheric sea surface temperature gradient in the Tropical Atlantic, *J. Clim.*, 9, 2010–2019, 1996.

Wu, L., and Z. Liu, Is tropical Atlantic variability driven by the North Atlantic Oscillation?, *Geophys. Res. Let.*, 13, doi:10.1029/2002GL014939, 2002.

Xie, P., and P. A. Arkin, Global precipitation: A 17-year monthly analysis based on gauge observations, satellite estimates, and numerical model outputs, *Bull. Am. Meteorol. Soc.*, 78, 2539–2558, 1997.

Xie, S.-P., A dynamic ocean–atmosphere model of the tropical Atlantic decadal variability, *J. Clim.*, 12, 64–70, 1999.

Xie, S.-P., and S. G. Philander, A coupled ocean–atmosphere model of relevance to the ITCZ in the eastern Pacific, *Tellus*, 46A, 340–350, 1994.

Xie, S.-P., and Y. Tanimoto, A pan-Atlantic decadal climate oscillation, *Geophys. Res. Lett.*, 25, 2185–2188, 1998.

Zebiak, S. E., Air–sea interaction in the equatorial Atlantic region, *J. Clim.*, 6, 1567–1586, 1993.

M. Barreiro, Program in Atmospheric and Oceanic Sciences, 205 Sayre Hall, Forrestal Campus, Princeton University, Princeton, New Jersey 08544–0710. (barreiro@princeton.edu)

P. Chang, Department of Oceanography, Texas A&M University, College Station, Texas 77843–3146. (ping@tamu.edu)

A. Giannini, International Research Institute for Climate Prediction, Columbia University, Palisades, New York 10964–8000. (alesall@iri.columbia.edu)

R. Saravanan, National Center for Atmospheric Research, Boulder, Colorado 80307–3000. (svn@ncar.ucar.edu)

Toward Understanding Tropical Atlantic Variability Using Coupled Modeling Surgery

Lixin Wu, Qiong Zhang, and Zhengyu Liu

Center for Climatic Research, University of Wisconsin-Madison, Madison, Wisconsin

The origins of tropical Atlantic variability (TAV) are investigated using a coupled modeling surgery approach. The coupled global circulation model captures the major TAV modes in the observations: the Atlantic Niño (ATL-Niño), the North Tropical Atlantic (NTA) and the South Tropical Atlantic (STA) modes. With the modeling surgery, it is found that these variability modes predominantly originate from local tropical Atlantic climate system, while remote ENSO and NAO forcing can enhance their variance and modulate their temporal evolution. Specifically, the interannual evolution of these modes is dictated predominantly by the remote ENSO, while the decadal evolution of these modes is dictated by extratropical-tropical interactions. Local ocean-atmosphere coupling is critical for the full development of the NTA and the ATL-Niño, but in general is not necessary for the STA mode. In the north tropical Atlantic and the subtropical south Atlantic (poleward of 20°S), the model simulated decadal SST anomalies are primarily generated by the surface heat flux and damped by the oceanic heat transport, but at the equator and in the south tropical Atlantic (equatorward of 20°S), they are primarily associated with oceanic heat transport through the anomalous vertical heat advection, and damped by the surface heat flux.

1. INTRODUCTION

Sea surface temperature in the tropical Atlantic Ocean exhibits variability on a variety of timescales and in broad spatial patterns [e.g., *Nobre and Shukla*, 1996]. The dominant patterns of variability are characterized by inter-hemispheric SST gradients, and modulations in the strength of the southeast and northeast trades as well as the position and intensity of the Intertropical Convergence Zone (ITCZ). Such climatic fluctuations thus have a profound impact on the rainfall in the surrounding landmasses, primarily northeastern Brazil and sub-Saharan West Africa [*Hastenrath*, 1978; *Moura and Shukla*, 1981; *Folland et al.*, 1986].

Earth's Climate: The Ocean-Atmosphere Interaction
Geophysical Monograph Series 147
Copyright 2004 by the American Geophysical Union
10.1029/147GM09

One of the key issues about the tropical Atlantic variability (TAV) concerns its origins [*Marshall et al.*, 2001; *Xie and Carton*, this volume]. Unlike the tropical Pacific, the tropical Atlantic is not dominated by any single process such as ENSO, rather it is subject to multiple, competing influences from both local and remote processes. Several mechanisms have been proposed to explain the origins of the TAV. Some studies suggest that the interannual to decadal variability in the tropical Atlantic arises from local coupled ocean-atmosphere interaction, which involves wind-evaporation-SST coupled feedback and ocean dynamics [e.g., *Chang et al.*, 1997]. Other studies, however, suggest that the TAV can be triggered by remote forcing from climate variability outside of the tropical Atlantic. Both observations and modeling studies suggest that ENSO in the tropical Pacific can exert a significant impact on the interannual variability over the tropical Atlantic [*Curtis and Hastenrath*, 1995; *Enfield and Mayer*, 1997; *Elliot et al.*, 2001; *Saravanan and Chang*, 2000; *Czaja et al.*, 2002; *Wu et al.*, 2002; *Huang et al., 2002*] via the anomalous atmospheric Walker circulation [e.g. *Klein et al., 1999; Chiang et al., 2000*] and the Pacific-North America (PNA) teleconnection [e.g., *Nobre and Shukla,* 1996]. Studies also suggest the ENSO can impact the variability in the equatorial Atlantic through changes of the equatorial easterlies. Within the Atlantic basin, the North Atlantic Oscillation (NAO) has also been found to have a significant impact on the tropical north Atlantic through oceanic [*Hansen and Bezdeck*, 1996; *Yang*, 1999; *Malanotte-Rizzoli et al.*, 2000; *Wu and Liu*, 2002], atmospheric [*Nobre and Shukla,* 1997; *Czaja et al.,* 2002] and coupled ocean-atmosphere processes [*Xie and Tanimoto,* 1998].

Key elements related to the origins of the low-frequency variability in the tropical Atlantic include how the changes of atmospheric circulation in response to local SST anomalies in the tropical Atlantic feed back on the ocean, and what role ocean dynamics plays. The latter is of particular concern for the cause of variability at decadal timescales. A recent study suggests that the north tropical Atlantic variability is largely a response to remote NAO and ENSO forcing without the need to invoke local air-sea coupling and oceanic dynamics [*Czaja et al., 2002*]. Other studies, however, suggest that the local air-sea coupled feedback is important for the full development of the NTA [e.g., *Chang et al.*, 2000; *Wu and Liu*, 2002]. At decadal timescales, modeling studies suggest that the change in ocean heat transport and upper-ocean heat content may impact SST anomalies, and can potentially allow the coupled system to oscillate [*Chang et al.*, 1997; *Huang and Shukla*, 1997], but these effects may be model-dependent [e.g., *Carton et al.*, 1996; *Dommenget and Latif*, 2000; *Seager et al.*, 2001]. In the south tropical Atlantic, some studies suggest that SST anomalies are predominantly generated by wind-driven oceanic circulation [*Häkkinen and Mo*, 2002], while others suggest that SST anomalies are associated with the internal atmospheric variability through changes of surface heat flux [*Sterl and Hazeleger*, 2003; *Barreiro et al.*, this volume].

In this paper, we will explore the origins of the TAV using coupled modeling surgery [CMS, *Wu et al.*, 2003] with focus on the roles of (1) remote ENSO forcing and extratropical-tropical interaction in the Atlantic, (2) local ocean-atmosphere coupling, and (3) oceanic dynamics. The paper is constructed as follows. The observational data and the coupled model used in this study will be described in section 2. In section 3, we will briefly describe the coupled modeling surgery. We then investigate, in section 4, the impacts of both ENSO and extratropical-tropical inter-action on the temporal evolution of the TAV (changes of SST variance forced by the ENSO and NAO have been studied in *Liu et al.* [2003a]) in section 4. In section 5 and 6, the roles of local ocean-atmosphere coupling and oceanic dynamics are studied, respectively. Conclusions and final discussions are in section 7.

2. DATA AND MODEL

The observational data we used in this study are the GISST SST [*Parker et al.*, 1995] from 1903 to 1994. All the data, including the model output discussed below, are seasonally averaged and detrended.

The model we used is the Fast Ocean-Atmosphere Model (FOAM1.0) [*Jacob*, 1997]. The AGCM is the fully parallel R15 version of the NCAR CCM2, but with 19 vertical levels, and with the atmospheric physics replaced by those of CCM3; the OGCM is developed following the GFDL MOM with a horizontal resolution of 2.8° longitude × 1.4° latitude × 16 vertical levels. The coupled FOAM simulation (the control run, CTRL) is integrated for about 350 years from the 456th year of a longer control simulation.

Without flux adjustment, the model captures most major features of the observed tropical climatology as in most state-of-the-art climate models [*Jacob*, 1997; *Liu et al.*, 2003b]. The most serious deficiency in the simulated tropical Atlantic climatology is the double ITCZ, with the ITCZ in boreal winter migrating into the southern hemisphere in FOAM, instead of staying north of the equator as in observations (see Figure 1 of *Liu et al.*, [2003b]). In addition, the model has a cold center emerging in the western equatorial Atlantic in fall and winter, resulting in a cold tongue in the western Atlantic in the annual mean SST. This is opposite to the observation, which has a cold tongue in the eastern equatorial Atlantic. These deficiencies appear to be common in many coupled models, including the NCAR CSM [*Boville and Gent*, 1998]. These deficiencies should be kept in mind in later discussions of the TAV, especially for those features in the equatorial and tropical south Atlantic.

In spite of the deficiencies in the model-simulated climatology, FOAM produces a reasonable TAV, comparable with some state-of-the-art coupled models [e.g., *Dommenget and Latif*, 2000; *Huang et al.*, 2002]. The two leading EOFs of the SSTA show a symmetric mode and a dipole mode [*Liu and Wu*, 2000], consistent with the observations [e.g., *Houghton and Tourre*, 1992]. The leading REOFs of SSTA show the North Tropical Atlantic Mode (Figure 1a1, NTA), the Atlantic Niño Mode (Figure 1a2, ATL-Niño) [*Zebiak*, 1993; *Ruiz-Barradas et al.*, 2000] and the South Tropical Atlantic Mode (Figure 1a3, STA). These modes are similar to the observations although the order of the modes is somewhat altered (Figure 1b1–b3). There are, of course, some deficiencies in the simulated TAV modes. The major deficiency of the FOAM TAV is its 2nd REOF, which is centered in the western equatorial Atlantic and has no counterpart in the observations (not shown). This western equatorial Atlantic mode appears to be caused by the deficient model equatorial annual cycle. In addition, the maximum SST anomaly in the NTA is shifted to the west instead of being near to the western African coast in the observations (Figure 1a1, b1), and variance of both the ATL-Niño (Figure 1a2, b2) and the STA modes (Figure 1a3, b3) is underestimated.

Both NTA and STA exhibit significant interannual variability with timescales ranging from 2 to 6 years, and decadal to interdecadal variability around 30 to 50 years (NTA, Figure 2a1) and 12 years (STA, Figure 2a3), respectively. In contrast to the NTA and STA, the ATL-Niño seems to be dominated by interannual variability (Figure 2a2). The spectrum of these three modes is broadly similar to the observations although the frequencies in the model are somewhat shifted (Figure 2b1–b3). For example, all three modes show substantial interannual variability with a common peak around 3.3 years in the observation and 4 years in the model, respectively. In addition, the observed NTA and STA also exhibit some remarkable multidecadal variability and decadal variability, respectively, although the statistical significance is weaker than those in the model.

In summary, FOAM captures major features of the observed TAV modes in spite of serious deficiencies of the tropical mean climatology in the model. In the following, we will explore the origins of the model-simulated TAV modes by using the coupled modeling surgery approach.

3. COUPLED MODELING SURGERY

The Coupled Modeling Surgery (CMS) broadly represents a set of modeling approaches that can be used to identify the origins and causes of a specific variability mode in the coupled climate system. Two CMS strategies have been implemented: the partial coupling (PC) and the partial blocking

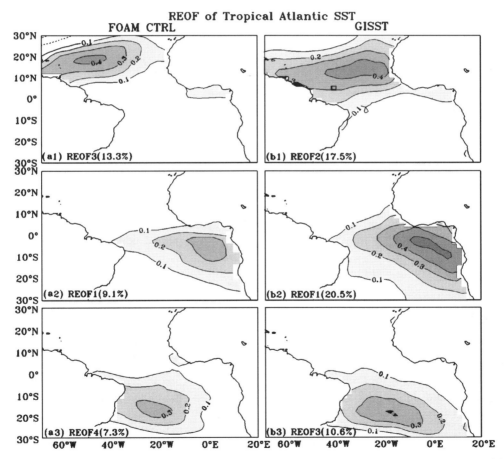

Figure 1. Rotated EOF modes of tropical Atlantic SST anomalies in the FOAM control simulation (left panel) and the observations (GISST 1903–1994, right panel). (a1, a2) the North Tropical Atlantic Mode (NTA), (b1,b2) the Atlantic Niño (ATL-Niño) mode, and (c1, c2) the South Tropical Atlantic Mode (STA). The data are averaged for each season and the linear trend is removed prior to the analysis. The rotated EOF is calculated over the domain (70°W, 20°E) × (30°S, 30°N). Contour interval is 0.1°C.

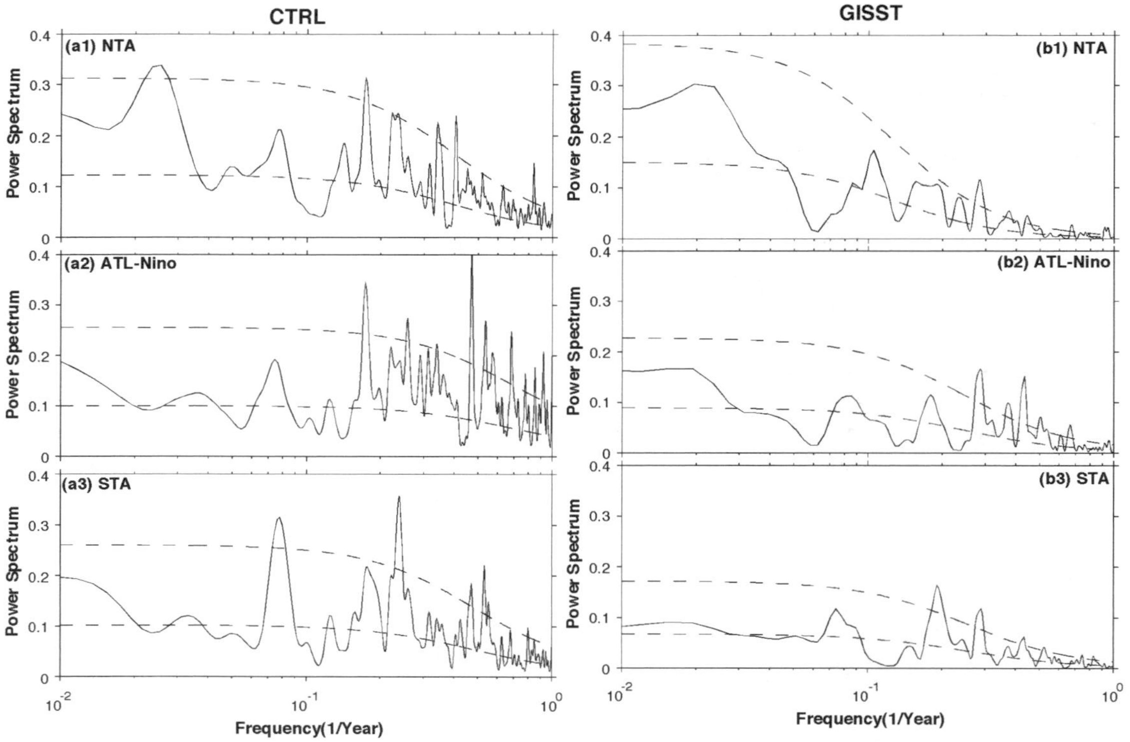

Figure 2. Power spectrum of SST indices corresponding to each TAV mode in the FOAM control simulation (left panel) and observations (right panel). SST indices for the NTA, ATL-Niño and STA are defined as SST anomalies averaged over (70°W, 20°W) × (5°N, 25°N), (30°W, 10°E) × (15°S, 5°N), and (40°W, 0°E) × (5°S, 25°S), respectively. The power spectrum is calculated using a multi-taper method (3 tapers) with 50% and 95% significance levels indicated (dashed lines).

(PB). In the PC experiments, the atmospheric model sees a prescribed annual cycle of SST that is obtained from the CTRL in a specified region (called the PC region) and sees the predicted SST from the full ocean-atmosphere coupling elsewhere. The ocean model is forced by the full atmosphere-ocean flux calculated by the atmospheric model over the entire domain. Over the PC domain, the surface fluxes that drive the ocean model are calculated using the SST predicted by the ocean model at each time step. Over the PC region, variability is generated in both the ocean and atmosphere by processes internal to the atmosphere. The PC approach may underestimate atmospheric stochastic variability due to the enhanced thermodynamic damping by the fixed SST boundary condition [*Barsugli and Batistti,* 1998]. Nevertheless, the fixed SST forcing of an AGCM is a frequently used approach in AGCM experiments, and therefore provides an useful reference to study the role of coupled ocean-atmosphere interaction in generating climate variability [e.g., *Chang et al.,* 2000]. In the PB experiments, "sponge walls" are placed at specified latitudinal bands of the ocean component of the coupled system, such that oceanic teleconnection between different latitudes is cut off. Within the sponge walls, temperature and salinity are restored towards the annual cycle of the CTRL run.

In the following, a series of coupled modeling surgery experiments are conducted to assess the role of remote forcing, local ocean-atmosphere coupling and oceanic dynamics in generation of the TAV modes. Unless explicitly specified, all the surgery experiments are integrated for 350 years from the same initial state as the CTRL experiment.

4. THE ROLE OF REMOTE FORCING

Observational studies have suggested the role of ENSO and the NAO [e.g. *Curtis and Hasterath*, 1995; *Nobre and Shukla*, 1996; *Saravanan and Chang*, 2000; *Czaja et al.*, 2002] in the generation of the TAV. Qualitatively, the observational results are also supported by coupled ocean-atmosphere modeling studies [e.g., *Elliot et al., 2001; Saravanan and Chang,* 2000; *Huang et al.,* 2002; *Wu et al.,* 2002; *Wu and Liu,* 2002]. Recently, *Liu et al.* [2003a] have quantified these two remote impacts using both statistical methods and model-aided dynamic methods. Both approaches suggest that the remote impact contributes to nearly half of the variance of the tropical Atlantic SST variability at interannual and decadal timescales. Here, we will focus on how ENSO and extratropical-tropical interaction modulate the temporal evolution of the TAV, which have not been documented in most of the previous studies. Some preliminary results have been reported in *Wu and Liu* [2002, ENSO].

4.1. The Role of ENSO

FOAM produces a reasonable ENSO [*Liu et al.*, 2000]. Over the Pacific, the leading EOF mode captures the major features of the observed ENSO, with a significant pattern correlation (0.82) and with an amplitude 75% of the observed. The power spectrum of the Niño-3 SST index shows dominant interannual variability around 4 years (not shown). The model also demonstrates a robust teleconnection between ENSO and TAV. The regression of the JFM (January-February-March) SSTA in the tropical Atlantic against the DJF (December-January-February) Niño-3 SST (Figure 3a) reproduces the major features of the observations as noted by *Enfield and Mayer* [1997], with broad warming in both hemispheres, although the warming in the south tropical Atlantic is somewhat stronger than the observations. The changes of SSTA, surface wind and surface turbulent heat flux (latent plus sensible) in the tropical Atlantic associated with ENSO are summarized in Figure 3b, which displays the zonal mean regression of these variables in the tropical Atlantic against the DJF Niño-3 SSTA. The ENSO-induced warming in the NTA

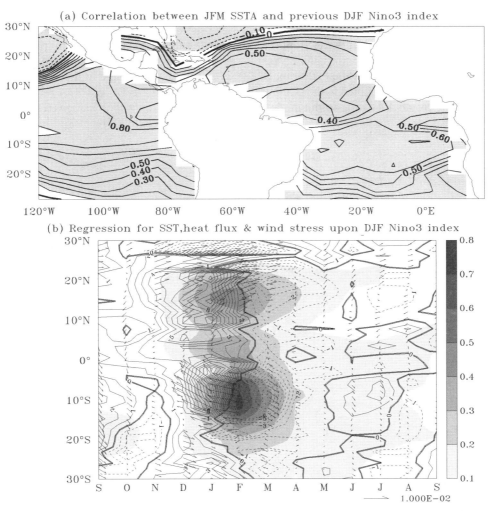

Figure 3. (a) Correlation of JFM (January-February-March) tropical Atlantic SSTA with previous winter (December-January-February) Niño-3 SSTA. (b) Zonally averaged regression of SSTA (shaded), surface heat flux (contours) and wind stress (arrows) with previous winter Niño-3 SSTA. Units for SST, heat flux to ocean and wind stress are °C, W/m^2 and N/m^2 per standard deviation of Niño-3 SSTA, respectively. Contour interval for the heat flux is 1.

matures in February in association with anomalous westerlies inducing heating of the ocean at a rate of 10 W/m^{2}K. After that, the anomalous wind is weakened and the heat flux changes the sign to damp the SST anomaly. The correlation of wintertime (JFM) sea level pressure (SLP) with DJF Niño-3 SST shows a SLP anomaly over the subtropical North Atlantic and an opposite anomaly in the tropical Atlantic (not shown). The former is suggested to be a part of PNA teleconnection, and the latter to be the reflection of an anomalous Walker circulation [e.g., *Covey and Hastenrath*, 1978; *Klein et al.*, 1999]. In the STA, ENSO-induced warming also matures on February. The downward heat flux is primarily associated with the warming of the troposphere induced by ENSO [*Chiang and Sobel*, 2002]. Relative to the NTA, wind anomalies in the STA decay faster.

Although the model captures the major features of the observed ENSO impact on the tropical Atlantic [*Enfield and Mayer*, 1997; *Czaja et al.*, 2002], some differences from the observations should be noted. In the observations, the ENSO induced warming tends to peak in MAM, while in the model it peaks in JFM. This may be partly attributed to the shallow mixed layer depth over the tropical Atlantic in the model. Also, the model overestimates the warming in the south tropical Atlantic, which tends to be associated with the deficiencies of the model climatology over that area.

The above analysis in the model control simulation provides some insight into the influence of ENSO on the TAV, but it can not answer whether the TAV is forced directly by ENSO or not, because the climate in the fully coupled GCM is the product of complex interaction involving feedbacks that are not easily separated.

To truly assess the contribution of the ENSO to the TAV, we perform a PC experiment, denoted as PC-TP, where ENSO is removed in the tropical Pacific by constraining the SST forcing of the atmosphere to the model mean annual cycle in the tropical Pacific (within 20° of the equator). By doing so, variance of ENSO variability in the tropical Pacific is reduced by over 80% and the pattern becomes incoherent. In contrast, the pattern of each TAV mode remains virtually identical to the corresponding mode in the control (Figure 4a1–a3). The temporal evolution of the NTA and STA modes, however, undergoes substantial change in the absence of ENSO. Without ENSO, both the NTA and STA modes do not exhibit any prominent interannual oscillations (Figure 4b1, b3), with vari-

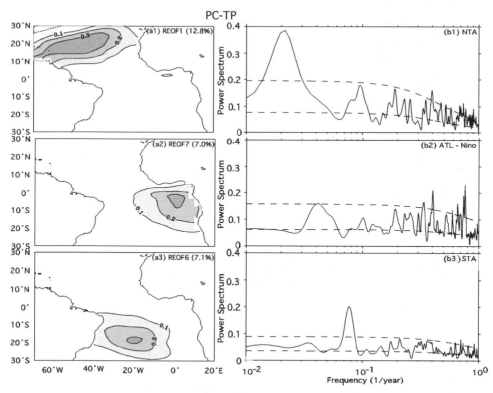

Figure 4. Rotated EOF modes of tropical Atlantic SST anomalies (left panel) and the power spectrum of corresponding SST indices (right panel) in experiment PC-TP, where ENSO is suppressed by deactivating air-sea coupling in the tropical Pacific. The calculation methods for both REOF and power spectrum are the same as those used in the control simulation.

ance between 2 to 6 years reduced by 40 to 50% (compared with the CTRL) for each mode [*Wu et al.*, 2002]. This suggests that the interannual variability of the NTA and STA modes is largely dictated by the ENSO. In contrast to the interannual variability, ENSO has less impact on the decadal variability of the tropical Atlantic. As in CTRL, both the NTA and STA mode remain significant multi-decadal or decadal variability, respectively (Figure 4b1, b3). It is noted that the suppression of ENSO tends to have less impact on the high frequency variability of these modes. Indeed, the ATL-Niño still retains substantial variability within timescales shorter than 3 years (Figure 4b2). In short, PC-TP experiment reveals that ENSO largely dictates the interannual variability of the TAV, but is not a necessary precondition for the TAV, and it also shows that decadal variability is not dictated by ENSO. In the next section, we will explore the role of extratropical-tropical interaction in the generation of the TAV.

4.2. The Role of Extratropical-Tropical Interaction

In the Atlantic basin, climate variability in the extratropics and high latitudes may affect the variability in the tropics through the oceanic [*Hansen and Bezdeck*, 1996; *Yang*, 1999; *Malanotte-Rizzoli et al.*, 2000; *Wu and Liu*, 2002], atmospheric [*Nobre and Shukla*, 1996; *Czaja et al.*, 2002] and coupled ocean-atmosphere [*Xie and Tanimoto*, 1998] processes. These extratropical-tropical pathways seem to be present in the FOAM control simulation. For example, the regression of SSTA with the NAO index shows a tripole-like structure over the northern hemisphere, and extratropical SSTA tends to show a tendency, although not significant, of equatorward movement [*Wu and Liu*, 2002].

To truly assess the impact of extratropical-tropical interaction on the TAV, we perform a sensitivity experiment, denoted as PBC-TAV. PBC-TAV is based on PC-TP, but with ocean-atmosphere coupling deactivated in the extratropical north and south Atlantic (poleward of 30°) by the PC surgery, as well as oceanic variability completely removed from the surface to the bottom by the PB surgery (also everywhere poleward of 30°). In this experiment, extratropical-tropical oceanic teleconnection is completely shut off, but the atmospheric internal variability in the extratropics remains. Statistical assessment shows that the projection of this extratropical atmospheric internal variability on the TAV is only about 10 to 20% [*Liu et al.*, 2003a]. Therefore, in PBC-TAV, SST variations in the tropical Atlantic, if significant, should arise predominantly from local ocean-atmosphere interaction. This experiment is integrated for 150 years.

In PBC-TAV, all three TAV modes including the NTA, STA and ATL-Niño modes remain largely similar to those in PC-TP, except some modest reduction of the magnitudes (about 20% for each mode) (Figure 5a1–a3 vs. Figure 4a1–a3). This readily tells us that the TAV originates predominantly from ocean-atmosphere interaction locally in the tropical Atlantic.

Extratropical-tropical interaction, although not necessary for the generation of the TAV, can lead to significant changes of the temporal evolution of the TAV modes, particularly at decadal timescales. In comparison with PC-TP, the multi-decadal variability of the NTA mode is suppressed substantially, while the interannual variability remains comparable (Figure 5b1 vs. Figure 4b1). A similar change is also observed for the STA mode, where the decadal peak around 12–14 years virtually disappears (Figure 5b3 vs. Figure 4b3). It is conceivable that the multidecadal variability of the NTA mode is mainly associated with the North Atlantic climate, perhaps as a reflection of the Atlantic Multidecadal Oscillation (AMO) as noted in the model [*Jacob*, 1997] and some observational studies [*Enfield et al.*, 2001]. Destroying the AMO by the PB surgery leads to the diminishment of the multidecadal variability of the NTA mode. Over the south Atlantic, the decadal variation of the STA mode tends to be associated with the south Atlantic basin-scale decadal to interdecadal variability (not shown), which has been documented in various observational studies [e.g. *Venegas et al.*, 1997]. For the ATL-Niño mode, the remarkable change seems to be the enhancement of variability around 8 years (Figure 5b2 vs. Figure 4b2), although the mechanism remains to be identified.

The PBC-TAV experiment suggests that the decadal variability in the north and south tropical Atlantic is predominantly associated with the extratropical-tropical teleconnection within the Atlantic. In other words, local ocean-atmosphere interaction alone cannot give rise to decadal oscillations. The detailed processes of how extratropical-tropical teleconnection affect the tropics will remain a topic for further study. In the next section, we will investigate the role of local ocean-atmosphere coupling.

5. THE ROLE OF LOCAL OCEAN-ATMOSPHERE COUPLING

The modeling surgery studies suggest that the TAV arises from local ocean-atmosphere interaction in the tropical Atlantic although remote ENSO forcing and extratropical-tropical interaction within the Atlantic can enhance the variance and modulate the temporal evolution of the TAV. Local forcing of the SSTA includes both atmospheric internal variability and ocean-atmosphere coupled feedback. Some studies suggest the local ocean-atmosphere coupling is critical for the full development of the TAV [*Chang et al.*, 1997, 2000; *Wu and Liu*, 2002], while others suggest that the coupling is not necessary to be invoked [*Czaja et al.*, 2002]. Recent studies show

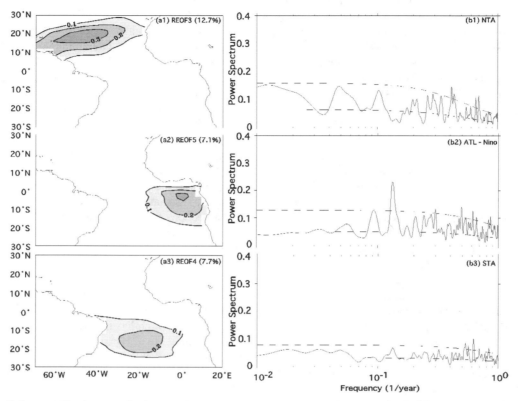

Figure 5. Same as Fig. 4, except for the experiment PBC-TAV, which is based on PC-TP, but further suppresses air-sea coupling and oceanic variability from the surface to the bottom in extratropical north and south Atlantic (>30°).

this positive feedback tends to occur in the deep tropics [*Saravanan and Chang*, 2000; *Chang et al.*, 2000].

To show the development of a NTA SST anomaly, we plot the regression of the surface wind, turbulent heat flux against the MAM NTA SSTA index (Figure 6). The surface wind anomalies are dominated by westerlies, which persist from earlier winter until middle summer. The anomalous westerlies from November to March are largely attributed to both NAO (NDJ) and ENSO (DJFM), and subsequently are likely the response to the warming SSTA. The westerly wind anomalies induce oceanic heating at a rate of about 15 to 20 W/m²K through reducing the evaporative heat loss. After April, the heating is replaced by cooling although the anomalous westerlies are still dominant. This is because the heating induced by the anomalous westerlies is overwhelmed by the enhanced evaporative heat loss due to the increase of specific humidity at the sea surface. In general, the model captures the major features of the development of a NTA event in the observations, although the feedback of the atmosphere to a NTA SST anomaly in the model tends to be over a broader latitudinal range rather than being geographically limited in the deep tropics [*Czaja et al.*, 2002].

To assess the role of local air-sea coupling on generation of the TAV, we perform a PC-experiment, denoted as PC-G, in which air-sea coupling is shut off everywhere including tropical Pacific, extratropical Atlantic and tropical Atlantic. Compared with PBC-TAV, the magnitude of the NTA mode is reduced substantially by about 65% (Figure 7a1). The message is clear: local air-sea coupling is important to the full development of the NTA mode, consistent with the above analysis and previous AGCM studies [e.g., *Chang et al.*, 2000]. In contrast, local air-sea coupling has virtually no contribution to the STA mode, suggesting that the STA mode may arise from atmospheric internal variability (Figure 7a3). The importance of the atmospheric internal variability on the STA mode tends to be consistent with other studies [*Sterl and Hazeleger*, 2003; *Barreiro et al.*, this volume]. For the ATL-Niño mode, local air-sea coupling contributes about 40% to the total variance (Figure 7a2), which is in contrast to ENSO in the tropical Pacific, where air-sea coupling contributes about 80% to the total variance [*Wu et al.*, 2003]. This suggests that both coupling and atmospheric stochastic forcing are important for the equatorial Atlantic variability [*Zebiak*, 1993].

To further show how ocean-atmosphere coupling affects the NTA and STA mode, we perform two additional experiments, PC-NTA and PC-STA, with each integrated for 150 years. Both experiments are based on PBC-TAV, but remove coupling in the latitudinal band (5°, 30°) of one hemisphere. In PC-STA (no coupling in the NTA), the NTA mode is substantially reduced as that in PC-G, but the ATL-Niño and STA modes remain virtually identical to those in PBC-TAV (not shown). In PC-NTA (no coupling in the STA), both the STA and the NTA modes remain virtually identical to those in PBC-TAV in spite of a substantial reduction of the ATL-Niño mode due to the absence of coupling in the south Atlantic (not shown). These two experiments provide a further support for the different role of air-sea coupling in each hemisphere, and also indicate that the contribution of interhemispheric (tropical) interaction to each mode is essentially negligible.

It is noted that the NTA in PC-G still exhibits some multidecadal variability, although the magnitude is much reduced and the frequency is somewhat shifted than that in CTRL and PC-TP (Figure 7b1). This is in contrast to PBC-TAV, where the NTA mode does not show any multidecadal peaks when the oceanic variability is also completely shut off over the North Atlantic. This leads us to conclude that the multidecadal variability over the North Atlantic originates essentially from oceanic process, which can impact the tropics through extratropical-tropical oceanic teleconnection.

6. THE ROLE OF OCEAN DYNAMICS

Previous studies have invoked oceanic dynamics to explain the decadal variability of the tropical Atlantic [e.g., *Huang and Shukla*, 1997; *Chang et al.*, 1997, 2001] because of the longer timescales associated with oceanic circulation, while other studies suggest that even the decadal SST signal can be primarily explained by surface flux alone [e.g., *Carton et al.*, 1996]. Recent OGCM studies by *Seager et al.* [2001] suggest that the role of ocean in the tropical Atlantic is largely passive and damping. Here, we will assess the role of oceanic dynamics in the generation of the decadal TAV based upon an

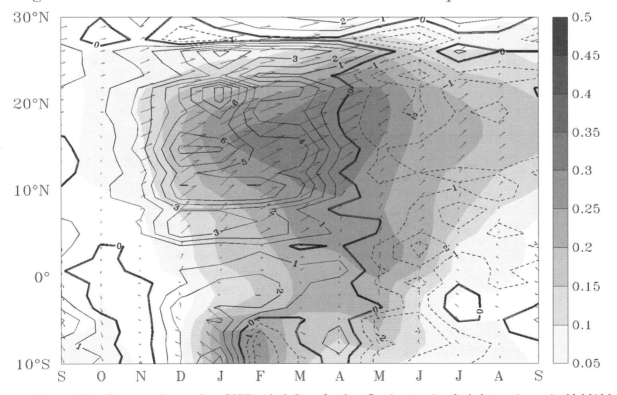

Figure 6. Zonally averaged regression of SSTA (shaded), surface heat flux (contours) and wind stress (arrows) with MAM NTA SSTA index. Units for SST, heat flux to ocean and wind stress are °C, W/m² and N/m² per standard deviation of the MAM NTA SSTA index, respectively. Contour interval for the heat flux is 1.

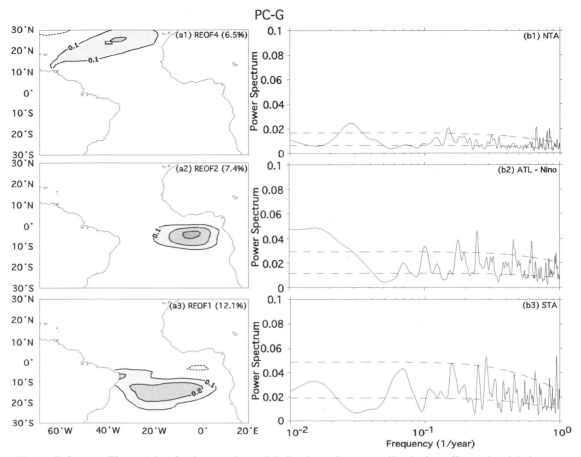

Figure 7. Same as Figure 4, but for the experiment PC-G, where air-sea coupling is shut off over the global ocean.

analysis of upper ocean heat budget and coupled modeling surgery experiments.

To examine the causes of decadal SST changes, we have calculated the regression of the zonal mean heat flux and oceanic heat transport in the upper 50 m with SST at different latitudinal bands. The data has been annually averaged and then low-pass filtered to retain the variability with frequency lower than 5 years. It can be seen that in the north tropical Atlantic, SST anomalies are predominantly driven by the surface heat flux, and damped by the oceanic heat transport; in contrast, in the equatorial and south tropical Atlantic (equatorward of 20°S), SST anomalies are predominantly generated by the oceanic heat transport, and damped by the surface heat flux (Figure 8a, b). Poleward of 20°S, SST anomalies tend to be generated by the surface heat flux again, and damped by the oceanic heat transport.

We further decompose the total ocean heat transport into the anomalous advection, mean advection as well as dissipation components. In the north tropical Atlantic, both the mean and anomalous meridional advection tend to damp SST anomalies (Figure 8c, d), consistent with the previous modeling studies [*Chang et al.*, 2001; *Seager et al.*, 2001]. Physically, this phenomenon can be interpreted as follows. Warm SST anomalies in the north tropical Atlantic are associated with anomalous southwest trades, which drive a southward Ekman flow to bring cold water from subtropics to the tropics (Figure 8c). The mean northward flow, albeit weaker in the model, tends to move the SST anomalies to the subtropics, thus also producing a cooling effect. In the equatorial and south tropical Atlantic (equatorward of 20°S), warming (cooling) is primarily generated by the anomalous advection of the mean vertical temperature gradient (Figure 8e) and damped by the mean advection of the anomalous vertical temperature gradient (Figure 8f) as well as surface heat flux (Figure 8a). Poleward of 20°S, SST anomalies are generated primarily by the surface heat flux, and damped by the anomalous vertical advection, which is somewhat similar to the north tropical Atlantic.

To further investigate the role of ocean dynamics in the generation of tropical Atlantic decadal variability, we perform an additional surgery experiment denoted as TAV-Wind. In this experiment, the wind stress in the tropical Atlantic is constrained to the model climatological annual cycle, but the surface wind speed and thus heat flux are allowed to change as in the CTRL. Outside of the tropical Atlantic, there are no constraints for any variables. Figure 9 displays the ratio of standard deviation of low-passed (>5 years) SST anomalies in

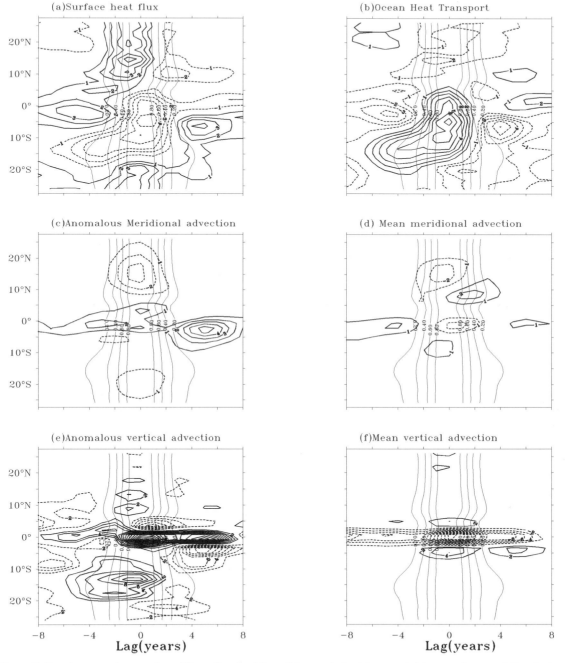

Figure 8. Zonal averaged regression of (a surface heat flux, (b) oceanic heat transport, (c) anomalous meridional advection, (d) mean meridional advection, (e) anomalous vertical advection, and (f) mean vertical advection on the zonal averaged SST at the same latitude in FOAM control simulation. All data have been annually averaged, and then low-passed filtered to retain variability lower than 5 years. Contour interval for each term is 1 W/m^{2}K, except for the anomalous vertical advection (2 W/m^{2}K). Solid lines in each figure are auto-regression of SST index at each latitude, with an interval of 0.2.

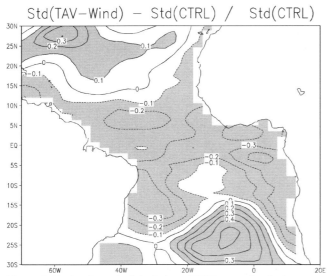

Figure 9. Ratio of standard deviation of SST anomalies in the experiment TAV-Wind relative to the control experiment. SST anomalies are annually averaged and then low-pass filtered to retain variability lower than 5 years. Areas with ratio exceeding ±0.1 are shaded.

TAV-Wind relative to the CTRL. It can be seen that SST variability is reduced substantially between roughly 20°S and 10°N, with the maximum reduction in the ATL-Niño and the STA region. Over the north tropical Atlantic and the south central Atlantic, SST variability is enhanced.

The change of SST variability in TAV-Wind can be readily interpreted in terms of the previous heat budget analysis. From 20°S and 10°N, SST anomalies are generated primarily by the anomalous vertical advection (Figure 8b), and therefore preventing the latter will suppress SST variations. In the north tropical Atlantic (north of 10°N) and subtropical south Atlantic (poleward of 20°S), oceanic dynamics tends to damp SST anomalies (Figure 8b). Specifically, the damping in the north tropical Atlantic is partly undertaken by the anomalous meridional advection, while in the subtropical south Atlantic by the anomalous vertical advection, and thus preventing these anomalous advections can enhance SST variations in these domains significantly.

7. SUMMARY AND DISCUSSION

In this paper, a series of coupled modeling surgery experiments have been performed to study the origins of the tropical Atlantic variability. The model captures the major observed TAV modes: the Atlantic Niño (ATL-Niño), the North Tropical Atlantic (NTA) and the South Tropical Atlantic (STA) modes. It is found that these variability modes predominantly originate from the local tropical Atlantic climate system, while the remote ENSO and NAO forcing can enhance their variance and modulate their temporal evolution. Specifically, the interannual evolution of these modes is predominantly dictated by the remote ENSO, while the decadal evolution of these modes is dictated by extratropical-tropical interaction. Local ocean-atmosphere coupling is critical for a full development of the NTA and the ATL-Niño, but is not necessary for the STA mode. In the north tropical Atlantic, the simulated decadal SST anomalies are primarily associated with surface heat flux and damped by oceanic heat transport, but in the equatorial and south tropical Atlantic, SST anomalies are primarily generated by oceanic heat transport through the vertical anomalous heat advection, and damped by the surface heat flux. In the subtropical south Atlantic (poleward of 20°S), the simulated decadal SST anomalies are also generated by the surface heat flux and damped by oceanic heat transport.

Our coupled modeling surgery study here highlights the different role of air-sea coupling in the north and south tropical Atlantic, and different role of ENSO and extratropical-tropical interaction on the interannual and decadal evolution of tropical Atlantic SST anomalies. The important role of oceanic dynamics in the tropical Atlantic is also highlighted. There are many issues left for future study. For example, we have not clearly assessed the role of each pathway of extratropical-tropical teleconnection in the generation of TAV, nor understood how the decadal oscillations in the north and south tropical Atlantic are established through extratropical-tropical interaction. We found that the STA originates from atmospheric forcing, but it is not clear that the forcing is from atmospheric internal stochastic processes or other external processes such as the adjacent land surface processes.

It should be noted that the conclusions regarding the origins of the south Atlantic climate variability may be physically consistent within the model context, however, the application to the observed south Atlantic variability may be limited because of the deficiencies of the model climatology in this area. Indeed, some deficiencies of the simulated tropical mean seasonal cycle appear even in AGCM simulations [*Wang and Carton*, 2003]. It is not clear to what extent the mean climatology will affect the tropical Atlantic climate variability. In our coupled model, in spite of the significant deficiencies of the mean climatology in the south Atlantic, the model still captures the observed SST modes in the south Atlantic (0°–50°S), for example, the leading two EOF modes show a monopole pattern extending over the entire south Atlantic and a dipole pattern straddling around 25°S, respectively [e.g., *Sterl and Hazeleger*, 2003].

Nevertheless, our coupled modeling study here sheds some light on the origins of the TAV at different timescales, and suggests the need for future studies on the mechanisms and predictability of the TAV. Similar modeling studies by other climate models are especially expected to foster a modeling intercomparison.

Acknowledgments. This study is supported by NASA and NOAA. Computational allocations from NCSA and NCAR are appreciated. Comments from two anonymous reviewers and Dr. John Kutzbach were helpful to improve the paper. This is CCR contribution 828.

REFERENCES

Barreiro, M., A. Giannini, P. Chang, and R. Saravanan, On the role of the southern hemisphere atmosphere circulation in tropical Atlantic variability, this volume.

Basugli, J. J., and D. S. Battisti, The basic effects of atmosphere-ocean thermal coupling on midlatitude variability, *J. Atmos. Sci.*, 55, 477–493, 1998.

Boville, B. A. and P. Gent, The NCAR Climate System Model, version one, *J. Clim.*, 11, 1115–1130, 1998.

Carton, J. A., X. Cao, B. S. Giese, and A. M. da Silva, Decadal and interannual SST variability in the tropical Atlantic, *J. Phys. Oceanogr.*, 26, 1165–1175, 1996.

Chang, P., L. Ji, and H. Li, A decadal climate variation in the tropical Atlantic Ocean from thermodynamic air-sea interactions, *Nature*, 385, 516–518, 1997.

Chang, P., R. Saravanan, L. Ji, and G. C. Hegerl, The effect of local sea surface temperature on atmosphere circulation over the tropical Atlantic sector, *J. Clim.*, 13, 2195–2216, 2000.

Chang, P., L. Ji, and R. Saravanan, A hybrid coupled model study of tropical Atlantic variability, *J. Clim.*, 14, 361–390, 2001.

Chiang, J. C. H., and A. H. Sobel: Tropical tropospheric temperature variations caused by ENSO and their influence on the remote tropical climate, *J. Clim.*, 15, 2616–2631, 2002.

Chiang, J. C. H., Y. Kushnir, and S. E. Zebiak, Interdecadal changes in eastern Pacific ITCZ variability and its influences on the Atlantic ITCZ, *Geophys. Res. Lett.*, 27, 3687–3690, 2000.

Covey, D. L., and S. Hastenrath, Pacific El Niño phenomena and Atlantic circulation, *Mon. Weather. Rev.*, 106, 1280–1287, 1978.

Curtis, S., and S. Hastenrath, Forcing of anomalous sea surface temperature evolution in the tropical Atlantic during Pacific warm events, *J. Geophys. Res.*, 100, 15835–15847, 1995.

Czaja, A., P. van der Vaart, and J. Marshall, A diagnostic study of the role of remote forcing in tropical Atlantic variability, *J. Clim.*, 15, 3280–3290, 2002.

Dommenget, D. and M. Latif, Interannual to decadal variability in the tropical Atlantic, *J. Clim.*, 13, 777–792, 2000.

Enfield, D. B., and D. A. Mayer, Tropical Atlantic sea surface temperature variability and its relation to the El Niño-Southern Oscillation, *J. Geophys. Res.*, 102, 929–945, 1997.

Enfield, D. B., A. M. Mestas-Nunez, and P. J. Trimble, The Atlantic multidecadal oscillation and its relation to rainfall and river flows in the continental U.S., *Geophys. Res. Lett.*, 28, 2077–2080, 2001.

Elliott, J. R., S. P. Jewson, and R. T. Sutton, The impact of the 1997/98 El Niño event on the Atlantic ocean, *J. Clim.*, 14, 1069–1077, 2001.

Folland, C. K., T. Palmer, and D. Parket, Sahel rainfall and worldwide sea temperatures: 1901–85, *Nature*, 320, 602–606, 1986.

Grötzner, A., M. Latif, and T. P. Barnett, A decadal climate cycle in the North Atlantic Ocean as simulated by the ECHO coupled GCM, *J. Clim.*, 11, 831–847, 1998.

Häkkinen, S., and K. Mo, Low-frequency variability of the tropical Atlantic, *J. Clim.*, 15, 237–250, 2002.

Hansen, D. V., and H. Bezdek, On the nature of decadal anomalies in North Atlantic sea surface temperature, *J. Geophys. Res.*, 101, 8749–8758, 1996.

Hastenrath, S., On modes of tropical circulation and climate anomalies, *J. Atmos. Sci.* 35, 2222–2231, 1978.

Houghton, R. W., and Y. M. Tourre, Characteristics of low-frequency sea surface temperature fluctuations in the tropical Atlantic, *J. Clim.*, 5, 765–771, 1992.

Huang, B., and J. Shukla, Characteristics of the interannual and decadal variability in general circulation model of the tropical Atlantic ocean, *J. Phys. Oceanogr.*, 27, 1693–1712, 1997.

Huang, B., P. S. Schopf, and Z. Pan, The ENSO effect on the tropical Atlantic variability: A regionally coupled model study, *Geophys. Res. Lett.*, 29, 2039–2042, 2002.

Klein, S. A., B. J. Soden, and N-C Lau, Remote sea surface temperature variations during ENSO: evidence for a tropical atmospheric bridge, *J. Clim.*, 12, 917–932, 1999.

Jacob, R., Low frequency variability in a simulated atmosphere ocean system, Ph.D. Thesis, University of Wisconsin-Madison, 1997.

Liu, Z. and L. Wu, Tropical Atlantic variability in a coupled GCM, *Atmospheric Science Letter*, 1, 26–36, 2000.

Liu, Z., J. Kutzbach and L. Wu, Modeling climatic shift of El Niño variability in the Holocene, *Geophys. Res. Lett.*, 27, 2265–2268, 2000.

Liu, Z., Q. Zhang, and L. Wu, Remote impact on tropical Atlantic climate variability: stochastical assessment and dynamic assessment, *J. Clim.*, in press, 2003a.

Liu, Z., B. Otto-Bliesner, J. Kutzbach, L. Li, and C. Shields, Coupled climate simulation of the evolution of global monsoons in the Holocene, *J. Clim.*, in press, 2003b.

Marshall, J., Y. Kushnir, D. Battisti, P. Chang, A. Czaja, R. Dickson, J. Hurrel, M. McCartney, R. Saravanan, and M. Visbeck: North Atlantic climate variability: Phenomena, impacts and mechanisms, *Int. J. Climatol.*, 21, 1863–1898, 2001.

Malanotte-Rizzoli, P., K. Hedstorm, H. Arango, and D. B. Haidvogel, Water mass pathways between subtropical and tropical ocean in a climatological simulation of the North Atlantic Ocean circulation, *Dyn. Atmos. Oceans.*, 32, 331–171, 2000.

Mann, M., and J. Lees, Robust estimation of background noise and signal detection in climate time series, *Climate Change*, 33, 409–445, 1996.

Moura, A. D., and J. Shukla, On the dynamics of droughts in northeast Brazil: Observations, theory and numerical experiments with a general circulation model, *J. Atmos. Sci.*, 38, 2653–2657, 1981.

Nobre, C., and J. Shukla, Variations of sea surface temperature, wind stress, and rainfall over the tropical Atlantic and South America, *J. Clim.*, *9*, 2464–2479, 1996.

Parker, D. E., C. K. Folland, A. Bevan, M. N. Ward, M. Jackson, and F. Maskell, Marine surface data for analysis of climate fluctuations on interannual to century time-scales, in *Natural Climate Variability on Decadal to Century Time Scales*, edited by D. G. Martinson et al., pp. 241–250, National Academy Press, 1995.

Ruiz-Barradas, A., J. A. Carton, and S. Nigam, Structure of interannual-to-decadal climate variability in the tropical Atlantic sector, *J. Clim.*, *13*, 3285–3297, 2000.

Saravanan, R., and P. Chang, Interaction between tropical Atlantic variability and El Niño-Southern Oscillation, *J. Clim.*, *13*, 2177–2194, 2000.

Seager, R., Y. Kushnir, P. Chang, N. Naik, J. Miller, and W. Hazeleger, Looking for the role of the ocean in tropical Atlantic decadal climate variability, *J. Clim.*, *14*, 638–655, 2001.

Sterl A., and W. Hazeleger, Coupled variability and air-sea interaction in the South Atlantic Ocean, *Clim. Dyn.*, in press

Venegas, S. A., L. A. Mysak, and N. Straub, Atmosphere-ocean coupled variability in the South Atlantic, *J. Clim.*, *10*, 2904–2920, 1997.

Wang, J., and J. Carton, Modeling climate variability in the tropical Atlantic, *J. Clim.*, *16*, 3858–3876, 2003.

Wu, L., and Z. Liu, Is Tropical Atlantic variability driven by the North Atlantic Oscillation?, *Geophys. Res. Lett*, *29*, doi: 10.1029/2002GL014939, 2002.

Wu, L., Q. Zhang, and Z. Liu, Searching for the role of ENSO in the tropical Atlantic variability, *CLIVAR Exchanges*, *25(7)*, 20–24, 2002.

Wu, L., Z. Liu, R. Gallimore, R. Jacob, D. Lee, and Y. Zhong, Pacific decadal variability: The tropical Pacific mode and the North Pacific mode, *J. Clim.*, *16*, 1101–1120, 2003.

Xie, S.-P., and J. A. Carton, Tropical Atlantic variability: Patterns, mechanisms, and impacts, this volume.

Xie, S.-P., and Y. Tanimoto, A pan-Atlantic decadal climate oscillation, *Geophys. Res. Lett.*, *25*, 2185–2188, 1998.

Yang, J., A linkage between decadal climate variations in the Labrador Sea and the tropical Atlantic Ocean, *Geophys. Res. Lett.*, *26*, 1023–1026, 1999.

Zebiak, S. E., Air-sea interaction in the equatorial Atlantic region, *J. Clim.*, *6*, 1567–1586, 1993.

L. Wu and Z. Liu, Center for Climatic Research, University of Wisconsin-Madison, 1225, West Dayton Street, Madison, Wisconsin 53706. (lixinwu@wisc.edu; zliu3@wisc.edu)

Q. Zhang, State Key Laboratory of Atmospheric Sciences and Geophysical Fluid Dynamics, Chinese Academy of Sciences, Beijing 100029, P.R. China. (zhq@lasg.lap.ac.cn)

Thermodynamic Coupling and Predictability of Tropical Sea Surface Temperature

R. Saravanan

National Center for Atmospheric Research, Boulder, Colorado

Ping Chang

Department of Oceanography, Texas A&M University, College Station, Texas

Air–sea coupling involves the exchange of both momentum and heat between the atmosphere and the ocean. Dynamic coupling, which relates to the momentum exchange, is believed to play the dominant role in the tropics, especially in phenomena such as the El Niño–Southern Oscillation in the tropical Pacific. However, thermodynamic heat exchange between the atmosphere and the ocean can also play a significant role in air–sea coupling. This is especially true in the tropical Atlantic, where dynamic coupling may be of secondary importance. In this study, the role of this thermodynamic air–sea coupling is studied using an atmospheric general circulation model coupled to a slab ocean model. Two thermodynamic feedback mechanisms are considered: the reduced thermal damping mechanism and the wind-evaporation-SST (WES) feedback. It is shown that thermodynamic coupling leads to amplification and increased persistence of surface wind variability in the deep tropical Atlantic region. This effect is anisotropic, being stronger in the meridional component than in the zonal component of the surface wind. These features cannot be explained by the isotropic reduced thermal damping mechanism, and indicates a possible role for the WES feedback. Predictability experiments using observed December sea surface temperature (SST) initial conditions were also carried out. These show that thermodynamic coupling can lead to forecasts of north tropical Atlantic SST that are significantly better than persistence forecasts during the boreal spring. These results mean that thermodynamic coupling certainly leads to a richer, more complex set of interactions than a local, Hasselmann-type of red-noise model would imply.

1. INTRODUCTION

Coupling between the atmosphere and the ocean is an important contributor to climate variability. A classic example of atmosphere–ocean interaction is the El Niño–Southern Oscillation (ENSO) phenomenon in the tropical Pacific [cf. *Wang and Picaut*, 2004]. ENSO primarily involves exchange of momentum between atmospheric surface winds and oceanic currents, which is sometimes referred to as "dynamic coupling", because thermodynamic processes such as surface heat flux exchange play a secondary role in this interaction. Dynamic coupling has been the subject of extensive studies [*Neelin et al.*, 1998], because of its crucial role in the ENSO mechanism [*Bjerknes*, 1969].

Another example of atmosphere–ocean interaction, which has received somewhat less attention, is one dominated by the heat flux exchange between the atmosphere and the ocean. We may refer to this as "thermodynamic coupling" to distinguish it from the predominantly dynamic coupling associated with El Niño. The canonical example of such interaction is the red-noise climate model of *Hasselmann* [1976]. However, more complex mechanisms of thermodynamic coupling have also been proposed in the literature: e.g., the reduced thermal damping effect [*Barsugli and Battisti*, 1998] and the Wind-Evaporation SST (WES) feedback mechanism [*Xie and Philander*, 1994; *Chang et al.*, 1997].

The reduced thermal damping effect can be explained as follows: when the atmosphere warms in response to a positive sea surface temperature (SST) anomaly, the resulting decrease in the air–sea temperature gradient leads to a decrease in the surface heat exchange and thus reduced thermal damping of the SST anomaly. Although many studies have emphasized the role of reduced thermal damping in the midlatitudes [e.g., *Bladé*, 1997], the same mechanism can work in the tropics as well [*Saravanan and Chang*, 1999]. The WES mechanism, which is inherently a tropical mechanism, works as follows: consider a positive SST anomaly north of the equator and a negative anomaly to the south. This induces a northward surface flow across the equator, which would be deflected westward by the Coriolis force in the southern hemisphere and eastward in the northern hemisphere. This would increase the windspeed over the negative southern SST anomaly, cooling it further through surface evaporation, and decrease the windspeed over the positive northern SST anomaly, warming it further. The net effect is a positive feedback on the original SST anomaly.

Although *Chang et al.* [1997] hypothesized a positive WES feedback associated with the "Atlantic dipole" mode [*Hastenrath and Heller*, 1977], subsequent studies have questioned the existence of both the dipole mode and the positive feedback [*Houghton and Tourre*, 1992; *Enfield and Mayer*, 1997; *Dommenget and Latif*, 2000; *Sutton et al.*, 2000a; *Czaja et al.*, 2002]. The evidence so far suggests that the meridionally-extended dipole mode, say with maxima at 15°S and 15°N, is most likely not a physical mode. This still leaves open the possibility that the cross-equatorial SST gradient mode is important and that the WES feedback operates in the deep tropics [cf. *Chang et al.*, 2000; *Okumura et al.*, 2001; *Barreiro et al.*, 2004]. How far away from the deep tropics the atmosphere is responsive to SST anomalies is still a matter of debate, and studies addressing this issue have thus far been inconclusive [e.g., *Watanabe and Kimoto*, 1999; *Sutton et al.*, 2000b].

In this study, we explore the role of thermodynamic coupling in tropical SST variability and predictability. We focus on the tropical Atlantic, because many studies have indicated that thermodynamic coupling may indeed play the dominant role in this region [e.g., *Chang et al.*, 1997; *Xie and Carton*, 2004]. We ask the following questions. How does thermodynamic coupling affect the amplitude and persistence of variability in the tropical Atlantic? How does the coupling affect predictability of SST? An earlier study by *Saravanan and Chang* [1999] suggested that the reduced thermal damping effect was indeed important. *Chang et al.* [2003] have recently suggested that thermodynamic coupling can contribute significantly to predictability on seasonal timescales. In this study, we use a hierarchical coupling approach [e.g., *Sutton et al.*, 2000b; *Huang et al.*, 2002; *Wu et al.*, 2004] to explore the dynamical underpinnings of this predictability and how it relates to the degree of coupling between the atmosphere and the ocean. In particular, we attempt to distinguish between the role of reduced thermal damping and the WES feedback. In Section 2, we describe the datasets and model configurations used in this study. Section 3 analyzes the role of thermodynamic coupling in atmospheric variability, by examining how the variance and persistence of a particular mode are affected by the coupling. A more practical assessment of predictability, using ensemble forecasts starting from observed initial conditions, is presented in Section 4, followed by concluding remarks in Section 5.

2. MODEL INTEGRATIONS AND DATASETS

Distinguishing between dynamic and thermodynamic coupling in the real atmosphere–ocean system is quite difficult because the two often occur together. Therefore, we turn to numerical models of the atmosphere–ocean system. We construct a coupled model that supports only thermodynamic coupling. Our experimental configuration consists of an atmospheric general circulation model coupled to a very simple mixed layer ocean model. The atmospheric model that we use is the Community Climate Model, Version 3.6.6 (CCM3) developed at the National Center for Atmospheric Research (NCAR). We shall refer to the coupled model using CCM3 as the CCM3-ML model.

CCM3 is an atmospheric general circulation model (AGCM) incorporating a comprehensive suite of physical parameterizations [*Kiehl et al.*, 1998]. Global-scale features of the climatology and variability of CCM3 are documented by *Hurrell et al.* [1998] and *Saravanan* [1998]. *Chang et al.* [2000] analyze the tropical Atlantic climatology of CCM3 integrations and find the simulations of surface fluxes and precipitation to be fairly realistic. The mixed layer ocean model is just a slab with spatially varying depth, governed by the simple equation:

$$\frac{\partial T_o}{\partial t} = \frac{F}{\rho_w C_w H} + Q, \qquad (1)$$

where T_o is the slab ocean temperature, F is the heat flux into the ocean, ρ_w is the density of sea water, C_w the heat capacity, and H is the spatially-varying mixed layer depth. Q is a seasonally-varying "q-flux" that represents oceanic heat transport, which is diagnosed from a control run of CCM3 with prescribed, seasonally-varying, SST values. Depth H of the mixed layer was specified from the annually-averaged observational estimates of mixed layer depth, which is derived from *Levitus* [1982] using a constant potential density difference criterion between the surface and the bottom of the mixed layer. Figure 1 shows the annual-mean mixed layer depth in the tropical region of interest to this study. Note that mixed layer depths in the deep tropical Atlantic region are quite shallow, ranging from 10 m to 40 m.

Since the slab ocean has no ocean dynamics, the CCM3-ML model allows only thermodynamic coupling, and no dynamic coupling. Two kinds of integrations were carried out using the CCM3-ML model: (i) a 100-year long control integration starting from an arbitrary initial condition; (ii) ensemble forecasts, each 9-months long, starting from the December 15th of each year from 1959 to 1997. For the ensemble forecasts, the AGCM was initialized with the observed atmospheric state for Dec. 15th of each year. An ensemble size of ten was used, with the atmospheric initial state being perturbed slightly to generate the ten different initial conditions for the forecasts [cf. *Shukla et al.*, 2000]. The slab ocean was initialized with observed December SST values for each of the years.

To compare the coupled variability in CCM3-ML with uncoupled variability, two additional integrations were also analyzed: a control uncoupled integration of CCM3 forced by the climatological annual cycle of SST and an AMIP-style uncoupled integration of CCM3 forced by observed SST from 1950–1999. For observed SST values, our source is the reconstructed dataset of *Smith et al.* [1996]. For observed atmospheric data, our source is the NCEP/NCAR reanalyis from 1958–1998.

We use monthly averages, and derived seasonal averages, throughout our analysis. The following is a list of names and summary descriptions of the various datasets used in this study:

OBS: Monthly-averaged observational data
ACYC: Control integration of CCM3, with repeating annual cycle of SST (100 years)
AMIP: AMIP-style integration of CCM3, forced by observed monthly-mean SST from 1950–1999 (50 yrs)
MIXL: Control integration of CCM3-ML, with repeating annual cycle of Q (100 years)
GIC: Ensemble forecasts using observed December SST initial conditions globally (10 member ensemble, each 9 months long, 1959–1997)
AIC: Ensemble forecasts using observed December SST initial conditions in the Atlantic basin, with climatological SST specified elsewhere

3. ATMOSPHERIC VARIABILITY

First, we consider how thermodynamic coupling can affect atmospheric variability in the tropics. In the Introduction, we described two possible mechanisms, reduced thermal damping and the WES feedback, that can amplify atmospheric variability through coupling. By comparing the characteristics of atmospheric variability in CCM3 integrations with different degrees of coupling, we can verify if this amplification is indeed present.

Studies of the WES feedback have suggested that the tropical Atlantic is perhaps the region with conditions most favorable for it to occur [*Chang et al.*, 1997]. Although earlier studies indicated that the WES feedback may occur over the whole tropical belt, subsequent studies have suggested that the feedback may be confined to the deep tropics [*Chang et al.*, 2000]. The feedback is also likely to be most noticeable in the meridional surface wind in this region. Therefore, we carry out empirical orthogonal function (EOF) and correlation analysis of the variability of 1000 hPa meridional wind (V1000) in the deep tropical Atlantic in the observations and in three different CCM3 integrations: ACYC, AMIP, and MIXL. These CCM3 integrations represent different degrees

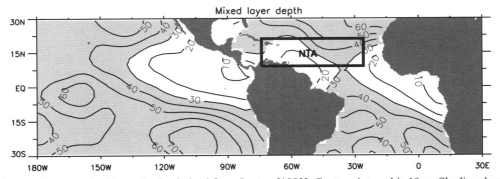

Figure 1. Annual-mean mixed layer depths derived from *Levitus* [1982]. Contour interval is 10 m. Shading denotes values > 30 m. The NTA region is outlined in bold lines.

of thermodynamic coupling, with ACYC being uncoupled, AMIP representing the partial effects of coupling through the specified SST variability, and MIXL exhibiting the effects of full thermodynamic coupling. One feature common to all CCM3 integrations is that the mean annual cycle of SST is the same as in the observations. In ACYC and AMIP, the annual cycle of SST is specified, and in the MIXL integration, our choice of seasonally and spatially varying Q in Equation (1) ensures that the SST climatology matches observations. Note that the MIXL integration does not represent ENSO, which is an important source of remote forcing in the tropical Atlantic. However, the ACYC integration also excludes ENSO, thus making comparisons between the MIXL and ACYC runs quite appropriate. We would have also liked to include fully coupled integrations, with dynamic coupling, in our analysis. Unfortunately, fully coupled models tend to have a rather poor simulation of the SST climatology in the tropical Atlantic [*Davey et al.*, 2002].

Figure 2 shows the dominant statistical mode of variability (EOF1) of V1000 in the deep tropical Atlantic (15°S–15°N) in observations and in the different CCM3 integrations. The most striking feature is that in all cases there is a monopolar mode of variability with maximum amplitude near the equator. The center of this mode is displaced slightly north of the equator in observations, but located right at the equator in all of the CCM3 integrations. Interestingly, even the uncoupled ACYC integration shows this mode of variability, although it is weaker than in the observations by about 30%. However, as the degree of coupling increases, the amplitude of the mode

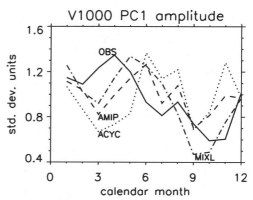

Figure 3. Root-mean-square amplitude of PC1 of Atlantic V1000 as a function of the calendar month (1 = January…): OBS (solid), ACYC (dotted), AMIP (dashed), MIXL (dash-dotted). Ordinate units are in standard deviations of the PC.

also increases. In the AMIP integration, the mode amplitude is somewhat stronger than in the observations, indicative of some model deficiencies. In the MIXL integration, the mode amplitude is even stronger than in the AMIP integration, presumably through the effect of coupling. This suggests that the MIXL model overestimates the oceanic feedback, and allowing dynamical coupling and active ocean dynamics could act as a damping effect. The zonal structure of the mode also changes somewhat with increased coupling, with western intensification along the equator.

To further check if the simulated mode of V1000 variability is realistic, we consider the root-mean-square amplitude of the principal component of EOF1 for each calendar month (Figure 3). We see that both the observations and simulations show a seasonal cycle in the amplitude of the mode. The minimum amplitudes occur during the Northern fall season. The maximum amplitude occurs during April in the observations and a month or two later in the simulations. This is broadly consistent with studies of the cross-equatorial gradient mode, which manifests itself in the Northern spring and summer seasons [e.g., *Chang et al.*, 2000; *Sutton et al.*, 2000a]. Except for some phase differences, the MIXL simulation also manages to capture a realistic seasonal amplitude variation of the mode.

To try and understand the seasonal variations in the V1000 variability, we analyze the climatological surface (1000 hPa) wind patterns during the Northern spring (March–April–May) and the Northern fall (September–October–November) seasons (Figure 4). During the MAM season, the surface wind convergence region straddles the equator, extending from about 5°S to 5°N, with a NE–SW tilt. The maximum variability of EOF1 of V1000 (Figure 2) is also approximately co-located with this region of convergence and low windspeeds. During the SON season, the region of surface wind convergence shifts

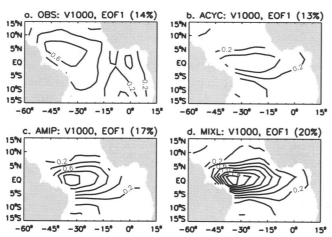

Figure 2. Dominant Empirical Orthogonal Function (EOF1) of meridional wind (V) at 1000 hPa in the deep Tropical Atlantic for different datasets: a. OBS (observations); b. ACYC (uncoupled atmosphere); c. AMIP (atmosphere forced by observed SST); d. MIXL (atmosphere coupled to mixed layer). The EOF patterns are normalized to one standard deviation. Contour interval is 0.2 m s^{-1}. Percentage variance explained by each EOF is shown in parentheses.

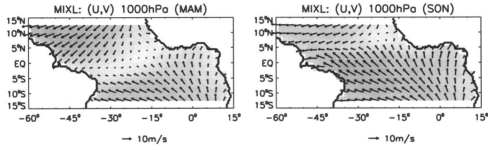

Figure 4. Climatological surface wind (vectors) and wind speed (shading) for the MIXL integration: a. March-April-May; b. September-October-November.

northward, and the mean windspeeds at the equator become considerably larger. Observed surface wind variability also shows a similar northward shift and colocation of the region of maximum wind variability with the region of minimum wind speed (not shown).

The absolute strength of the WES feedback would depend only upon the magnitude of the anomalous windspeed. However, strong mean winds, acting upon the anomalous air–sea surface temperature difference, would tend to damp SST anomalies, and thus act to counter the WES feedback [*Saravanan and Chang*, 2000; *Chang et al.*, 2000]. Therefore, what is of interest is the relative strength of the WES feedback, which would depend upon the ratio of the anomalous windspeed to the mean windspeed. This means that the WES feedback would be most effective where the mean wind is of the same magnitude as the amplitude of wind variations. The amplitude of seasonally-averaged V1000 variations is about 1 m/s (Figure 2), whereas the mean wind speed between 15°S and 15°N in the tropical Atlantic can be up to 10 m/s (Figure 4). This suggests that the relative impact of the WES feedback would be strongest in regions where the mean wind speed is on the order of 1 m/s, i.e., in the vicinity of the convergence regions. The WES feedback would also be stronger when the convergence region is closer to the equator, because the positive surface flux feedback tends to be confined to the deep tropics [*Chiang et al.*, 2002].

Next we consider the temporal persistence characteristics of V1000 variability. Figure 5 shows the autocorrelation of the time series of V1000 EOF1 (i.e., PC1) at different lags. As expected, the ACYC integration shows very little autocorrelation for V1000 even after one month, due to the fact that atmospheric memory is rather short. However, the observed autocorrelation is fairly high at one month lag (about 0.7), and the AMIP and the MIXL integrations exhibit similar autocorrelation values. This corresponds to a decay timescale of about 3 months. In the AMIP case, this can be explained through the persistence of the observed SST anomalies, but in the MIXL case, the increased autocorrelation is clearly due to thermodynamic coupling. The MIXL integration is also able to simulate quite well the decay of the observed autocorrelation function at longer lag times.

So far we have focused on the variability of the meridional wind in the deep tropics. To determine if the effects of the feedback are isotropic, we next consider the variability of zonal wind at 1000 hPa (U1000). Figure 6 shows the structure of EOF1 of U1000 for the different cases, which again tends to be monopolar in the deep tropics. The ACYC integration shows a deep tropical mode of U1000 variability that is considerably weaker than in the observations, with the center of action shifted northward. The AMIP integration shows variability quite similar in amplitude to the observations, although the maximum variability occurs slightly to the south of the equator, unlike in the observations. The MIXL integration accentuates this difference when compared to the observations, with fairly strong variability centered around 5°S. The maximum U1000 in the MIXL integration is about 50% larger than in the observations, which is somewhat weaker than the factor of two difference seen for V1000. By comparing U1000 variability in the MIXL and ACYC integrations, it is clear that coupling leads to enhanced variability in the equatorial region, although the effect is not as dramatic as for V1000. We also computed the combined vector EOFs of U1000 and

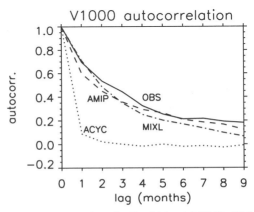

Figure 5. Lag autocorrelation of PC1 of Atlantic V1000: OBS (solid), ACYC (dotted), AMIP (dashed), MIXL (dash-dotted).

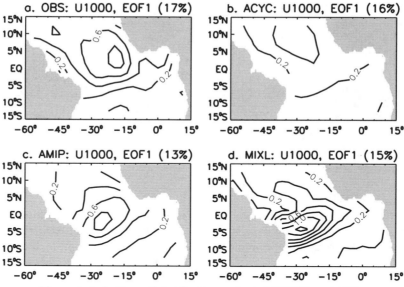

Figure 6. As in Figure 2, but for the zonal wind (U1000) variability.

V1000 (not shown), where we find that the dominant EOFs show significantly stronger meridional orientation in the MIXL run as compared to the ACYC run.

Like the differences in the amplification of variability, the temporal persistence characteristics of U1000 are also somewhat different when compared to V1000, as we see in Figure 7. The figure compares the lag autocorrelation of EOF1 of V1000 and U1000, and we see that V1000 variability is considerably more persistent than that of U1000. The decay timescale for V1000 variability is about 3 months, whereas it is only about 1.5 months for U1000 variability.

Overall, we see that thermodynamic coupling tends to amplify atmospheric variability in the deep tropics, perhaps to unrealistically large amplitudes. This means that thermodynamic coupling provides less of a feedback for U1000 vari-

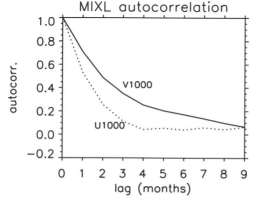

Figure 7. Lag autocorrelation of PC1 in the Atlantic region for the MIXL integration: V1000 (solid), U1000 (dashed).

ability, as compared to V1000 variability. One can conclude from Figure 2 that thermodynamic coupling can significantly affect atmospheric variability in the deep tropics in the Atlantic.

4. SST PREDICTABILITY

Previously, we described how thermodynamic feedback from the ocean results in increased amplitude and persistence of atmospheric variability. In this section, we present evidence that this feedback is not just of academic interest, and that it can lead to practical predictability in the tropical Atlantic region. We have carried out ensemble forecasts using the CCM3-ML model as described in Section 2. Two sets of forecasts were carried out, using observed December SST initial conditions globally (GIC), and in the Atlantic region only (AIC).

Using 16 years of data from the GIC and AIC experiments, *Chang et al.* [2003] found that the CCM3-ML model showed significant forecast skill in the tropical Atlantic region on seasonal timescales, even though dynamical oceanic feedbacks were excluded. We have repeated the calculation using 39 years (1959–1997) of data to assess the predictability of SST in a North Tropical Atlantic (NTA) box (76°W–26°W, 9°N–21°N; see Figure 1). Figure 8 shows the statistically-averaged skill associated with the forecasts as measured by the correlation between the observed and predicted SST anomalies in the NTA region for the GIC and AIC forecasts. Also shown in the figure is the skill of the persistence forecast, where the initial SST anomaly is assumed to persist for all time. Note that the GIC forecasts easily beat persistence, showing considerable skill all the way up to 6–7 months of lead time. Part of this impressive skill comes from the ability of the CCM3-ML model to capture the remote influence of ENSO [e.g., *Enfield and Mayer*, 1997; *Huang et al.*, 2002], when initialized with the global observed SST. This is true even though the forecasted evolution of the ENSO SST anomaly in the tropical Pacific develops errors after about three months of lead time, due to the lack of dynamical coupling (not shown). As it takes 2–3 months for errors in tropical Pacific SST anomalies to affect tropical Atlantic SST anomalies, the Atlantic forecasts still remain skillful for the first six months or so.

When we consider the AIC case, where the remote influence of ENSO is absent, the CCM3-ML forecasts still manage to beat persistence at long lead times. Although the decay of anomaly correlation with increasing lag does not fit an exponential curve very well, the effective decay timescale for the AIC forecasts is about 8 months. Note that the only coupling effect present in the AIC experiments is thermodynamic coupling local to the Atlantic basin. This means that all of the predictive skill in the AIC forecasts is attributable to thermodynamic coupling.

Different types of feedbacks can be associated with thermodynamic coupling: e.g., the reduced thermal damping effect, the WES feedback, and possibly others. Which of these feedbacks is responsible for the improved skill of the AIC experiment relative to persistence seen in Figure 8? The improved skill cannot be attributed to the reduced thermal damping effect, because the optimal forecast for a *local* red-noise model of atmosphere–ocean interaction [*Hasselmann*, 1976; *Barsugli and Battisti*, 1998] is simply damped persistence, where the initial SST anomaly monotonically decays to zero. In the correlation measure that we have used, the red-noise model will have the same skill as the persistence forecast. Furthermore, the ensemble-averaged SST anomalies do not always evolve monotonically in the AIC forecasts, as would be expected from a red-noise model.

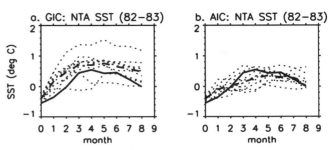

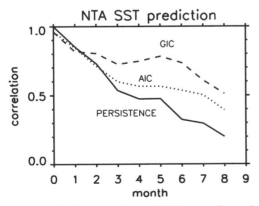

Figure 8. Correlation between observed SST anomalies and ensemble-average predicted SST anomalies in the NTA region: persistence forecast (solid), GIC (dashed) and AIC (dotted). The NTA region is defined in Figure 1.

Figure 9. Evolution of NTA SST anomaly during the 1982–83 warm ENSO event as a function of forecast lead time: observed (solid), ensemble-average (dashed), and ensemble members (dotted). (a) GIC forecasts; (b) AIC forecasts. Month 0 represents December.

To illustrate the non-monotonic evolution of predicted SST anomalies, we consider how CCM3-ML forecasts the evolution of Tropical Atlantic anomalies associated with the strong El Niño event of 1982–83. Figure 9a shows the observed SST anomaly (solid), the ensemble-average forecast (dashed), and the individual ensemble members (dotted), for the GIC forecasts. The observed anomaly in the NTA region starts out with an amplitude of about –0.5°C in December 1982 and increases rapidly over the next 3–4 months to reach a value of +0.5°C, followed by a slow decay. Presumably, this Atlantic warming is a result of the warming in the equatorial Pacific, transmitted via the anomalous Walker circulation and the associated tropical tropospheric warming [e.g., *Saravanan and Chang*, 2000]. The ensemble-averaged GIC prediction tracks the observed SST evolution rather well for this case. Although the individual ensemble members show some spread, the overall evolution is quite consistent.

Interestingly, the AIC forecasts (Figure 9b) also capture this Atlantic warming, albeit with a weaker amplitude. This simulated SST evolution cannot be explained by local, Hasselmann-type of red-noise models, because they can predict the decay of an initial SST anomaly to zero, but not the change of sign. This suggests that *nonlocal* thermodynamic interactions (within the tropical Atlantic) are responsible for the change of sign of the simulated SST anomaly. The WES feedback could well play a role in such interactions. In the WES feedback, the wind speed change does not simply respond to local SST, but to the SST gradient that brings a nonlocal effect to thermodynamic interactions. *Chang et al.* [2004] recently explored this effect in a theoretical framework. Further mechanistic studies are needed to confirm whether the WES feedback is indeed playing a role in the coupled integrations.

5. SUMMARY AND DISCUSSION

Research in tropical atmosphere–ocean interaction has typically been focused on *full* coupling between these two components of the climate system, involving both dynamic and thermodynamic feedbacks. Although there have been several studies of thermodynamic coupling between the atmosphere and a motionless slab ocean, these have been more focused on the middle latitudes [cf. *Alexander et al.*, 2002]. The tropics present a different parameter regime from the middle latitudes for two reasons: the oceanic mixed layer is much shallower, and the atmosphere is more responsive to SST anomalies [*Saravanan and Chang*, 1999]. In this study, we have explored the role of thermodynamic coupling in the tropics.

We have analyzed variability in the tropical Atlantic region in a comprehensive atmospheric GCM coupled to a slab ocean model. This coupled model only supports thermodynamic coupling, or surface heat exchange, between the atmosphere and the ocean. It does not support dynamic coupling, i.e., surface momentum exchange. We analyzed the surface wind variability in the deep tropics, and found that the dominant mode of variability straddles the equator, and is co-located with the region of minimum climatological wind speed. By comparing the variability in the coupled model to uncoupled atmosphere-only integrations, we are able to demonstrate that coupling leads to significant amplification of surface wind variability. We also find that the thermodynamic feedback from the ocean leads to increased persistence of surface wind variability. Furthermore, this effect is not isotropic. The meridional surface wind component exhibits significantly more persistence as compared to the zonal surface wind component.

We find that the thermodynamic ocean–atmosphere feedback also leads to significant forecast skill in the tropical Atlantic for lead times of 6–8 months, beating out persistence forecasts. The CCM3-ML model is able to simulate not just the local decay of SST anomalies in the North Tropical Atlantic, but also their growth due to nonlocal effects. To quantitatively analyze the effects of thermodynamic coupling, we compute the decay time of SST anomalies in the tropics, assuming that there is no thermodynamic coupling, i.e., no reduced thermal damping or WES feedbacks. We approximate the surface heat flux F from Equation (1) as follows:

$$F = \kappa(T_a - T_o) \qquad (2)$$

where T_a is the surface atmospheric temperature, and κ is an exchange coefficient relating the heat flux to the local air–sea temperature difference. Estimates of κ are quite uncertain and scale-dependent. We choose a reference value of 40 W m^{-2} K^{-1}, based on linearizations of bulk formulae for air–sea flux exchange [e.g., *Haney*, 1971]. Note that the mixed layer depth in the NTA region (Figure 1) is quite shallow, on the order of a few tens of metres. For a reference mixed layer depth of 30 m (say), and the reference value of κ, the *e*-folding decay time of SST anomalies is about 1.5 months. In midlatitude estimates of the reduced damping effect, κ is effectively reduced to a value of about 20 W m^{-2} K^{-1} [e.g., *Saravanan*, 1998], which would imply an *e*-folding timescale of about 3 months for a 30 m mixed layer. Our coupled forecasts of SST anomalies in the tropical Atlantic, on the other hand, are skillful for periods of more than six months. This suggests that the effective κ in this region is substantially lower, i.e., on the order of 10 W m^{-2} K^{-1}, which is consistent with the observational estimates of *Czaja et al.* [2002].

The equatorial location of the coupled amplification of surface wind variability, its anisotropy and non-local growth characteristics suggest that this amplification is not due to reduced thermal damping, but rather, due to the WES feedback. If the reduced thermal damping effect were responsible for the amplification of surface wind variability in the western Tropical Atlantic, one would expect the feedback to be strongest where the mixed layer is shallowest, i.e., north of the equator in the western Tropical Atlantic, and its effects would be more isotropic because this mechanism simply amplifies existing variability. On the other hand, the WES feedback would be most effective close to the equator, where the Coriolis effect is small and an SST gradient would generate a surface wind response directed along the gradient, not perpendicular to it. The typical scenario for the WES feedback in the tropical Atlantic involves meridional flow across the equator [*Chang et al.*, 1997], which would imply a weaker impact of this feedback on the zonal flow very close to the equator. The WES feedback can also explain growth in SST anomalies seen in the predictability experiments, unlike the reduced thermal damping mechanism.

Although this study suggests that thermodynamic coupling can result in a positive air–sea feedback, this does not necessarily imply that the *fully* coupled system will also exhibit positive feedbacks. There could be dynamic oceanic feedbacks that are negative and oppose any positive thermodynamic feedbacks. Furthermore, thermodynamic feedbacks cannot give rise to timescales much longer than the oceanic mixed layer timescale, which is on the order of several months. Indeed, negative oceanic feeedbacks such as meridional advection [*Chang et al.*, 1997, 2001] and Ekman transport [*Xie*, 1999] have been proposed to explain the origin of decadal timescales in tropical Atlantic variability.

To summarize, we have shown that thermodynamic atmosphere–ocean interaction can play an important role in the tropical Atlantic. It can lead to amplification and increased persistence of atmospheric variability in the deep tropics. It can also explain much of the seasonal timescale predictability in this region. These results mean that thermodynamic coupling leads to a richer, more complex set of interactions than a local, Hasselmann-type of red-noise model would imply. Although we have focused on the tropical Atlantic, we have also found evidence that thermodynamic coupling is active even in the tropical Pacific, although its strength is substantially weaker than the dynamic air–sea interaction associated with ENSO. We also plan to carry out further mechanistic studies that would allow us to better distinguish between the WES feedback and other types of thermodynamic feedback in this region. Since the deep tropics seems to play an important role in thermodynamic coupling, we also hope to explore how improving the meridional resolution affects the model simulations.

Acknowledgments. We would like to acknowledge the contribution of Link Ji, who carried out the model integrations. This work was partially supported by NASA InterAgency Agreement W-19,750, NOAA Grant NA16GP1575, and NSF Grant ATM-0337846. Ping Chang also acknowledges the support from the National Natural Science Foundation of China (NSFC) through Grant 40128003.

REFERENCES

Alexander, M. A., I. Bladé, M. Newman, J. R. Lanzante, N.-C. Lau, and J. D. Scott, The Atmospheric Bridge: the influence of ENSO teleconnections on air–sea interaction over the global oceans, *J. Clim., 15,* 2205–2231, 2002.

Barreiro, M., A. Giannini, P. Chang, and R. Saravanan, On the role of the South Atlantic atmospheric circulation in Tropical Atlantic Variability, 2004.

Barsugli, J. J., and D. S. Battisti, The basic effects of atmosphere–ocean thermal coupling on midlatitude variability, *J. Atmos. Sci., 55,* 477–493, 1998.

Bjerknes, J., Atmospheric teleconnections from the equatorial Pacific, *Mon. Weather Rev., 97,* 163–172, 1969.

Bladé, I., The influence of midlatitude ocean–atmosphere coupling on the low–frequency variability of a GCM. Part I: No tropical SST forcing, *J. Clim., 10,* 2087–2106, 1997.

Chang, P., L. Ji, and H. Li, A decadal climate variation in the tropical Atlantic ocean from thermodynamic air–sea interactions, *Nature, 385,* 516–518, 1997.

Chang, P., R. Saravanan, L. Ji, and G. C. Hegerl, The effect of local sea–surface temperatures on atmospheric circulation over the tropical Atlantic sector, *J. Clim., 13,* 2195–2216, 2000.

Chang, P., L. Ji, and R. Saravanan, A hybrid coupled model study of tropical Atlantic variability, *J. Clim., 14,* 361–390, 2001.

Chang, P., R. Saravanan, and L. Ji, Tropical Atlantic seasonal predictability: the roles of El Niño remote influence and thermodynamic air–sea feedback, *Geophys. Res. Lett., 30,* 1501–1504, 2003.

Chang, P., R. Saravanan, T. DelSole, F. Wang, and L. Ji, Predictability of linear coupled systems. Part II: An application to a simple coupled model of Tropical Atlantic variability, *J. Clim.,* in press, 2004.

Chiang, J. C. H., Y. Kushnir, and A. Giannini, Deconstructing Atlantic ITCZ variability: Influence of the local cross-equatorial SST gradient, and remote forcing from the eastern equatorial Pacific, *J. Geophys. Res., 107*(D1), doi:10.1029/2000JD000307, 2002.

Czaja, A., P. van der Vaart, and J. Marshall, A diagnostic study of the role of remote forcing in tropical Atlantic variability, *J. Clim., 15,* 3280–3290, 2002.

Davey, M. K., et al., STOIC, 2002: A study of coupled model climatology and variability in tropical ocean regions, *Clim. Dyn., 18,* 403–420, 2002.

Dommenget, D., and M. Latif, Interannual to decadal variability in the tropical Atlantic, *J. Clim., 13,* 777–792, 2000.

Enfield, D. B., and D. A. Mayer, Tropical Atlantic sea surface temperature variability and its relation to the El Niño–Southern Oscillation, *J. Geophys. Res., 102,* 929–945, 1997.

Haney, R. L., Surface thermal boundary condition for ocean circulation models, *J. Phys. Oceanogr., 1,* 241–248, 1971.

Hasselmann, K., Stochastic climate models: Part I. Theory, *Tellus, 28,* 473–485, 1976.

Hastenrath, S., and L. Heller, Dynamics of climatic hazards in northeast Brazil, *Q. J. R. Meteorol. Soc., 103,* 77–92, 1977.

Houghton, R. W., and Y. Tourre, Characteristics of low frequency sea surface fluctuations in the tropical Atlantic, *J. Clim., 5,* 765–771, 1992.

Huang, B., P. S. Scopf, and Z. Pan, The ENSO effect on the tropical Atlantic variability: A regionally coupled model study, *Geophys. Res. Lett., 29,* doi:10.1029/2002GL01014872, 2002.

Hurrell, J. W., J. J. Hack, B. A. Boville, D. L. Williamson, and J. T. Kiehl, The dynamical simulation of the NCAR Community Climate Model version 3, *J. Clim., 11,* 1207–1236, 1998.

Kiehl, J. T., J. J. Hack, G. B. Bonan, B. A. Boville, D. L. Williamson, and P. J. Rasch, The National Center for Atmospheric Research Community Climate Model: CCM3, *J. Clim., 11,* 1131–1149, 1998.

Levitus, S., Climatological Atlas of the World Oceans, *NOAA Prof Pap no 13,* US Government Printing Office, Washington, DC, 1982.

Neelin, J. D., D. S. Battisti, A. C. Hirst, F.-F. Jin, Y. Wakata, T. Yamagata, and S. E. Zebiak, ENSO theory, *J. Geophys. Res., 103,* 14261–14290, 1998.

Okumura, Y., S.-P. Xie, A. Numaguti, and Y. Tanimoto, Tropical Atlantic air–sea interaction and its influence on the NAO, *Geophys. Res. Lett., 28,* 1507–1510, 2001.

Saravanan, R., Atmospheric low frequency variability and its relationship to midlatitude SST variability: Studies using the NCAR Climate System Model, *J. Clim., 11,* 1386–1404, 1998.

Saravanan, R., and P. Chang, Oceanic mixed layer feedback and tropical Atlantic variability. *Geophys. Res. Lett., 26,* 3629–3632, 1999.

Saravanan, R., and P. Chang, Interaction between tropical Atlantic variability and El Niño–Southern Oscillation, *J. Clim., 13,* 2177–2194, 2000.

Smith, T. M., R. W. Reynolds, R. E. Livezey, and D. C. Stokes, Reconstruction of historical sea surface temperatures using empirical orthogonal functions, *J. Clim., 9,* 1403–1420, 1996.

Shukla, J., et al., Dynamical seasonal prediction, *Bull. Amer. Meteorol. Soc., 81,* 2593–2606, 2000.

Sutton, R. T., S. P. Jewson, and D. P. Rowell, The elements of climate variability in the tropical Atlantic, *J. Clim., 13,* 3261–3284, 2000a.

Sutton, R. T., W. A. Norton, and S. P. Jewson, The North Atlantic

Oscillation–What role for the ocean? *Atmospheric Science Letters, 1,* 89–100, 2000b.

Wang, C., and J. Picaut, An overview of El Niño–Southern Oscillation understanding, 2004.

Watanabe, M., and M. Kimoto, Tropical–extratropical connection in the Atlantic atmosphere–ocean variability, *Geophys. Res. Lett., 26,* 2247–2250, 1999.

Wu, L., Q. Zhang, and Z. Liu, Toward understanding the tropical Atlantic variability using coupled modeling surgery, 2004.

Xie, S.-P., A dynamic ocean–atmosphere model of the tropical Atlantic decadal variability, *J. Clim., 12,* 64–70, 1999.

Xie, S.-P., and S. G. H. Philander, A coupled ocean–atmosphere model of relevance to the ITCZ in the eastern Pacific, *Tellus, 46A,* 340–350, 1994.

Xie, S.-P., and J. A. Carton, Tropical Atlantic variability: patterns, mechanisms, and impacts, 2004.

P. Chang, Department of Oceanography, Texas A&M University, College Station, Texas 77843-3146. (ping@ocean.tamu.edu)

R. Saravanan, National Center for Atmospheric Research, Boulder, Colorado 80307-3000. (svn@ncar.ucar.edu)

Internal Variability of the Tropical Atlantic Ocean

Markus Jochum[1], Raghu Murtugudde[2], Paola Malanotte-Rizzoli[1], and Antonio J. Busalacchi[2]

A 100 year integration of an eddy resolving numerical model of the tropical Atlantic is analyzed to quantify the interannual variability caused by internal variability of ocean dynamics. It is found that, except for the spring position of the SST maximum, the strength of internal variability in the tropical Atlantic is comparable to published mid-latitude values but is dwarfed by the strength of the seasonal cycle. During spring however, the equatorial meridional SST gradient is very weak, and internal oceanic variability causes a variability in the position of the SST maximum that is comparable to its observed variability. It is shown that these variations in the SST are due to tropical instability waves whose strength varies from year to year, even under climatological forcing. The results suggest that in winter, the predictability of the location of the tropical SST maximum is limited to the persistence time of SST anomalies which is approximately 100 days.

1. INTRODUCTION

The coupled ocean-atmosphere system varies on many timescales; however, it is the variability on multi-year and decadal timescales that currently receives the most attention. Understanding these long term variabilities will improve climate forecasts and the interpretation of historical climate records. Variability in the ocean-atmosphere system can be attributed to external forcing (e.g., ice ages), ocean-atmosphere coupling (e.g., El Niño), internal atmospheric variability (e.g., North Atlantic Oscillation) and, the focus of the present study, internal oceanic variability (e.g., Kuroshio path). Because of the relatively low ocean temperature in higher latitudes, ocean-atmosphere coupling is thought to be stronger in the tropics, whereas the relatively low tropical ratio between background velocity and planetary wave speed suggests that the tropical oceans are governed by linear dynamics, thereby restricting internal variability to higher latitudes. Observational evidence for internal oceanic variability in mid-latitudes has been reported by Taft [1972] who shows that the Kuroshio switches back and forth between a large and a small meander state. Both states can persist for several years and the transitions between them occur within a couple of months [*Kawabe*, 1986]. Observations of the Gulf Stream also show a weak bimodality of the path [*Bane and Dewar*, 1988]. High resolution ocean general circulation models (OGCMs) are able to reproduce the observed bimodalities [*Schmeits and Dijkstra*, 2001].

Several authors demonstrated that the observed internal variability of the western boundary currents can be understood within the framework of dynamical systems theory [*Jiang et al.*, 1995; *Primeau*, 1998; *Meacham*, 2000; *Simmonet et al.*, 2003; and references therein]. These studies typically use a one or two layer high resolution OGCM set in a rectangular basin with a mid-latitude double gyre. For certain ranges of Rossby and Ekman numbers the solutions exhibit chaotic or limit cycle behaviour which is usually tied to the strength of the inertial recirculation gyres near the western boundary. The not very comforting picture that emerges from these studies is that the nature of the ocean circulation could be sensitive to parameters that are not well known (e.g., friction or boundary conditions). The present authors are not aware of a study that shows this internal mid-latitude varibility to have an impact on large scale climate, but it can be speculated that it affects the water mass properties and the heat budget of the mid-latitude oceans.

---

[1]Massachusetts Institute of Technology, Cambridge, Massachusetts.
[2]University of Maryland, College Park, Maryland.

Earth's Climate: The Ocean-Atmosphere Interaction
Geophysical Monograph Series 147
Copyright 2004 by the American Geophysical Union
10.1029/147GM11

In the tropical Atlantic, there is a large interannual variability of the seasonal march of the Inter-Tropical Convergence Zone [ITCZ, *Chiang et al.*, 2002] with disastrous consequences for the Brazilian and West African population. Although the mechanisms behind this variability are not entirely clear [*Xie and Carton*, 2004], internal oceanic variability has to our knowledge not yet been suggested. There are no reports of internal interannual variability in the tropical oceans; in fact, it would be difficult to observe considering the strength of El Niño and the seasonal cycle in the tropics. Furthermore, the success of Cane et al. [1986] in predicting El Niño suggested that "the tropical ocean response on interannual timescales is reasonably well captured by linear or weakly nonlinear approximations to the ocean dynamics" [*Neelin et al.*, 1998]. However, we, the authors, recently concluded several studies that show that at least in the Atlantic nonlinear dynamics are a major part of the tropical circulation: the barotropically unstable North Equatorial CounterCurrent (NECC) generates rings which carry South Atlantic water and potential vorticity into the subtropical gyre [*Jochum and Malanotte-Rizzoli*, 2003] and the unstable Equatorial UnderCurrent–South Equatorial Current (EUC-SEC) system generates tropical instability waves [TIWs; *Jochum et al.*, 2004b; JMB hereafter] that drive the Atlantic Tsuchiya jets [*Jochum and Malanotte-Rizzoli*, 2004]. TIWs have a period of approximately 20 to 40 days and, due to their nonlinear nature, are of varying strength every year, even under seasonal forcing (JMB). Since they are a major contributor to the equatorial mixed layer heat and momentum budget [*Hansen and Paul*, 1984] one can infer that the equatorial temperature and velocity fields change from year to year. The present study quantifies, with the help of an OGCM, to what extent the nonlinear effects in the tropical Atlantic ocean contribute to its observed interannual variablity. It is not attempted, like in the aforementioned mid-latitude studies, to determine the particular kind of nonlinear regime that dominates the tropical Atlantic. The OGCM used here has in comparison to JMB a more realistic oceanic and atmospheric mixed layer model and is numerically more efficient, so that high resolution runs lasting 100 years are feasible.

Van der Vaart and Dijkstra [1998] showed that in idealized coupled tropical ocean-atmosphere models intraseasonal instabilities in conjunction with seasonal forcing can lead to interannual variability. Whether instabilities can generate interannual variability without atmospheric coupling is investigated here. One important difference between the tropical Atlantic and the aforementioned idealized mid-latitude studies is the fate of the eddy kinetic energy (EKE). In the latter, the EKE is focused on a small region along the western boundary by two converging western boundary currents and the westward group speed of Rossby waves; on the other hand, in the tropical Atlantic the orientation of the Brazilian coastline allows the North Brazil Current rings to carry away the EKE generated by the NECC from the equatorial Atlantic into the Caribbean Sea. TIWs, the other large carrier of EKE, are mostly dissipated before they can reach the western boundary (JMB). Thus, instead of the basin wide growth and collapse of recirculation gyres which can be observed in idealized mid-latitude studies, the internal variability in the tropical Atlantic is likely to produce more subtle changes due to local instability processes. Nevertheless, the present study shows that these instabilities can contribute a significant portion to the observed interannual variability of the location of the SST maximum.

The following section describes the numerical model, Section 3 describes and quantifies the internal interannual variability and discusses mechanisms for this variability. Section 4 summarizes the results.

2. MODEL DESCRIPTION

The OGCM employed for this study is the reduced gravity, primitive equation, sigma-coordinate model of Gent and Cane [1989]. The OGCM is coupled to an advective atmospheric mixed layer model which computes surface heat fluxes without any restoring boundary conditions or feedbacks to observations [*Seager et al.*, 1995; *Murtugudde et al.*, 1996]. A variable depth oceanic mixed layer represents the three main processes of oceanic turbulent mixing, namely, the entrainment/detrainment due to wind and buoyancy forcing, the gradient Richardson number mixing generated by the shear flow instability, and the convective mixing related to static instabilities in the water column [*Chen et al.*, 1994].

In previous studies this model has demonstrated its ability to reproduce the observed SST and circulation in the tropical Atlantic [*Murtugudde et al.*, 1996; *Inui et al.*, 2002]. In this study, the model is initialized with Levitus [1994] temperature and salinity fields, driven by seasonal Hellerman and Rosenstein [1983] winds, has a 0.25° degree horizontal resolution and 8 layers in the vertical. At the meridional boundaries at 25°S and 25°N, temperature and salinity are restored to Levitus [1994]. The model is spun up for 20 years and the presented results are taken from the subsequent 100 years of simulation. Years 21 and 22 of the data are saved as 5 day snapshots to compute the strength of the eddy field whereas the remainder is saved as monthly means to limit the amount of data to a managable size. Most of the analysis is done with anomaly fields for which the seasonal cycle has been removed. The correlation analysis was performed after smoothing the anomalies with an 11 month Hanning smoother [*Press et al.*, 1992] so that only the integral effects of the high frequency

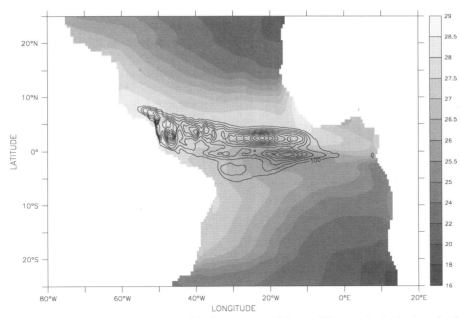

Figure 1. Annual mean of the SST; superimposed is the variance of the meridional velocity in the mixed layer (contour lines: 100 cm^2/s^2, the maximum is 1800 cm^2/s^2). The seasonal cycle has been removed from the velocity before computing the variance.

oscillations remain. It should be pointed out that the separation between mean, seasonal, intra-, and interannual variability is mainly mathematical. Because TIWs are nonlinear, they will contribute not only to intraseasonal variability but also to the mean, the seasonal cycle and, as shown here, to interannual variability. Thus, the seasonal cycle is the result of seasonal changes in the direct surface forcing and the forcing due to seasonally varying instabilities.

The pattern and strength of the variability is, apart from the southward shift of the NBC/NECC retroflection, consistent with altimeter observations [*Stammer*, 1997; Figure 1]. The southward shift of the retroflection can be explained by the model's weak meridional overturning circulation of 2 Sv [*Fratantoni et al.*, 2000; *Jochum and Malanotte-Rizzoli*, 2001]. Accordingly, the simulated EKE along the Brazilian coast at 4°N and 8°N is about 40% weaker than observed [*Johns et al.*, 1990, 1998], whereas the simulated EKE for the TIWs and for the NBC at the equator matches the observed values [*Weisberg and Weingartner*, 1988; *Schott et al.*, 1993].

As motivated in the introduction, internal variability of the SST is most likely to arise in regions with strong instability processes. Figures 1 and 2 suggest that for the present study the most interesting region is the central basin along the equator, where high eddy activity is combined with a strong meridional temperature gradient. The next section will describe and quantify the internal variability in this experiment.

3. INTERNAL VARIABILITY

An energy analysis connects this study with the aforementioned idealized mid-latitude studies: near the equator, variations of the kinetic energy are an order of magnitude larger than variations in the potential energy [*Pedlosky*, 1979; *Weisberg and Weingartner*, 1988], therefore only kinetic energy is considered here, specifically the kinetic energy integrated over all model layers of the most energetic part of the tropics (10°S to 15°N). The solution's energy levels (after removing the seasonal cycle) are narrowly distributed around a mean and are slightly skewed towards states of higher energy (Figure 3). This is different from the results of the idealized mid-latitude experiment of McCalpin and Haidvogel [1996] who find two distinct peaks in their energy distribution, a low level and a high level state which correspond to two different states of the circulation. In an experiment similar to McCalpin and Haidvogel [1996], Meacham [2000] shows that a low viscosity simulation can develop strong aperiodic oscillations which destroy the limit cycle behaviour observed by McCalpin and Haidvogel [1996]. The energy distribution of the present simulation is similar to the low viscosity case in Meacham [2000]: There are no multiple prefered states of the circulation, but the irregular TIW activity (Figure 2) causes a spread of energy around the mean. Apart from the annual harmonics, there is no spectral peak in the energy time series and, as in the altimeter observations by Stammer [1997], the spectrum is white

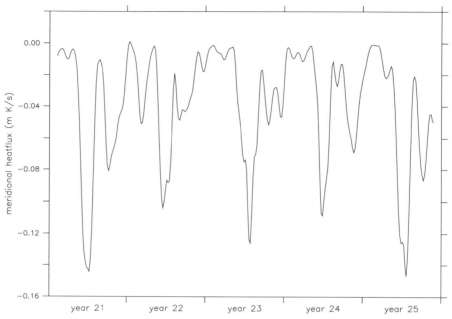

Figure 2. Meridional heatflux on the equator at the surface at 20°W. Note the difference in strength from year to year.

on timescales longer than 100 days (not shown). Since the seasonal cycle of the tropics is considerably larger than that at mid-latitudes, the contribution of the internal variability to the energetics of the tropics tends to be small despite the internal variability in the tropics being comparable to that in the mid-latitudes.

Although the internal variability in the kinetic energy is negligible in the tropical Atlantic ocean, it is not obviously so for the SST. The model's internal interannual SST variability rarely exceeds 0.2 K anywhere, but there are areas where this signal can be as strong as 30% of the seasonal signal of the SST (Figure 4). Two different SST indices are thought to be important for the tropical Atlantic: the Atlantic Niño and the Atlantic Gradient [*Servain*, 1991; *Zebiak*, 1993]. The first index uses the SST anomalies averaged over an equatorial domain (here and in *Zebiak* [1993]: 3°S to 3°N and 20°W to 0°W) to determine the presence of an Atlantic coupled mode similar to the Pacific ENSO: the Atlantic Niño. The Reynolds and Smith [1994] climatology (RS from here on) and results from a numerical model driven by NCEP reanalysis winds [J. Kroeger, manuscript in preparation, 2004] show that, after removing the seasonal cycle, the Atlantic Niño index has a standard deviation (std) of 0.5 which is approximately ten times larger than the internal variability produced in the model (not shown). Hence, internal variability in the tropical Atlantic ocean is unlikely to have a significant impact on the Atlantic Niño.

The situation is very different for the gradient index which defines a meridional SST gradient. Near the equator the wind is directed from low to high SST, therefore the position of the ITCZ is connected to the position of the SST maximum [*Lindzen and Nigam*, 1987; *Ruiz-Barradas et al.*, 2000]. The interannual variability of the ITCZ position manifests itself primarily as a meridional displacement from its mean position and leads to severe rainfall anomalies in West Africa and Northeast Brazil [*Nobre and Shukla*, 1996]. Thus, a small change in the seasonal cycle of the SST maximum can have a large impact on the climate of these areas.

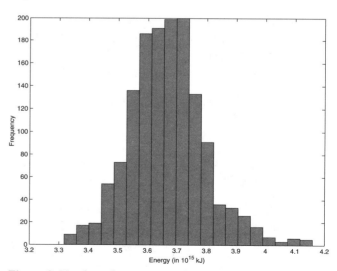

Figure 3. Number of occurences of different energy states after removal of the seasonal cycle. For example, there are 200 months in which the energy is between 3.70 and $3.75 \cdot 10^{15}$ kJ.

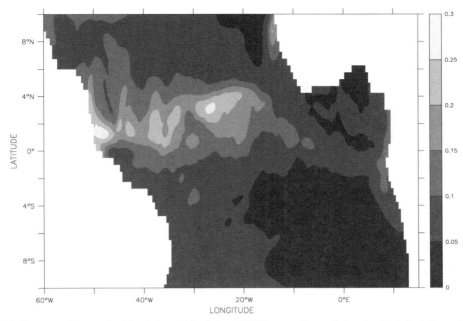

Figure 4. Ratio between the standard deviation of the internal SST anomalies and the standard deviation of the seasonal cycle.

In the present study, the internal variability in the model can account for a signifcant part of the observed interannual variability in the SST gradient in a narrow band centered on the equator (Figure 5). Basing the index on areas further polewards will reduce the relative contribution of the internal variability; for example, if instead of the area between 1°S and 1°N we choose the area between 5°S and 5°N the simulated internal variability is only about 50% of the observed interannual variability in the summer and 15% in the spring. For the area between 20°S and 20°N the values are 10% and 4%, respectively. The seasonal cycle of the internal variability reflects the seasonality of the TIWs: weak in the early spring, sudden increase in May, and slow decay during fall and winter.

Even though the relative contribution of the internal variability to the interannual variability is largest in the summer and fall, the position of the SST maximum is most sensitive to disturbances in spring (April to June) when the SST gradient is weak (Figure 6). In spring the insolation maximum returns from the southern hemisphere and crosses the equator. During the austral summer the waters south of the equator are heated, the cold tongue vanishes and the meridional temperature gradient is very weak [*Mitchell and Wallace*, 1992]. Due to this weak gradient, small disturbances caused by internal variability are sufficient to cause the spring maximum in the present study to be sometimes north and sometimes south of the equator which leads to a large std of the yearly southernmost position of the SST maximum of 3.5° (mean: 3.3°S). Based on the RS climatology the observational value is 2.0°S ± 3.2°. During the boreal summer the contributions of the internal variability to the position of the SST maximum are negligible because of the strong meridional temperature gradient: The std of the northernmost latitude of the SST maximum is 0.2° as opposed to the observed 2.3°. We conclude that, although internal variability in the tropical Atlantic is largely

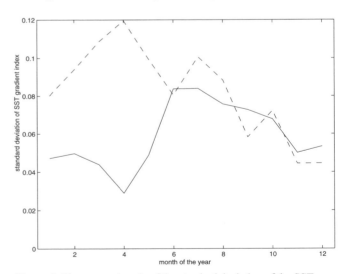

Figure 5. The seasonal cycle of the standard deviation of the SST gradient index for the observations (dashed line) and the model (solid line). The index is here computed as the difference between the zonally averaged (35°W to 15°W) SST between the equator and 1°N and the equator and 1°S.

neglible because of the strong seasonal cycle in the tropics, it causes a significant variability in the position of the SST maximum during spring and therefore must be accounted for in studies of tropical climate variability. The question now is whether this SST variability is generated locally, or remotely and subsequently advected to the surface.

The internal SST variability is limited to the equatorial band which suggests that it is either generated locally by TIWs or advected by the EUC and upwelled into the mixed layer. A correlation analysis rules out the advection hypothesis and supports the TIW hypothesis: The SST anomaly at the equator at 20°W is correlated with the temperature anomaly in the rest of the domain. This particular point is chosen because it has both, strong TIW activity and upwelling. Since water in the EUC travels about 2000 km in a month, the analysis should yield positive values to the west of 20°W and below the surface if the anomalies were advected. This is clearly not the case (Figure 7, right); instead the correlation pattern reflects the typical structure of TIWs: high meridional coherence and anticorrelation between surface and thermocline [JMB, Figure 7, left].

4. SUMMARY AND DISCUSSION

A high resolution OGCM with climatological forcing has been used to study the internal variability of the tropical Atlantic. It is found that because of the strong seasonal cycle (and not because of the absence of nonlinear effects) the contribution of internal variability to the interannual variability of the kinetic energy is mostly negligible. However, near the equator internal variability can modulate the seasonal cycle of SST from year to year and causes interannual variability. Most importantly it can change the position of the spring position of the SST maximum, whose internal variability in the model is comparable to its observed interannual variability. The main sources of internal variability in the tropical Atlantic are the NECC and EUC. Their instabilities generate high frequency waves that have interannual variations in strength. In the case of the EUC, the instabilities are in a region with a strong meridional temperature gradient which leads to interannual SST variabilities. Thus, nonlinear and aperiodic TIW generation leads to a significant variance of the spring time position of the SST maximum: It can be north as well as south of the equator. This has important implications for the predictability of tropical Atlantic climate, because it means that the spring time position of the SST maximum *cannot* be known before the preceding winter, limiting the forecast range to the persistence time of temperature anomalies of the equatorial mixed-layer which is approximately 100 days [*Kessler et al.*, 1996; see also *Saravannan and Chang*, 2004, for a discussion on predictability]. The importance of accurate TIW representation might also explain the problems that coupled models have in reproducing eastern tropical Atlantic climate [*Davey et al.*, 2000].

The seasonal cycle of the ITCZ is observed to be very uncertain in the spring. Similar to the modeled SST maximum its spring time position can be north or south of the equator [*Chiang et al.*, 2002]. While theory and observations suggest that the ITCZ position depends on the position of the SST maximum, it is not clear how strong the internal variability influences the

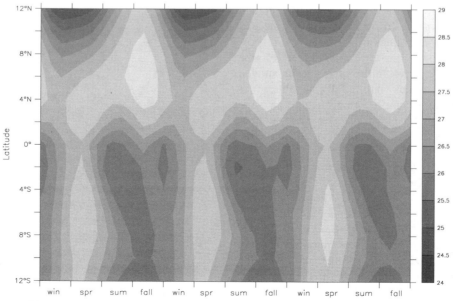

Figure 6. 3 years of SST averaged from 35°W to 15°W. Note the strong SST gradient around the SST maximum at 6°N during summer and the weak gradient during spring, especially the reversal of the SST gradient in the third spring.

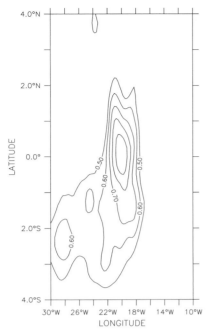
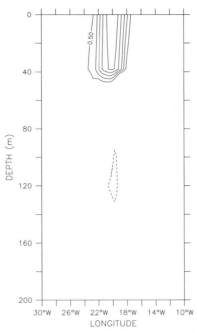

Figure 7. Correlation of the SST anomalies at 20°W/0°N with the surrounding SST anomalies (left) and with the temperature anomalies at 0°N (right). Only correlations with absolute values of 0.5 or higher are shown.

ITCZ position. Hashizume et al. [2001] provide observational evidence for the Atlantic and Pacific that the TIWs directly affect the local wind field. However, how much the internal SST variability will perturb the seasonal cycle of the ITCZ position has to be investigated with an OGCM that is coupled to an atmospheric GCM which will be part of our future research.

Since the TIWs are an important part of the equatorial heat budget [JMB] and are shown here to have the potential to perturb the seasonal cycle of the ITCZ during spring, it is worthwhile asking to what extent they provide a positive or negative feedback for anomalies in the cross equatorial temperature gradient. Chang et al. [1997] propose a coupled ocean-atmosphere mode for the Atlantic in which anomalies in the cross equatorial wind lead to anomalies in the cross equatorial temperature gradient which amplify again the initial wind anomalies. The TIW heat flux depends on the meridional temperature gradient [Hansen and Paul, 1984], therefore an increased northward temperature gradient would lead to an increased southward heat flux, resulting in a negative feedback. Furthermore, Philander and Delecluse [1983] showed that an increased northward wind across the equator leads to a strengthening of the EUC which should lead to a stronger TIW activity and an increased southward heat flux, again a negative feedback. There is observational evidence for this: Compared to 1983, 1984 has a weaker meridional SST gradient (RS) *and* weaker TIWs [Weisberg and Weingartner, 1988, their Figure 7]. This suggests that the positive feedback cycle of Chang et al. [1997] could be damped by TIWs. Numerical studies are currently under way to support this proposed negative TIW feedback.

Acknowledgments. This research was funded with NOAA grant Nr NA16GP1576 and NASA grant Nr NAG5-7194 at MIT and ESSIC/UMD.

REFERENCES

Bane, J. M, and W. K. Dewar, Gulf Stream bimodality and variability downstream of the Charleston bump, *J. Geophys. Res.*, *93*, 6695–6710, 1988.

Cane, M. A., S. C. Dolan, and S. E. Zebiak, Experimental forecasts of the 1982/83 El Nino, *Nature*, *321*, 827–832, 1986.

Chang, P., L. Ji, and H. Li, A decadal climate variation in the tropical Atlantic ocean from thermodynamic air-sea interactions, *Nature*, *385*, 515–518, 1997.

Chen, D., L. Rothstein, and A. J. Busalacchi, A hybrid vertical mixing scheme and its applications to tropical ocean models, *J. Phys. Oceanogr.*, *24*, 2156–2179, 1994.

Chiang, J., Y. Kushnir, and A. Giannini, Deconstructing Atlantic ITCZ variability, *J. Geophys. Res.*, *107*, doi: 10.1029/2000JD000307, 2002.

Davey, M., M. Huddleston, and K. Sperber, A study of coupled model climatology and variability in tropical ocean regions, *CLIVAR Exchanges*, *17*, 2000.

Fratantoni, D. M., W. E. Johns, T. Townsend, and H. Hurlburt, Low-latitude circulation and mass transport in a model of the tropical Atlantic ocean, *J. Phys. Oceanogr.*, *30*, 1944–1966, 2000.

Gent, P., and M.A. Cane, A reduced gravity, primitive equation model of the upper equatorial ocean, *J. Comput. Phys.*, *81*, 444–480, 1989.

Hansen, D., and C. Paul, Genesis and effects of long waves in the equatorial Pacific, *J. Geophys. Res.*, *89*, 10,431–10,440, 1984.

Hashizume, H., S.-P. Xie, T. Liu, and K. Takeuchi, Local and remote atmospheric response due to tropical instability waves: A global view from space, *J. Geophys. Res.*, *106*, 10,173–10,185, 2001.

Hellerman, S., and M. Rosenstein, Normal monthly wind stress over the world ocean with error estimates, *J. Phys. Oceanogr.*, *13*, 1093–1104, 1983.

Inui, T., A. Lazar, P. Malanotte-Rizzoli, and A. Busalacchi, Wind stress effects on the Atlantic tropical-subtropical circulation, *J. Phys. Oceanogr.*, *32*, 2257–2276, 2001.

Jiang, S., F.-F. Jin, and M. Ghil, Multiple equilibra, periodic and aperiodic solutions in a wind-driven, double-gyre, shallow-water model, *J. Phys. Oceanogr.*, *25*, 764–786, 1995.

Jochum, M., and P. Malanotte-Rizzoli, On the influence of the meridional overturning circulation on the tropical-subtropical pathways, *J. Phys. Oceanogr.*, *31*, 1313–1323, 2001.

Jochum, M., and P. Malanotte-Rizzoli, On the generation of North Brazil Current Rings, *J. Mar. Res.*, *61*, 147–162, 2003.

Jochum, M., and P. Malanotte-Rizzoli, A new driving mechanism for the South Equatorial Undercurrent, *J. Phys. Oceanogr.*, *7*, 755–771, 2004.

Jochum, M., P. Malanotte-Rizzoli, and A. Busalacchi, Tropical Instability Waves in the Atlantic Ocean, *Ocean Modelling*, *7*, 145–163, 2004.

Johns, W. E., T. Lee, F. Schott, R. Zantropp, and R. Evans, North Brazil Current retroflection: seasonal structure and eddy variability, *J. Geophys. Res.*, *95*, 22,103–22,120, 1990.

Johns, W.E., T. Lee, R. Beardsley, J. Candela, R. Limeburner, and B. Castro, Annual cycle and variability of the North Brazil Current, *J. Phys. Oceanogr.*, *28*, 103–128, 1998.

Kawabe, M., Transition regimes between the three typical paths of the Kuroshio, *J. Oceanogr. Soc. Jpn.*, *25*, 3103–3117, 1986.

Kessler, W. S., M. Spillane, M. McPhaden, and D. Harrison, Scales of variability in the equatorial Pacific inferred from the TAO buoy array, *J. Clim.*, *9*, 2999–3024, 1996.

Levitus, S., Climatological Atlas of the World Ocean, *NOAA Prof. Paper 13*, 1994.

Lindzen, R., and S. Nigam, On the role of SST gradients in forcing low level winds and convergence in the tropics, *J. Atmos. Sci.*, *44*, 2418–2436, 1987.

McCalpin, J., and D. Haidvogel, Phenomenology of the low-frequency variability in a reduced-gravity, quasigeostrophic double-gyre model, *J. Phys. Oceanogr.*, *26*, 739–752, 1996.

Meacham, S., Low-frequency variability in the wind-driven circulation, *J. Phys. Oceanogr.*, *30*, 269–293, 2000.

Mitchell, T., and J. Wallace, The annual cycle in equatorial convection and sea surface temperature, *J. Clim.*, *5*, 1140–1156, 1992.

Murtugudde, R., R. Seager, and A. J. Busalacchi, Simulation of the tropical oceans with and ocean GCM coupled to an atmospheric mixed layer model, *J. Clim.*, *9*, 1796–1815, 1996.

Neelin, J., D. Battisti, A. Hirst, F.-F. Jin, Y. Wakata, T. Yamagata, and S. Zebiak, ENSO theory, *J. Geophys. Res.*, *103*, 14,261–14,290, 1998.

Nobre, P., and J. Shukla, Variations of SST, wind stress and rainfall over the tropical Atlantic and South America, *J. Clim.*, *9*, 2464–2479, 1996.

Pedlosky, J., *Geophysical Fluid Dynamics*, Springer, 1979.

Philander, S. G. H., and P. Delecluse, Coastal currents in low latitudes, *J. Mar. Res.*, *30*, 887–902, 1983.

Press, W., S. Teukolsky, W. Vetterling, and B. Flannery, Numerical Recipies, *Cambridge University Press*, 1992.

Primeau, F., Multiple equilibria of a double-gyre ocean model with super-slip boundary conditions, *J. Phys. Oceanogr.*, *28*, 2130–2147, 1998.

Reynolds, R., and T. Smith, Improved global SST analyses using optimal interpolation, *J. Clim.*, *7*, 929–948, 1994.

Ruiz-Barradas, A., J. A. Carton, and S. Nigam, Structure of interannual to decadal climate variability in the tropical Atlantic sector, *J. Clim.*, *13*, 3285–3297, 2000.

Saravanan, R., and P. Chang, Thermodynamic coupling and predictability of tropical sea surface temperature, this volume.

Schmeits, M., and H. Dijkstra, Bimodal behaviour of the Kuroshio and the Gulf Stream, *J. Phys. Oceanogr.*, *31*, 3435–3456, 2001.

Schott, F., J. Fischer, J. Reppin, and U. Send, On mean and seasonal currents and transports at the western boundary of the equatorial Atlantic, *J. Geophys. Res.*, *98*, 14,353–14,368, 1993.

Seager, R., M. Blumenthal, and Y. Kushnir, An advective atmospheric mixed layer model for ocean modeling purposes: global simulation of atmospheric heat fluxes, *J. Clim.*, *8*, 1951–1964, 1995.

Servain, J., Simple climatic indecies for the tropical Atlantic ocean and some applications, *J. Geophys. Res.*, *96*, 15,137–15,146, 1991.

Simmonet, E., M. Ghil, K. Ide, R. Temam, and S. Wang, Low-frequency variability in shallow-water models of the wind-driven ocean circulation. Part I and II, *J. Phys. Oceanogr.*, *33*, 712–752, 2003.

Stammer, D., Global characteristics of ocean variability estimated from TOPEX/POSEIDON altimeter measurements, *J. Phys. Oceanogr.*, *27*, 1743–1769, 1997.

Taft, B., Characteristics of the flow of the Kuroshio south of Japan, in: *Physical aspects of the Japan Current. University of Washington Press*, 1972.

van der Vaart, P., and H. Dijkstra, The nonlinear evolution of unstable coupled equatorial ocean-atmosphere modes, *Nonlinear Processes in Geophysics*, *5*, 39–52, 1998.

Weisberg, R.H., and T.J. Weingartner, Instability Waves in the equatorial Atlantic Ocean, *J. Phys. Oceanogr.*, *18*, 1641–1657, 1988.

Xie, S.-P., and J.A. Carton, Tropical Atlantic Variability: Patterns, mechanisms and impacts, this volume.

Zebiak, S. E., Air-sea interaction in the tropical Atlantic, *J. Clim.*, *6*, 1567–1586, 1993.

A. J. Busalacchi, University of Maryland, College Park, Maryland 20742. (ajb@essic.umd.edu)

M. Jochum, Massachusetts Institute of Technology, Cambridge, Massachusetts 02139. (mjochum@mit.edu)

P. Malanotte-Rizzoli, Massachusetts Institute of Technology, Cambridge, Massachusetts 02139. (rizzoli@ocean.mit.edu)

R. Murtugudde, University of Maryland, College Park, Maryland 20742. (ragu@essic.umd.edu)

Coupled Ocean-Atmosphere Variability in the Tropical Indian Ocean

Toshio Yamagata[1,2], Swadhin K. Behera[1], Jing-Jia Luo[1], Sebastien Masson[1], Mark R. Jury[3], and Suryachandra A. Rao[1]

The Indian Ocean Dipole (IOD) is a natural ocean–atmosphere coupled mode that plays important roles in seasonal and interannual climate variations. The coupled mode locked to boreal summer and fall is distinguished as a dipole in the SST anomalies that are coupled to zonal winds. The equatorial winds reverse their direction from westerlies to easterlies during the peak phase of the positive IOD events when SST is cool in the east and warm in the west. In response to changes in the wind, the thermocline rises in the east and subsides in the west. Subsurface equatorial long Rossby waves play a major role in strengthening SST anomalies in the central and western parts. The SINTEX-F1 coupled model results support the observational finding that these equatorial Rossby waves are coupled to the surface wind forcing associated with IOD rather than ENSO. The ENSO influence is only distinct in off-equatorial latitudes south of 10°S. Although IOD events dominate the ocean–atmosphere variability during its evolution, their less frequent occurrence compared to ENSO events leads the mode to the second seat in the interannual variability. Therefore, it is necessary to remove the most dominant uniform mode to capture the IOD statistically. The seasonally stratified correlation between the indices of IOD and ENSO peaks at 0.53 in September–November. This means that only one third of IOD events are associated with ENSO events. Since a large number of IOD events are not associated with ENSO events, the independent nature of IOD is examined using partial correlation and pure composite techniques. Through changes in atmospheric circulation and water vapor transport, a positive IOD event causes drought in Indonesia, above normal rainfall in Africa, India, Bangladesh and Vietnam, and dry as well as hot summer in Europe, Japan, Korea and East China. In the Southern Hemisphere, the positive IOD causes dry winter in Australia, and dry as well as warm conditions in Brazil. The identification of IOD events has raised a new possibility to make a real advance in the predictability of seasonal and interannual climate variations that originate in the tropics.

[1] Frontier Research System for Global Change, Yokohama, Kanagawa, Japan.
[2] Department of Earth and Planetary Science, The University of Tokyo, Tokyo, Japan.
[3] Environmental Science Department, University of Zululand, South Africa.

Earth's Climate: The Ocean-Atmosphere Interaction
Geophysical Monograph Series 147
Copyright 2004 by the American Geophysical Union
10.1029/147GM12

1. INTRODUCTION

Tropical oceans play major roles in the natural variability of the world climate. Anomalous coupled ocean–atmosphere phenomena generated in the tropical oceans produce global atmospheric and oceanic circulation changes that influence regional climate conditions even in remote regions. On the interannual time scale, the El Niño-Southern Oscillation (ENSO) of the tropical Pacific Ocean is known as one typical example of such phenomena and has so far received worldwide attention because of the enormous societal impact.

In contrast to the Pacific, the interannual variability originated in the tropical Indian Ocean has not been paid much attention. This is mainly because the variability in the basin is overwhelmed by seasonal winds. The southwest monsoon winds that dominate the annual cycle produce strong upwelling along the Somali coast of the western Indian Ocean. Weakening of these winds during the boreal fall after the summer monsoon period gives rise to warmer SST mainly owing to weakened upwelling. During this time, the equatorial westerly winds become stronger and generate the strong equatorial currents known as the *Yoshida-Wyrtki* jet, which transports the warm waters to the east. In some years such as 1994, however, the *Yoshida-Wyrtki* jet did not evolve in such a normal way. This was a puzzle [e.g., *Vinayachandran et al.*, 1999].

It has turned out that the 1994 anomalous event in the basin is due to an ocean–atmosphere coupled phenomenon. This coupled mode is now widely called the Indian Ocean Dipole (IOD) [*Saji et al.*, 1999; *Yamagata et al.*, 2003]; it is sometimes referred to as the Indian Ocean Zonal (IOZ) mode. During these IOD/IOZ events, an east–west dipole pattern in SST anomalies evolves in tropical Indian Ocean. The changes in SST are found to be closely associated with changes in surface winds; equatorial winds reverse direction from westerlies to easterlies during the peak phase of the positive IOD events when SST is cool in the east and warm in the west. Changes in surface winds are associated with a basin-wide anomalous Walker circulation [*Yamagata et al.*, 2002]. These ocean and atmosphere conditions imply that a Bjerknes-type [*Bjerknes*, 1969] feedback mechanism is responsible for the IOD evolution.

The dipole pattern is not restricted only to SST anomalies. The thermocline rises in the east and deepens in the central and western parts in response to anomalous equatorial winds during the IOD events, thus giving rise to a subsurface dipole [*Rao et al.*, 2002a]. Since the seasonal southeasterly winds along the Java coast are also strengthened during the positive IOD events, the anomalous coastal upwelling causes further SST cooling in the east [*Behera et al.*, 1999]. The dipole pattern related to IOD is thus observed in heat content/sea level anomalies [*Rao et al.*, 2002a]. The atmospheric component of IOD is seen in OLR anomalies [*Behera et al.*, 1999, 2003a; *Yamagata et al.*, 2002] and sea level pressure anomalies [*Behera and Yamagata*, 2003]. Therefore, IOD indices are derived for various ocean–atmosphere variables: SST anomalies from GISST, zonal wind anomaly from NCEP-NCAR reanalysis data, sea surface height anomaly from simple ocean data assimilation (SODA) products, sea level anomaly from TOPEX/POSEIDON data and satellite derived OLR anomaly [see Figure 1 of *Yamagata et al.*, 2003].

Several other studies also discussed various aspects of this Indian Ocean coupled phenomenon [*Webster et al.*, 1999; *Murtugudde et al.*, 2000; *Feng et al.*, 2001; *Li and Mu*, 2001; *Rao et al.*, 2002b; *Vinayachandran et al.*, 1999; 2002; *Xie et al.*, 2002; *Saji and Yamagata*, 2003a; *Guan et al.*, 2003; *Masson et al.*, 2003a; *Ashok et al.*, 2003a; *Annamalai et al.*, 2003; *Shinoda et al.*, 2003] using observed data and ocean/atmosphere model simulations. We also note that coupled general circulation models (CGCMs) are now successful in reproducing the IOD events [*Iizuka et al.*, 2000; *Gualdi et al.*, 2003; *Behera et al.*, 2003b; *Lau and Nath*, 2004; *Cai et al.*, 2003].

As expected from the practice of ENSO, the impact of the IOD is not limited only to the equatorial Indian Ocean. Through the changes in the atmospheric circulation, IOD influences the world climate [e.g., *Saji and Yamagata*, 2003b]. For example, the IOD influences the Southern Oscillation in the Pacific [*Behera and Yamagata*, 2003], rainfall variability during the Indian summer monsoon [*Behera et al.*, 1999; *Ashok et al.*, 2001], the summer climate condition in East Asia [*Guan and Yamagata*, 2003; *Guan et al.*, 2003], the African rainfall [*Black et al.*, 2003; *Clark et al.*, 2003; *Behera et al.*, 2003b; *Rao et al.*, 2004], the Sri Lankan Maha rainfall [*Lareef et al.*, 2003] and the Australian winter climate [*Ashok et al.*, 2003b].

Discovery of the IOD has stimulated exciting research in other disciplines of science such as paleoclimate, marine biology and atmospheric chemistry. In a recent paper, *Abram et al.* [2003] reported that the scattered particulates from severe wildfires in the Indonesian region during the 1997 IOD event caused exceptional coral bleaching in the Mentawai Island (off Sumatra) reef ecosystem. They also traced the IOD signal back to the mid-Holocene period using the fossil coral records from the region, revealing the first evidence of paleo-IOD. In another context, *Fujiwara et al.* [1999] found that the variability in tropospheric ozone distribution over Indonesia is related to the IOD phenomenon.

Since the above new concept of IOD was introduced in 1999, several interesting issues related to its existence/nonexistence and dependence/independence of ENSO have been raised mostly on the basis of statistics [e.g., *Allan et al.*, 2001; *Dommenget and Latif*, 2002; *Baquero-Bernal et al.*, 2002; *Hastenrath*, 2002]. Although simple statistical analyses can capture physical modes, more sophisticated methods sometimes could be misleading if not supported by our state-of-art understanding of dynamics. In the present study, we address these issues using multiple datasets including results from a coupled model simulation. The long time series of the data derived from the coupled model simulation on the Earth Simulator improves the statistical confidence in identifying the independent evolution of IOD and its global teleconnections.

2. MODEL AND DATA

The model results used in the study are obtained from an ocean–atmosphere–land coupled general circulation model (CGCM) simulation. The CGCM known as SINTEX-F1 (SINTEX-FRSGC] is an upgraded version [*Masson et al.,* 2003b; *Luo et al.,* 2003] of the original SINTEX (Scale Interaction Experiment of EU project) model described in *Gualdi et al.* [2003]. The model is also modified in coding to adapt to the unique new-generation machine called the Earth Simulator. In this model, the atmospheric component ECHAM-4 [*Roeckner et al.,* 1996] is coupled to the ocean component OPA 8.2 [*Madec et al.,* 1998] through the coupler OASIS 2.4 [*Valcke et al.,* 2000]. The atmosphere model has a spectral triangular truncation of T106 with 19 vertical levels. The ocean model OPA8.2 of the ORCA2 grid uses the Arakawa C-grid with a finite mesh of 2° x 2° *cosine* (latitude) with increased meridional resolutions to 0.5° near the equator. The model's finite mesh is designed in a way that the North Pole is replaced by two node points over land, one on North America and the other on Asia. The model has 31 levels in the vertical. The details of the coupling strategy are reported in *Guilyardi et al.* [2001] and the model's skill to reproduce the Indian Ocean variability is found in *Gualdi et al.* [2003]. We note here that the present version of the SINTEX-F1 model differs in a number of ways from the SINTEX model reported in the *Gualdi et al.* [2003]. The spectral truncation in the atmospheric component here is T106 in contrast to T42. The ocean component includes a free surface, improved runoff parameterization, and a few enclosed seas previously absent in the model geometry [*Masson et al.,* 2003b]. In addition, a number of small changes boost up the performance of SINTEX-F1 [*Luo et al.,* 2003]. A complete intercomparison study of the two versions of the model along with other available coupled models is now under preparation. We use here the last 200 years monthly data derived from the 220 years simulation to compare with the observational data.

The observed data used in the analysis are from 1958 to 1999. Monthly anomalies of atmospheric fields are derived from the NCEP-NCAR reanalysis data [*Kalnay et al.,* 1996]. SST anomalies are computed from the GISST 2.3b dataset [*Rayner et al.,* 1996]. Other ocean variables like the sea surface height and heat content are derived from the SODA [*Carton et al.,* 2000]. The rainfall anomalies are derived from the gridded precipitation data [*Willmott and Matsuura,* 1995]. Following *Saji et al.* [1999], the Dipole Mode Index (DMI) is defined as the SST anomaly difference between western (50°E–70°E, 10°S–10°N) and eastern (90°E–110°E, 10°S–Eq) tropical Indian Ocean. Niño-3 index for the eastern Pacific is derived from the GISST data.

Besides simple linear statistical tools such as a composite technique and a correlation method, a partial correlation technique [e.g., *Yule,* 1907] is used to show a partial relationship between two variables while excluding influences arising from another independent variable. For example, the partial correlation between DMI and global precipitation anomalies, while excluding the influence due to the correlation between Niño-3 and precipitation anomalies [cf. *Saji and Yamagata,* 2003b], is defined as follows:

$$r_{13,2} = (r_{13} - r_{12} \cdot r_{23}) / \sqrt{(1 - r^2_{12})} \sqrt{(1 - r^2_{23})},$$

where r_{13} is the correlation between DMI and global precipitation anomalies, r_{12} is the correlation between DMI and Niño-3 index and r_{23} is the correlation between Niño-3 and global precipitation anomalies. Similarly, the partial correlation can also be obtained for Niño-3 and precipitation anomalies while excluding the influence due to the correlation between IOD and precipitation anomalies. Statistical significance of the correlation coefficients is determined by a 2-tailed "*t*-test".

We note that the low frequency variabilities of periods longer than 7 years are removed from all the datasets.

3. DOMINANT MODES OF THE INDIAN OCEAN SST VARIABILITY

3.1. The Basin-wide Mode

A basin-wide SST anomaly of almost uniform polarity is present as the most dominant interannual mode (Figure 1a) in the Indian Ocean [*Cadet,* 1985; *Klein et al.,* 1999]. We note that this basin-wide uniform mode shows a high correlation with the eastern Pacific SST anomalies. The peak correlation coefficient of 0.8 is found when the Niño-3 index leads the basin-wide uniform mode by 4 months (Figure 2). It is no wonder that most of the previous Indian Ocean studies mainly focused on the ENSO influence on SST variability in the basin [cf. *Latif and Barnett,* 1995; *Tourre and White,* 1997; *Venzke et al.,* 2000]. The surface fluxes are identified as the major cause of SST changes during the Pacific ENSO events [*Venzke et al.,* 2000]. The basin-wide warming (cooling) is observed after an El Niño (La Niña) peak owing to the reduced (enhanced) cloud cover and increased (decreased) solar insolation. Reduction (enhancement) in the wind speed also contributes to the warming (cooling) through changes in the latent heat flux. Therefore, it is clear that the basin-wide uniform mode is a consequence of ENSO forcing. A simple estimate shows that a change of 10 Wm^{-2} in the surface fluxes could lead to a change of 0.5°C in the SST over a period of 4 months for a typical mixed-layer of 50 m. This

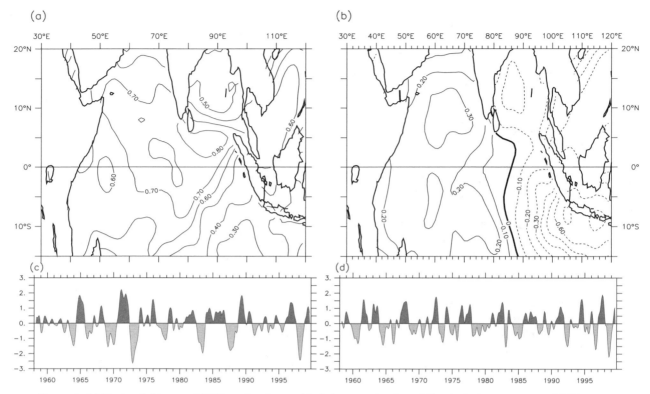

Figure 1. (a) First and (b) second EOF modes along with their respective (c and d) principal components of SST anomalies derived from GISST data.

simple estimate explains the 4 months lag in the peak correlation.

3.2. The Dipole Mode

Although IOD appears as a major signal during some years, its less frequent occurrence compared to the basin-wide uniform mode linked closely to ENSO provides IOD the second most dominant seat in the EOF analysis (Figure 1b). We also note that the period of basin-wide uniform mode is longer compared to that of the IOD mode. It is rare until now for climate dynamists to discuss the second mode of variability. This is why some researchers felt difficulties in accepting the new concept of the IOD [cf. *Allan et al.*, 2001; *Hastenrath*, 2002]. Since the basin-wide uniform mode dominates the SST variability on interannual time scale, we need to filter out the externally forced mode to find the signal related to IOD in a linear statistical analysis [*Yamagata et al.*, 2003; *Behera et al.*, 2003a]. In the wavelet spectra of raw SST anomalies in the eastern pole (10°S–Eq., 90°E–110°E) and western pole (10°S–10°N, 50°E–70°E), we do not find much coherence [Figure 2 in *Yamagata et al.*, 2003]; we are apt to be misled to a denial of the dipole. However, as shown in *Yamagata et al.* [2003], a remarkable seesaw is found between the two boxes after removing the external ENSO effect (readers are referred to Figure 3 of their article). This shows quite a contrast to other major oscillatory modes such as the Southern Oscillation and the North Atlantic Oscillation. Because those are the first dominant modes even statistically, a negative correlation is observed between poles of those two modes in raw data. Since the IOD appears as the second mode statistically in SST variability, we need to remove the first dominant mode to detect its sea-saw mode statistically. This is the basic reason why some statistical analyses fail to capture the IOD signal [cf. *Dommenget and Latif*, 2002; *Hastenrath*, 2002] even if the IOD appear as a see-saw mode dramatically in a physical space during event years. The above subtlety is demonstrated mathematically in *Behera et al.* [2003a].

3.3. Ocean–atmosphere Coupling During IOD Events

The dipole mode in the SST anomalies is found to be coupled with subsurface temperature variability as well as atmospheric variability in the Indian Ocean. In fact, the first dominant mode of subsurface temperature variability, in contrast to the second dominant mode of SST variability, is characterized by a dipole related to the IOD [*Rao et al.*, 2002a]. To show a close link between the surface signal and the sub-

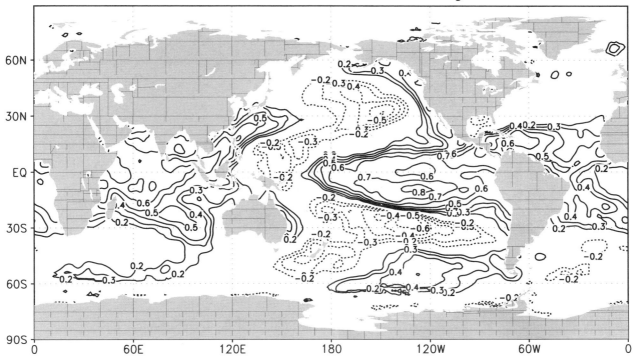

Figure 2. Correlation of the Niño-3 index with the global SST anomalies, when the former is leading the latter by 4 months. Values higher than 99% confidence limit using a 2-tailed t-test are contoured.

surface signal, we calculate an index by taking the area average of the anomalies of upper 125m heat content obtained from SODA. The area used to compute the heat content index is from 60°E–70°E and from 5°S to 5°N. This box falls within the domain of the western pole of the DMI in SST anomalies. Figure 3 shows the correlation between this heat content index and SST and wind stress anomalies during August–November. The high correlation of the heat content index with SST anomalies in the western Indian Ocean along with its simultaneous inverse correlation in the eastern part demonstrates the basin-wide coupling between the surface signal and the subsurface signal during the IOD season. Also, the correlation with equatorial wind anomalies shows the close coupling between the ocean and the atmosphere. Thus the dominant mode of subsurface variability, which is coupled with surface winds, provides a necessary feedback to SST during the IOD events.

The above is confirmed in the region of high correlation between 60°E and 80°E, where a signature of the coupled Rossby waves excited by the IOD-related winds in the equatorial region is prominent [*Rao et al.*, 2002a; *Xie et al.*, 2002]. *Rao et al.* [2002a] found that the evolution of the dominant dipole mode in the subsurface is controlled by equatorial ocean dynamics forced by zonal winds in the equatorial region. The subsurface dipole provides the delayed time required to reverse the phase of the surface dipole in the following year through propagation of oceanic Rossby/Kelvin waves. This is further confirmed from a recent coupled model study [*Gualdi et al.*, 2003]. Thus, the turnabout of the phase of the subsurface dipole leads to the quasi-biennial oscillation (QBO) of the tropical Indian Ocean and may play an important role for the QBO in the Indo–Pacific sector [cf. *Meehl*, 1987].

Rossby waves with interannual periodicity (from 3 to 5 years) are reported in the southern Indian Ocean by *Perigaud and Delecluse* [1993], *Masumoto and Meyers* [1998], *Chambers et al.* [1999] and *White* [2000]. In particular, *Masumoto and Meyers* [1998] concluded that these waves are primarily forced by the wind stress curl along the Rossby wave paths. *Xie et al.* [2002], along this context, have suggested that Rossby waves in the southern Indian Ocean play a very important role in air–sea coupling in the region, claiming also that these Rossby waves are dominantly forced by ENSO. In a more elaborate study, however, *Rao et al.* [2004] have distinguished two regions based on the major difference of forcing. The IOD dominates the forcing of the equatorial Rossby waves in the equatorial waveguide north of 10°S. In higher latitudes, i.e. south of 10°S, the ENSO influence prevails as discussed by *Xie et al.* [2002] and *Jury and Huang* [2003]. In the following,

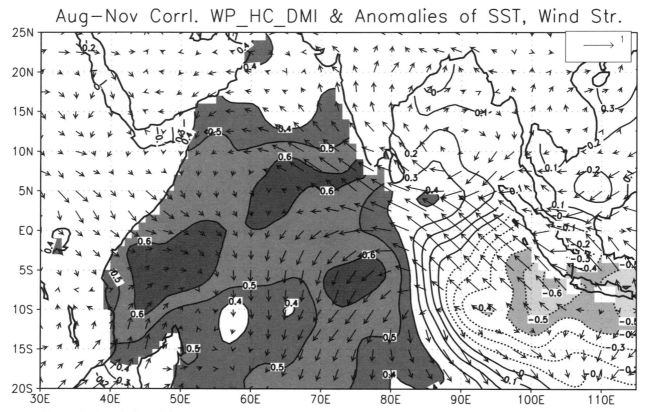

Figure 3. Correlation of the heat content anomaly index from the western box with the SST anomalies from tropical Indian Ocean and the wind stress anomalies. Contour interval is 0.1 and negative (positive) values are indicated by dashed (solid) contours. The contours that exceed ±0.4, which correspond to statistical significance at 99% with a 2-tailed t-test, are shaded.

we further verify this interesting difference in the forcing of Rossby waves using a coupled model simulation.

3.4. Evidence of Coupling in a CGCM Simulation

We here discuss the simulation results from SINTEX-F1 CGCM. Just like the SINTEX [*Gualdi et al.*, 2003], the SINTEX-F1 has shown a very high skill in simulating IOD events as well as ENSO events [*Behera et al.*, 2003b]. Putting aside a comprehensive intercomparison between the SINTEX and SINTEX-F1 simulations, which is beyond the scope of the present paper, we report here some of our analyses that support the observational results. Since the eastern pole of the SST anomaly in the model extends to the central part of the basin, the western box (40°–60°E, 10°S–10°N) used in deriving the model DMI is slightly different from that for the observation (50°–70°E, 10°S–10°N) [*Saji et al.*, 1999]. The standard deviation of the model DMI is 0.5°C that is slightly higher compared to the observed DMI [*Saji et al.*, 1999], whereas the model Niño-3 standard deviation is 0.8°C that is similar to the observation.

As in the observation, we find that the model SST variability is closely coupled to the subsurface variability in the eastern and southern tropical Indian Ocean (Figure 4). We here focus our attention on the air–sea coupling [*Rao et al.*, 2004] in the southern tropical Indian Ocean as it has interesting implications in the African rainfall variability [*Black et al.*, 2003; *Saji and Yamagata*, 2003b; *Behera et al.*, 2003b; *Jury*, 2002]. Seasonal indices of DMI (September–November) and Niño-3 (October–December) are used in the analyses. Figure 5 shows the correlation coefficients between the SSH anomaly at different latitude bands and either the DMI or the Niño-3 index. We note that the correlation in case of the DMI is higher near the equatorial region. In contrast, the correlation with the Niño-3 index is higher in the off-equatorial region. The correlation with the ENSO is not surprising. As discussed in the following section, the analysis is affected by the 28% co-occurrence of IODs with ENSOs. Therefore, we have separated their unique interactions using the partial correlation method here (Figure 6). As evident in Figure 6, the correlation partial to DMI did not change much but the

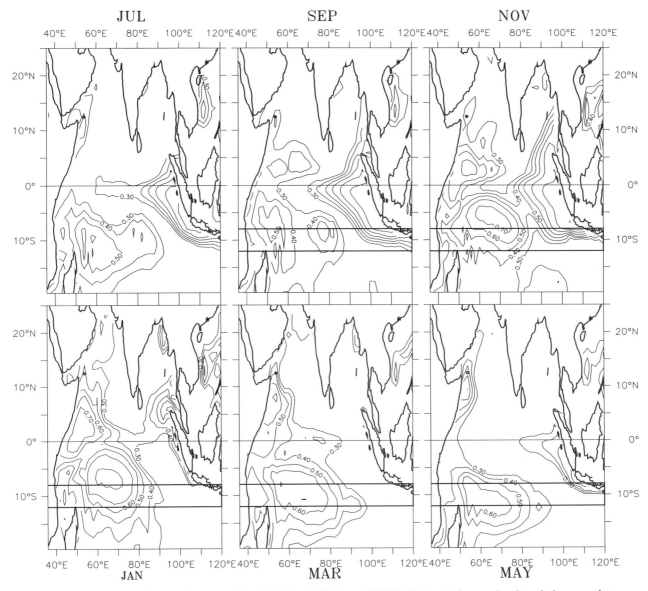

Figure 4. Correlation between the anomalies of SST and SSH from SINTEX-F1 simulation results. Correlations are plotted for alternate months starting from July to the May of the following year. Values higher than 0.2 are statistically significant at 99% level using a 2-tailed t-test.

corresponding correlation partial to Niño-3 weakened appreciably in the equatorial regions. It is interesting to note here that the correlation peaks during boreal fall in the former case, whereas the correlation peaks in spring of the following year in the latter case. Therefore, we must be careful in interpreting relations between the Indian Ocean signals and influences of Pacific climate variations [cf. *Xie et al.*, 2002]; the interannual SSH anomalies in the equatorial waveguide north of 10°S is dominated by IOD, whereas that in the off-equatorial region south of 10°S is more influenced by ENSO.

The westward propagation of the long Rossby waves is evident in the correlation patterns; it is particularly clear in the higher latitude bands as the phase speed decreases with the increasing latitude. We note that the theoretical phase speed of the first baroclinic Rossby mode is about 0.7 m s^{-1}, 0.4 m s^{-1}, 0.2 m s^{-1} and 0.1 m s^{-1} for the latitude bands of 2°S–4°S, 6°S–8°S, 10°S–12°S and 14°S–16°S, respectively. Since the phase speed is faster near the equator, the westward orientation is not noticeable in the correlation pattern. We also note that the actual phase propagation can differ from that of the free

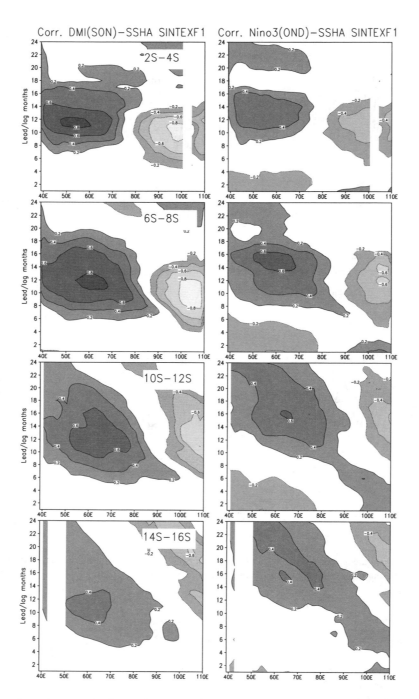

Figure 5. Correlation between the September–November DMI and the south Indian Ocean SSH anomalies (left panels) for different latitude bands in SINTEX-F1 simulation. The corresponding correlation for the October–December Niño-3 index is shown on the right panels. Contour interval is 0.2 and negative (positive) values are indicated by dashed (solid) contours. The contours that exceed ±0.2, which correspond to statistical significance at 99% with a 2-tailed t-test, are shaded.

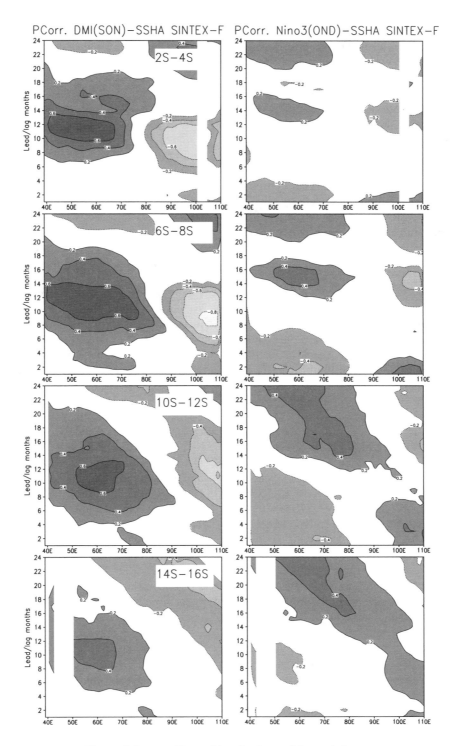

Figure 6. Same as Figure 5 but for the partial correlation.

modes since the waves in the southern Indian Ocean are much influenced by air–sea coupling [e.g., *White*, 2000]. All those from the model simulation are in good agreement with the observations [*Rao et al.*, 2004].

In order to understand the unique nature of subsurface coupling with winds in the model, we have calculated the partial correlation between the wind stress curl anomaly and either the DMI or the Niño-3 index (Figure 7). The correlation partial to DMI becomes significant from July. In particular, high values of correlation are observed in south-central tropical Indian Ocean region during the peak IOD season of September–November. The correlation partial to Niño-3 is insignificant during this time. It becomes significant only in December in a small region south of 10°S. Therefore, it is clear that the ocean–atmosphere conditions related to the IOD is essential for the coupled Rossby waves in the southern equatorial regions. The wind stress curl associated with the IOD forces the westward propagating downwelling long Rossby waves which increase the heat content of the upper layer in the central and western parts of the basin. The heat content anomaly maintains the SST anomaly, which is, in turn, tied to the wind stress anomalies, thereby completing the feedback loop. The ENSO correlation with the southern Indian Ocean is not very clear. One possibility is the ENSO signal through the Indonesian throughflow. The oceanic anomaly of the Pacific origin may propagate westward and enhance local air–sea coupling south of 10°S, thus generating wind anomalies necessary to excite the Rossby waves in the higher latitude [cf. *Masumoto and Meyers*, 1998].

4. IOD AS AN INHERENT COUPLED MODE

As already mentioned, the IOD events sometimes co-occur with the Pacific ENSO events in both the present model and the observation. Therefore, one concern about IOD is whether it is independent from ENSO. A simple correlation between DMI and Niño-3 index is 0.33 in the observation [*Saji et al.*, 1999]. Since the IOD evolution is locked to seasons, it is important to introduce the seasonal stratification in the statistical analysis as pointed out by *Nicholls and Drosdowsky* [2000] and *Allan et al.* [2001]. For the peak IOD season of September–November, the correlation between DMI and Niño-3 index amounts to 0.53. Based on this significant correlation, one is apt to conclude in a straightforward way that IOD events occur as a part of ENSO [*Allan et al.*, 2001; *Baquero-Bernal et al.*, 2002]. Another way in interpreting this statistics is that it merely reflects the fact that one third of the positive IOD events co-occur with El Niño events. The latter view is based on the fact that the non-orthogonality of two time series does not necessarily mean that the two phenomena are always connected in a physical space. Observation of clear independent occurrence of IOD in some years as discussed in the following tilts the discussion in favor of the latter interpretation.

Several positive IOD events actually evolved in absence of El Niño in certain years such as 1961, 1967 and 1994. This is further analyzed using a composite technique. In addition to composite pictures based on all IOD and ENSO events, we have prepared composites for pure events. A positive (negative) pure IOD event is identified as an event when El Niño (La Niña) does not co-occur (Table 1). In total, 9 independent IOD events and 10 independent ENSO events are used to prepare composite pictures for pure events. The total number assures us of robustness of the results. Figure 8 shows such composites of SST anomalies for all and pure IOD events during the IOD evolution period. Both composites look very similar. We find cold SST anomalies near the Sumatra coast and in the eastern Indian Ocean and warm SST anomalies in the regions of central and western Indian Ocean. It may be noted that the SST anomalies are stronger in the pure IOD composite. To understand the characteristics of the ENSO-related anomalies during those months, we have plotted in Figure 9 the corresponding composites of SST anomalies. Although a dipole-like pattern emerges in all El Niño composite, it almost disappears in a pure El Niño composite (right panel in Figure 9) in which we have removed the co-occurring IOD years. In the pure El Niño composite, we find that the cold SST anomalies near the Java coast propagate along the west coast of Australia. This is understood on the basis of the oceanic finding that the mature ENSO signal in the western Pacific intrudes into the eastern Indian Ocean through the coastal wave-guide around the Australian continent [*Clarke and Liu*, 1994; *Meyers*, 1996]. The SST in the eastern Indian Ocean near the west coast of Australia during the boreal fall and winter is thus influenced by ENSO. This is known as the *Clarke-Meyers effect*. The changes in the SST may cause local air–sea interaction in this region [*Hendon*, 2003; *Tozuka and Yamagata*, 2003]. This phenomenon appears to be different from the cooling off Sumatra related to the basin-wide IOD phenomenon that involves the equatorial ocean dynamics as discussed earlier. However, it apparently enhances the IOD-ENSO correlation during the boreal fall.

As discussed earlier, the variability of the IOD and ENSO is simulated realistically in the SINTEX-F1 CGCM. The correlation between DMI and Niño-3 is 0.4 for the whole year and 0.54 for the boreal fall season, which is very similar to the observation. Therefore, the long time series of the model simulation gives us an opportunity to verify the above independence issue of IOD with better statistical confidence. We have shown the composite of anomalies of model SST and SSH in Figures 10 and 11. Here we only show the composites for pure IOD and pure ENSO events. In total, we find 25 pure IOD events and 15 pure ENSO events in the model simulation.

Figure 7. Same as Figure 5 but for the partial correlations with the tropical Indian Ocean wind stress curl anomalies for the months starting from September to December. Contoured values are statistically significant at 99% level using a 2-tailed t-test and positive values > 0.3 are shaded.

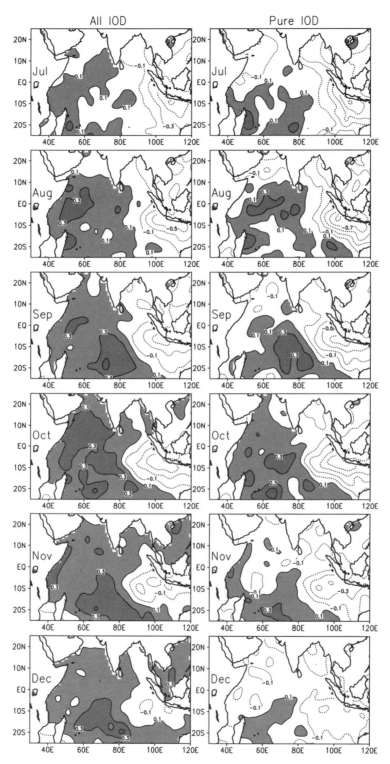

Figure 8. Composite ((positive−negative)/2) plots of SST anomalies for months starting from July to December. Left panels show the composite for the IOD years that also include ENSO events. Right panels are the composites for the pure IOD years. Contour interval in 0.2^0C and positive values $> 0.1^0C$ are shaded.

Table 1. Years of IOD and ENSO events considered in the composite analyses. The asterisk denotes pure events, i.e. no El Niño (La Niña) during a positive (negative) IOD event.

	Years of Positive IOD	Years of Negative IOD	Years of El Niño	Years of La Niña
1	**1961***	**1958***	1963	1964
2	1963	**1960***	**1965***	**1967***
3	**1967***	1964	**1969***	1970
4	1972	1970	1972	**1971***
5	**1977***	**1989***	**1976***	**1973***
6	1982	**1992***	1982	**1975***
7	**1994***	**1996***	**1986***	**1988***
8	1997	-	**1991***	-
9	-	-	1997	-

The evolution of IOD events is clearly seen in the pure IOD composites but the dipole-like variability cannot be seen in the pure ENSO composites. Rather, we notice the basin-wide uniform mode. These model results, together with the observed data, demonstrate the existence of the independent evolution of IOD.

The zonal Walker circulation is another process that may link the IOD events in the Indian Ocean with ENSO events in the Pacific. *Yamagata et al.* [2002] have shown that the atmospheric bridge between the two basins is apparent when we consider all IOD events that include the co-occurring ENSO events. However, they have demonstrated that an anomalous Walker cell exists only in the Indian Ocean during pure IOD events [Figure 4 of *Yamagata et al.*, 2003]; this also confirms the independent evolution of some IOD events. To avoid misunderstanding, we repeat that this linear analysis does not exclude completely the possibility of physical interaction between the two climate signals in some cases. From a case study of the 1997–98 El Niño event, *Ueda and Matsumoto* [2000] suggested that the changes in the Walker circulation related to the El Niño could influence the evolution of IOD through changes in the monsoon circulation. Conversely, *Behera and Yamagata* [2003] showed that IOD modulates the Darwin pressure variability, i.e., the Southern Oscillation.

The precondition for IOD evolution is yet another issue for deliberation. Several studies indicate presence of a favorable mechanism in the eastern Indian Ocean [e.g., *Saji et al.*, 1999; *Behera et al.*, 1999] that combines cold SST anomalies, strengthening of southeasterlies and suppression of convection into a feedback loop. However, recent studies offer a few alternatives: atmospheric pressure variability in the eastern Indian Ocean [e.g., *Gualdi et al.*, 2003; *Li et al.*, 2003], favorable changes in winds in relation to the Pacific ENSO and Indian monsoon [e.g., *Annamalai et al.*, 2003] and influences from the southern extratropical region [e.g., *Lau and Nath*, 2004]. All those studies fall short in more than one occasion to answer the failure (success) in IOD evolution in spite of favorable (unfavorable) precondition. For example, *Gualdi et al.* [2003] reported the failure of their proposed favorable mechanism to excite the IOD event in 1979. We also find several instances (e.g. the aborted 2003 event) when an IOD event is aborted at its premature stage although air–sea conditions are apparently favorable for its complete evolution. This indicates the evolution mechanism of the IOD is more complex than we expect now and we need further studies. Processes that need immediate attention are 1) mechanisms that connect/disconnect subsurface signals with surface ones including the barrier layer structure [*Masson et al.*, 2003a], and 2) roles of intraseasonal disturbances in both atmosphere and ocean. The IOD's unique influence on the global climate as reviewed in the following section provides motivation for further intensive field and modeling efforts.

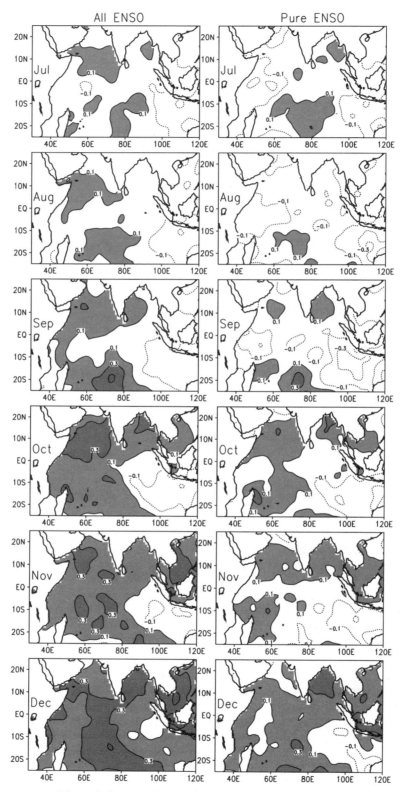

Figure 9. Same as Figure 8 but for all and pure ENSO years.

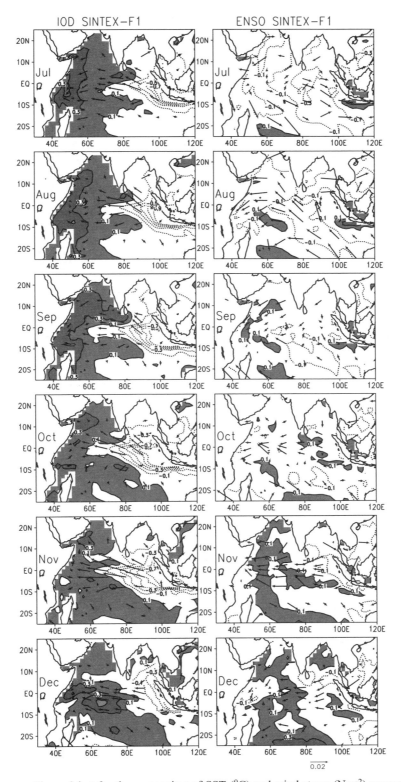

Figure 10. Same as Figure 8 but for the composites of SST (^{0}C) and wind stress (Nm^{-2}) anomalies from SINTEX-F1 simulation results for pure IOD events (left panels) and pure ENSO events (right panels).

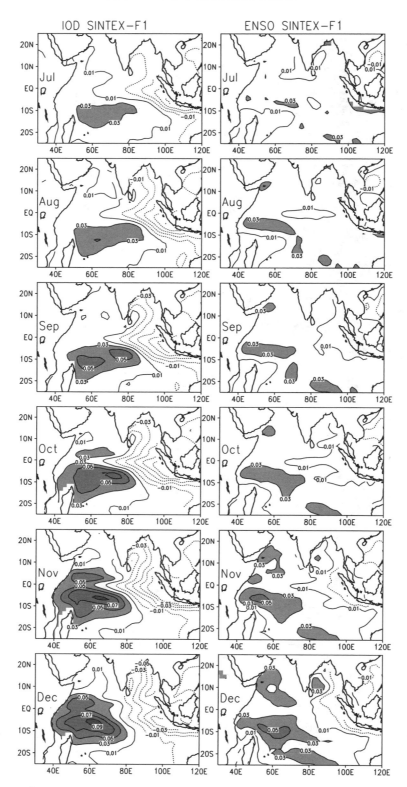

Figure 11. Same as Figure 10 but for model SSH anomalies. Contour interval is 0.02 m and positive values > 0.03 m are shaded.

5. IOD TELECONNECTION

One of the most important directions of the IOD research from the societal viewpoint is the identification of its unique teleconnection bringing regional climate variability in various parts of the globe. Since IOD and ENSO indices are not completely orthogonal, IOD influences must be carefully appreciated. Now that the degeneracy is almost resolved using the partial correlation analysis, relative influences caused by these two major tropical phenomena are becoming clear. *Behera and Yamagata* [2003] showed that the IOD influences the Darwin pressure variability, i.e., Southern Oscillation. Positive IOD and El Niño have similar impacts in the Indonesian region owing to anomalous atmospheric subsidence, and thereby induce drought there. Interested readers are referred to *Saji and Yamagata* [2003b], where they have shown for the first time, using the partial correlation analysis, the IOD teleconnection pattern over the globe.

Here, we show rainfall variabilities associated with either IOD or ENSO using a composite technique which is similar to but much simpler than the partial correlation analysis. Plate 1 shows the composite plots for two consecutive seasons. Notice that the anomalies in the East African region are unlike those from a conventional view. Several previous studies suggested enhanced short rains during El Niño events [*Ropelewski and Halpert*, 1987; *Ogallo*, 1989; *Hastenrath et al.*, 1993; *Mutai and Ward*, 2000]. Recent statistical analyses of observed data, however, have revealed that IOD rather than El Niño is more responsible for the enhancement of the East African short rains [*Black et al.*, 2003; *Saji and Yamagata*, 2003b]. Plate 1 supports clearly this new view. In addition, *Behera et al.* [2003b], after deriving an index for the East African short rains, have shown that SINTEX-F1 simulation results reproduces an east–west dipole in the correlation between the short rain index and the SST anomalies in the Indian Ocean. They have also found a high simultaneous correlation of the zonal wind anomalies, emphasizing the existence of air–sea coupling during the IOD's influence on the East African short rains. The slow propagation of the air–sea coupled mode in the western Indian Ocean provides a scope for the predictability of the IOD-induced short rains at least a season ahead [*Rao et al.*, 2004]. The anomalous westward low-level winds in response to the anomalous zonal gradient of SST increase the moisture transport to the western Indian Ocean and enhance atmospheric convection in East Africa. The model DMI has a high correlation coefficient of 0.65 with that for East African short rains. In contrast, the model Niño-3 index has a correlation coefficient of only 0.28 with that for East African short rains. In Plate 2, we show the partial correlation of the model rainfall anomalies with either DMI or Niño-3. The correlation patterns are consistent with the observed variation of rainfall anomalies in the East African region; positive IOD (El Niño) events are related to enhanced (reduced) rainfall in East Africa. Interestingly, the current coupled model captures even the higher impact of IOD on the Sri Lankan Maha rainfall as observed by *Lareef et al.* [2003]. In the Indonesian regions, the model rain anomaly shows higher negative partial correlation with the IOD index as compared to that of ENSO.

From the rainfall composite plots in Plate 1 and similar composites for surface temperature (figures not shown), we find that the positive IOD and El Niño have opposite influences in the Far East, including Japan and Korea; positive IOD events give rise to warm and dry summer, while negative IOD events lead to cold and wet summer [see *Saji and Yamagata*, 2003b for more details]. For example, the record-breaking hot and dry summer during 1994 (just like 1961) in East Asia was actually linked to the IOD [*Guan and Yamagata*, 2003]. It is well known that the summer climate condition over East Asia is dominated by activities of the East Asian summer monsoon system. Since the East Asian summer monsoon system is one subsystem of the Asian Monsoon [*Wang and Fan*, 1999], it interacts with another subsystem, the Indian summer monsoon, via variations of the Tibetan high and the Asian jet [*Rodwell and Hoskins*, 1996; *Enomoto et al.*, 2003]. The precipitation over the northern part of India, the Bay of Bengal, Indochina and the southern part of China was enhanced during the 1994 positive IOD event [*Behera et al.*, 1999; *Guan and Yamagata*, 2003; *Saji and Yamagata*, 2003b]. Using the NCEP/NCAR reanalysis data [*Kalnay et al.*, 1996] from 1979 through 2001 and the CMAP precipitation data from 1979 through 1999, *Guan and Yamagata* [2003] analyzed the summer conditions of 1994 and found that the equivalent barotropic high pressure system known as the Bonin High was strengthened over East Asia (Figure 12). The anomalous pressure pattern bringing the hot summer is well known to Japanese weather forecasters as a whale tail pressure pattern. The tail part (the Bonin High) is equivalent barotropic in contrast to the larger baroclinic head part the Pacific High). The IOD-induced summer circulation changes over East Asia are understood through a triangular mechanism as shown schematically in Plate 3. One process is that a Rossby wavetrain is excited in the upper troposphere by the IOD-induced divergent flow over the Tibetan Plateau [*Sardeshmukh and Hoskins*, 1988]. This wave train propagates northeastward from the southern part of China. This is quite similar to Nitta's Pacific–Japan pattern [*Nitta*, 1987] although the whole system is shifted a little westward. Another process is that the IOD-induced diabatic heating around India excites a long atmospheric Rossby wave to the west of the heating. The latter reminds us of the monsoon–desert mechanism that connects the circulation changes over the Mediterranean Sea/Sahara region with the heating over India [*Rodwell and Hoskins*, 1996]. Interestingly, this monsoon–desert mecha-

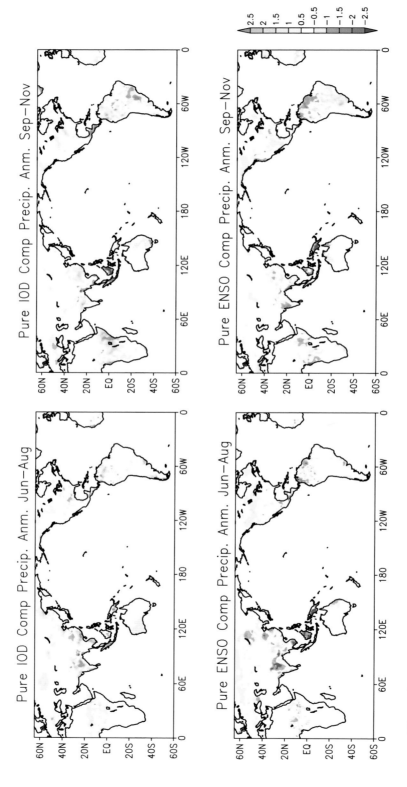

Plate 1. Composites of gridded precipitation anomalies (mm/day) for pure IOD (upper) and ENSO (lower) events. The left panels are for June–August season and the right panels are for September–November season.

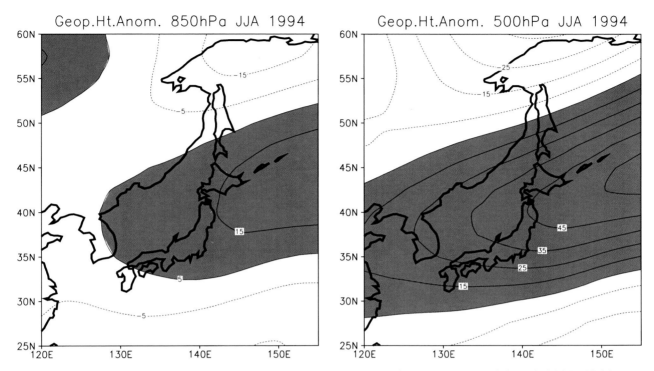

Figure 12. June–August geopotential height anomalies during summer of 1994 for 850 hPa (left) and 500 hPa (right).

nism was introduced by examining the anomalous summer condition of 1994 prior to the discovery of IOD [cf. *Hoskins*, 1996]. The westerly Asian jet acts as a waveguide for eastward propagating tropospheric disturbances to connect the circulation change around the Mediterranean Sea with the anomalous circulation changes over East Asia. This mechanism called the Silk Road process may contribute to strengthening the equivalent barotropic Bonin High in East Asia [*Enomoto et al.*, 2003]. The scenario was confirmed by calculating the wave activity flux [cf. *Plumb*, 1986; *Takaya and Nakamura*, 2001] by *Guan and Yamagata* [2003].

In the Southern Hemisphere, the impact of the IOD is remarkable in the southwestern part of Australia [*Saji et al.*, 2003b; *Ashok et al.*, 2003b] and Brazil [*Saji et al.*, 2003b]; positive IOD events cause warm and dry conditions and negative events cause cold and wet conditions (Plates 1 and 2). The IOD teleconnection in the winter hemisphere is more like a Rossby wave train.

6. SUMMARY

Using various ocean and atmosphere data, we have shown that the IOD is a natural ocean–atmosphere coupled mode in the Indian Ocean. This important tropical ocean–atmosphere coupled phenomenon has been overlooked for a long period as the ENSO-forced basin-wide mode dominates statistically the SST variability in the basin. The SINTEX-F1 CGCM simulates successfully the IOD as an ocean–atmosphere coupled mode and confirms the importance of oceanic dynamics in the evolution of IOD.

Although the IOD emerges statistically as the second major mode in observed SST anomalies, it shows up as a remarkable event in some years, just like a normal mode in classical dynamics, and induces climate variations in many parts of the world. The year of 1994 was such a case and the dramatic impact on summer conditions in East Asia actually led the authors to shed light on this important climate signal as a synthesis of the phenomenon described in part in the past. We have discussed here how the IOD event influences summer conditions in East Asia mostly on the basis of our recent work. The abnormally hot and dry summer in 1994 was induced by the IOD-related rainfall anomalies in South Asia through a triangular mechanism (Plate 3). On the eastern side of the triangle, the mechanism is similar to the Nitta's Pacific–Japan pattern with slight westward shift. The combination of monsoon–desert mechanism and the Silk Road process on the other two sides of the triangle strengthens the Bonin High over Japan as a whale tail pattern.

The air–sea coupling in the western Indian Ocean related to the IOD events produces anomalously active short rains in East Africa. During a positive event the wind stress curl anomalies in the off-equatorial region of the central part, related

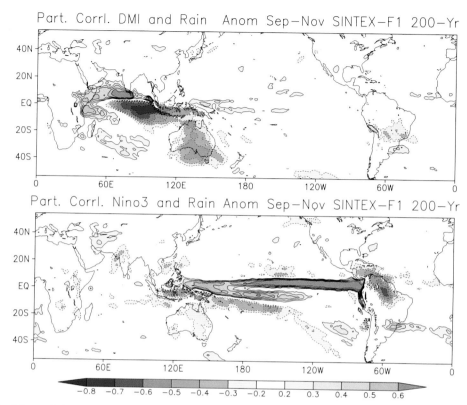

Plate 2. Partial correlation of model DMI with rainfall anomalies. The upper panel shows the partial correlation of the model DMI with the rainfall anomalies (where the Niño-3 influence is removed from the correlation) and the lower panel shows the corresponding partial correlation for the Niño-3 index (where the DMI influence is removed from the correlation). Shaded values are statistically significant at 99% level using a 2-tailed t-test.

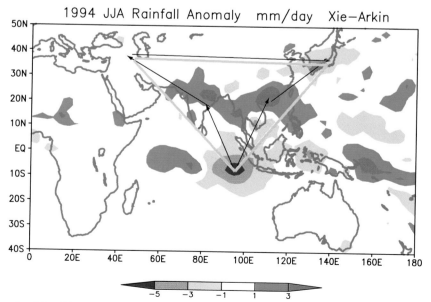

Plate 3. Schematic of the 1994 IOD event influence on the East Asia summer conditions. The triangular mechanism is shown over the shaded rainfall anomalies.

to the changes in the atmospheric convection near Sumatra, produces warmer SST. This, together with the zonal gradient in the SST anomalies in the tropical region, enhances moisture transport to the East Africa and thereby induces higher rainfall there. This is verified here by the CGCM simulation. The influence of the IOD is also remarkable in the Southern Hemisphere, particularly over Australia and Brazil. Interestingly, this observed behavior (Plate 1) is also well simulated by the SINTEX-F1 CGCM (Plate 2).

Understanding the teleconnection patterns related to either IOD or ENSO and their positive/negative interference during years of co-occurrence is very important from a societal viewpoint because people at large suffer from their regional influences rather than the coupled phenomena themselves. We have shown here that the SINTEX-F1 model simulates such climate signals and their influences reasonably well. Predicting the regional derivatives of climate variations is becoming a challenge [cf. *Philander*, 2004]. Despite the complexity in the coupled system, recent progress in high-resolution CGCM development along with the increased interest in setting up the Indian Ocean observing network composed of TRITON buys, ARGO floats and proposed equatorial moorings will enhance our understanding and predictability of the coupled variability in the Indian Ocean.

Acknowledgments. Discussions with Drs. Ashok, Delecluse, Gualdi, Guan, Meyers, Navarra, Philander, and Saji were very helpful in preparing this review from our new viewpoint. Two anonymous reviewers provided useful suggestions.

REFERENCES

Abram, N. J., M. K. Gagan, M. T. McCulloch, J. Chappell, and W. S. Hantoro, Coral reef death during the 1997 Indian Ocean Dipole linked to Indonesian wildfires, *Science, 301,* 952–955, 2003.

Allan, R., D. Chambers, W. Drosdowsky, H. Hendon, M. Latif, N. Nicholls, I. Smith, R. Stone, and Y. Tourre, Is there an Indian Ocean dipole, and is it independent of the El Niño–Southern Oscillation? *CLIVAR Exchanges, 6,* 18–22, 2001.

Annamalai, H., R. Murtugudde, J. Potemra, S. P. Xie, P. Liu, and B. Wang, Coupled dynamics over the Indian Ocean: Spring initiation of the zonal mode, *Deep-Sea Res. II, 50,* 2305–2330, 2003.

Ashok, K., Z. Guan, and T. Yamagata, Impact of the Indian Ocean Dipole on the Decadal relationship between the Indian monsoon rainfall and ENSO, *Geophys. Res. Lett., 28,* 4499–4502, 2001.

Ashok, K., Z. Guan., and T. Yamagata, A look at the relationship between the ENSO and the Indian Ocean Dipole, *J. Meteorol. Soc. Jpn., 81,* 41–56, 2003a.

Ashok, K., Z. Guan, and T. Yamagata, Influence of the Indian Ocean Dipole on the Australian winter rainfall, *Geophys. Res. Lett., 30,* doi:10.1029/2003GL017926, 2003b.

Baquero-Bernal, A., M. Latif, and S. Legutke, On dipole-like variability in the tropical Indian Ocean, *J. Clim., 15,* 1358–1368, 2002.

Behera, S. K., R. Krishnan, and T. Yamagata, Unusual ocean–atmosphere conditions in the tropical Indian Ocean during 1994, *Geophys. Res. Lett., 26,* 3001–3004, 1999.

Behera, S. K., and T. Yamagata, Influence of the Indian Ocean Dipole on the Southern Oscillation, *J. Meteorol. Soc. Jpn., 81,* 169–177, 2003.

Behera, S. K., S. A. Rao, H. N. Saji, and T. Yamagata, Comments on "A cautionary note on the interpretation of EOFs", *J. Clim., 16,* 1087–1093, 2003a.

Behera, S. K., J.-J. Luo, S. Masson, T. Yamagata, P. Delecluse, S. Gualdi, and A. Navarra, Impact of the Indian Ocean Dipole on the East African short rains: A CGCM study, *CLIVAR Exchanges, 27,* 43–45, 2003b.

Bjerknes, J., Atmospheric teleconnections from the equatorial Pacific, *Mon. Weather Rev., 97,* 163–172, 1969.

Black, E., J. Slingo, and K. R. Sperber, An observational study of the relationship between excessively strong short rains in coastal East Africa and Indian Ocean SST, *Mon. Weather Rev., 31,* 74–94, 2003.

Cadet, D. L., The Southern Oscillation over the Indian Ocean, *J. Clim., 5,* 189–212, 1985.

Cai, W., H. Hendon, and G. Meyers, Indian Ocean dipole-like variability in the CSIRO Mark 3 coupled climate model, *J. Clim.,* submitted, 2003.

Carton, J. A., G. Chepurin, X. Cao, and B. S. Giese, A simple ocean data assimilation analysis of the global upper ocean 1950–1995, Part 1: methodology, *J. Phys. Oceanogr., 30,* 294–309, 2000.

Chambers, D. P., B. D. Tapley, and R. H. Stewart, Anomalous warming in the Indian Ocean coincident with El Niño, *J. Geophys. Res., 104,* 3035–3047, 1999.

Clark, C. O., P. J. Webster, and J. E. Cole, Interdecadal variability of the relationship between the Indian Ocean Zonal Mode and East African coastal rainfall anomalies, *J. Clim., 16,* 548–554, 2003.

Clarke, A. J., and X. Liu, Interannual sea level in the northern and eastern Indian Ocean, *J. Phys. Oceanogr., 24,* 1224–1235, 1994.

Dommenget, D., and M. Latif, A cautionary note on the interpretation of EOFs, *J. Clim., 15,* 216–225, 2002.

Enomoto, T., B. J. Hoskins, and Y. Matsuda, The formation of the Bonin high in August, *Q. J. R. Meteorol. Soc., 587,* 157–178, 2003.

Feng, M., G. Meyers, and S. Wijffels, Interannual upper ocean variability in the tropical Indian Ocean, *Geophys. Res. Lett., 28,* 4151–4154, 2001.

Fujiwara, M., K. Kita, S. Kawakami, T. Ogawa, N. Komala, S. Saraspriya, and A. Suripto, Tropospheric ozone enhancements during the Indonesian forest fire events in 1994 and 1997 as revealed by ground-based observations, *Geophys. Res. Lett., 26,* 2417–2420, 1999.

Gualdi, S., E. Guilyardi, A. Navarra, S. Masina, and P. Delecluse, The interannual variability in the tropical Indian Ocean as simulated by a CGCM, *Clim. Dyn., 20,* 567–582, 2003.

Guan, Z., K. Ashok, and T. Yamagata, Summer-time response of the tropical atmosphere to the Indian Ocean dipole sea surface temperature anomalies, *J. Meteorol. Soc. Jpn., 81,* 531–561, 2003.

Guan, Z., and T. Yamagata, The unusual summer of 1994 in East Asia: IOD Teleconnections, *Geophys. Res. Lett., 30,*

doi:10.1029/2002GL016831, 2003.
Guilyardi, E., P. Delecluse, S. Gualdi, and A. Navarra, The role of lateral ocean physics in the upper ocean thermal balance of a coupled ocean–atmosphere GCM, *Clim. Dyn., 17,* 589–599, 2001.
Hastenrath, S, A. Nicklis, and L. Greischar, Atmospheric–hydrospheric mechanisms of climate anomalies in the western equatorial Indian Ocean, *J. Geophys. Res., 98 (C11),* 20219–20235, 1993.
Hastenrath, S., Dipoles, Temperature Gradient, and Tropical Climate Anomalies, *Bull. Am. Meteorol. Soc., 83,* 735–738, 2002.
Hendon, H. H., Indonesian Rainfall Variability: Impacts of ENSO and Local Air–Sea Interaction, *J. Clim., 16,* 1775–1790, 2003.
Hoskins, B. J., On the existence and strength of the summer subtropical anticyclones, *Bull. Am. Meteorol. Soc., 77,* 1287–1292, 1996.
Iizuka, S., T. Matsuura, and T. Yamagata, The Indian Ocean SST dipole simulated in a coupled general circulation model, *Geophys. Res. Lett., 27,* 3369–3372, 2000.
Jury, R. M., Economic impacts of climate variability in South Africa and development of resource prediction models, *J. Appl. Meteorol., 41,* 46–55, 2002.
Jury, R. M., and B. Huang, The Rossby wave as a key mechanism of Indian Ocean climate variability, *Deep-Sea Res.,* submitted, 2003.
Kalnay, E., and coauthors, The NCEP/NCAR 40 year Reanalysis Project, *Bull. Am. Meteorol. Soc., 77,* 437–471, 1996.
Klein, S. A., B. J. Soden, and N. C. Lau, Remote sea surface temperature variations during ENSO: Evidence for a tropical atmospheric bridge, *J. Clim., 12,* 917–932, 1999.
Lareef, Z., S. A. Rao, and T. Yamagata, Modulation of Sri Lankan Maha rainfall by the Indian Ocean Dipole, *Geophys. Res. Lett., 30,* doi:10.1029/2002GL015639, 2003.
Latif, M., and T. P. Barnett, Interactions of the tropical oceans, *J. Clim., 8,* 952–968, 1995.
Lau, N.-C., and M. J. Nath, Coupled GCM simulation of atmosphere–ocean variability associated with zonally asymmetric SST changes in the tropical Indian Ocean, *J. Clim., 17,* 245–265, 2004.
Li, C., and M. Mu, Influence of the Indian Ocean dipole on Asian monsoon circulation, *CLIVAR Exchanges, 6,* 11–14, 2001.
Li, T., B. Wang, C.P. Chang, and Y. Zhang, A theory for the Indian Ocean Dipole–Zonal Mode, *J. Atmos. Sci., 60,* 2119–2135, 2003.
Luo J.-J., S. Masson, S. Behera, S. Gualdi, A. Navarra, P. Delecluse, and T. Yamagata, The updating and performance of the SINTEX coupled model on the Earth Simulator: Atmosphere part, *EGS Meeting, Nice, France,* 2003.
Madec, G, P. Delecluse, M. Imbard, and C. Levy, OPA version 8.1 ocean general circulation model reference manual, *Technical Report/Note 11,* LODYC/IPSL, *Paris, France,* pp 91, 1998.
Masson, S., J-P Boulanger, C. Menkes, P. Delecluse, and T. Yamagata, Impact of salinity on the 1997 Indian Ocean dipole event in a numerical experiment, *J. Geophys. Res.,* in press, 2003a.
Masson, S., J.-J. Luo, S. Behera, S. Gualdi, E. Guilyardi, A. Navarra, P. Delecluse, and T. Yamagata, The updating and performance of the SINTEX coupled model on the Earth Simulator: Ocean part, *EGS Meeting, Nice, France,* 2003b.
Masumoto, Y., and G. Meyers, Forced Rossby waves in the southern tropical Indian Ocean, *J. Geophys. Res., 103,* 27589–27602, 1998.
Meehl, G. A., The annual cycle and the interannual variability in the tropical Pacific and Indian Ocean regions, *Mon. Weather Rev., 115,* 27–50, 1987.
Meyers, G., Variation of Indonesian throughflow and El Niño–Southern Oscillation, *J. Geophys. Res., 101,* 12255–12263, 1996.
Murtugudde, R. G., J. P. McCreary, and A. J. Busalacchi, Oceanic processes associated with anomalous events in the Indian Ocean with relevance to 1997–1998, *J. Geophys. Res., 105,* 3295–3306, 2000.
Mutai, C. C., and M. N. Ward, East African rainfall and the tropical circulation/convection on intraseasonal to interannual timescales, *J. Clim., 13,* 3915–3939, 2000.
Nicholls, N., and W. Drosdowsky, Is there an equatorial Indian Ocean SST dipole, independent of the El Niño–Southern Oscillation? *Symp. on Climate Variability, the Oceans, and Societal Impacts,* Albuquerque, NM, Amer. Met. Soc., *17–18,* 2000.
Nitta, T., Convective activities in the tropical western Pacific and their impact on the Northern Hemisphere summer circulation, *J. Meteorol. Soc. Jpn., 65,* 373–390, 1987.
Ogallo, L., The spatial and temporal patterns of the East African seasonal rainfall derived from principal component analysis, *Int. J. Climatol., 9,* 145–167, 1989.
Perigaud, C., and P. Delecluse, Interannual sea-level variations in the tropical Indian Ocean from Geosat and shallow water simulations, *J. Phys. Oceanogr., 23,* 1916–1934, 1993.
Philander, S. G. H., Our Affair with El Niño, *Princeton Univ. Press,* in press, 2004.
Plumb, R.A., Three-dimensional propagation of transient quasi-geostrophic eddies and its relationship with the eddy forcing of the time-mean flow, *J. Atmos. Sci., 43,* 1657–1678, 1986.
Rao, S. A., S. K. Behera, Y. Masumoto, and T. Yamagata, Interannual variability in the subsurface tropical Indian Ocean with a special emphasis on the Indian Ocean Dipole, *Deep-Sea Res. II, 49,* 1549–1572, 2002a.
Rao, S. A, V. V. Gopalkrishnan, S. R. Shetye, and T. Yamagata, Why were cool SST anomalies absent in the Bay of Bengal during the 1997 Indian Ocean Dipole Event? *Geophys. Res. Lett., 29,* doi:10.1029/2001GL014645, 2002b.
Rao, S. A., S. K. Behera, and T. Yamagata, Subsurface feedback on the SST variability during the Indian Ocean Dipole events, *Dyn. Atmos. Ocean,* submitted, 2004.
Rayner, N. A., E. B. Horton, D. E. Parker, C. K. Folland, and R. B. Hackett, *Clim. Res. Tech. Note, 74,* UK Met. Office, Bracknell, 1996.
Rodwell, M. J., and B. J. Hoskins, Monsoons and the dynamics of deserts, *Q. J. R. Meteorol. Soc., 122,* 1385–1404, 1996.
Roeckner, E., and coauthors, The atmospheric general circulation model ECHAM4: model description and simulation of present day climate, *Max Plank Institute fur Meteorologie Rep., 218,* Hamburg, Germany, pp90, 1996.
Ropelewski, F., and M. S. Halpert, Global and regional scale precipitation patterns associated with the El Niño /Southern Oscillation, *Mon. Weather Rev., 115,* 1602–1626, 1987.
Saji, N. H., B. N. Goswami, P. N. Vinayachandran, and T. Yamagata, A dipole mode in the tropical Indian Ocean, *Nature, 401,* 360–363, 1999.
Saji, N. H., and T. Yamagata, Structure of SST and surface wind

variability during Indian Ocean Dipole Mode events: COADS observations, *J. Clim.*, *16*, 2735–2751, 2003a.

Saji, N. H., and T. Yamagata, Possible impacts of Indian Ocean Dipole Mode events on global climate, *Clim. Res.*, *25*, 151–169, 2003b.

Sardeshmukh, P. D., and B. J. Hoskins, The generation of global rotational flow by steady idealized tropical divergence, *J. Atmos. Sci.*, *45*, 1228–1251, 1988.

Shinoda, T., M. A. Alexander, and H. H. Hendon, Remote response of the Indian Ocean to interannual SST variations in the tropical Pacific, *J. Clim.*, in press, 2003.

Takaya, K., and H. Nakamura, A formulation of a phase-independent wave-activity flux for stationary and migratory quasigeostrophic eddies on a zonally varying basic flow, *J. Atmos. Sci.*, *58*, 608–627, 2001.

Tourre, Y. M., and W. B. White, ENSO signals in the global upper-ocean temperature, *J. Phys. Oceanogr.*, *25*, 1317–1332, 1995.

Tozuka, T., and T. Yamagata, Annual ENSO, *J. Phys. Oceanogr.*, *33*, 1564–1578, 2003.

Ueda, H., and J. Matsumto, A possible trigerring process of east–west asymmetric anomalies over the Indian Ocean in relation to 1997/98 El Niño, *J. Meteorol. Soc. Jpn.*, *78*, 803–818, 2000.

Valcke S., L. Terray, and A. Piacentini, The OASIS coupler user guide version 2.4, *Technical Report TR/CMGC/00-10*, CERFACS, Toulouse, France, pp 85, 2000.

Venzke, S., M. Latif, and A. Villwock, The coupled GCM ECHO-2. Part II: Indian Ocean response to ENSO, *J. Clim.*, *13*, 1371–1383, 2000.

Vinayachandran, P. N., N. H. Saji, and T. Yamagata, Response of the equatorial Indian Ocean to an anomalous wind event during 1994, *Geophys. Res. Lett.*, *26*, 1613–1616, 1999.

Vinayachandran, P. N., S. Iizuka, and T. Yamagata, Indian Ocean Dipole mode events in an ocean general circulation model, *Deep-Sea Res. II*, *49*, 1573–1596, 2002.

Wang, B., and Z. Fan, Choice of South Asian summer monsoon indices, *Bull. Am. Meteorol. Soc.*, *80*, 629–638, 1999.

Webster, P. J., A. Moore, J. Loschnigg, and M. Leban, Coupled ocean–atmosphere dynamics in the Indian Ocean during 1997–98, *Nature*, *40*, 356–360, 1999.

White, W. B., Coupled Rossby waves in the Indian Ocean on interannual time scales, *J. Phys. Oceanogr.*, *30*, 2972–2989, 2000.

Willmott, C. J., and K. Matsuura, Smart interpolation of annually averaged air temperature in the United States, *J. Appl. Meteorl.*, *34*, 2557–2586, 1995.

Xie, S.-P., H. Annamalai, F. Schott, and J. P. McCreary, Structure and Mechanisms of south Indian Ocean climate variability, *J. Clim.*, *15*, 864–878, 2002.

Yamagata, T., S. K. Behera, S. A. Rao., Z. Guan, K. Ashok, and H. N. Saji, The Indian Ocean dipole: a physical entity, *CLIVAR Exchanges*, *24*, 15–18, 2002.

Yamagata, T., S. K. Behera, S. A. Rao, Z. Guan, K. Ashok, and H. N. Saji, Comments on "Dipoles, Temperature Gradient, and Tropical Climate Anomalies", *Bull. Am. Meteorol. Soc.*, *84*, 1418–1422, 2003.

Yule, G. U., On the Theory of Correlation for any Number of Variables, Treated by a New System of Notation, *Proc. Royal Soc. Series A*, *79*, 182–193, 1907.

S. K. Behera, J.-J. Luo, S. Masson, S. A. Rao, and T. Yamagata, Frontier Research System for Global Change, Yokohama, Kanagawa 236–0001, Japan. (yamagata@eps.s.u-tokyo.ac.jp)

M. R. Jury, Environmental Science Department, University of Zululand, South Africa.

Role of the Indian Ocean in Regional Climate Variability

H. Annamalai

IPRC/SOEST, University of Hawaii, Honolulu, Hawaii

Raghu Murtugudde

ESSIC/Meteorology, University of Maryland, College Park, Maryland

The role of the Indian Ocean in regional climate variability has been studied for a long time. Whether the Indian Ocean plays an active role or simply responds passively to the wind and heat flux variability generated elsewhere remains somewhat of an open question. Here we attempt a fairly comprehensive review of the literature relating to the Indian Ocean variability at all time-scales and to the current understanding of the role of this tropical ocean in the coupled climate system. Despite an investigative history of more than a century, it is fair to say that the role of the Indian Ocean in regional monsoon variability, the most important climate process in this sector, remains to be fully understood and remains an active area of diagnostic and modeling research. This translates into limited success of the state-of-the-art monsoon forecast systems, statistical and dynamical. Much attention in the last few years has been focused on the east–west mode of variability, referred to as the dipole or the zonal mode. While there is incontrovertible evidence that the Indian Ocean plays an active role in some of these dipole/zonal mode events, the self-sustainability of this mode and its impact on regional and global climate is still a matter of debate. The most significant impediment to improving the predictive understanding of the region is the fact that at all time-scales sea surface temperature variability of the Indian Ocean is most often in the range of observational errors. The need for coordinated and sustained observations and diagnostic studies along with continued investigations using a hierarchy of models has been well-recognized, because monsoons form an integral part of the global climate system and the ENSO cycles. It is thus hoped that the role of the Indian Ocean in the coupled climate system will be much better quantified in the coming years, leading to significant improvements in regional coupled climate forecasts.

1. INTRODUCTION

1.1. Background

Anomalous events in the Indian Ocean (IO) during 1994 and 1997 have generated much renewed interest in all aspects of the coupled climate system over this sector. The purpose of this study is to review the literature on IO dynamics and thermodynamics and place these studies in the context of the role of the IO in regional climate variability. Most earlier studies of the IO were typically motivated by the unique aspect of the region, namely, the seasonally reversing monsoon forcing [*Lighthil*, 1969; *Leetmaa*, 1972; *Anderson and Rowlands*, 1976; see *Schott and McCreary*, 2001, for a review]. Numerous studies reported so far address the impact of monsoonal forcing on the physical, biological, and biogeochemical variability of the IO basin. The majority relate to climatological features as opposed to the Pacific basin, where the number of studies addressing the interannual to decadal variability far outnumber the climatological studies [see TOGA Special Issue of J. Geophys. Res., 1998].

Plate 1a shows the standard deviation (a measure of interannual variability) of sea surface temperature (SST). It is evident that, unlike the tropical Pacific [e.g., *McPhaden*, 1999], the SST variance in the IO is relatively small. This may account for the lack of motivation to address the heat budgets that control SST variability, the main variable of interest for atmospheric forcing. Although small in magnitude, the SST variability, particularly that over the Southern IO, is tightly coupled to thermocline variability (Plate 1b), and therefore expected to influence the regional climate variability. Yet, many of the studies investigating the heat budgets for SST variability in the IO did not seek thermocline/SST feedbacks [*Shetye*, 1986; *Molinari et al.*, 1986; *McCreary and Kundu*, 1989; *McCreary et al.*, 1993; *Anderson and Carrington*, 1993; *Behera et al.*, 2000]. Recently, such feedback is being investigated to find the role of IO in regional monsoons or the IO response to El Niño–Southern Oscillation (ENSO) forcing [*Hastenrath and Greischar*, 1993; *Murtugudde and Busalacchi*, 1999; *Schiller et al.*, 2000; *Loschnigg and Webster*, 2000; *Rao and Sivakumar*, 2000; *Xie et al.*, 2002; *Meehl et al.*, 2003].

Some atlases, which provide an excellent background for any IO studies have been constructed for the IO [e.g., *Hastenrath and Greischar*, 1989; *Rao et al.*, 1989]. Recently, *Schott and McCreary* [2001] have provided an excellent review of the physical oceanography of the IO. Unlike the numerous volumes of literature, review articles and books on ENSO and its global impacts [e.g., *Philander*, 1990; TOGA Special issue of J. Geophys. Res., 1998], a comprehensive review on the role of IO on regional climate variability is unavailable. After the anomalous IO events of 1994 and 1997, there is a surge of interest to understand the interannual variability of the IO dipole/zonal mode (hereafter IODZM). In light of this recent interest in the IO, this review is appropriate and timely. In the following subsections, we briefly highlight the characteristics of the subregions of the IO. This will lay the foundation for the role of these regional SST anomalies on the regional climate variability, particularly on the Asian–African–Australian monsoon systems.

1.2. Subregions of the Indian Ocean

1.2.1. Arabian Sea. The IO variability is fundamentally affected by the northern land boundary that consists of the Asian continent and the Indian subcontinent, which bisects the Northern IO into the Arabian Sea and the Bay of Bengal. The response to the reversing monsoonal forcing is most dramatic in the Somali upwelling region in the western boundary of the Arabian Sea with the Somali Current directed equatorward during the Northeast (winter) Monsoon and poleward during the Southwest (summer) Monsoon. It is known that most of the SST variability in the Arabian Sea can be attributed to changes in surface heat fluxes [*McCreary and Kundu*, 1989; *Rao et al.*, 1989; *Murtugudde and Busalacchi*, 1999; *Rao and Sivakumar*, 2000].

1.2.2. Bay of Bengal. The Bay of Bengal receives freshwater input from several major rivers such as Irrawaddy, Ganga, and Brahmaputra in addition to significant amounts of rainfall during the summer monsoon season [*Shetye et al.*, 1996; *Vinayachandran et al.*, 2002]. While synoptic observational estimates show the formation of a freshwater induced barrier layer [*Lukas and Lindstrom*, 1991; *Sprintall and Tomczak*, 1992; *Vinayachandran et al.*, 2002], buoy observations and modeling studies indicate that the impact of freshwater input on regional SST is not as obvious [*Sengupta and Ravichandran*, 2001; *Han et al.*, 2001; *Howden and Murtugudde*, 2001]. The higher mean SST in the head Bay (Plate 2) is instrumental for the maintenance of the higher mean precipitation during the monsoon season (Plate 3b) [*Shenoi et al.*, 2002]. Yet, the SST variations over the Bay of Bengal are negligible (Plate 1a) except at intraseasonal time scales during the summer monsoon season (Section 2.6).

1.2.3. Equatorial region. Unlike the other two oceans, the IO has no significant upwelling in the eastern equatorial region. The upwelling is, however, stronger in the western equatorial region and the seasonal cycle of the SST propagates eastward unlike the equatorial Atlantic and Pacific Oceans where it propagates westward. The IO is the only tropical ocean where the annual-mean winds on the equator are westerly, resulting in a deeper thermocline in the eastern equatorial region [*Reverdin*, 1987].

The most prominent feature of the equatorial IO is the semi-annual equatorial Kelvin waves and the associated eastward jets [*Wyrtki*, 1973; *O'Brien and Hurlburt*, 1974; *Yamagata et al.*, 1996; *Sprintall et al.*, 2000]. As a consequence, all year around, the SST in the eastern equatorial IO

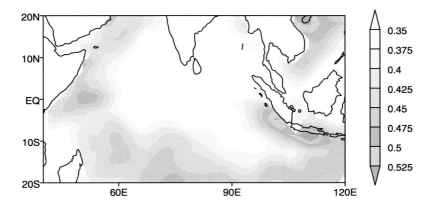

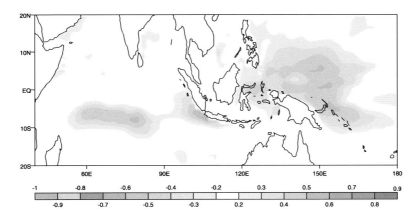

Plate 1. (a) Standard deviation of interannual SST, and (b) simultaneous correlation between anomalous SST and thermocline as measured by the depth of the 20°C isotherm.

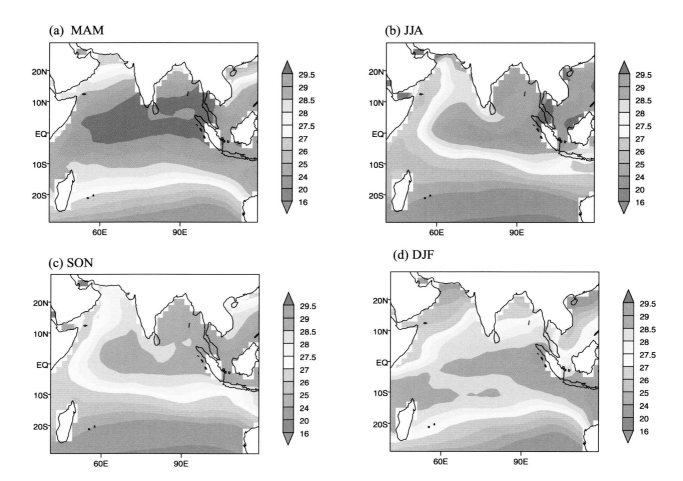

Plate 2. Seasonal SST climatology over the tropical Indian Ocean, (a) March–May, (b) June–August, (c) September–November, and (d) December–February. The figure illustrates the seasonal cycle of SST.

in the vicinity of (90°E–110°E, 10°S–0) is warmer than 27.5°C (Plate 2), the threshold required for deep convection in the tropics [e.g., *Gadgil et al.*, 1984]. The eastern equatorial IO experiences heavy rainfall throughout the year (Plate 3) resulting in a barrier layer [*Sprintall and Tomczak*, 1992; *Masson et al.*, 2002], and the barrier layer variability there may play a role in the interannual variability of the tropical IO [*Murtugudde and Busalacchi*, 1999; *Murtugudde et al.*, 2000]. Despite these constraints, occasionally there is significant upwelling of cold thermocline waters off the coasts of Sumatra and Java (as indicated by thermocline feedback in Plate 1b), leading to a cold tongue for a few months.

1.2.4. Southwest Indian Ocean (SWIO). The Southern IO is influenced by predominant southeasterlies to the south of about 10°S and seasonally reversing southwest and northeast monsoons to the north of 10°S. Thus the deep tropics between the equator and about 10°S is subject to seasonally reversing Ekman pumping, which leads to a doming thermocline structure, especially in the west [*Reverdin*, 1987; *Murtugudde and Busalacchi*, 1999; *Xie et al.*, 2002]. As suggested by Plate 1b, there is a strong linkage between SST and thermocline variations over the SWIO in the vicinity of 8°S, 60°E. *Xie et al.* [2002] and *Huang and Kinter* [2002] showed that the SWIO thermocline variability is influenced by Rossby waves, whose signal may extend into the equatorial region [*Tsai et al.*, 1992; *Hastenrath et al.*, 1993; *Murtugudde et al.*, 2000; *Vinayachandran et al.*, 2003]. The importance of SWIO SST variability on the regional climate will be addressed in Section 2.

1.2.5. Southeast Indian Ocean and Indonesian throughflow. One major factor in the variability of the Southern IO is the introduction of Pacific waters through the Indonesian Seas via the Indonesian throughflow (ITF) which has been studied extensively [see *Godfrey*, 1996, for a review]. The effect of upwelling on SST variability off Java is reduced due to the spreading of the thermocline by the ITF [*Murtugudde et al.*, 1998b]. The west coast of Australia has persistent alongshore winds which are upwelling favorable. However, the steric gradient set up by the ITF drives a coastal current poleward into the prevailing winds leading to the Leeuwin current complex [*Godfrey and Ridgeway*, 1985; *Godfrey and Weaver*, 1991]. The SST variability over the subtropical Southeast IO (poleward of 20°S) is largely due to heat fluxes [*Xie et al.*, 2002].

In summary, most of the observed SST variance in the Northern IO is due to heat fluxes [e.g., *Klein et al.*, 1999; *Rao and Sivakumar*, 2000], while ocean dynamics contribution is substantial over the Southern IO, particularly in the neighborhood of the equatorial zone [e.g., *Murtugudde et al.*, 2000; *Xie et al.*, 2002; *Huang and Kinter*, 2002].

1.3. ENSO and Non-ENSO Variability in the Indian Ocean

1.3.1. ENSO signature. The IO response to ENSO is well known to be a basin scale surface warming/cooling with a lag of about a season [*Nigam and Shen*, 1993; *Tourre and White*, 1995; *Klein et al.*, 1999; *Venzke et al.*, 2000], whereas the subsurface response appears to be simultaneous, forced by ENSO related wind anomalies [*Reverdin et al.*, 1986; *Hastenrath et al.*, 1993; *Shen and Kimoto*, 1999; *Murtugudde and Busalacchi*, 1999; *Chambers et al.*, 1999; *Webster et al.*, 1999; *Schiller et al.*, 2000; *Hastenrath*, 2002; *Krishnamurthy and Kirtman*, 2003; *Masson et al.*, 2003; *Allan et al.*, 2001]. In the basin-wide structure the maximum SST variance is observed over SWIO (Plate 1a). Analysis of in-situ measurements and model-assimilated dataset reveals a strong influence of subsurface thermocline variability on SST over the SWIO, and ENSO is the dominant forcing for thermocline variability there [*Xie et al.*, 2002; *Huang and Kinter*, 2002]. Some recent studies, based on short records, argue that subsurface temperature variability in the IO is predominantly quasi-biennial and is independent of ENSO [*Rao et al.*, 2002].

1.3.2. Indian Ocean dipole/zonal mode. A new wave of interest has emerged in understanding the air–sea interactions in the equatorial Indian Ocean, known as IODZM [*Reverdin et al.*, 1986; *Kapala et al.*, 1994; *Hastenrath et al.*, 1993; *Murtugudde et al.*, 1998a; *Murtugudde and Busalacchi*, 1999; *Saji et al.*, 1999; *Webster et al.*, 1999; *Anderson*, 1999; *Yu and Rienecker*, 1999, 2000]. Of particular importance is the SST variability off Java/Sumatra and its relationship to thermocline (Plate 1). Probably for the first time, *Murtugudde et al.* [1998a] and *Murtugudde and Busalacchi* [1999] reported that this was one of the regions in the IO where Bjerknes [1969] feedbacks occur due to interactions between the thermocline and SST. Some of the interpretations of the surface and subsurface variability associated with the IODZM have been reported to be independent of ENSO [*Saji et al.*, 1999; *Iizuka et al.*, 2000; *Rao et al.*, 2002] but the debate continues [*Nicholls and Drosdowsky*, 2001; *Baquero–Bernal et al.*, 2002; *Hastenrath*, 2002; *Huang and Kinter* 2002; *Krishnamurthy and Kirtman* 2003; *Li et al.*, 2003; *Murtugudde et al.*, 2003; *Yamagata et al.*, 2003]. We return to the IODZM in Section 3.

The review article is organized as follows. In Section 2, we provide a comprehensive review of the various studies that explore the role of the IO in the monsoon variability and Tropospheric Biennial Oscillation (TBO). Recent spike of interest in the IO region has been sparked by the anomalous events of 1994 and 1997 and the role of this mode on the regional climate variability are discussed in Section 3. Section 4 addresses some outstanding issues in the IO variability. We summarize our review in Section 5.

2. INDIAN OCEAN AND MONSOON VARIABILITY

The tropical IO and the land mass around its rim experiences one of the most energetic components of the Earth's climate system, the Asian–Australian–African monsoon system. In fact, deep convection/intense precipitation occurs during all seasons around the IO rim (Plate 3). The Asian summer monsoon peaks during June–September (Plate 3b), the African monsoon establishes in October–November (short-rains, Plate 3c), and the Australian monsoon occurs during December–February (Plate 3d). During boreal winter and spring the southwestern IO extending into Southern Africa experiences deep convection (Plate 3a, d).

The variability of the monsoons, at both intraseasonal and interannual time scales exert considerable influence on socio-economic aspects in many regions such as south Asia, Africa, and Australia [e.g., *Webster et al.*, 1998]. There is increasing evidence that the diabatic heating associated with the monsoon system influences the global climate [e.g., *Rodwell and Hoskins*, 1996]. Owing to both the scientific and social importance, a large volume of literature that describes the annual cycle and variability of the monsoons is available in the form of review articles [e.g., *Shukla*, 1987; *Hastenrath*, 1988; *Webster et al.*, 1998; *Holland*, 1986; *Krishnamurthy and Kinter*, 2003] and books [e.g., *Lighthill and Pearce*, 1981; *Fein and Stephens*, 1987; *Chang and Krishnamurti*, 1987; *Pant and Rupakumar*, 1997]. Despite these sustained efforts, simulating the annual cycle and the interannual variability of the monsoons remain one of the toughest challenges for the modeling community [*Sperber and Palmer*, 1996; *Gadgil and Sajani*, 1998].

In this section, first we present the arguments for why SST anomalies are important for the interannual variability of the monsoon (Section 2.1), then we review the observational and modeling studies that focused on the role of IO SST on the Indian summer monsoon (ISM, Section 2.2), the African monsoon (AFRM, Section 2.3), and the Australian monsoon (AUSM, Section 2.4). Based on the above linkages, we review the studies that implicate the IO in the TBO (Section 2.5). Recent studies suggest that the statistics of intraseasonal variability can influence interannual climate variability, particularly over the monsoon regions. Therefore, we conclude the section with a review of the effects of IO SST on the summer and winter season intraseasonal variability (Section 2.6). Each subtopic is closed with a summary of our current understanding of the issue discussed, while what is remaining to be done is presented in Section 4.

2.1. Effect of Lower Boundary Forcing on Monsoon Variability

The land–sea thermal contrast between the Asian land mass and tropical IO plays an important role during the onset and developing stages of the ISM [e.g., *Shukla*, 1987; *Webster et al.*, 1998]. The north–northwestward migration of convection associated with the annual cycle is tightly linked to the poleward migration of maximum SST over the Asian–Australian monsoon region [*Meehl*, 1987; *Webster et al.*, 1998] (Plates 2, 3). *Joseph* [1990] hypothesized that the warm pool of the northern IO in May–June forces the onset of the ISM. From a modeling study the instrumental role of the seasonal cycle of IO SST on the monsoon evolution was demonstrated by *Shukla and Fennessy* [1994].

A successful prediction of the seasonal mean rainfall and circulation is based on the premise that the monsoon is dynamically a stable system and its seasonal mean rainfall and circulation and their interannual variability are largely governed by slowly varying boundary conditions such as SST, snow cover, soil moisture, etc. [*Charney and Shukla*, 1981; *Shukla*, 1981]. Since the seminal works of *Walker and Bliss* [1932], many observational [e.g., *Sikka*, 1980; *Rasmusson and Carpenter*, 1983; *Webster and Yang*, 1992; *Slingo and Annamalai*, 2000; *Miyakoda et al.*, 2002] and modeling studies [e.g., *Keshvamurthy*, 1982; *Palmer et al.*, 1992; *Ju and Slingo*, 1995; *Soman and Slingo*, 1997; *Nigam*, 1994; *Lau and Nath*, 2000] have confirmed the strong influence of the contemporaneous SST anomalies in the central–eastern Pacific on the ISM. Similarly, the variations of the AUSM [e.g., *Nicholls*, 1984] and AFRM [e.g., *Farmer*, 1988; *Hastenrath et al.*, 1993] are also linked to ENSO.

At interannual time scales, despite the above evidence that ENSO dominates the monsoon variations, the SST variability in the IO can also influence the monsoons, particularly in years when the local SST anomalies are strong and/or years when the SST anomalies in the tropical Pacific are small and insignificant. Such a premise is possible since much of the central–eastern tropical IO lies in a warm pool region (SST > 28°C, Plate 2b), and due to the non-linearity of the Clausius–Clapeyron equation a small variation in SST (~ 0.5°C) could have larger impact on moisture availability, and therefore on the tropical convection [*Soman and Slingo*, 1997; *Zhang*, 1993]. From a diagnostic study, *Lau and Wu* [1999] note that about 19% of the Asian–Australian monsoon variability is due to local coupled processes.

2.2. Indian Ocean and the Indian Summer Monsoon

2.2.1. Observational studies. During the pre-monsoon season (March–May) the warmest SST in the global oceans

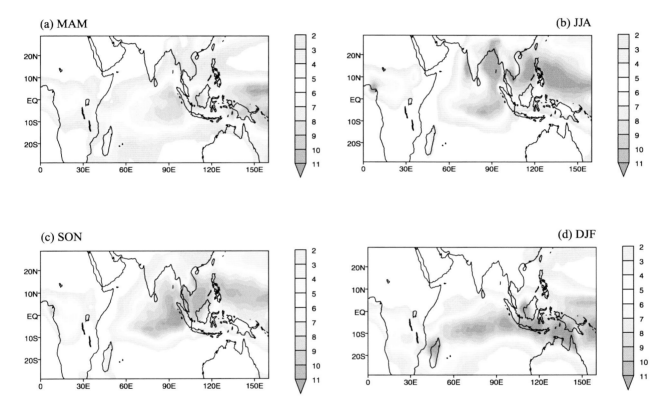

Plate 3. Observed precipitation (mm/day) climatology over the Indian Ocean for (a) March–May, (b) June–August, (c) September–November, and (d) December–February.

is observed over the Northern IO [e.g., *Joseph*, 1990; *Sengupta et al.*, 2002] (Plate 2a). Therefore, a moderate fluctuation in SST over the Northern IO during spring and early summer is expected to have a larger impact on the ensuing ISM primarily through changes in evaporation and moisture convergence. With the availability of reliable SST observations/analyses for the last 4–5 decades, many observational studies indicate a causal relationship between Arabian Sea SST anomalies and ISM. *Ellis* [1952] carried out the first case study and noted that during a strong (weak) ISM year the SST anomalies were above (below) normal over the Arabian Sea. This case study probably motivated the subsequent observational studies since the physics involved was rather appealing: warm SST can produce more evaporation and therefore stronger ISM rainfall [*Shukla*, 1987]. Using long observations along particular ship tracks, many studies examined the correlations between ISM and SST over the Arabian Sea [e.g., *Shukla and Misra*, 1977; *Weare*, 1979]. *Rao and Goswami* [1988] find that SST anomalies during March–April in the region of 5°N–10°N, 60°E–75°E to be significantly correlated with the ISM. *Clark et al.* [2000] propose predictive relationships between Arabian Sea SST over autumn and winter to ISM in the following year while *Terray* [1995] found quasi-biennial and ENSO time-scale SST modes in the IO that were related to ISM. *Lau et al.* [2000] showed that boreal spring warming in the northern Arabian Sea has a significant impact on the ISM. *Li et al.* [2001a] conclude that IO SST do influence ISM on the TBO time-scale through local moisture convergence.

The first observational study to examine the importance of Arabian Sea water vapor flux on the ISM was done by *Pisharoty* [1976]. He reported that the in situ flux of water vapor over the Arabian Sea was much larger than that obtained from the Southern IO and underscores the importance of local SST anomalies. The research that followed these findings, however, obtained contradictory results. The calculations of *Saha and Bavadekar* [1973], *Saha* [1974], *Cadet and Reverdin* [1981], and *Cadet and Diehl* [1984] suggested that the Southern IO is the main moisture source region for the ISM. Yet, there is no demonstrative relationship between ISM and SST over the Southern IO except for the correlation between the SST gradient, Somali Jet, and ISM reported by *Murtugudde et al.* [1998a] and *Murtugudde and Busalacchi* [1999]. The latter appears to be driven by the monsoonal heating rather than being the cause for it.

Nicholls [1983, 1995] proposed that SST of northwest Australia (5°S–10°S, 120°E–160°E) has a predictive value for ISM since warm SST during April there was correlated with stronger ISM in the following summer. In a related study, *Sadhuram and Wells* [1999] find a significant correlation between ISM and SST over 0°N–5°N, 80°E–85°E during November of the previous year.

2.2.2. Modeling studies. Motivated by the observational studies mentioned above, *Shukla* [1975] carried out the first ever-modeling study to understand the impact of Arabian Sea SST on the ISM. His atmospheric general circulation model (AGCM) results corroborated the observational findings in that when cold SST anomalies over the Arabian Sea were imposed, the model simulated rainfall over India and adjoining oceanic regions was significantly reduced. Results from a different AGCM with similar SST forcing produced a conflicting response over the ISM region [*Washington et al.*, 1977]. The differences in the response primarily depend on the individual AGCM's capability in simulating the mean monsoon [*Shukla*, 1984]. Consistent with the results of *Shukla* [1975, 1984], in a recent modeling study, *Arpe et al.* [1998] re-emphasized the importance of Northern IO SST anomalies on the ISM.

Recently, some modeling studies focused on the impact of equatorial central–eastern IO SST anomalies on the ISM. In an idealized case, *Chandrasekar and Kitoh* [1999] imposed warm (cold) SST anomalies over the equatorial IO and their model simulated suppressed (enhanced) rainfall over the Indian sub-continent. In the absence of an El Niño during the summer of 2000, *Krishnan et al.* [2003] attributed the deficient rainfall over India in that year to the warm SST anomalies in the IO. From model simulations they suggested that the persistent warm SST anomalies favored anomalous convection over the equatorial IO that subsequently resulted in an extended monsoon break over India. Both these studies indicated that SST anomalies in the equatorial IO modulate the local Hadley circulation, and thereby the monsoon precipitation over India.

Lau and Nath [2000] carried out a series of model experiments to assess the role of remote versus local SST on the ISM. In the run called TOGA (Tropical Ocean–Global Atmosphere), the AGCM was forced by ENSO SST variability while in the run termed TOGA–ML, the ENSO forcing from the Pacific is included with mixed layer–atmosphere coupling in the IO. In another run, GOGA (Global Ocean–Global Atmosphere), the SST variability over the global oceans is prescribed. The simulated precipitation and winds between TOGA–ML minus TOGA would provide information about air–sea coupling outside the tropical Pacific. In a similar way, the difference between GOGA and TOGA runs would indicate the role of SST outside the tropical Pacific. The difference charts in all the four panels of Plate 4 suggest enhanced southwesterly flow and precipitation over India. The inference is that the basin-wide air–sea interaction over the tropical IO opposes the remote ENSO forcing on the ISM.

Schubert and Wu [2001] examined the predictability of the 1997 and 1998 ISM low-level winds. Using an AGCM, they made a 10–member ensemble simulation with prescribed SST anomalies over the tropical Indo–Pacific regions. Their model simulations suggested that the 1998 monsoon is considerably more predictable than the 1997 monsoon. They found that during May and June of 1998 the predictability of the low-level wind anomalies is largely associated with a local response to anomalously warm IO SST. If not in a statistically significant sense, case studies for selective years imply the importance of IO SST anomalies on the ISM.

In summary, a synthesis of all the correlation studies is that the significant area of correlation between ISM and IO SST is observed only over small regions. It is now well accepted that the ISM is a large-scale phenomena. The questions of how the SST over such small regions would persist to influence seasonal rainfall over such a large region have not been fully explained. It has been suggested that part of the controversy may be related to the lack of sufficient accuracy in the observed SST [*Shukla*, 1987; *Rao and Goswami*, 1988; *Terray*, 1994]. Most current state-of-art AGCMs have serious problems in simulating the mean monsoon [e.g., *Sperber and Palmer*, 1996] and the AGCM's response to prescribed SST anomalies crucially depends on the mean monsoon of the model [e.g., *Shukla*, 1987]. Due to this inherent limitation with the AGCMs, fully coupled models too face serious limitations in simulating the monsoons. Three constraints, small magnitude in SST variance (Plate 1a), small areas of significant correlations and lack of simulation of mean monsoon in models diminish the value of any IO SST in predicting the ISM. Despite small in magnitude and area, if the anomalous SST persist over the warm pool region of IO then its effect on the ISM needs to be understood.

2.3. Indian Ocean and the African Monsoon

2.3.1. Observational studies. The African monsoon consists of two subcomponents, the East African monsoon and South African monsoon. There is increasing evidence that different regional SST anomalies within the tropical IO influence each of the subcomponents.

2.3.1.1. Indian Ocean and East Africa. The seasonally reversing Asian monsoon circulation also affects large parts of East Africa producing two seasons of rainfall during boreal spring and fall seasons [*Ogallo*, 1988; *Farmer*, 1988; *Hutchinson*, 1992; *Nicholson and Kim*, 1997]. The rainfall during spring (April–May) is abundant, while it is more variable during fall (October–November, Figure 5a) [*Hastenrath et al.*, 1993]. As for the Asian monsoons, the majority of the studies exploring the variability of the East African rainfall have looked to ENSO as a remote driver [*Farmer*, 1988; *Ogallo*, 1988; *Hutchinson*, 1992; *Hastenrath et al.*, 1993; *Nicholson and Kim*, 1997]. But *Hastenrath et al.* [1993] pointed out that the mechanisms for the precipitation anomalies in the two rainy seasons are distinct with only the October–November rainfall anomalies showing a strong correlation with the Southern Oscillation. They also pointed to the role of the SST anomalies over western IO in suppressing convection during fall, particularly during the high phase of the Southern Oscillation (La Niña) and affecting East African rainfall (Figure 1). The SST anomalies themselves, however, appear to be driven by wind changes associated with ENSO that is consistent with *Nicholson and Kim* [1997], and the recent findings of *Allan et al.* [2001].

2.3.1.2. Indian Ocean and Southern Africa. There is observational evidence that the South African rainfall variability is forced remotely by SST anomalies in the western tropical IO [*Mason and Jury*, 1995; *Jury*, 1996; *Goddard and Graham*, 1999]. The variability of summer rainfall over Zimbabwe is related to tropical IO SST [*Makarau and Jury*, 1997]. *Landman and Mason* [1999] argue that the relation between IO SST and summer rainfall over South Africa has changed since the late 1970s, coincident with hypothesized changes in the ENSO characteristics [*Trenberth and Hoar*, 1996; *Rajagopalan et al.*, 1999]. Warmer SST in the western equatorial IO were associated with drier conditions over Southern Africa prior to the 1970s whereas they are associated with wetter conditions since the late 1970s, even though the ENSO influence appeared to remain the same [*Landman and Mason*, 1999]. Recently, *Behera and Yamagata* [2001] identified a dipole-like variability in SST over the subtropical Southern IO that is correlated with rainfall over Southern Africa.

The seasonal mean rainfall can be significantly affected by the number of tropical cyclone days. *Jury et al.* [1999] show that the SST over central and southern IO has an impact on the tropical cyclone days in the SWIO and has a predictive value with a lead of a few months. *Xie et al.* [2002] showed a similar predictive value based on the Rossby waves associated with ENSO and also with the IODZM and the associated SST anomalies (Plate 5).

2.3.2. Modeling studies. Since the number of modeling studies on the African monsoon variability is meagre we will combine the AGCM studies that address the East and South African monsoon subcomponents together in this subsection. *Latif et al.* [1999] used an AGCM to show that East African rainfall anomalies that led to severe flooding during December–January of 1998 were forced largely by the western IO warming. Plate 6 taken from *Latif et al.* [1999] indicates the direct role of IO SST anomalies on the rainfall over South

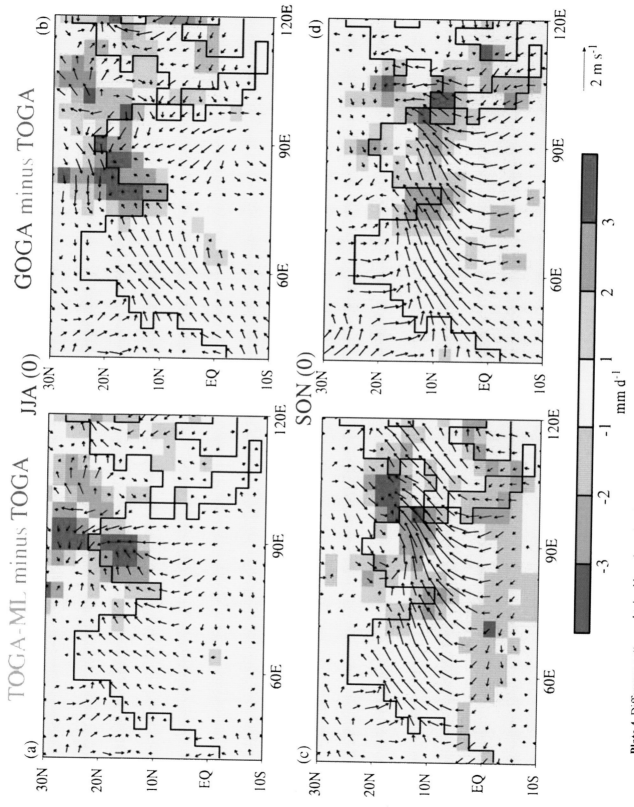

Plate 4. Difference patterns obtained by subtracting the warm minus cold composites for TOGA from the corresponding composites for TOGA–ML (left panels), and by subtracting the composites for TOGA from those for GOGA (right panels). Results are shown for the surface wind vector (arrows; see scale at bottom right) and precipitation (shading; see scale bar at bottom; units: mm/day) during [(a), (b)] JJA, and [(c), (d)] SON [from *Lau and Nath*, 2000].

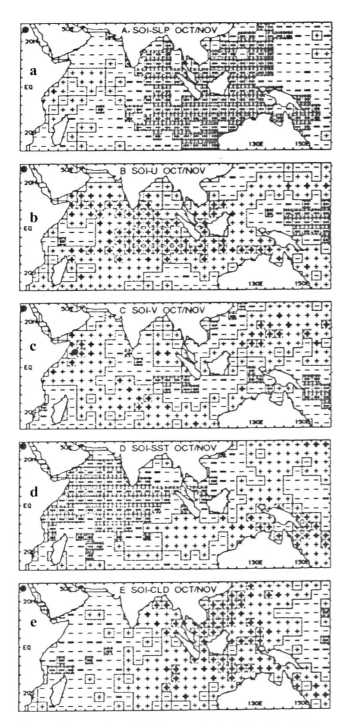

Figure 1. Patterns of correlations between October–November rainfall over equatorial Africa and (a) SLP, (b) zonal winds, (c) meridional winds, (d) SST, and (e) clouds for October–November. Plus and minus symbols denote the sign of the correlation, and boldface type indicates values with local significance at 5% level. Shading represents absolute values bigger than 0.4. Base period is 1948–87 [from *Hastenrath et al.*, 1993].

Africa. Similar AGCM experiments by *Goddard and Graham* [1999] show that Southern African rainfall anomalies were also forced by the western IO SST anomaly.

In sharp contrast to the uncertainty on the role of IO SST on the ISM, observational studies clearly indicate that the effect of IO SST on the East African monsoon is more significant and robust. One primary reason is that, in the mean, the equatorial IO is swept by a zonal circulation [*Hastenrath*, 2000] maintained by the deep convection over the eastern IO (Plate 3c). The changes in this zonal circulation are largely determined by local SST anomalies [e.g., *Hastenrath et al.*, 1993; *Clark et al.*, 2003] that directly account for the variations in East African rainfall. For reasons unknown, very few modeling studies have been carried out to understand the response of the East African monsoon to IO SST anomalies.

2.4. Indian Ocean and the Australian Monsoon

2.4.1. Observational studies. The Australian monsoon is dominated by seasonal rainfall during austral winter with a well-known ENSO driven interannual variability [*Quinn et al.*, 1978; *McBride and Nicholls*, 1983; *Nicholls*, 1988; *Webster et al.*, 1998]. While much of Australia tends to be drier during warm ENSO events, the correlation is not perfect similar to the case of the ISM [*Nicholls*, 1989].

Nicholls [1989] investigated the role of Indo–Pacific SST on the Australian winter monsoon variability. From an EOF analysis, he identified two different spatial patterns of rainfall variability. The correlations between the Principal Components (PCs) and Indo–Pacific SST revealed some interesting results. The dominant spatial pattern (EOF1) was significantly correlated with the SST difference between south–central IO (10°S–20°S, 80°–90°E) and the Indonesian Sea (0–10°S, 120°–130°E) while EOF2 was correlated with ENSO-related SST anomalies in the equatorial Pacific. He applied partial correlation analysis that removes the effects of ENSO. Again, EOF1 was highly correlated with SST in the south–central IO and Indonesian Seas (Figure 2). Based on this statistical analysis *Nicholls* [1989] concluded that the SST anomalies in the south–central IO may be independent of ENSO, and has a strong predictive skill for the winter monsoon over Australia. Even though causality was not evident, it was clear that the IO SST were important for Australian winter rainfall.

Indonesian Seas are subject to the reversing seasonal monsoons albeit with a much weaker seasonal contrast than the Indian subcontinent [*Nicholls*, 1981; *Hackert and Hastenrath*, 1986]. *Braak* [1919] first suggested that high local atmospheric pressure over Indonesia in the first half of the year was an indicator of a drier second half of the year with a late onset of the wet season, and vice versa. *Nicholls* [1981] found evi-

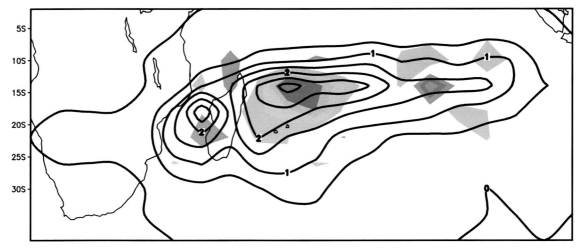

Plate 5. Climatological mean tropical cyclone days (contours) in Dec–Apr, and the difference (colors) between years of anomalously deep and years of anomalously shallow thermocline over 8°S–12°S, 50°E–70°E [from *Xie et al.*, 2002].

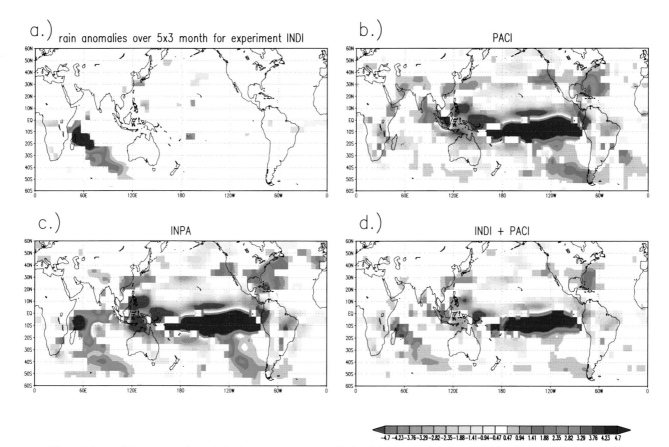

Plate 6. Ensemble mean and precipitation responses (mm/day) of the atmosphere model ECHAM3 (T21) to the SST anomaly patterns; (a) response to the Indian Ocean SST anomaly, (b) response to the Pacific Ocean SST anomaly, (c) response to the complete Indo–Pacific SST anomaly, and (d) linear superposition of the responses shown in (a) and (b). Shown are the anomalies (relative to a 50-year control run with climatological SST) that exceed 95% significance level according to a t-test [from *Latif et al.*, 1999].

dence to support this hypothesis based on a positive feedback in the air–sea interactions causing the initial SST anomalies to persist from austral winter into spring. *Nicholls* [1978] also pointed out that the air–sea interactions in the Indonesian Seas and Northern Australia are dependent on the seasonally reversing relation between SST and pressure anomalies.

Hackert and Hastenrath [1986] analyzed ship and land observations in the IO region to investigate the mechanisms of rainfall anomalies over Indonesia and arrived at the conclusion that the interannual anomalies are largely a modulation of the annual cycle. Wet (dry) years are thus associated with stronger (weaker) Northeast monsoon with coincident cloudiness and SST anomalies although the ocean–atmosphere interactions are distinct in the austral spring and fall seasons, which was noted to be crucial for reversing the SST anomalies between the two seasons. The analyses of *Hackert and Hastenrath* [1986] clearly captured the correlation between Java rainfall and eastern and central IO SST. The most dominant interannual signal over Indonesia is of course the ENSO related drought [*Hendon,* 2003] that not only leads to severe agricultural devastation but also forest fires and haze [*Nichol,* 1997, 1998].

2.4.2. Modeling studies. Motivated by the correlation statistics, many AGCM studies have investigated the impact of Pacific and IO SST anomalies on the Australian rainfall variability [e.g., *Voice and Hunt,* 1984; *Simmonds and Trigg,* 1988; *Simmonds et al.,* 1989; *Simmonds,* 1990; *Simmonds and Rocha,* 1991; *Frederiksen and Balgovind,* 1994; *Frederiksen and Frederiksen,* 1996; *Frederiksen et al.,* 1999]. Most of these studies confirm the role of the SST gradient between the Indonesian archipelago and the south–central IO on the winter rainfall variability.

The possible role of IO SST anomalies on the Australian winter rainfall has been investigated while ENSO-related SST anomalies in the Pacific dominate its summer rainfall variability. Despite the fact that the local winter seasonal climatological rainfall over Australia is small (Plate 2b) its interannual variability has significant agro-economical consequences [*Nicholls,* 1989].

2.5. Indian Ocean SST, Monsoons and the TBO

The processes within the tropical IO have been proposed to play an "active" role in the life cycle of the TBO, one of the major climate signals in the tropics [e.g., *Meehl,* 1987; *Li et al.,* 2001b; *Meehl et al.,* 2003]. A dry ISM is followed by a spatially large warm SST anomaly over the tropical IO that is observed to persist up to the following monsoon [*Joseph and Pillai,* 1984]. Such a large warm SST anomaly is found to result in a wet monsoon in the following year producing a biennial type oscillation [*Joseph,* 1981; *Joseph and Pillai,* 1986]. In the context of the Asian–Australian monsoon, the TBO is defined as the tendency for a relatively strong monsoon to be followed by a relatively weak one, and vice-versa [*Meehl and Arblaster,* 2002a]. Therefore, it has been suggested that the TBO is not so much an oscillation, but a tendency for the system to flip-flop or transition from year to year. There is compelling observational evidence that a strong ISM is followed by a strong AUSM, and the strong AUSM is followed by a weak ISM [e.g., *Meehl,* 1987; *Chang and Li,* 2000; *Yu et al.,* 2003]. An analysis of observed data shows that the TBO encompasses most ENSO years (with their well-known biennial tendency) as well as additional years that contribute to biennial transitions [*Meehl and Arblaster,* 2002b]. Thus it has been hypothesized that the TBO is a fundamental feature of the coupled climate system over the entire Indian–Pacific regions [see *Li et al.,* 2001b for a review on TBO theories].

Figure 3, taken from *Yu et al.* [2003], shows the lagged correlation coefficients calculated between monthly ISM and AUSM rainfall anomalies. Two large correlation coefficients, a positive one (in-phase) with ISM leading the AUSM by about 2 seasons, and a negative one (out-of-phase) with the AUSM leading the ISM by about 2 seasons, are readily apparent. *Yu et al.* [2003] examined the role of Indo–Pacific oceans for the TBO from a series of coupled model integrations. The coupled model run, when both the Indian and Pacific Oceans are interactive produces both the transitions (Figure 3b). The run, with an interactive Pacific (Indian) Ocean produces the in-phase (out-of-phase, Fig. 3d) transition (Fig.

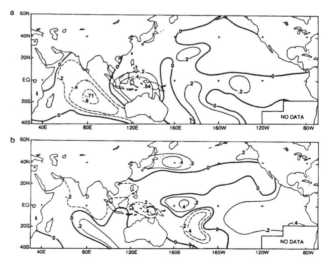

Figure 2. Partial correlations between SST and (a) the rotated principal component of Australian rainfall after removal of ENSO effects, and (b) the second rotated principal component of Australian rainfall after removal of ENSO effects [from *Nicholls,* 1989].

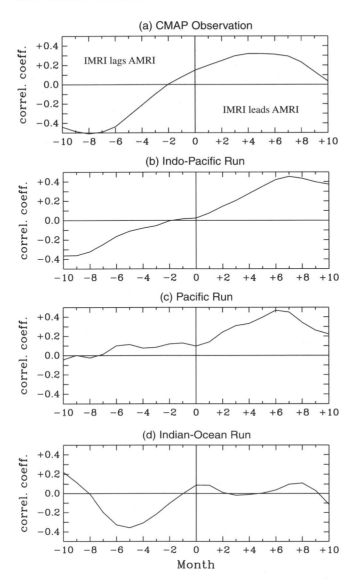

Figure 3. Time-lagged correlation coefficients between monthly Indian monsoon rainfall index and Australian monsoon rainfall index calculated from (a) CMAP observations, (b) Indo–Pacific CGCM run, (c) Pacific CGCM run, and (d) Indian Ocean CGCM run [from *Yu et al.*, 2003].

3c). These model simulations, in agreement with observations [e.g., *Meehl and Arblaster*, 2002a], suggest that the IO plays an "active" role in the life cycle of the TBO. *Meehl and Arblaster* [2002b] and *Loschnigg et al.* [2003] argue that the IODZM events are large amplitude excursions of the TBO in the tropical IO.

The possible role of the IO on the TBO, if fully established, will have immense consequences on the tropical climate variability. Over the period 1871–2000, the ISM is dominated by the quasi-biennial variability and all ENSO indices themselves have a strong biennial component. In the real system, ENSO, TBO, monsoons and IODZM are all mutually interactive components and it is very difficult to isolate the sole effect of IO on the TBO. Careful experiments with realistic coupled models will possibly reveal if the IO plays an active rather than a passive role in the TBO life cycle.

2.6. Indian Ocean and the Intraseasonal Variability

Both during summer and winter seasons, convection associated with the intraseasonal oscillations (ISOs), typically with a time scale of 30–50 days, originate over the equatorial IO [e.g., *Madden and Julian*, 1971, 1972; *Yasunari*, 1979; *Sikka and Gadgil*, 1980; *Lau and Chan*, 1986; *Annamalai and Slingo*, 2001]. Although the ISOs are inherent to the atmosphere, both observational and coupled modeling studies clearly demonstrate the crucial role played by IO SST in the organization, intensification, and propagation of the convection and circulation associated with the ISOs [e.g., *Krishnamurti et al.*, 1988; *Li and Wang*, 1994; *Hendon and Glick*, 1997; *Flatau et al.*, 1997; *Shinoda et al.*, 1998; *Waliser et al.*, 1999; *Woolnough et al.*, 2000; *Sengupta and Ravichandran*, 2001; *Sengupta et al.*, 2002; *Kemball-Cook et al.*, 2002; *Fu et al.*, 2003; *Waliser et al.*, 2003, 2004].

SST observations from moored buoys in the North Bay of Bengal during July–August of 1998 indicate that the amplitude variation of SST at ISO time scale could be as large as 2°C [*Sengupta et al.*, 2001; *Bhat et al.*, 2001]. From satellite derived SST data, *Vecchi and Harisson* [2002] also noted similar magnitudes in SST fluctuations at ISO time scales. BOBMEX, the Bay of Bengal Monsoon Experiment was carried out under the auspices of the Indian Climate Research Program during the boreal summers of 1999–2001 [*Bhat et al.*, 2001]. The SST observations from BOBMEX further indicate large amplitude variations in SST at ISO time scales over the North Bay of Bengal, and *Bhat et al.* [2001] suggest that the SST variations have direct impact on the genesis of monsoon depressions.

The phase of the ISOs is linked to the onset, and active/break phases of the ISM [e.g., *Joseph et al.*, 1994; *Sikka and Gadgil*, 1980], and the AUSM [*Hendon and Liebmann*, 1990]. The large spatial scale convection and circulation associated with the ISOs modulates the mean monsoon circulation [e.g., *Krishnamurthy and Shukla*, 2000; *Sperber et al.*, 2000; *Goswami and Ajayamohan*, 2001] and also the genesis of synoptic systems [e.g., *Liebmann et al.*, 1994]. Therefore, it has been hypothesized that the statistical properties of the ISOs can modulate the interannual variability of the monsoons [e.g., *Palmer*, 1994; *Sperber et al.*, 2000; *Krishnamurthy and Shukla*,

2000; *Goswami and Ajayamohan*, 2001]. Therefore, a better representation of the ISOs in the models may be pre-requisite for addressing the interannual variability.

The AGCMs have limitations in the simulation of ISOs [e.g., *Slingo et al.*, 1996]. Motivated by the observational evidence, some recent studies indicate that an AGCM coupled to either a simple or a comprehensive ocean model has a better representation of the ISOs [e.g., *Waliser et al.*, 1999; *Kemball–Cook et al.*, 2002]. Plate 7, taken from *Fu et al.* [2003], suggests that at ISO time scales, SST anomalies from the coupled model leads the convection by about 10 days. This lead–lag relationship between SST and convection at ISO time scale appears to be crucial for the organization, and propagation of ISOs [e.g., *Fu et al.*, 2003].

Recent observational and coupled modeling studies indicate the importance of SST variability at ISO time scale on the atmospheric ISOs. But the stumbling block is to understand how the ISO SST feeds back onto the organization and intensification of atmospheric convection. In addition, it is unclear what physical processes are responsible for the preferred timescale of the ISO. Also, the role of the suppressed phase of convection on the life-cycle of the ISO needs to be established.

3. IMPACT OF THE DIPOLE/ZONAL MODE ON CLIMATE

Even though the IODZM events with significant climatic impacts occurred only during 1961, 1994, and 1997, they produce the largest SST anomalies in the IO, and clearly involve an active role by the ocean. The event of 1961 that led to historic floods over parts of East Africa was noted in several studies [e.g., *Reverdin et al.*, 1986] and even prompted the terminology IO El Niño [*Kapala et al.*, 1994]. Considering the extensive literature on all aspects of IODZM, we restrict ourselves to a very brief discussion of the coupled nature of this coupled climate mode and its potential impacts in the regional climate variability.

3.1. A Coupled Mode in the Indian Ocean

As pointed out by *Hastenrath* [2002], the correlation between the zonal SST gradient, equatorial winds, and Southern Oscillation Index (SOI) was recognized before the 1994 event [*Hastenrath et al.*, 1993]. A few years later, he noted the zonal circulation in the equatorial IO to be highly correlated with the SOI [*Hastenrath*, 2000]. While investigating the sea level and thermocline variability in the IO, *Murtugudde et al.* [1998a] rediscovered the east–west variability to be an ocean dynamic mode and not just an SST mode. As noted earlier, they argued that the equatorial SST gradient has a significant correlation with the Somali Jet that in turn has a strong correlation with the ISM [also see *Murtugudde and Busalacchi*, 1999]. While their simple forced atmospheric model experiments led them to conclude that the zonal SST gradient was a passive response to the ISM variability, they showed that there is significant correlation between thermocline and SST anomalies off Sumatra. This is a clear indication of *Bjerknes* [1969] feedback that can lead to a coupled climate mode within the IO.

The anomalous cooling off Sumatra during 1994 and 1997 has led several investigators to study the unusual physical and biological conditions in the IO [e.g., *Webster et al.*, 1999; *Saji et al.*, 1999; *Behera et al.*, 1999; *Murtugudde et al.*, 1999; *Chambers et al.*, 1999; *Yu and Rienecker*, 1999; *Vinayachandran et al.*, 1999]. The SST gradient defined by *Murtugudde et al.* [1998a] was employed by *Saji et al.* [1999] to propose that a physical dipole existed in the IO that evolves during boreal spring, mature in the fall before decaying abruptly in the winter [*Yamagata et al.*, 2003]. Identifying a number of IODZM events, they allowed room for the fact that the trigger for such events may not be entirely independent of ENSO. *Webster et al.* [1999] also speculated that at least during 1997–98, the anomalous conditions in the IO may have been independent of ENSO.

It has now been confirmed by a number of studies that during the strong IODZM events, there is indeed a coupling between the thermocline depth anomalies and SST anomalies in the eastern IO which leads to the suppression of precipitation and the growth of the upwelling-alongshore wind structure from its onset during the boreal spring to its maturity during fall [*Saji et al.*, 1999; *Webster et al.*, 1999; *Yamagata et al.*, 2003]. What remains under intense debate is the question of the trigger for the onset of the IODZM. As in the tropical Atlantic, more and more studies suggest that the coupled mode in the IO is weak on its own and does not appear to be self-sustained [*Annamalai et al.*, 2003; *Loschnigg et al.*, 2003; *Gualdi et al.*, 2003; *Lau and Nath*, 2003; *Li et al.*, 2003]. This raises the possibility of an external trigger for the initiation of the IODZM.

The diagnostics carried out by *Annamalai et al.* [2003] showed that in the IO there is a natural air–sea coupled mode with weak cooling off Sumatra. During a window from spring to early summer, this natural cooling can intensify, should El Niño-like conditions prevail in the central Pacific. This is because the anomalous precipitation during spring over Sumatra/Java is largely controlled by SST anomalies over the equatorial central Pacific. The changes in convection in the central Pacific induce subsidence and suppress the precipitation off Sumatra/Java. The heating anomaly associated with this deficient rainfall forces an anticyclone at low-level as a Rossby wave response. The increased alongshore winds

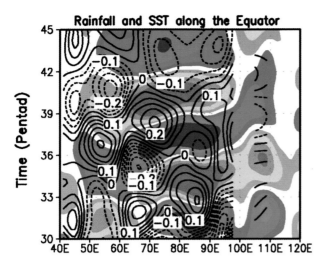

Plate 7. Longitude–time Hovemueller diagram of intraseasonal variability of rainfall rate (positive in yellow/red; negative in green/blue; shading interval 2 mm/day) and SST (contours; negative values in dotted lines; contour interval is 0.05°C) along the equator. Pentad 30 corresponds to 01 June [from *Fu et al.*, 2003].

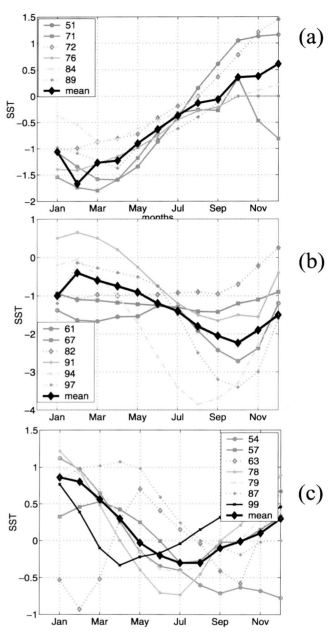

Plate 8. SST anomalies in the eastern equatorial Indian Ocean standardized and averaged over 90°E-110°E, 10°S-Equator to define (a) aborted, (b) strong, and (c) weak zonal mode years. The monthly s.t.d of SST is 0.3°C [from *Annamalai et al.*, 2003].

in turn enhance oceanic upwelling and surface cooling. *Ueda and Matsumoto* [2000] also suggested that the IODZM in 1997 is triggered by El Niño of that year. Based on the intensity and persistence of SST and thermocline variations off Java and Sumatra, *Annamalai et al.* [2003] defined IODZM events during the last 50 years (Plate 8). As seen from Plate 8a, there are occasions when SST anomalies are colder than normal in boreal spring but change sign abruptly thereafter. Therefore, depending on the proximity of the thermocline to the surface, initial cooling off Java/Sumatra can grow into a fully coupled IODZM.

Two of the issues intensely debated in the last three years are the term "dipole" and the relationship between IODZM and ENSO. As noted by several studies [*Annamalai et al.*, 2003; *Huang and Kinter*, 2002; *Feng and Meyers*, 2003; *Hendon*, 2003], the cooling off Java/Sumatra is robust in all strong IODZM years while the location of western warming varies from event to event. An example of the potential deficiencies of the IOD definition of *Saji et al.* [1999] is that during years such as 1972 where the gradient definition implies a strong dipole, no notable cooling occurred off Sumatra during the summer and fall months (Plate 8a). Over the western IO, observed SST variance has a local maximum around 50°E, 0°N (Plate 1a) but with no appreciable thermocline feedback (Plate 1b). Therefore, SST anomalies in the western equatorial IO appear to be flux driven and more work needs to be done in understanding their effect during IODZM events. Analyzing XBT measurements, *Feng and Meyers* [2003] showed that SST in the western IO leads the eastern IO by about 4–6 months, and also indicated that there are years when the eastern cooling was not accompanied by western warming.

Xie et al. [2002], among others, noted that the simultaneous correlation between IODZM and NINO3.4 SST indices during fall is very high. This observational fact led many to argue that most IODZM events are forced by ENSO [e.g., *Baquero–Bernal et al.*, 2002; *Annamalai et al.*, 2003] and maintained by local air–sea interaction. *Wang et al.* [2003] proposed that the atmospheric Rossby wave–SST dipole feedback is involved in maintaining the air–sea interaction in the southern IO during the developing phase of ENSO. On the other hand, during non-IODZM years, the ENSO related cooling off Java/Sumatra appears to be surface heat flux driven [*Wolter and Hastenrath*, 1989; *Yu and Reinecker*, 1999; *Krishnamurthy and Kirtman*, 2003; *Hendon*, 2003] while subsurface oceanic processes are actively involved during strong IODZM years [*Murtugudde et al.*, 2000].

Setting aside the debates about the exact nature of the triggers for IODZM, or the lack of well-defined warm SST anomalies in the western IO, or its relationship to ENSO, or why it terminates abruptly, it is nonetheless important to note that the impact of IO on regional climate may in fact occur through this intrinsic IO coupled mode since the SST anomalies during strong IODZM years are substantial.

3.2. Consequences of the IODZM on Monsoon Variability

Since the SST anomalies associated with the IODZM, in particular off Sumatra/Java, are above the observational noise, there is potential for IODZM to influence the climate around the IO rim. In this subsection, we review how this intrinsic coupled mode influences the Asian–Australian–African monsoon systems.

3.2.1. IODZM and the Asian monsoons. *Slingo and Annamalai* [2000] investigated the unusual response of the ISM

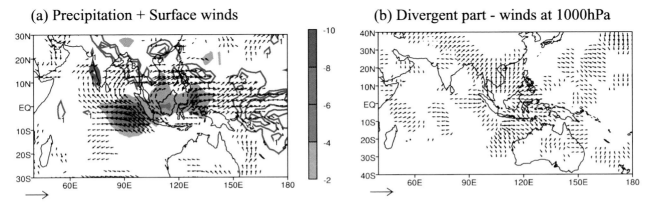

Figure 4. July–August composite of (a) precipitation and surface winds. Negative (positive) values are shaded (contoured). The contour interval is 1 mm/day while the shading interval is 2mm/day. (b) Divergent winds at 1000 hPa. Reference vector is shown in both panels. [from *Annamalai et al.*, 2003].

230 INDIAN OCEAN AND CLIMATE

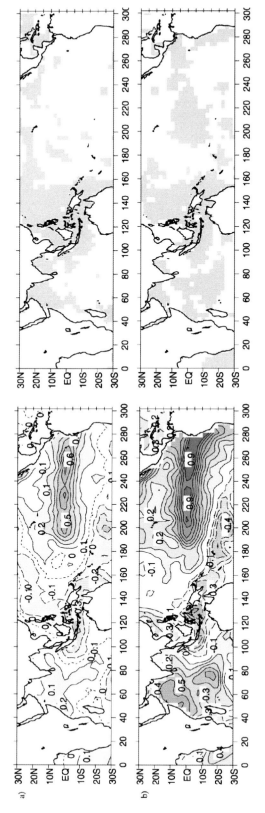

Plate 9. SST anomaly patterns associated with heavy East African short rains based on SON rainfall from Figure 5. Extreme rainfall years are based on anomalies greater than one standard deviation from the mean: (a) equatorial region (2.5°S–2.5°N), and (b) southern region (10°S–12.5°S). SST anomalies are calculated relative to the 1900–97 climatological mean. The left panels show the SST anomalies, and contour interval is 0.1 K. Gray shading in the right panels indicates regions where anomalies are significant at the 95% level [from *Black et al.*, 2003].

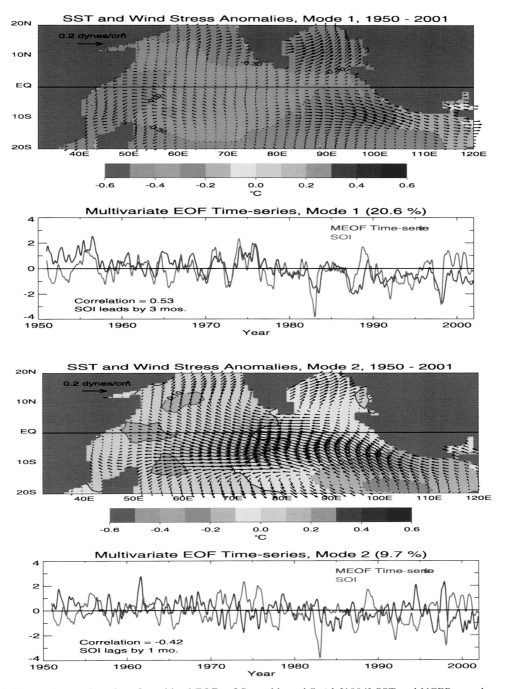

Plate 10. First and second modes of combined EOFs of *Reynolds and Smith* [1994] SST and NCEP reanalyses winds.

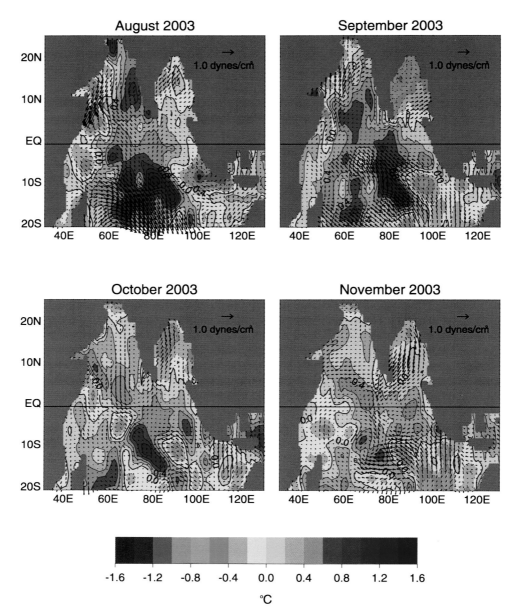

Plate 11. Anomalies of *Reynolds and Smith* [1994] sea surface temperatures and NCEP reanalyses winds for Aug–Nov of 2003 show a surface cooling off Java during boreal summer months which failed to grow into a fully coupled dipole/zonal mode.

during the 1997–98 El Niño. By analyzing observations and reanalysis products they concluded that the relatively prolonged cooling of SST over the Maritime Continent and eastern equatorial IO was able to generate changes in the local Hadley circulation such that the ISM was above normal. *Behera et al.* [1999] suggested that the abnormal ISM during 1994 was due to the strong IODZM. *Ashok et al.* [2001] extended this idea for other IODZM years. Figure 4 taken from *Annamalai et al.* [2003], shows the composites of anomalous precipitation and surface winds during strong July–August of IODZM events. The negative precipitation anomalies in the eastern equatorial IO are surrounded by positive precipitation anomalies along the entire monsoon trough, from the Indian sub-continent extending eastwards into the tropical northwest Pacific. The surface winds indicate that the convergence over the Bay of Bengal is due not only to the westerly anomalies but also due to southerly flow originating off Sumatra. The inference from the Figure 4b is that the north–south heating gradient (precipitation) over the eastern IO favors a local meridional circulation [*Gill*, 1980], which is important for transporting moisture towards the monsoon trough. Therefore, if IODZM events co-occur with El Niño then the regional coupled processes can counteract the remote suppressing effect of El Niño on the ISM. On the other hand, the north–south circulation strengthens along-shore winds off Sumatra and therefore, the ISM too has a role in the stronger upwelling and in the intensity of IODZM events.

Encouraged by the observed features, *Annamalai and Liu* [2004] carried out a series of AGCM experiments with ECHAM5 to isolate the role of Pacific and IO SST anomalies on the ISM. Their results are broadly consistent with *Lau and Nath* [2000] in that the remote forcing from the Pacific dominates the ISM variability while the local forcing from the IO adds to the details in the ISM response in terms of precipitation anomalies over the Indian subcontinent. The authors note that prescribing SST anomalies associated with the IODZM events alone do not significantly influence the precipitation over the Indian subcontinent. Similar conclusions were arrived by *Wu and Kirtman* [2004] using the COLA AGCM.

During the winter monsoon season, *Rasmusson and Carpenter* [1983] found a strong correlation between ENSO and Sri Lankan rainfall during October–December. However, employing correlation analyses with over a century long data, *Zubair et al.* [2003] conclude that the anomalous convergence associated with the warm SST anomalies over the western IO associated with the IODZM events strongly modulate the September–December rainfall over Sri Lanka. The differing views in the studies of *Rasmusson and Carpenter* [1983] and *Zubair et al.* [2003] suggest the importance of separating ENSO-induced vs. IODZM-induced warm SST anomalies in the western IO to clearly demonstrate the influence of IODZM on Sri Lankan rainfall.

Despite the fact that observed correlations indicate IODZM possibly influences the ISM, it is unclear why AGCMs fails to simulate above normal rainfall when forced by IODZM SST anomalies. This may be related to the limitations in the simulation of mean monsoon precipitation by the AGCMs or to the small "spatial scale" associated with the SST anomalies off Sumatra. On a positive note, results from these model experiments indicate that the SST anomalies over the IO are very effective when Pacific SST anomalies are also included. Therefore, the SST anomalies over the entire tropical Indo–Pacific region need to be considered for a realistic simulation of the intensity of the ISM. However, if the SST anomalies themselves are influenced by the strength of the anomalous ISM then realistic coupled models are the only option for monsoon prediction.

3.2.2. IODZM and the East African monsoon. Birkett et al. [1999] suggested that the abnormal East African rainfall during fall 1997 that resulted in rapid rise in regional lake levels was due to IO SST. *Black et al.* [2003] examined in detail the role of IODZM on the African monsoon system during boreal fall. Their analysis focused separately on the rainfall variations over the equatorial region and the region slightly south of it, and they noted that the standard deviation of rainfall over the former region is twice as strong compared to the latter region (Figure 5). The SST anomalies associated with heavy rainfall for both regions are shown in Plate 9 [from *Black et al.*, 2003]. It is clear that high rainfall is associated with a warming of the western IO near the East African coast, and another robust signal is the cooling off the Sumatran and Australian coast. In addition, there is a pronounced warming in the equatorial Pacific. Of all the features, the SST anomalies off Sumatra show high statistical significance (right panels in Plate 9). Despite the lack of robustness, it should be noted that the average SST anomaly over the Pacific is strongly positive, which supports the hypothesis that El Niño conditions are associated with high rainfall in equatorial Africa. In a similar study, *Clark et al.* [2003] also note that correlations of East African rainfall with IO SST are not only significant over the western IO but also off Sumatra/Java in the eastern IO and they also find high simultaneous correlation between boreal fall East African rainfall and ENSO indices.

Behera et al. [2003] analyze a long-run of a coupled model and conclude that the East African rainfall is largely influenced by IODZM. All the above studies are related to *Hastenrath et al.* [1993], one of the original works to propose that the east–west SST gradient in the IO influences regional climate over East Africa.

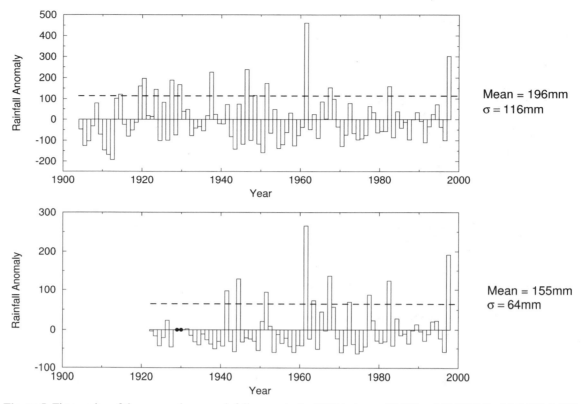

Figure 5. Time series of the seasonal mean rainfall anomaly for SON between 37.5°E and 41.25°E for (a) 2.5°S–2.5°N, and (b) 10°S–12.5°S. Black circles indicate missing data [from *Black et al.*, 2003].

The effect of IO SST anomalies on the East African rainfall has been further demonstrated with the knowledge gained in recent years. Although encouraging, to what degree the IO SST anomalies that directly influence the East African rainfall were driven by ENSO vs. local air–sea interactions remains an open issue. For example, what is intriguing is that the rainfall during 1994, one of the strongest IODZM events, is not anomalous (Figure 5a). In addition, the effect of IODZM on the East African rainfall is asymmetric. In most positive IODZM events the rainfall is deficient (Figure 5a) while during negative IODZM years the relationship is not obvious [*Slingo*, 2004 personal communication]. As mentioned earlier in Section 2.3.2, more modeling studies are necessary to understand the influence (or lack thereof) of IODZM events on the East African monsoon rainfall.

3.2.3. IODZM and the Indonesian–Australian rainfall. The importance of IO SST gradient on Indonesian rainfall variability can be inferred from the analyses of *Reverdin et al.* [1986] in the context of anomalous events in the IO, especially 1961. *Hendon* [2003] revisited the Indonesian rainfall variability in the context of the 1997/98 drought associated with the ENSO and IODZM events. The cooling in the eastern IO during the dry season grows and persists due to local air–sea interactions and ocean dynamics with remote ENSO forced winds acting to enhance the cooling. The zonal SST gradient in the IO is found to be related to ENSO and prominent in explaining the dry season rainfall variability. Even during years when IODZM does not cause the strong cooling off Sumatra, ENSO related cooling due to heat flux anomalies still appears to drive similar interactions with Indonesian rainfall variability [*Hendon*, 2003].

Ashok et al. [2003] suggest through AGCM experiments that the anomalous anticyclonic circulation in the lower troposphere forced by cooling off Sumatra extends over much of the western Australia influencing the winter rainfall anomalies there. Although the results of *Ashok et al.* [2003] are appealing, the observed July–August rainfall anomaly composite during IODZM events (Fig. 4a), however, does not show appreciable rainfall anomalies over western Australia. *Xie et al.* [2002] showed that the SST variability off northwest Australia during boreal fall is primarily due to surface fluxes. From correlation analysis, the authors showed that the SST variability in this region is highly correlated with the SST variations off Sumatra associated with IODZM events. The possible role of the southeastern IO SST anomalies on the Australian summer monsoon needs to be investigated.

3.2.4. IODZM and global climate anomalies. Partial correlation techniques were used over a combination of observational and reanalysis products to quantify the possible impacts of the IODZM on global climate. *Behera and Yamagata* [2003] used partial correlation method and showed that IODZM can influence the Darwin pressure and therefore ENSO. To understand such a possibility, *Annamalai et al.* [2004] conducted sensitivity experiments with an AGCM and a simple atmospheric model, the latter used to identify basic processes. Their model results showed that when basin-wide warm SST anomalies are present in the IO, an atmospheric Kelvin wave with easterly wind flow over the equatorial western–central Pacific is generated. The easterly wind anomalies then weaken the westerly anomalies associated with El Niño. In contrast, during IODZM years the east–west contrast in SST and convective anomalies does not generate a significant Kelvin-wave response, and there is little effect on the El Niño-induced westerlies. Based on the model and observational results the authors concluded that the strength of El Niño is amplified during IODZM years.

The heating anomalies associated with above normal ISM force descending long Rossby waves to the west over the Mediterranean [*Rodwell and Hoskins,* 1996]. Recognizing the observational fact that IODZM influences ISM, *Guan and Yamagata* [2003] suggest that the indirect forcing through modulation of the wave-guide over Sahara region were responsible for the hot and dry summer of 1994 over East Asia. On the other hand, *Park and Schubert* [1997] use assimilated model products to conclude that the anomalous circulation and the 1994 summer drought over East Asia were mainly driven by orographic forcing associated with the zonal wind changes over the Tibetan plateau. This alternate explanation indicates that further studies are required before any conclusions can be drawn about the role of IODZM in forcing anomalous summer conditions over East Asia.

Saji and Yamagata [2003] find large partial correlations between land temperatures and IODZM events from South Africa and South America to North America and Europe, from Papua New Guinea to Iran, mostly during boreal summer and fall seasons. Partial correlation techniques do allow extraction of causalities but the processes responsible for the teleconnection between IODZM and global climate remain to be explored and documented. For example, the theories proposed by *Hoskins and Karoly* [1981] and *Simmons et al.* [1983] are fundamental for explaining the observed correlations between ENSO and North American climate. These were further substantiated by AGCM experiments [e.g., *Shukla and Wallace,* 1983]. Similar idealized modeling studies should be carried out to understand the processes involved in the possible impact of IODZM on global climate anomalies.

4. OUTSTANDING ISSUES

Much of the 1980s were dominated by extensive investments in the tropical Pacific to understand and forecast the ENSO phenomenon [e.g., *McPhaden et al.,* 1998]. One of the achievements under the auspices of TOGA includes the successful ENSO forecasts with simple coupled models, which led *Cane* [1991] to claim that ENSO was a simple, robust and large-scale phenomenon. However, no such claim can be made about any aspect of the IO sector climate variability, especially the monsoons or the IODZM. In this section, we summarize the most important scientific issues that need to be addressed in the future to advance our knowledge on the role of IO on regional climate variability.

4.1. Separable and Non-separable SST Signals

The exact processes by which ENSO events modulate the IO SST can only be studied in models since in the real world, local variability often co-occurs making it difficult to separate the remote forcing form the local forcing. In doing so, the ENSO related wind anomalies versus the ENSO induced heat flux anomalies remain the most difficult to quantify since most models remain deficient in simultaneously capturing the monsoon–ENSO correlations in terms of wind and radiative forcing [e.g., *Lau and Nath,* 2003].

In a statistical sense, the combined EOFs of observed SST and reanalysis winds capture the wind driven SST variability in the IO. The first EOF (Plate 10) explains nearly 21% of the combined variance and the SST pattern depicts a basin wide mode with a local maximum in loading near 10°S, 60°E which is a region of domed thermocline. The wind patterns capture the alongshore component off Java and Sumatra. Note that the correlation of PC1 with SOI is 0.53 (significant at the 99% level) with SOI leading by 3 months.

The second combined EOF explains nearly 10% of the variance and the SST shows a east–west pattern with strong alongshore winds in the eastern IO. The peaks in the time-series clearly capture the 1961, 1994, and 1997 events. The correlation of PC2 with SOI is –0.42 still significant at 95% level but with the SOI lagging by 1 month. This mode obviously captures the intrinsic IO mode, namely the IODZM. It is rather intriguing to note that both EOFs, independent of each other in a statistical sense, capture wind variability off Sumatra and Java, and their PCs have significant correlation with SOI. The message is that it is not always easy or possible to separate ENSO forced variability especially when remotely induced anomalies can generate local feedbacks. Sustained observations such as moored buoys may provide opportunities to design process studies around them based on ENSO forecasts so that wind and heat flux variabilities in the IO can be recorded in detail during an ENSO cycle. Careful experiments with coupled models need to be carried out to determine the effect of remote forcing on local feedbacks.

4.2. Need for Accurate Measurements of Indian Ocean SST

The greatest challenge in the IO sector is to be able to capture the anomalous variability of SST that is at present in the range of observational errors at all time-scales. A detailed comparison of the NCEP–NCAR and ECMWF reanalysis products demonstrates that the reanalysis products have systematic errors over the tropical IO during the ISM season [*Annamalai et al., 1999*]. The primary reason is that the tropical IO is a data void region and therefore the analyses depend heavily on the first guess supplied by the forecast model; this in turn will be sensitive to the diabatic heating distribution produced by the physical parameterizations used in the model. In this subsection we highlight the pressing need for accurate measurements of SST over the IO for better representation of the mean and variability of the monsoon systems.

Most AGCMs fail to simulate the observed precipitation maximum over the Bay of Bengal. The dynamics and thermodynamics of the Bay of Bengal are only beginning to be understood at intraseasonal and longer time-scales. It is clear that this region is remarkably different in terms of surface fluxes and oceanic structure and response from the western Pacific warm pool region despite many similarities. Intense field experiments similar to those like TOGA–COARE can be applied to the Bay of Bengal to understand role of surface fluxes and convective parameterizations that could improve the simulation of the mean monsoon. The new experiment ARMEX (Arabian Sea Monsoon Experiment) will shed insight into the air–sea processes that control the precipitation maximum off the west coast of India. Such field studies should be given top priority so that the data gathered can easily be translated into state variables represented in predictive models

Simple and complex atmospheric modeling studies demonstrate that local SST anomalies over the eastern IO and Maritime Continent are the primary factors in determining the local precipitation anomalies during boreal summer and fall, when IODZM events are expected to influence regional climate variability. The SST and subsurface variability over the eastern IO need to be carefully monitored to evaluate the growth and intensification of IODZM events. In addition, accurate measurements of SST over the equatorial western IO will benefit the prediction of East African rainfall. This will require the extension of TAO array into the equatorial IO.

The role of the annual cycle of wind stress curl and thermocline doming in the Southern tropical IO has been noted repeatedly as discussed in the previous sections (Plates 1 and 5). As in the eastern IO, the SWIO experiences large interannual variability in thermocline and SST, and subsequently in convection and in the number of tropical cyclone days. Recent simple atmospheric modeling studies indicate that the circulation anomalies forced by SWIO convective anomalies modulate the tropical West Pacific convective anomalies during boreal winter [*Watanabe and Jin, 2002, 2003*]. Apart from the interannual oceanic Rossby waves, the thermocline and SST variability in SWIO is most likely related to the ITF variability at all timescales. Therefore, sustained efforts are necessary to measure and monitor surface and subsurface variability over the SWIO.

Even if the seasonal mean rainfall over India is normal, the unusual space–time evolution of the monsoon within the season can cause hardships to agriculture. Therefore, understanding and predicting the space–time evolution of the ISM is paramount. The recent knowledge gained clearly indicates an active role of SST for monsoon intraseasonal variability [*Webster et al., 2002*]. The need for new data was easily justified leading to more recent observational studies such as BOBMEX and JASMINE [*Bhat et al., 2001; Webster et al., 2002*] that clearly demonstrate the importance and usefulness of monitoring the ocean–atmosphere conditions over the tropical IO in understanding the various phases of the monsoon. The recent availability of satellite based SST observations from TRMM (Tropical Rain Measuring Mission) Microwave Imager (TMI) has provided a unique opportunity to quantify the SST variations at ISO timescales. Since the SST variations lead the variations in atmospheric convection by about 10 days on average, monitoring and measuring the SST accurately over the Bay of Bengal will offer a predictive skill on the intraseasonal variability of the ISM.

During the ISM season westward propagating synoptic disturbances known as monsoon depressions are the main rain-bearing systems. Most of the monsoon depressions form and intensify over the Bay of Bengal and to a certain degree their genesis is modulated by the phase of the intraseasonal variability. It was learned from MONEX–79 that warm SST and high heat contents in the Bay of Bengal provided the necessary conditions for the genesis of low-pressure systems and monsoon depressions [*Rao, 1987*]. After the passage of these systems, the SST was cooled by 0.2°C to 3°C, depending on the strength of these events. Since the temporal sampling of SST by space–borne sensors is not frequent enough, continuous monitoring by buoys are necessary. The role of the diurnal cycle of the oceanic mixed layer in determining SST variability at ISO time scales and the rectification of lower frequencies by the diurnal cycle are potentially important for understanding the role of air–sea interactions in regional atmospheric convection. Moored arrays in the IO can help us understand the diurnal cycle in the upper ocean. In summary, more meticulous observational efforts, by both in-situ and space-borne sensors, will likely be transferred to improve monsoon forecasts, particularly at intraseasonal timescales.

Comparison of *in-situ* and satellite-derived SST products such as *Reynolds and Smith* [1994] can differ by several degrees at ISO time-scales pointing to the need for sustained coverage by instruments such as the TMI at high spatial and temporal resolution. *Reynolds and Smith* [1994] remains the most popular SST dataset for most AGCM studies and this product may potentially be improved substantially with TMI-quality products in the future since extensive spatial coverage can only be accomplished with remote sensing platforms.

4.3. Process Studies Using Ocean Model to Understand SST Variability

The coupled dynamics in the IO differ significantly from that in the equatorial Pacific. In this subsection, we present some issues that can be addressed by ocean models to improve our knowledge of the role of IO in regional climate variability.

Ocean models can also quantify the oceanic mass and heat budgets associated with strong and weak monsoons. The mechanisms of transporting waters across the equator are only beginning to be addressed in models. The subsurface and surface variability during strong and weak monsoons may have a direct bearing on the TBO in the region. For example, during the period 1977–2002 the observed ISM has a strong biennial component while ENSO is dominated by a low-frequency (4–5 years) component. In addition, more frequent IODZM events developed during 1977–2002. Carefully designed ocean model experiments can possibly reveal the memory required for the TBO signal in the IO sector during this period.

Stand-alone ocean models can address several issues such as the role of the ITF in the variability of the IO at all time-scales. The Southern IO has a gyre structure that is unique compared to the Atlantic and Pacific southern gyres since the South Equatorial Current in the IO is never a part of the equatorial current system. The injection of the ITF occurs into this gyre and may modulate the low-frequency variability of the gyres. With the availability of atmospheric reanalysis and multidecadal flux products, it has become routine to carry out long simulations of the IO. At this juncture, it is important to analyze the ocean model simulations carefully to determine if the decadal and longer time-scale variability in the IO is crucially dependent on the Pacific through the ITF and if so, what is the best way to model the impact of the ITF on the IO. The basin scale impacts of the ITF have typically been studied in model simulations with the ITF closed [*Hirst and Godfrey,* 1993; *Schneider,* 1998; *Wajsowicz and Schneider,* 2001]. The role of the heat carried in via the ITF into the IO on regional climate variability remains to be fully understood.

Satellite based SST measurements reveal that at ISO time scales, the peak to peak amplitude of SST over the Bay of Bengal can be as large as 3°C. However, the SST variability at ISO time scales is largely a response to atmospheric wind forcing and it is unclear what effects are manifested on the oceanic processes. As pointed out by *Webster et al.* [2002] and *Waliser et al.* [2003], heat transports and currents display significant intraseasonal variability and yet no direct impact on SST has been quantified other than inferential connections. Fate of the freshwater input into the Bay of Bengal during the ISM season and its impact on the upper ocean hydrological cycle has to be revisited especially in terms of its impact on air–sea interactions at intraseasonal time-scales.

Ocean modelers lag far behind the atmospheric modelers in intercomparison studies that have proven to be instrumental in focusing community efforts in improving AGCMs [e.g., *Slingo et al.,* 1996; *Sperber and Palmer,* 1996; *Gadgil and Sajani,* 1998]. The need for multiple surface fluxes to force ocean models is a great inhibitor of ocean model intercomparison studies but these issues must be addressed if real progress is to be made in understanding the role of the IO in regional climate variability. As discussed in Section 2, numerous observational and modeling evidence exist pointing to the role of IO heat content variability in regional climate variability including the TBO. However, no systematic studies exist to consistently connect heat content anomalies to SST anomalies.

4.4. Atmospheric Model Studies Over the Indian Ocean

The large interannual SST anomalies over the eastern and southwestern IO (Plate 1a) exert substantial influence on the local precipitation anomalies. Over the open oceans the precipitation anomalies are inherently tied to local SST anomalies and less sensitive to changes in atmospheric initial conditions [*Charney and Shukla,* 1981]. *Farrara et al.* [2000] forced their AGCM with the boreal winter IO SST anomalies associated with the 1997/98 El Niño and noted the generation of a Rossby-wave source over East Asia. Continued efforts to perform sensitivity experiments with AGCMs that have realistic climatology will reveal the importance of these large IO SST anomalies on the atmosphere, local and remote. A sufficiently large ensemble member is needed to assess the simulated precipitation to changes in atmospheric initial conditions. Since ocean dynamics is involved in the SST variations over the eastern and southwestern IO (Plate 1b) realistic ocean models can be used to predict the SST which in turn can be used in AGCMs to predict the seasonal climate over the IO. If this feasibility is demonstrated then one can hope for dynamical seasonal prediction of climate over the IO.

More AGCM experiments need to be carried out to understand why the East African rainfall anomalies are sensitive to only one phase of the IODZM events. In addition, the observed correlations between monsoon variations and ENSO have changed sign and/or have become insignificant after the 1976–77 climate shift in the Pacific. Systematic AGCM studies are required to understand the possible role of local SST anomalies in the monsoon variations after the Pacific climate shift. Sensitivity experiments with simple atmospheric models forced either with SST or diabatic heating anomalies need to be conducted to understand the basic dynamics involved in the teleconnections, particularly for the SST anomalies associated with the IODZM.

Inter-comparison of simulated winter time MJOs in AGCMs revealed the importance of the oceanic contribution in the life cycle of MJOs [e.g. *Slingo et al.*, 1996]. Such a systematic effort to inter-compare summer time intraseasonal variability is beginning to be considered. The SST anomalies at intraseasonal timescales play an important role in the initiation of convection in the equatorial western IO during summer and winter. To understand the possible role of SST anomalies in the life-cycle of monsoon intraseasonal variability diagnostics have to be performed in stand-alone AGCMs and in their coupled versions. Careful experiments need to be carried out to pin down whether the coupling or the changes in the basic state improve the simulation of monsoon intraseaseal variability. Modeling studies are necessary to understand the processes responsible for the lead/lag relationship between SST and convection at intraseasonal timescales. Since the vertical structure of the atmospheric heating determines the vertical structure of the atmospheric circulation, it is very important to design model experiments to understand the role of intraseasonal SST on atmospheric convection and on the vertical structure of divergence, moisture and vertical motion.

4.5. Modeling the Modes of Variability in the Indian Ocean

The recent attention to the IO based on IODZM variability has led to numerous modeling studies. However, even when the coupled models generate IODZM-like variability, they are not always accompanied by realistic monsoon and/or ENSO variability and in particular, the mean thermocline in the EEIO simulated by most of the coupled models is unrealistic. More stringent and careful experiments need to be done in each component of the coupled models before strong claims such as "IODZM is independent of ENSO" can be made. A coupled model intercomparison focused on IODZM may be called for if more events such as those during 1994 and 1997 occur in the coming years. We outline a few problems that could be addressed by the modeling community.

One basic problem in most models is that the annual cycle of precipitation over the eastern IO is solely determined by the annual cycle in local SST. This results in an unrealistic simulation of precipitation over the eastern and western IO. The resulting precipitation gradient drives mean easterlies in the equatorial IO during boreal fall that subsequently result in a much shallow mean thermocline in the coupled models. In contrast, in observations the SST–convection relationship in the eastern IO is out-of-phase in the annual cycle [*Annamalai et al.*, 2003]. Careful experiments with AGCMs need to be first conducted to understand what processes control the annual cycle in precipitation over this very important region for IODZM events.

Unlike the equatorial eastern Pacific, the mean thermocline in the eastern IO is deep. Yet, subsurface oceanic processes play an active role during strong IODZM events. Observations indicate strong IODZM events are clustered in certain decades with a preferred quasi-biennial tendency. Some IODZM events co-occur with ENSO while others do not. The following relevant questions can be addressed with realistic ocean models: (i) What is the role of oceanic Kelvin and Rossby-waves during the life-cycle of an IODZM event?, (ii) What is the role of the spring Wyrkti Jet in the initiation of an IODZM?, (iii) Why are strong IODZM events clustered in certain decades and what controls the quasi-biennial tendency?, and (iv) What oceanic processes control the SST variability over the equatorial western IO in the neighborhood of 50°E, 0°N (Plate 1a)?.

Simple hybrid coupled models which proved enormously useful for understanding all aspects of ENSO variability could be constructed for the IODZM if the monsoon variability in the region can be represented appropriately in terms of their feedbacks to IODZM. The basic questions about the self-sustainability of IODZM in-light of its apparent episodic nature or its other possible incarnations such as the TBO may be better understood based on simpler process oriented models [e.g., *Li et al.*, 2003].

The role of the IO in ENSO variability continues to be investigated since the early works of *Barnett* [1983] and *Anderson and McCreary* [1985] using more sophisticated coupled ocean–atmosphere models [e.g., *Nagai et al.*, 1995; *Yu et al.*, 2002; *Wu and Kirtman*, 2003; *Annamalai et al.*, 2004]. The conclusion of the studies points to an interactive role for the IO with a stronger and more realistic ENSO when the coupling in the IO is explicitly represented. Apart from the direct impact of IO SST on the Pacific, potential impacts could also occur via the monsoon [*Kirtman and Shukla*, 2000; *Goswami and Jayevelu* 2001]. *Landsea and Knaff* [2000] presented limitations of the state-of-art coupled models in predicting the intensity of the 1997–98 El Niño. The inference that strong IODZM events influence the ISM and the equatorial wind over the central Pacific suggests that a proper representation of IO SST anomalies in coupled models will contribute to a more realistic simulation of ENSO and therefore the global climate variability. Existing coupled model runs need to be examined to determine if the intensity of ENSO is stronger or not during IODZM events. One could further address questions such as: Does the realistic simulation of IODZM in coupled models lead to better ENSO–monsoon correlations?

Rather appropriately, at the time of this writing, the IO underwent significant cooling off the coasts of Java and Sumatra during the boreal summer months of 2003. Even as claims were being made about the arrival of another IODZM event, the cold SST anomalies dissipated by the end of September 2003 (see Plate 11). The Pacific started out with a La Niña-like precursors followed by predictions of an impending La Niña but by the end of 2003, the NINO3 index was near neutral to slightly warm. Obviously, we still lack complete understanding of the processes involved in the growth of the IODZM events. The cooling of 2003 can at best be classified as a weak event according to the definition of *Annamalai et al.* [2003].

5. CONCLUDING REMARKS

A fairly comprehensive review of the published literature on the role of the IO in regional climate variability has been attempted. It is clear that despite a lack of coordinated observational and modeling attempts to quantify the role of the IO in the coupled climate variability, the predictive understanding has made considerable progress in the past decade as a result of the recognition of monsoons as an integral part of the ENSO cycle. Despite very focused efforts towards ENSO prediction over the last decade or so, many issues remain in the understanding and prediction of the ENSO cycle. Monsoon prediction attempts can be traced back to more than a century, yet the dynamic prediction of the monsoon is in various stages of development with much of the gain having been accomplished as an ENSO teleconnected climate process. It has become more and more evident that the earlier conjectures on the role of the intraseasonal variability in determining the seasonal mean monsoons have been confirmed by more recent observational and modeling activities.

Despite the enormity of the task of accurately modeling and predicting intraseasonal to interannual variability in the IO region as a coupled system, the convergence of ideas on observational and modeling requirements for achieving such a goal provides hope for future success. With observational systems such as Argo and Atlas/Triton buoys achieving their critical mass in the IO, the knowledge of IO variability and its role in regional climate variability will improve greatly. The need for understanding the intrinsic variability as opposed to IO response to ENSO will lead to significant progress towards monsoon forecasts especially in terms of improving forecasts at intraseasonal time-scales. With the growth of the observational database in the coming years, it is hoped that our understanding of the IO variability will improve at all spatial and time-scales. We look forward to the future with great expectations that more attention will be paid in understanding the IO climate variability.

Acknowledgments. As always, RM would like to express his deep gratitude to James Beauchamp for his excellent assistance in the analyses of model output and also literature search. RM is also grateful NASA Salinity, TOPEX/JASON, and Indian Ocean Biogeochemistry grants. We acknowledge the NOAA OGP/Pacific Program for partial support. Discussions with Clara Deser, Shang-Ping Xie, Jay McCreary, and Tony Busalacchi were very illuminating and helped provide a better focus. We thank Dr. Joseph for bringing to our attention some earlier works on IO and monsoon variability published in Mausam. We also thank the American Meteorological Society and American Geophysical Union for permitting to reproduce some of the figures. Yuko Okumura and Justin Small are appreciated for editorial assistance. IPRC is partly funded by the Frontier System for Global Change. This is IPRC contribution 270 and SOEST contribution 6364.

REFERENCES

Allan, R. J., D. Chambers, W. Drosdowsky, H. Hendon, M. Latif, N. Nicholls, I. Smith, R. Stone, and Y. Tourre, Is there an Indian Ocean dipole independent of the El Niño–Southern Oscillations?, *CLIVAR Exchanges, 6*, 4–8, 2001.

Anderson, D. L. T., Extremes in the Indian Ocean, *Nature, 401*, 337–339, 1999.

Anderson, D. L. T., and D. J. Carrington, Modeling interannual variability in the Indian Ocean using momentum fluxes from the UKMO and ECMWF, *J. Geophys. Res., 98*, 12,483–12,499, 1993.

Anderson, D. L. T., and J. P. McCreary, On the role of the Indian Ocean in a coupled ocean–atmosphere model of El Niño and the Southern Oscillation, *J. Atmos. Sci., 42*, 2439–2444, 1985.

Anderson, D. L. T., and P. B. Rowlands, The Somali Current response to the monsoon: The relative importance of local and remote forcing, *J. Mar. Res. 34*, 395–417, 1976.

Annamalai, H., and J. M. Slingo, Active/break cycles: diagnosis of the intraseasonal variability of the Asian summer monsoon, *Clim. Dyn., 18*, 85–102, 2001.

Annamalai H., J. M. Slingo, K. R. Sperber, and K. Hodges, The mean evolution and variability of the Asian summer monsoon: Comparison of ECMWF and NCEP/NCAR reanalysis, *Mon. Weather Rev., 127*, 1157–1186, 1999.

Annamalai, H., R. Murtugudde, J. Potemra, S.-P. Xie, and B. Wang, Coupled dynamics over the Indian Ocean: spring initiation of the Zonal Mode, *Deep Sea Res., Part II, 50*, 2305–2330, 2003.

Annamalai, H., and P. Liu, Response of the Asian summer monsoon to changes in ENSO properties, *Q. J.R. Meteorol. Soc.,* submitted, 2004.

Annamalai, H, S.-P. Xie, J. P. McCreary, and R. Murtugudde, Impact of Indian Ocean sea surface temperature on developing El Niño, *J. Clim,* accepted, 2004.

Arpe, K., L. Dümenil, and M. A. Giorgetta, Variability of the Indian Monsoon in the ECHAM3 model: Sensitivity to sea surface temperature, soil moisture, and the stratospheric quasi-biennial oscillation, *J. Clim., 11,* 1837–1858, 1998.

Ashok, K., Z. Guan, and T. Yamagata, Impact of the Indian Ocean dipole on the relationship between the Indian monsoon rainfall and ENSO, *Geophys. Res. Lett., 28,* 4499–4502, 2001.

Ashok, K., Z. Guan, and T. Yamagata, Impact of Indian Ocean dipole on the Australian winter rainfall, *Geophys. Res. Lett., 30,* 1821, doi:10.1029/2003GL017926, 2003.

Baquero-Bernal, A., M. Latif, and M. Legutke, On dipole like variability of sea surface temperature in the tropical Indian Ocean, *J. Clim., 15,* 1358–1368, 2002.

Barnett, T. P., Interaction of the monsoon and Pacific trade wind system at interannual time scales. Part I: The equatorial zone, *Mon. Weather Rev., 111,* 756–773, 1983.

Behera, S.K., and T. Yamagata, Subtropical SST dipole events in the southern Indian Ocean, *Geophys. Res. Lett., 28(2),* 227–230, 10.1029/2000GL011461, 2001.

Behera, S. K., and T. Yamagata, Influence of the Indian Ocean dipole on the Southern Oscillation, *J. Meteorol. Soc. Jpn., 81,* 169–177, 2003.

Behera, S. K., R. Krishnan, and T. Yamagata, Unusual ocean–atmosphere conditions in the tropical Indian Ocean during 1994, *Geophys. Res. Lett., 26,* 3001–3004, 1999.

Behera, S. K., P. S. Salvekar, and T. Yamagata, Simulation of interannual SST variability in north Indian Ocean, *J. Clim., 13,* 3487–3499, 2000.

Behera, S. K., J. Luo, S. Masson, T. Yamagata, P. Delecluse, S. Gualdi, and A. Navarra, Impact of the Indian Ocean Dipole on the East African short rains: A CGCM study, *CLIVAR Exchanges, 27,* 1–4, 2003.

Bhat, G. S. , et al. BOBMEX: The Bay of Bengal Monsoon Experiment, *Bull. Am. Meteorol. Soc., 82,* 2217–2243, 2001.

Birkett, C., R. Murtugudde, and T. Allan, Indian Ocean climate event brings floods to East Africa's lakes and the Sudd marsh, *Geophys. Res. Lett., 26,* 1031–1034, 1999.

Bjerknes, J., Atmospheric teleconnections from the equatorial Pacific, *Mon. Weather Rev. 97,* 163–172, 1969.

Black, E., J. M. Slingo, and K. R. Sperber, An observational study of the relationship between excessively strong short rains in coastal Africa and Indian Ocean SST, *Mon. Weather Rev., 131,*74–94, 2003.

Braak, C., Atmospheric variations of short and long duration in the Malay Archipelago, *Meded. Verh. Koninklijk Magnetischen Meterologisch Observatorium te Batavia,* No. 5, 57 pp., (available from Royal Netherlands Meteorological Institute, De Bilt, 3730 AE, Netherlands.), 1919.

Cadet, D., and B. C. Diehl, Interannual variability of surface fields over the Indian Ocean during recent decades, *Mon. Weather Rev., 112,* 1921–1935, 1984.

Cadet, D., and G. Reverdin, The Monsoon over the Indian Ocean during summer 1975. Part I: Mean fields, *Mon. Weather Rev., 109,* 148–158, 1981.

Cane, M. A., Forecasting El Niño with a geophysical model, in *Teleconnection Connecting World-Wide Climate Anomalies,* edited by M. Glantz, R. Katz, and N. Nicholls, pp. 345–369, Cambridge University Press, 1991.

Chambers, D. P, B. D. Tapley, R. H. Stewart, Anomalous warming in the Indian Ocean coincident with El Niño, *J. Geophys. Res., 104,* 3035–3047, 1999.

Chandrasekar, A., and A. Kitoh, Impact of localized sea surface temperature anomalies over the equatorial Indian Ocean on the Indian summer monsoon, *J. Meteorol. Soc. Jpn., 76,* 841–853, 1999.

Chang, C.-P., and T. N. Krishnamurti, *Monsoon modeling,* Cambridge University Press, 1987.

Chang, C.-P., and T. Li, A theory for the tropospheric biennial oscillation, *J. Atmos. Sci., 57,* 2209–2224, 2000.

Charney, J., and J. Shukla, Predictability of monsoons, in *Monsoon Dynamics,* pp. 99–110, Cambridge University Press, 1981.

Clark C. O., C. Oelfke, J. E. Cole, and P. J. Webster, Indian Ocean SST and Indian summer rainfall: Predictive relationships and their decadal variability, *J. Clim., 13,* 4452–4452, 2000.

Clark, C. O., P. J. Webster, and J. E. Cole., Interdecadal variability of the relationship between the Indian Ocean Zonal Mode and East African coastal rainfall anomalies, *J. Clim., 16,* 548–554, 2003.

Ellis, R. S., A Preliminary study of a relation between surface temperature of the North Indian Ocean and precipitation over India, M.S. thesis, Department of Meteorology, Florida State University, Tallahassee, 1952.

Farmer, G., Seasonal forecasting of the Kenya coast short rains, *J. Climatol., 8,* 489–497, 1988.

Farrara, J. D., C. R. Mechoso, and A.W. Robertson, Ensembles of AGCM two-tier predictions and simulations of the circulation anomalies during winter 1997–98, *Mon. Weather Rev., 128,* 3589–3604, 2000.

Fein, J. S., and P. L. Stephens (Eds), *Monsoons,* 632 pp., Wiley-Interscience Publication, John Wiley and Sons, New York, 1987.

Feng, M., and G. Meyers, Interannual variability in the tropical Indian Ocean: a two-year time-scale of Indian Ocean dipole, *Deep Sea. Res., Part II, 50,* 2263–2284, 2003.

Flatau, M., P. J. Flatau, P. Phoebus, and P. P. Niiler, The feedback between equatorial convection and local radiative and evaporative processes: The implications for intraseasonal oscillations, *J. Atmos. Sci., 54,* 2373–2386, 1997.

Frederiksen, C. S., and R. C. Balgovind, The influence of the Indian Ocean Indonesian SST gradient on the Australian winter rainfall and circulation in atmospheric GCM, *Q. J. R. Meteorol. Soc., 120,* 923–952, 1994.

Frederiksen, C. S., and J. S. Frederiksen, A theoretical model of Australian northwest cloud band disturbances and southern hemisphere storm tracks: The role of SST anomalies, *J. Atmos. Sci., 53,* 1410–1432, 1996.

Frederiksen, C. S., D. P. Rowell, R. C. Balgovind, and C. K. Folland, Multidecadal simulations of Australian rainfall variability: The role of SST, *J. Clim., 12,* 357–379, 1999.

Fu, X., B. Wang, T. Li, and P. J. McCreary, Coupling between northward-propagating intraseasonal oscillations and sea surface temperature in the Indian Ocean, *J. Atmos. Sci., 60,* 1733–1753, 2003.

Gadgil, S., and S. Sajani, Monsoon precipitation in the AMIP runs, *Clim. Dyn., 14,* 659–689, 1998.

Gadgil, S., P. V. Joseph, and N. V. Joshi, Ocean atmosphere coupling over monsoon regions, *Nature, 312,* 141–143, 1984.

Gill, A. E., Some simple solutions for heat-induced tropical circulation, *Q. J. R. Meteorol. Soc.,106,* 447–462, 1980.

Goddard, L., and N. E. Graham, Importance of the Indian Ocean for simulating rainfall anomalies over eastern and southern Africa, *J. Geophys. Res., 104,* 19,099–19,116, 1999.

Godfrey, J. S, The effect of the Indonesian throughflow on ocean circulation and heat exchange with the atmosphere: A review, *J. Geophys. Res., 101,* 12,217–12,238, 1996.

Godfrey, J. S., and K. R. Ridgeway, The large scale environment of the poleward flowing Leeuwin Current, western Australia: Alongshore steric height gradients, wind stress, and geostrophic flow, *J. Phys. Oceanogr., 15,* 481–495, 1985.

Godfrey, J. S., and A. J. Weaver, Is the Leeuwin Current driven by Pacific heating and winds?, *Prog. Oceanogr., 27,* 225–272, 1991.

Goswami, B. N. and R. S. Ajaymohan, Intraseasonal oscillations and interannual variability of the Indian summer monsoon, *J. Clim., 14,* 1180–1198, 2001.

Goswami, B. N., and V. Jayavelu, On possible impact of the Indian summer monsoon on the ENSO, *Geophys. Res. Lett., 27,* 571–574, 2001.

Gualdi, S., E. Guilyardi, A. Navarra, S. Masina, and P. Delecluse, The interannual variability in the Indian Ocean as simulated by a CGCM, *Clim. Dyn., 20,* 567–582, 2003.

Guan, Z., and T. Yamagata, The unusual summer of 1994 in east Asia: IOD teleconnections, *Geophys. Res. Lett., 30,* 1544, doi: 10.1029/2002GL016831, 2003.

Hackert, E. C., and S. Hastenrath, Mechanisms of Java rainfall anomalies, *Mon. Weather Rev., 114,* 745–757, 1986.

Han, W., J. P. McCreary, and K. E. Kohler, Influence of precipitation minus evaporation and Bay of Bengal rivers on dynamics, thermodynamics, and mixed layer physics in the upper Indian Ocean, *J. Geophys. Res., 106,* 6895–6916, 2001.

Hastenrath, S., Prediction of Indian monsoon rainfall: Further exploration, *J. Clim., 1,* 298–305, 1988.

Hastenrath, S., Zonal circulations over the equatorial Indian Ocean, *J. Clim., 13,* 2746–2756, 2000.

Hastenrath, S., Dipoles, temperature gradients, and tropical climate anomalies, *Bull. Am. Meteorol. Soc., 83,* 735–740, 2002.

Hastenrath, S., and L. Greischar, Upper-ocean structure, in *Climatic Atlas of the Indian Ocean*, 273 pp., University of Wisconsin Press, 1989.

Hastenrath, S., A. Nicklis, and L. Greishar, Atmospheric hydrospheric mechanics of climate anomalies in the western equatorial Indian Ocean, *J. Geophys. Res., 98,* 20,219–20,235, 1993.

Hendon, H. H., Indonesian rainfall variability: Impact of ENSO and local air–sea Interaction, *J. Clim.,* 16, 1775–1790, 2003.

Hendon H. H., and J. Glick, Intraseasonal air–sea interaction in the tropical Indian and Pacific Oceans, *J. Clim., 10,* 647–661, 1997.

Hendon, H. H., and B. Liebmann, A composite study of onset of the Australian summer monsoon, *J. Atmos. Sci., 47,* 2227–2240, 1990.

Hirst, A. C., and J. S. Godfrey, The role of Indonesian throughflow in a global ocean GCM, *J. Phys. Oceanogr., 23,* 1057–1086, 1993.

Holland, G. J., Interannual variability of the Australian summer monsoon at Darwin–1952–82, *Mon. Weather Rev., 114,* 594–604, 1986.

Hoskins, B. J., and D. J. Karoly, The steady linear response of a spherical atmosphere to thermal and orographic forcing, *J. Amos. Sci., 38,* 1179–1196, 1981.

Howden, S., and R. Murtugudde, Effects of river inputs into the Bay of Bengal, *J. Geophys. Res., 106,* 19,285–19,483, 2001.

Huang, B. H., and J. L. Kinter, The interannual variability in the tropical Indian Ocean and its relations to El Niño–Southern Oscillation, *J. Geophys. Res., 107,* 3199, doi: 10.1029/2001JC 001278, 2002.

Hutchinson, P., The Southern Oscillation and prediction of 'Der' season rainfall in Somalia, *J. Clim., 5,* 525–531, 1992.

Iizuka, S., T. Matsuura, and T. Yamagata, The Indian Ocean SST dipole simulated in a coupled general circulation model, *Gephys. Res. Lett., 27,* 3369–3372, 2000.

Joseph, P. V., Ocean–atmosphere interaction on a seasonal scaleover north Indian Ocean and Indian monsoon rainfall and cyclone tracks–a preliminary study, *Mausam, 32,* 237–246, 1981.

Joseph, P. V., Monsoon variability in relation to equatorial trough activity over India and west Pacific Oceans, *Mausam, 41,* 291–296, 1990.

Joseph, P. V., and P. V. Pillai, Air–sea interaction on a seasonal scale over north Indian Ocean, Part I: Interannual variability of sea surface temperature and Indian summer monsoon rainfall, *Mausam, 35,* 323–330, 1984.

Joseph, P. V., and P. V. Pillai, Air–sea interaction on a seasonal scale over north Indian Ocean, Part II: Monthly mean atmospheric and oceanic parameters during 1972 and 1973, *Mausam, 37,* 159–168, 1986.

Joseph, P. V., J. K. Eischeid, and R. J. Pyle, Interannual variability of the onset of the Indian summer monsoon and its association with atmospheric features, El Niño and sea surface temperature anomalies, *J. Clim., 7,* 81–105, 1994.

Ju, J., and J. M. Slingo, The Asian summer monsoon and ENSO, *Q. J. R. Meteorol. Soc., 122,* 1133–1168, 1995.

Jury, M. R., Regional teleconnection patterns associated with summer rainfall over South Africa, Namibia, and Zimbabwe, *Int. J. Climatol., 16,* 135–153, 1996.

Jury, M. R., B. Pathack, and B. Parker, Climatic determinants and statistical prediction of tropical cyclone days in the Southwest Indian Ocean, *J. Clim., 12,* 1738–1746, 1999.

Kapala, A., K. Born, and H. Flohn, Monsoon anomaly or an El Niño event at the equatorial Indian Ocean? Catastrophic rains of 1961/62 in East Africa and their teleconnections, paper presented at Int. Conf. Monsoon Variab. and Pred., World Meteorol. Org., Trieste, Italy, 1994.

Kemball-Cook, S., B. Wang, and X. Fu., Simulation of the intraseasonal oscillation in ECHAM-4 model: the impact of coupling with an ocean model, *J. Atmos. Sci, 59,* 1433–1453, 2002.

Keshvamurthy, R. N., Response of the atmosphere to sea surface temperature anomalies over the equatorial Pacific and the teleconnections of the southern oscillation, *J. Atmos. Sci.*, *39*, 1241–1259, 1982.

Kirtman, B., and J. Shukla, Influence of the Indian summer monsoon on ENSO, *Q. J. R. Meteorol. Soc.*, *126*, 213–239, 2000.

Klein, S. A., B. J. Soden, and N.-C. Lau, Remote sea surface temperature variations during ENSO: Evidence for a tropical atmospheric bridge, *J. Clim.*, *12*, 917–932, 1999.

Krishnamurthy, V., and J. Kinter, The Indian Monsoon and its relation to global climate variability, in *Global Climate: Current Research and Uncertainties in the Climate System*, edited by Rodo and Comin, Springer, 2003.

Krishnamurthy, V., and B. Kirtman, Variability of the Indian Ocean: Relation to monsoon and ENSO, *Q. J. R. Meteorol. Soc.*, *129*, 1623–1646, 2003.

Krishnamurthy, V., and J. Shukla, Intraseasonal and interannual variability of rainfall over India, *J. Clim.*, *13*, 4366–4377, 2000.

Krishnamurti, T. N., D. K. Oosterhof, and A. V. Mehta, Air–sea interaction on the time scale of 30 to 50 days, *J. Atmos. Sci.*, *45*, 1304–1322, 1988.

Krishnan, R., M. Mujumdar, V. Vaidya, K. V. Ramesh, and V. Satyan, The abnormal Indian summer monsoon of 2000, *J. Clim.*, *16*, 1177–1194, 2003.

Landman, W. A., and S. J. Mason, Change in the association between Indian Ocean sea–surface temperatures and summer rainfall over South Africa and Namibia, *Int. J. Climatol.*, *19*, 1477–1492, 1999.

Landsea, C. W., and J. A. Knaff, How much skill was there in forecasting the very strong 1997–98 El Niño, *Bull. Am. Meteorol. Soc.*, *81*, 2107–2119, 2000.

Latif, M., D. Dommenget, M. Dima, and A. Grötzner, The role of Indian Ocean sea surface temperature in forcing east African rainfall anomalies during December–January 1997/98, *J. Clim.*, *12*, 3497–3504, 1999.

Lau, K.-M., and P. H. Chan, Aspects of the 40–50 day oscillation during the northern summer as inferred from outgoing longwave radiation, *Mon. Weather Rev.*, *114*, 1354–1367, 1986.

Lau, K.-M., and H.-T. Wu, Assessment of the impacts of the 1997–98 El Niño on the Asian–Australian monsoon, *Geophys. Res. Lett.*, *26*, 1747–1750, 1999.

Lau, K.-M., K. M. Kim, and S. Yang, Dynamical and boundary forcing characteristics of regional components of Asian summer monsoon, *J. Clim.*, *13*, 2461–2482, 2000.

Lau, N.-C., and M. J. Nath, Impact of ENSO on the variability of the Asian–Australian monsoons as simulated in GCM experiments, *J. Clim.*, *13*, 4287–4309, 2000.

Lau, N.-C., and M. J. Nath, Atmosphere–ocean variations in the Indo–Pacific sector during ENSO episodes, *J. Clim.*, *16*, 3–20, 2003.

Leetma, A., The response of the Somali Current to the southwest monsoon of 1970, *Deep Sea Res.*, *19*, 397–400, 1972.

Li, T., and B. Wang, The influence of sea surface temperature on the tropical intraseasonal oscillation: A numerical experiment, *Mon. Weather Rev.*, *122*, 2349–2362, 1994.

Li, T., Y. Zhang, C.-P. Chang, and B. Wang, On the relationship between Indian Ocean sea surface temperature and Asian summer monsoon, *Geophys. Res. Lett.*, *28*, 2843–2846, 2001a.

Li, T., C.-W. Tham, and C.-P. Chang, A coupled air–sea–monsoon oscillator for the tropospheric biennial oscillation, *J. Clim.*, *14*, 752–764, 2001b.

Li, T., B. Wang, C.-P. Chang, and Y. Zhang, A theory for the Indian Ocean dipole–zonal mode, *J. Atmos. Sci.*, *60*, 2119–2135, 2003.

Liebmann, B., H. H. Hendon, and J. D. Glick, The relationship between tropical cyclones of the western Pacific and Indian Oceans and Madden–Julian Oscillation, *J. Meteorol. Soc. Jpn.*, *72*, 401–412, 1994.

Lighthill, M. J., Dynamic response of the Indian Ocean to the onset of southwest monsoon, *Philos. Trans. Soc. London, A265*, 45–92, 1969.

Lighthill, M. J., and R. P. Pierce, *Monsoon Dynamics*, Cambridge University Press, 1981.

Loschnigg, J., and P. J. Webster, A coupled ocean–atmosphere system of SST modulation for the Indian Ocean, *J. Clim.*, *13*, 3342–3360, 2000.

Loschnigg, J., G. A. Meehl, P. J. Webster, J. M. Arblaster, and G. P. Compo, The Asian monsoon, the tropospheric biennial oscillation, and the Indian Ocean Zonal Mode in the NCAR CSM, *J. Clim.*, *16*, 1617–1642, 2003.

Lukas, R., and E. Lindstrom, The mixed layer of the western equatorial Pacific, *J. Geophys. Res.*, *93*, 3343–3357, 1991

Madden, R. A., and P. R. Julian, Detection of a 40–day oscillation pressures and zonal winds in tropical Pacific, *Bull. Am. Meteorol. Soc.*, *52*, 789–799, 1971

Madden, R. A., and P. R. Julian, Description of global-scale circulation cells in the tropics with a 40–50 day period, *J. Atmos. Sci.*, *29*, 1109–1123, 1972.

Makaru, A., and M. R. Jury, Predictability of Zimbabwe summer rainfall, *Int. J. Climatol*, *17*, 1421–1432, 1997

Mason, S. J., and M. R. Jury, Climatic variability and change over southern Africa: A reflection on underlying processes, *Prog. Phys. Geogr.*, *21*, 23–50, 1997.

Masson, S., P. Delecluse, J.-P. Boulanger, and C. Menkes, A model study of the variability and formation mechanisms of the barrier layer in the eastern equatorial Indian Ocean, *J. Geophys. Res.*, *107*, 8017, doi:10.1029/2001JC000832, 2002.

Masson, S., C. Menkes, P. Delecluse, and J.-P. Boulanger, Impacts of salinity on the eastern Indian Ocean during the termination of the fall Wyrtki Jet, *J. Geophys. Res.*, *108*, 3067, doi: 10.1029/2001JC000833, 2003.

McBride, J. L., and N. Nicholls, Seasonal relationship between Australian rainfall and the Southern Oscillation, *Mon. Weather Rev.*, *111*, 1998–2004, 1983.

McCreary, J. P., and P. K. Kundu, A numerical investigation of sea surface variability in the Arabian Sea, *J. Geophys. Res.*, *94*, 16,097–16,114, 1989.

McCreary, J. P., P. K. Kundu, and R. L. Molinari, A numerical investigation of dynamics, thermodynamics and mixed-layer processes in the Indian Ocean, *Prog. Oceanogr.*, *31*, 181–244, 1993.

McPhaden, M. J., Genesis and evolution of the 1997–98 El Niño, *Science*, *283*, 950–954, 1999.

McPhaden, M. J., et al., The tropical ocean global atmosphere observing system: A decade of progress, *J. Geophys. Res.*, *103*, 14,169–14,240, 1998.

Meehl,, G. A., The annual cycle and interannual variability in the tropical Pacific and Indian Ocean regions, *Mon. Weather Rev.*, *115*, 27–50, 1987.

Meehl, G. A., and J. M. Arblaster, The tropospheric biennial oscillation and Asian–Australian monsoon rainfall, *J. Clim.*, *15*, 722–744, 2002a.

Meehl, G. A., and J. M. Arblaster, Indian monsoon GCM sensitivity experiments testing tropospheric biennial oscillation transition conditions, *J. Clim.*, *15*, 923–944, 2002b.

Meehl, G. A., J. M. Arblaster, and J. Loschnigg, Coupled ocean atmosphere dynamical processes in the tropical Indian and Pacific Oceans and the TBO, *J. Clim.*, *16*, 2138–2158, 2003.

Miyakoda, K., J. L. Kinter, and S. Yang, The role of ENSO in the south Asian monsoon and pre-monsoon signals over the Tibetan Plateau, *COLA Tech. Rep. 113*, Center for Ocean–Land–Atmosphere Studies, Calverton, MD, 2002.

Molinari, R. L., J. Swallo, and J. F. Festa, Evolution of the near-surface thermal structure in the western Indian Ocean during FGGE, 1979, *J. Mar. Res.*, *44*, 739–762, 1986.

Murtugudde, R., and A. Busalacchi, Interannual variability of the dynamics and thermodynamics of the tropical Indian Ocean, *J. Clim.*, *12*, 2300–2326, 1999.

Murtugudde, R., B. N. Goswami, and A. J. Busalacchi, Air–sea interactions in the southern Indian Ocean and its relation to interannual variability of the monsoon over India, *proceedings of the International Conference on Monsoon and Hydrological Cycle*, Kyongju, Korea, pp. 184–188, 1998a.

Murtugudde, R., A. J. Busalacchi, and J. Beauchamp, Seasonal to interannual effects of the Indonesian throughflow on the tropical Indo–Pacific basin, *J. Geophys. Res.*, *103*, 21,425–21,441, 1998b.

Murtugudde, R., S. Signorini, J. Christian, A. Busalacchi, and C. McClain, Ocean color variability of the tropical Indo–Pacific basin observed by SeaWiFS during 1997–98, *J. Geophys. Res.*, *104*, 18,351–18,366, 1999.

Murtugudde, R., J. McCreary, and A. Busalacchi, Oceanic processes associated with anomalous events in the Indian Ocean, *J. Geophys. Res.*, *105*, 3295–3306, 2000.

Murtugudde, R., A. Busalacchi, and J. P. McCreary, Comments on 'Dipoles' temperature gradients, and tropical climate anomalies, *Bull. Am. Meteorol. Soc.*, *84*, 1422–1423, 2003.

Nagai, T., Y. Kitamura, M. Endoh, and T. Tokioka, Coupled atmosphere–ocean model simulations of El Niño/Southern Oscillation with and without an active Indian Ocean, *J. Clim.*, *8*, 3–14, 1995.

Nichol, J., Bioclimatic impacts of the 1994 smoke haze event in Southeast Asia, *Atmos. Environ.*, *31*, 1209–1219, 1997.

Nichol, J., Smoke haze in Southeast Asia: A predictable recurrence, *Atmos. Environ.*, *32*, 14–15, 1998.

Nicholls, N., Air–sea interaction and quasi-biennial oscillation, *Mon. Weather Rev.*, *106*, 1505–1508, 1978.

Nicholls, N., Air–sea interactions and the possibility of long-range weather prediction in the Indonesian archipelago, *Mon. Weather Rev.*, *109*, 2435–2443, 1981.

Nicholls, N., Predicting Indian monsoon rainfall from sea–surface temperature in the Indonesia North Australia area, *Nature*, *306*, 576–577, 1983.

Nicholls, N., A system for predicting the onset of the North Australian wet season, *J. Climatol.*, *4*, 425–435, 1984.

Nicholls, N., El Niño–Southern Oscillation impact prediction, *Bull. Am. Meteorol. Soc.*, *69*, 173–176, 1988.

Nicholls, N., Sea surface temperatures and Australian winter rainfall, *J. Clim.*, *2*, 965–973, 1989.

Nicholls, N., All-India summer monsoon rainfall and sea surface temperatures around Northern Australia and Indonesia, *J. Clim.*, *8*, 1463–1472, 1995.

Nicholls, N., and W. Drosdowsky, Is there an equatorial Indian Ocean SST dipole, independent of the El Niño–Southern Oscillation?, paper presented at 81st American Meteorological Society, Annual meeting, Albuquerque, NM, USA, 2001.

Nicholson, S. E., and J. Kim, The relationship of the El Niño–Southern Oscillation to African rainfall, *Int. J. Climatol.*, *17*, 117–135, 1997.

Nigam, S., On the dynamical basis for the Asian summer monsoon rainfall–El Niño relationship, *J. Clim.*, *7*, 1750–1771, 1994.

Nigam, S., and H.-S. Shen, Structure of oceanic and atmospheric low-frequency variability over the tropical Pacific and Indian Oceans. Part I: COADS observations, *J. Clim.*, *6*, 657–676, 1993.

O'Brien, J. J., and H. E. Hurlburt, Equatorial jet in the Indian Ocean: Theory, *Science*, *184*, 1075–1077, 1974.

Ogallo, L. J., Relationships between seasonal rainfall in East Africa and the Southern Oscillation, *J. Climatol.*, *8*, 31–43, 1988.

Palmer, T. N., Chaos and predictability in forecasting the monsoons, *Proc. Indian Natn. Sci. Acad., Part A*, *60*, 57–66, 1994.

Palmer, T. N., C. Brankovic, P. Viterbo, and M. J. Miller, Modeling interannual variability of summer monsoons, *J. Clim.*, *5*, 399–417, 1992.

Pant, G. B., and K. Rupakumar, *Climates of south Asia*, 320 pp., John Wiley Sons, 1997.

Park, C.-K., and S. D. Schubert, On the nature of 1994 East Asian summer drought, *J. Clim.*, *10*, 1056–1070, 1997.

Philander, G. H., *El Niño, La Niña and Southern Oscillation*, 280 pp., Academic Press, New York, 1990.

Pisharoty, P. R., Forecasting droughts in the subcontinent of India, *Proc. Indian Natn. Sci. Acad., Part A*, *42*, 220–223, 1976.

Quinn, W. H., D. O. Zopf, K. S. Short, and R. T. W. K. Yang, Historical trends and statistics of Southern Oscillation, El Niño, and Indonesian droughts, *Fish. Bull.*, *76*, 663–678, 1978.

Rajagopalan, B., U. Lall, and M. A. Cane, Anomalous ENSO occurrences: An alternate view, *J. Clim.*, *10*, 2351–2357, 1997.

Rao, K. G., and B. N. Goswami, Interannual variations of sea surface temperature over the Arabian Sea and the Indian Monsoon: A new perspective, *Mon. Weather Rev.*, *116*, 558–568, 1988.

Rao, R. R., Further analysis on the thermal response of the upper Bay of Bengal to the forcing of pre-monsoon cyclonic storm and summer monsoon onset during MONEX–79, *Mausam*, *38*, 147–156, 1987.

Rao, R. R., and R. Sivakumar, Seasonal variability of near-surface thermal structure and heat budget of the mixed layer of the Indian Ocean from a new global ocean temperature climatology, *J. Geophys. Res.*, *105*, 995–1015, 2000.

Rao, R. R., R. L. Molinari, and J. F. Festa, Evolution of the climatological near-surface thermal structure of the tropical Indian Ocean. 1. Description of mean monthly mixed layer depth, and sea surface temperature, surface current, and surface meteorological fields, *J. Geophys. Res., 94,* 10,801–10,815, 1989.

Rao, S. A., S. K. Behera, Y. Masumoto, and T. Yamagata, Interannual subsurface variability in the tropical Indian Ocean with a special emphasis on the Indian Ocean Dipole, *Deep Sea Res., Part II, 49,* 1549–1572, 2002.

Rasmusson, E. M., and T. H. Carpenter, The relationship between eastern equatorial Pacific sea surface temperature and rainfall over India and Sri Lanka, *Mon. Weather Rev., 111,* 517–528, 1983.

Reverdin, G., The upper equatorial Indian Ocean. The climatological seasonal cycle, *J. Phys. Oceanogr., 17,* 903–927, 1987.

Reverdin, G., D. L. Cadet, and D. Gutzler, Interannual displacements of convection and surface circulation over the equatorial Indian Ocean, *Q. J. R. Meteorol. Soc., 112,* 43–67, 1986.

Reynolds, R.W., and T.M. Smith, Improved global sea surface temperature analyses using optimal interpolation. *J. Clim, 7,* 929–948, 1994.

Rodwell, M. J., and B. J. Hoskins, Monsoons and the dynamics of deserts, *Q. J. R. Meteorol. Soc., 122,* 1385–1404, 1996.

Sadhuram, Y., and N. C. Wells, Role of the Indian Ocean on the Southern Oscillation, atmospheric circulation indices and monsoon rainfall over India, *Glob. Atmos. Ocean Sys., 7,* 47–72, 1999.

Saha, K. R., Some aspects of the Arabian sea summer monsoon, *Tellus, 26,* 464–476, 1974.

Saha, K. R., and S. N. Bavadekar, Water vapor budget and precipitation over the Arabian Sea during the northern summer, *Q. J. Roy. Meteor. Soc., 99,* 273–278, 1973.

Saji, N. H., and T. Yamagata, Possible impacts of Indian Ocean dipole mode events on global climate, *Clim. Res., 25,* 151–169, 2003.

Saji, N. H., B. N. Goswami, P.N. Vinayachandran, and T. Yamagata, A dipole mode in the tropical Indian Ocean, *Nature, 401,* 360–363, 1999.

Schiller, A., J. S. Godfrey, P. C. McIntosh, G. Meyers, and R. Fiedler, Interannual dynamics and thermodynamics of the Indo–Pacific oceans, *J. Phys. Oceanogr., 30,* 987–1012, 2000.

Schneider, N., The Indonesian throughflow and the global climate system, *J. Clim., 11,* 676–689, 1998.

Schott, F., and J. P. McCreary, The monsoon circulation of the Indian Ocean, *Prog. Oceanogr., 51,* 1–123, 2001.

Schubert, S. D., and M. L. Wu, Predictability of the 1997 and 1998 south Asian monsoon low-level winds, *J. Clim., 14,* 3173–3191, 2001.

Sengupta, D., and M. Ravichandran, Oscillations of the Bay of Bengal sea surface temperatures during 1998 summer monsoon, *Geophys. Res. Lett., 28,* 2033–2036, 2001.

Sengupta, D., B. N. Goswami, and R. Senan, Coherent intraseasonal oscillations of the ocean and atmosphere during the Asian summer monsoon, *Geophys. Res. Lett., 28,* 4127–4130, 2001.

Sengupta, D., P. K. Ray, and G. S. Bhat, Spring warming of the eastern Arabian Sea and Bay of Bengal from buoy data, *Geophys. Res. Lett., 29,* 1734–1737, 2002.

Shen, X., and M. Kimoto, Influence of El Niño on the 1997 Indian summer monsoon, *J. Meteorol. Soc. Jpn., 77,* 1023–1037, 1999.

Shenoi, S.S.C., D. Shankar, and S.R. Shetye, Differences in heat budgets of the near-surface Arabian Sea and Bay of Bengal: Implications for the summer monsoon, *J. Geophys. Res., 107,* doi:10.1029/2001JC000679, 2002.

Shetye, S. R., A model study of the seasonal cycle of the Arabian Sea sea surface temperature, *J. Mar. Res., 44,* 521–542, 1986.

Shetye, S. R., A. Gouveia, D. Shankar, S. Shenoi, P. Vinayachandran, D. Sundar, G. Michael, and G. Nampoothiri, Hydrography and circulation in the western Bay of Bengal during the northeast monsoon, *J. Geophys. Res, 101,* 14,011–14,025, 1996.

Shinoda, T., H. H. Hendon, and J. D. Glick, Mixed layer modeling of intraseasonal sea surface temperature variability in the tropical western Pacific and Indian Ocean, *J. Clim., 11,* 2668–2685, 1998.

Shukla, J., Effects of Arabian Sea sea–surface temperature anomaly on Indian summer monsoon: Numerical experiment with GFDL model, *J. Atmos. Sci., 32,* 503–511, 1975.

Shukla, J., Dynamical predictability of monthly means, *J. Atmos. Sci., 38,* 2547–2572, 1981.

Shukla, J., Predictability of time averages: Part II. The influence of the boundary forcing, in *Problems and prospects in long and medium range weather forecasting*, edited by D.M. Burridge and E. Kallen, pp. 155–206, Springer-Verlag, London, 1984.

Shukla, J., Interannual variability of monsoons, in *Monsoons*, edited by J. S. Fein and P. L. Stephens, pp. 399–464, John Wiley and Sons, New York, 1987.

Shukla, J., and M. Fennessy, Simulation and predictability of monsoons, *Proc. Int. Conf. on Monsoon Variab. and Pred.*, Tech. Rep. WCRP–84, pp. 567–575, World Climate Research Programmer, Geneva, Switzerland, 1994.

Shukla, J., and B. M. Misra, Relationship between sea–surface temperature and wind speed over central Arabian Sea and monsoon rainfall over India, *Mon. Weather Rev., 105,* 998–1002, 1977.

Shukla, J., and J. M. Wallace, Numerical simulation of the atmospheric response to equatorial Pacific sea surface temperature anomalies, *J. Atmos. Sci., 40,* 1613–1630, 1983.

Sikka, D., Some aspects of the large-scale fluctuations of summer monsoon rainfall over India in relation to fluctuations in the planetary and regional scale circulation parameters, *Proc. Indian Acad. Sci. (Earth and Planetary Science), 89,* 179–195, 1980.

Sikka, D., and S. Gadgil, On the maximum cloud zone and the ITCZ over the Indian longitudes during the southwest monsoon, *Mon. Weather Rev., 108,* 1840–1853, 1980.

Simmonds, I., A modeling study of winter circulation and precipitation anomalies associated with Australian region ocean temperatures, *Aust. Meteorol. Mag., 38,* 151–162, 1990.

Simmonds, I., and A. Rocha, The association of Australian winter climate with ocean temperatures to the west, *J. Clim., 4,* 1147–1161, 1991.

Simmonds, I., and G. Trigg, Global circulation and precipitation changes induced by sea surface temperature anomalies to the north of Australia in a general circulation model, *Math. Comput. Simul., 30,* 99–104, 1988.

Simmonds, I., M. Dix, P. Rayner, and G. Trigg, Local and remote response to zonally uniform sea surface temperature in a July general circulation model, *Int. J. Climatol., 9,* 111–131, 1989.

Simmons, A. J., J. M. Wallace, and G. W. Branstator, Barotropic wave propagation and instability, and atmospheric teleconnection patterns, *J. Atmos. Sci., 40,* 1363–1392, 1983.

Slingo, J. M., and H. Annamalai, 1997: The El Niño of the century and the response of the Indian summer monsoon, *Mon. Weather*

Rev., 128, 1778–1797, 2000.
Slingo, J. M., et al., Intraseasonal oscillations in 15 atmospheric general circulation models: Results from an AMIP diagnostic subproject, *Clim. Dyn., 12,* 325–357, 1996.
Soman, M. K., and J. M. Slingo, Sensitivity of the Asian summer monsoon to aspects of sea–surface–temperature anomalies in the tropical Pacific, *Q. J. R. Meteorol. Soc., 123,* 309–336, 1997.
Sperber, K. R., and T. N. Palmer, Interannual tropical rainfall variability in general circulation model simulations associated with the Atmospheric Model Intercomparison Project, *J. Clim., 9,* 2727–2750, 1996.
Sperber, K. R., J. M. Slingo, and H. Annamalai, Predictability and relationship between subseasonal and interannual variability during the Asian summer monsoon, *Q. J. R. Meteorol. Soc.,126,* 2545–2574, 2000.
Sprintall, J., and M. Tomczak, Evidence of the barrier layer in the surface layer of the tropics, *J. Geophys. Res., 97,* 7305–7316, 1992.
Sprintall, J., A. Gordon, R. Murtugudde, and D. Susanto, A semi-annual Indian Ocean forced Kelvin wave observed in the Indonesian Seas in May 1997, *J. Geophys. Res, 105,* 17,217–17,230, 2000.
Terray, P., An evaluation of climatological data in the Indian Ocean area, *J. Meteorol. Soc. Jpn. 2,* 359–385, 1994.
Terray, P., Space–time structure of monsoon interannual variability, *J. Clim., 8,* 2595–2619, 1995.
Tourre, Y. M., and W. B. White, ENSO signals in global upper ocean temperature, *J. Phys. Oceanogr., 25,* 1317–1332, 1995.
Trenberth, K. E., and T. J. Hoar, The 1990–1995 El Niño Southern Oscillation event: Longest on record, *Geophys. Res. Lett., 23,* 57–60, 1996.
Tsai, P. T. H., J. J. O'Brien, and M. E. Luther, The 26-day oscillation observed in the satellite sea surface temperature measurements in the equatorial western Indian Ocean, *J. Geophys. Res., 97,* 9605–9618, 1992.
Ueda, H., and J. Matsumoto, A possible triggering process of east–west asymmetric anomalies over the Indian Ocean in relation to 1997/98 El Niño, *J. Meteorol. Soc. Jpn., 7,* 8803–8818, 2001.
Vecchi, G. A., and D. E. Harrison, Monsoon breaks and subseasonal sea surface temperature variability in the Bay of Bengal, *J. Clim., 15,* 1485–1493, 2002.
Venzke, S., M. Latif, and A. Villwock, The coupled GCM ECHO–2. Part II: Indian Ocean response to ENSO, *J. Clim., 13,* 1371–1383, 2000.
Vinayachandran, P. N., N. H. Saji, and T. Yamagata, Response of the equatorial Indian Ocean to an unusual wind event during 1994, *Geophys. Res. Lett., 26,* 1613–1616, 1999.
Vinayachandran, P. N., V. S. N. Murty, and V. Ramesh Babu, Observations of barrier layer formation in the Bay of Bengal during summer monsoon, *J. Geophys. Res., 107,* 8018, doi: 10.1029/2001JC000831, 2002.
Vinayachandran, P. N., S. Iizuka, and T. Yamagata, Indian Ocean dipole mode events in an ocean general circulation model, *Deep Sea Res., Part II,* 49, 1573–1596, 2003.
Voice, M. E., and B. G. Hunt, A study of the dynamics of drought initiation using a global general circulation model, *J. Geophys. Res., 89,* 9504–9520, 1984.

Wajsowicz, R. C., and E. K. Schneider, The Indonesian throughflow's effect on global climate determined from the COLA coupled climate system, *J. Clim., 14,* 3029–3042, 2001.
Waliser, D. E., K.-M. Lau, and K. M. Kim, The influence of coupled sea surface temperature on the Madden–Julian Oscillation: A model perturbation experiment, *J. Atmos. Sci., 56,* 333–358, 1999.
Waliser, D. E., R. Murtugudde, and L. E. Lucas, Indo–Pacific Ocean response to atmospheric intraseasonal variability: 1. Austral summer and Madden–Julian Oscillation, *J. Geophys. Res.,* 108, 3160, doi: 10.1029/2002JC001620, 2003.
Waliser, D. E., R. Murtugudde, and L. E. Lucas, Indo–Pacific Ocean response to atmospheric intraseasonal variability: 2. Boreal summer and the intraseasonal oscillation, *J. Geophys. Res.,* in press, 2004.
Walker, G. T., and E. W. Bliss, World weather V, *Mem. R. Meteorol. Soc., 4,* 53–84, 1932.
Wang, B., R. Wu, and T. Li., Atmosphere–warm ocean interaction and its impact on Asian–Australian monsoon variation, *J. Clim., 16,* 1195–1211, 2003.
Washington, W. M., R. M. Chervin, and G. V. Rao., Effects of a variety of Indian Ocean surface temperature anomaly patterns on the summer monsoon circulation: Experiments with the NCAR general circulation model, *Pure Appl. Geophys.,* 1335–1356, 1977.
Watanabe, M., and F.-F. Jin, Role of the Indian Ocean warming in the development of Philippine Sea anticyclone during ENSO, *Geophys. Res. Lett.,* doi: 10.1029/2001GL014762, 2002.
Watanabe, M., and F.-F. Jin, A moist linear baroclinic model: Coupled dynamical–convective response to El Niño, *J. Clim., 16,* 1121–1139, 2003.
Weare, B. C., A statistical study of the relationships between ocean surface temperatures and the Indian monsoon, *J. Atmos. Sci.,* 36, 2279–2291, 1979.
Webster, P. J., and S. Yang, Monsoon and ENSO: Selectively interactive systems, *Q. J. R. Meteorol. Soc., 118,* 877–926, 1992.
Webster, P. J., V. Magaña, T. Palmer, J. Shukla, R. Tomas, M. Yanai, and T. Yasunari, Monsoon: processes, predictability, and the prospects for prediction, *J. Geophys. Res., 103,* 14,451–14,510, 1998.
Webster, P. J., A. M. Moore, J. P. Loschnigg, and R. R. Leben, Coupled ocean–atmosphere dynamics in the Indian Ocean during 1997–98, *Nature, 401,* 356–360, 1999.
Webster, P. J., et al., The JASMINE pilot study, *Bull. Am. Meteorol. Soc., 83,* 1603–1630, 2002.
Wolter, K., and S. Hastenrath, Annual cycle and long-term trends of circulation and climate variability over the tropical oceans, *J. Clim., 2,* 1329–1351, 1989.
Woolnough, S., J. Slingo, and B. J. Hoskins, The relationship between convection and sea surface temperature on intraseasonal timescales, *J. Clim., 13,* 2086–2104, 2000.
Wu, R., and B. Kirtman, Understanding the impact of the Indian Ocean on ENSO variability in a coupled GCM, *COLA Technical Report 151,* 33 pp., (available at http://www.iges.org), 2003.
Wu, R., and B. Kirtman, Impacts of Indian Ocean on the Indian summer monsoon–ENSO relationship, *J. Clim.,* 2004 (in press).
Wyrtki, K., An equatorial jet in the Indian Ocean, *Science, 181,* 262–264, 1973.

Xie, S.-P., H. Annamalai, F. A. Schott, and J. P. McCreary, Structure and mechanisms of south Indian Ocean climate variability, *J. Clim.*, *15*, 867–878, 2002.

Yamagata, T., K. Mizuno, and Y. Masumoto, Seasonnal variations in the equatorial Indian Ocean and their impact on the Lombok throughflow, *J. Geophys. Res.*, 12,465–12,473, 1996.

Yamagata, T., N. H. Saji, and S. K. Behera, Comments on Indian Ocean dipole, *Bull. Am. Meteorol. Soc.*, *84*, 1440–1442, 2003.

Yasunari, T., Cloudiness fluctuations associated with the northern hemisphere summer monsoon, *J. Meteorol. Soc. Jpn.*, *57*, 227–242, 1979.

Yu, J.-Y., C. R. Mechoso, J. C. McWilliams, and A. Arakawa, Impacts of the Indian Ocean on the ENSO cycle, *Geophys. Res. Lett.*, *29*, 46.1–46.4, 2002.

Yu, J.-Y., S. P. Weng, and J. D. Farrara, Ocean roles in the TBO transitions of the Indian–Australian monsoon system, *J. Clim.*, *16*, 3072–3080, 2003.

Yu, L. S., and M. M. Rienecker, Mechanisms for the Indian Ocean warming during the 1997–98 El Niño, *Geophys. Res. Lett.*, *26*, 735–738, 1999.

Yu, L. S., and M. M. Rienecker, Indian Ocean warming of 1997–1998, *J. Geophys. Res.*, *105*, 16,923–16,393, 2000.

Zhang, C., Large-scale variability of deep convection in relation to sea surface temperature in the tropics, *J. Clim.*, *6*, 1898–1913, 1993.

Zubair, L., S. A. Rao, and T. Yamagata, Modulation of Sri Lanka Maha rainfall by the Indian Ocean dipole, *Geophys. Res. Lett.*, *30*, 1063, doi: 10.1029/2002GL015639, 2003.

H. Annamalai, IPRC/SOEST, University of Hawaii, 1680 East West Road, POST Bldg 401, Honolulu, Hawaii 96822. (hanna@hawaii.edu)

R. Murtugudde, ESSIC/UMD, CSS Bldg, Room 2207, College Park, Maryland 20742. (ragu@essic.umd.edu)

Interannual Indian Rainfall Variability and Indian Ocean Sea Surface Temperature Anomalies

Gabriel A. Vecchi[1]

JISAO, University of Washington, Seattle, Washington

D. E. Harrison

NOAA/Pacific Marine Environmental Laboratory, Seattle, Washington

It is shown that interannual variations in Indian continental rainfall during the southwest monsoon can be usefully represented by two regional rainfall indices. Indian rainfall is concentrated in two regions, each with strong mean and variance in precipitation: the Western Ghats (WG) and the Ganges-Mahanadi Basin (GB) region. Interannual variability of rainfall averaged over each of the two regions (WG and GB) is uncorrelated; however, the rainfall over these two regions together explains 90% of the interannual variance of All-India rainfall (AIR). The lack of correlation between WG and GB rainfall suggests that different mechanisms may account for their variability. During the period 1982-2001, rainfall variability over each of these two regions exhibits distinct relationships to Indian Ocean SST: warm SSTA over the western Arabian Sea at the monsoon onset is associated with increased WG rainfall (r = 0.77), while cool SSTA off of Java and Sumatra is associated with increased GB rainfall (r = -0.55). The connection between SSTA and AIR is considerably weaker, and represents the superposition of that associated with each region. We find the relationship with WG rainfall is robust, while that with GB results from a single exceptional year. Each region also exhibits distinct relationships to El Niño SSTA indices.

1. INTRODUCTION

The southwest monsoon provides most of the annual rainfall in India. This monsoon exhibits non-trivial variability of precipitation over India on interannual timescales [*Shukla,* 1976; *Webster et al.,* 1998]. This variability has a profound effect on Indian economic and agricultural output, and impacts strongly the over 1 billion people in the region [*Mooley et al.,* 1981; *Parthasarathy et al.,* 1988a; *Webster et al.,* 1998]. Predicting the variability of annual rainfall over India has been a goal for the climate community for decades.

Much work on Indian continental precipitation has focused on large-scale indicators of interannual rainfall variability, often the total over India [*Shukla,* 1976; *Parthasarathy et al.,* 1988b, 1992]. A commonly used index is the area-weighted integral of the rainfall measured by a national network of rain gauges, usually referred to as "All-India Rainfall" (AIR), records of which exist into the nineteenth century. Integrated indices of Indian continental precipitation have been found to connect with a variety of Indian economic and agricultural indicators [*Mooley et al.,* 1981; *Parthasarathy et al.,* 1988a; *Webster et al.,* 1998].

---

[1]Now at NOAA/Geophysical Fluid Dynamics Laboratory, Princeton, New Jersey.

Earth's Climate: The Ocean-Atmosphere Interaction
Geophysical Monograph Series 147
Copyright 2004 by the American Geophysical Union
10.1029/147GM14

Using subdivisional Indian rainfall data [*Parthasarathy et al.*, 1995], it has been established that there is considerable spatial structure to rainfall variability over India. Interannual and decadal variability, as well as long-term trends, in Indian rainfall have been found to exhibit significant spatial variability [*Parthasarathy*, 1984; *Rassmusson and Carpenter*, 1983; *Shukla* 1987; *Subbaramayya and Naidu*, 1992; *Gadgil et al.*, 1993]. Also, the relationships between Indian continental rainfall variability and large-scale climate parameters have been found to be spatially variable [*Mooley and Parsarathy*, 1983, 1984; *Parthasarathy and Pant*, 1984; *Mooley at al.*, 1986; *Parthasarathy et al.*, 1990, 1991].

We here show that interannual variability of total Indian rainfall can be described effectively using indices of variability over two distinct regions. The time series of averages over the regions are uncorrelated to each other. Their sum explains a large fraction of All-India rainfall variability—90% of the variance over the past 23 years. The first region encompasses the Western Ghats coastal region and the other the Ganges and Mahanadi River Basins. That rainfall anomalies over the two regions are uncorrelated suggests that different processes may account for the interannual variability of precipitation in each region. We find significant and distinct simultaneous and leading sea surface temperature anomaly (SSTA) patterns associated with each regional index. However, we are unable to find similarly large-scale connections to All-India Rainfall.

In the following Section we describe the datasets used in this analysis. In Section 3 we present our parsing of Indian precipitation into the two regional Indices. In Section 4 we describe the correlations between the regional precipitation indices and Indian Ocean SSTA. Section 5 offers a summary of the results and discussion of their implications.

2. DATA AND METHODS

We use the Reynolds/NCEP Version 2 weekly 1°×1° optimally interpolated SST product [*Reynolds and Smith*, 1994; *Reynolds et al.*, 2002]. A monthly climatology of SST was generated by averaging, 1982 through 2001; anomalies were computed from this monthly climatology. Though EOF-based reconstructions of SST extending to the beginning of the century are available, we restrict our analysis of SSTs to the 20-year period 1982–2001, because from 1982 onwards there is a relative homogeneity of the data sources (satellites and in situ) and the analysis techniques for the gridded dataset.

We use the pentad CPC Merged Analysis of Precipitation (CMAP) on a global 2.5°x2.5° grid, which combines satellite estimates of precipitation with in situ rain gauge measurements [*Xie and Arkin*, 1996, 1997; available online from NOAA/CIRES/CDC at http://cdc.noaa.gov/]. Two monthly climatologies of precipitation were generated by averaging, based on the years 1979 through 2001, and 1982 through 2001; anomalies were computed from these monthly climatologies. Each climatology is applied depending on the time-period being considered; for example, in analyses spanning 1979–2001, the 1979 through 2001 climatology was used.

We remove the trend and long-term mean from SST and precipitation anomaly data prior to our analysis. The trends are significant, over the period 1982–2001 SSTA warmed by an average of 0.2–0.6°C over most of the Indian Ocean (except for the southwest Indian Ocean, which cooled by 0.2–0.4°C); these trends represent between 50% and 250% of the local standard deviation of monthly mean SSTA. The trends in Indian precipitation are more modest, with a general tendency for drying of 0.2–0.8 mm day^{-1} for the 20 year period (1982-2001); these trends represent 20%-60% of the standard deviation of the local monthly mean precipitation anomaly. The inclusion or removal of the trends impacts our principal results only in detail, and not in substance.

To confirm the robustness of our analysis of the spatial structure of Indian rainfall we use rain gauge-based datasets with a considerably longer record than that of CMAP. We use a global gridded rain gauge-based, land precipitation dataset [*Hulme*, 1992, 1994; *Hulme et al.*, 1998; data are available online from Dr. Mike Hulme at http://www.cru.uea.ac.uk/~mikeh/datasets/global), on a monthly, 5° x 5° grid, from 1900 through 1998. We also make use of the seasonal 1871-2000 rain-gauge based sub-divisional Indian rainfall dataset produced at the Indian Institute of Tropical Meteorology by Dr. B. Parthasarathy [*Parthasarathy et al.*, 1995; data are available online at http://grads.iges.org/india/partha.subdiv.html] we refer to this data as Parth95.

3. STRUCTURE OF INDIAN RAINFALL

We performed a grid-box by grid-box analysis of the CMAP precipitation anomalies of Indian precipitation, and found that it is dominated by two regions of high mean and interannual variance in southwest monsoon (June-September) precipitation. We develop two regional-average indices, and find that they are uncorrelated to each other. Their sum is correlated with All-India Rainfall (AIR) at 0.95 over the period 1979-2001, and at 0.88 over the period 1900-1998.

Plate 1 shows maps of the mean (Plate 1a) and the standard deviation (Plate 1b,c) of the monthly mean anomalies of SW monsoon land precipitation anomalies from the CMAP dataset. In each panel of Plate 1 we indicate the two regions which will be used to generate precipitation indices. Plate 1c shows the actual grid cells that are used to construct the land precipitation indices. There are two primary regions of strong

VECCHI AND HARRISON 249

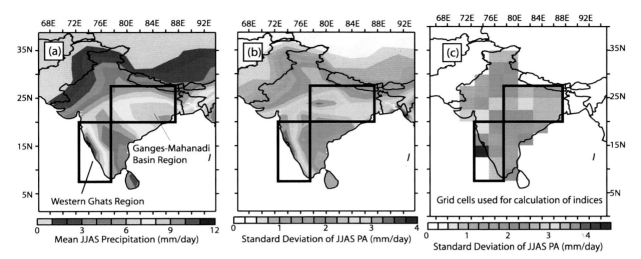

Plate 1. Filled maps of the (a) mean CMAP precipitation over June-September, (b) standard deviation of the June-September monthly-mean CMAP precipitation anomaly. Shaded map (c) of the standard deviation of monthly-mean June-September CMAP precipitation anomaly, indicating the actual grid-cells used in the computation of the indices. The figures are based on the 1979-2001 CMAP precipitation data. Units are mm/day.

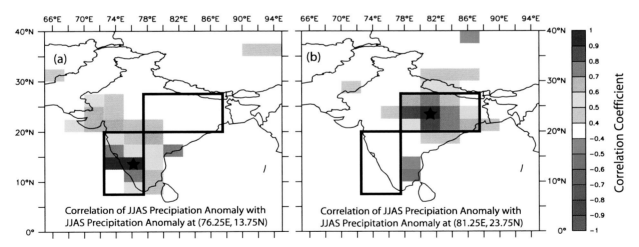

Plate 2. Shaded maps of the correlation coefficient of the mean June-September CMAP precipitation anomaly in each land grid-cell with mean June-September precipitation anomaly at (a) (76.25°E, 13.75°N) and (b) (81.25°E, 23.75°N). Correlations are based on the 1979-2001 CMAP data, only values greater than 0.4 are shaded (representing statistical significance at the 92.5% level). In each panel the areas that are used to define the regional indices are indicated by black boxes. Notice that the precipitation anomalies are correlated only locally within each region.

precipitation and strong precipitation variability: the Western Ghats, and the Ganges and Mahanadi River Basins. The Western Ghats have the strongest mean and variability in precipitation, concentrated along the coast west of the Western Ghat Mountains. The Ganges-Mahanadi Basin has strong mean and variability extending inland from the Bay of Bengal.

To explore relationships between the two regions we compute the correlation between mean SW monsoon anomaly at each CMAP grid box in South Asia and the precipitation anomaly at points in each region, examples are shown in Plate 2a-b. The mean SW monsoon precipitation anomaly for points within each region is significantly correlated (at the 95% level) with that of the other points within the region, but not between the regions. Because the precipitation variability is locally coherent in each region, it is sensible to define regional precipitation indices. The Western Ghats Index (WGI) is defined as the spatial average of continental precipitation anomaly in the region (72.5°E-77.5°E, 7.5°N-20°N), and the Ganges-Mahanadi Basin Index (GBI) is defined as the spatially averaged continental precipitation anomaly over the region (77.5°E-87.5°E, 20°N-27.5°N). We define our All-India Rainfall Index as the precipitation anomaly averaged over all the grid cells shown in Figure 1c.

Figure 1a shows time series of the mean WGI and GBI during the SW monsoon. The two indices are not correlated; for the 23-year period, 1979-2001, the correlation coefficient between the two regional indices is 0.2, which is not statistically significant even at the 85% level. Further, if the year 1980 is removed from the analysis, the correlation between the two time series is 0.08; thus the small nominally-positive correlation between the two time series arises principally from one extreme year. Each regional index is significantly and positively

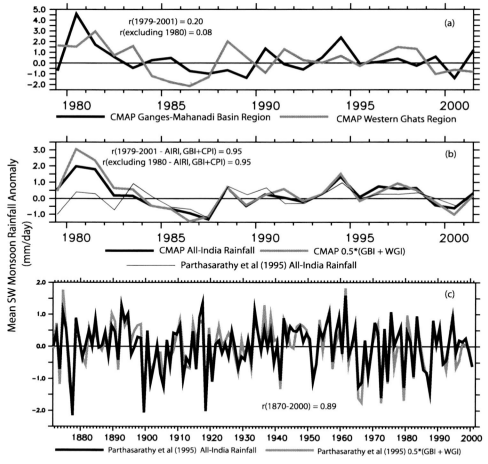

Figure 1. (a) Time series of the mean CMAP June-September precipitation anomaly over the two regions: Ganges-Mahanadi Basin (black line) and Western Ghats (gray line). (b) Time series of the mean CMAP June-September precipitation anomaly over All-India (black line), the average of GBI and WGI (gray line), and the *Parthasarathy et al.* [1995] AIRI (thin line). (c) Time series of the mean *Parthasarathy et al.* [1995] AIRI (black line) and the average of GBI and WGI (gray line). Units are mm/day. Correlation coefficients between the time series are indicated in the panels.

correlated with AIRI; the correlation between WGI and AIRI is 0.76 and between GBI and AIRI it is 0.72.

Figure 1b shows time series of the AIRI and of the average of the two regional indices. The sum of these two independent regional indices explains 90% of the interannual variance of Indian subcontinent precipitation in the CMAP precipitation dataset. This relationship is not altogether surprising, since GBI and WGI represent the two regions of largest mean and variability in Indian rainfall, and they are uncorrelated. Moderate changes in the definition of the regions affects only the details of the results.

Extrema in All-India rainfall result sometimes from one index or the other, and sometimes by a combination of the two. For example, the years 1984-86 showed negative precipitation over All-India; however, all of India wasn't dry: while the Western Ghats were extremely dry, the Ganges-Mahanadi Basin was close to normal. In 1988, All-India indicates a wetter than normal year, which is made up principally of enhanced precipitation over the Western Ghats; the Ganges-Mahanadi Basin was drier than normal. In 1994 the All-India signal is dominated by the extreme precipitation anomaly in the Ganges-Mahanadi Basin. The converse occurs in 1997-98, which was wetter in the Western Ghats but near-normal in the Ganges-Mahanadi Basin. And sometimes the precipitation anomalies in both regions are in phase, such as 1980-1981 (wet), 1986-1987 (dry) and 2000 (dry).

The CMAP dataset is based on both satellite and rain gauge estimates of precipitation, and thus spans only the period 1979-present. To confirm that the results based on the recent decades were evident in the past century, we used the Hulme gridded precipitation (HGP) dataset over the period 1900-1998. Again, there are two variance and mean precipitation maxima in the Indian subcontinent, corresponding well with those found in CMAP. Further, the correlation coefficient between the two regional indices (defined as above, but computed using the HGP data) is 0.15, which is not statistically significant at the 90% level, and the correlation of the average of the two indices with AIRI is 0.88. Thus, over the entire 20th century the two regional indices are uncorrelated, and explain a large part of the variance of AIRI.

The Indian sub-divisional rainfall data of (Parth95) is one of the most commonly used in studies of Indian monsoon variability, thus it is of interest to see how it compares with the CMAP rainfall data. The AIRI index computed from the Parth95 data is shown in Figure 1b over the period 1979-2000, for comparison with the CMAP estimate. The disagreement over the period 1979-1981 between the two estimates of AIRI is quite striking, but the two time series follow each other relatively well over the post-1981 period—with correlation coefficient of 0.76. The differences between the two estimates of continental rainfall may arise from any of a variety of (non-exclusive) factors, including the relatively coarse grid of the CMAP data, the distribution of the rain gauges in the network, or the CMAP algorithm.

Though the temporal evolution of the indices differs in the Parth95 and the CMAP data, the spatial parsing seen in the CMAP data is evident in the Parth95 data. There are two spatial maxima in the Parth95 JJAS rainfall and rainfall variability, one in the Western Ghats and the other in the Ganges-Mahanadi Basin. We generate indices for the Western Ghats and Ganges-Mahanadi Basin regions using the Parth95 data. The Parth95 rainfall WGI is computed as the area-weighted sum of rainfall in: Konkan; Madhya Mharashtra; Marathawada; Coastal, Northern Interior and Southern Interior Karnataka; and Kerala sub-divisions. The Parth95 rainfall GBI is computed as the area-weighted sum of rainfall in: Orissa, Bihar Plateau, Bihar Plains, East Uttar Pradesh and East Madhya Pradesh sub-divisions. The regional indices computed from CMAP and from the Parth95 data are relatively well correlated over the entire record (0.70 for GBI and 0.80 for WGI), with the period 1979-1981 standing out. When only the post-1981 period is considered the correlations become stronger: 0.77 for GBI and 0.83 for WGI. The strong disagreement between CMAP and Parth95 rainfall during the period 1979-1981 bears further examination.

The relationship between the two regional indices and All-India rainfall is the same in the Parth95 data as in CMAP. The correlation between the two regional indices computed using the Parth95 data over the period 1871-2000 is not significant ($r = 0.04$); while their sum is explains most of the variability in All-India rainfall ($r = 0.89$) (see Figure 1c). Thus, the relationship found in recent decades using the merged satellite/rain gauge dataset is a characteristic of the interannual rainfall variability over India.

4. SSTA/RAINFALL CORRELATIONS

In this section we describe the correlations between Indian Ocean SSTA and WGI, GBI and All-India Rainfall, using the rainfall indices based on the CMAP precipitation dataset. The analysis of SSTA-Rainfall relationships is done over the period 1982-2001. Since the greatest discrepancies between the CMAP and Parth95 rainfall data are in the period 1979-1981, the character of the SSTA correlation features is largely the same when computed using rainfall indices computed with the 1982-2000 Parth95 rainfall data than with CMAP, and differs only in magnitude.

Plate 3 shows the correlations of GBI and WGI, respectively, precipitation anomaly during JJAS and Indian Ocean SSTA in different two-month averages, starting from the June-July (JJ) at lag zero (Plate 3a, 3e) and moving back in time to the September-October (SO) preceding the southwest monsoon (Plate

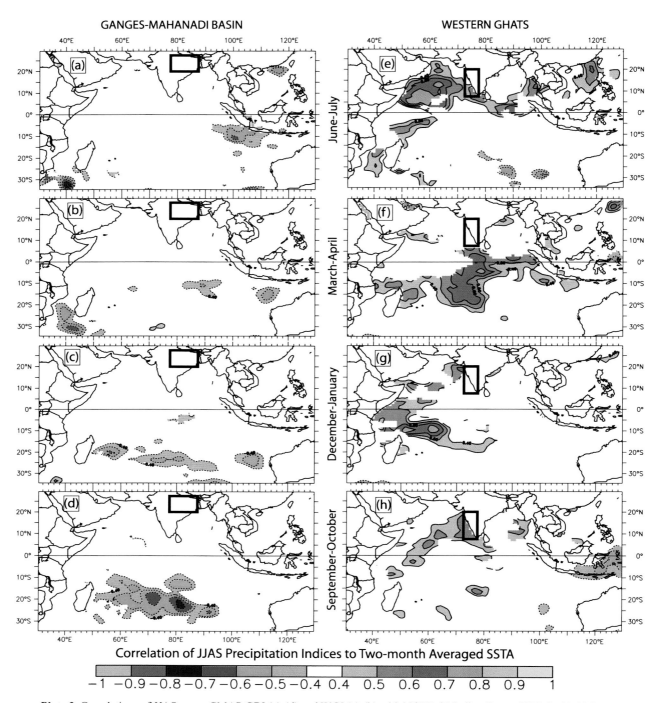

Plate 3. Correlations of JJAS-mean CMAP GBI (a)-(d) and WGI (e)-(h) with NCEP-OI Indian Ocean SSTA in (a),(e) June-July, (b),(f) March-April, (c),(g) December-January, (d),(h) September October. Only those correlations whose amplitude exceeds 0.4 are shown (representing statistical significance at the 92.5% level), and only over those regions where the local SSTA standard deviation (SSTA first averaged over the months to be correlated) exceeds 0.3°C. Correlations are computed over the period 1982-2001.

3d, 3h). Only correlations greater than 0.4 (statistically significant at roughly the 92.5% level, assuming a Normal distribution and that each year is an independent sample); only correlations in areas where the standard deviation of SSTA exceeds 0.3°C are shown (assuming a Normal distribution, if the standard deviation is less than 0.3°C, one expects less than 10% of the samples to have an amplitude exceeding 0.5°C).

Plate 3a shows that moderate simultaneous negative correlations between GBI and SSTA off the coasts of Sumatra and Java exist (peak values <0.6). These correlations are evident from the beginning of the SW monsoon, but disappear following that. In the inter-monsoon period there are scattered negative correlations between GBI and SSTA in the southeastern Indian Ocean (Plate 3b). There are more extensive negative correlations with southern Indian Ocean SSTA in the preceding NE monsoon (Plate 3c). The negative correlations between SW monsoon GBI and southern Indian Ocean SSTA in the preceding SW to NE inter-monsoon period are stronger and more extensive than any of the instantaneous correlations.

We note that in the correlations between JJ SSTA and JJAS CMAP-based GBI there is an area of negative correlation in the northern Arabian Sea which is just below the threshold for significance, with values between 0.3-0.4. When the SSTA correlations are computed using the GBI based on the Parth95 Indian rainfall this area of negative correlation exceeds the significance threshold of 0.4. As the SST record lengthens this relationship should be revisited.

Plate 3e shows a strong simultaneous positive correlation between WGI and SSTA in the western Arabian Sea; this correlation exceeds 0.7 over a large area. There is also positive correlation between SSTA in the southwestern Indian Ocean, south of the Bay of Bengal and in the South China Sea, though of more moderate amplitudes. Relatively strong positive correlations are evident at the beginning of the southwest monsoon in the western Arabian Sea (MJ), but not prior to that. However, the southern hemisphere correlations continue through the inter-monsoon period (Plate 3f) and strengthen during the NE monsoon preceding the enhanced precipitation. In fact, in DJ they exceed 0.6 in the southwest Indian Ocean, and locally exceed 0.7 (Plate 3g). Prior to DJ the southern hemisphere correlations are smaller and not significant (Plate 3h). In SO prior to the SW monsoon there are negative correlations between SSTA in the western Pacific Ocean, and positive SSTA correlations over the eastern Arabian Sea.

The percentage of grid-points which are statistically significant in the maps of correlation between GBI and SSTA (Plate 3a-d) are, in order: 7.8%, 5.9%, 10%, and 13%. These percentages are comparable to the null expected values of 7.5% from a 92.5% threshold, thus it appears that the GBI correlation patterns may be spurious. However, the percentage of grid-points which pass the threshold of statistical significance in the maps of correlation between GBI and SSTA shown in Plate 3e-h is considerably larger than the 7.5% expected; the percentages are 23% (3e), 25% (3f), 17% (3g) and 16% (3h). This suggests that the SSTA patterns associated with GBI may not be spurious.

To determine whether the statistically significant patterns of correlation we find are representative of the entire 20-year record or whether they result from extrema in the record we examine more carefully the three principal correlation patterns for each precipitation index. For the GBI we explore the June-July Sumatra/Java, and the December-January and September-October southern Indian Ocean signals. For the WGI we explore the June-July Western Arabian Sea, the March-April central south Indian Ocean and the December-January western south Indian Ocean signals.

Figures 2 and 3 show scatter plots of each SSTA index with the corresponding JJAS precipitation Index, each panel shows the correlation coefficient. Only for the June-July Sumatra/Java SSTA index correlation with GBI (Figure 2a) does the removal of an outlier fundamentally change the correlation coefficient; in the absence of the extreme value in 1994 the correlation coefficient drops from –0.55 to –0.21; thus, the relationship between June-July Sumatra/Java SSTA and JJAS GBI is not robust. However, for all other SSTA indices, the relationship appears to be robust. The strongest correlation coefficient is that between western Arabian SSTA and WGI at 0.77 (Figure 3a).

We explore the correlations of Indian Ocean SSTA to All-India precipitation in Plate 4; we shade those values whose amplitude exceeds 0.4 and whose local standard deviation of SSTA exceeds 0.3°C, and we contour without shading those correlations which are not significant but whose local standard deviation of SSTA exceeds 0.3°C. Notice first that for none of the month-pairs are there large-scale areas of significant correlation akin to those evident in the regional index correlations (Plate 3). Also, in all the months, the correlation patterns are a superposition of the WGI and GBI correlation patters, with reduced amplitude; this is to be expected since GBI and WGI are uncorrelated and their sum represents a large part of the variance of AIRI.

Starting in DJ through the beginning of the SW monsoon there is an indication of an east-west SST dipole associated with AIRI, whose poles have limited—if any—statistical significance. However, the dipole pattern arises from two independent correlation patterns and does not represent a single dipole mode. The positive correlations in the western part of the basin arise from the WGI contribution to AIRI, while the negative correlations in the eastern part of the basin arise from the GBI contribution to AIRI.

We have computed correlations between AIRI, WGI and GBI and two commonly used El Niño SSTA indices (NIÑO3:

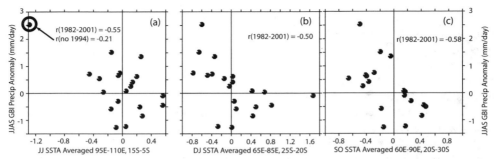

Figure 2. Scatter plots of Ganges-Mahanadi Basin CMAP Precipitation anomaly vs. NCEP-OI SSTA averaged over (a) June-July, 95°E-110°E, 15°S-5°S; (b) December-January, 65°E-85°E, 25°S-20°S; (c) September-October, 60°E-90°E, 30°S-20°S. These are the months/regions with which GBI exhibits large-scale SSTA correlations. Indicated in each panel is the correlation coefficient between the two quantities. Notice that the correlation with June-July 1994 arises from a single extreme point.

SSTA averaged 150°W-90°W, 5°S-5°N; NIÑO3.4: SSTA averaged 170°E-125°W, 5°S-5°N). With our chosen data sets and over our analysis period, we find no significant correlation between the CMAP-based All-India Rainfall Index and NIÑO3 or NIÑO3.4 SSTA, at any lag between –25 and 25 months. There is a modest correlation (r = 0.42) between GBI and NIÑO3.4, with GBI leading NIÑO3.4 by ~4 months, but this correlation is due principally to 1994 and is thus not robust. However, we are able to find significant lead correlations between El Niño SSTA indices and WGI (r = 0.43 with NIÑO3, r up to 0.6 with NIÑO3.4), with equatorial Pacific SSTs leading WGI by 15-20 months.

5. DISCUSSION

On interannual time scale there are two distinct regions of Indian continental summer rainfall, which we have labeled "Western Ghats" and "Ganges-Mahanadi Basin". Over the two decades of satellite data used here, and over 130 years using land station data, the two regional-average time series of rainfall, WGI and GBI, are uncorrelated at 90%. Further the sum of the two time series accounts between 80-90% of the interannual variability of rainfall averaged over all of India, using "All-India" rainfall indices. To a very useful degree then, these two regions may be used to study interannual rainfall variability of total rainfall over India.

A commonly used method for decomposing the variability of a geophysical field is Empirical Orthogonal Function (EOF) analysis. We performed an EOF analysis of the 20-year record of rainfall, which produces substantially the same result as other EOF analysis of regional Indian precipitation anomalies [*Rassmusson and Carpenter*, 1983; *Shukla*, 1987]. The EOFs indicate that there are two principal modes of variability: a continent-wide "breathing" mode (~30% of the variance) and a dipole-like mode (where the Ganges-Mahanadi Basin varies inversely with the Western Ghats; ~11% of the variance). On the surface, the EOF description of the modes of precipitation variability over India appears to differ with our two-region decomposition. However, these EOFs are the expected statistical descriptions for a system whose principal dynamical modes are two independently varying centers of strong variance [*Domenget and Latif*, 2002]. Thus, despite

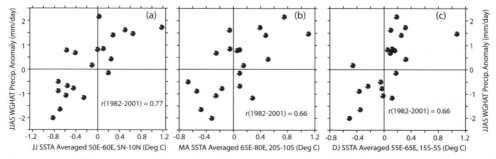

Figure 3. Scatter plots of Western Ghats CMAP Precipitation anomaly vs. NCEP-OI SSTA averaged over (a) June-July, 50°E-60°E, 5°N-10°N; (b) March-April, 65°E-80°E, 20°S-10°S; (c) September-October, 55°E-65°E, 15°S-5°S. These are the months/regions with which GBI exhibits large-scale SSTA correlations. Indicated in each panel is the correlation coefficient between the two quantities. Notice that the correlations are all relatively robust.

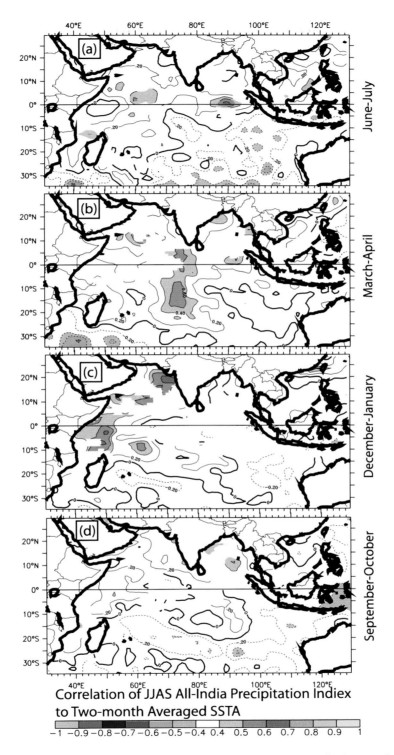

Plate 4. As in Plate 3, but for CMAP-based AIRI. Only those correlations whose amplitude exceeds 0.4 are shaded, and only over those regions where the local NCEP-OI SSTA standard deviation exceeds 0.3°C. Contours indicate values where the amplitude of the correlation is less that 0.4, contour interval in non-shaded regions is 0.2, in shaded regions it is 0.1. Correlations are computed over the period 1982-2001.

the apparent dissimilarity, the EOF description is consistent with our result that Indian precipitation is made up primarily of two independently varying centers of action.

Our results also at first appear to contradict previous work that has described widespread correlation of net summer rainfall over India. *Hastenrath* [1987, Figure 3] shows that the correlation coefficient between summer mean local precipitation anomalies and the All-India Rainfall total is positive throughout India. *Goswami et al.* [1999], using the same precipitation dataset we have used (but with fewer years), find that the precipitation anomaly at all points in an extended Indian monsoon region is positively correlated with the average over the entire region. *Goswami et al.* [1999] argue that the interannual variability of precipitation of the Indian monsoon is largely spatially coherent because "positive correlations over the entire... region establishes that the interannual variability of the seasonal summer mean precipitation variability is indeed coherent over the... region" [*Goswami et al.*, 1999, p 615].

However, it does not follow, from modest widespread local correlation with a large-scale average, that there is widespread coherence between individual locations. Our results provide a clear counterexample. While it is true that both the WGI and GBI time series are correlated with All-India rainfall, it is also true that these two time series are uncorrelated to each other. Thus, there is no inconsistency between our description of the system and the results of *Hastenrath* [1987] and *Goswami et al.* [1999].

Recognition of the independence of the summer averages of these two primary regions of Indian rainfall invites a new perspective, one in which it is not surprising that there are different useful statistical relationships between large scale environmental conditions and the individual regional time series. It appears likely that different mechanisms may be responsible for the interannual variability of precipitation in each of the two regions.

In hindsight, this result is not altogether unexpected: India is a very large country with orographic features that divide it, and the moisture pathways to each of the two regions are different. During the SW Monsoon, there is strong moisture transport onto India and the rest of southern Asia, some of which is picked up by evaporation from the surrounding waters (the Arabian Sea and Bay of Bengal) and some of which is transported from the southern Hemisphere [*Findlater*, 1969; *Saha*, 1970b; *Cadet and Greco*, 1987a,b]. The moisture flux across the west coast of India over the Arabian Sea is a principal source of the precipitation over the Western Ghat Mountains, often resulting from periods of heavy orgraphically forced rainfall [*Saha and Bavadekar*, 1977; *Rakhecha and Pisharoty*, 1996]. Meanwhile the rainfall over the Ganges-Mahanadi River Basins is associated with tropical depressions formed over the Bay of Bengal which propagate northwestward onto continental India, and whose effect is principally limited to the Ganges-Mahanadi River Basins [*Murakami*, 1976; *Rakhecha and Pisharoty*, 1996]. Thus, the rain that falls in the two regions, Ganges-Mahanadi Basin and Western Ghats, results from different synoptic circulation structures and is also liable to be directly influenced by two different adjacent seas, the Bay of Bengal and the Arabian Sea respectively.

There is a strong positive correlation, at the onset of the monsoon, between Arabian SST anomaly and WGI. The robustness of this correlation (Section 4) encourages its use in short term forecasting of the coming SW monsoon net rainfall over the Western Ghats. Others have noted a relationship between Indian continental rainfall and SST in the Arabian Sea. *Saha* [1970a, 1974] suggested that the SST in the Arabian Sea could be an important aspect of the monsoon circulation and its rainfall over India. Many data analysis studies have found that the SW monsoon Indian rainfall is positively correlated to the SSTs in the Arabian Sea, with the correlation being strongest with western India precipitation [*Shukla and Misra*, 1977; *Joseph and Pillai*, 1984; *Rao and Goswami*, 1988; *Ramesh Kumar and Sastry*, 1990]. Modeling studies [*Shukla*, 1975; *Druyan et al.*, 1983; *Krenshaw*, 1988] have found that Arabian Sea SST anomalies influence monsoon circulation and rainfall over India, and have identified mechanisms for the connection. Increased Arabian Sea SSTs are associated with enhanced evaporation and atmospheric circulation, which increases rainfall in the western part of India. In this perspective, the factors that control SST anomalies in the Arabian Sea also significantly affect the net summer rainfall over the Western Ghats (and a significant component of All-India Rainfall).

There is also a zero-lag negative correlation between SSTA in the Java-Sumatra region and GBI (Section 3). This zero lag result is not at all robust; it results primarily from a single year (1994) with very large SST and rainfall anomalies. There is an area of borderline significance in the northern Arabian Sea which exhibits negative correlation between JJ SSTA and GBI, this relationship crosses the threshold for significance when the Parth95 rainfall dataset is used for an index. We suggest that, as the record lengthens, this relationship be re-examined.

There are also statistically significant positive correlations with WGI (and negative correlations with GBI) and SSTA leading the onset of the southwest monsoon by up to nine months (Plate 3). The relationships are relatively robust, although less strongly so than the zero lag western Arabian Sea connection to WGI (Figures 2, 3). No simple ideas have occurred to us to rationalize these patterns, and it is puzzling that the lead relationships of six and nine months with GBI should be stronger than the simultaneous correlations. However, since these correlations are both significant and relatively robust they may be useful in SW monsoon rainfall forecasts. Data and modeling studies should be pursued to understand the mechanisms, if any, behind these observed relationships.

Somewhat to our surprise, we find no correlations were found between GBI and Bay of Bengal SST anomalies. Most of the synoptic systems that bring rainfall to that part of India originate over the Bay of Bengal, and there is a strong relationship between Bay of Bengal SST and subseasonal monsoon breaks [*Vecchi and Harrison*, 2002]. While it is possible that there is, in fact, no systematic interannual relationship between Bay of Bengal SST and GBI rainfall, it is also possible that this lack of correlation on interannual timescales arises from the fact that the Bay of Bengal is particularly poorly sampled for SST over much of the year [*Reynolds and Smith*, 1994; *Reynolds et al.*, 2002]. On subseasonal timescales, the widely used NCEP SST analysis seriously underestimates the SSTA variability in the Bay of Bengal during the southwest monsoon [*Premkumar et al.*, 2000; *Sengupta and Ravichandran*, 2001; *Vecchi and Harrison*, 2002]. Thus, it is possible that interannual SST variations in the Bay of Bengal are also poorly represented in the NCEP SST product. Aliasing of the large amplitude (1-2°C) subseasonal variability could alone easily impact the utility of this product for low frequency climate studies. We do not wish to single out the NCEP SST analysis for concern here; any modern blended SST analysis that depends on infrared satellite as well as in situ data will be limited in this region.

We also looked for relationships between Indian Ocean SST and the first two principal components (EOFs) of Indian rainfall, and between All-India Rainfall. In neither case were we able to find large scale SST anomaly patterns with significant correlation at the 92.5% level. Relative to AIRI, we find correlation patterns consistent with the union of our WGI and GBI results, but the correlations fail to reach our 92.5% significance threshold over a large area. Interestingly, the structure of the non-significant correlation between AIRI and Indian Ocean SST between December-January and June-July is reminiscent of the recently identified Indian Ocean Dipole or Zonal Mode [*Saji et al.*, 1999; *Webster et al.*, 1999]. However, this pattern arises because each of the two uncorrelated components of AIRI (GBI and WGI) is associated with a different pattern in SSTA (cold SST in the east, warm SST in the west, respectively). This suggests caution in interpreting relationships between the IODM/IOZM and AIRI.

Other relationships between AIRI and Indian Ocean SST, besides the Arabian Sea connections, have been found. We do not wish to do a comprehensive summary, but offer some examples of relationships described in the literature. *Nicholls* [1995] finds that AIRI is correlated positively to April SSTA in the Australia/Indonesia region. *Sadhuram et al.* [1997] find significant positive correlation between November-December SSTA in the eastern equatorial Indian Ocean and AIRI. *Goswami et al.* [1999], defining a "broad-scale circulation index" for the Indian monsoon, find significant positive correlations with southern Indian Ocean SSTA. In an extensive analysis of the relationship between Indian Ocean SSTA and AIRI, *Clark et al.* [2000] find significant correlation between AIRI and SSTA in a variety of locations in the Indian Ocean; these include positive correlations between the eastern Arabian Sea and central Indian Ocean in September-November (for the years 1977-1995), the northern Arabian Sea and Indian Ocean northwest of Australia in December-February (for the years 1945-1995).

From this brief summary, and the results presented in this paper, it is evident that a wide range of somewhat conflicting results have been described between Indian Ocean SST anomalies and Indian continental rainfall anomalies. Every study finds one or more correlations (although few discuss the robustness of their correlations) but the relationships often involve different regions at different leads. We believe that this uncomfortable situation may partly arise because of uncertainties in our knowledge of Indian Ocean SST.

A specific example usefully illustrates the present situation. *Clark et al.* [2000] have recently described an SST anomaly pattern that leads All-India rainfall by 9 months. They worked with roughly the same time period that we have used here, but used an early version of the Hadley Centre SST analysis instead of the NCEP SST analysis that we used. We are unable to reproduce their result using NCEP SST, even when we use a rain gauge-based All-India Rainfall index. The region in which *Clark et al.* [2000] found correlation has very small SST anomaly variance. Further, there is strong SST variability on subseasonal timescales in that region [*Harrison and Vecchi*, 2001; *Vecchi and Harrison*, 2002]; the SST analysis 'signal to noise' is large there. It is possible that the large relative uncertainties associated with the SSTA in this region are responsible for the different reported SSTA/AIRI relationships.

Much of the Indian Ocean is not well observed during various periods of the year, either by infrared satellites or by ships and drifting buoys [*Reynolds and Smith*, 1994; *Reynolds et al.*, 2002]. *Harrison and Vecchi* [2001] and *Vecchi and Harrison* [2002] discuss some aspects of how different the Indian Ocean SST variability appears to be when it is observed via the TRMM Microwave Imager rather than the NCEP weekly SST analysis. In the presence of significant persistent cloud, storminess and high winds, there typically are little data available for the preparation of fields of SST. It is possible that our entire view of Indian Ocean SST will change as we accumulate multi-decadal microwave SST information. We note that the international climate community is working to improve the in situ Indian Ocean observing system [*Smith and Koblin-*

sky, 2001]. Sustained deployments of surface drifters and moorings in the region should substantially improve SST analysis products.

While we are on the topic of limitations in this study, we wish to acknowledge that use of only the most recent 20 years of data may affect our SSTA correlation results. Many studies have noted that statistical relations can be different in these modern decades than they were in earlier decades [*Hastenrath*, 1987; *Torrence and Webster*, 1999; *Clark et al.*, 2000; *Krishnamurthy and Goswami*, 2000]. But it is only over these recent decades that relatively homogeneous (in the sense that they were produced from similar sensors and consistent analyses) and complete SST and precipitation data sets are available. As the satellite data records increase in duration, all of these relationships should be revisited.

There are distinct connections between the El Niño-Southern Oscillation (ENSO) phenomenon on WGI and GBI. A connection has been found between deficient rainfall in India in the SW monsoon and El Niño conditions [*Rassmusson and Carpenter*, 1983; *Mooley and Parthasarathy*, 1983, 1984; *Yu and Slingo*, 1995]. This relationship is not robust, however, with the major 1997-98 El Niño standing as a notable exception; nor is it stationary, there have been interdecadal changes in the relationship [*Torrence and Webster* 1999, *Krishnamurthy and Goswami* 2000]. We find that each regional index has its own relationship to El Niño indices, with the strongest connection being between WGI and NIÑO3.4 ($r \sim 0.6$ at 15-20 months). Thus, equatorial Pacific temperatures may prove useful in forecasts of WGI.

That Indian rainfall can be parsed into two uncorrelated regions of high variability suggests that there are distinct mechanisms controlling the variability on interannual timescales in each region, whose superposition controls most of the rainfall over all of India. The potential utility of this new parsing of the system is evident in our SSTA correlation analysis over the Indian Ocean, in which the correlations between each region were distinct and more significant than any found with AIRI. The utility of these correlations for forecasts of monsoon precipitation over India should be investigated. We suggest that the environmental associations and mechanisms controlling the interannual variability of precipitation in each region be examined separately, and combined to gain understanding of the variability over India as a whole.

Acknowledgments. This publication is funded by the Joint Institute for the Study of the Atmosphere and Ocean (JISAO), under NOAA Cooperative Agreement No. NA17RJ11232, Contribution #943, by NOAA (OAR HQ and OGP) through UW/PMEL Hayes center, and by NASA's physical oceanography program. Analysis and graphics done using freeware package Ferret (http://ferret.wrc.noaa.gov/). This is PMEL contribution 2518. The global gridded rain gauge precipitation data (Version 1.0) is constructed and supplied by Dr Mike Hulme at the Climatic Research Unit, University of East Anglia, Norwich, UK; whose work has been supported by the UK Department of the Environment, Transport and the Regions (Contract EPG 1/1/85). Indian sub-divisional rainfall is made available by Dr. B. Parthasarathy, Indian Institute of Tropical Meteorology, Pune 411 008, India. Helpful comments, discussion and suggestions from N. Bond, J. Callahan, K. McHugh, and S. Ilcane.

REFERENCES

Cadet, D.L., and S. Greco, Water vapor transport over the Indian Ocean during the 1979 summer monsoon, Pt. 1, Water vapor fluxes, *Mon. Weather Rev.*, *115*(2), 653-663, 1987a.

Cadet, D.L., and S. Greco, Water vapor transport over the Indian Ocean during the 1979 summer monsoon, Pt. 2, Water vapor budgets, *Mon. Weather Rev.*, *115*(10), 2358-2366, 1987b.

Clark, C.O., J.E. Cole, and P.J. Webster, Indian Ocean SST and Indian Summer Rainfall: Predictive Relationships and Their Decadal Variability, *J. Clim.*, *13*, 2503-2519, 2000.

Dommenget, D., and M. Latif, A cautionary note on the interpretation of EOFs, *J. Clim.*, *15*, 216–225, 2002.

Druyan, L.M., J.R. Miller, and G.L. Russell, Atmospheric General Circulation Model Simulations With an Interactive Ocean: Effects of Sea Surface Temperature Anomalies in the Arabian Sea, *Atmos.-Ocean*, *21*(1), 94-106, 1983.

Findlater, J., Interhemispheric transport of air in the lower troposphere over the western Indian Ocean, *Q. J. R. Meteorol. Soc.*, *95*, 400-403, 1969.

Gadgil, S., Yadumani, and N.V. Joshi, Coherent Rainfall Zones of the Indian Region, *Int. J. Climatol.*, *13*, 547-566, 1993.

Goswami, B.N., V. Krishnamurthy, and H. Annamalai, A broad-scale circulation index for interannual variability of the Indian summer monsoon, *Q. J. R. Meteorol. Soc.*, *125*, 611-633, 1999.

Harrison, D.E., and G.A. Vecchi, January 1999 Indian Ocean cooling event, *Geophys. Res. Lett.*, *28*(19), 3717–3720, 2001.

Hastenrath, S., On the Prediction of India Monsoon Rainfall Anomalies, *J. Climate and Applied Meteorol.*, *26*, 847-857, 1987.

Hulme, M., A 1951-80 global land precipitation climatology for the evaluation of General Circulation Models, *Clim. Dyn.*, *7*, 57-72, 1992.

Hulme, M., Validation of large-scale precipitation fields in General Circulation Models, in *Global precipitations and climate change* (eds.) Desbois, M. and Desalmand, F., NATO ASI Series, Springer-Verlag, Berlin, pp. 387-406, 1994.

Hulme, M., T.J. Osborn, and T.C. Johns, Precipitation sensitivity to global warming: Comparison of observations with HadCM2 simulations, *Geophys. Res. Lett.*, *25*, 3379-3382, 1998.

Joseph, P.V., and P.V. Pillai, Air-Sea interaction on a seasonal scale over north Indian Ocean–Part I: Inter-annual variations of sea surface temperature and the Indian summer monsoon rainfall, *Mausam*, *35*(3), 323-330, 1984.

Kershaw, R., The effect of a sea surface temperature anomaly on a prediction of the onset of the south-west monsoon over Indian, *Q. J. R. Meteorol. Soc.*, *114*, 325-345, 1988.

Krishnamurthy, V., and B.N. Goswami, Indian Monsoon-ENSO Relationship on Interdecadal Timescales, *J. Clim.*, 13, 579-595, 2000.

Mooley, D.A., B. Parthasarathy, N.A. Sontakke, and A.A. Munot, Annual Rain-Water over India, its Variability and Impact on the Economy, *J. Climatology*, 1, 167-186, 1981.

Mooley, D.A., and B. Parthasarathy, Indian Summer Monsoon and El Nino, *Pageoph*, 121(2), 339-352, 1983.

Mooley, D.A., and B. Parthasarathy, Indian Summer Monsoon and the East Equatorial Pacific Sea Surface Temperature, *Atmosphere-Ocean*, 22(1), 23-35, 1984.

Murakami, M., Analysis of Summer Monsoon Fluctuations over India, *J. Meteorol. Soc. Jpn*, 54(1), 15-31, 1976.

Nicholls, N., All-Indian Summer Monsoon Rainfall and Sea Surface Temperatures around Northern Australia and Indonesia, *J. Clim.*, 8, 1463-1467, 1995.

Parthasarathy, B., Interannual and long-term variability of Indian summer monsoon rainfall, *Proc. Indian Acad. Sci.*, 93(4), 371-385, 1984.

Parthasarathy, B., and G.B. Pant, The spatial and temporal relationships between Indian summer rainfall and the Southern Oscillation, *Tellus*, 36A, 269-277, 1984.

Parthasarathy, B., A.A. Munot, and D.R. Kothawale, Regression Model for Estimation of Indian Foodgrain Production from Summer Monsoon Rainfall, *Agricultural and Forest Meteorol.*, 42, 167-182, 1988a.

Parthasarathy, B., H.F. Diaz, and J.K. Eischeid, Prediction of All-India Summer Monsoon Rainfall With Regional and Large-Scale Parameters, *J. Geophys. Res.*, 93(D5), 5341-5350, 1988b.

Parthasarathy, B., K. Rupa Kumar, and D.R. Kothawale, Indian summer monsoon rainfall indices: 1871-1990, *Meteorol. Mag.*, 121, 174-186, 1991.

Parthasarathy, B., A.A. Munot, and D.R. Kothawale, Monthly and seasonal rainfall series for All-India homogeneous regions and meteorological subdivisions: 1871-1994. *Contributions from Indian Institute of Tropical Meteorology*, Research Report RR-065, Pune 411 008 India, August 1995

Premkumar, K., M. Ravichandran, S. R. Kalsi, D. Sengupta, and S. Gadgil, First results from a new observational system over the Indian seas, *Curr. Sci.*, 78, 323–330, 2000.

Rakhecha, P.R., and P.R. Pisharoty, Heavy rainfall during monsoon season: Point and spatial distribution, *Curr. Sci.*, 71(3), 179-186, 1996.

Ramesh Kumar, M.R., and J.S. Sastry, Relationships Between Sea Surface Temperature, Southern Oscillation, Position of the 500 mb Ridge along 75°E in April and the Indian Monsoon Rainfall, *J. Meteorol. Soc. Jpn*, 68(6), 741-745, 1990.

Rao, K.G., and B.N. Goswami, Interannual Variations in Sea Surface Temperature of the Arabian Sea and the Indian Monsoon: A New Perspective, *Mon. Weather Rev.*, 116, 558-568, 1988.

Rasmusson, E.M., and T.H. Carpenter, The relationship between eastern equatorial Pacific sea surface temperature and rainfall over India and Sri Lanka, Mon. Weather Rev., 11, 517-528, 1983.

Reynolds, R.W., and T.M. Smith, Improved global sea surface temperature analyses using optimum interpolation, J. Clim., 7, 1195-1202, 1994.

Reynolds, R.W, N.A. Rayner, T.M. Smith, D.C. Stokes, W. Wang, An Improved In Situ and Satellite SST Analysis for Climate, *J. Clim.*, 15(13), 1609–1625, 2002.

Sadhuram, Y., Predicting monsoon rainfall and pressure indices from sea surface temperature, *Curr. Sci.*, 72(3), 166-168, 1997.

Saha, K., Zonal Anomaly of sea surface temperature in equatorial Indian Ocean and its possible effect upon monsoon circulation, *Tellus*, XXII, 403-409, 1970a.

Saha, K., Air and water vapor transport across the equator in western Indian Ocean during northern summer, *Tellus*, XXII, 681-687, 1970b.

Saha, K., Some aspects of the Arabian Sea summer monsoon, *Tellus*, XXVI, 464-476, 1974.

Saha, K.R., and S.N. Bavadekar, Moisture flux across the west coast of India and rainfall during the southwest monsoon, *Q. J. R. Meteorol. Soc.*, 103(436), 370-374, 1977.

Sengupta, D., and M. Ravichandran, Oscillations of Bay of Bengal sea surface temperature during the 1998 summer monsoon, *Geophys. Res. Lett.*, 28, 2033–2036, 2001.

Shukla, J., Effect of Arabian Sea-Surface Temperature Anomaly on Indian Summer Monsoon: A Numerical Experiment with the GFDL Model, *J. Atmos. Sci.*, 32, 503-511, 1975.

Shukla, J., and B.M. Misra, Relationships Between Sea Surface Temperature and Wind Speed Over the Central Arabian Sea, and Monsoon Rainfall Over India, *Mon. Weather Rev.*, 105, 998-1002, 1977.

Shukla, J., Interannual variability of monsoons, in "Monsoons", edited by J.S. Fein and P.L. Stephens, 399-464. John Wiley and Sons, 1987.

Subbaramayya, I., and C.V. Naidu, Spatial Variations and Trends in the Indian Monsoon Rainfall, *Int. J. Climatol.*, 12, 597-609, 1992.

Torrence, C., and P.J. Webster, Interdecadal Changes in the ENSO-Monsoon System, *J. Clim.*, 12, 2679-2690, 1999.

Vecchi, G.A., and D.E. Harrison. Monsoon Breaks and sub-seasonal sea surface temperature variability in the Bay of Bengal, *J. Clim.*, 15(12), 1485-1493, 2002.

Webster, P.J. et al., Monsoons: Processes, predictability and prospects for prediction, *J. Geophys. Res. (TOGA Special Issue)*, 103, 14,451-14,510, 1998.

Xie, P. and P.A. Arkin, Analyses of Global Monthly Precipitation Using Gauge Observations, Satellite Estimates, and Numerical Model Predictions, *J. Clim*, 9, 840-858, 1996.

Xie, P., and P.A. Arkin, Global precipitation: a 17-year monthly analysis based on gauge observations, satellite estimates, and numerical model outputs, *Bull. Amer. Meteorol. Soc.*, 78(11), 2539-2558, 1997.

Yu, J., and J. Slingo, The asian summer monsoon and ENSO, *Q. J. R. Meteorol. Soc.*, 121, 1133-1168, 1995.

D.E. Harrison, Pacific Marine Environmental Laboratory, NOAA, 7600 Sand Point Way NE, Seattle, Washington 98115. (D.E.Harrison@noaa.gov)

Gabriel A. Vecchi, Geophysical Fluid Dynamics Laboratory, NOAA, Princeton Forrestal Campus Rte. 1, P.O. Box 308, Princeton, New Jersey 08542-0308. (Gabriel.A.Vecchi@noaa.gov)

Shallow Overturning Circulations of the Tropical-Subtropical Oceans

Friedrich A. Schott

IFM-GEOMAR Leibniz Institut für Meereswissenschaften an der Universität Kiel, Kiel, Germany

Julian P. McCreary, Jr.

International Pacific Research Center, University of Hawaii, Honolulu, Hawaii

Gregory C. Johnson

NOAA/Pacific Marine Environmental Laboratory, Seattle, Washington

The Subtropical Cells (STCs) of the Pacific and Atlantic Oceans connect the subtropical subduction regions of both hemispheres to the eastern, equatorial upwelling regimes by equatorward thermocline and poleward surface flows. In the Indian Ocean, where equatorial upwelling is absent, a cross-equatorial cell (CEC) connects the southern-hemisphere subduction regime with upwelling regions north of the equator, and it is closed by southward, cross-equatorial Ekman/Sverdrup transport at the ocean surface. We review here the theory explaining the mean features of the STCs and CEC, the observational evidence for their various branches, and results of realistic model simulations. A topic of particular interest is the partition of the equatorward STC branch between interior and western-boundary pathways. Observational results are only now beginning to reveal the structure of the interior pathways, and model results of these flows vary with model type and the wind forcing applied. We also review studies of STC variability, which has been hypothesized to play a role in climate variability. Existing work indicates that wind-driven STC transport variations ($\mathbf{v'}\overline{T}$ processes) are more important than advection of subducted temperature anomalies by the mean STC currents (the $\overline{\mathbf{v}}T'$ processes) in generating equatorial sea-surface temperature anomalies and, hence, climate variability.

1. INTRODUCTION

The Subtropical Cells (STCs) are shallow overturning circulations confined to the upper 500 m. In the Atlantic and Pacific Oceans, they connect subduction zones of the eastern, subtropical ocean with upwelling zones in the tropics. In addition to the dominant equatorial upwelling regimes,

Earth's Climate: The Ocean-Atmosphere Interaction
Geophysical Monograph Series 147
Copyright 2004 by the American Geophysical Union
10.1029/147GM15

upwelling in off-equatorial regions (e.g., the Costa Rica Dome and the Peruvian coast in the Pacific, and the Guinea and Angola domes in the Atlantic) also has to be considered as a potential STC driver; so far, very little is known about these upwelling regimes, but offshore Ekman transports suggest that their contribution may not be negligible. The subsurface STC branches carry thermocline water to the equator either in western boundary currents after circulating across the basin in the Subtropical Gyres or directly in the ocean interior. They are closed by poleward surface currents, largely Ekman transports, that return the upwelled waters to the subtropics (e.g.,

McCreary and Lu [1994]; *Liu et al.* [1994]; *Lu et al.* [1998]; *Johnson and McPhaden* [1999]; *Malanotte-Rizzoli et al.* [2000]).

In contrast to the other two oceans, the Indian Ocean does not possess an equatorial upwelling zone because its annual-mean equatorial winds are westerly. As a result, its shallow cells differ markedly from those in the other two oceans. One cell, the Cross-Equatorial Cell (CEC), has its descending branches in the eastern subtropics of the southern hemisphere but its upwelling branches in the northern hemisphere, thereby requiring cross-equatorial flow [*Schott et al.*, 2002a; *Miyama et al.*, 2003]. Another, the Indian Ocean STC, is associated with upwelling driven by the Ekman divergence from 2–12°S at the northern edge of the Southeast Trades [*McCreary et al.*, 1993; *Murtugudde et al.*, 1999; *Webster et al.*, 1999], and hence is confined to the southern hemisphere.

The STCs provide the cool subsurface water that is required to maintain the tropical thermocline. For this reason, STC variability has been hypothesized to be important for the decadal modulation of ENSO and for Pacific decadal variability, and it may affect Atlantic equatorial SST as well. In the Indian Ocean, a recently identified climate anomaly, the Indian Ocean Dipole (IOD) or Zonal Mode (IOZM), has been associated with changes in the 2–12°S upwelling regime and, hence, variations in the Indian Ocean STC [*Xie et al.*, 2002; *Feng and Meyers*, 2003].

STC pathways are complicated by their interaction with the other ocean currents, particularly with the interocean circulations in each basin, namely, the northward flow of warm water in the Atlantic by the Meridional Overturning Circulation (MOC) with a transport of about 15 Sv [*Ganachaud and Wunsch*, 2001; *Lumpkin and Speer*, 2003], and the circulation driven by the Indonesian Throughflow (ITF) in the Pacific with a transport of 10–15 Sv [*Ganachaud et al.*, 2000; *Sloyan et al.*, 2003]. One result of these interactions is that the southern STC is stronger than the northern one in both oceans [*Johnson and McPhaden*, 1999; *Sloyan et al.*, 2003; *Zhang et al.*, 2003; *Fratantoni et al.*, 2000; *Lazar et al.*, 2002].

The STCs also interact with even shallower overturning cells confined to the tropics. In the Atlantic and Pacific, the shallow Tropical Cells (TCs) are associated with downwelling driven by the decrease of the poleward Ekman transport 4–6° off the equator [*Liu et al.*, 1994; *McCreary and Lu*, 1994; *Molinari et al.*, 2003]. Interestingly, although the TCs are strong in zonal integrations along constant depths, they are much diminished in integrations carried out along isopycnal layers, indicating that they have little influence on heat transport [*Hazeleger et al.*, 2000]. Their existence implies that any measure of STC strength must be defined poleward of the TC convergences, that is, closer to the dynamical intersection between the tropics and subtropics (near 8–10° say). In the Indian Ocean, the CEC interacts with a shallow equatorial roll, analogous to the TCs (or to the Equatorial Cell defined by *McCreary and Lu* [1994]) but driven by meridional winds; it distorts the southward, cross-equatorial pathway of the CEC, requiring that it takes place about 50 m below the surface [*Schott et al.*, 2003; *Miyama et al.*, 2003].

Finally, the off-equatorial undercurrents in the Atlantic and Pacific have to be considered when discussing STCs. In the Pacific, they are referred to as North and South Subsurface Countercurrents (NSCC and SSCC) or simply Tsuchiya Jets [*Tsuchiya*, 1972], whereas in the Atlantic they are called North and South Equatorial Undercurrents (NEUC and SEUC). They are attached to the EUC in the western Pacific [*Rowe et al.*, 2000], but are clearly separated from the EUC in the western Atlantic [*Schott et al.*, 1998; *Bourles et al.*, 1999b]. The fate of these currents is not clear. They diverge poleward toward the east [*Johnson and Moore*, 1997], and hence may upwell in the aforementioned off-equatorial regions or partially recirculate in the other mid-depth equatorial currents [*Rowe et al.*, 2000]. Although most of their transport is located deeper than the EUC, their shallow portions overlap with the deeper part of the EUC. Thus, they can potentially impact the STCs by partially blocking the equatorward transport of thermocline water to the equator, instead carrying water eastward where it may upwell along the eastern boundary or in domes. Even more fundamentally, they can perhaps be considered part of the STCs themselves, a deep off-equatorial branch.

Our review is organized as follows. In Section 2, we discuss the theoretical concepts that explain the basic properties of mean STC pathways. Next, we review observations and model simulations for the individual oceans in Sections 3–5. In Section 6, we present observational and modeling evidence on STC variability, and discuss its potential role in climate variability. We close (Section 7) with a summary of our conclusions and a brief outlook on needed research.

2. STEADY-STATE THEORY

In this section, we discuss basic properties of steady-state STCs (their existence, strength, equatorward extension, and subsurface pathways) using a $2\frac{1}{2}$-layer model, the simplest system that can represent all the STC branches. Its advantage is that analytic solutions can be obtained in which basic STC properties are clearly expressed [*Luyten et al.*, 1983; *Pedlosky*, 1987, 1988, 1991; *Pedlosky and Samelson*, 1989; *McCreary and Lu*, 1994; *Liu*, 1994], allowing their underlying dynamics to be readily identified and their sensitivity to model parameters assessed. Among these studies, only *McCreary and Lu* [1994] obtained solutions in a closed basin, allowing the existence of a complete STC. Here, then, we focus on results from that paper.

As noted here and discussed in greater detail in Sections 3–5, similar dynamics appear to be at work both in the real ocean and in solutions to oceanic GCMs. Also see *Liu et al.* [1994] for a discussion of basic STC properties from a GCM perspective.

2.1. Model Overview

The model equations are

$$f\mathbf{k} \times \mathbf{v}_i + \nabla p_i = \delta_{1i} \tau^x(y) \mathbf{i}/h_i,$$
$$\nabla \cdot (h_i \mathbf{v}_i) = -(-1)^i w_1,$$ (1)

where $i=1,2$ is a layer index, h_i and $\mathbf{v}_i = (u_i, v_i)$ are the thickness and velocity fields of layer i, the pressures are $p_1 = g'_{12}h_1 + g'_{23}h$ and $p_2 = g'_{23}h$, $h = h_1 + h_2$, $g'_{ij} = g(\rho_j - \rho_i)/\rho_o$, $\mathbf{i}$ and $\mathbf{k}$ are unit vectors in the zonal and vertical directions, and δ_{1i} is the Kronecker symbol–δ (that is, $\delta_{11} = 1$ and $\delta_{12} = 0$). Water can transfer between layers at the rate w_1, allowing for the existence of upwelling and downwelling STC branches. To allow for analytic solutions, w_1 is restricted to the values $\pm\infty$ or 0. Horizontal mixing is not explicitly included in (1), but is assumed to be present in boundary layers.

McCreary and Lu [1994] obtained analytic solutions to (1) in the rectangular basin shown in Figure 1 with $y_n = L = 50°$, subject to the constraints that h_1 and h_2 are fixed to constant values H_1 and H_2, respectively, along the eastern boundary of the basin. The solutions were forced by zonally independent wind fields, $\tau^x(y)$, with peak westerlies at 37.5°N, peak easterlies at 12.5°N, and hence a region of negative curl for 12.5°N $< y <$ 37.5°N (as in the thick $\tau^x(y)$ curve in Figure 1).

Figure 1 also indicates the different dynamical regimes of the solution. In Region 1 ($y > y_d = 18°$), h_1 is kept fixed to H_1. In this region, then, layer 1 behaves like a constant-thickness mixed layer, in which water instantly transfers between the layers (i.e., $w_1 = \pm\infty$) in response to Ekman pumping; in particular, when $w_{ek} < 0$ water instantly subducts from layer 1 to layer 2. In Region 2 ($\delta_e < y < y_d$), no entrainment or detrainment is allowed ($w_1 = 0$), so that h_1 is allowed to adjust freely; thus, there is no subduction in the model equatorward of y_d, consistent with its observed weakening at low latitudes. In Region 3 ($y < \delta_e$), the equatorial boundary region, there is upwelling but no detrainment ($w_1 > 0$). The lines $y = y_n - \delta_n$ and $x = \delta_w$ designate edges of frictional boundary currents attached to the northern and western boundaries, respectively.

With the above specifications, the physical situation in Region 2 is essentially the same as in the *Luyten et al.* [1983] study with y_d corresponding to their "ventilation latitude," the two systems differing only in that *Luyten et al.* [1983] assumed $H_1 = 0$ (i.e., an infinitesimally thin mixed layer in Region 1). Regions 1 and 2 are the regions considered by *Liu*

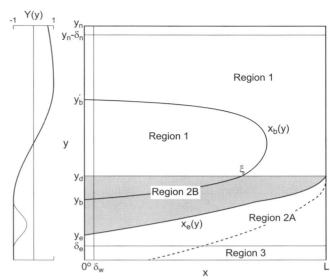

Figure 1. A schematic diagram illustrating the structure of the analytic solution of *McCreary and Lu* [1994]. Two wind profiles, $Y(y)$, are plotted at the left margin, where $\tau^x(y) = \tau_o Y(y)$ and $\tau_o = 0.6$. Parameter values for the solid x_e and x_b curves are $g'_{12} = 1.96$ cm s^{-2} and $g'_{23} = 3.675$ cm s^{-2}, $H_1 = 50$ m, and $H_2 = 200$ m, whereas for the dashed x_e curve $g'_{12} = 1.23$ cm s^{-2} and $H_2 = 150$ m.

[1994]. Regions 2 and 3 correspond to the domain considered by *Pedlosky* [1987, 1988, 1991] and *Pedlosky and Samelson* [1989], who investigated pathways by which subducted water flows to the equator to join the EUC.

2.2. STC Existence and Strength

The net upper-layer transport that flows across y_d is given by

$$M_1(y_d) = -L\frac{\tau^x_d}{f_d}\left(1 - \frac{2H_1}{D + \sqrt{D^2 - 2L\tau^x_d/g'_{23}}}\right)$$ (2)

where $D = H_1 + H_2$, $\tau^x_d = \tau^x(y_d)$, and $f_d = f(y_d)$. It is determined by the Ekman drift across y_d, $-L\tau^x_d/f_d$, and by the upper-layer geostrophic flow. The geostrophic transport always tends to counteract the Ekman transport, but is never strong enough to reverse it since $\tau^x_d > 0$ and hence

$$\sqrt{D^2 - 2L\tau^x_d/g'_{23}} > D \geq H_1.$$

Indeed, for realistic parameter choices and forcing, the geostrophic contribution is relatively small so that $M_1(y_d)$ is largely determined by the Ekman transport [*McCreary and Lu*, 1994].

Because $w_1 = 0$ in Region 2, it follows from the upper-layer continuity equation that $M_1(y) = M_1(y_d)$ *everywhere* in Region

2, so that $M_1(y_d)$ measures the strength of the upper branch of the model's STC. Mass conservation also requires that the corresponding lower-layer transport is $M_2(y) = -M_1(y_d)$, so that (2) also measures the strength of the STC's lower branch. Thus, $M_1(y_d)$ is really the driving force for subsurface equatorward flow: By draining upper-layer water from the tropics, it requires a compensating subsurface flow.

It is rather surprising that $M_1(y_d)$ is determined *entirely* by properties along $y = y_d$, and does not depend on details of the forcing inside either Region 1 or Region 2. In particular, it does not depend on the strength of the wind curl in Region 1, as might be expected: If there is no change in τ_d^x, an increase (decrease) in wind curl in Region 1 only leads to a strengthening (weakening) of the Subtropical Gyre. This result clearly depends on the model property that subduction cuts off along a *single* latitude, y_d. Nevertheless, it is supported in several studies that do not adopt this restriction. For example, in *Liu and Philander's* [1995] idealized GCM solutions, STC strength is almost unchanged in solutions forced by wind fields in which the midlatitude wind curl differs by a factor of two. *Nonaka et al.* [2002] concluded that the STC strength in their GCM solutions was set by zonal winds in the bands 17.5°S < y < 7.5°S and 7.5°N < y < 22.5°N.

2.3. Subsurface Equatorward Flow

In response to negative Ekman pumping ($y \leq 37.5°N$), upper-layer water in Region 1 subducts into layer 2, and, according to (2), some of it is driven across y_d into Region 2. In Region 2, equation (1) can be solved for a single equation in h,

$$\bar{v}_g h_y + (\bar{u}_g - c_r) h_x = 0, \quad (3)$$

where

$$\bar{u}_g = -\frac{1}{h}\Psi_y, \quad \bar{v}_g = \frac{1}{h}\left(\Psi_x + \frac{\tau^x}{f}\right) = \frac{f}{h\beta}w_{ek}, \quad (4)$$

are the depth-averaged, zonal and meridional geostrophic currents in both layers,

$$\Psi = (1/\beta) \int_L^x \operatorname{curl} \tau \, dx$$

is the Sverdrup transport streamfunction, $c_r = (\beta/f^2)\, g'_{12} (h_1 h_2/h)$, and $\beta = f_y$ [*Luyten et al.*, 1983; *Luyten and Stommel*, 1986]. According to (3), isolines of h are parallel to the characteristic curves, $x_c(s)$ and $y_c(s)$, obtained by integrating the equations

$$\frac{dx_c}{ds} = \bar{u}_g - c_r, \quad \frac{dy_c}{ds} = \bar{v}_g, \quad (5)$$

where s is a time-like variable, the integration requiring that h is specified on some boundaries of the domain. Since, according to (1), isolines of h are also identifiable with geostrophic streamlines of layer-2 flow, so are the characteristics.

Region 2 is divided into two subregions, Regions 2A and 2B (Figure 1), by the lower-layer characteristic, $x_e(y)$, that emanates from the point (L, y_d),

$$x_e(y) = L - \frac{g'_{12}\beta}{f\tau_y^x - \beta\tau^x}\left(1 - \frac{f}{f_d}\right)\left[\frac{1}{2}\left(1 - \frac{f}{f_d}\right)D^2 \right.$$
$$\left. + \frac{f}{f_d}DH_1 - \frac{1}{2}\left(1 + \frac{f}{f_d}\right)H_1^2\right]. \quad (6)$$

East of x_e (Region 2A), h is determined by its value along the eastern boundary, D, and this constant value is carried into the interior of Region 2A along characteristics. As a result, $h = D$ throughout Region 2A and there is no layer-2 flow there, the "shadow zone" of the *Luyten et al.* [1983] solution. West of x_e (Region 2B), h is specified by its value along y_d, which increases to the west since $w_e(y_d) < 0$ for the wind profiles considered here. This westward increase of h is carried into the interior of Region 2B, ensuring that there is a region of equatorward flow in layer 2 that extends well into the tropics.

McCreary and Lu [1994] obtained an expression similar to (6) for the lower-layer streamline, $x_b(y)$, that extends to the bifurcation point of the model's low-latitude western boundary current, $(0, y_b)$ (Figure 1). It divides Region 2B into two parts, with water east of x_b flowing to the equator to participate in the STC, and water west of it returning to the subtropics in the western boundary current. It also separates the subduction region in Region 1 into two parts, with only water that subducts east of x_b participating in the STC.

The interesting property of the Region-2 response is that there is *any* penetration of layer-2 water into the tropics at all. This property is entirely due to the nonlinear terms in (1), that is, the terms that involve h_i. (In a linear model, the variable thicknesses h_i in the nonlinear terms are replaced by externally specified constants H_i, typically assumed to be uniform throughout the basin.) The nonlinearities are the source of the term proportional to $\bar{v}_g$ in (3), which, according to (5), accounts for the meridional displacement of characteristics. Consistent with this result, $x_e(y)$ becomes a straight line along $y = y_d$ in the limit that the forcing, and hence the expression $\left(f\tau_y^x - \beta\tau^x\right)$ tends to zero.

2.4. Interior and Western-Boundary Pathways

In Figure 1, all of layer-2 first flows across the basin before moving to the equator in a western boundary current, but it can also flow to the equator in the interior ocean. The equatorial limit of (6) is,

$$x_e(0) = L + \frac{g'_{12}}{2\tau^x(0)} H_2(H_2 + 2H_1), \quad (7)$$

which is well defined even though the equator lies outside the range of validity of (1). Since there are no lower-layer currents in Region 2A, a necessary condition for lower-layer water to flow to the equator in the interior ocean is that $x_e(0) > 0$, that is, the shadow-zone boundary intersects the equator.

The value of $x_e(0)$ can be positive or negative for realistic parameter choices, with $x_e(0)$ being positive for larger L and $|\tau^x(0)|$ and smaller g'_{12} and H_2 (as in the dashed curve in Figure 1). The dependency on L suggests that the existence of an interior pathway is more likely in the Pacific Ocean than the Atlantic. The dependency on the latter two parameters suggests that, in the real ocean, water from shallower thermocline levels is more likely to move to the equator in the interior ocean. As we shall see, these dependencies are consistent with both observed and modelled properties (Sections 3–5).

Provided there is an Intertropical Convergence Zone (ITCZ, as in the thin τ^x curve in Figure 1), however, the inequality $x_e(0) > 0$ is not a sufficient condition for the existence of an interior pathway. If the ITCZ is sufficiently weak, $x_e(y)$ first bends more sharply to the west on the northern flank of the ITCZ where $|w_{ek}|$ is small, and then bends back to the east on the southern flank where it is large. If the ITCZ is sufficiently strong, $x_e(y)$ intersects the western boundary *before* it can loop back, and so layer-2 flow follows the western-boundary pathway to the equator even though $x_e(y) > 0$. If the ITCZ is strong enough for $w_{ek} > 0$, then no interior pathway is possible.

The idea of the ITCZ being a barrier for equatorward flow in the Pacific Ocean was discussed by *Lu and McCreary* [1995], since in their solution to a 2½-layer model the interior pathway was completely eliminated by the ITCZ. As discussed in Section 3, however, interior pathways exist at shallow levels in GCM solutions [*Liu et al.*, 1994; *Blanke and Raynaud*, 1997; *Rothstein et al.*, 1998]. *Liu and Huang* [1998] suggested that the lack of an interior pathway in the *Lu and McCreary* [1995] solution was their value of H_2 being somewhat too thick, an idea supported by *Lu et al.* [1998; see Figure 7].

3. PACIFIC OCEAN

The STCs have been better documented and quantified in the Pacific Ocean than in the other two oceans, mainly because their equatorward branches are located within the region occupied by the ENSO signal, which is relatively well sampled by moored arrays and repeat shipboard sections. In this section (and in Sections 4 and 5 as well), we first describe key properties of the wind field that drives the STCs, then review observations of their individual (subduction, equatorward, upwelling, and poleward) branches, and conclude with a discussion of several modeling studies designed to simulate them.

3.1. Wind-stress Fields and Ekman Transport Divergences

The annual-mean wind-stress field from the NCEP reanalysis (Figure 2a) shows the tradewind circulations in each hemisphere with the ITCZ just north of the equator. A band of positive (or weakened negative) Ekman pumping, $w_{ek} = \text{curl}(\tau/f)$ (Figure 2b) associated with these winds extends across the interior ocean under the ITCZ. As discussed in Section 2.4 and below, it acts to inhibit the possible interior STC pathways in the northern hemisphere. Note, however, that values of w_{ek} remain negative within the band from about the dateline to 140°W, providing a "window" for the existence of interior pathways; details of the window depend on the wind-stress climatology used, but it is always there. In the southern hemisphere, there is no region of positive w_{ek} in the eastern and central ocean, allowing the possibility for interior pathways over a broader longitudinal range. On the other hand, details of the curl structure and its role for STC pathways are still evolving. For example, *Kessler et al.* [2003] recently derived w_{ek} near the equator on a finer scale from scatterometer winds, finding a conspicuous band of positive w_{ek} along about 1°N between 150–100°W; the presence of this band results in a more realistic Sverdrup circulation than for existing wind-stress and reanalysis products, yielding a stronger EUC and northern branch of the South Equatorial Current (SEC).

As discussed in Section 2.2, the STCs are forced by the poleward transport of near-surface, equatorial waters across the latitudes where subduction cuts off in each hemisphere, roughly the equatorward boundaries of the Subtropical Gyres; moreover, that divergence is composed primarily of Ekman transport. The location of the subduction cut-off latitudes is not precisely known in the real ocean. They must be chosen poleward of the Tropical Cell convergence, about 4–6° north and south of the equator, and we define them here to be 10°N/10°S.

The annual-mean value of the poleward NCEP Ekman transports divergence across these latitudes is 54 Sv (and similarly, 58 Sv for the ERS-1,2 scatterometer stresses), providing an estimate of the upper limit for equatorial upwelling discussed next. The longitudinal distribution of the meridional Ekman transports along 10°N has its maximum contribution to the total divergence in the center of the basin, while at 10°S the maximum southward contributions are further to the east, as shown in Figure 2b.

The area integral of positive w_{ek} in Figure 2b in the region $x > 100°W$, $5°N > y > 20°N$ plus the line integral of offshore

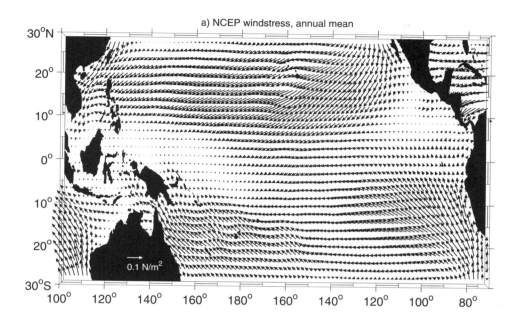

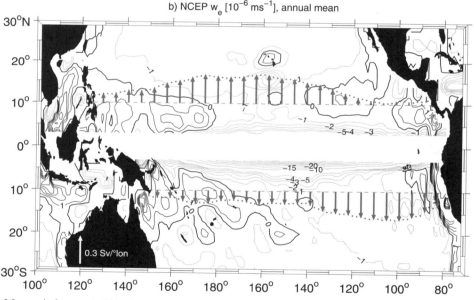

Figure 2. a) Mean wind stress and b) Ekman pumping (outside ± 3° latitude belt) for Pacific from NCEP reanalysis 1990-1999; also shown are meridional Ekman transports (vector at lower left = 0.3 Sv/degree longitude) across 10°N and 10°S.

Ekman transport along the coast, is 2.2 Sv, providing an upper limit for off-equatorial upwelling in the Costa Rica Dome. Similarly, the area integral of positive w_{ek} for $x > 85°W$, $20°S > y > 5°S$ plus the coastal Ekman divergence driven by the alongshore winds is 7.7 Sv, yielding a total eastern boundary Ekman divergence of about 10 Sv. Some amount of this eastern upwelling occurs within the 10°S to 10°N latitude range and thus contributes to the Ekman divergence across 10°N/10°S; thus, this amount must be subtracted from the 10°N/10°S divergence to estimate open-ocean equatorial upwelling.

The variations of the Ekman divergence across 10°N and 10°S for the Pacific (Plate 1) shows variations of about 5 Sv amplitude at interannual and longer timescales, which do not

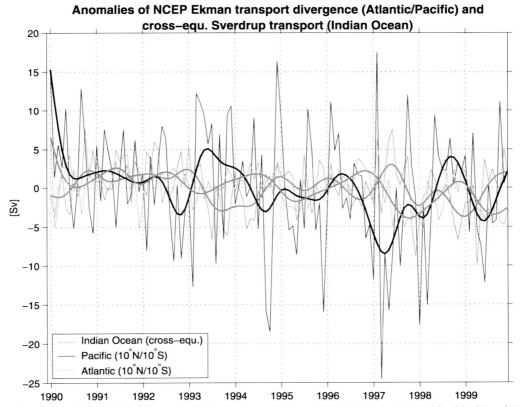

Plate 1. Time series 1990–1999 of equatorial Ekman transport divergence (in Sv) for the Pacific (black) and Atlantic (blue), calculated for the band 10°S–10°N and coast to coast, and of cross-equatorial Sverdrup transport for the Indian Ocean (red) from NCEP reanalysis wind stresses. Thin curves are for monthly means, heavy curves for low-passed interannual variability.

occur in any obvious association with El Niño/La Niña events; however, if the divergence is calculated closer to the equator, at 3°N/3°S say, a relationship to ENSO appears (see Section 6).

3.2. Observations

For convenience, we organize the discussion in this section to follow the subsurface branch of the North and South STCs from their sources (subduction) to their sinks (upwelling). We conclude by noting key properties of their surface return flows.

3.2.1. Subduction. Estimating subduction rates is an involved process that requires knowledge of oceanic density structure and air-sea fluxes, and these estimates have been limited to the annual mean [*Marshall et al.*, 1993]. Analyses suggest that subduction rates are fairly high in a band across the subtropical North Pacific, but highest in the west [*Qiu and Huang*, 1995]. These high subduction rates lead to an enhanced presence of certain density classes within the pycnocline, namely, regions of reduced stratification known as "mode waters". The most eastward and equatorward of these mode waters [*Hautala and Roemmich*, 1998] are the ones that flow westward and equatorward to participate in the STCs. Using *Hellermann and Rosenstein* [1983] wind stresses and hydrographic climatology from *Levitus* [1982], *Qiu and Huang* [1995] estimated a total subduction of 35 Sv for the North Pacific. By calculating geostrophic pathways of the subducted waters, they showed that most of the subducted water returns northward through the NEC (north of the bifurcation in the Mindanao Current) and then via the Kuroshio, but the authors did not estimate the southward transfer into the STC.

In the South Pacific, the highest subduction rates are at the equatorward and eastward edges of the subtropics, extending even into the tropics [*Huang and Qiu*, 1998; *Karstensen and Quadfasel*, 2002], and they are associated with a prominent low-latitude mode water [*Wong and Johnson*, 2003]. Using the mean hydrographic data base of *Levitus and Boyer* [1994a,b] and Southampton Oceanographic Center (SOC) air-sea fluxes, *Karstensen and Quadfasel* [2002] determined a total water-mass formation of 44 Sv for the South Pacific. Of this total, 27 Sv are inserted into densities lighter than 26.5 kg m^{-3} that can participate in tropical upwelling (Section 3.2.4). How much of this subduction reaches the East Australian Current and turns southward versus how much enters the STC has not been estimated.

3.2.2. Subsurface equatorward flow.

3.2.2.1. Western-boundary pathways: In the South Pacific, subsurface waters flow to the equator in the New Guinea Coastal Undercurrent (NGCU). Water flows to the NGCU primarily through two channels: the Vitiaz Strait between Papua New Guinea and New Britain, and the St. Georges Channel between New Britain and New Ireland [*Butt and Lindstrom*, 1994]. The inverse model result of *Sloyan et al.* [2003] yielded a transport of 14.7±1.5 Sv westward for the NGUC. They further concluded that the NGUC overshoots the equator and retroflects back southeastward to supply the EUC (Plate 2), as suggested earlier by *Tsuchiya et al.* [1989]. At 143°E, the EUC transport in the density range 25.5 kg m^{-3} < σ_θ < 26.3 kg m^{-3} essentially balances with the NGUC supply.

In the northern hemisphere, the Mindanao Current (MC) flows southward along the Philippine Islands. Analysis of eight CTD/ADCP surveys off Mindanao along 8°N [*Wijffels et al.*, 1995] showed a southward MC transport of 23±4 Sv above σ_θ = 26.7 kg m^{-3} (around 350 m); 15 Sv of this transport falls into the density range σ_θ = 23–26.2 kg m^{-3} [*Liu and Philander*, 2001] and is used here in our schematic diagram (Plate 2). Just south of Mindanao (about 5°N), 9±3 Sv of the MC flows into the Celebes Basin to provide much of the water for the ITF [*Gordon et al.*, 1999], the rest (14±5 Sv) turning east to feed the NECC and eventually joining the EUC (*Johnson and McPhaden* [1999]; Plate 2). It is not likely that all of this water reaches the EUC, however, as some of the denser MC waters contribute a relatively fresh signature that is advected eastward in the northern part of the NSCC [*Johnson and McPhaden*, 1999], which itself carries about 7 Sv across 165°E [*Sloyan et al.*, 2003], and some of the lighter MC waters are too light to supply the EUC.

3.2.2.2. Interior pathways: While there are substantial difference between different wind stress products [*Kessler et al.*, 2003], the possibility of an interior pathway is allowed by the property that w_{ek} remains generally negative in the central Pacific (Figure 2b). Such pathways with interior flow to the equator in the northern hemisphere were first suggested by tracer studies [*Fine et al.*, 1987]. Moreover, distributions of salinity, acceleration potential, and potential vorticity (Figure 3) all point toward the presence of a circuitous, southward, interior flow, with a strong westward component north of the ITCZ in the North Equatorial Current (NEC), an eastward component in latitude band of the NECC, and westward flow in the northern branch of the SEC. By contrast, southern-hemisphere distributions all show a direct northwestward route toward the equator within the southern branch of the SEC.

From historical CTD data, *Johnson and McPhaden* [1999] estimated the relative contributions of the equatorward geostrophic flows in the interior ocean above a neutral density of about γ_n = 26 kg m^{-3} to be 5±1 Sv and 15±3 Sv in the North and South Pacific, respectively (Figure 4). In that study, the bulk of the equatorward thermocline transport

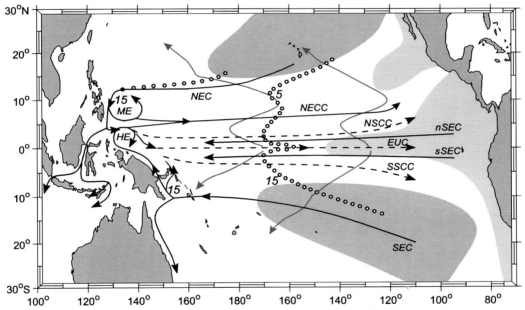

Plate 2. Schematic representation of the Pacific circulation branches, subduction (blue) and upwelling (green) zones that participate in the STC. Current branches participating in STC flows are NEC = North Equatorial Current; nSEC, sSEC = South Equatorial Current north and south of the equator; NECC = North Equatorial Countercurrent; EUC = Equatorial Undercurrent; NSCC, SSCC = North and South Subsurface Countercurrent; ME = Mindanao Eddy; HE = Halmahera Eddy. Interior equatorward thermocline pathways dotted, with interior equatorward transport estimates (in Sv) marked; selected surface poleward pathways for the central basin (from Figure 6) marked by thin, magenta lines; see text for details.

across 7°S occurs from 160–90°W, whereas the flow between the NEC and the NECC is distributed from 130–180°W (Figure 4). Near-equatorial pathways, however, apparently shift further westward in both hemispheres before merging into the EUC (Plate 2). *Sloyan et al.* [2003] obtained equatorward transports for the density range $\sigma_\theta = 23.0–26.3$ kg m^{-3}, finding 5.8 Sv across 8°N from 156°E–170°W and 13.6 Sv across 8°S from 170–95°W, confirming the earlier estimates. These results show that about half of the total equatorial Pacific upwelling is supplied by interior meridional flows at the upper thermocline.

3.2.3. Equatorial currents

3.2.3.1. EUC: The EUC shoals from a core depth of 200 m at 143°E to 80 m at 110°W (Figure 5; *Johnson et al.* [2002]). At the western end, across 156°E, its transport is estimated to be 19.8±1.1 Sv. It attains its maximum transport of 30.5±2.0 Sv in the eastern central Pacific at 125°W, and then drops to 16.2±1.9 Sv at 95°E [*Sloyan et al.*, 2003]. Most of its transport is upwelled before it reaches the eastern boundary, but some likely feeds into the Peru-Chile Undercurrent [*Lukas*, 1986].

3.2.3.2. NSCC and SSCC: The North and South Subsurface Countercurrents (NSCC, SSCC) flow eastward in the density range 26.3 kg m^{-3} < σ_θ < 26.9 kg m^{-3} in the western and central equatorial Pacific so that they are located underneath the interior equatorward flow discussed above, but rise across isopycnals on their way east (Figure 5). They are supplied mostly from western boundary currents. In the east, they likely supply water for the off-equatorial upwelling regions along the South American coast and in the Costa Rica Dome. Thus, they should probably be viewed as a deep branch of the STC system. Connections with westward-flowing tropical currents in this density range have also been suggested [*Rowe et al.*, 2000].

The pathways of the subsurface countercurrents shift polewards toward the east (Figure 5; *Rowe et al.* [2000]), as discussed kinematically by *Johnson and Moore* [1997] and dynamically by *McCreary et al.* [2002]. The SSCC transport was estimated to be 6.7±2.2 Sv at 165°E with its core at 2.5°S, and 6.1±1.9 Sv at 110°W with its core at 5.5°S, with a suggestion of a reduction to 3.8±2.4 Sv at 95°W due to upwelling across the $\sigma_\theta = 26.3$ kg m^{-3} isopycnal [*Rowe et al.*, 2000; *Sloyan et al.*, 2003]. The NSCC has similar strength, with a transport of 7.6±1.7 Sv at 165°E centered near 2.5°N and 6.4±2.2 Sv at 110°W near 4.5°N. In contrast to the SSCC, the NSCC is only slightly weakened at 95°W, perhaps indicating a larger role of the NSCC for supplying eastern Pacific upwelling.

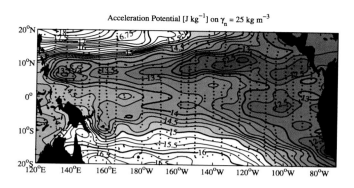

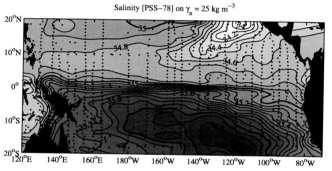

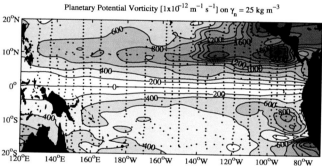

Figure 3. Objective maps of properties on $\gamma_n = 25.0$ kg m^{-3}, within the tropical Pacific pycnocline made from binned mean hydrographic profiles (dots), a) acceleration potential relative to 900 db, contour intervals of 0.25 J kg^{-1}; b) salinity, contour intervals of 0.1; c) planetary potential vorticity, contour intervals of 200×10^{-12} m^{-1}s^{-1}. (After *Johnson and McPhaden* [2001].)

3.2.3.3. Tropical cells: Analysis of shipboard sections from 95–170°W taken during the 1990's, mostly during maintenance of the Tropical Atmosphere Ocean (TAO) project [*McPhaden et al.*, 1998] moorings, shows equatorward subsurface flow beneath poleward near-surface Ekman flow, the latter with off-equatorial convergence occurring mostly in the NECC about 7°N [*Johnson et al.*, 2001]. Drifter data suggest a more symmetric off-equatorial convergence with maxima near ±4° having an amplitude about half that of the equatorial divergence [*Johnson*, 2001]. In their inverse model study, *Sloyan et al.* [2003] estimated the overall strength of the TCs to be 15±13 Sv and 20±11 Sv for the northern and southern TCs, respectively. This diverse evidence of off-equatorial convergence likely indicates the downwelling limbs of the Tropical Cells predicted by models (e.g., *Stommel* [1960]; *Philander and Pacanowski* [1980]; *McCreary* [1985]; *Liu et al.* [1994]; *McCreary and Lu* [1994]). The existence of the TCs implies that the strengths of the STCs' poleward branches need to be evaluated at least poleward of the TC convergences.

3.2.4. Upwelling. The existence of annual-mean equatorial easterlies ensures that most of the upwelling in the tropical ocean occurs along the equator, due to the strong Ekman divergence there. Some water also upwells along the coast of Peru, and there is potential for upwelling in the Costa Rica Dome where thermocline, and even subthermocline, waters are brought very close to the surface.

3.2.4.1. Source depth of upwelled waters: From tracers (^{14}C and CO_2) incorporated into a box model, *Quay et al.* [1983] set an upper limit for the density of water that upwells along the equator to be $\sigma_\theta = 26.5$ kg m^{-3} (a depth of about 225 m). In their recent inverse model study, *Sloyan et al.* [2003] merged CTD data and mean zonal currents from shipboard ADCP data [*Johnson et al.*, 2002] with air-sea flux estimates to obtain a consistent picture of the equatorial Pacific circulation and upwelling.

They determined for the central equatorial Pacific that diapycnal upwelling existed only for $\sigma_\theta < 24.0$ kg m^{-3}, with upward velocities in the depth range 100–200 m occurring mostly along isopycnals. For the eastern equatorial Pacific, however, some diapycnal upwelling was found across isopycnals as dense as $\sigma_\theta = 26.3$ kg m^{-3}, corresponding to a depth of about 200 m.

3.2.4.2. Equatorial upwelling: Estimates of upwelling transports for the zonally sloping EUC differ widely depending on whether they are determined across horizontal or isopycnal surfaces. *Wyrtki* [1981] made an early calculation of the upwelling transport across 50 m in an equatorial box extending from 170°E to 100°W and from 5°S to 5°N. Based on estimates of the horizontal-transport convergence across the sides of the box (determined from existing equatorial current measurements, geostrophy, and net Ekman divergence), he reported a range of estimates for the upwelling transport with 51 Sv being his preferred value. *Bryden and Brady* [1985] carried out a similar analysis in a smaller box from 110–140°W for the same latitude range, determining an upwelling transport of 22 Sv across 62.5 m. They also noted that because the thermocline and EUC shoal toward the east, only 7 Sv of that amount actually crossed 23°C to upwell into the surface layer.

Meinen et al. [2001] determined equatorial upwelling and its variability (see Section 6.1.1) across 50 m in an equatorial

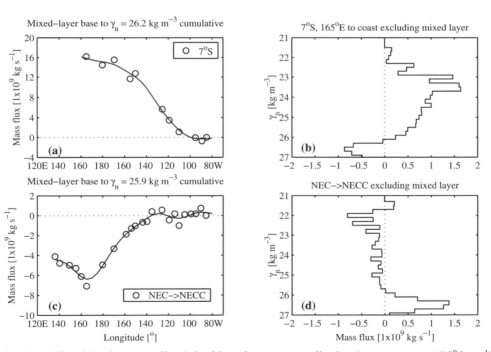

Figure 4. Quasi-meridional interior pycnocline (mixed-layer base to pycnocline base) mass transport (10^9 kg s^{-1}) zonally accumulated from the Americas westward at (a) 7°S and (c) between the North Equatorial Current (NEC) and North Equatorial Countercurrent (NECC) in the Pacific Ocean. The same quantity summed in 0.1 kg m^{-3} γ_n bins at (b) 7°S from 165°E to the Americas and (d) between the NEC and NECC from 135°E to the Americas. (After *Johnson and McPhaden* [1999].)

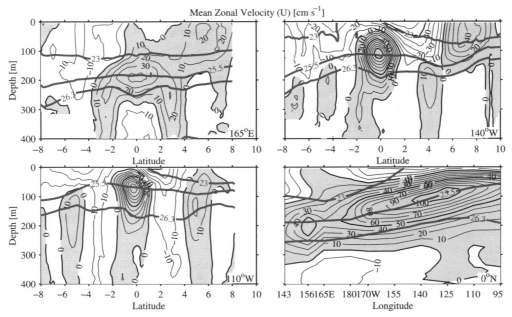

Figure 5. Sections of mean zonal velocity, estimated at 3 different longitudes and along the equator in the Pacific Ocean. Locations marked in the bottom right corner of each section. Contour interval is 10 cm s^{-1}, with enhanced contours at 50 cm s^{-1} intervals; eastward velocities are shaded. Heavy lines correspond to isopycnal surfaces σ_θ = 23.0, 25.5 and 26.3 kg m^{-3} (After Johnson et al. [2002].)

box extending from 5°S to 5°N and from 155°E to 95°W. They estimated the upwelling transport by the horizontal transport divergence out of the box, utilizing TAO-mooring and shipboard data, and applying geostrophy and Ekman dynamics. *Meinen et al.* [2001] obtained 24±3 Sv of upwelling across 50 m for that box. Cross-isopycnal transports were significantly smaller, 15 Sv across the 24°C isotherm. The inverse calculation of *Sloyan et al.* [2003] yielded 24±4 Sv of upwelling across the σ_θ = 23.0 kg m^{-3} isopycnal surface in the central Pacific (2°S–2°N, 170–125°W) and 13±4 Sv across σ_θ = 24.0 kg m^{-3} in the eastern equatorial Pacific (125–95°W). A direct velocity-based upwelling estimate [*Johnson et al.*, 2001] is larger than all of these box model estimates, but arguably more uncertain.

3.2.4.3. Peru-Chile coastal upwelling: The EUC surfaces in the east and what remains of it at the Galapagos Islands turns southeastward toward the west coast of South America [*Lukas*, 1986]. It then presumably feeds the poleward-flowing Peru-Chile Undercurrent (PCUC), but conclusive evidence supporting a continuity between these currents has not been provided yet. The PCUC transport at 10°S has been estimated from several hydrographic sections to be 1 Sv, sufficient to supply the coastal upwelling to at least 15°S [*Huyer et al.*, 1991]. This estimate is considerably smaller than the Ekman transport-derived upper bounds mentioned previously for the region. On the other hand, water properties of the PCUC [*Blanco et al.*, 2001] suggest that its source may just as likely be the SSCC, which could reach the eastern boundary south of 10°S.

3.2.4.4. Costa Rica Dome upwelling: Finally, the Costa Rica Dome (near 89°W, 9°N) provides another possible location for upwelling. The top of an extremely sharp thermocline is often located within 10 m of the sea surface within this dome [*Wyrtki*, 1964]. Relative to surrounding waters, surface waters within the dome are slightly cooler, saltier, undersaturated in oxygen, and nutrient-rich, all properties indicative of upwelling. A recent analysis yielded an upwelling transport of about 3.5 Sv [*Kessler*, 2002]. *Johnson and McPhaden* [1999] speculated that Costa Rica Dome upwelling might be associated with northward flow of the NSCC, and *Kessler* [2002] built a stronger case for this connection (see Plate 2). *McCreary et al.* [2002] provided theoretical justification for the idea.

3.2.5. Poleward surface flow. Surface drifter trajectories (Figure 6) clearly show the poleward component of the Ekman flow driven by the easterly trade winds in both hemispheres [*Reverdin et al.*, 1994; *Johnson*, 2001]. The poleward Ekman flows are superimposed on the geostrophic currents, which are predominantly zonal and, except for the NECC and the weaker SECC (not shown in Plate 2), are directed westward.

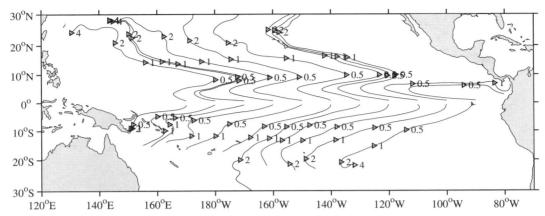

Figure 6. Near-surface trajectories calculated from mean surface drifter velocities; numbers are mean time in years to reach the location from starting points at ±0.75° latitude from equator. (From *Johnson* [2001].)

Shipboard ADCP data give some indication of the vertical distribution of poleward Ekman flow, and it appears that these flows reach below the mixed layer [*Johnson et al.*, 2001].

Interestingly, most of the trajectories for the surface flow reach the subtropics in the western half of the basin (*Johnson* [2001]; Figure 6). Since subduction of the water that participates in the STCs occurs predominantly in the eastern subtropics, there is not an obvious direct and complete closure of the surface and subsurface STC branches within the subtropics (Plate 2), suggesting that the STCs are linked to circulations farther poleward. See *McCreary and Lu* [1994], *Liu et al.* [1994], and *Lu et al.* [1998] for discussions of these linkages in ocean models; in particular, the latter two studies note the interaction of the North Pacific STC with the Subpolar Cell (SPC), a shallow overturning circulation that connects the subtropical and subpolar oceans. In the eastern basin, where the EUC and off-equatorial undercurrents surface, the drifters can also travel eastward (Figure 6).

A cautionary note has to be added here regarding drifter representations of particle pathways for the waters that are upwelled at the equator. Real water parcel followers would have to downwell partially at the TC convergence and thus be removed from the poleward Ekman transports for some time, thereby extending further westward than indicated by the trajectories of Figure 6 or Plate 2.

3.3. Models

A number of modeling studies have investigated the Pacific STCs. Here, we briefly summarize four of them [*Lu et al.*, 1998; *Blanke and Raynaud*, 1997; *Rothstein et al.*, 1998; *Huang and Liu*, 1999] that span a variety of model types. The solutions all capture the overall STC structure, but differ in the strength and location of interior pathways. None of the solutions reproduces the subsurface countercurrents at 3–5° latitude. This shortcoming may not be significant for understanding STCs, however, since the observed equatorward thermocline flow appears to mostly pass over them, but more research on the subject is needed.

Lu et al. [1998] obtained solutions to a $3\frac{1}{2}$-layer model, the three active layers representing tropical, thermocline, and upper-intermediate (subthermocline) waters. As for the simpler layer model discussed in Section 2, water is allowed to transfer between layers to parameterize the processes of upwelling, subduction, and diapycnal mixing, and subduction is cut off equatorward of 18°S and 18°N. Solutions are forced by climatological annual-mean winds and by the ITF, which is prescribed as an outflow of 10 Sv from layers 1 and 2 along the western boundary from 1–6°N and a compensating inflow into layer 3 across the open southern boundary.

Figure 7 plots the v_2 field from their main-run solution, together with a number of streamlines that divide the flow field into several subregions and shaded areas indicating where water subducts from layer 1 into layer 2. Streamlines Sz and Bi correspond to the shadow-zone and bifurcation streamlines, x_e and x_b, in Section 2, and the others are defined in the caption to Figure 7. The fate of subducted water varies considerably depending on where it subducts. Only the water that subducts in the two darkest-shaded regions in each hemisphere flows to the equator to join the EUC and, hence, to participate in the STCs.

The layer-3 water imported into the South Pacific first recirculates within the deepest portion of the South Pacific Subtropical Gyre, and then flows north to the equator in a western-boundary current, consistent with the observed flow of subthermocline water. Most of it turns eastward to flow along the equator as the deepest part of the EUC, where it upwells into layers 2 and 1 in the central and eastern oceans. The addition of this water, together with the draining of water

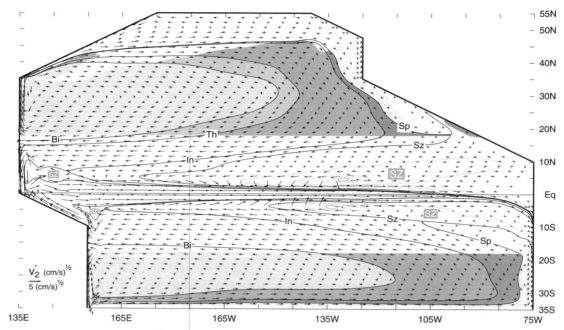

Figure 7. STC-circulation in the $3\frac{1}{2}$-layer model of *Lu et al.* [1998]. Horizontal distributions of v_2 with regions where $w_s \neq 0$ shaded. Current arrows are of the vector field $v'_i = v_i/|v_i|^{\frac{1}{2}}$, which has the same direction as v_i but an amplitude of $|v_i|^{\frac{1}{2}}$; this modification enhances the strength of weak flows relative to stronger ones, allowing them to be more visible in each plot. The shading indicates where layer-1 water subducts into layer 2. The flow field is divided into subregions by streamlines of v_2. Streamline Sp defines the southern boundary of the Subpolar Gyre in layer 2. Streamlines Sz and Bi correspond to the shadow-zone and bifurcation streamlines, x_e and x_b, in Section 2. Streamline In bifurcates in the western basin at 150°E, 6°N, with one branch extending on to the western boundary. Streamline Th intersects the western boundary at the southern edge of the Indonesian passage, so that water between streamlines Th and Bi leaves the basin in the throughflow region. Analogous streamlines are also plotted in the southern hemisphere (the throughflow streamline only existing in the northern hemisphere).

from layers 1 and 2 by the ITF, ensures that the EUC is composed mostly of water of southern-hemisphere origin (63%). In a subsequent study with a different parameterization of layer-2 entrainment, *McCreary et al.* [2002] concluded that the layer-3 equatorial branch upwells rather in the Costa Rica Dome and along the Peruvian coast, thereby generating the NSCC and SSCC, respectively.

Blanke and Raynaud [1997] investigated the sources and sinks of the Pacific EUC, using an OGCM with 30 vertical levels (a 10-m resolution in the top 150 m) and a horizontal resolution of 0.33° near the equator, decreasing to 1.5° poleward of ±47°. The model was forced by *Hellermann and Rosenstein* [1983] wind stresses and *Esbensen and Kushnir* [1981] heat fluxes, and salinity was relaxed to the *Levitus* [1982] surface climatology. There was, however, no explicit representation of the ITF. Pathways of water parcels backtracked from the central equatorial Pacific show source regions for EUC water similar to those discussed in previous sections. Particles reach the EUC via the equatorward western boundary currents but also through windows in the interior ocean via circuitous pathways with large westward excursions before reaching the equatorial interior Pacific.

Rothstein et al. [1998] studied STC pathways, using a σ-coordinate model of the Pacific domain with 11 layers in the upper ocean (roughly the top 700 m) beneath a surface mixed layer and with a horizontal resolution of 0.3° near the equator, increasing to 1° at 50° latitude. The forcing was similar to that for the *Blanke and Raynaud* [1997] study, and the model also did not include the ITF. Pathways of subsurface flow are identified by the Bernoulli function and current vectors (Figure 8). They again show a westward excursion of the equatorward thermocline flow in both hemispheres. In the northern hemisphere, only the waters that subduct in the northeastern subtropics (region III) follow pathways that extend directly to the equator in the interior ocean; the model equatorward transport across 10°N through the 140–160°E window is 3 Sv, of the same order as observational numbers quoted earlier.

Waters that subduct west of about 170°W (region I, Figure 8) do not reach the equator at all but rather bend northward at the western boundary to remain in the subtropical gyre. Particles subducted in the central subtropical North Pacific at about 170–120°W (region II) travel to the western boundary. Almost all of them bend southward there and then eastward within the NECC to join the EUC in the central ocean. Note the existence of an additional pathway (the 12.5 isoline) that first flows to the equator and then retroflects to join the NECC. In the southern hemisphere, the lack of the ITCZ simplifies the trajectories. Subduction west of 140°W and north of 15°S reaches the EUC via the western boundary, whereas subduction from further east can supply interior pathways.

A further development in Pacific STC studies is the evaluation of the NCEP Pacific Ocean reanalysis product, in which observed temperature data are assimilated into an ocean model (*Huang and Liu* [1999]; see also *Liu and Philander* [2001]). The model has 28 vertical layers, its longitudinal resolution is 1.5°, and its meridional resolution decreases from $1/3°$ from 10°S–10°N to 1° at midlatitudes. Forcing is by annual-mean NCEP momentum and heat fluxes. The authors presented Lagrangian trajectories obtained by releasing particles at a depth of 50 m along 24°N and 24°S in the North and South Pacific (Figs. 9a,b), confirming the existence of the exchange windows (Figs. 9c,d) discussed above.

Some of the particles released at 24°N take the western-boundary pathway to the equator. Interestingly, they do not join the EUC at the boundary, but rather travel eastward with the NECC until mid-basin and only then extend to the equator (Figure 9a). Water subducted east of about 135°W takes the interior exchange window, first travelling westward with the NEC and then eastward in the NECC before extending to the equator. Similarly, particles released at 24°S have two exchange windows. Water subducted east of about 100°W reaches the equator by the interior exchange window, and the western-boundary exchange window is supplied out of longitude range 170–100°W. Time scales for particles to reach the equatorial zone range from less than 2 years near the western boundary to 15 years in the central subtropics.

Overall, *Huang and Liu* [1999] confirm that the EUC is dominantly supplied out of the South Pacific (where the role of the ITF in comparison to observations needs to be taken into account), and they find that the southern exchange window reaches much further poleward than the northern one (Figs. 9c,d). *Huang and Liu* [1999] also calculated the meridional transports across 10°S and 10°N, obtaining values for the western boundary and northern interior thermocline transports similar to those discussed in the previous sections and marked in Plate 2. Their southern interior thermocline transport is weaker (12 Sv) than estimated from observations (15 Sv), perhaps a consequence of the model not including the ITF.

4. ATLANTIC OCEAN

Only in the western Atlantic (near 35°W) are shipboard ADCP observations sufficiently dense to be comparable to the coverage in the equatorial Pacific and hence to allow quantitative circulation estimates for the STC mean flows [*Schott*

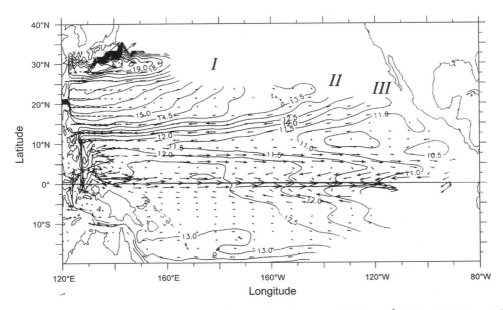

Figure 8. Bernoulli function and current vectors on isopycnal surface $\sigma_\theta = 24.0$ kg m^{-3} in the *Rothstein et al.* [1998] σ-coordinate model simulation.

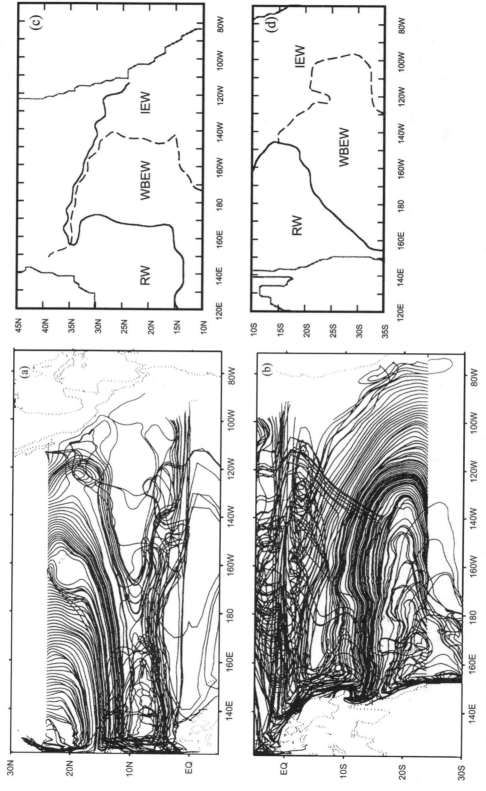

Figure 9. Trajectories of particles released at a) 24°N, b) 24°S in NCEP model study; c) corresponding thermocline water exchange windows in the North Pacific and d) South Pacific (RW = recirculation window, WBEW = western boundary exchange window, IEW = interior exchange window). (From *Huang and Liu* [1999].)

et al., 2003; *Molinari et al.*, 2003]. Farther east, interpretation of observations gets more speculative as the coverage becomes increasingly sparse. Nevertheless, it is clear from the available data that the Atlantic STCs have many features in common with those in the Pacific. One prominent difference, though, is the narrowness of the Atlantic basin, which limits the contribution of interior thermocline pathways to the equatorward transport (Section 2.3).

Another difference is the strong influence of the Atlantic MOC (analogous to the ITF-driven circulation in the Pacific) on the structure of the weaker Atlantic STCs. It carries an estimated 16±3 Sv of warm and intermediate waters from the South to the North Atlantic via the NBC [*Ganachaud and Wunsch*, 2001]. The northward flow reaches northern subtropical latitudes by two routes: 1) by ring shedding from the NBC retroflection; and 2) by eastward detours in the EUC, NEUC, and NECC, with some of the MOC water upwelling to shallower levels along the equator. Recent NBC ring studies [*Garzoli et al.*, 2003; *Johns et al.*, 2003] have shown that the rings contribute 8–9 Sv of South Atlantic NBC water to interhemispheric exchange, about half of the total and a much larger fraction than previously thought. Among other things, the MOC prevents much of the subsurface branch of the North Atlantic STC from reaching the equator.

4.1. Wind-Stress Fields and Ekman-Transport Divergences

The Atlantic ITCZ shows a marked seasonal migration (Figs. 10a,b), moving from its most equatorial position during February to its most northern location near 10°N during August. The Ekman transports are directed poleward in both hemispheres with a zonally integrated, annual-mean divergence between 10°S and 10°N of 23 Sv for the NCEP reanalysis stresses (and 21 Sv for ERS-1,2 scatterometer stresses). The mean longitudinal distribution of the meridional NCEP Ekman transports across 10°N and 10°S (Figure 10c) shows that they are strongest in the central and western parts of the respective sections. The time series of the Ekman transport divergence across 10°N/10°S (Plate 1) shows variations of about 2 Sv amplitude at interannual but also longer time scales.

The Ekman-pumping velocity field, w_{ek}, for the annual mean winds (Figure 10c) is similar to the North Pacific, with a band of positive (or weakened negative) w_{ek} that extends nearly across the basin in the North Atlantic, limiting the possibility of interior pathways in the northern hemisphere. There is also an area of positive w_{ek} in the eastern tropical-subtropical South Atlantic, allowing a larger longitude range of interior exchange with the tropics. The area integral of positive w_{ek} in Figure 10 in the region $x > 30°W$, $5°N > y > 22°N$ plus the line integral of offshore Ekman transport along the coast, is 4.4 Sv, providing an upper limit for off-equatorial upwelling in the

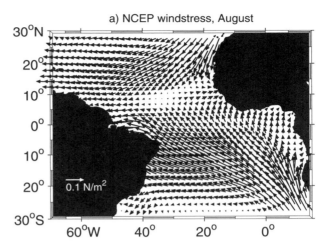

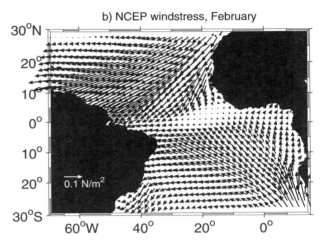

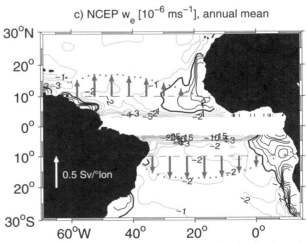

Figure 10. Wind stresses (NCEP) over the tropical and subtropical Atlantic for a) August and b) February, c) annual-mean Ekman pumping (outside ±3° latitude belt); also shown, in c), are meridional Ekman transports (vector at lower left = 0.5 Sv/degree longitude) across 10°N and 10°S.

Guinea Dome and offshore of Northwest Africa. Similarly, the area integral of positive w_{ek} for $x > 5°W$, $20°S > y > 5°S$ plus the coastal Ekman divergence driven by the alongshore winds yields an upper estimate on coastal and Angola-Dome upwelling of 5.5 Sv. The total eastern-boundary Ekman divergence for the Atlantic is therefore about 10 Sv, similar to the eastern Pacific and a substantial fraction of the equatorial 10°N/10°S divergence (23 Sv). As stated in Section 3.1, some amount of this eastern upwelling occurs from 10°S–10°N and so contributes to the Ekman divergence across 10°N/10°S; thus, this amount must be subtracted from the divergence to estimate equatorial open-ocean upwelling.

From recent satellite observations, an ITCZ has also been identified in the western tropical South Atlantic. It extends eastward from the Brazilian coast in the latitude range 3–10°S during boreal summer [*Grodsky and Carton*, 2003], and is associated with wind convergence, high SST, reduced surface salinities, and increased precipitation. Its possible effects on the structure of the southern-hemisphere STC, if any, have not yet been investigated.

4.2. Observations

4.2.1. Subduction. For the North Atlantic, *Qiu and Huang* [1995] estimated an annual-mean subduction rate of 27 Sv. By calculating geostrophic pathways of the subducted waters, they further showed that most of the subducted water returns northward within the NEC and Gulf Stream, but they did not specifically estimate the southward transfer that would contribute to the northern STC. For the South Atlantic, *Karstensen and Quadfasel* [2002] estimated a total subduction of 22.5 Sv south of about 10°S, of which 18.7 Sv was inserted into density classes lighter than 26.8 kg m^{-3} that can upwell in the eastern tropical Atlantic (see Section 4.2.4). Much of the subducted water that is introduced into the southern SEC, however, reaches the western boundary south of the bifurcation latitude (12–15°S; Plate 3), returning southward within the Brazil Current.

4.2.2. Subsurface equatorward flow.

4.2.2.1. Western-boundary pathways: In the South Atlantic, the SEC carries thermocline waters subducted in the southeastern ocean toward the northwest. The bifurcation of the SEC occurs at 12–15°S, and from there the North Brazil Undercurrent (NBUC) transports the bulk of the STC waters equatorward. The NBUC has a strong maximum of more than 50 cm s^{-1} at a depth of about 250 m, with weak, or even reversed, surface currents due to the Ekman transports driven by the Southeast Trades (Figure 11a). It turns westward after passing Cape San Roque (at 5°S) and is augmented by the

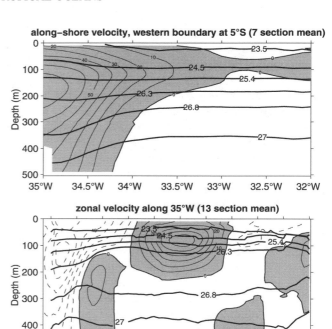

Figure 11. Mean current distributions for a) meridional flow of the North Brazil Undercurrent across 5°S (7 sections), and the zonal equatorial current system b) at 35°W (13 sections), c) at 23°W (1 section); also shown (heavy lines) are several relevant isopycnals. (After *Schott et al.* [2002b, 2003].)

shallower inflow of the low-latitude SEC. This additional flow causes the NBUC to lose its undercurrent character, and it continues to flow northward as the surface-intensified North Brazil Current (NBC; Plate 3 and Figure 11b).

Based on 7 shipboard current-profiling sections at 5°S, *Schott et al.* [2002b] estimated that the NBUC transports 25 Sv northward across that latitude (Figure 11a). This transport is a superposition of the MOC, the South Atlantic STC, and a recirculation of the southward interior Sverdrup transport, the latter estimated to be about 10 Sv near 5°S [*Mayer et al.*, 1998]. From 13 sections along 35°W, *Schott et al.* [2003] estimated an average NBC transport of 32 Sv northwest of Cape

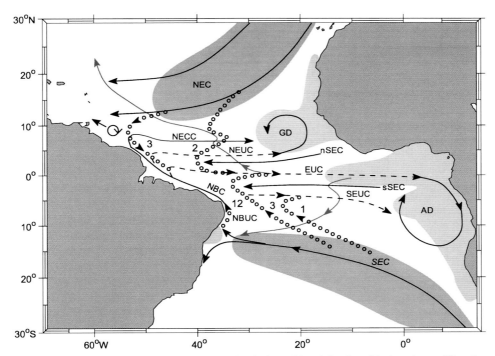

Plate 3. Schematic representation of the Atlantic STC circulation with subduction (blue) and upwelling (green) zones that participate in the STC. Current branches participating in STC flows are NEC, SEC, NECC and EUC as in Plate 2; NEUC, SEUC = North and South Equatorial Undercurrent; NBC, NBUC = North Brazil Current and Undercurrent; GD, AD = Guinea and Angola domes. Interior equatorward thermocline pathways dotted, transport estimates marked for interior and western boundary pathways; surface poleward pathways for the central basin (from drifter tracks, after *Grodsky and Carton* [2002]) marked by thin, magenta line; see text for details.

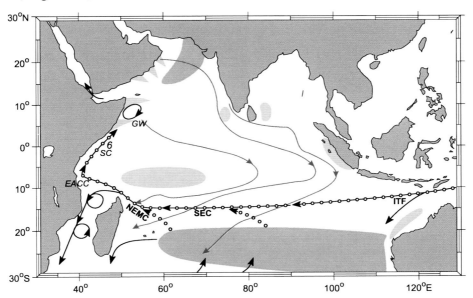

Plate 4. Schematic representation of the Indian Ocean Cross-Equatorial Cell (CEC) with subduction (blue) and upwelling (green) zones that participate in the CEC; ITF = Indonesian Throughflow, SEC = South Equatorial Current, NEMC = Northeast Madagascar Current, EACC = East African Coast Current, SC = Somali Current, GW = Great Whirl; also shown are model surface trajectories (magenta) of southward CEC return flow originating from upwelling sites off Somalia, Oman and west of India; see text for details. (After *Schott et al.* [2002a] and *Miyama et al.* [2003].)

San Roque with another 7 Sv of SEC flow just to its north, yielding a total of 39 Sv flowing across 35°W. *Schott et al.* [1998] reported 35 Sv of this amount crossing the equator at 44°W, after some loss to the South Intermediate Countercurrent (SICC). The NBC overshoots the equator [*Johns et al.,* 1998; *Schott et al.,* 1998] and, after passing through a retroflection zone known for its intense eddy-shedding activity [*Garzoli et al.,* 2003], most of it merges into various zonal currents in the interior ocean. When evaluating the NBUC currents at 5°S from the ADCP observations of *Schott et al.* [2002b] for the STC density range of $\sigma_\theta = 23.4–26.2$ kg m^{-3} (Figure 11a), a northward transport of 12 Sv results, as marked in Plate 3.

In the North Atlantic, thermocline water in the NEC is carried equatorward by the Guyana Undercurrent, which bends eastward to join the NECC at 5–8°N and the NEUC at 3–5°N. It is a very weak flow, transporting only about 3 Sv [*Wilson et al.,* 1994; *Bourles et al.,* 1999b; *Schott et al.,* 1998]. Model studies suggest that it may penetrate further southward during boreal winter to supply the EUC [*Schott and Boening,* 1991]. *Zhang et al.* [2003] estimate a western boundary undercurrent transport of 3.3±1.0 Sv, in agreement with the earlier observational estimates. The weakness of the northern equatorward STC flow compared to the southern one is of course a consequence of the Atlantic MOC, transporting about 15 Sv of warm water across the equator (see Section 4.3.2).

4.2.2.2. Interior pathways: In the tropical-subtropical Atlantic, the water masses subducted in the eastern subtropics and inserted into the thermocline by Ekman pumping are characterized by a potential-vorticity (PV) minimum. The Ekman upwelling associated with the ITCZ, however, brings stratified waters with higher PV into the density range of the subducted waters, causing them to make a westward detour around this barrier on their way south. This path is clearly seen in the PV distribution ($f/\Delta h$, where Δh is layer thickness) on the isopycnal surface $\sigma_\theta = 25.4$ kg m^{-3}, as determined by *Zhang et al.* [2003] from climatological hydrographic data (Figure 12a). Trajectories of the subducted waters on the $\sigma_\theta = 25.4$ kg m^{-3} surface show that there are indeed interior pathways in the South Atlantic, facilitated by the narrowness of the PV barrier there (Figure 12b). For the South Atlantic, *Zhang et al.* [2003] determined the geostrophic transport across 6°S, between the African coast and 34°W for various density classes (Figure 12c), obtaining a total of 4.0±0.5 Sv for the density range $\sigma_\theta = 23.5–26.3$ kg m^{-3}. For the North Atlantic, they estimated 2 Sv of equatorward interior flow across 10°N, concentrated in the density range $\sigma_\theta = 23.5–26.0$ kg m^{-3}. A similar calculation was carried out by *Lazar et al.* [2002], also yielding interior pathways. It should be noted that the lightest portions of the equatorward flow ($< \sigma_\theta = 24.0$ kg m^{-3}) are subducted at quite low latitudes and therefore do not really

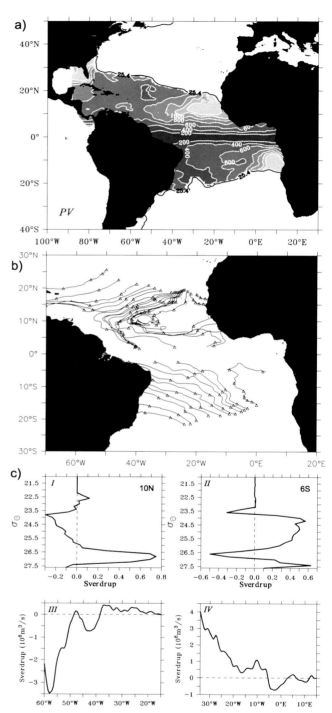

Figure 12. a) Distribution of potential vorticity ($f/\Delta h$, with Δh layer thickness, in units of 10^{-12}m^{-1}s^{-1}); b) geostrophic currents on the isopycnal surface $\sigma_\theta = 25.4$ kg m^{-3} for the tropical Atlantic, based on climatological hydrographic data; c) interior net meridional transports by density layers across 10°N (I: African coast to 60°W) and 6°S (II: African coast to 35°W) as well as accumulated layer transports across 10°N (II) and across 6°S (IV). (From *Zhang et al.* [2003].)

qualify as part of the STC, because they are not participating in a subtropical-tropical exchange.

4.2.3. Equatorial currents.

4.2.3.1. EUC: The EUC draws most of its water from the NBC retroflection (Plate 3). *Schott et al.* [2002b] reported that the undercurrent layer ($\sigma_\theta = 24.5–26.8$ kg m^{-3}) is supplied by a northward flow of 13.4±2.7 Sv at 5°S, and estimated that about 80% of this flow enters the EUC. The EUC has a total eastward transport of 21.9±3.5 Sv in the mean section at 35°W, of which 8.6 Sv occur in the near-surface layer above $\sigma_\theta = 24.5$ kg m^{-3} [*Schott et al.*, 2003]. The surface-layer eastward flow occurs predominantly during the spring when the ITCZ migrates close to the equator (Figure 10a,b), and weakened easterly, or even westerly, wind stresses drive the near-surface flow [*Bourles et al.*, 1999a; *Schott et al.*, 1998].

Farther east at 0°E, the transport is reduced to about 6 Sv [*Bourles et al.*, 2002]. These authors showed that during boreal summer the Atlantic EUC can even surface and terminate away from the boundary. The fate of the mean EUC at the eastern boundary and its possible supply of the Gabon-Congo Undercurrent or Angola Dome is still uncertain [*Stramma and Schott*, 1999].

4.2.3.2. North and South Equatorial Undercurrents: The NEUC and SEUC seem to be weak in the west and to strengthen into the central ocean. Their potential role in the STCs is to provide, from their shallower density layers, water for the off-equatorial eastern upwelling regimes along the coasts and in the Guinea and Angola domes. The SEUC is recognizable (Figure 11b) at a depth of about 100–400 m at 3–4°S in the density range 26.2–27.0 kg m^{-3}; its transport at 35°W is only 3 Sv, but at 23°W (Figure 11c) it has strengthened to about 10 Sv. Although a supply of the SEUC out of the NBC retroflection regime has not been established from ship surveys, the SEUC is distinguishable from the low-oxygen interior thermocline waters of the tropical South Atlantic by a relative oxygen maximum [*Arhan et al.*, 1998]. An interpretation is that it is supplied by a mixture of NBUC waters with interior SEC recirculations. Mean transports from sections further east still have to be composed from the sparse data base.

Since the equatorward flow of subducted water in the interior ocean occurs in the density range $\sigma_\theta = 23.5–26.3$ kg m^{-3} [*Zhang et al.*, 2003], most of it can pass over the SEUC in the western basin (Figure 11b) but not at 23°W in the central basin where isopycnals $\sigma_\theta = 25.4$ kg m^{-1} and deeper pass through its upper part (Figure 11c). Thus, the denser part of the subducted water (Figure 12c) cannot reach the EUC. *Zhang* [2003; pers. comm.] estimated that about 1 Sv of the interior equatorward thermocline flow is trapped by the SEUC in this way and carried eastward with it.

4.2.3.3. Tropical cells: From their ADCP section analysis, *Molinari et al.* [2003] inferred the existence of a North Atlantic TC, consisting of shallow upwelling at the equator and downwelling from 3–6°N. For the downwelling branch, *Grodsky and Carton* [2002] estimated a transport of 42 Sv for 35°W–10°E from drifter convergences. The existence of a TC in the tropical South Atlantic has not yet been quantified. In any event, as for the Pacific, the presence of Atlantic TCs has to be taken into account when evaluating the wind forcing and net meridional upper-layer STC transports, that is, the relevant quantifications need to be carried out poleward of about 6° latitude.

4.2.4. Upwelling.

4.2.4.1. Source depth of upwelled waters: The core of the Atlantic EUC lies in the density range $\sigma_\theta = 24.5–26.8$ kg m^{-3}, and the SEUC and NEUC extend to densities of the order $\sigma_\theta = 27.0$ kg m^{-3}. Since the $\sigma_\theta = 26.25$ kg m^{-3} isopycnal within the EUC reaches the surface at the Greenwich meridian [*Bourles et al.*, 2002], only densities near and less than this value can upwell to the surface at the equator [*Snowden and Molinari*, 2003]. The fate of the denser water is unclear, but it may upwell in the off-equatorial areas.

4.2.4.2. Equatorial upwelling: From an evaluation of early direct current sections across the equator, *Gouriou and Reverdin* [1992] estimated a divergence of 15 Sv for the 4–35°W band, and concluded that upwelling into the surface layer was confined to the upper part of the EUC. From an average of 12 cross-equatorial western Atlantic ADCP sections extending for 10° of longitude centered near 35°W, *Molinari et al.* [2003] estimated an upwelling transport of about 11 Sv. These estimates are comparable to upwelling transports obtained from inverse-model analyses of the divergence across off-equatorial sections. *Lux et al.* [2001], for example, derived an upwelling transport of 7.5 Sv out of the thermocline water layer ($\sigma_\theta = 24.58–26.75$ kg m^{-3}) for a box closed by zonal sections at 7.5°N and 4.5°S. Earlier, *Roemmich* [1983] obtained an upwelling transport of 6–10 Sv across $\sigma_\theta = 26.2$ kg m^{-3} for basin-wide sections at 10°N and 10°S, the spread of values depending on model assumptions. When comparing these estimates with the Ekman divergence of about 25 Sv across 10°N/10°S, it has to be considered that diapycnal upwelling velocities have a sharp profile and maximum upwelling values may not be reached when averaging over density ranges.

4.2.4.3. Off-equatorial upwelling: Off-equatorial upwelling happens along the eastern coasts and in two cyclonic domes, namely, the Guinea and Angola domes in the northern and southern hemispheres, respectively (Plate 3), which are driven by the regions of strong positive w_{ek} in the eastern tropical ocean (Figure 10). Upwelling estimates based on observations have not been reported for these regions, but, as noted above (Section 4.1), their combined Ekman upwelling effect (10 Sv) is not negligible. Their effects, while not known in detail, are nevertheless included in aforementioned inverse studies for those contributions that fall inside the northern and southern boundaries of their respective analysis domains.

4.2.5. Poleward surface flow. Surface-drifter pathways for the tropical Atlantic were determined by *Grodsky and Carton* [2002]. As for the Pacific (Section 3.2.5; Figure 6), they indicate that the STC return flow is carried westward by the zonal currents, except for an eastward portion in the NECC. Consequently, much of the upwelled water returns to the subtropics west of the subduction sites, as indicated in Plate 3 for two trajectories from the central equatorial basin. Again, it has to be noted that these drifter trajectories are not the tracks of real water parcels since they cannot follow downward motions at TC convergences; hence real trajectories would be deflected even more westward than those of Plate 3. This property ensures that much of the STC water does not return directly to the equator, so that the STCs are primarily closed indirectly after one or more recirculations in the Subtropical Gyre, or even an excursion into the subpolar region. Similar to the Pacific, pathways in the eastern basin also extend eastward to be trapped in the Gulf of Guinea.

4.3. Models

4.3.1. STC pathways and exchange windows. A variety of model studies have addressed Atlantic STC pathways and exchange windows between the subtropical subduction regions and the tropical and eastern upwelling regions. *Malanotte-Rizzoli et al.* [2000] used an intermediate-resolution (non-eddy resolving) OGCM, driven by COADS climatology and obtained PV distributions that looked qualitatively similar to those observed (Figure 12a). Regarding the exchange windows, they find an interior exchange zone in the northern hemisphere, whereas all the waters subducted in the eastern South Atlantic take the western-boundary pathway. As expected, the seasonal effect of the northern PV barrier is most pronounced in late boreal summer and much reduced in winter, so that the annual mean is dominated by the summer situation. They attributed the stronger than observed equatorward northern thermocline flow in their simulation to their model's having a somewhat weak MOC.

It is obvious that there must be a dependence of the exchange windows on the patterns and intensity of the wind stress climatology, since that determines upwelling and the PV barrier. *Inui et al.* [2002] studied the differences in STC pathways between the commonly used *Hellermann and Rosenstein* [1983] and *da Silva et al.* [1994] forcing fields. They found that for the stronger *Hellerman and Rosenstein* [1983] forcing the interior exchange window is much reduced in comparison to the weaker *da Silva et al.* [1994] forcing. This calculation was confirmed by a similar model study with both wind stress climatologies carried out by *Lazar et al.* [2002].

4.3.2. Effect of the MOC. Several investigators have studied the interaction of the Atlantic MOC with the STC. *Fratantoni et al.* [2000] compared two solutions to a six layer isopycnal model of the tropical Atlantic, one forced by wind alone and the other by winds and a 15-Sv MOC imposed at the model's northern and southern boundaries. A similar calculation was carried out by *Boening* [2002; pers. comm.] using an Atlantic-basin GCM. Both studies found that the STCs were nearly symmetric about the equator in the solution forced only by winds. In contrast, for the solution forced by wind and the MOC, the northern cell was so weak that there was only about 2 Sv of equatorward thermocline flow in the northern hemisphere, similar to the observational evidence presented above.

The transports of solutions forced by both winds and the MOC are approximately a linear superposition of the transports from solutions forced by MOC and wind forcing alone. This linearity, however, is not at all true for the mesoscale variability. The eddy kinetic energy at the western boundary is greatly enhanced with the addition of the MOC. *Fratantoni et al.* [2000] attribute this to the combined effects of increased current shear, advection of potential vorticity from the equatorial waveguide by the strengthened NBC and enhanced potential vorticity gradient. In their model, NBC eddy shedding only occurs in the combined forcing case.

5. INDIAN OCEAN

In the Indian Ocean, the Southeast Trades do not extend to the equator, and the annual-mean equatorial winds have a slight westerly component (Figure 13c). As a result, there is no annual-mean, equatorial upwelling as in the other oceans. Instead, there are prominent upwelling regions in the northern hemisphere off Somalia, Oman, and India during the summer monsoon. Since subduction occurs predominantly in the southeastern, subtropical Indian Ocean, the shallow overturning circulation associated with these upwelling regions involves interhemispheric flow, forming the Cross-equatorial Cell (CEC). There are also regions of open-ocean upwelling in the southern tropical Indian Ocean and off northwestern Australia that generate intra-hemispheric

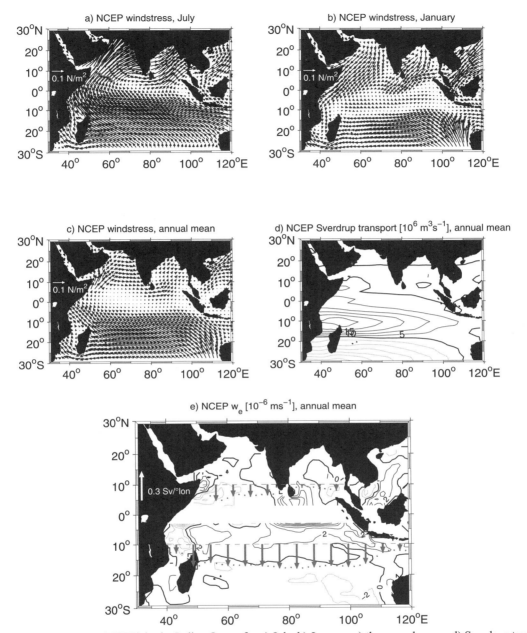

Figure 13. Wind stresses (NCEP) in the Indian Ocean for a) July, b) January, c) the annual mean; d) Sverdrup transport function, and e) annual mean of Ekman pumping (outside ±3° latitude belt), with meridional Ekman transports (vector at left = 0.3 Sv/degree longitude) across 10°N and 10°S.

overturning cells. Superimposed on the basin circulation is the ITF, allowing for complex pathways relating the CEC and hemispheric cells with the other oceans.

5.1. Wind Stress

5.1.1. Monsoon variability. The wind-stress fields for July and January show the drastic changes of the monsoons (Figs. 13a,b). During the northern summer, the Southeast Trades attain their seasonal maximum and extend almost to 5°S, their most northerly position. At this time, they flow into the Arabian Sea over the western basin, forming a narrow atmospheric jet with strong northward cross-equatorial stresses (the Findlater Jet; Figure 13a). As a result, in the Arabian Sea there is Ekman convergence and mixed-layer warming southeast of the jet, and Ekman divergence and cooling northwest of it. North of the equator, the summer-monsoon winds have an eastward component (Figure 13a) and south of it (in the domain of the Southeast Trades) they have a westward component, that is, they drive southward Ekman transports on both sides of the equator. The equatorial winds are directed primarily northward *against* the off-equatorial Ekman transports in both hemispheres (Figure 13a), and they are strongest in the west.

During the winter monsoon, the Southeast Trades are confined south of 10°S, and there is a belt of eastward wind stress between 10°S and the equator (Figs. 13b). Ekman divergence and thermocline doming along the northern edge of the Trades are largest at this time. The zonal-mean stresses are eastward north of the equator and westward south of it. Thus, the Ekman transport is directed northward on both sides of the equator, the opposite situation from the summer. On the equator, the meridional stress is (weakly) southward, again against the Ekman transport.

The equatorial wind stress is eastward during the intermonsoon seasons (April–June and October–December), so that there is Ekman convergence on the equator. These winds drive semiannual eastward surface jets, leading to a deep, annual-mean thermocline in the eastern equatorial ocean [*Wyrtki*, 1973].

5.1.2. Sverdrup and Ekman transports. The annual-mean wind stress distribution is dominated by the summer-monsoon pattern, having anticyclonic stresses over the Arabian Sea (Figure 13c). The resulting mean Sverdrup transport function (Figure 13d) shows the subtropical and SEC circulation of the South Indian Ocean, with a weak cell in the Somali Current area. Using the NCEP climatology for the time period 1990–98, there is a mean southward transport of 6.5±0.9 Sv, based on the mean values for individual years.

As noted by *Godfrey et al.* [2001] and *Miyama et al.* [2003], the zonal component of the wind stress nearly vanishes at the equator during both monsoons, and it is roughly proportional to the distance from the equator on either side. For such a wind field ($\tau^x \propto y$), the Ekman pumping velocity, $-(\tau^x/f)_y$, vanishes completely. It follows that no pressure gradients and, hence, no geostrophic currents are generated, so that the flow field is composed entirely of Ekman drift. For this wind field, then, the concept of Ekman flow is valid all the way to the equator (i.e., in the limit $y \to 0$). Consistent with this property, $-\tau^x_y/\beta = -\tau^x/f$ so that the Sverdrup and Ekman transports are equal, a relation that is valid even at the equator. As a consequence, the cross-equatorial Ekman/Sverdrup flow is very shallow, providing the driving force for the CEC.

The near-equivalence of Ekman and Sverdrup transports has been confirmed by *Schott et al.* [2002a]. For the NCEP winds averaged from 1990–98, the annual-mean Ekman transports across 3°N and 3°S are −6.9±3.6 Sv and −6.0±3.9 Sv, respectively, and the cross-equatorial Sverdrup transport is −6.5±0.9 Sv. The same averages for the ERS-1/2 scatterometer winds yield Ekman transports of 7.4±4.0 Sv and 9.3±6.4 Sv with an across-equatorial Sverdrup transport of −6.4±2.6 Sv.

The longitudinal distributions of the annual-mean Ekman transports across 10°N and 10°S are sketched in Figure 13e. They underscore the completely different behavior of the Indian Ocean in comparison to the Atlantic (Plate 3) and Pacific (Plate 2) Oceans, with (weak) mean southward Ekman transports across 10°N. South of about 8°S, the Southeast Trades are present throughout the year. They are associated with a band of intensified w_{ek} from 2–12°S and 55–90°E (Figure 13e). Ekman divergence estimates for the region are almost 10 Sv [*Schott et al.*, 2002a], and this divergence drives the Indian Ocean's southern-hemisphere STC.

As discussed above, the main driver of the Indian Ocean cross-equatorial cell (CEC) is the cross-equatorial Sverdrup transport. Time series of this field show intraseasonal variations of 5 Sv or more and year-to-year differences of 2–3 Sv. These transport variations are comparable to those in the other two oceans, but they amount to a much larger fraction for the mean CEC than for the mean STCs in the Pacific and Atlantic Oceans.

5.1.3. Subduction. Subduction in the Indian Ocean occurs predominantly in the southeastern subtropical Indian Ocean. Based on their global analysis of climatological data, *Karstensen and Quadfasel* [2002] estimated that a total of 36 Sv is subducted into the 23–27 kg m^{-3} density range in that region. Of this amount, 12.2 Sv enters into densities less than 25.7 kg m^{-3}, corresponding to an upwelling depth off Somalia of about 150 m. A small amount of subduction, estimated to be about 0.5 Sv in the annual mean by *Karstensen* [2003; pers. comm.], also happens in the northern Arabian Sea during the winter monsoon in density classes that can upwell locally (Plate 4). A similar amount of surface water is transformed into thermocline waters in the Red Sea and Arabian Gulf, but into density classes of 27.2 and 26.6 kg m^{-3}, respectively, too dense for upwelling [*Schott et al.*, 2002a].

5.1.4. Southern-hemisphere subsurface flow. Given the structure of the southern-hemisphere winds (Figure 13c), we can expect that south of the equator the Indian Ocean's shallow overturning circulation has a structure like that in the other oceans, with subsurface water following the western-boundary pathway to the equator since $\tau^x(0) \approx 0$ so that $x_e(0) \to -\infty$ [Eq. (7)]. Indeed, this structure is suggested by the distributions of salinity and nutrients in the thermocline (*Schott et al.*, 2002a). Subducted water masses are carried

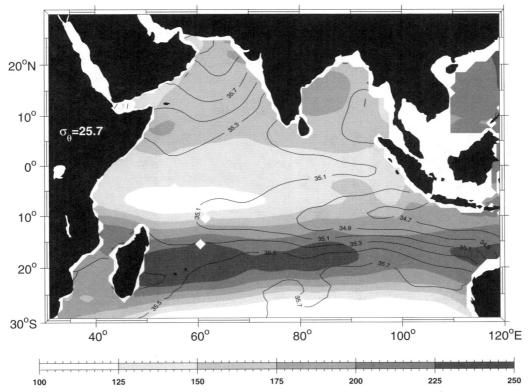

Figure 14. Depths (shading) of and salinities (contours) on potential density surface 25.7 kg m^{-3} for the annual-mean distributions, showing spreading of low-salinity ITF waters and doming northeast of Madagascar.

westward within the South Equatorial Current (SEC). The SEC waters bifurcate at the Madagascar coast, with part turning southward and the rest eventually joining the East African Coastal Current (EACC) to flow to the equator. The denser waters do not participate in the CEC, either retroflecting southeast of Madagascar or flowing out of the basin within the Agulhas Current.

Salinity (Figure 14) and nutrient distributions further indicate a connection to the ITF region. In a recent inverse modeling study, *Ganachaud et al.* [2000] emphasized the connection of the Throughflow to the southern Indian Ocean via the Mozambique Current. Evidence from drifter trajectories and earlier regional budget studies [*Swallow et al.*, 1991] indicates that the near-surface SEC water that arrives north of Madagascar mostly continues to flow northwestward toward the EACC and Somali Current (Plate 4); it is only the deeper thermocline flow that supplies the Mozambique Current.

5.1.5. Cross-Equatorial Cell.

5.1.5.1. Subsurface cross-equatorial flow: *Schott et al.* [1990] observed the cross-equatorial western-boundary flow with moored stations. They found that the annually reversing part of the Somali Current was restricted to be not much deeper than the upper 100 m (Figure 15a,b), with the stronger northward flow during the summer monsoon resulting in an annual northward mean of 3.5 Sv from the surface to 100 m, while from 100–400 m the flow was northward in both seasons. *Schott et al.* [2002a] derived a total of 6.3 Sv for the 50–300 m layer, which carries water in the density ranges that upwell off Somalia. The deeper portions of the northward flow, about 1.6 Sv from 300–500 m must also participate in the northern-hemisphere upwelling, since there is no southward return flow at these densities elsewhere along the equator. (Presumably, all subsurface cross-equatorial flow occurs near the western boundary where mixing and strong nonlinearities can change the sign of its potential vorticity.)

5.1.5.2. Upwelling: During the Southwest Monsoon, coastal upwelling off Somalia typically occurs in wedge-shaped areas, formed by offshore advection along the northern flanks of two prominent summertime gyres, the "Southern Gyre" at 3–5°N and the "Great Whirl" at 8–10°N [*Schott and McCreary*, 2001]. Although upwelling densities as high as $\sigma_\theta = 6.8$ kg m^{-3} have been observed off Somalia, the typical upwelled water is lighter than 25.7 kg m^{-3}.

286 OVERTURNING CIRCULATIONS OF THE TROPICAL-SUBTROPICAL OCEANS

From shipboard sections off Somalia, the Southern Gyre outflow of upwelled water was estimated by *Schott et al.* [2002a] to be about 9 Sv. The Great-Whirl outflow has two portions, one through the passage between Socotra and the African continent (Plate 4) and another by offshore and southeastward recirculation, both branches taking up heat and losing density. *Schott et al.* [2002a] estimated a combined upwelling transport of 12.5 Sv for the summer monsoon and 4.2 Sv for the annual mean. Along the Omani coast, coastal upwelling associated with filaments connected to topographic features

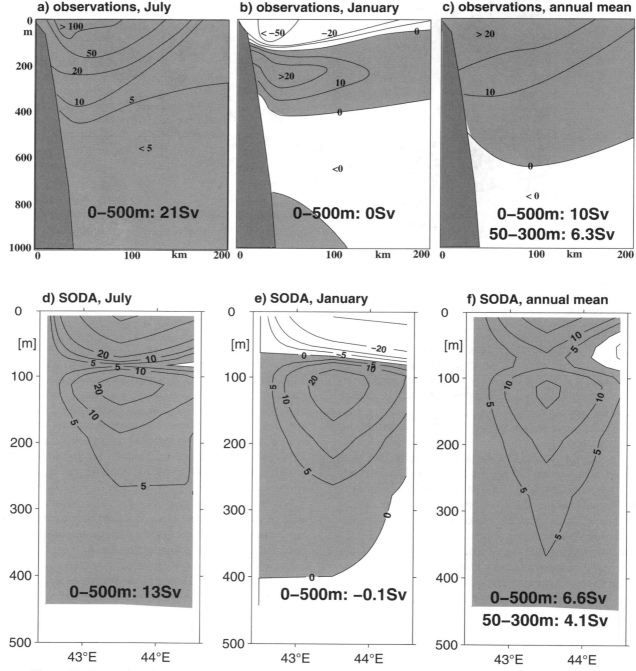

Figure 15. Observed meridional currents and transports of the Somali Current at the equator: a) summer monsoon, b) winter monsoon, c) mean; d)–f) same as a)–c) but from SODA model.

carry the upwelled waters far into the interior Arabian Sea. From various pieces of evidence, *Schott et al.* [2002a] estimated this upwelling to be about 3 Sv for the monsoon season or 1 Sv annually.

There is also open-ocean upwelling in the northern Indian Ocean during the summer monsoon. It has been emphasized in several modeling studies with regard to cyclonic domes east and west of southern India and Sri Lanka [*McCreary et al.*, 1993; *Vinayachandran and Yamagata*, 1998; *Miyama et al.*, 2003], which are driven by the positive Ekman pumping there (Figure 13e). Presumably, open-ocean upwelling also occurs in the Arabian Sea west of the Findlater Jet; however, it has been discounted as a significant upwelling source, likely because its effects are overwhelmed by the nearby Somali and Omani coastal upwelling and by local mixed-layer deepening [*McCreary et al.*, 1993; *McCreary et al.*, 1996; *Lee et al.*, 2000].

5.1.5.3. Surface cross-equatorial flow and equatorial roll: As discussed in Section 5.1, the wind-stress distribution suggests that the CEC is closed by a surface, southward Ekman/Sverdrup transport in the interior ocean. *Schott et al.* [2002a] noted the agreement between the observed northward Somali Current in depth range 50–300 m and the Sverdrup transport, both carrying about 6 Sv (Plate 4). Drifters provide observational evidence for this interior flow, indicating southward flow on either side of the equator to within ±3°. Close to the equator, however, the flow is masked by the equatorial roll, as discussed next.

Solutions forced by meridional winds at the equator [*Wacongne and Pacanowski*, 1996; *Miyama et al.*, 2003] suggest that there should be a vertical-meridional overturning circulation within a narrow band around the equator, with northward (southward) surface currents on the equator during the summer (winter) monsoon and subsurface counterflow underneath. This counterflow connects the southward (northward) Ekman transports from one side of the equator to the other during the summer (winter) monsoon. Based on shipboard ADCP sections taken during the summer monsoon in the western basin where strong northward winds exist on the equator (Figure 13a), *Schott et al.* [2002a] reported observational evidence for the subsurface flow of the roll at speeds exceeding 20 cm s^{-1}. Evidence for the reverse circulation during the winter monsoon was also found, but the roll was not as well developed at that time.

Whether the rolls can cause diapycnal fluxes depends on whether or not they penetrate through the bottom of the mixed layer. In the observations, however, the evidence points to the roll being dominantly restricted to the surface-mixed layer or at least uppermost part of the thermocline, leading *Schott et al.* [2002a] to the conclusion that the equatorial roll has small diapycnal effects and is therefore of little consequence for the meridional heat transport. This conclusion was confirmed by the model study of *Miyama et al.* [2003] (see Section 5.3).

Estimation of near-surface mean flows from drifter trajectories is much more difficult than for the other two oceans because of the seasonal reversal of the monsoon circulation north of about 8°S and the existence of the equatorial roll, the latter preventing drifters from being directly advected across the equator by the Ekman transports in both hemispheres. Therefore, likely mean surface trajectories have been inserted into Plate 4 from the model study of *Miyama et al.* [2003]. Three trajectories are shown from a simulation based on forcing by annual-mean winds, which emanate from the upwelling regimes of Somalia, the eastern Arabia Peninsula and west of India. They all slant to the southeast in the northern hemisphere and down to about 5°S, then reverse to slant to the southwest. As in the other oceans, much of the surface return flow does not return to the southeastern subduction regime, again requiring indirect STC recirculations with the subtropical or subpolar Indian Ocean, or even an interocean transfer into the Atlantic.

5.1.6. Southern-hemisphere STC. There are indications of open-ocean upwelling within the band of intensified Ekman pumping from 5–12°S noted in Section 5.1. Direct evidence of this upwelling, including transport estimates from ocean measurements, is still lacking. It is suggested, however, by the doming of the thermocline along the northern edge of the Southeast Trades (Figure 14). Although not obvious in SST maps because it is masked by the general meridional temperature increase in that region, it is also suggested by satellite color images that show a chlorophyll maximum in that region [*Murtugudde et al.*, 1999]. As determined from numerical simulations, the region has an annual-mean upwelling transport of 5–10 Sv (*McCreary et al.* [1993]; *Ferron and Marotzke* [2003]; *Miyama et al.* [2003]; Figure 16 below), consistent with the amplitude of the Ekman divergence there (Section 5.1.2).

This upwelling regime points toward the existence of a second overturning circulation, the Indian Ocean STC. The subsurface pathways associated with this cell have not been determined from observations. Possible sources for the thermocline water that supplies the upwelling are subduction in the southeastern Indian Ocean, recirculation from the western Indian Ocean, and the ITF. In the *McCreary et al.* [1993] solution, all the water comes from the second source: Just south of the equator part of the subsurface EACC bends offshore to eastward into the upwelling band (see their Figure 3).

5.1.7. Other overturning cells. There is another regime of southern-hemispheric upwelling along the northwest Aus-

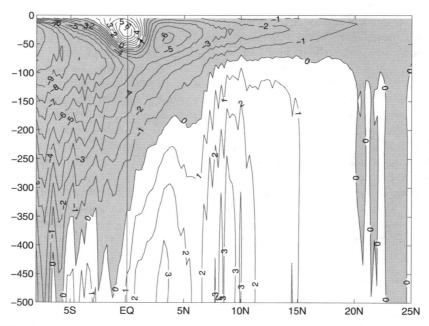

Figure 16. Meridional overturning function of the JAMSTEC model of *Miyama et al.* [2003] for the upper 500 m of the Indian Ocean.

tralian shelf and Arafura Sea, generating another overturning cell. *Godfrey and Mansbridge* [2000] established mass budgets for the combined region based on Ekman and geostrophic currents, estimating an annual-mean upwelling transport of 1.4 Sv (Plate 4), so that cell is weak.

Finally, IOD episodes are marked by westward wind anomalies along the equator, leading to significant uplifting of the thermocline and SST reduction in the eastern equatorial ocean and off Sumatra [*Saji et al.*, 1999; *Feng and Meyers*, 2003], but their contribution to mean upwelling is estimated to be small [*Schott et al.*, 2002a].

5.2. Models

A number of modeling studies have addressed the Indian Ocean overturning circulations [*McCreary et al.*, 1993; *Wacongne and Pacanowski*, 1996; *Garternicht and Schott*, 1997; *Ferron and Marotzke*, 2003; *Miyama et al.*, 2003]. Here, we review results from *Miyama et al.* [2003], who compared solutions to a variety of ocean models: a $1\frac{1}{2}$-layer model and a linear, continuously stratified model in order to explore the dynamics of cross-equatorial flows; a $2\frac{1}{2}$-layer model without the IT; a $4\frac{1}{2}$-layer model with the throughflow included; and the Japan Marine Science and Technology Center (JAMSTEC) global GCM in order to investigate circulation pathways. We also discuss output from the Simple Ocean Data Assimilation (SODA) product of *Carton et al.* [2000] that was analyzed by *Schott et al.* [2002a]. In SODA, a variety of ocean measurements (altimetry, surface temperatures, XBT track-lines, hydrographic stations, etc.) are assimilated into an ocean model, with a horizontal resolution of 1° × 1° at midlatitudes and 1° × 0.45° (longitude–latitude) in the tropics, 20 vertical levels, and a 15-m resolution near the sea surface.

5.2.1. Pathways. Flow pathways in both the model solutions and SODA compare well with the observed currents. The mean surface circulation of SODA shown by current vectors for the upper 15 m (Plate 5) indicates the dominant effect of the summer monsoon, with a northward Somali Current and anticyclonic circulation across the Arabian Sea. The structure of SODA's cross-equatorial Somali Current (Figure 15d–f) compares well with the observed one (Figure 15a–c). On the other hand, in SODA the total cross-equatorial Somali Current transport above 500 m (6.6 Sv) and the thermocline transport in the 50–300 m range (4.1 Sv) are weaker than observational estimates, while its upwelling corresponds to the numbers given above (see below).

The meridional overturning circulation in the JAMSTEC solution (Figure 16) has a cross-equatorial transport of 5 Sv. Of course, the overturning circulation illustrated by this 2-d streamfunction is deceptively simple; for example, the solution's subsurface northward branch occurs mostly in the Somali Current, whereas its near-surface southward branch occurs across the interior of the basin primarily as an Ekman/Sverdrup flow.

Note that the JAMSTEC streamfunction has a well-developed equatorial roll. The existence of the roll is confirmed in model surface trajectories, which sometimes carry out several orbits within the roll before extending into the southern hemisphere. A consequence of the northward surface flow associated with the roll is that model surface drifters only cross the equator in the far-eastern ocean, a property that can be inferred from the structure of the near-equatorial flow field

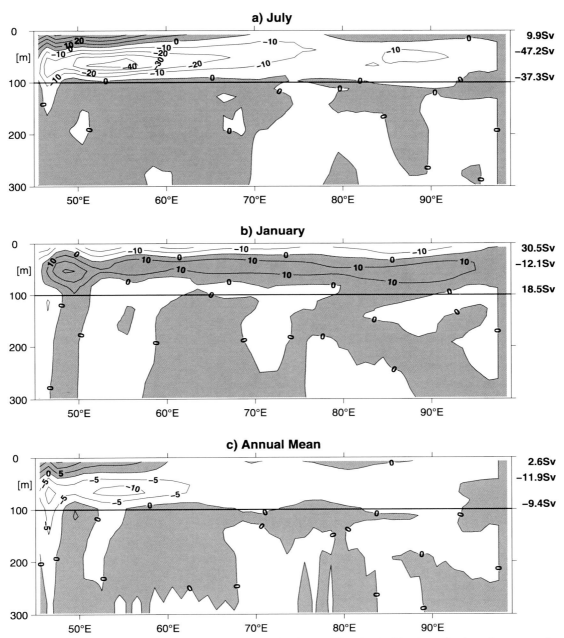

Figure 17. The equatorial roll in the SODA model, shown by the meridional velocities (cms^{-1}), along the equator for a) July, b) January and c) annual mean. Positive (shaded) is northward. Also shown (right margin) are northward, southward and net transports in 0–100-m layer. (From *Schott et al.* [2002a].)

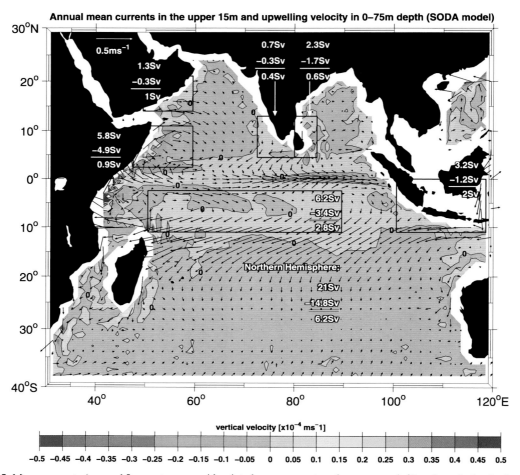

Plate 5. Mean currents (upper 15 m, not every grid point shown; current scale see upper left) and vertical velocity at 75-m depth of SODA model. Shown for marked boxes are upwelling (positive), downwelling (negative) and net transports in Sv across 75-m level. (From *Schott et al.* [2002a].)

(Plate 5). The equatorial roll in the SODA product is developed during the summer monsoon in response to the strong northward winds over the western equatorial ocean, with northward surface flow and southward subsurface flow (Figure 17a); it has a reversed rotation during the winter monsoon (Figure 17b), and is dominated by the summer conditions in the annual mean (Figure 17c).

Miyama et al. [2003] investigated the sources of waters that upwell in the northern hemisphere by tracing particle trajectories backwards from individual upwelling zones. In the JAMSTEC solution, water came from three regions: flow into the basin from south of 30°S, subduction in the southeastern Indian Ocean north of 30°S, and from the Pacific via the ITF. Particles from the Throughflow could follow either a shallow, fast (<4 years) pathway to the upwelling regions or a deep, slow (>10 years) one that involved subduction (Plate 6). The authors provided transports for each of the sources (9.7, 4.7, and 5.1 Sv, respectively, in their Table 2), but were not able to quantify precisely the contribution of each type to the 6-Sv cross-equatorial transport.

5.2.2. Northern-hemisphere upwelling. The model solutions and SODA all produce similar values for the net annual-mean upwelling in the northern hemisphere, a consequence of its being determined by the cross-equatorial Ekman/Sverdrup transport and, hence, solely by the winds. In contrast, the solutions differ markedly in the division of transport between *individual* upwelling regions, indicating that local upwelling is a highly model-dependent process.

Off Somalia, the annual-mean SODA upwelling across 75 m is 5.8 Sv (Plate 5), a value between the low and high estimates of *Schott et al.* [2002a]. This value, however, is nearly compensated by 4.9 Sv of downwelling in the region, yielding an annual-mean upwelling of only 0.9 Sv, much too small in comparison to observed estimates. In the three solutions analyzed by *Miyama et al.* [2003], the net annual-mean upwelling in the region varied from 1.3–3.1 Sv, still somewhat small compared to observations. Off Oman, the net upwellings in SODA and the *Miyama et al.* [2003] solutions vary from 1.0–1.9 Sv, in reasonable agreement with observations. Offshore from southern India and Sri Lanka, there is a total upwelling of 3.0 Sv (1.0 Sv net) in the SODA model (Plate 5), mostly east of 80°E where there is Ekman suction driven by large and positive wind curl. Similarly, the net upwelling transports in the *Miyama et al.* [2003] solutions vary from 1.2–1.8 Sv.

5.2.3. Southern-hemisphere upwelling. For the upwelling band northeast of Madagascar (2–12°S, 50–90°E), *Miyama et al.* [2003] obtained upwelling transports of 5–8 Sv, depending on model type. In SODA, the average upwelling transport across 75 m from 1990–99 is 6.2 Sv, but there was also downwelling of 3.4 Sv across that level, leaving a net upwelling of 2.8 Sv, small compared to the Ekman divergences and model simulations.

6. VARIABILITY

Interest in understanding STC variability stems from its possible influence on equatorial upwelling and SST, and hence climate. So far, observational evidence of STC variability is restricted to the Pacific because of its exceptional coverage of the near-equatorial zone over the past two decades. While modes of variability involving subtropical-tropical SST and ocean advection have also been determined for the Atlantic (e.g., *Huang and Shukla* [1997]) and Indian (e.g., *Xie et al.* [2002]) Oceans, observational evidence for the involvement of the STCs or CEC (Indian Ocean) in climate anomalies in these oceans is so far not available.

6.1. Observations

6.1.1. Pacific. In the North Pacific, one manifestation of Pacific decadal variability is a contrast between anomalously cold SSTs in the center of the basin with anomalously warm SSTs around the eastern boundary, and vice versa [*Mantua et al.*, 1997]. Analysis of subsurface temperature data suggests that when these anomalous SSTs subduct, they gradually propagate downward and southwestward to cool or warm the subtropical thermocline [*Deser et al.*, 1996]. Without salinity data, however, it is difficult to determine whether these anomalies result from changes in the temperature-salinity (T-S) properties of the subducted water or from wind-forced changes in the pycnocline depth.

Considerably less work has been done on the variability of air-sea fluxes and subduction in the South Pacific. This lack is due partly to the relative dearth of oceanic and atmospheric observations there, and partly to the perception that the region exhibits little low-frequency variability; however, a study of the NCEP reanalysis product did show low-frequency variability in the southern hemisphere with a character and magnitude similar to that seen in the north [*Garreaud and Battisti*, 1999]. *Kessler* [1999] provides one of the few pertinent studies of South Pacific variability, focusing on T-S properties of the core isopycnal in the salty tongue of subtropical water that reaches the equator to feed the EUC, a shallow feature that overlies the mode water formed just to its south [*Wong and Johnson*, 2003]. He showed that there is considerable interannual variability along 5–10°S, attributing it to advection and changes in the SEC associated with El Niño. Additionally, there is an interdecadal trend over the record length from 1983–1997, with

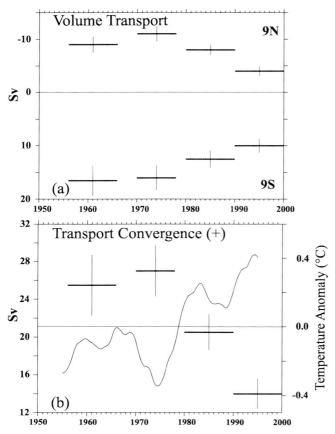

Figure 18. STC variability in the Pacific during 1955–2000, documented by decadal means of a) interior equatorward geostrophic thermocline transports across 9°N and 9°S and b) their sum, showing decrease of STC thermocline convergence from 27 Sv in the 70's to 14 Sv in the 90's; simultaneously, a warming of the equatorial Pacific (curve in b) by about 0.6°C is observed. (From *McPhaden and Zhang* [2002].)

strength, since the edges of the box are located well inside the tropics, not near the subtropical/tropical boundary.

Evidence for interdecadal variability of the strength of the Pacific STCs was recently documented by *McPhaden and Zhang* [2002]. They determined the interior, geostrophic, equatorward convergences across 9°S and 9°N during the past 50 years (Figure 18a), finding that the total convergence reduced from 27±1 Sv during the decade before the mid-1970's regime shift [*Mantua et al.*, 1997] to only 14±1.5 Sv in the 1990's (Figure 18b). There were comparable reductions in estimates of the same quantities when the Sverdrup relation was applied to various wind products, as well as in the poleward Ekman transports. They inferred a weakening of both the North and South STCs over this time period, and hence of the equatorial upwelling. In support of this idea, they noted that there was a gradual increase in NINO3 SST during the same time span.

6.1.2. Atlantic. Aspects of the Atlantic STCs exhibit interannual-to-decadal variability. There are equatorial SST anomalies at interannual periods, corresponding to an Atlantic ENSO, and also at decadal periods [*Carton et al.*, 1996; *Huang and Shukla*, 1997]. The Ekman divergence between 10°N and 10°S has amplitudes of several Sverdrups at these time scales (Plate 1), and temperature variability at thermocline levels has been documented in several studies. The transfer of South Atlantic thermocline waters by North Brazil Current rings (Section 4) also undergoes longer period variations [*Goni and Johns*, 2003], which likely has consequences for the distribution of water masses and STC pathways. To date, though, there is no Atlantic counterpart to the *McPhaden and Zhang* [2002] Pacific STC variability study that would point toward these individual variations being collectively associated with Atlantic STC variability.

6.1.3. Indian Ocean. In the Indian Ocean, interannual variability occurs in both the Cross-equatorial Cell (CEC; Section 5.2.3) and in the STC of the southern hemisphere (Section 5.1.6). Variability of the cross-equatorial heat transport was suggested by *Loschnigg and Webster* [2000] to affect northern hemisphere SST and thereby to influence the following monsoon. In the southern hemisphere, tropical and subtropical anomalies associated with IOD events [*Saji et al.*, 1999; *Feng and Meyers*, 2003] have been found to affect SST strongly at interannual time scales in the upwelling zone northeast of Madagascar. The upwelling is interrupted by Rossby waves arriving from the east that increase upper-layer thickness and SST, causing increased atmospheric convection and enhanced cyclogenesis [*Webster et al.*, 1999; *Xie et al.*, 2002]. Since anomalous Ekman downwelling is involved, this phenomenon may be considered to be an anomaly of the Indian Ocean's southern-hemisphere STC. Similarly, the upwelling

a tendency toward a warmer, saltier tongue, for which he could find no satisfactory explanation.

In their upwelling study (see Section 3.2.4), *Meinen et al.* [2001] also examined the seasonal-to-interannual variability of equatorial upwelling across 50 m in an equatorial box extending from 5°S to 5°N and from 155°E to 95°W. On seasonal time scales, the divergence is mostly determined by variations in zonal geostrophic and meridional Ekman transports. During the 1997–1998 El Niño, the meridional geostrophic convergence was nearly eliminated as the thermocline flattened and the Ekman divergence was reduced even more, halving the upwelling across 50 m. During La Niña phases, they found increased Ekman transports and divergences due to the strengthened trades. It is not clear, however, that these changes represent variations in STC

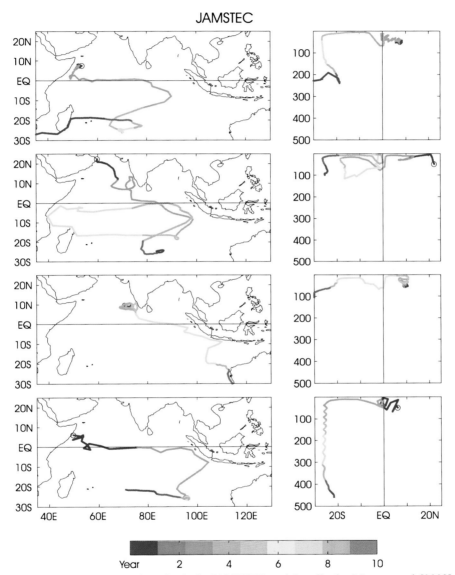

Plate 6. Typical Indian Ocean particle trajectories in the JAMSTEC model studies by *Miyama et al.* [2003], traced backward from the upwelling to the subduction zones, showing x–y and y–z views in corresponding left and right panels.

off Sumatra undergoes substantial variations [*Susanto and Gordon*, 2001; *Feng and Meyers*, 2003], also suggesting shallow cell anomalies.

6.2. Models

A number of recent modeling studies have been carried out to investigate the role of STC variability in climate variability. Two processes have been proposed for this influence. In one process, SST anomalies are subducted into the thermocline in the eastern subtropics, are advected to the equatorial ocean at thermocline levels with a time lag of the order of a decade, and influence equatorial SST when they upwell there (the $\overline{v}T'$ hypothesis, first proposed by *Gu and Philander* [1997]). In the other, STC strength responds to interannual-to-decadal variability of the winds, thus altering the amount, rather than temperature, of the water that is upwelled in the equatorial cold tongue (the $v'\overline{T}$ hypothesis; *Kleeman et al.* [1999]).

6.2.1. $\overline{v}T'$ processes. Existing solutions suggest that $\overline{v}T'$ processes are not adequate for generating equatorial SST anomalies of realistic amplitude at decadal time scales. In the Pacific, *Schneider et al.* [1999] reported a weak response in their wind-forced GCM solution, noting that northern-hemisphere isopycnal-thickness anomalies (assuming a climatological T-S relationship) were advected equatorward by the mean circulation, but only as far south as 18°N; equatorward of that latitude, temperature anomalies resulted from changes in the pycnocline depth forced by local wind variability. *Nonaka and Xie* [2000], *Shin and Liu* [2000], and *Hazeleger et al.* [2001] also noted a weak equatorial response in their GCM solutions forced by realistic midlatitude temperature anomalies.

Schneider [2000] discussed the potential climatic influence of "spiciness" anomalies in the Pacific region of his coupled-model solution. Spiciness anomalies are temperature anomalies accompanied by salinity anomalies such that there is no density change. He noted that such anomalies were subducted into the thermocline in the subtropics and subsequently advected to the equator by the STCs, where they perturbed tropical climate. Essentially, spiciness anomalies behave like "passive" tracers, which do not feedback to affect the pressure field and, hence, the circulation. (In contrast, a temperature anomaly alone is an "active" tracer that does affect pressure.) *Liu and Shin* [1999], *Nonaka et al.* [2000], and *Nonaka and Xie* [2000] have considered the propagation of both active and passive tracers, the latter study concluding that uncompensated temperature anomalies propagate less readily to the equator than a passive tracer.

In the Atlantic, *Lazar et al.* [2001] inserted a warm temperature anomaly (2.2°C) into the eastern subtropical subduction area of their OGCM, and followed its propagation across the South Atlantic and to the equator for a period of 10 years; during years 6–8, there was an increase in equatorial temperature along the $\sigma_\theta = 25.3$ kg m^{-3} surface, but only by 0.2°C. Noting an increase in the anomaly across the South Atlantic, the authors conclude that they are not just passively advected along with the mean thermocline flow but are partly salinity-compensated (a partial spiciness anomaly), indicating the need to observe both salinity and temperature variability in STC studies. They also concluded that, in addition to mean advection, wave processes associated with such dynamic anomalies need to be taken into account.

6.2.2. $\overline{v}T'$ processes. Concerning $\overline{v}T'$ processes, *Klinger et al.* [2002] used a $3\frac{1}{2}$-layer ocean model (a thermodynamic version of the *Lu et al.* [1998] model discussed in Section 3.3) to investigate the influence of switched-on and periodic, off-equatorial winds on the Pacific STCs and equatorial SST. They showed that decadal wind anomalies along the tropical-subtropical boundary can alter the STC transport, leading to significant changes in equatorial SST, whereas wind anomalies at midlatitudes do not. They also showed that the various STC branches respond at quite different time scales, with the surface branch adjusting much more rapidly than the subsurface one. As a result, only at periods considerably longer than a decade can STC variability be assumed to be quasi-steady. It follows that quasi-steady theories of climate variability (e.g., *Huang and Pedlosky* [1999]) should be viewed with caution.

Nonaka et al. [2002] confirmed the importance of tropical-subtropical wind anomalies in causing STC variability in their GCM solution forced by NCEP reanalysis winds. They showed that at decadal time scales weaker (stronger) STC transports are associated with warmer (cooler) equatorial SST anomalies (Plate 7a) and more (less) heat content (Plate 7b). (The green-dashed line in Plate 7a is a trend in STC strength, removed from the green curve. It indicates that the model STCs weakened by 6 Sv from 1965–1990, consistent with but weaker than the observed weakening noted by *McPhaden and Zhang* [2002].) In contrast, there was no such relationship at interannual time scales, pointing toward fundamentally different dynamics for interannual and decadal variability (see their Figure 2). They also noted the importance of forcing by near-equatorial, decadal winds, concluding that the model's decadal variability was initiated by near-equatorial wind anomalies and subsequently maintained by STC variability. Finally, they noted that there was a phase difference between two different measures of STC strength, with a measure based on heat transport V' leading another based on the two-dimensional streamfunction $\Delta\psi'$ (compare blue and green curves in Plate 7a), pointing toward the need for a definition of STC variability that takes into account the different spinup times of the various STC branches.

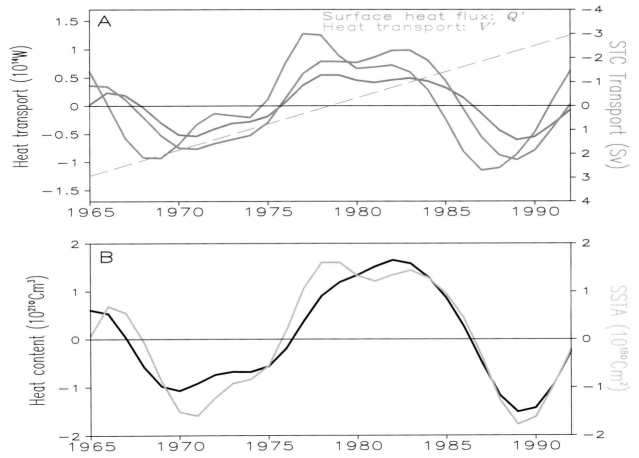

Plate 7. a) Decadal anomalies of heat transport V' across 10°S and 15°N (blue line; left axis), surface heat flux Q' integrated over the area extending across the Pacific and from 10°S and 15°N basin A (red line; left axis), and STC strength $\Delta\psi'$ (green solid line; right axis) for a solution forced by winds without equatorial anomalies (Solution NoEQ). b) Decadal anomalies of heat content from 10°S to 15°N and from the surface to 500 m H' (black line; left axis), and SST anomalies averaged over area A (purple line; right axis) for solution NoEQ. Trends are removed from all curves, with the trend of $\Delta\psi'$ plotted in the green dashed line in a). The $\Delta\psi'$ curves are reversed for easy comparison with other curves. (From *Nonaka et al.* [2002].)

Kleeman et al. [1999] and *Solomon et al.* [2003] developed an intermediate coupled model to explore the potential influence of STC variability. Their solutions developed two modes of oscillation: an ENSO-like interannual mode (~4 year) and a decadal one (~13 year). The decadal mode was generated in northern midlatitudes, and was similar in structure to the *Latif and Barnett* [1994] mode. It was found to influence the equatorial cold tongue by altering the transport of the ocean model's northern STC ($v'\overline{T}$ hypothesis), thereby causing ENSO Decadal Variability (EDV). The STC transport was determined largely by the Ekman transport across the edge of the tropical warm pool, thereby linking subtropical wind variability to equatorial SST anomalies. *Solomon et al.* [2003] noted that for weaker midlatitude coupling, the decadal mode was damped but could still be maintained by tropical air-sea interactions (the ENSO mode) through tropical-to-midlatitude atmospheric teleconnections. In this parameter range, then, both tropical and midlatitude processes are required to cause PDV and EDV, a property consistent with the conclusions of *Nonaka et al.* [2002].

Latif and Lohmann [2003, pers. comm.] discussed Atlantic SST anomalies in a solution to the MPI OPYC coupled model. They found that, although Atlantic anomalies existed at decadal time scales, they had small amplitudes in the tropics compared to those in the Pacific. Since the Atlantic STC anomalies last for decadal time scales, though, they may still be climatically significant.

7. SUMMARY AND CONCLUSIONS

We have reviewed observational evidence and model results for the STCs in all the oceans and for the CEC in the Indian Ocean. Here, we summarize our findings and comment on outstanding issues.

7.1. Theory

Basic STC dynamics can be understood within the framework of a $2\frac{1}{2}$-layer model (Section 2). In this model, the existence and strength of the STCs is determined by the tradewinds across the subtropical/tropical boundary: They cause poleward Ekman drift that drains near-surface water out of the tropics by Ekman drift, thereby forcing a compensating return flow (Section 2.2). The existence of subsurface, equatorward currents in the interior ocean results from the nonlinear terms in Eqs. (1), which require that subsurface flow follows equatorward-bending characteristics (Section 2.3; *Luyten et al.* [1983]). Two key characteristics are the shadow-zone x_e [Eq. (6)] and bifurcation x_b streamlines (Figure 1): Only subsurface water located between these two streamlines flows to the equator to participate in the STC, the water west of x_b recirculating poleward within the subtropical gyre. Thus, it is primarily thermocline water subducted in the eastern subtropics that participates in the STCs. Thermocline water can flow directly to the equator in the interior ocean (the interior pathway) provided that $x_e(0) > 0$ (as for the dashed curve in Figure 1; Section 2.4). The ITCZ, however, can distort or even block interior pathways, by bending characteristics more to the west on its northern flank where $|w_e|$ is small. If the ITCZ is sufficiently strong, $x_e(y)$ intersects the western boundary before it can approach the equator even if $x_e(0) > 0$. For a weaker ITCZ, it first bends westward but then returns eastward to extend to the equator, creating a sinuous interior pathway, for example, as in Figs. 8 and 9. According to (5), the ITCZ's influence on characteristics has a convenient interpretation in terms of wind-driven currents, with characteristics bending westward in the NEC and eastward in the NECC. Another interpretation is that the ITCZ creates a "PV barrier" by uplifting stratified waters with high PV, thereby forcing low-PV subducted waters to take a westward detour around these zones.

7.2. Pacific and Atlantic Oceans

7.2.1. Subduction. Available estimates of subduction rates were obtained by combining Ekman transports determined from climatological winds with referenced geostrophic currents from the mean hydrographic data base. Using this approach, estimates of the total subduction rates in both hemispheres are large, 35 Sv (27 Sv) and 44 Sv (23 Sv) for the North and South Pacific (Atlantic), respectively [*Huang and Qiu*, 1998; *Karstensen and Quadfasel*, 2002]. Only fractions of these totals, however, are subducted in regions with pathways that extend to the equator and density classes that upwell there. It should also be noted that subduction is not restricted to the subtropics, a consequence of the lateral-induction term that results from near-surface flow across a negative gradient of mixed-layer thickness; it can happen even in low latitudes where the Ekman pumping is weak or absent [*Karstensen and Quadfasel*, 2002]. By definition, however, tropical subduction should not be considered part of the STC.

7.2.2. Interior and western-boundary pathways. In the Pacific and Atlantic Oceans, thermocline water moves to the equator via western-boundary and interior pathways (Plates 2 and 3). In the North (South) Pacific, about 14 Sv (15 Sv) of thermocline water takes the western-boundary pathway, with the northern branch first feeding the NECC before joining the EUC and the southern branch moving directly to the equator in the NGUC. Actually, the northern western boundary current carries 23 Sv southward, but about 9 Sv of this North Pacific water feeds the Indonesian Throughflow near the equator.

In the North (South) Atlantic, the western-boundary pathway carries about 3 Sv (12 Sv) to the EUC, the northern branch being nearly blocked by the MOC (see Section 7.4). In the Pacific, the transports of the interior pathways were recently quite well determined by *Sloyan et al.* [2003], who applied a box inverse model to the multitude of direct current observations compiled along cross-sections by the TAO project. They found 5 Sv for the northern interior pathway and 15 Sv for the southern one (Plate 2). In the Atlantic, preliminary estimates of STC transports were derived by *Zhang et al.* [2003] based on geostrophic currents derived from the climatological hydrographic data base, yielding 2 Sv for the northern and 4 Sv for the southern interior STC transport (Plate 3). The interior equatorward STC pathways (dotted in Plates 2 and 3) are deflected eastward (by the NECC) and westward (by the SEC), and therefore particles have to describe wide excursions before joining the EUC (Figs. 8 and 9).

7.2.3. Equatorial currents. The Pacific and Atlantic EUCs are the primary equatorial branches of the STCs. In the western regions, they are both predominantly supplied from the southern hemisphere by the NGUC and NBUC, respectively (Plates 2 and 3), a consequence of interactions with interocean circulations (see Section 7.4 below). The EUCs in both oceans slant upwards across isopycnals to the east (e.g., Figure 5d) with maximum transports of about 40 Sv and 20 Sv, respectively. They seem to terminate near the eastern boundary, and a clear connection to the eastern-boundary circulations has not been established. Water-mass properties, however, suggest that EUC water flows southward, rather than northward, near the eastern boundary.

There are off-equatorial undercurrents in each ocean, located 3–5° on either side of the equator, the NSCC and SSCC (or Tsuchiya Jets) in the Pacific and the NEUC and SEUC in the Atlantic. Source waters for these flows appear to be western-boundary currents, with the exception of the SEUC for which the connection to the NBC retroflection region is not clear [*Arhan et al.*, 1998; *Schott et al.*, 1998]. They diverge poleward [*Johnson and Moore*, 1997; *Rowe et al.*, 2000] as they flow eastward, as shown schematically in Plates 2 and 3. They are merged with the EUC in the western Pacific (Figure 5), but are clearly separate from the EUC in the western Atlantic (Figure 11b). The role of these currents in the STCs, if any, is still a matter of debate. For one thing, they occupy a certain density range that can block interior STC pathways (Figs. 4, 5). Indeed, the SEUC is estimated to block about 1 Sv out of 4 Sv interior thermocline flow across 6°S (*Zhang* [2003, pers. comm.]; dotted line in Plate 3). They can perhaps be viewed as a deep STC branch that upwells in off-equatorial regions. At present, though, their relation to eastern upwelling has not been quantitatively documented.

There is a region of Ekman convergence on either side of the equator, a consequence of the poleward increase of $|f|$. In response, water appears to downwell there and return to the equator at the top of the thermocline or even within the surface mixed layer, to form Tropical Cells (called Equatorial Cells in the latter case by *McCreary and Lu* [1994]). This tendency is particularly prominent at the northern edge of the Southeast Trades (4–6°N), where there is strong convergence associated with weaker winds in the ITCZ. For the Pacific, *Sloyan et al.* [2003] estimated the combined northern and southern TCs to have a strength of 35 Sv (with large error bars), whereas for the Atlantic *Grodsky and Carton* [2003] estimated the transport of the northern TC to be only 4 Sv based on drifter convergences. The TCs are also prominent features in numerical solutions; however, because they occur at shallow levels with little diapycnal exchange, their strength is much reduced when calculated in density coordinates [*Hazeleger et al.*, 2001]. The existence of the TCs requires that any measure of STC strength based on surface transports occurs poleward of the TC convergences (about 7°).

7.2.4. Upwelling. Because upwelling velocities are so small (~1 m/day), upwelling transports are typically estimated indirectly, by calculating the horizontal transport divergence out of a closed box surrounding the region of interest. A complicating factor with this approach is that estimates depend on the chosen depth of the box, so that the calculation should be repeated for a range of depths to determine the maximum upwelling. In addition, the EUC shoals to the east, so the method typically yields higher values than the actual, across-isopycnal transport.

Applying these procedures to the central and eastern Pacific, *Sloyan et al.* [2003] estimated 37±4 Sv of equatorial upwelling in their box inverse model analysis, which is the best available estimate. For the Atlantic, with much less complete observational coverage, only local estimates of upwelling transports have been possible. It should be noted that equatorial upwelling alone is not a measure of STC strength, it has to be corrected for the contribution of the TCs, which never reaches the subtropics. There are no quantitative estimates of eastern-boundary and dome upwelling in either ocean that are based on modern methods.

Upper limits for upwelling transports can be determined from wind-stress divergences. For the Pacific, the total Ekman divergence across 10°S and 10°N is 54 Sv (for NCEP reanalysis stresses) and the coastal and dome upwelling is 10 Sv (Figure 2b). For the Atlantic, the divergence across 10°N and 10°S is 23 Sv; the divergence along coasts and in domes is also 10 Sv (Figure 10c), a significant amount in comparison to the Atlantic's equatorial upwelling.

7.2.5. Poleward surface flows. There is poleward Ekman drift of the surface flow field from the equator to the subtropics, the basic driving force of the STCs (Figs. 2b, 10c). At the same time, surface water is deflected westward by the NEC and SEC and eastward by the NECC on its way north. Therefore, surface pathways (Figure 6, Plate 2) are as sinuous as subsurface interior pathways (Figs. 8, 12), and much of the poleward return flow of the STC ends up westward of the subduction sites (Plates 2, 3). Direct closure within the STC is thus not possible for this flow, leading to interactions with the subtropical gyre and Subpolar Cell (SPC).

7.3. Indian Ocean

The Indian Ocean is unique among the three oceans, in not possessing an eastern upwelling zone and cold tongue. Instead, the mean equatorial winds are westerly, making the equator a downwelling regime so that the upwelling branches of the overturning cells all occur off the equator. In the annual mean, almost all subduction occurs in the southern hemisphere. Some of the subducted water upwells in the southern hemisphere, but a significant portion upwells north of the equator, thereby forming the Indian Ocean's distinctive CEC.

Almost all subduction occurs in the southern hemisphere. It is estimated to be 36 Sv, out of which 12 Sv enter density classes that can upwell in the northern hemisphere [*Karstensen and Quadfasel*, 2002]. The subducted waters flow westward in the SEC alongside waters originating from the ITF and joined by northward recirculation from the southern subtropics (Plate 4), so that it is not clear just what is the final mix of waters that supply the CEC. In the numerical solutions reported by *Miyama et al.* [2003], subsurface water also flows across 30°S to participate in the CEC. There is also a small amount of subduction in the northern Arabian Sea, that might feed the upwelling there (Plate 4).

One measure of the strength of the CEC is the southward, near-surface transport across the equator (analogous to $M_1(y_d)$ in (2) for the STCs). For the particular mean wind field of the Indian Ocean, in which τ^x is roughly proportional to y near the equator, the cross-equatorial Sverdrup transport is nearly equal to the Ekman transport, a consequence of the property that

$$\lim_{y \to 0} \tau^x/f = \tau^x_y/\beta$$

for $\tau^x \propto y$ [*Godfrey et al.*, 2001; *Miyama et al.*, 2003]. *Schott et al.* [2002a] demonstrate this near-equality, finding that the winds drive a southward Ekman-Sverdrup transport of about 6 Sv. Consistent with this value, the CEC is closed by northward flow of the Somali Current at the thermocline level, which is also close to 6 Sv in western boundary observations (Figure 15). This strength is also consistent with the overturning streamfunction of the JAMSTEC GCM solution and other high-resolution models (Figure 16).

During the Southwest Monsoon, upwelling occurs in the northern hemisphere, mostly along the coasts of Somalia and Oman where it is driven by the intense Findlater Jet. It extends offshore from Somalia in wedges between large quasi-stationary gyres (the Southern Gyre and Great Whirl) and from Oman in filaments that extend well into the interior of the northern Arabian Sea (Plate 4). Model studies also suggest that there is open-ocean upwelling south of India and Sri Lanka, but observational evidence is not available to quantify it.

In addition to the CEC, the Indian Ocean appears to have another prominent overturning circulation confined to the southern hemisphere, the Indian Ocean STC. It is associated with upwelling driven by Ekman pumping along the northern edge of the Southeast Trades from 5–10°S (Plate 4). This cell was first reported in the modeling study of *McCreary et al.* [1993]. *Murtugudde et al.* [1999] noted its influence on ocean color.

7.4. Influence of Interocean Circulations

In the Pacific and Atlantic Oceans, one might expect the STCs to be nearly symmetric about the equator based on the structure of the wind fields that drive them (with poleward Ekman transports roughly equal in both hemispheres), but they are not. In the Atlantic, the reason for the asymmetry is the northward transport of 15–20 Sv by the warm limb of the Atlantic MOC. It strengthens the southern STC, but weakens the northern one so much that only small equatorward thermocline flow is observed in the northern hemisphere (Plate 3). In the Pacific, the asymmetry is due to the ITF with an estimated transport of 10–15 Sv [*Ganachaud et al.* 2000; *Sloyan et al.*, 2003]. The ITF is supplied by thermocline and intermediate water from the South Pacific, which must cross the equator and upwell to shallower levels before eventually merging into the ITF in the Makassar Strait (Plate 2). Thus, it also acts to strengthen the southern STC and weaken the northern one. Its effects, however, are less apparent than those of the Atlantic's MOC because of the much larger transports of the Pacific STCs.

In the Indian Ocean, the ITF alters the mix of waters that participate in the CEC. Water-mass properties indicate that a portion of ITF water participates in the cross-equatorial Somali Current but the ITF influence on CEC pathways is not clear. As the (model) surface pathways in Plate 4 indicate, the cross-equatorial return CEC flow from the northern upwelling sites can arrive west of the subduction regime, thus feeding into the gyre return flow or even into exchanges with the Atlantic.

7.5. STC Variability and Climate

Because the STCs account for a significant amount of poleward heat transport [*Talley*, 2003], their variability has been hypothesized to be involved in climate. In one hypothesis ($\mathbf{v'}\overline{T}$ processes), temperature anomalies subducted into the thermocline at midlatitudes are advected to the equator, where they upwell to affect SST [*Gu and Philander*, 1997]. In the other $\mathbf{v'}\overline{T}$ processes, variability in the strength of the STCs leads to SST anomalies [*Kleeman et al.*, 1999].

There is little observational evidence to support either hypothesis. In support of the former, *Deser et al.* [1996] reported the downward and equatorward propagation of temperature anomalies, but they did not follow them to the equator. In support of the latter, *McPhaden and Zhang* [2002] reported a decrease in the equatorward transport convergence across 9°N and 9°S in the Pacific Ocean from 27 Sv during the 1970's to about half that value during the 1990's. There was a similar decrease in Ekman transport divergence across these latitudes during the same time period, as well as an increase of equatorial SST by about 0.6°C (Figure 18b). A recent update of this calculation to 1998–2003 reveals a rebound of the STC strength associated with an SST cooling [McPhaden and Zhang, pers. comm., 2004].

Modeling studies generally suggest that $\overline{v}T'$ processes do not generate significant equatorial SST anomalies [*Schneider et al.*, 1999; *Nonaka and Xie*, 2000; *Hazeleger et al.*, 2001; *Lazar et al.*, 2001], although there is an indication of the potential influence of spiciness anomalies [*Schneider*, 2000]. In contrast, ocean-only models forced by wind anomalies along the subtropical/tropical lead to changes in STC strength that, in turn, cause variations in equatorial SST that are consistent with observed decadal variability [*Klinger et al.*, 2002; *Nonaka et al.*, 2002], and intermediate coupled models have demonstrated the potential feedback of STC-driven SST anomalies to the atmosphere [*Kleeman et al.*, 1999; *Solomon et al.*, 2003]. A noteworthy result of the ocean-model studies is that the STC response is not quasi-steady even at quite low frequencies (periods of 10 years and longer), the surface branch responding more rapidly than the subsurface one.

7.6. Conclusions

We are only beginning to understand the role of STCs in ocean circulation and climate. Even their mean transports and pathways are still poorly determined from observations, and there are significant differences in their representation in different model types. In the Pacific our quantitative knowledge is fairly advanced [*Meinen et al.*, 2001; *Sloyan et al.*, 2003; *McPhaden and Zhang*, 2002] due to the continuous observational efforts related to the El Niño Southern Oscillation (ENSO) observing system [*McPhaden et al.*, 1998]. In contrast, the existing Atlantic data base is much weaker: Only in the western tropical Atlantic have continuous observations of western boundary STC components been started [*Molinari et al.*, 2003; *Schott et al.*, 2003].

There is a particular need for observational work to determine pathways of thermocline flow between identified subduction regions and the EUCs, as well as the potential participation of the off-equatorial undercurrents (NSCC and SSCC in the Pacific; NEUC and SEUC in the Atlantic) in the STCs. One method of pursuing STC pathways involves deploying Autonomous Profiling Explorers (APEX) that drift at shallow levels to provide velocity information there, but take temperature and salinity profiles to deeper levels at preselected time intervals to provide water mass information. As shown for the tropical Atlantic [*Schott et al.*, 2002b], however, floats at constant levels are of reduced usefulness as tracers for STC pathways since they cross density surfaces. Instead, isopycnic RAFOS floats are needed, which can be analyzed as true density-surface followers similar to model Lagrangian studies (e.g., Figure 9a).

Regarding the surface circulation, the Atlantic is lacking not only in data on subsurface zonal currents but also in surface drifter data. In the Pacific, surface drifter data are sufficiently dense to quantify mean STC pathways, equatorial divergence, and off-equatorial TC convergence [*Johnson*, 2001]. In contrast, surface drifter data are much more sparse in the tropical Atlantic, especially in the east, making such estimates uncertain there [*Grodsky and Carton*, 2002; *Molinari et al.*, 2003].

Large uncertainties also surround subduction and upwelling by the STCs. So far, estimates of mean subduction rates have been limited to the evaluation of historical data sets. When fully implemented, the Array for Realtime Geostrophic Oceanography (Argo) Program [*Argo Science Team*, 1998] will provide sufficient CTD data from profiling floats to estimate nearly global fields of large-scale interior geostrophic circulation and mixed-layer properties. These data, combined with satellite scatterometer winds fields, should allow large-scale estimates of the temporal evolution of subduction from season to season. In addition, time series are required to understand the influence of high-frequency small-scale forcing in subduction. Regarding upwelling, we noted that the relation between the various zonal equatorial currents and the eastern coastal and dome upwelling zones is still largely unknown, although Ekman transport divergences hint at some importance of these zones for the STCs. Focused process studies are needed to fill these knowledge gaps.

For determining potential STC predictability, transports by the STC branches need to be measured continuously at key locations, and then related to indicators of STC strength (such as Ekman divergence at 10°S/10°N), to equatorial SST variability, and to model simulations. Key locations are the low-latitude western boundary currents and cross-equatorial sections in the interior ocean, to measure the transports of the western-boundary and interior STC pathways, respectively. The interior observations, which are already underway in the Pacific, need to be extended meridionally beyond the latitudes of the TC convergences. While seasonal ship sections might provide sufficient temporal coverage for the STC variability time scale (Plate 7), aliasing by Tropical Instability Waves (and possibly their role in STC currents and stratification) may be substantial, requiring longer-term moored stations. Several such studies are presently being discussed within the international Climate Variability and Predictability studies (CLIVAR) program.

Regarding modeling studies, it is important to determine the causes (model resolution, mixing parameterizations, forcing, etc.) for the considerable differences in mean STC transports and pathways among models. It is equally important to understand the processes that govern STC variability and its influence on equatorial SST in greater detail. Finally, models can be used to develop a comprehensive measure for STC strength and variability; there are indications that variability of the meridional streamfunction, which implicitly assumes that surface and subsurface branches respond at the same time scale, is not adequate.

In conclusion, we have shown here that STCs and the CEC play important roles in the mean circulation of all three tropical-subtropical oceans, and that there is evidence for their participation in climate anomalies. We are, however, only at the beginning to develop a quantitative understanding of the overturning circulations and their variability. Further progress requires new dedicated observational activities and modeling efforts.

Acknowledgments. We thank Rena Schoenefeldt, Verena Hormann and Jens Schafstall (all of IFM-GEOMAR) for help with the graphic presentations; J. Karstensen (IFM-GEOMAR) for supplying subduction transports for the three southern oceans, and Lothar Stramma (IFM-GEOMAR) for critical readings of several versions of the manuscript. Fritz Schott acknowledges support from the German CLIVAR program through Contract 03F0246A. Julian McCreary acknowledges support from the Frontier Research System for Global Change through its sponsorship of the IPRC. Gregory Johnson acknowledges support from the NOAA Office of Oceanic and Atmospheric Research. This manuscript is SOEST contribution No. 6350, IPRC contribution No. 265, and PMEL contribution No. 2623.

REFERENCES

Argo Science Team, On the design and implementation of Argo: An initial plan for a global array of profiling floats, *International CLI-VAR Project Office Report 21*, 32 pp., 1998.

Arhan, M., H. Mercier, Y. Gouriou, and B. Bourlès, Hydrographic sections across the Atlantic Ocean at 730'N and 430'S, *Deep Sea Res., Part I*, *45*, 829–872, 1998.

Blanco, J. L., A. C. Thomas, M.-E. Carr, and P. T. Strub, Seasonal climatology of hydrographic conditions in the upwelling region off northern Chile, *J. Geophys. Res.*, *106* (C6), 11,451–11,467, 2001.

Blanke, B., and S. Raynaud, Kinematics of the Pacific equatorial undercurrent. An Eulerian and Lagrangian approach from GCM Results, *J. Phys. Oceanogr.*, *27*, 1038–1053, 1997.

Bourlès, B., Y. Gouriou, and R. Chuchla, On the circulation in the upper layer in the western equatorial Atlantic, *J. Geophys. Res.*, *104*(C9), 21,151–21,170, 1999a.

Bourlès, B., R.L. Molinari, E. Johns, W.D. Wilson, and K.D. Leaman, Upper layer currents in the western tropical Atlantic (1989–1991). *J. Geophys. Res.*, *104*(C1), 1361–1375, 1999b.

Bourlès, B., M. D. D'Orgeville, G. Eldin, Y. Gouriou, R. Chuchla, Y. DuPenhoat, and S. Arnault, On the evolution of the thermocline and subthermocline eastward currents evolution in the equatorial Atlantic, *Geophys. Res. Lett.*, *29*(16), doi:10.1029/2002GL015098, 2002.

Bryden, H. L., and E. C. Brady, Diagnostic model of the three-dimensional circulation in the upper equatorial Pacific Ocean, *J. Phys. Oceanogr.*, *15*, 1255–1273, 1985.

Butt, J., and E. Lindstrom, Currents off east coast of New Ireland, Papua-New Guinea, and their relevance to regional undercurrents in the western equatorial Pacific Ocean, *J. Geophys. Res.*, *99*, 12,203–12,514, 1994.

Carton, J. A., X. Cao, B. S. Giese, and A. M. da Silva, Decadal and interannual SST variability in the tropical Atlantic Ocean, *J. Phys. Oceanogr.*, *26*, 1165–1175, 1996.

Carton, J. A., G. Chepurin, X. Cao, and B. Giese, A simple ocean data assimilation analysis of the global upper ocean 1950–95. Part I: Methodology. *J. Phys. Oceanogr.*, *30*, 294–309, 2000.

da Silva, A. M., C. C. Young, and S. Levitus, *Atlas of Surface Marine Data 1994*. Vol. 3: Anomalies of heat and momentum fluxes; Vol. 4: Anomalies of fresh water fluxes NOAA Atlas NESDIS 8, U.S. Department of Commerce, NOAA, NESDIS, 413 pp., 1994.

Deser, C., A. Alexander, and M. S. Timlin, Upper-ocean thermal variations in the North Pacific during 1970–1991, *J. Clim.*, *9*, 1840–1855, 1996.

Esbensen, S. K., and V. Kushnir, The heat budget of the global ocean: An atlas based on estimates from marine surface observations. *Climatic Research Institution Rep. 29*, 27 pp., 1981.

Feng, M., and G. Meyers, Interannual variability in the tropical Indian Ocean: a two-year time-scale of Indian Ocean dipole, *Deep Sea Res., Part II*, *50*, 2263–2284, 2003.

Ferron, B., and J. Marotzke, Impact of 4D-variational assimilation of WOCE hydrography on the meridional circulation of the Indian Ocean, *Deep Sea Res., Part II*, *50*, 2005–2021, 2003.

Fine, R. A., W. H. Peterson, and H. G. Ostlund, The penetration of the tritium into the tropical Pacific, *J. Phys. Oceanogr.*, *17*, 553–564, 1987.

Fratantoni, D. M., W. E. Johns, T. L. Townsend, and H. E. Hurlburt, Low-latitude circulation and mass transport pathways in a model of the tropical Atlantic Ocean, *J. Phys. Oceanogr.*, *30*, 1944–1966, 2000.

Ganachaud, A., and C. Wunsch, Improved estimates of global ocean circulation, heat transport and mixing from hydrographic data, *Nature*, *408*, 453–457, 2001.

Ganachaud, A., C. Wunsch and J. Marotzke, The Meridional Overturning and Large-Scale Circulation of the Indian Ocean, *J. Geophys. Res.*, *105*, 26,117–26,134, 2000.

Garreaud, R. D., and D. S. Battisti, Interannual (ENSO) and interdecadal (ENSO-like) variability in the southern hemisphere tropospheric circulation, *J. Clim.*, *12*, 2113–2123, 1999.

Garzoli, S. L., A. Ffield, and Q. Yao, North Brazil current rings and the variability in the latitude of retroflection, in: *Interhemispheric water exchange in the Atlantic Ocean*, Goni, G.J., and P. Malanotte-Rizzoli (Editors), Elsevier Oceanography Series, pp.357–373, 2003.

Garternicht, U., and F. Schott, Heat fluxes of the Indian Ocean from a global eddy-resolving model, *J. Geophys. Res.*, *102*, 21,147–21,159, 1997.

Godfrey, J. S., G. C. Johnson, M. J. McPhaden, G. Reverdin, and S. Wijffels, The tropical ocean circulation. In: *Ocean circulation and climate*, Church, J., J. Gould, and G. Siedler (Editors), Academic Press, London, pp. 215–245, 2001.

Godfrey, J. S. and J. V. Mansbridge, Ekman transports, tidal mixing, and the control of temperature structure in Australia's northwest waters, *J. Geophys. Res.*, *105*(C10), 24,021–24,044, 2000.

Goni, G. J. and W. E. Johns, Synoptic study of warm rings in the North Brazil Current retroflection region using satellite altimetry, in: *Interhemispheric water exchange in the Atlantic Ocean*, Goni, G.J., and P. Malanotte-Rizzoli (Editors), Elsevier Oceanography Series, pp. 335–356, 2003.

Gordon, A. L., R. D. Susanto, and A. Ffield, Throughflow within Makassar Strait, *Geophys. Res. Lett.*, *26*, 3325–3329, 1999.

Gouriou, Y., and G. Reverdin, Isopycnal and diapycnal circulation of the upper equatorial Atlantic Ocean in 1983–1984, *J. Geophys. Res.*, *97*(C3), 3543–3572, 1992.

Grodsky, S.A. and J.A. Carton, Surface drifter pathways originating in the equatorial Atlantic cold tongue, *Geophys. Res. Lett.*, *29*(23), 62–65, 2002.

Grodsky, S. A., and J. A. Carton, The intertropical convergence zone in the South Atlantic and the equatorial Cold Tongue, *J. Clim.*, *16*(4), 723–733, 2003.

Gu, D., and S. G. H. Philander, Interdecadal climate fluctuations that depend on exchanges between the tropics and extratropics, *Science*, *272*, 805–808, 1997.

Hautala, S., and D. Roemmich, Subtropical mode water in the northeast Pacific basin, *J. Geophys. Res.*, *103*, 13,055–13,066, 1998.

Hazeleger, W., P. de Vries, and G. J. van Oldenborgh, Do tropical cells ventilate the Indo-Pacific equatorial thermocline?, *Geophys. Res. Lett.*, *28*, 1763–1766, 2000.

Hazeleger, W., M. Visbeck, M. Cane, A. Karspack, and N. Naik, Decadal upper ocean temperature variability in the tropical Pacific, *J. Geophys. Res.*, *106*, 8971–8988, 2001.

Hellerman, S., and M. Rosenstein, Normal monthly wind stress over the world ocean with error estimates, *J. Phys. Oceanogr.*, *13*, 1093–1105, 1983.

Huang, B., and Z. Liu, Pacific subtropical-tropical thermocline water exchange in the National Centers for Environmental Prediction ocean model, *J. Geophys. Res.*, *104*, 11,065–11,076, 1999.

Huang, R. X., and J. Pedlosky, Climate variability inferred from a layered model of the ventilated thermocline, *J. Phys. Oceanogr.*, *29*, 779–790, 1999.

Huang, R. X., and B. Qiu, The Structure of the wind-driven circulation in the subtropical south Pacific Ocean. *J. Phys. Oceanogr.*, *28*, 1173–1186, 1998.

Huang, B., and J. Shukla, Characteristics of the interannual and decadal variability in a general circulation model of the tropical Atlantic Ocean, *J. Phys. Oceanogr.*, *27*(8), 1693–1712, 1997.

Huyer, A., M. Knoll, T. Paluskiewicz, and R. L. Smith, The Peru Undercurrent: a study in variability, *Deep Sea Res.*, *Suppl. 1*, *38*, 247–271, 1991.

Inui, T., A. Lazar, P. Malanotte-Rizzoli, and A. Busalacchi, Wind stress effects on subsurface pathways from the subtropical to tropical Atlantic, *J. Phys. Oceanogr.*, *32*(8), 2257–2276, 2002.

Johns, W. E., T. N. Lee, R. C. Beardsley, J. Candela, R. Limeburger, and B. Castro, Annual cycle and variability of the North Brazil Current, *J. Phys. Oceanogr.*, *28*, 103–128, 1998.

Johns, W. E., R. Zantopp, and G. J. Goni, Cross-gyre transport by North Brazil Current rings, in: *Interhemispheric water exchange in the Atlantic Ocean*, Goni, G.J., and P. Malanotte-Rizzoli (Editors), Elsevier Oceanography Series, pp. 411–441, 2003.

Johnson, G. C., The Pacific Ocean subtropical cell surface limb, *Geophys. Res. Lett.*, *28*, 1771–1774, 2001.

Johnson, G. C., and M.J. McPhaden, Interior pycnocline flow from the subtropical to the equatorial Pacific Ocean, *J. Phys. Oceanogr.*, *29*(12), 3073–3089, 1999.

Johnson, G. C., M. J. McPhaden, and E. Firing, Equatorial Pacific Ocean horizontal velocity, divergence, and upwelling, *J. Phys. Oceanogr.*, *31*(3), 839–849, 2001.

Johnson, G. C., and D.W. Moore, The Pacific Subsurface Countercurrents and an inertial model, *J. Phys. Oceanogr.*, *27*(11), 2448–2459, 1997.

Johnson, G. C., B. M. Sloyan, W. S. Kessler, and K.E. McTaggart, Direct measurements of upper ocean currents and water properties across the tropical Pacific during the 1990s, *Prog. Oceanogr.*, *52*, 31–61, 2002.

Karstensen, J. and D. Quadfasel, Formation of southern hemisphere thermocline waters: water mass conversion and subduction, *J. Phys. Oceanogr.*, *32*(11), 3020–3038, 2002.

Kessler, W. S., Interannual variability in the subsurface high-salinity tongue south of the equator at 165E, *J. Phys. Oceanogr.*, *29*(8), 2038–2049, 1999.

Kessler, W. S., Mean three-dimensional circulation in the northeast tropical Pacific, *J. Phys. Oceanogr.*, *32*, 2457–2471, 2002.

Kessler, W. S., G. C. Johnson, and D. W. Moore, Sverdrup and nonlinear dynamics of the Pacific Equatorial Currents. *J. Phys. Oceanogr.*, *33*, 994–1008, 2003.

Kleeman, R., J. P. McCreary Jr., and B. A. Klinger, A mechanism for generating ENSO decadal variability, *Geophys. Res. Lett.*, *26*(12), 1743–1746, 1999.

Klinger, B. A., J. P. McCreary Jr., and R. Kleeman, The relationship between oscillating subtropical wind stress and equatorial temperature, *J. Phys. Oceanogr.*, *32*, 1507–1521, 2002.

Latif, M., and T. P. Barnett, Causes of decadal climate variability over the north Pacific and North America. *Science*, *266*, 634–637, 1994.

Lazar, A., T. Inui, P. Malanotte-Rizzoli, A. J. Busalacchi, L. Wang, and R. Murtugudde, Seasonality of the ventilation of the tropical Atlantic thermocline in an ocean general circulation model, *J. Geophys. Res.*, *107*(C8), 3104, doi:10.1029/2000JC000667, 2002.

Lazar, A., R. Murtugudde, and A. J. Busalacchi, A model study of temperature anomaly propagation from the subtropics to tropics within the South Atlantic thermocline, *Geophys. Res. Lett.*, *28*, 1271–1274, 2001.

Lee, C. M., B. H. Jones, K. H. Brink, and A.S. Fischer, The upper ocean response to monsoonal forcing in the Arabian Sea, seasonal and spatial variability, *Deep Sea Res., Part II*, *47*, 1177–1226, 2000.

Levitus, S., Climatological atlas of the world ocean. *NOAA Profess. Paper No 13*, US. Dept. of Commerce, Rockville, MD, 173 pp., 1982.

Levitus, S., and T. Boyer, World Ocean Atlas 1994, Vol. 3: Salinity. *NOAA Atlas NESDIS 3*, U.S. Government Printing Office, Wash., D.C., 93 pp., 1994a.

Levitus, S., and T. Boyer, World Ocean Atlas 1994, Vol. 4: Temperature. *NOAA Atlas NESDIS 4*, U.S. Government Printing Office, Wash., D.C., 117 pp., 1994b.

Liu, Z., A simple model of the mass exchange between the subtropical and tropical ocean, *J. Phys. Oceanogr.*, *24*, 1153–1165, 1994.

Liu, Z., and S. G. H. Philander, How different wind stress patterns affect the tropical-subtropical circulations of the upper ocean, *J. Phys. Oceanogr.*, *25*, 449–462, 1995.

Liu, Z. and G. Philander, Tropical-extratropical oceanic exchange pathways, in: *Ocean Circulation and Climate: Observing and Modeling the Global Ocean*, G. Siedler et al. (Editors), pp. 247–254, 2001.

Liu, Z., S. G. H. Philander, and R.C. Pacanowski, A GCM study of the tropical-subtropical upper-ocean exchange, *J. Phys. Oceanogr.*, *24*, 2606–2623, 1994.

Liu, Z., and B. Huang, Why is there a tritium maximum in the central equatorial Pacific thermocline, *J. Phys. Oceanogr.*, *28*, 1527–1533, 1998.

Liu, Z., and S. Shin, On thermocline ventilation of active and passive tracers, *Geophys. Res. Lett.*, *26*, 357–360, 1999.

Loschnigg, J., and P.J. Webster, A coupled ocean-atmosphere system of SST modulation for the Indian Ocean. *J. Clim.*, *13*(19), 3342–3360, 2000.

Lu, P., and J. P. McCreary Jr., Influence of the ITCZ on the flow of the thermocline water from the subtropical to the equatorial Pacific Ocean, *J. Phys. Oceanogr.*, *25*, 3076–3088, 1995.

Lu, P., J. P. McCreary Jr., and B. A. Klinger, Meridional circulation cells and the source waters of the Pacific equatorial undercurrent, *J. Phys. Oceanogr.*, *28*, 62–83, 1998.

Lukas, R., The termination of the equatorial undercurrent in the eastern Pacific, *Prog. Oceanogr.*, *16*, 63–90, 1986.

Lumpkin, R. and K. Speer, Large-scale vertical and horizontal circulation in the North Atlantic Ocean, *J. Phys. Oceanogr.*, *33*, 1902–1920, 2003.

Lux, M., H. Mercier, and M. Arhan, Interhemispheric exchanges of mass and heat in the Atlantic Ocean in January–March 1993, *Deep Sea Res., Part I*, *48*, 605–638, 2001.

Luyten, J. R., J. Pedlosky, and H. Stommel, The ventilated thermocline, *J. Phys. Oceanogr.*, *13*, 292–309, 1983.

Luyten, J. R., and H. Stommel, Gyres driven by combined wind and buoyancy flux, *J. Phys. Oceanogr.*, *16*, 1551–1560, 1986.

Malanotte-Rizzoli, P. K. Hedstrom, H. Arango, and D.B. Haidvogel, Water mass pathways between the subtropical and tropical ocean in a climatological simulation of the North Atlantic ocean circulation, *Dyn. Atmospheres and Oceans*, *32*, 331–371, 2000.

Mantua, N. J., S. R. Hare, Y. Zhang, J. M. Wallace, and R. C. Francis, A Pacific interdecadal climate oscillation with impacts on salmon production, *Bull. Am. Meteorol. Soc.*, *78*, 1069–1079, 1997.

Marshall, J. C., A. J. G. Nurser, and R. G. Williams, Inferring the subduction rate and period over the North Atlantic, *J. Phys. Oceanogr.*, *23*, 1315–1329, 1993.

Mayer, D. A., R. L. Molinari, and J. F. Fiesta, The mean and annual cycle of upper layer temperature fields in relation to Sverdrup dynamics within the gyres of the Atlantic Ocean, *J. Geophys. Res.*, *103*, 18545–18566, 1998.

McCreary, J. P., Jr., Modeling equatorial ocean circulation, *Ann. Rev. Fluid Mech.*, *17*, 359–409, 1985.

McCreary, J. P., Jr., P. K. Kundu, and R. L. Molinari, A numerical investigation of dynamics, thermodynamics and mixed-layer processes in the Indian Ocean, *Prog. Oceanogr.*, *31*, 181–244, 1993.

McCreary, J. P., Jr., and P. Lu, Interaction between the subtropical and equatorial ocean circulations: the subtropical cell, *J. Phys. Oceanogr.*, *24*(2), 466–497, 1994.

McCreary, J. P., Jr., K. E. Kohler, R. R. Hood, and D. Olson, A four-component model of biological activity in the Arabian Sea. *Prog. Oceanogr.*, *37*, 193–240, 1996.

McCreary, J.P., Jr., P. Lu, and Z. Yu, Dynamics of the Pacific Subsurface Countercurrents, *J. Phys. Oceanogr.*, *32*, 2379–2404, 2002.

McPhaden, M. J., A.J. Busalacchi, R. Cheney, J. R. Donguy, K. S. Gage, D. Halpern, M. Ji, P. Julian, G. Meyers, G. T. Mitchum, P. P. Niiler, J. Picaut, R.W. Reynolds, N. Smith, and K. Takeuchi, The Tropical Ocean-Global Atmosphere (TOGA) observing system: A decade of progress, *J. Geophys. Res.*, *103*, 14,169–14,240, 1998.

McPhaden, M. J., and D. Zhang, Slowdown of the meridional overturning circulation in the upper Pacific Ocean, *Nature*, *415*, 603–608, 2002.

Meinen, C. S., M. J. McPhaden, and G. C. Johnson, Vertical velocities and transports in the equatorial Pacific during 1993–1999, *J. Phys. Oceanogr.*, *31*, 3230–3248, 2001.

Miyama, T., J. P. McCreary Jr., T. G. Jensen, J. Loschnigg, S. Godfrey, and A. Ishida, Structure and dynamics of the Indian-Ocean cross-equatorial cell, *Deep Sea Res., Part II*, *50*, 2023–2047, 2003.

Molinari, R. L., S. Bauer, D. P. Snowden, G. C. Johnson, B. Bourles, Y. Gouriou, and H. Mercier, A Comparison of kinematic evidence for tropical cells in the Atlantic and Pacific oceans, in: *Interhemispheric Water Exchange in the Atlantic Ocean*, Goni, G.J., and P. Malanotte-Rizzoli (Editors), Elsevier Oceanographic Series, pp.

269–286, 2003.

Murtugudde, R., S. Signorini, J. Christian, A. Busalacchi, C. McClain, and J. Picaut, Ocean color variability of the tropical Indo-Pacific basin observed by SeaWiFS during 1997–98, *J. Geophys. Res.*, *104*, 18,351–18,366, 1999.

Nonaka, M., S.-P. Xie, and K. Takeuchi, Equatorward spreading of a passive tracer with application to North Pacific interdecadal temperature variations, *J. Oceanogr.*, *56*, 173–183, 2000.

Nonaka, M., and S.-P. Xie, Propagation of North Pacific interdecadal subsurface temperature anomalies in an ocean GCM, *Geophys. Res. Lett.*, *27*, 3747–3750, 2000.

Nonaka M., S.-P. Xie, and J. P. McCreary Jr., Decadal variations in the Subtropical Cells and equatorial Pacific SST, *Geophys. Res. Lett.*, *29*(7), 1116, doi:10.1029/2001GL013-717, 2002.

Pedlosky, J., An inertial theory of the equatorial undercurrent, *J. Phys. Oceanogr.*, *17*, 1978–1985, 1987.

Pedlosky, J., Entrainment and the termination of the equatorial undercurrent, *J. Phys. Oceanogr.*, *18*, 880–886, 1988.

Pedlosky, J., The link between western boundary current and equatorial undercurrent, *J. Phys. Oceanogr.*, *21*, 1553–1558, 1991.

Pedlosky, J., and R.M. Samelson, Wind forcing and the zonal structure of the equatorial undercurrent, *J. Phys. Oceanogr.*, *19*, 1244–1254, 1989.

Philander S. G. H., and R. C. Pacanowski, The generation of equatorial currents, *J. Geophys. Res.*, *85*, 1123–1136, 1980.

Qiu, B. and R. X. Huang, Ventilation of the North Atlantic and North Pacific: Subduction Versus Obduction, *J. Phys. Oceanogr.*, *25*, 2374–2390, 1995.

Quay, P. D., M. Stuiver, and W. S. Broeker, Upwelling rates for the equatorial Pacific Ocean derived from the bomb 14C distribution, *J. Mar. Res.*, *41*, 769–792, 1983.

Reverdin, G., C. Frankignoul, E. Kestenare, and M. J. McPhaden, Seasonal variability in the surface currents of the equatorial Pacific, *J. Geophys. Res.*, *99*, 20323–20344, 1994.

Roemmich, D., The balance of geostrophic and Ekman transports in the tropical Atlantic Ocean, *J. Phys. Oceanogr.*, *13*, 1534–1539, 1983.

Rothstein, L. M., R.-H. Zhang, A. J. Busalacchi, and D. Chen, A numerical simulation of the mean water pathways in the subtropical and tropical Pacific Ocean, *J. Phys. Oceanogr.*, *28*, 322–343, 1998.

Rowe, G., E. Firing, and G.C. Johnson, Pacific equatorial subsurface countercurrent velocity, transport, and potential vorticity, *J. Phys. Oceanogr.*, *30*, 1172–1187, 2000.

Saji, N. H., B. N. Goswami, P. N. Vinayachandran, and T. Yamagata, A dipole in the tropical Indian Ocean, *Nature*, *401*, 360–363, 1999.

Schneider, N., A decadal spiciness mode in the tropics, *Geophys. Res. Lett.*, *27*, 257–260, 2000.

Schneider, N, S. Venzke, A. J. Miller, D. W. Pierce, T. P. Barnett, C. Deser, and M. Latif, Pacific thermocline bridge revisited, *Geophys. Res. Lett.*, *26*, 1329–1332, 1999.

Schott, F., J. C. Swallow and M. Fieux, The Somali current at the equator: annual cycle of currents and transports in the upper 1000 m and connection to neighboring latitudes, *Deep Sea Res.*, *37*, 1825–1848, 1990.

Schott, F. and C.W. Böning, The WOCE model in the western equatorial Atlantic: upper-layer circulation, *J. Geophys. Res.*, *96*(C4), 6993–7004, 1991.

Schott, F., J. Fischer and L. Stramma, Transports and pathways of the upper-layer circulation in the western tropical Atlantic, *J. Phys. Oceanogr.*, *28*(10), 1904–1928, 1998.

Schott, F., and J. P. McCreary Jr., The monsoon circulation of the Indian Ocean, *Prog. Oceanogr.*, *51*, 1–123, 2001.

Schott, F., Dengler, M., and R. Schoenefeldt, The shallow thermohaline circulation of the Indian Ocean, *Prog. Oceanogr.*, *53*, 57–103, 2002a.

Schott, F., P. Brandt, M. Hamann, J. Fischer, and L. Stramma, On the boundary flow off Brazil at 5-10°S and its connection to the interior tropical Atlantic, *Geophys. Res. Lett.*, *29*(17), 1840, doi:10.1029/2002 GL014786, 2002b.

Schott, F., M. Dengler, P. Brandt, K. Affler, J. Fischer, B. Bourlès, Y. Gouriou, R. L. Molinari, and M. Rhein, The zonal currents and transports at 35W in the tropical Atlantic, *Geophys. Res. Lett.*, *30*(7), 1349, doi:10.1029/2002GL016849, 2003.

Shin, S., and Z. Liu, Response of equatorial thermocline to extratropical buoyancy forcing, *J. Phys. Oceanogr.*, *30*, 2883–2905, 2000.

Sloyan, B. M., G. C. Johnson, and W. S. Kessler, The Pacific cold tongue: A pathway for interhemispheric exchange, *J. Phys. Oceanogr.*, *33*(5) 1027–1043, 2003.

Snowden, D. P. and R .L. Molinari, Subtropical cells in the Atlantic Ocean: An observational summary, in: *Interhemispheric Water Exchange in the Atlantic Ocean*, Goni, G.J., and P. Malanotte-Rizzoli (Editors), Elsevier Oceanographic Series, pp. 287–312, 2003.

Solomon, A., J. P. McCreary Jr., R. Kleeman, and B.A. Klinger, Interannual and decadal variability in an intermediate coupled model of the Pacific region, *J. Clim.*, *16*, 383–405, 2003.

Stommel, H., Wind-drift near the equator, *Deep Sea Res.*, *6*, 298–302, 1960.

Stramma, L., and F. Schott, The mean flow field of the tropical Atlantic Ocean, *Deep Sea Res., Part II*, *46*, 279–303, 1999.

Susanto, D. R., and A .L. Gordon, Upwelling along the coasts of Java and Sumatra and its relation to ENSO, *Geophys. Res. Lett.*, *28*, 1599–1602, 2001.

Swallow, J. C., F. Schott, and M. Fieux, Structure and transport of the East African Coastal Current, *J. Geophys. Res.*, *96*, 22,254–22,267, 1991.

Talley, L., Shallow, intermediate, and deep overturning components of the global heat budget, *J. Phys. Oceanogr.*, *33*, 530–560, 2003.

Tsuchiya, M., A subsurface north equatorial countercurrent in the eastern Pacific Ocean, *J. Geophys. Res.*, *77*, 5981–5986, 1972.

Tsuchiya, M., R. Lukas, R. A. Fine, E. Firing, and E. Lindstrom, Source waters of the Pacific equatorial undercurrent, *Prog. Oceanogr.*, *23*, 101–147, 1989.

Vinayachandran, P. N., and T. Yamagata, Monsoon response of the sea around Sri Lanka: generation of thermal domes and anticyclonic vortices, *J. Phys. Oceanogr.*, *28*, 1946–1960, 1998.

Wacongne, S. and R. Pacanowski, Seasonal heat transport in a primitive equations model of the tropical Indian Ocean, *J. Phys. Oceanogr.*, *26*(12), 2666–2699, 1996.

Webster, P. J., A. M. Moore, J.P. Loschnigg, and R.R. Leben, Coupled ocean-atmosphere dynamics in the Indian Ocean during 1997–98, *Nature*, *401*, 356–359, 1999.

Wijffels, S., E. Firing, and J. Toole, The mean structure and variability of the Mindanao Current at 8N, *J. Geophys. Res.*, *100*, 18,421–18,435, 1995.

Wilson, W. D., E. Johns, and R. L. Molinari, Upper layer circulation in the western tropical North Atlantic Ocean during August 1989, *J. Geophys. Res.*, 99, 22,513–22,523, 1994.

Wong, A. P. S., and G. C. Johnson, South Pacific eastern subtropical mode water, *J. Phys. Oceanogr.*, *33*, 1493–1509, 2003.

Wyrtki, K., Upwelling in the Costa Rica Dome, *U.S. Fish and Wildlife Service Fishery Bull.*, *63*, 355–372, 1964.

Wyrtki, K., An estimate of equatorial upwelling in the Pacific, *J. Phys. Oceanogr.*, *11*, 1205–1214, 1981.

Wyrtki, K., An equatorial jet in the Indian Ocean, *Science*, *181*, 262–264, 1973.

Xie, S.-P., H. Annamalai, F. Schott, and J. P. McCreary, Jr., Origin and predictability of South Indian Ocean climate variability, *J. Clim.*, *15*(8), 864–874, 2002.

Zhang, D., M. J. McPhaden and W. E. Johns, Observational evidence for flow between the subtropical and tropical Atlantic: The Atlantic Tropical Cells, *J. Phys. Oceanogr.*, *33*, 1783–1797, 2003.

Gregory C. Johnson, NOAA/Pacific Marine Environmental Laboratory, Seattle, Washington. (Gregory.C.Johnson@noaa.gov)

Julian P. McCreary, Jr., International Pacific Research Center, University of Hawaii, POST Bldg. 401, 1680 East West Road, Honolulu, Hawaii 96822. (jay@hawaii.edu)

Friedrich A. Schott, IFM-GEOMAR Leibniz Institut für Meereswissenschaften an der Universität Kiel, Düsternbrooker Weg 20, D-24105 Kiel, Germany. (fschott@ifm-geomar.de)

Seasonal Variation of the Subtropical/Tropical Pathways in the Atlantic Ocean From an Ocean Data Assimilation Experiment

Meyre P. Da Silva and Ping Chang

Department of Oceanography, Texas A&M University, College Station, Texas

An ocean data assimilation product from the Geophysical Fluid Dynamics Laboratory (GFDL) was used to investigate the seasonal variability of the ocean circulation in the northern tropical Atlantic. Analyses indicate that the seasonal variation of the circulation plays an important role in mediating meridional exchange between the subtropics and equatorial zone. In particular, seasonal changes in the Northern Equatorial Countercurrent (NECC) and in the cyclonic recirculation of the Guinea dome directly affect the trajectory of a subducted water parcel. Furthermore, seasonal variation of the zonal slope of the thermocline in the interior of the basin controls, to a large extent, the amount of water participating in the equatorial circulation. These changes can be explained in terms of a simple dynamical model where local Ekman pumping dominates thermocline variation in the western part of the basin, and Rossby wave adjustment comes into playing in the eastern basin.

1. INTRODUCTION

In the Ekman convergence region in the mid latitudes surface waters are subducted into the thermocline. After subduction they move equatorward predominantly along isopycnals, and at the equator a fraction of these waters is incorporated into the Equatorial Undercurrent (EUC). The waters of the EUC are characterized by a salinity maximum, and are referred as the subtropical underwater (STUW) [*O'Connor*, 2002]. In the Atlantic the EUC is fed mainly by waters from the southern subtropics [*Metcalf and Stalcup*, 1967], an asymmetry caused by the interaction between the wind-driven gyres and the meridional overturning circulation (MOC) [*Jochum and Malanotte-Rizzoli*, 2001; *Fratantoni et al.*, 2000].

Bryden and Brady [1985] provided evidence that water returning to the surface in the equatorial region is also predominantly along isopycnals, implying that the density structure of the equatorial thermocline must be determined in analogy with the midlatitude thermocline. Inspired by this result *Pedlosky* [1987] extended the concept of ventilated thermocline of extratropics to tropics.

Following Pedlosky's work, numerical and analytical models have been used to study the processes that control the tropical/subtropical water exchange, and identify the pathways of this exchange. Focus has been given primarily to the Pacific Ocean [*McCreary and Lu*, 1994; *Liu*, 1994; *Liu et al.*, 1994; *Liu and Philander*, 1995; *Lu and McCreary*, 1995; *Rothstein et al*, 1998] and recently to the Atlantic Ocean [e. g., *Harper*, 2000; *Malanotte-Rizzoli et al*, 2000]. The results of these studies emphasize the link between the Equatorial Undercurrent (EUC) and the subtropical thermocline, providing basic support to Pedlosky's theory. However, the use of models with more complete dynamics and realistic wind forcing in more recent studies has produced conflicting results.

At the heart of the controversy is a subject not considered in Pedlosky's theoretical work, that is, the impact of Ekman suction induced by the ITCZ on the pathways of waters from the Northern Hemisphere. Does the ITCZ act as a barrier to interior exchange water between the equator and northern subtropics, as concluded by *Lu and McCreary* [1995]? Results from a high resolution general circulation model (GCM) [*Harper*, 2000] indicate that there is no total barrier, and "the thermal structure under the ITCZ merely redirects the pathways of the thermocline ventilation". Based on a reduced-gravity GCM of the Atlantic Ocean, *Inui et al.* [2002] suggested that on seasonal time scales the Ekman suction region permits interior flow towards the equatorial region during the winter (February) and prevents communication during the summer (August).

Traditionally, numerical studies of the tropical/subtropical pathways are based on annual-mean circulation fields, implying that the water pathways after subduction are not significantly affected by seasonal variation of the circulation. Even when dealing with seasonal variations, it is often assumed that the ocean is in equilibrium with the winds. These assumptions are under question for the northern tropical Atlantic because in this region the pronounced migration of the ITCZ causes considerable seasonal variation in the wind stress curl, and the oceanic adjustment to these wind fluctuations takes at least a few months. In addition to pumping the thermocline up and down, changes in the wind stress curl also cause changes in the thermocline slope and hence in the geostrophic meridional flow. In order to understand the extent to which the ITCZ can have an effect on the water pathways, it is necessary to take into consideration the oceanic response to seasonal changes in the wind stress. Furthermore, a realistic simulation of the thermal structure within the thermocline and its variability is imperative.

Although the equatorial thermocline is mostly supplied from the southern Atlantic, the water pathway in the northern hemisphere is likely to be affected strongly by seasonality [*Hazeleger et al.*, 2003; *Lazar et al.*, 2002]. Thus, in this study we will assess the seasonal variation of the northern tropical Atlantic circulation using a global ocean data assimilation (ODA) product. Our main objective is to diagnose the sensitivity of the relation between the tropical/subtropical pathways and seasonal changes of the wind stress caused by the annual migration of the ITCZ. A description of the ocean assimilation system is given in section 2. In section 3 simulated fields are compared with available observations. The main characteristics of simulated mean and seasonal circulation of the northern tropical Atlantic are presented in section 4, as well as details of the oceanic adjustment to fluctuations in the wind stress curl. In section 5 seasonal changes in the tropical/subtropical pathways are diagnosed using a Lagrangian trajectory analysis. Finally, section 6 summarizes the results.

2. THE OCEAN ASSIMILATION SYSTEM

The results presented in this paper are based on the ODA analysis carried out at the Geophysical Fluid Dynamics Laboratory. The Ocean Data Assimilation system consists of three components: the quality controlled observational data, a statistical interpolation algorithm and a numerical ocean model.

Surface and subsurface temperature data incorporated into the ocean model were taken from the World Ocean data Bank-94 [*Levitus and Boyer*, 1994], the Global Temperature and Salinity Profile Project and TOGA/TAO moorings. Additional data from Global SST data sets were provided by the National Center for Environmental Prediction (NCEP). Quality control checks included: a basin dependent range check for temperatures, standard deviation checks, and gravitational stability checks. The univariate variational optimal interpolation scheme developed by *Derber and Rosati* [1989] was used to assimilate the observed data into the numerical model.

The numerical model was based on the GFDL Modular Ocean Model version 3 (MOM3) [*Pacanowski*, 1999]. The model domain covers the global ocean basin from 78°S to 65°N, with a uniform zonal resolution of 1°. The meridional resolution is 1/3° between 10°S–10°N, and gradually increases to 1° polewards of ±20°. In the vertical, the model has 40 levels with a constant 10 m resolution in the upper 210 m. For horizontal mixing of momentum and tracers the model employed a *Smagorinsky* [1993] non-linear scheme and a *Gent-McWilliams* [1990] lateral tracer diffusivity scheme. In the vertical, a non-local K-Profile Parameterization (KPP) scheme was used together with a *Bryan-Lewis* [1979] tracer mixing scheme. The model was forced with climatological surface wind stresses derived from NASA/ DAO/SSMI analysis superimposed on which were daily wind stress anomalies from the NCEP Reanalysis for the period from 1980 to 2000. Restoring boundary conditions to both weekly Reynolds OISST product and Levitus salinity were used at the ocean surface.

3. COMPARISON WITH OBSERVATION

The existing observations in the tropical Atlantic do not have sufficient spatial and temporal resolutions to readily compare with the simulated fields. In order to evaluate the model performance, the time series from the Pilot Research Moored Array in the Tropical Atlantic (PIRATA), which cover the period from the end of 1997 up to the present, were combined with estimates and measurements from earlier observational studies and data sets.

We begin the model validation by examining the transports of the Equatorial Undercurrent (EUC), the North Equatorial Countercurrent (NECC) and the North Brazil Current (NBC). Table 1 provides monthly, seasonal and annual mean transport values from the model and available observation from 1980 to 1994, corresponding to the first 15 years of the assimilated experiment. To minimize the differences in sampling size and resolution between data and model, the comparison was performed using, whenever possible, the specific period, section

Table 1. Observed and Simulated Transports of the EUC, the NECC and the NBC in Sverdrups.

Current	Section	Period	Obs.	ODA
EUC (0m - 26.8 σ_θ)	35°W	mean (1990-94)	22.3 Sv	23.0 Sv
		10/90	26.3 Sv	24.2 Sv
		06/91	43.6 Sv	28.7 Sv
		11/92	19.5 Sv	20.3 Sv
		03/94	21.2 Sv	18.8 Sv
NBC (0 m - 26.8 σ_θ)	10°S	mean (1992-94)	11.1 Sv	10.2 Sv
		11/92	12.5 Sv	10.8 Sv
		03/94	9.6 Sv	9.5 Sv
	5°S	mean (1990-94)	14.6 Sv	19.3 Sv
		10/90	13.5 Sv	20.3 Sv
		06/91	21.6 Sv	20.2 Sv
		11/92	10.8 Sv	18.9 Sv
Schott et al., 1998 (direct measurements)		03/94	13.2 Sv	18.0 Sv
NBC (0 - 300 m)	44°W	mean (1990-91)	23.8 ± 4.6 Sv	19.2 Sv
	0° - 1.5°N	mean 12/90 - 02/91	22.0 ± 3.2 Sv	20.6 Sv
Schott et al., 1993 (direct measurements)		mean 06/91 - 08/91	27.8 ± 4.4 Sv	19.5 Sv
NBC (0 - 800 m)	4°N	mean (1989-91)	26.1 Sv	26.6 Sv
		04-05/89	13.0 ± 6.0 Sv	15.1 Sv
Johns et al., 1998 (direct measurements)		07-08/89	36.0 ± 6.0 Sv	40.7 Sv
EUC ($\geq$ 20 cm/s)	23°W	mean (1982-84)	15.0 Sv	15.2 Sv
Hisard and Henin, 1987 (direct measurements)		Std. dev.	7.0 Sv	4.1 Sv
NECC (4.5° - 10.5°N)	28°W	mean (1980-85)	9.0 Sv	9.2 Sv
		Fall	12.0 Sv	13.4 Sv
Richardson et al., 1992 (indirect estimates)		Spring	5.0 Sv	4.2 Sv

and current definition as given by each observational work listed in Table 1.

Overall, the simulated transport values for the EUC and the NECC are in reasonable agreement with observations. Since, in some cases, we are comparing instantaneous measurements with monthly mean output fields, simulated transport variations are weakened by time average. For instance, in summer 1991 a maximum transport of 43.6 Sv was measured at 35°W for the EUC [*Schott et al.*, 1998]. Even though the simulated transport is high (28.7 Sv) in this period relative to the other period, it is about 34% lower than observation. This atypical transport value was not used for the mean EUC in Table 1.

The NBC transports at 10°S, 5°S, 4°N, and at the western equatorial region along 44°W were computed for comparison with transports estimates based on direct measurements by *Schott et al.* [1998], *Johns et al.* [1998] and *Schott et al.* [1993], respectively. While the simulated transport compares well to the observed transport at 10°S and 4°N, the agreement is not completely satisfactory at 5°S and 44°W. The 5°S-simulated transports are up to 8 Sv higher than the observed ones (see Table 1). Such an overestimate might be due to the additional inflow received by the western boundary current from the east through the South Equatorial Current (SEC). Whereas the observation only reveals a mean inflow of 3.5 Sv from the east between 5° and 10°S, the ODA renders mean SEC inflow of about 9 Sv. At 44°W the NBC simulated seasonal cycle is slightly different from the observed cycle found by *Schott et al.* [1993]. In the observation the NBC has high transports (>26.0 Sv) from June to August and two transport minima, one in January (<20.0 Sv) and another in May ($\approx$20.0 Sv). The simulated NBC shows maximum transports in August (24.6 Sv) and one minimum in November (19.0 Sv) and another in May (14.4 Sv).

To assure that the variation of the thermocline depth is realistically simulated, the simulated depth of the 20°C isotherm was evaluated against two independent data sets: For the period prior to 1994 we compared the model results with observed XBT data compiled by *White* [1995], while from 1997 to 2000 the PIRATA buoys at 15°N–38°W, 0°–35°W and 10°S–10°W were used. Due to the coarse resolution of the White data set, a time series of thermocline depth averaged over 54.5°W–22.5°W and 5°N–20°N was used to represent the northern tropics. Correspondingly, the southern tropics was represented by a time series averaged over 38.5°W–11.5°E and 15°S–0°S.

Figures 1a to 1e display both observed and simulated time series of the thermocline depth. The main difference identified between simulated and observed time series is a mean offset, with the simulated thermocline depth being too shallow in the north and too deep in the south. The agreement between simulated and observed thermocline depth is better in the northern tropics, where the correlation between the two series is 0.82 (Fig. 1a). In the southern tropics, the correlation has a lower value of 0.67 (Fig 1b). The simulated and PIRATA time series also display a better agreement for those buoys located on and north of the equator (Figs. 1c and 1d). In both regions the correlations are higher than 0.80 which is at 99% significance level. By contrast, the 0.37 correlation between the simulated and PIRATA time series at 10°S, 10°W) is slightly below the 99% significance level (Fig. 1e).

Taking into the consideration of the better quality of the observation in the Northern Atlantic, it is perhaps not surprising that the simulated thermocline depth in this region agrees better with observations than the southern Atlantic where the sparsity of data is a major concern. In sum, we conclude that the assimilated product gives a satisfactory overall description of the seasonal variation of the tropical Atlantic Ocean circulation. In the following, we focus on a more detailed analysis of important circulation features based on the assimilated data.

4. SIMULATED CIRCULATION

4.1. Annual Mean

The analysis focuses first on the mean flow below the mixed layer, which in this study is defined as the depth at which the density difference from the sea surface is $0.125\sigma_\theta$ [*Monterey and Levitus*, 1997]. The flow in the thermocline will be described in terms of horizontal velocity vectors and streamlines of geostrophic currents projected onto an isopycnal surface which gives a good representation of the seasonal thermocline variation. This analysis provides a view of adiabatic circulation pattern (the cross isopycnal velocity is zero) so that the result can be compared to the theoretical picture of Pedlosky's analytical model. The superposition of streamlines of geostrophic currents to the total velocity, on the other hand, gives an indication of those regions in the model where the flow departs from geostrophy.

The streamlines of geostrophic currents Ψ on an isopycnal surface are given by

$$\Psi = \alpha' \mathbf{P} + \Phi', \qquad (1)$$

where α' is the specific volume anomaly, $\mathbf{P}$ the pressure, and Φ' the geopotential anomaly. A computational method to derive Ψ using equation (1) is given in *Kessler* [1999]. The ref-

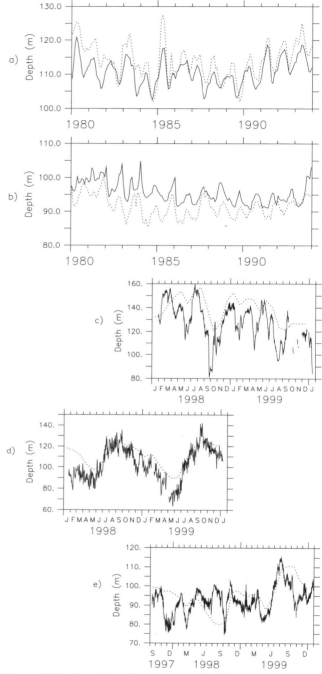

Figure 1. Simulated (dotted) and observed (solid) thermocline depth: a) northern tropics (54°–22°W and 5°–20°N); b) southern tropics (38°–11°E and 5°S–0°N); c) at 15°N and 38°W; d) at 0°N and 35°W; e) at 10°S and 10°W.

erence level to calculate Ψ was chosen at 500 m. For the tropical Atlantic region, this is approximately the depth of $27.15\sigma_\theta$, which is considered to be the transition layer between thermocline and intermediate waters [*Lux et al., 2000*], and thus is a logical choice to represent the geostrophic flow within the thermocline.

Plate 1a shows contours of Ψ, and current vectors on the $25.0\sigma_\theta$ surface. Shaded areas indicate the regions of positive Ekman pumping computed according to $W_e = curl(\frac{\tau}{f\rho})$, where τ is the wind stress used to force the model, f is the Coriolis parameter and ρ is density. The main features seen in the circulation on the $25.0\sigma_\theta$ surface are representative of the flow in the thermocline. In the interior off-equatorial regions, the flow is predominantly geostrophic as indicated by Plate 1a. In agreement with Pedlosky theory, three types of streamlines can be identified in both hemispheres: 1) streamlines connecting to the subtropical gyres, 2) western boundary streamlines feeding subducted waters to the EUC, and 3) interior streamlines linking the subduction zone directly to the EUC.

An additional type of streamline is seen in Plate 1a connecting one part of the outcrop zone to another part without interacting with either the western boundary or the equatorial region. This type of circulation is not considered in Pedlosky's theory, but appears in other recent modeling results [*Lazar et al., 2002; Malanotte-Rizzoli et al., 2000; Inui et al., 2002*]. They occupy the shadow zone described by the well known ventilated theory of *Luyten et al.* [1983], and form a region of closed geostrophic contours which is apparently caused by the upwelling and shoaling of the thermocline driven by Ekman suction. The shoaling of the thermocline results in higher potential vorticity, thus no net southward flow can go over this potential vorticity barrier, and water can only reach the equator to the west of this region.

Plate 1b displays a zonal section of annual mean meridional velocity (*cm/s*) and potential density (σ_θ) at 10°N, which lies approximately in the center of the region of closed geostrophic contours. In this region the isopycnals present a dome-like shape which is associated with the cyclonic flow pattern. This feature has been identified as the Guinea dome [*Rossignol and Meyrueis, 1964*]. Consistent with geostrophy, the isopycnals slope up to the east in the region of southward flow and in the opposite direction in the region of northward flow. However, there is some evidence of ageostrophic flow in the upper thermocline between 25° and 20°W and in the subthermocline close to the African coast. The meridional velocities associated with both southward and northward branches reach values as large as 3 *cm/s*. West of 40°W, a relatively strong vertical shear occurs where the northward surface Ekman drift overlies the southward geostrophic flow. Even further west (at about 60°W) the upper western boundary current, with velocities higher than 10 *cm/s*, is fed largely by subtropical waters that impinge on to the coast of South America and deflect northward.

The total southward meridional transport of subtropical waters crossing 10°N, between the base of mixed layer and $26.8\sigma_\theta$ is about 8 Sv. Among them, approximately 2.6 Sv join the western boundary current and return back to the subtropics, while 3.8 Sv are captured by the cyclonic gyre and return to the surface off the coast of West Africa. Thus, only 1.6 Sv from the northern subtropics remain to eventually join the EUC. Net northward flows occur in the mixed layer (7.0 Sv), and in the subthermocline (3.4 Sv) and intermediate (5.6 Sv) layers. Thus, the mean warm return flow of the MOC in the ODA is about 16 Sv which is consistent with values from the literature [e. g., *Schmitz and McCartney, 1993*].

4.2. Seasonal Circulation

During the boreal spring the ITCZ is situated close to the equator, while during the boreal fall it reaches its extreme northward position at about 10°N. The wind stress curl associated with the annual migration of the ITCZ forces the thermocline to adjust, changing its meridional slope such that the strength of the NECC is altered. The simulated NECC is well developed during the boreal fall and diminishes almost completely in the west during the boreal spring (see Plate 2 and Table I).This seasonal change in the meridional slope of the thermocline and in the NECC is one of the best examples of oceanic response to atmospheric forcing and has been well studied. In the following, we discuss the relevant oceanic dynamics in response to the ITCZ annual migration with a particular emphasis on the changes in the zonal extent of the Ekman pumping and its effect on the change in the thickness and zonal slope of the thermocline.

Shown in Plate 2 is the annual cycle of horizontal velocity and geostrophic streamlines on $25.0\sigma_\theta$, shaded areas indicate the regions of positive Ekman suction. The cyclonic gyre in the northern tropics is present throughout the year. With the exception of fall, its center always is located northward of the region of Ekman suction. While the center of the gyre hardly moves meridionally, its western edge expands considerably, from near 38°W in boreal spring to about 50°W during fall and early winter. This implies that as the season progresses from spring to winter the exchange between tropics and northern subtropics becomes more and more confined to the western part of the basin.

Figure 2a illustrates the seasonal changes in the zonal distribution of Ekman pumping velocity along 10°N. The annual mean has been removed at each grid point. The meridional migration of the ITCZ results in changes in the wind stress curl and hence Ekman pumping velocity that are predominantly taking place in the zonal direction. For the first part of the year,

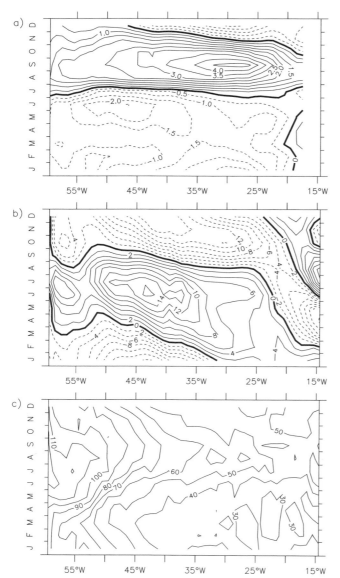

m/s) in the west. Also worth noting is a semi-annual cycle of Ekman pumping close to the eastern boundary.

Simulated seasonal variations in the thermocline depth at 10°N after removing the annual mean are shown in Figure 2b. Just as with the Ekman pumping velocity, the displacement of the thermocline along 10°N has a semiannual character in the east while in the central and western part of the basin is dominated by the annual signal. However, the simulated thermocline depth anomalies exhibit westward propagation that is not seen in the annual cycle of the Ekman pumping velocity (Fig. 2a). During the late boreal fall and early winter the thermocline is shallower than normal almost over the entire basin, except close to the African coast. At 40°W the thermocline depth anomaly reaches minimum value (<−14 m) in December, about three months after the positive Ekman pumping anomaly reaches its maximum. Towards the boreal spring the thermocline gradually deepens in the center of the basin and rises in the east. The deepening progresses in the west until July when positive anomalies reach a maximum value of 14 m.

Seasonal changes of the southward flow along 10°N are associated with variations of the thermocline thickness in the interior of the basin. Figure 2c displays the annual cycle of thermocline thickness, defined as the height from the base of the mixed layer to the thermocline depth. Figure 3 shows the transport along 10°N which has been decomposed into four components: 1) total southward flow (from 62°W to 14°W), 2) interior southward flow (east of 50°W), 3) western boundary southward flow (west of 50°W), and 4) northward flow in the eastern part of the basin. Each of these components was computed between the base of the mixed layer and $26.8\sigma_\theta$ layer.

The analysis of meridional transports along 10°N yields low transport values (≈ 6 Sv) for the total southward flow from February to April and high transports of about 10 Sv from July to November. The seasonality of western boundary transports is not very pronounced, and the interior flow

Figure 2. a) Ekman pumping/suction velocity anomalies (in m/s, scaled by 10^6) along 10°N. Solid contours indicate anomalous Ekman suction, dashed contours anomalous Ekman pumping. b) Thermocline depth anomalies (in m) at 10°N. Solid contours indicate deep anomalies, dashed contours shallow anomalies. c) Thermocline thickness (in m) at 10°N.

Ekman pumping anomalies are positive in the far east (1×10^{-6} m/s) and negative in the west ($<-2.0 \times 10^{-6}$ m/s). After the boreal spring, as the ITCZ retreats to its northern latitudes, an area of positive anomalies appears over the whole basin and reaches its maximum (>4.0×10^{-6} m/s) during boreal fall. A transitional phase can be seen from late boreal fall to early winter when the anomalous Ekman pumping velocities are negative ($<-2.0 \times 10^{-6}$ m/s) in the east and positive (≈1.0×10^{-6}

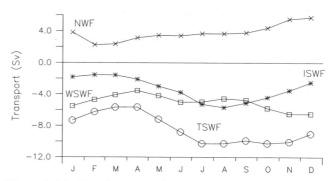

Figure 3. Meridional transport (in Sv) at 10°N decomposed into 4 components: total southward flow (TSWF), interior southward flow (ISWF), northward flow (NWF), and western boundary southward flow (WSWF).

takes into account most of the seasonal changes in the total southward flow. The northward flow in the eastern part of the basin which is a good indicator of the strength of the cyclonic gyre shows a low value of about 2 Sv in February/March and a high value of almost 6 Sv in November/December. Comparing Figs. 2c and 3, it is clear that, in the interior, the maximum transport corresponds with a period of strong thermocline thickness gradient and thick thermocline layer (about 75 m averaged between 50° and 25°W). The minimum transport is associated with small thermocline thickness (about 35 m averaged between 50° and 25°W) and a weak thermocline thickness gradient.

To examine how changes in the thermocline thickness affect the interior meridional flow, we selected three months that exhibit very distinctive characteristics: 1) March when both interior and western boundary southward flow are weak; 2) August when interior flow has maximum southward transport and the southward flow in the western boundary is relatively weak; and 3) December when the interior southward transport is weak and the western boundary is at its maximum.

In March (Plate 3a) the mixed layer reaches its maximum depth, and the thermocline begins to deepen (Fig. 2b). The thickness between the base of the mixed layer and thermocline is more affected by the mixed layer depth, resulting in a relatively thin thermocline layer. Two weak cores of southward flow are seen in the interior, and the northward branch of the recirculation is relatively weak and confined to the east in the upper thermocline.

In August (Plate 3b) the mixed layer is shallow while the thermocline is still deep in the interior of the basin (Fig. 2b), the combined effect of these two processes results in a maximum thermocline thickness. The southward flow in the interior intensifies and extends eastward. An eastern boundary current with velocities higher than 5 *cm/s* develops in the upper thermocline.

In December (Plate 3c) the mixed layer begins to deepen, while the thermocline is upwelled in most parts of the basin (Fig. 2b). Since the upwelling is not homogeneous across the basin, the thermocline thickness tends to be thinner in the eastern part of the basin. The southward flow with maximum velocities up to 5 *cm/s* is confined to the region west of 40°W, extending westward to the Brazilian coast and forming a southward western boundary current from about 100 to below 300 m depth. A diffused and weak (< 3 *cm/s*) northward flow is observed in the eastern half of the basin.

The thermocline thickness as defined here is dependent on both mixed layer and thermocline depth. Thinking in terms of potential vorticity barrier as outlined in *Lu and McCreary* [1995], thickening/thinning of the layer bounded by the base of the mixed layer and the thermocline would permit/restrict interior southward flow. However, an inspection of Plate 3 and Figure 2c indicates that the strength of the interior southward flow is controlled by changes in the gradient of the thermocline thickness (pressure gradient) and not by the absolute value of the thermocline thickness. For instance, during August the southward flow extends almost over the entire basin (Plate 3b) which corresponds to a strong zonal pressure gradient in the interior (Fig. 2c). In contrast, during March, the zonal gradient is considerably weakened in the interior, which results in a weak southward flow. From Plate 3, it is clear that the seasonal changes in the mixed layer are almost uniform across the interior of the basin, therefore changes in the zonal slope of the thermocline are basically responsible for changes in the pressure gradient and thus the magnitude of the southward flow and westward extension of the cyclonic gyre.

4.3. Adjustment of Thermocline to Seasonal Winds

The above analysis shows that seasonal variations in the southward flow are directly related to changes in the zonal slope of the thermocline in the interior of the basin. To better understand the processes that control the displacement of the northern tropical thermocline, we use in this section a simple model of thermocline response to the changes in winds based on the quasi-geostrophic approximation:

$$\overset{A}{\frac{\partial h}{\partial t}} - \overset{B}{Cr\frac{\partial h}{\partial x}} + \overset{C}{\gamma h} = \overset{D}{-W_e} \qquad (2)$$

where $Cr = \beta g'H_0/f^2$ is the phase speed propagation for non-dispersive long Rossby wave, h and H_0 are the depth and mean depth of the thermocline, β is the derivative of the Coriolis parameter f, γ damping rate, W_e is the Ekman pumping/suction velocity. Solutions to (2) consist of a local oceanic response to the Ekman pumping (balance between A and D) and a remote response through non-dispersive long Rossby waves (balance between B + C and D).

Similar or modified versions of the equation (2) have commonly been used in observational studies for the Pacific [*Meyers*, 1979; *Kessler*, 1990] and for the Atlantic [*Garzoli and Katz*, 1983; *Katz*, 1987]. The results of these earlier studies suggest that in the southern boundary of the NECC, Rossby waves contribute dominantly to the annual cycle of thermocline. Near 10°N, local response to Ekman pumping is the primary source for annual thermocline depth variations.

Following the approach by *Kessler* [1990] and *Meyers* [1979], displacements of the thermocline (h) were first calculated using only the Ekman pumping velocity with the boundary condition that $h_b(t) = \int -W_e dt$. The solutions are then compared to those calculated according to (2) using

Rossby wave speeds (Cr) that give the best fit in phase to the simulated annual cycle of 20°C isotherm at each latitude. A maximum correlation value of 0.9 between the simple model hindcast and the ODA was found at 10°N for Cr value of 16.0 cm/s. This best-fit value is higher than the theoretical phase speed of 7.0 cm/s given by $Cr = \beta g' H_0 / f^2$.

Observational studies also indicate that linear theory underestimates the phase speed of baroclinic Rossby waves [e.g., *Chelton and Schlax*, 1996]. However, for a latitude as low as 10°N the discrepancy between the theoretical and observation estimated wave speeds is normally much less than a factor of 2. Limitations of the simple approach used here may be responsible for such large difference. When finding a best-fit value for Cr, it was assumed that a single Rossby wave propagates exactly westward along a given latitude. In reality, this assumption may not be accurate because the waves that affect the oceanic adjustment at 10°N may consist of many wave modes of different propagating speed and have a significant meridional group velocity depending on frequency and wavelength [*Philander*, 1990]. In addition, as shown in the previous section (see Plate 2), the forcing has also westward propagation which can contribute to the high value of Cr found by the fitting process.

The thermocline depth at the eastern boundary h_e was taken from the ODA product and used as a boundary condition to (2) $h_b(t) = h_e(t)$. In order to reproduce amplitudes comparable with simulation, a simple Newtonian damping term (γh) was added to the vorticity equation (2) to take into consideration effect of dissipation. Solutions with a damping rate varying from 1/2 to 3 $year^{-1}$ were calculated for each latitude. Along 10°N a decay rate of about 1 $year^{-1}$ gives the best agreement with simulation. The vorticity equation (2) was solved numerically using a leap-frog scheme with a time step $\Delta t = 3$ days and spatial resolution $\Delta x = 1°$. The annual cycle of the Ekman pumping W_e and thermocline displacement at the eastern boundary h_e were represented by six Fourier harmonics.

Figure 4 shows thermocline displacements at 10°N from the assimilated experiment (Fig. 4a), and the hindcasted solutions based on (2) with all dynamical processes (Fig. 4b) and with only Ekman pumping (Fig. 4c). The results suggest that the oceanic response at 10°N can be divided into three parts: western part, interior and eastern part. For the region west of 45°W, the local Ekman response closely follows the fully assimilated response, indicating that local Ekman dynamics are most important. In the interior, between 45° and 25°W, solution to (2) agrees better with the assimilated thermocline variation. In particular, the westward propagation of the thermocline depth anomalies in the eastern part of the basin is successfully reproduced. This indicates that long Rossby waves play an essential role in the oceanic adjustment in this region.

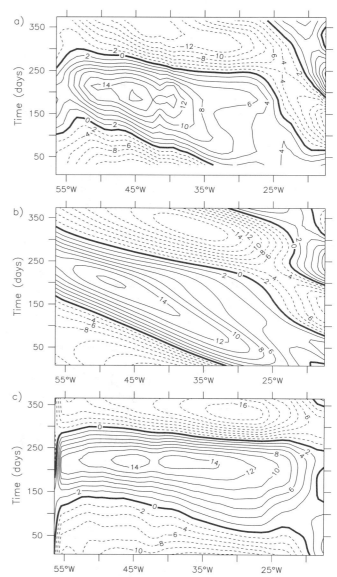

Figure 4. a) Simulated variations in the thermocline depth (in m) at 10°N. Hindcasted thermocline depth anomalies at 10°N using b) equation (2), c) Ekman pumping alone.

East of 25°W the thermocline displacements are also better described by solution to (2), which is not a surprise when taking into account the boundary condition imposed. However, whereas in the interior the oceanic response is predominantly an annual cycle, in the eastern boundary the response has a semiannual component. This semiannual signal is in part related with the semiannual changes in the wind stress curl near the African coast, and in part with semiannual fluctuations of the thermocline depth at the eastern boundary which can be traced back to the equatorial region. *Yamagata and Iizuka* [1995] found a similar propagation in their model results,

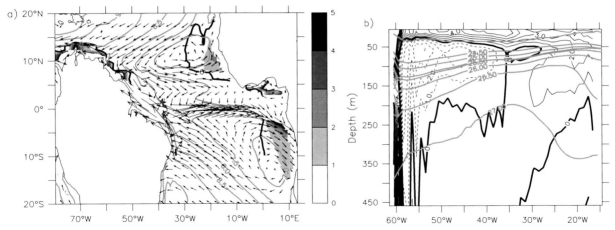

Plate 1. a) Annual mean geostrophic streamlines (in m^2/s^2) in red and velocity vectors (in cm/s) on 25.0 σ_θ. Shaded areas indicate regions of Ekman suction (in m/s scaled by 10^6). b) Zonal section of annual mean meridional velocity (in cm/s) and potential density along 10°N. Solid contours represent northward flow and dashed contours southward flow.

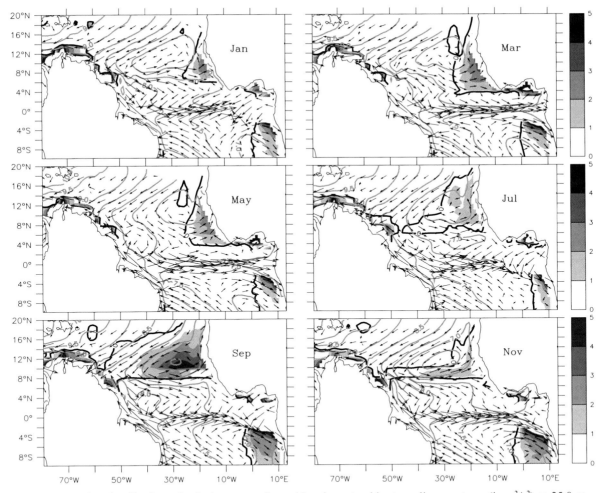

Plate 2. Annual cycle of horizontal velocity vectors (in cm/s) and geostrophic streamlines contours (in m^2/s^2) on 25.0 σ_θ. Shaded areas indicate regions of Ekman suction (in m/s scaled by 10^6).

314 SEASONALITY OF THE ATLANTIC PATHWAYS

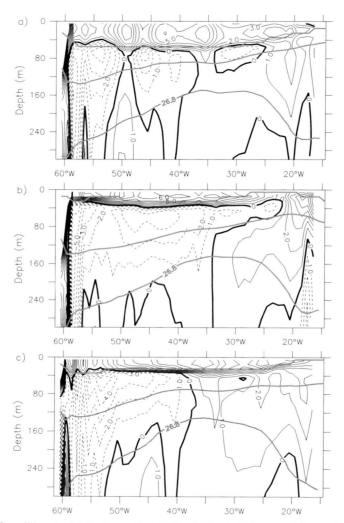

Plate 3. Zonal section of meridional velocity (in cm/s) at 10°N: a) March, b) August, c) December. Red lines indicate the mixed layer depth, the thermocline depth, and the 26.8 σ_θ. Dashed lines indicate southward flow.

which they attributed to intrusion of coastal Kelvin waves excited in the equator due to the semiannual relaxation of the trade winds east of 30°W [*Philander and Pacanowski*, 1986].

Thus the zonal slope of the thermocline at 10°N results from the response to local Ekman pumping in the western part of the basin and wave adjustment in the east. In particular, the thermocline response in the region of meridional southward flow is mainly governed by local process. Since in this region the time rate of change of the thermocline depth is given by the local wind stress curl, the response of the thermocline depth lags the local winds by a quarter of the forcing period [*Philander*, 1990]. As the wind stress curl at 10°N has a strong annual signal, it will take approximately three months for the thermocline depth to respond to the seasonal changes of the wind, which is in agreement with the results of the previous section. The absence of Rossby waves in this region may be attributed to the fact that the waves have a significant meridional group velocity due to the existence of the NECC, which causes wave energy to propagate equatorward and be absorbed by critical layer [*Chang and Philander*, 1988]. Thus, waves excited in the east are difficult to reach the western part of the basin.

5. LAGRANGIAN ANALYSIS

The tropical/subtropical water exchange is not modulated only by the strength of the southward meridional transport in the ITCZ region, seasonal variation in the northern cyclonic gyre and the NECC also play an important role. In this section we explore how these changes in the circulation affect the water exchange and pathways between tropics and subtropics by using a Lagrangian trajectory analysis. To track water particle movement along a representative isopycnal surface under adiabatic assumption, horizontal velocities were projected on the $25.0\sigma_\theta$ layer. The flow within the mixed layer, where diabatic processes are important, is masked out. Particles were released along 12°N and from 60° to 20°W and integrated forward in time using the seasonally varying velocity fields, which were decomposed into the six Fourier harmonics in time to provide continuous time series.

Plates 4a, 4b and 4c display trajectories for particles deployed on the $25.0\sigma_\theta$ in March, August and December over 3 years period. In these plots the trajectories color are coded such that the color changes every half year. In March, the meridional velocity in the interior is weak (Plate 3a), the cyclonic gyre is also weak and confined to the region east of 42°W, retracting further east in the next 2 months (Plate 2). Plate 4a illustrates that particles released west of 50°W flow towards the western boundary and deflect north. Particles released in the interior of the basin take 5-6 months to reach 9°N, arriving just when the gyre starts to strengthen and shifts westwards. Thus, in the following 4 months these particles are carried eastward until the gyre weakens again and particles in the southern limit of the gyre escape southward joining the EUC. Only those particles released between 48° and 43°W are able to reach the equator, taking about 15 months to up to 3 years.

The condition of the tropical circulation in August is almost opposite to that in March. The southward flow in the interior of basin strengthens (Plate 3b) and the cyclonic gyre is expanding westward (Plate 2). With the intensification of the gyre from July to December the interior flow south of 10°N has a more pronounced southeasterly component and particles that do not make it to the western boundary are brought eastward (Plate 4b). After 7 months, particles released between 52° and 44°W reach the northern boundary of the NECC (≈8°N) just as both the NECC and the cyclonic gyre start to weaken. Thus, they move equatorward along trajectories that have a more southward component, while particles arriving to the same latitudes 6–8 months later are carried further east before joining the EUC. Most of particles reach the equatorial region in about 15 months and only a few take more than 2 years.

Plate 4c shows that particles released in December favor the western boundary pathway to reach the equator. This is a result of several factors: First, during the time when the particles are released, the western edge of the cyclonic gyre is in its westernmost position (Plate 2). As a consequence, southward flow is confined to the western part of the basin and has considerable strength close to the western boundary (Plate 3c). Thus particles tend to move very fast southwestward in the beginning. Then just 2 months later the gyre weakens and starts shrinking eastward, forcing the particles released between 54° and 44°W to assume a more southward trajectory. Finally, when these particles arrive at about 7°N, five months later, the NECC is weak and the cyclonic gyre is confined to the east, thereby favoring particles to continue traveling close along the western boundary until they reach the equatorial region. Most particles arrive at the equator in a record time of only 9 months.

The expression "exchange window" is used here to indicate the longitudinal band along 12°N where particles released eventually join the equatorial circulation. It is included in the above definition the initial longitude of those particles that after years following the cyclonic gyre, when the conditions are favorable, are able to escape and move equatorward. In general, the exchange window has a large longitudinal extension from late boreal summer through winter, reaching its maximum in November when extends from 54° to 43°W. By contrast, the exchange window is restricted to the region between 48° and 43°W in the boreal spring.

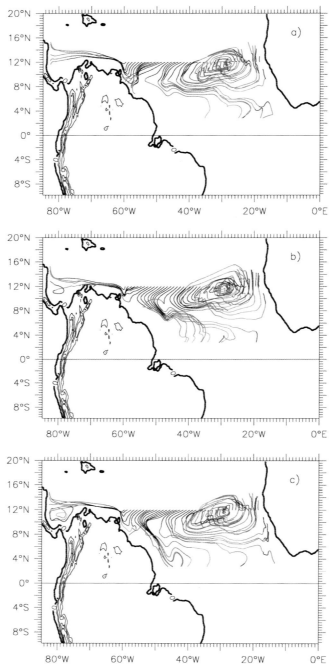

Plate 4. Lagrangian trajectories for particles released on the 25.0 σ_θ along 12°N and from 60° to 20°W over 3 years period. Particles released in a) March; b) August; c) December. Trajectories color are coded such that the color changes every half year.

6. SUMMARY AND DISCUSSION

Based on a state-of-the-art assimilated GFDL run from 1980–2000, the seasonal variability of the Northern Tropical Atlantic circulation is investigated. Most of the circulation features revealed by our analysis agrees qualitatively and quantitatively with previous observational and modeling works, except for a few noted differences. Among these differences is the importance of seasonal variation of the circulation in mediating meridional exchange process between the subtropics and equatorial zone in the Northern Tropical Atlantic. While some of the previous modeling studies suggest that the exchange process is determined mainly by the mean circulation and is not affected strongly by its annual cycle, our results suggest that the seasonal change in the Northern Tropical Atlantic Circulation not only can affect the amount of water participating in the equatorial circulation, but also the pathways taken by the subtropical water toward the equator.

The annual migration of the ITCZ causes changes in the wind stress curl that are not homogeneous across the basin, and neither is the oceanic response to these changes. Using a simple model it is shown that the thermocline displacement along 10°N is mainly controlled by local Ekman pumping in the western part of the basin, while in the east Rossby wave adjustment is also an important part of the oceanic response. This difference between the oceanic responses in the eastern and western sets the thermocline slope in the interior of the basin, which controls, to a large extent the seasonal variation of the southward meridional transport toward the equator. The total southward flow has high transports of more than 9 Sv from boreal summer to late fall, when the thermocline slope is well developed. Low transports of less than 6 Sv are observed from February to April when the thermocline slope relaxes in the central part of the basin.

According to the Lagrangian analysis, more particles tend to join the EUC when released during the period when the total southward flow is strong. However, the fate of the subducted water depends also on the seasonal variation of the tropical circulation in the period following the release of particles. In particular, seasonal changes in the NECC and in the cyclonic gyre affect directly the destination of a subducted water parcel. These variations also allow exchanges between the cyclonic gyre and interior pathway which are not observed when the time-mean flow is used.

While the northern cyclonic gyre is present all year around, both its zonal extent and strength undergo a well noted seasonal variation. In February/March, the gyre is weak ($\approx$2 Sv) and confined to the region east of 38°W. In November, with a transport value of about 5 Sv, its western edge shifts to approximately 50°W, following the seasonal change in the zonal slope of the thermocline. However, there appears to be no significant meridional movements in the gyre location. Although the center of this gyre coincides with the observed cyclonic circulation feature related to the Guinea Dome, its dimension appears to be much larger and more related with counterclockwise flow fields composed of the eastward NECC (upper thermocline) and the westward NEC (see Figs. 4 and 5 of [*Stramma and Schott*, 1999]).

Our results indicate that the role of the cyclonic gyre in the tropical/subtropical pathways is twofold. On the one hand, it prevents particles released east of 42°W from influencing in the equatorial circulation. Although the western edge of gyre moves to east of 38°W during the boreal spring, the meridional velocities in the interior of basin are so weak that particles are not able to move away from the sphere of action of the gyre before it strengthens again. On the other hand, for particles released west of 42°W, the cyclonic gyre acts to reinforce the communication between tropic/subtropics. The flow in its southwest edge is predominantly eastward, hence particles under its area of influence tend to move towards the interior of basin, instead of reaching the western boundary and deflecting north. The subsequent relaxation of the gyre occurs when the particles have entered the NECC. Once entrained into the NECC, the likehood of water particles being recaptured by the recirculation is low. An interior pathway will be favored when particles are released during the strengthening of the gyre (June/July), and a western boundary pathway is favored when the gyre is well developed (November/December).

Acknowledgments. We thank Matt Harrison and Tony Rosati for providing us the GFDL ODA product. We also thank two anonymous reviewers for their constructive comments which helped to improve the paper. The work is supported by NOAA, NASA and NSF through research grants: NA16GP1572, NAG5-9409 and ATM-0337846. MPDS acknowledges the support from the CNPq - Conselho Nacional de Desenvolvimento Cientifico e Tecnologico of the Ministry for Science and Technology of Brazil. PC acknowledges the support from the National Natural Science Foundation of China (NSFC) through grant 40128003.

REFERENCES

Bryan, K., and L. J. Lewis, A water mass model of the world ocean, *J. Geophys. Res*, 84, 2503–2517, 1979.

Bryden, H. L., and E. C. Brady, Diagnostic Model of the three-dimensional circulation in the upper equatorial Pacific Ocean, *J. Phys. Oceanogr.*, 15, 1255–12773, 1985.

Chang, P., and S. G. H. Philander, Rossby wave packets in baroclinic currents, *Deep Sea Res.*, 36, 17–37, 1988.

Chelton, D. B., and M. G. Schlax, Global observations of oceanic Rossby waves, *Science, 272*, 234–238, 1996.

Derber, J., and A. Rosati, A global oceanic data assimilation system. *J. Phys. Oceanogr., 19*, 1333–1347, 1989.

Fratantoni, D. M., W. E. Johns, T. L. Townsend, and H. E. Hulburt, Low-latitude circulation and mass transport pathways in a model of the tropical Atlantic ocean, *J. Phys. Oceanogr., 30*, 1944–1966, 2000.

Garzoli, S. L., and E. J. Katz, The forced annual reversal of the Atlantic north equatorial countercurrent, *J. Phys. Oceanogr., 13*, 2082–2090, 1983.

Gent, P. R., and J. C. McWilliams, Isopycnal mixing in ocean circulation models, *J. Phys. Oceanogr., 20*, 150–155, 1990.

Harper, S., Thermocline ventilation and pathways of tropical-subtropical water mass exchange, *Tellus, 52A*, 330–345, 2000.

Hazeleger, W., P. de Fries, and Y. Fiocourt, Sources of the Equatorial Undercurrent in the Atlantic in a high resolution ocean model, *J. Phys. Oceanogr., 33*, 677–693, 2003.

Hisard, P., and C. Hénin, Response of equatorial Atlantic Ocean to the 1983–1984 wind from the Programme Franais Océan et Climat Dans l'Atlantique Equatorial Cruise data set, *J. Geophys. Res., 92*, 3759–3768, 1987.

Inui, T., P. Malanotte-Rizzoli, and A. Busalacchi, Wind stress sffects on the Atlantic Subtropical-Tropical circulation, *J. Phys. Oceanogr., 32*, 2257–2276, 2002.

Jochum, M., and P. Malanotte-Rizzoli, Influence of the meridional overturning circulation on tropical/subtropical pathways, *J. Phys. Oceanogr., 31*, 1313–1323, 2001.

Johns, W. E., T. N. Lee, R. C. Beardsley, J. Candela, R. Limeburner, and B. Castro, Annual Cycle and variability of the North Brazil Current, *J. Phys. Oceanogr., 28*, 103–128, 1998.

Katz, E. J., Seasonal response of the sea surface to the wind in the equatorial Atlantic, *J. Geophys. Res., 92*, 1885–1893, 1987.

Kessler, W. S., and J. P. McCreary, Observations of long Rossby wave in the nothern tropical Pacific, *J. Phys. Oceanogr., 95*, 5183–5217, 1990.

Kessler, W. S., Interannual variability of the subsurface high salinity tongue south of the equator at 165°E, *J. Geophys. Res., 29*, 2038–2049, 1999.

Lazar, A., T. Inui, A. J. Busalacchi, P. Malanotte-Rizzoli, and L. Wang, Seasonality of the ventilation of the tropical Atlantic thermocline, *J. Geophys. Res., 107*, (18)1–17, 2002.

Levitus, S., and T. Boyer, World Ocean Atlas 1994, Vol. 4: Temperature, *NOAA Atlas NESDIS 4*, 117pp, U.S. Gov. Print. Off., Washington, D.C., 1994.

Liu, Z., A simple model of the mass exchange between the subtropical and tropical ocean, *J. Phys. Oceanogr., 24*, 1153–1165, 1994.

Liu, Z., and S. G. H. Philander, How different wind stress patterns affect the tropical-subtropical circulations of the upper ocean, *J. Phys. Oceanogr., 25*, 449–462, 1995.

Liu, Z., S. G. H. Philander, and R. Pacanowski, A GCM study of tropical-subtropical upper-ocean exchange, *J. Phys. Oceanogr., 24*, 2606–2623, 1994.

Lu, P., and J. P. McCreary, Influence of the ITCZ on the flow of thermohaline water from the subtropical to the equatorial Pacific Ocean, *J. Phys. Oceanogr., 25*, 3076–3088, 1995.

Lux, M., H. Mercier, and M. Arhan, Interhemispheric exchanges of mass and heat in the Atlantic Ocean in January–March 1993, *Deep Sea Res., 48*, 605–638, 2000.

Luyten, J. R., J. Pedlosky, and H. Stommel, The ventilated thermocline, *J. Phys. Oceanogr., 13*, 292–309, 1983.

Malanotte-Rizzoli, P., K. Hedstrom, H. Arango and D. B. Haidvogel, Water mass pathways between the subtropical and tropical ocean in a climatological simulation of the North Atlantic ocean circulation, *Dyn. Atmos. Oceans, 32*, 331–371, 2000.

Meyers, G., On the annual Rossby wave in the tropical North Pacific Ocean, *J. Phys. Oceanogr., 9*, 885–891, 1979.

McCreary, J. P., and P. Lu, On the interaction between the subtropical and equatorial ocean circulation: the subtropical cell, *J. Phys. Oceanogr., 24*, 466–497, 1994.

Metcalf, W. G., and M. C. Stalcup, Origin of the Atlantic Equatorial Undercurrent, *J. Geophys. Res., 72*, 4959–4875, 1967.

Monterey, G, and S. Levitus, Seasonal variability of mixed layer depth for the world ocean, *NOAA Atlas, NESDIS 14*, 96pp, U.S. Gov. Print. Off., Washington, D.C., 1997.

O'Connor, B. M., Formation rates of subtropical underwater in the Pacific Ocean, *Deep Sea Res. Part I, 49*, 1571–1590, 2002.

Pacanowski, R. C., MOM 3.0 Manual, *The GFDL Ocean Group Technical Report*, 668 pp, Geophysical Fluid Dynamics Laboratory, Princeton, N. J., 1999.

Pedlosky, J., A inertial theory of the equatorial undercurrent, *J. Phys. Oceanogr., 17*, 1978–1985, 1987.

Philander, S. G. H., and R. C. Pacanowski, A model of the seasonal cycle in the tropical Atlantic Ocean, *J. Geophys. Res., 91*, 14,192–14,206, 1986.

Philander, S. G., Oceanic adjustment: I, In *El Niño, La Niña, and the Southern Oscillation*, edited by R. Dmowska and J. R. Holton, pp. 103–157, Academic Press, London, 1990.

Richardson, P. L., S. Arnault, S. Garzoli, and J. G. Bruce, Annual cycle of the Atlantic North Equatorial Countercurrent, *Deep Sea Res., 39*, 997–1014, 1992.

Rothstein, L. M., R. -H. Zhang, A. J. Busalacchi, and D. Chen, A numerical simulation of the mean water pathways in the subtropical and tropical Pacific Ocean, *J. Phys. Oceanogr., 11*, 794–812, 1998.

Rossignol, M., and A. M. Meyrueis, Campagnes océanographiques du Gérard-Tréca, 53pp, Cent. Oceanogr. Dakar-Thiaroye, ORSTOM, Dakar, Senegal, 1964.

Schott, F. A., J. Fischer, J. Reppin, and U. Send, On mean and seasonal currents and transports at the western boundary of the equatorial Atlantic, *J. Geophys. Res., 98*, 14,353–14,368, 1993.

Schott, F. A., J. Fischer, and L. Stramma, Transports and Pathways of the upper-layer circulation in the western tropical, *J. Phys. Oceanogr., 28*, 1904–1928, 1998.

Schmitz Jr., W. J., and M. S. McCartney, On the North Atlantic Circulation, *Rev. Geophys., 31*, 29–49, 1993.

Smagorinsky, J., Some historical remarks on the use of nonlinear viscosities, In *Large eddy simulation of complex engineering and geophysical flows*, edited by B. Galperin and S. A. Orszag, Cambridge University Press, 1993.

Stramma, L., and F. A. Schott, The mean flow field of the tropical Atlantic ocean, *Deep Sea Res., 46*, 279–303, 1999.

White, W. B., Design of a global observing system for gyre-scale upper ocean temperature variability, *Prog. Oceanogr., 36*, 169–217, 1995.

Yamagata, T., and S. Iizuka, Simulation of the tropical thermal domes in the Atlantic: a seasonal cycle, *J. Phys. Oceanogr., 25*, 2129–2140, 1995.

P. Chang and M. P. Da Silva, Texas A&M University, Department of Oceanography, College Station, Texas 77843. (ping@tamu.edu; meyre@ocean.tamu.edu)

Gyre-connected Variations Inferred From the Circulation Indices in the Northern Pacific Ocean

Dongxiao Wang, Yun Liu, and Dejun Gu

LED, South China Sea Institute of Oceanonolgy, Chinese Academy of Sciences, Guangzhou, China

Three circulation indices are defined for the tropical, subtropical and subpolar gyres in the northern Pacific (hereafter, TG, STG and SPG) based on the Simple Ocean Data Assimilation product. A singular spectrum analysis of these three gyre-indices reveals that the temporal evolution of the SPG, STG and TG are dominated by different timescales. A possible mechanism is proposed that the observed interannual to interdecadal variations in the upper ocean temperatures involves the adjustment of different gyres. Results show that the North Pacific multidecadal variability tends to be associated with the SPG adjustment, while the decadal variability tends to be associated with both the STG and SPG adjustment. The tropical–extratropical interaction on the interannual timescale can be also explained in the view of changes of the basin scale gyre's intensity.

1. INTRODUCTION

Recent analyses of historical data have revealed substantial longer timescale variations in the ocean- atmosphere system of the Pacific basin [*Graham*, 1994; *Deser et al.*, 1996; and *Barnett et al.*, 1999]. These variations are characterized by a broad band of frequencies ranging from decadal to multidecadal timescales [*Minobe*, 1999].

Decadal temperature fluctuations in the Pacific Ocean have a significant impact on the climate and ecosystems of Northern Hemisphere [*Trenberth and Hurrell*, 1994; *Mantua et al.*, 1997; *Tourre et al.*, 1999]. The physical mechanisms of these fluctuations, however, remain poorly understood [see the review of *Miller and Schneider*, 2000]. Some theories ascribe a central role to the extratropical- tropical interaction [*Gu and Philander*, 1997], while others suggest the midlatitude ocean-atmosphere coupled feedback and oceanic gyre adjustment play a dominant role [*Latif and Barnett*, 1994, 1996]. Some other studies also suggest the role of reddening of atmospheric stochastic forcing by the oceans [*Jin*, 1997; *Frankignoul et al.*, 1997; and *Saravanan and McWilliams*, 1998].

On the decadal timescales, it is conceivable that dynamic processes within the oceans play important role due to its large heat capacity. Among these processes, the anomalous temperature advection by mean currents ($U \cdot \Delta T'$) and the mean temperature by anomalous currents ($U' \cdot \Delta T$) have both been paid much attention over the recent decades. In the North Pacific, both modeling and observations suggest that the adjustment of the subpolar/subtropical gyres to the anomalous wind stress can affect SST and thermocline in the Kuroshio Extension region [*Miller et al.*, 1998; *Deser et al.*, 1999; *Seager et al.*, 2001; and *Schneider et al.*, 2002]. The subtropical-tropical meridional shallow overturning cell has been suggested to play a central role for the decadal variability in the subtropical and tropical Pacific Oceans [*Kleeman et al.*, 1999]. Observations show that the shallow overturning circulation has been slowed down since the 1970s, which tends to be associated with the tropical Pacific SST change [*McPhaden and Zhang*, 2002]. In addition to the change of subtropical-tropical cell, the subduction of the extratropical warm (cold) anomalies may also affect the tropical SST, especially in the southern Pacific.

Earth's Climate: The Ocean-Atmosphere Interaction
Geophysical Monograph Series 147
Copyright 2004 by the American Geophysical Union
10.1029/147GM17

In this note, we will explore a potential connection of the adjustment of the northern Pacific oceanic gyres with the North Pacific climate variability by using a simple gyre-index analysis. Our study suggests the different roles of northern Pacific subpolar and subtropical gyres in the decadal and multidecadal variability.

2. DATA AND METHOD

This work utilizes a 50-year global retrospective analysis of upper-ocean temperature, salinity and currents, namely the Simple Ocean Data Assimilation (SODA) product [*Carton et al.*, 2000]. SODA provides a monthly global data from January 1950 to December 2000. The horizontal resolution is 0.45° latitude by 1.0° longitude in the tropics. The latitudinal resolution decreases in poleward direction to 1.0° in the midlatitude. There are 20 vertical levels distributed on a stretched grid.

The large-scale circulation of the Pacific Ocean is characterized by two great anticyclonic subtropical gyres, two high-latitude cyclonic gyres, two westward flows along 10° to 15° north and south, and an eastward flow that takes place just north of the equator at the surface and at about 500m, but lies along the equator at all other depths. This observational pattern is roughly symmetric about the equator [*Reid*, 1997].

To characterize the variations of three large-scale gyres over the northern Pacific, we define three gyre indices by integrating the upper 500 m velocity along the each gyre path (approximated by the idealized rectangles, Figure 1) as:

$$\vec{U} = \int_{-500}^{0} \vec{u} \cdot dz.$$

Then the indices are obtained by summing up along the idealized rectangles with direction matching the climatological current patterns (see Figure 1 and Table 1),

$$I = \oint_{\Sigma} \vec{U} \cdot dl.$$

We use SODA 2001 version to calculate the indices. All three indices are converted into their anomalies by removing their 50-year averaged seasonal cycles. A positive (negative) anomaly indicates gyre intensification enhancing (weakening). Physically, the three indices represent the regionally-integrated Ekman pumping variations (figure omitted).

Additionally, heat storage anomalies of the upper 400m layer [*White*, 1995] are used here to study the oceanic thermal variations. Three key regions with intense signatures are selected, namely, the central North Pacific (180°E–140°W, 36°N–20°N), the Kuroshio Extension

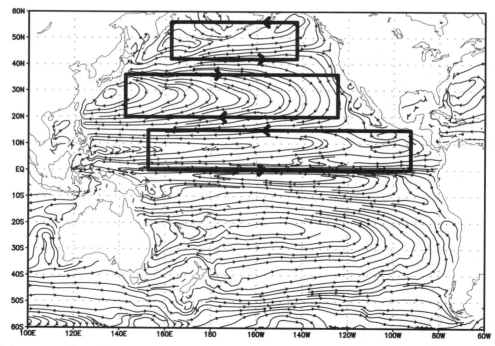

Figure 1. Long-term climatological current streamlines (averaged over upper 500 m) derived from the 1950–1999 SODA product. Three rectangles respectively represent idealized TG, STG, and SPG along which indices are integrated in the directions indicated by arrows.

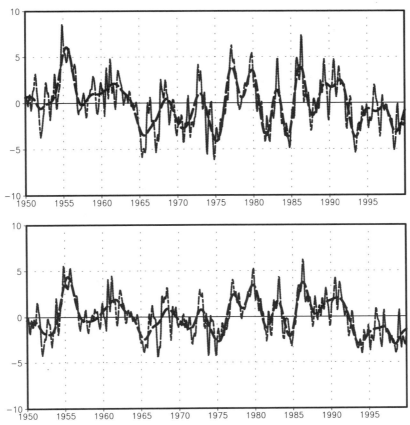

Figure 2. STG's indices integrated over different layers. Upper panel, the integrated index in the upper 200 m; lower panel, the subsurface one integrated between 200 m and 500 m. Short dashed line is the raw data, long dashed line for the reconstructed one based on the ten leading SSA modes. Unit: 10^8 m^3/sec.

(150°E–170°W, 36°N–20°N), and the Niño3 region (150°W–90°W, 5°N–5°S).

We use the Singular Spectrum Analysis–Multi-taper Method (SSA-MTM) toolkit, which consists of a set of programs to perform detailed spectral analyses and decompositions on a univariate input time series [*Ghil and Vautard*, 1991]. The toolkit contains the following procedures: (a) estimating the spectrum of a time series, (b) decomposing the time series into components as trends, oscillations, and noise, and (c) reconstructing the contributions of selected components (or, modes). The SSA method is used here to separate the different intrinsic modes that vary from interannual to interdedacadal timescales.

3. GENERAL FEATURES

We will only focus on the northern Pacific because in the southern Pacific there are not sufficient historical observations and data assimilation products like SODA may provide unrealistic ocean circulation. It should be also noted that neither of the STG and SPG box extends far enough west to encompass the western boundary currents, since SODA simulation of the western boundary currents is not so good.

It's not trivial to determine a specific depth that the major momentum of the wind-driven circulation is restrained. In general, the thermocline is deeper in the high latitudes than that in the tropics and subtropics.

Here we describe a simple validation for our choice of least depth of 500 m for definition of the gyre index for the STG. We compute two different indices, one for the upper 200 m layer and another for the 200 m to 500 m layer. The upper layer transport is of the same order and shows similar low frequency vari-

Table 1. Definition of the circulation indices

Location	west	east	north	south
Subpolar gyre (SPG)	162.5°E	142.5°W	56°N	42°N
Subtropical gyre (STG)	142.5°E	124.5°W	36°N	20°N
Tropical gyre (TG)	152.5°E	92.5°W	15°N	0°N

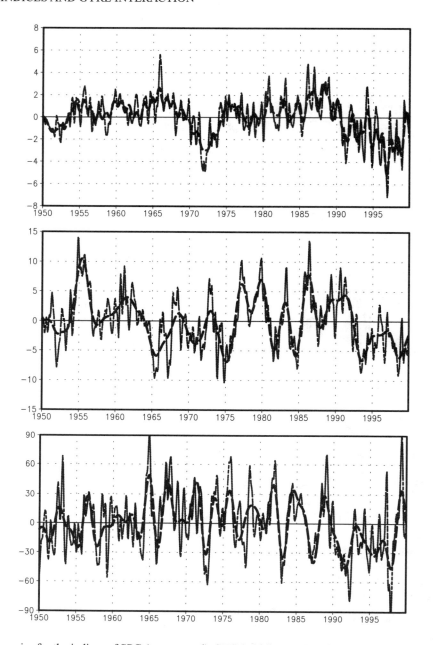

Figure 3. Time series for the indices of SPG (upper panel), STG (middle panel) and TG (lower panel). Long dashed lines are the reconstructed one based on the ten leading SSA modes. Accounting percentages of individual modes for three indices are showed in Table 2. Note that a pair of modes with almost same percentages indicates a periodical oscillation, according to SSA method.

ability as the lower layer transport. This indicates that both the upper layer (upper 200 m, Figure 2a) and the subsurface layer (200–500 m, Figure 2b) are equally important.

Figure 3 shows the time series for the three indices we compute over the SPG, STG, and TG. In order to remove the high frequency noise, we performed SSA analysis upon these time series and reconstructed a filtered version based on the leading ten modes. Since each mode explains a significant percentage of the variance (see Table 2), the reconstructed time series essentially captures the main features for each index. Generally, we can find there are different predominant timescales for variability in each circulation index.

Table 2. Variance percentages explained by the individual modes

Modes	1st	2nd	3rd	4th	5th	6th	7th	8th	9th	10th
SPG	17.0	16.2	6.2	5.4	4.4	3.5	3.3	2.6	2.5	2.4
STG	11.3	9.7	8.5	8.5	7.1	6.9	5.3	5.0	3.8	2.7
TG	8.3	6.1	6.1	5.3	5.0	5.0	4.5	4.5	4.0	4.0

3.1. SPG Index

The SPG index is characterized predominantly by interdecadal variations (Figure 3a). The gyre tends to be intensified from the middle 1950s to the end of 1960s as well as from 1980 to 1990. A significant weakening appears from 1970 to 1980. There are also some interannual and decadal variations, but in general much weaker than the multidecadal variations. Such decadal to interdecadal anomalies will be analyzed further in the next section.

3.2. STG Index

In contrast to the SPG, the STG is dominated by decadal variability with a timescale of about one or two decades (Figure 3b). The STG tends to be intensified roughly in 1954–1964, 1976–1981, and 1986–1992, and weakened in 1965–1976, and 1992–2000, as well as a short period of 1982–1986. The intensification from the 1970s to 1980s seems to be consistent with other previous observational studies [*Qiu and Joyce*, 1992; *Deser et al.*, 1999]. This transition tends to occur several years after the 1976–77 climate shift, probably as a delayed response to the interior wind change. We will return to this issue in a latter section.

3.3. TG Index

As the circulation is close to the equator, the TG variations are strongly associated with the ENSO. It can be seen that the TG is characterized predominantly by interannual variations (Figure 3c). This is in contrast to the STG and SPG. A decade-long weakening can also be seen in the 1990s, which tends to be linked to the persistent warming in the tropical Pacific.

In summary, the gyre-index defined here clearly reveals distinctive temporal characteristics of the northern Pacific basin-scale circulation variability. For the subpolar gyre, the variability is dominated by multidecadal timescale, while in the subtropical and tropical gyres, the variability is dominated by decadal and interannual timescales, respectively. In the next section, we will explore how the oceanic gyre variations are associated with SST and atmospheric circulation changes. To do so, we simply use the Pacific decadal oscillation (PDO) index [*Mantua et al.*, 1997], the North Pacific Index (NPI, defined as area-averaged SLP anomalies in the region of 30°–65°N, 160°E–140°W, *Trenberth and Hurrell* [1994]), and ENSO index (here using Niño3 index). A 9-month running mean is applied to all the indices before the analysis.

4. NORTH PACIFIC DECADAL VARIABILITY INFERRED FROM GYRE-INDEX

As seen from the 9-month running mean evolution in Figure 4a, all five indices, namely, STG, TG, PDO, NPI, and Niño3, show signals of interannual to interdecadal timescales. The correlation between these indices is shown in Figures 4b and 4c. For the STG, all PDO, NPI and Niño3 indices lead the subtropical gyre variation. This phase leading could be supported by the observational evidence that the Rossby waves induced by wind curl in the central subtropical Pacific can affect the gyre circulation, say, via modifying the interior Sverdrup transport balance and thus subsequently alternating western boundary current a couple of years later. For the TG, the correlation with the PDO and the NPI is relatively weak, in contrast, it is significantly correlated with the Niño3 index with one or two months lag.

One of the possible mechanisms of Pacific decadal variability is the unstable air-sea interaction between the subtropical gyre circulation in the North Pacific and the Aleutian Low system. *Latif and Barnett* [1994, 1996] suggested a delayed negative feedback of the interdecadal oscillations in the North Pacific based on wind-stress curl forced Rossby wave adjustment and the subsequent heat transport of the subtropical gyre western boundary current. This tends to be supported by *Deser et al.* [1999], who found that the Kuroshio Current Extension is intensified from the 1970s to the 1980s as a delayed response to the anomalous interior wind stress curl. This delayed negative feedback, however, is not supported from other modeling studies [*Seager et al.*, 2001; *Schneider et al.*, 2002].

The delayed response of the subtropical and subpolar gyres to the anomalous wind stress can also be seen in our index analysis. The correlation between the STG and NPI indices is asymmetric with significantly negative correlation when the NPI leads (Figure 4b). Physically, a negative NPI index represents a strengthening of the Aleutian Low, and thus a positive curl anomaly in the subtropics, which subsequently spins up the subtropical gyre.

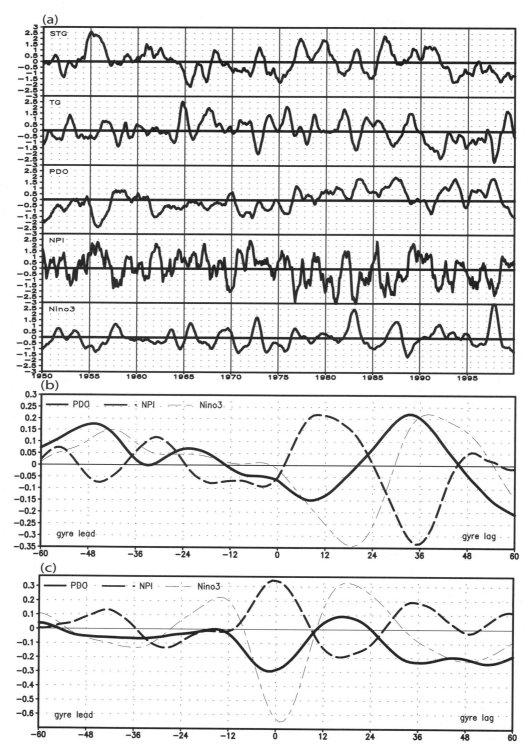

Figure 4. Comparison of time series of SPG, STG, PDO indices, NPI, and Niño3. (a) The 9-month running mean time series, (b) the lagged/leading correlation coefficient of STG vs. PDO index, NPI, and Niño3 index, (c) same as (b) but for TG. X-axis is the leading or lagged months, positive for the gyre lagged. Thick solid line for PDO index, thick long dashed line for NPI, and thin long short dashed line for Niño3 index.

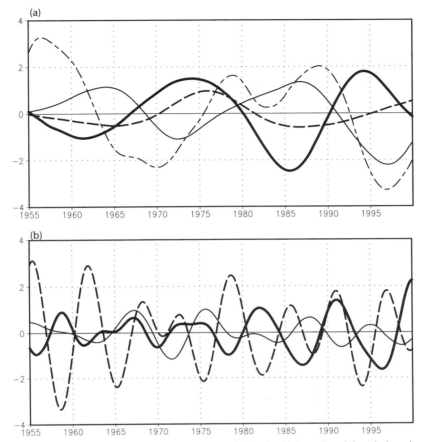

Figure 5. Decadal to interdecadal variations in the reconstructed signals. (a) Interdecadal variations in the reconstructed signals of the first two leading SSA modes (thin line for the SPG, accounting for 33.2% of the total variance; long short dashed line for the STG, accounting for 21.0% of the total variance; long dashed line for the heat storage in the central North Pacific (CNP), accounting for 53.3%; thick line for the heat storage in the Kuroshio Extension (KE), accounting for 36.5%). (b) Decadal variations in the reconstructed signals of the sum of SSA Mode 3 and 4 (long dashed line for the STG index, accounting for 17.0% of the total variance; thin line for the SPG index, accounting for 11.6%; thick line for KE heat storage, accounting for 18.1%).

We may examine the mechanism of the advection of the mean temperature gradients by the gyre transport fluctuations at different timescales. As shown in Figure 5a, at the interdecadal timescales, the SPG tends to lead the heat storages both in the central North Pacific and in the Kuroshio Extension region, whereas, the STG is orthogonal to these interdecadal thermal variations. If the subpolar gyre dominates, heat storage in the Kuroshio Extension region appears to increase when the SPG spins up. Physically, it is conceivable that the SPG dominates the interdecadal variation of the heat storage because the adjustment of the SPG in response to the wind stress change takes a longer time, which is usually one or two decades because of the slow phase speed of Rossby waves at these latitudes.

For decadal variability, say, in the heat storage in the Kuroshio Extension region, both the SPG and the STG show evidence of possible contribution (see Figure 5b). The STG index and heat storage in the Kuroshio Extension region are out of phase, except in several years around 1990. Recalling the above mentioned Rossby wave mechanism, the STG may lead the heat storage in the Kuroshio Extension region by 4–5 years. Though the SPG decadal signal is weaker than that of the STG, it seems to lead the heat storage in the Kuroshio Extension region by 2–3 years. Note that the enhanced SPG would result in negative anomalies of the heat storage in the Kuroshio Extension region. Therefore, the thermal decadal variation in the Kuroshio Extension can be contributed by the both the SPG and STG in response to the interior wind stress change, which may shift the gyre boundary or strengthen (reduce) the gyre transport. If the subtropical gyre dominates, heat storage in the Kuroshio Extension region should appear to increase when the STG spins up. Therefore, these strong signals can indicate that the subtropical gyre may be more important than the subpolar gyre in the observed regime shifts in the North Pacific such as the 1976–77 and 1989–90 cases.

In short, our index analysis tends to suggest that the 1976–77 climate shift is associated with the strengthening of the SPG. The result shows heat storage in the Kuroshio Extension region is associated with the SPG at multi-decadal timescale, but is more correlated with the STG at decadal timescale.

5. INTERACTION OF TROPICAL-EXTRATROPICAL GYRES

McPhaden and Zhang [2002] pointed out that the shallow overturning circulation has been slowed down since the 1970s. The subsequent reduction in equatorial upwelling is associated with a rise in equatorial sea surface temperatures. The enhanced and reduced TG index periods for 1961–1975 and 1981–1999 are indicated in Figure 3c, respectively. It seems that the decadal shift of the TG index occurs around 1976–77, while the so-called regime shift occurs in the North Pacific. The reason of such a correspondence is to be investigated in the future.

It is very important to investigate the water exchange between subtropical and tropical Pacific Oceans. The mechanism and implication of the observed tritium maximum in the central equatorial Pacific have been explored using a trajectory method [*Liu and Huang*, 1998], and the interior pathway of exchange water in the North Pacific Ocean was revealed.

We now describe the tropical-extratropical interaction in terms of gyre-related slow-down and spin-up. Generally, the STG and TG indices are out of phase in the past 50 years with the exception of several years around 1965, when their

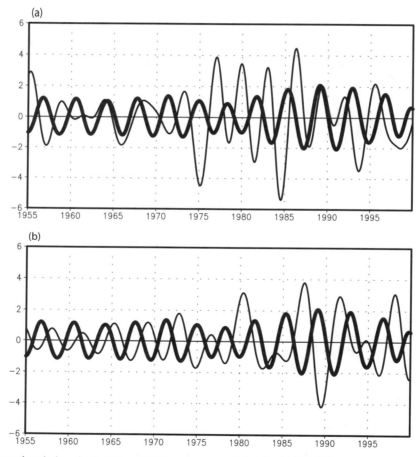

Figure 6. Interannual variations in the TG and STG reconstructed signals. (a) Thick line represents the reconstructed TG index based on its pair of interannual SSA modes (Modes 2 and 3, accounting for 12.2%); thin line represents the reconstructed STG index based on its pair of interannual SSA modes (Modes 5–10, jointly accounting for 30.8%). (b) Thin line represents the reconstructed Niño3 heat storage based on its interannual SSA modes (Modes 4–7, jointly accounting for 25.2% of total variance); thick line same as in (a).

leading interannual variations are concerned (Figure 6a). Note that 12.2% of total variance is explained by the TG's SSA modes 2 and 3 and 30.8% by the STG's modes 5–10. The North Equatorial Current bifurcation leads to a low-latitude western boundary current and Kuroshio [*McCreary and Lu*, 1994], both of them weight much in the TG and STG indices, respectively. If the total transport carried by the North Equatorial Current is fixed, an enhanced low-latitude western boundary current would be usually accompanied by a weakened Kuroshio, and *vice versa*.

On the interannual timescale, the TG index and the heat storage in the Niño3 region are out of phase significantly. With the enhanced (or reduced) TG, there is much more (less) subtropical cold water entering the tropics and will be upwelled mostly at the eastern equatorial Pacific.

Figure 6b also shows the changes of the amplitude and period over different decades associated with the TG index and El Niño events. Beginning at the late 1970s, the period of the TG index changed from 3–4 years to 5–6 years with increasing amplitude. So does the heat storage oscillations in the Niño3 region. The probable explanation is that the TG's larger amplitudes mean larger heat exchanges between the equatorial and off-equatorial zones, and larger anomalies in the heat storage in the Niño3 region lead to a longer oscillating period for relaxing to the normal state.

6. DISCUSSIONS AND REMARKS

The above indices analysis reveals an oceanic gyre connection with interannual to interdecadal variability in the northern Pacific. Our simple gyre indices clearly reveal different timescales of the SPG, STG and TG adjustment over the North Pacific. We hypothesize that the interdecadal variability of the North Pacific tends to be associated with the adjustment of the SPG, while the decadal variability tends to be associated with the adjustment of the STG.

Although wind as an external forcing is not considered in this study, it is possible that atmosphere-ocean interaction over different regions and on different timescales should be responsible for the above-mentioned gyre-connected variations in the Pacific Ocean.

It is also found the decadal shift of the TG index occurred around 1976–77 when its interannual amplitude is concerned, while the regime shift occurred in the North Pacific. However, the possible linkage between two phenomena remains unknown, although it has been shown that the STG associated to poleward heat transport leads the generation of heat anomalies in North Pacific on the interdecadal timescales [*Pierce et al.*, 2000].

Some results from this note need to be further addressed. First, the SPG is associated with the multidecadal variability, while both the STG and SPG are associated with the decadal variability in the northern Pacific. Second, the tropical-extratropical interaction can be explained in view of changes of the basin scale gyre's intensity. The STG and TG indices were out of phase in the past 50 years. On the interannual timescale, the TG index and the heat storage in the Niño3 region are out of phase. The results also show that the interdecadal variation modulates the interannual amplitude and period of the TG and El Niño almost simultaneously. Beginning in the late 1970s, the periods of them change from 3–4 years to 5–6 years with gradually enhanced amplitude.

Further studies are needed. We will extend our study to the southern Pacific Ocean. Our previous observational analysis further identifies the South Pacific as a major source of influence of recent decadal variability of equatorial thermocline [*Wang and Liu*, 2000].

Some OGCM and CGCM experiments are to be performed to test the above hypothesis. Coupled ocean-atmosphere models would be designed to isolate the role of the basin-scale gyres on the global atmosphere from the more local equatorial circulation, because it was found heat fluxes and wind forcing might play different impact on the oceanic decadal-interdecadal variations over different latitudes [*Wang et al.*, 2003]. The experimental strategy for answering the fundamental question of the roles of the STG, TG and SPG in the low frequency climate variability is our concern.

Acknowledgments: We thank Dr. Jim Carton for providing the SODA dataset. Comments from two anonymous reviewers, Prof. Rui Xin Huang at WHOI and the editor, Dr. Chunzai Wang were constructive to improve the paper substantially. This work is supported by the NSFC (Grant 40136010) and the CAS (Grant ZKCX2-SW-210).

REFERENCES

Barnett, T. P., D. W. Pierce, R. Saravanan, N. Schneider, D. Dommenget, and M. Latif, Origins of the midlatitude Pacific decadal variability, *Geophys. Res. Lett.*, 26, 1453–1456, 1999.

Carton, J. A., G. Chepurin, X. Cao, and B. Giese, A simple ocean data assimilation analysis of the global upper ocean 1950–1995. Part I: methodology, *J. Phys. Oceanogr.*, 30, 294–309, 2000.

Deser, C., M. A. Alexander, and M. S. Timlin, Upper ocean thermal variations in the North Pacific during 1970–1991, *J. Clim.*, 9, 1840–1855, 1996.

Deser, C., M. A. Alexander, and M. S. Timlin, Evidence for a wind-driven intensification of the Kuroshio Current Extension from the 1970s to the 1980s, *J. Clim.*, 12, 1697–1706, 1999.

Frankignoul, C., P. Muller, and E. Zorita, A simple model of the decadal response of the ocean to stochastic wind stress forcing, *J. Phys. Oceanogr.*, 27, 1533–1546, 1997.

Ghil, M., and R. Vautard, Interdecadal oscillations and the warming trend in global temperature time series, *Nature*, 350, 324–327, 1991.

Graham, N. E., Decadal scale variability in the 1970s and 1980s: observations and model results, *Clim. Dyn.*, 10, 135–162, 1994.

Gu, D., and S. G. H. Philander, Interdecadal climate fluctuations that depend on exchanges between the tropics and extratropics, *Science*, 275, 805–807, 1997.

Jin, F. F., A theory of interdecadal climate variability of the North Pacific ocean-atmosphere system, *J. Clim.*, 10, 1821–1835, 1997.

Kleeman, R., J. P. McCreary, and B. A. Klinger, A mechanism for generating ENSO decadal variability, *Geophys. Res. Lett.*, 26, 1743–1746, 1999.

Latif, M., and T. P. Barnett, Causes of decadal climate variability over the North Pacific and North America, *Science*, 266, 634–637, 1994.

Latif, M., and T. P. Barnett, Decadal climate variability over the North Pacific and North America: dynamics and predictability, *J. Clim.*, 9, 2407–2423, 1996.

Liu, Z. and B. Huang, Why is there a tritium maximum in the central equatorial Pacific thermocline? *J. Phys. Oceanogr.*, 28, 1527–1533, 1998.

Mantua, N. J., S. R. Hare, Y. Zhang, J. M. Wallace, and R. C. Francis, Pacific interdecadal climate oscillation with impacts on salmon production, *Bull. Am. Meteorol. Soc.*, 78, 1069–1079, 1997.

McCreary, J. P., and P. Lu, Interaction between the subtropical and equatorial ocean circulation—the subtropical cell, *J. Phys. Oceanogr.*, 24, 466–497, 1994.

McPhaden M. J., and D. Zhang, Slowdown of the meridional overturning circulation in the upper Pacific Ocean, *Nature*, 415, 603–608, 2002.

Miller, A. J., D. R. Cayan, and W. B. White, A westward intensified decadal in the North Pacific thermocline and gyre-scale circulation, *J. Clim.*, 11, 3112–3127, 1998.

Miller, A. J., and N. Schneider, Interdecadal climate regime dynamics in the North Pacific Ocean: theories, observations and ecosystem impacts, *Prog. Oceanogr.*, 47, 355–379, 2000.

Minobe, S., Resonance in bidecadal and pentadecadal climate oscillations over the North Pacific: role in climatic regime shifts, *Geophys. Res. Lett.*, 26, 855–858, 1999.

Overland, J. E., J. M. Adams, and N. A. Bond, Decadal variability of the Aleutian Low and its relation to high-latitude circulation, *J. Clim.*, 12, 1542–1548, 1999.

Pierce, D. W., T. P. Barnett, and M. Latif, Connections between the Pacific Ocean tropics and midlatitudes on decadal time scales, *J. Clim.*, 13, 1173–1194, 2000.

Qiu, B., and T. M. Joyce, Interannual variability in the mid- and low-latitude western North Pacific, *J. Phys. Oceanogr.*, 22, 1062–1079, 1992.

Reid, J. L., On the total geostrophic circulation of the Pacific Ocean: flow patterns, tracers, and transports, *Prog. Oceanogr.*, 39, 263–352, 1997.

Seager, R., Y. Kushnir, N. H. Naik, M. A. Cane, and J. Miller, Wind-driven shifts in the latitude of the Kuroshio–Oyashio Extension and generation of SST anomalies on decadal timescales, *J. Clim.*, 14, 4249–4265, 2001.

Saravanan, R., and J. C. McWilliams, Advective ocean–atmosphere interaction: an analytical stochastic model with implications for decadal variability, *J. Clim.*, 11, 165–188, 1998.

Tourre, Y., W. B. White, and Y. Kushnir, Evolution of interdecadal variability in sea level pressure, sea surface temperature, and upper ocean temperature over the Pacific Ocean, *J. Phys. Oceanogr.*, 29, 1528–1541, 1999.

Trenberth, K. E., and J. W. Hurrell, Decadal atmosphere–ocean variations in the Pacific, *Clim. Dyn.*, 9, 303–319, 1994.

Wang, D., and Z. Liu, 2000. The pathway of the interdecadal variability in the Pacific Ocean, *Chin. Sci. Bull.*, 45(17), 1555–1561.

Wang, D., J. Wang, L. Wu, and Z. Liu, Relative importance of wind and buoyancy forcing for interdecadal regime shifts in the Pacific Ocean, *Sci. China, Ser. D*, 46(5): 417–427, 2003.

White, W. B., Design of a global observing system for gyre scale upper ocean temperature variability, *Prog. Oceanogr.*, 36, 169–217, 1995.

Dejun Gu, Yun Liu, and Dongxiao Wang, 164 West Xingang Road, Guangzhou 510301, LED, South China Sea Institute of Oceanology, Chinese Academy of Sciences, China. (dxwang@scsio.ac.cn)

Observed Associations Among Storm Tracks, Jet Streams and Midlatitude Oceanic Fronts

Hisashi Nakamura[1,2], Takeaki Sampe[1], Youichi Tanimoto[2,3], and Akihiko Shimpo[4]

An association is discussed among a midlatitude storm track, a westerly polar-front jet stream and an underlying oceanic frontal zone. Their close association is observed when a subtropical jet stream is weak, as in the Southern Hemisphere summer or in the North Atlantic. Along a near-surface baroclinic zone that tends to be anchored around a frontal zone, storm track activity is enhanced within a well-defined polar-front jet with modest core velocity. This eddy-driven jet exhibits a deep structure with the strong surface westerlies maintained mainly through a poleward eddy heat flux. The westerly wind stress exerted along the frontal zone acts to maintain it by driving the oceanic current system, suggestive of a feedback loop via midlatitude atmosphere–ocean interaction. It is argued that the context of this feedback must be included in interpreting the tropospheric general circulation and its variability. In fact, decadal-scale sea–surface temperature anomalies observed in the North Pacific subarctic frontal zone controlled the anomalous heat release to the atmosphere. Seemingly, the local storm track responded consistently to the decadal-scale shift of the frontal axis, acting to reinforce basin-scale flow anomalies. Over the North and South Pacific, the association is disturbed in winter by an intensified subtropical jet that traps eddy activity into its sharp core. The trapping impairs baroclinic interaction of upper-level eddies with the surface baroclinicity along a midlatitude oceanic front, leading to the suppression of eddy activity as observed in midwinter over the North Pacific.

1. INTRODUCTION

Synoptic-scale baroclinic eddies migrating along mid-latitude storm tracks not only influence daily weather but also play a crucial role in the climate system by systematically transporting heat, moisture and angular momentum. Seasonal variations of synoptic-scale eddies have been examined for the Northern Hemisphere (NH) [*Petterssen*, 1956; *Klein*, 1958; *Whittaker and Horn*, 1984; *Rogers*, 1990] and the Southern Hemisphere (SH) [*Sinclair*, 1994, 1995; *Simmonds and Keay*, 2000], by tracking the centers of individual moving cyclones (or anticyclones) at the surface. The "Lagrangian-type" approach based on cyclone tracks ("storm tracks" in this framework) is a straightforward application of weather chart analysis. *Hoskins and Hodges* [2002] applied this tracking method to upper-level fields of other variables from which planetary-scale signals had been removed.

In addition to the "synoptic" viewpoint, another approach has been adopted, where emphasis is placed on propagation behavior of wavy disturbances and their ensemble feedback on

---

[1]Department of Earth and Planetary Science, University of Tokyo, Tokyo, Japan.
[2]Frontier Research System for Global Change, Yokohama, Japan.
[3]Graduate School of Environmental Earth Science, Hokkaido University, Sapporo, Japan.
[4]Climate Prediction Division, Marine and Climate Department, Japan Meteorological Agency, Tokyo, Japan.

Earth's Climate: The Ocean-Atmosphere Interaction
Geophysical Monograph Series 147
Copyright 2004 by the American Geophysical Union
10.1029/147GM18

the time-mean flow in which they are embedded. This "Eulerian-type" approach is based on high-pass filtering of daily time series at individual grid points, for extracting subweekly fluctuations associated with migratory synoptic-scale eddies [*Blackmon et al.*, 1977, 1984]. In this "wave dynamic" approach, regions of large variance in geopotential height or meridional wind velocity or of a strong poleward eddy heat flux are called "storm tracks", and "storm track activity" signifies the magnitude of the variance or heat flux. *Wallace et al.* [1988] discussed the relationship between cyclone and anticyclone tracks in the synoptic framework and storm tracks in the wave dynamic viewpoint. Though critically argued recently [*Held*, 1999], this approach has an advantage that local correlation between high-pass-filtered time series of air temperature and meridional wind velocity or vertical motion gives a measure for baroclinic structure of migratory eddies. The high positive correlation indicates baroclinic structure of those eddies that allows efficient energy conversion from the time-mean flow for their growth. Climatological seasonal variations observed in storm tracks were documented in this framework by *Trenberth* [1991] and *Nakamura and Shimpo* [2004, hereafter *NS04*] for the SH and by *Nakamura* [1992, hereafter *N92*] for the NH. Some of the related dynamical issues are discussed by *Cai* [2004].

As reviewed by *Chang et al.* [2002], recent studies have substantiated a notion of downstream development, in recognition of group-velocity propagation of synoptic eddies along storm tracks [*Chang*, 1993, 1999; *Lee and Held*, 1993; *Swanson and Pierrehumbert*, 1994; *Orlanski and Chang*, 1995; *Berbery and Vera*, 1996; *Chang and Yu*, 1999; *Rao et al.*, 2002]. The notion requires us to interpret eddy statistics in relation to cyclogenesis from a viewpoint of an initial value problem. This type of cyclogenesis has been known as the "B-type cyclogenesis" [*Petterssen and Smebye*, 1971] or "coupling development" [*Takayabu*, 1991], to which *Hoskins et al.* [1985] added further elucidation from a potential-vorticity (PV) perspective. In the "PV thinking", baroclinic eddy growth is interpreted as mutual reinforcement between PV anomalies at the tropopause and those in the form of temperature anomalies at the surface. In the downstream development, the thermal anomalies are triggered by wind fluctuations across a surface baroclinic zone induced by a propagating upper-level vortex. Thus, surface temperature gradient is of particular significance in baroclinic instability. Nevertheless, in most of the studies from the wave dynamic perspective, storm tracks have been regarded as a pure atmospheric issue.

Forecast experiments have shown the importance of heat and moisture supply from the warm ocean surface of the Gulf Stream or Kuroshio in individual events of rapid cyclone development [*Nuss and Kamikawa*, 1990; *Kuo et al.*, 1991; *Reed et al.*, 1993; *Neiman and Shapiro*, 1993]. A regional-model experiment by *Xie et al.* [2002] indicates that cyclone development is sensitive to a fine frontal structure in a sea–surface temperature (SST) field between the Kuroshio and the shallow East China Sea. Climatologically, rapid cyclone development over the NH is most likely along the Gulf Stream and Kuroshio [*Sanders and Gyakum*, 1980]. Over the SH, maritime cyclogenesis is most frequent around an intense oceanic frontal zone in the Indian Ocean [*Sinclair*, 1995]. These observational tendencies suggest the oceanic influence on storm track formation. At the same time, storm tracks can in turn influence the underlying ocean. By means mainly of their poleward heat flux, eddies migrating along a storm track transfer the mean-flow westerly momentum downward, acting to sustain the surface westerlies [*Lau and Holopainen*, 1984]. In fact, *Hoskins and Valdes* [1990, hereafter *HV90*] considered a storm track could be self-maintained under the heat and moisture supply from a nearby warm ocean current that is driven by the eddy-maintained surface westerlies. Those eddies also supply fresh water to the ocean along the storm track, influencing the stratification in the midlatitude upper ocean [*Lukas*, 2001].

The main purpose of this paper is to further discuss the importance of the atmosphere–ocean coupling via storm tracks in the tropospheric circulation system and its long-term variability from the wave dynamic viewpoint, based on observational statistics. Our argument may be viewed as an extension of *HV90*, but unlike in *HV90*, we put emphasis on oceanic frontal zones associated with major oceanic currents. As the surface air temperature over the open ocean is linked to SST underneath, maritime surface baroclinic zones tend to be anchored along oceanic fontal zones [*NS04*]. Though acting as thermal damping for the evolution of individual eddies, heat exchange with the underlying ocean, on longer time scales, can act to restore atmospheric near-surface baroclinicity against the relaxing effect by atmospheric eddy heat transport, as evident in sharp meridional contrasts in upward turbulent heat fluxes observed climatologically across midlatitude frontal zones [*Oberhuber*, 1988]. Some observations are shown in section 2 to suggest that SST anomalies in a midlatitude frontal zone can likely play a more active role in the air–sea interaction than act to damp atmospheric anomalies thermally. In section 3, we discuss associations among storm tracks, polar–frontal (or subpolar) jet streams and underlying oceanic frontal zones over the two hemispheres. In section 4, we then discuss how such an association can be disturbed in winter by the intensification of a subtropical jet stream. In the final section, we propose a working hypothesis through which our understanding might be deepened on the observed tropospheric circulation system and its variability.

2. IMPORTANCE OF STORM TRACKS AND OCEANIC FRONTAL ZONES IN EXTRATROPICAL COUPLED OCEAN–ATMOSPHERE VARIABILITY

2.1. Atmospheric Forcing Over Central/Eastern Basins

The interaction between the midlatitude ocean and storm tracks is by no means a new concept. The importance has been emphasized in the notion of the "atmospheric bridge", through which the effect of tropical Pacific SST anomalies (SSTAs), associated with the El Niño/Southern Oscillation (ENSO), is transferred into midlatitudes to drive SSTAs remotely with the opposite sign [*Lau and Nath,* 1994, 1996, 2001; *Lau,* 1997; *Alexander et al.,* 2002, 2004; *Hoerling and Kumar,* 2002]. A similar mechanism must be operative also in decadal SST variability over the North Pacific driven by tropical variability [*Nitta and Yamada,* 1989; *Trenberth,* 1990]. Pacific decadal variability is reviewed by *Seager et al.* [2004]. Once a stationary atmospheric teleconnection pattern forms in response to tropical SSTAs with equivalent barotropic anomalies at midlatitudes, local storm track activity and associated poleward heat transport are altered [*Trenberth,* 1990; *Hoerling and Ting,* 1994]. It is this anomalous heat flux through which anomalous upper-level westerly momentum is transferred to the surface. Surface wind anomalies thus enhanced drive SSTAs locally by changing surface turbulent heat fluxes, entrainment at the oceanic mixed-layer bottom, and a cross-frontal Ekman current [*Frankignoul and Reynolds,* 1983; *Frankignoul,* 1985; *Alexander,* 1992; *Miller et al.,* 1994].

The ocean–atmosphere interaction in the "atmospheric bridge" paradigm is thus primarily one-way forcing by atmospheric anomalies on the upper ocean. Thus, local correlation must be *negative* between SST and upward turbulent flux anomalies [*Cayan,* 1992ab; *Hanawa et al.,* 1995; *Tanimoto et al.,* 1997; *Alexander et al.,* 2002], and so is the local correlation between a SSTA and anomalous surface wind speed. Midlatitude SSTAs thus generated tend to have large horizontal extent, reflecting the spatial scale of atmospheric anomalies that have forced them [*Namias and Cayan,* 1981; *Wallace and Jiang,* 1987]. The one-way nature is consistent with the fact that most of the atmospheric general circulation models (AGCMs) fail to generate systematic response to prescribed midlatitude SSTAs [*Kushnir et al.,* 2002]. Several experiments, however, in each of which an AGCM is coupled thermally with a slab ocean mixed layer model [*Lau and Nath,* 1996, 2001; *Watanabe and Kimoto,* 2000] showed that midlatitude SSTAs can reinforce atmospheric anomalies that have driven them. This weak local feedback is called "reduced thermal damping" [*Kushnir et al.,* 2002], as elucidated by *Barsugli and Battisti* [1998] in a linearized one-dimensional coupled model.

2.2. Oceanic Influence From Western-basin Frontal Zones

As discussed above, the atmospheric forcing dominates in the coupled variability over the vast central and eastern domains of a basin. SSTA formation, however, cannot be interpreted solely with local exchanges of heat and momentum through the surface around western boundary currents, where the oceanic thermal advection is substantial in the upper-ocean heat budget [*Qiu and Kelly,* 1993; *Qiu,* 2000, 2002; *Kelly and Dong,* 2004]. Thus, the role of SSTAs in air–sea interaction can be more than the "reduced thermal damping". In fact, *Nonaka and Xie* [2003] found the SST–wind correlation in satellite data to be *positive* along the Kuroshio and its extension [*Xie,* 2004], indicative of modification in near-surface stratification by underlying SSTAs. Analyzing wintertime shipboard measurements compiled on a high-resolution grid over the North Pacific, *Tanimoto et al.* [2003] found that turbulent heat flux anomalies are *positively* correlated with SSTAs in the subarctic frontal zone located in the Kuroshio-Oyashio Extension [*Yasuda et al.,* 1996; *Yuan and Talley,* 1996], and that the positive correlation is stronger when the SSTAs lead the flux anomalies. Confined to a meridionally narrow region along the Kuroshio or frontal zone, the signal of this oceanic thermal forcing would hardly be captured in data complied on a coarse resolution grid (with ~5° latitudinal intervals) or through a statistical method that preferentially extracts basin-scale anomaly patterns such as a singular-value decomposition used by *Deser and Timlin* [1997] and others.

A close association has been found in the extratropics between frontal zones and decadal SST variability [*Nakamura et al.,* 1997a; *Nakamura and Yamagata,* 1999; *Nakamura and Kazmin,* 2003]. *Schneider et al.* [2002] argued that the Kuroshio Extension could be the key region for oceanic feedback on the atmosphere for the Pacific decadal variability, although the associated frontal zone is unlikely to be resolved in their model. *Tanimoto et al.* [2003] argued how SSTAs observed in the frontal zone with decadal variability inherent to the North Pacific [*Deser and Blackmon,* 1995; *Nakamura et al.,* 1997a; *Nakamura and Yamagata,* 1999; *Xie et al.,* 2000; *Tomita et al.,* 2001] can reinforce associated stationary atmospheric anomalies. Their results are summarized in Figure 1, Plates 1 and 2. In the presence of warm (cool) SSTAs in early part of winter (Plate 1a), latent heat release is enhanced (reduced) along the frontal zone (Figure 1; Plate 2a). Linearization of heat flux anomalies [*Halliwell and Mayer,* 1996] reveals that the enhanced (reduced) heat release is attributed to the effect of the local SSTAs (Figure 1; Plate 2b), part of which is offset by a contribution from surface air-temperature (and moisture) anomalies (Figure 1; Plate 2c). A contribution from wind anomalies is negligible in the subarctic

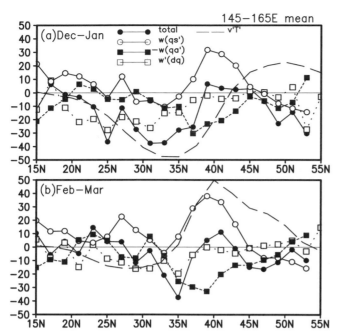

Figure 1. Meridional profiles of anomalous upward latent heat flux (W m^{-2}; solid line with closed circles; upward positive) at the surface, as an average between 145°E and 165°E, for (a) Dec.–Jan. and (b) Feb.–Mar. The subarctic frontal zone is between 36°N and 44°N. Superimposed are individual contributions to the total anomalous flux solely from (and proportional to) local SSTAs (solid line with open circles), anomalous air temperature (dotted line with closed squares) and anomalous surface wind speed (dotted line with open squares). Anomalous 850-hPa poleward heat flux associated with subweekly eddies (unit: 10 K m s^{-1}; long dashed line), as an average over the same longitudinal sector, is also superimposed. These profiles are based on the anomalies shown in Plate 2.

frontal zone (Figure 1; Plate 2d). As shown in Plate 2e, the local storm track is displaced poleward (equatorward), probably in response to changes in near-surface baroclinicity, as consistent with the observed decadal shift of the frontal axis [*Nakamura and Kazmin*, 2003]. This tendency is obvious particularly in early winter, as the eddy heat flux is enhanced (reduced) where the anomalous SST gradient is enhanced (relaxed) as in Figure 1a. In the upper troposphere, the anomalous storm track activity exerts anticyclonic (cyclonic) forcing over the midlatitude North Pacific through anomalous vorticity transport (Plate 2f), reinforcing the pre-existing stationary anticyclonic (cyclonic) anomalies. Consistent with this eddy forcing, a wave-activity flux of stationary Rossby waves [*Takaya and Nakamura*, 2001] is strongly divergent from the anomalies that resemble the Pacific/North American (PNA) pattern [*Wallace and Gutzler*, 1981], which is regarded as a preferred mode of variability in the exit of the North Pacific jet [*Simmons et al.*, 1983; *Peng and Robinson*, 2001]. As in the "atmospheric bridge", anomalous westerly momentum associated with the PNA pattern is transferred downward by eddies, to reinforce the anomalous surface Aleutian low (Plate 1e). Surface wind anomalies thus reinforced exert thermal forcing upon the upper ocean over the central and eastern North Pacific, acting to extend warm (cool) SSTAs downstream of the frontal zone and drive cool (warm) SSTAs off western Canada, in a manner consistent with the observed tendency in SSTAs to late winter (Plates 1a-c).

Kushnir et al. [2002] have postulated a similar mechanism as a paradigm for the coupling between a meridional dipole of atmospheric stationary anomalies and dipolar SSTAs as typically observed in the North Atlantic. Again, a critical factor in forcing the atmospheric anomalies is anomalous storm track activity in response to changes in surface baroclinicity associated with the SSTAs. They considered a particular situation where the SSTAs have been generated by the atmospheric anomalies, as in the "atmospheric bridge". In contrast, over the decadal SSTAs observed in the Pacific subarctic frontal zone, surface wind anomalies are weak, especially in early winter (Plate 1d), indicative of greater importance of oceanic processes [*Xie et al.*, 2000]. It has been suggested that SST variations around the Kuroshio Extension are strongly influenced by changes in oceanic condition [*Tomita et al.*, 2002; *Qiu*, 2003; *Kelly and Dong*, 2004], including the gyre adjustment to atmospheric forcing exerted far to the east [*Schneider et al.*, 2002]. Once zonally elongated SSTAs form in a frontal zone through oceanic processes, they would act to modify the surface baroclinicity locally.

Owing to the two-way interactive nature, a more convincing argument on the oceanic influence on atmospheric anomalies requires modeling studies. Part of the mechanisms argued by *Tanimoto et al.* [2003] is essentially the same as what *Peng and Whitaker* [1999] suggested from their careful diagnosis of an AGCM response to warm SSTAs in the Pacific subarctic frontal zone. They revealed the critical importance of a local storm track in yielding a PNA-like stationary atmospheric response in barotropic structure. A similar suggestion was made by *Watanabe and Kimoto* [2000] for the North Atlantic variability. *Peng and Whitaker* [1999] showed that their AGCM response is sensitive to subtle differences in the model time-mean flow. The sensitivity stems from how effectively the barotropic response is excited under the storm track feedback from near-surface anomalies as the robust direct response to the SSTAs. Since the SSTA pattern given in the model was taken from the observation, a mismatch could happen in their positions between the model storm track and the direct thermal response. The model sensitivity suggests the potential importance of their association, though may not be quite robust, between the frontal zone and storm track in reinforcing the PNA-like anomalies. The association may be part of

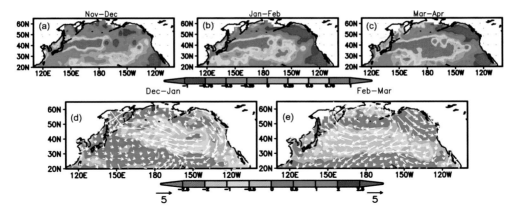

Plate 1. Difference maps of bi-monthly SSTA (°C; reddish and bluish colors for warm and cool anomalies, respectively, as shown below the panels) associated with the North Pacific decadal variability between 4-year composites for 1968/69~ 1971/72 and 1982/83~1985/86 (i.e., "warm" minus "cold"). Based on the Comprehensive Ocean–Atmosphere Data Set (COADS) complied on a 2° × 2° lat.–lon. grid for (a) Nov.–Dec., (b) Jan.–Feb. and (c) Mar.–Apr. (d) As in (a), but for surface wind velocity (arrows with scaling below the panels) and scalar wind speed (m s^{-1}; bluish and reddish colors for stronger and weaker winds, respectively, as shown below the panels) for Dec.–Jan. (e) As in (d), but for Feb.–Mar. After *Tanimoto et al.* [2003].

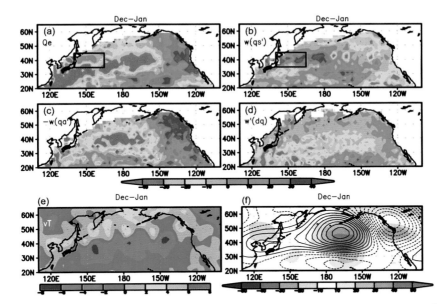

Plate 2. (a) As in Plate 1, but for Dec.–Jan. total upward latent heat flux anomalies at the surface (W m^{-2}; bluish and reddish colors for enhanced and reduced heat release, respectively, from the ocean). The subarctic frontal zone is indicated with a rectangle. (b) As in (a), but for a contribution only from local SSTAs, based on linearization applied to the total anomalous flux in (a). (c) As in (b), but for a contribution only from local air temperature anomalies. (d) As in (b), but for a contribution only from local wind speed anomalies. (e) As in (a), but for storm track activity measured by 850-hPa poleward heat flux associated with subweekly eddies (K m s^{-1}; reddish and bluish colors for the flux enhancement and reduction, respectively), based on the NCEP/NCAR reanalyses. (f) As in (e), but for 250-hPa height (contoured with 20-m intervals), superimposed on feedback forcing from anomalous storm track activity measured by 250-hPa geopotential height tendency (m day^{-1}; bluish and reddish colors for the cyclonic and anticyclonic tendencies, respectively) due only to eddy vorticity flux convergence [*Nakamura et al.*, 1997b]. After *Tanimoto et al.* [2003]. Coloring conventions are shown below the panels.

a feedback loop that is likely operative in the decadal variability inherent to the North Pacific.

3. CLIMATOLOGICAL ASSOCIATIONS AMONG STORM TRACKS, POLAR-FRONT JETS AND MIDLATITUDE OCEANIC FRONTS

3.1. Southern Hemisphere (SH)

Figure 2 shows the SH climatology of storm track activity, westerly wind speed and SST gradient. A prototype example of a close association among a subarctic frontal zone, midlatitude storm track and polar-front jet can be found around 50°S especially in austral summer [Figures 2d-f; *NS04*]. In winter, the association is still close over the South Atlantic and Indian Ocean (Figures 2a-c). There the low-level storm track activity is stronger than over the South Pacific, which seems in correspondence with tighter SST gradient across the Antarctic Polar Frontal Zone (APFZ) [*Colling*, 2001], a subarctic frontal zone along the Antarctic Circumpolar Circulation (ACC), over the former oceans. Along that frontal zone, a strong baroclinic zone forms near the surface (Fig-

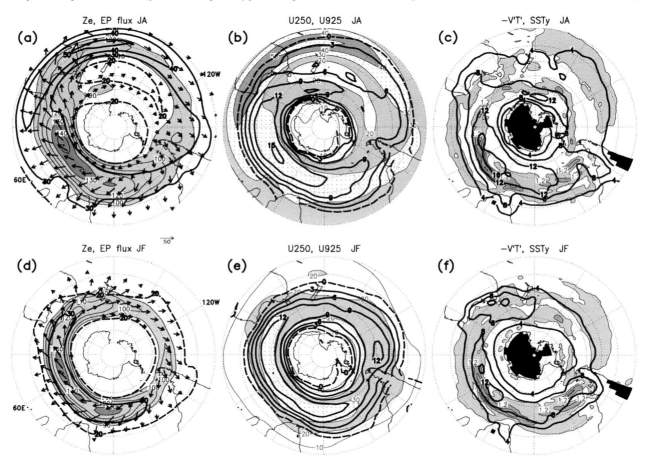

Figure 2. (a) Climatological Jul.–Aug. distribution for the upper-level SH storm track activity (stippling) and horizontal component of 250-hPa extended E-P flux (arrows indicating eddy transport of mean-flow easterly momentum; scaling at the bottom: unit: m² s⁻²), with 250-hPa westerly wind speed (U: m s⁻¹; heavy solid lines for 30, 40, 50 and 60; heavy dashed line for 20). Light and heavy stippling is applied where amplitude of subweekly fluctuations in geopotential height (Z_e: m) at the 250-hPa level is between 90 and 130 and above 130, respectively, with thin lines for every 10. Based on the NCEP (National Centers for Environmental Prediction) reanalyses. (b) As in (a) but for 925-hPa U (m s⁻¹; heavy lines for 3, 6, 9, 12 and 15; dashed for $U = 0$) and 250-hPa U (m s⁻¹; light stippling for 20~30; heavy stippling for 40~50). (c): As in (a), but for 850-hPa poleward heat flux associated with subweekly eddies (heavy lines for 4, 8, 12 and 16 K m s⁻¹). Light and heavy stippling indicates oceanic frontal zones where meridional SST gradient (°C/110 km) exceeds 0.6 and 1.2, respectively (thin lines are drawn for every 0.6), based on satellite and shipboard data complied by *Reynolds and Smith* [1994]. Dark shading indicates data-void regions. (d–f) As in (a–c), respectively, but for Jan.–Feb.

ures 3d-e). Both in the upper and lower troposphere (Figure 2), the storm track core forms in the southwestern Indian Ocean, almost coinciding with the core of the APFZ. In fact, *Sinclair* [1995] found that the most frequent cyclogenesis in the SH occurs around this APFZ core. There, in the course of the seasonal march, the low-level storm track activity exhibits high positive correlation with baroclinicity for a layer just above the surface. *NS04* showed that the correlation is even higher than that with the baroclinicity near the steering (700~850 hPa) level of subweekly eddies, which is also the case for the South Atlantic. Meridional sections in Figures 3d-e show a deep structure of the storm track over the Atlantic and Indian Ocean. The structure reflects the pronounced baroclinic eddy growth above the intense surface baroclinic zone and the downstream development of eddies along the upper-level polar-front jet that acts as a good waveguide for baroclinic wavepackets (Figures 3a-b). In fact, the extended Eliassen-Palm (E-P) flux [*Trenberth*, 1986] has a strong eastward component in the core of the upper-level storm track [*NS04*]. The jet is the sole westerly jet in summer. Even in winter when a subtropical jet intensifies, the storm track core over the South Indian Ocean remains preferentially along the polar-front jet (Figure 2).

The SH storm track core is collocated with the core of the surface westerly jet (Figure 2) as part of the deep polar-front jet (Figures 3a-b) maintained mainly by the downward transport of mean-flow westerly momentum via eddy heat fluxes. The fact that the strongest annual-mean wind stress within the world ocean is observed around the SH storm core [*Trenberth et al.*, 1990] suggests the importance of the storm track activity in driving the ACC and associated APFZ. As shown in Figures 4b and 4d, the annually averaged surface westerly acceleration induced as the feedback forcing through heat and vorticity transport by subweekly eddies is indeed strong along or slightly poleward of the surface westerly axis, and it is strongest near the core of the APFZ. The slight poleward displacement of that axis relative to the APFZ (Figure 2) seems consistent with a tendency for surface upward turbulent heat

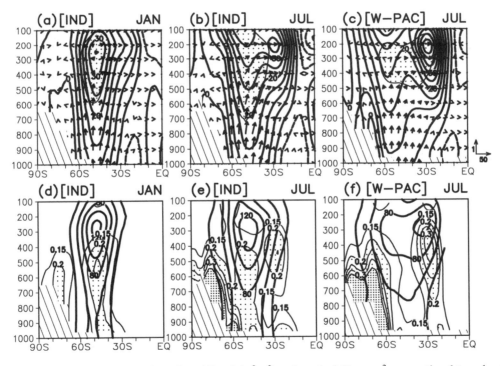

Figure 3. (a) Climatological Jan. section of meridional ($m^2 s^{-2}$) and vertical (Pa m s^{-2}; proportional to poleward eddy heat flux) components of the extended E-P flux (arrows; scaling at the lower-right corner), and U (contoured for every 5 m s^{-1}; dashed for the easterlies), both for the South Indian Ocean (50°~90°E). Based on the NCEP reanalyses. (b) As in (a) but for July. (c) As in (a) but for Jul. in the Australian sector (120°~160°E). Hatching indicates topography. (d–f) As in (a–c), respectively, but for eddy amplitude in geopotential height (Z_e; unit: m; heavy lines for every 20 from 40) and local baroclinicity (G; thin lines for every 0.05 from 0.15; light stippling for 0.2~0.35 and heavy stippling for above 0.35). Here, $G = |g/f_0| \cdot |\nabla \theta/(\theta N)|$, where N denotes the Brunt-Väisälä frequency, θ potential temperature, g the acceleration of gravity, f the Coriolis parameter and $f_0 = f(45°S)$. In linear theories of baroclinic instability for the zonally uniform westerlies, the maximum growth rate of the most unstable mode is proportional to G. In (a–c), stippling for $Z_e > 80$ (m). After *NS04*.

fluxes, wind stirring effect on the oceanic mixed layer, and Ekman velocity to be all maximized along the wind velocity axis. Consistent with an evaluation by *Lau and Holopainen* [1984] for the NH, a contribution from eddy heat transport is stronger than that from eddy vorticity transport, but their contributions are more comparable (not shown).

Over the South Pacific, the association among a midlatitude storm track, polar-front jet and subarctic frontal zone is less robust than over the Atlantic and Indian Ocean [*NS04*]. Though vulnerable to the seasonal intensification of a subtropical jet, their close association can still be found in austral summer and autumn when the jet is diminished. In these seasons, the Pacific storm track at the upper and lower levels is part of a well-defined circumpolar storm track along the ~50°S circle, accompanied consistently by the deep polar-front jet (Figures 2d-f). The low-level eddy activity gradually weakens downstream across the Pacific, as the SST gradient relaxes eastward along the APFZ (Figure 2f). The close association was observed also in a very unusual winter at the beginning of the 1998 La Niña event, in the absence of the intense subtropical jet due to the marked interannual variability. In that winter, the upper-level westerly bifurcation was much less apparent than in the climatology, which marks a sharp contrast with a distinct double-jet structure in the previous winter, as in other El Niño winters [*Chen et al.*, 1996]. As well inferred from a difference map in Figure 5a, no well-defined storm track formed over the subtropical South Pacific in the 1998 winter, under the extremely weakened subtropical jet. Instead, eddy activity over the South Pacific was enhanced at midlatitudes and organized into a single storm track along the polar-front jet at ~55°S throughout the troposphere (Figure 5), which indeed resembled the summertime situation (Figure 2). In 1998, the midlatitude westerlies were stronger not only in the upper troposphere but also near the surface (Figure 5), consistent with coherent vertical structure of the midlatitude storm track. In that winter, upper-level wave activity was dispersed strongly equatorward from the enhanced subpolar storm track in the central and eastern Pacific, through which the westerly momentum was transported poleward. Its downward transfer by eddies sustained the strong surface westerlies. In a macroscopic view, the Pacific APFZ remained similar between the two winters, seemingly to keep anchoring the low-level storm track and polar-front jet (not shown).

3.2. Northern Hemisphere (NH)

Figure 6 shows the NH climatology of storm track activity, westerly wind speed and SST gradient. Over each of the ocean basins, a major storm track extends eastward from an intense surface baroclinic zone anchored along a subarctic frontal zone off the western boundary of the basin (Figure 6b), where warm and cool boundary currents are confluent. In a macroscopic view, the storm track is along the boundary between subtropical and subpolar gyres. In addition, the thermal contrast between a warm boundary current (the Gulf Stream or Kuroshio) and its adjacent cooler landmass also influences the storm track activity in winter [*Dickson and Namias*, 1976; *Gulev et al.*, 2003]. Over the North Atlantic, a belt of the surface westerlies between the Icelandic Low and Azores High is situated slightly to the south of the storm track axis. Over the wintertime North Pacific, the poleward displacement of the low-level storm track relative to the surface westerly axis is more apparent. The latter is closer to the subtropical jet axis aloft especially over the western Pacific, although the poleward secondary branch of the surface westerlies is close to the storm track.

Despite the modest intensity of the local upper-level westerly jet (Figure 6a), midwinter storm track activity is stronger

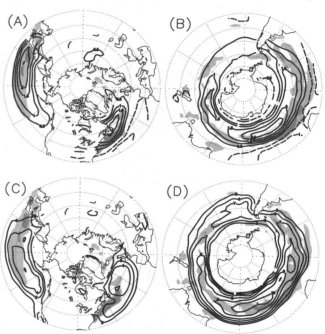

Figure 4. (a) Climatological annual-mean westerly acceleration (solid lines at 0.5 m s^{-1} day^{-1} intervals with zero lines omitted; dashed lines for easterly acceleration) at the 1000-hPa level over the NH, as the feedback forcing from storm tracks evaluated in the same manner as in *Lau and Holopainen* [1984] but based on 8-day high-pass-filtered NCEP reanalysis data for 1979~98. Light and heavy stippling indicates oceanic frontal zones where climatological annual-mean meridional SST gradient (°C/110 km) is 0.8~1.6 and above 1.6, respectively, based on the data by *Reynolds and Smith* [1994]. (b) As in (a) but for the SH. (c) As in (a) but for the 1000-hPa westerly wind speed for the NH (solid lines for 2, 3, 4 and 5 m s^{-1}). (d) As in (a) but for the 1000-hPa westerly wind speed for the SH (solid lines for 5, 6, 7 8, 9 and 10 m s^{-1}).

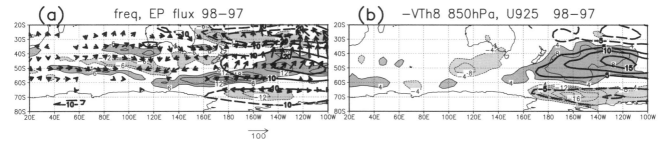

Figure 5. Difference maps over the South Indian and Pacific Oceans for Jul.~Aug. between 1997 and 1998 (1998 minus 1997). (a) Horizontal component of 250-hPa extended E-P flux (arrows; scaling at the bottom; unit: m^2 s^{-2}) associated with subweekly eddies, 250-hPa U (heavy lines for 10, 20 and 30 m s^{-1}; dashed for the anomalous easterlies) and 250-hPa storm tracks (stippling). Light and heavy stippling is applied where decrease and increase, respectively, in the frequency of an eddy amplitude maximum passing through a given data point with 2.5° intervals on a given meridian, defined as the number of days over a 62-day period, exceed 6 (thin lines for every 6). (b) 925-hPa U (heavy lines for 5, 10 and 15 m s^{-1}; dashed for the anomalous easterlies) and 850-hPa poleward eddy heat flux (K m s^{-1}; light contours for every 4; light and heavy stippling for positive and negative values). Based on the NCEP reanalyses.

over the North Atlantic than over the North Pacific (Figure 6b). The low-level storm track axis is closer to a subarctic frontal zone in the Atlantic than in the Pacific, and the cross-frontal SST gradient is substantially stronger in the Atlantic than in the Pacific (Figure 6b). While its main surface axis extends along the Oyashio Extension at ~42°N, the North Pacific subarctic frontal zone at the surface is meridionally broader, including the Interfrontal Zone in the Kuroshio-Oyashio Extension [*Lin and Talley,* 1996; *Yasuda et al.,* 1996; *Nakamura and Kazmin,* 2003]. The North Atlantic subarctic frontal zone is shaper and more intense, contributing to the more pronounced local eddy growth and perhaps to the stronger eddy activity.

Another factor that contributes to the Atlantic-Pacific difference in wintertime storm track activity is latitudinal displacement between a storm track and subarctic frontal zone. In the course of its seasonal march, the North Atlantic storm track stays to the north of the subarctic frontal zone, and it is nearest to the front in midwinter when eddy activity peaks (not shown). The westerly jet axis closely follows the underlying frontal zone, especially downstream of the jet core (Figure 6). The North Pacific storm track undergoes larger seasonal migration in its latitudinal position [*N92*], and eddy activity tends to be suppressed in midwinter when the storm track axis stays to the south of the Pacific subarctic frontal zone [*Nakamura and Sampe,* 2002; hereafter *NS02*]. *NS02* found that upper-level eddies traveling from the Asian continent tend to propagate above the surface baroclinic zone along the frontal zone when the storm track activity peaks in spring and late fall [*N92*]. In those seasons, the upper-level westerly jet core is substantially weaker than in midwinter and located somewhat poleward [*NS02*]. The suppression occurs despite the fact that the tropospheric baroclinicity peaks in midwinter. *NS02* pointed out that midwinter eddy activity has enhanced substantially since the late 1980s, as the Pacific storm track tends to stay over the subarctic frontal zone under the decadal weakening of the subtropical jet. They found that, for most of the time during the recent midwinter periods, the eddy amplitude maximum stayed at the midlatitude tropopause right above the frontal zone (Figure 7a), which allowed eddies efficient baroclinic growth through their interaction with a surface baroclinic zone along the frontal zone, as in fall and spring. In fact, eddies exhibited a deeper structure with vigorous poleward heat transport (Figure 7a). In each of these situations over either the Atlantic or Pacific, the extended E-P flux is strongly divergent in the upper troposphere out of the storm track core (not shown). Thus, a westerly jet with modest core velocity bears an eddy-driven nature, a characteristic of a polar-front jet [*Lee and Kim,* 2003]. These results suggest that the association with the underlying frontal zones contributes to the enhancement of the NH storm track activity.

Despite pronounced seasonal march in the axial position and intensity of the NH storm tracks, especially over the Pacific, the annually averaged surface westerly acceleration induced as the feedback forcing from the storm tracks is strongest along the poleward flank of a subarctic frontal zone over each of the ocean basins (Figure 4a), driving oceanic gyres. In the winters of enhanced eddy activity (Figure 8a), the surface westerly axis was situated along the northern fringe of the subarctic frontal zone in the western Pacific, and it was systematically below the upper-level storm track axis over the eastern Pacific. *N92* showed that, in the course of the seasonal march, the axis of the low-level westerlies tends to follow the upper-level storm track over the eastern Pacific, indicative of the reinforcement of the westerlies by the storm track.

4. INFLUENCE OF A SUBTROPICAL JET ON A STORM TRACK AND ITS ASSOCIATION WITH AN OCEANIC FRONT

4.1. Southern Hemisphere (SH)

In the SH climatology (Figure 2), the influence of the seasonal evolution of a subtropical jet stream on storm track activity is apparent only over the South Pacific [*NS04*]. Its wintertime intensification disturbs the association among a midlatitude storm track, polar-front jet and subarctic frontal zone observed over the South Pacific in austral summer and autumn. In the presence of double-jet structure [*Karoly et al.*, 1998; *Bals-Elsholz et al.*, 2001], upper-tropospheric storm track activity bifurcates from the core region into the main branch along the strong subtropical jet and the sub-branch along the weaker polar-front jet (Figure 2a). Thus, the westerlies and storm track are no longer circumpolar. The intense velocity core of the subtropical jet confined to the tropopause (Figure 3c) acts as an excellent waveguide for synoptic-scale eddies. In fact, the extended E-P flux associated with sub-weekly eddies is consistently eastward along the jet (Figures 2a). Located above a surface subtropical high-pressure belt, however, the jet does not favor baroclinic eddy growth, despite the modest surface baroclinicity across the underlying subtropical frontal zone (Figure 2c). Consistently, the subtropical jet does not accompany the strong westerlies at the surface (Figure 2b), thus yielding no significant contribution to the local mechanical driving of the ocean circulation. Over the extratropical SH, the annual-mean surface westerly acceleration induced as eddy feedback forcing is weakest over the South Pacific (Figure 4b), due to the winter-spring breakdown of the well-defined midlatitude storm track.

In winter and spring, the main branch of the low-level storm track is still along the polar-front jet (Figures 2b-c), though displaced poleward above an enhanced low-level baroclinic zone that forms along the seasonal sea-ice margin (Figure 3f). The low-level storm track forms despite the upper-level wave activity from upstream core region is mostly dispersed toward the subtropical jet (Figures 2a and 3b), suggestive of the importance of surface baroclinicity in the storm track formation.

4.2. Northern Hemisphere (NH)

A factor that contributes to the Atlantic-Pacific difference in storm track activity is the midwinter eddy-activity minimum (suppression) in the North Pacific [*N92*]. As opposed to linear theories of baroclinic instability [*Charney*, 1947; *Eady*, 1949], this unique aspect of the seasonal cycle occurs despite the local westerly jet is strongest in midwinter. *Bosart* [1999] speculated critically that the minimum might merely be an artifact due to the sampling by *N92* on the 250-hPa surface that tends to be above the tropopause only in midwinter. However, his speculation is inconsistent with the activity minimum also observed at the lower levels [*N92*; *Nakamura et al.*, 2002]. The minimum has been reproduced in AGCMs [*Christoph et al.*, 1997; *Zhang and Held*, 1999; *Chang*, 2001]. In reanalysis data, *Nakamura et al.* [2002] found the activity

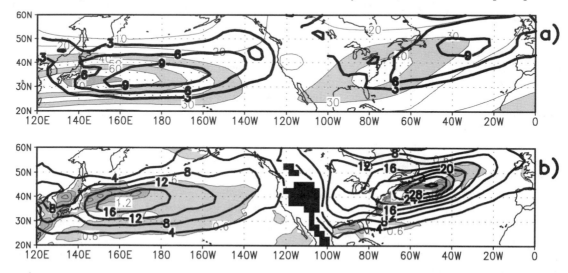

Figure 6. (a) Climatological Jan.~Feb. distribution of 925-hPa U (heavy lines for every 3 m s^{-1}) and 250-hPa U (light and heavy stippling for 30~40 and 50~60 m s^{-1}, respectively), based on the NCEP reanalyses. (b) As in (a) but for 850-hPa poleward eddy heat flux (heavy lines for every 4 K m s^{-1}). Light and heavy stippling indicates oceanic frontal zones where meridional SST gradient (°C/110 km) is 0.6~1.2 and above 1.2, respectively (with thin lines for every 0.6), based on the data by *Reynolds and Smith* [1994].

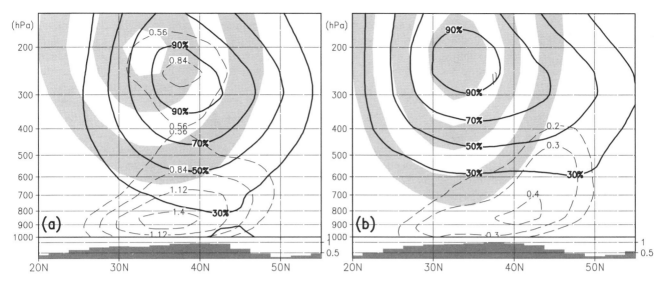

Figure 7. Meridional structure of a typical baroclinic eddy in the North Pacific storm track (170°E~170°W). Based on subweekly fluctuations in geopotential height (Z') field regressed linearly on 300-hPa Z' at [47°N, 105°E] with a 2-day lag, for (a) five Jan.~Feb. periods in 1979~95 with the weakest suppression in eddy activity and for (b) other five periods with the most distinct suppression. Reflecting the decadal weakening in the winter monsoon, the winters for (a) were all since 1987, whereas those for (b) were mostly before 1987. Eddy amplitude in Z' is normalized by its maximum (30, 50 70 and 90%). Associated poleward heat flux based on the regression (K m s^{-1}; density adjusted) is plotted with dashed lines for 0.56, 0.84, 1.12 and 1.40 in (a) and 0.2, 0.3 and 0.4 in (b). Note that eddy amplitude is larger in (a) by 67%. The westerly jet is indicated with stippling (U: 20~30, 40~50 and 60~70 m s^{-1}), and meridional SST gradient is plotted at the bottom (°C/110 km). Based on the NCEP reanalyses. After *NS02*.

minimum, which had been found by *N92* for 1965~84, has disappeared since the late 1980s, under the decadal weakening of the East Asian winter monsoon and associated relaxing of the subtropical jet. This modulation has been confirmed in *Chang's* [2003] analysis of unassimilated aircraft and rawinsonde data. As the mechanism of the activity suppression, *Chang* [2001] argued that enhanced precipitation in out-breaking cold air behind individual cyclones in midwinter does not favor the generation of eddy available potential energy. Alternatively, we argue in the following that the suppression can be interpreted as the dynamical influence of a seasonally intensified subtropical jet.

In the wintertime Far East, the low-level monsoonal northerlies and the enhanced subtropical jet aloft, as observed before the late 1980s, are associated with the marked deepening of a planetary-wave trough, and a polar-front jet tends to merge itself into the subtropical jet [*Mohri*, 1953]. By the northerly component behind the trough, upper-level eddies are driven strongly toward the intensified subtropical jet and then trapped into its core at ~32°N at the 200-hPa level. The core is ~12 km in altitude, ~3 km higher than the midlatitude tropopause (300 hPa) at which eddies have been propagating through the polar-front jet over the Asian continent. In fact, the storm track underwent greater equatorward excursion from its annual-mean position in five midwinter periods with the most distinct eddy-activity minimum than in five other midwinter periods without the minimum [*NS02*]. Trapped by the subtropical jet core, eddies were lifted up by ~3 km and then staying 500~800 km away from the surface baroclinic zone above the subarctic frontal zone at ~40°N (Figure 7b). Thus, eddy interaction with the surface baroclinic zone tended to be impaired, while eddies underwent substantial distortion in their structure. The coherency is thus lowered between subweekly fluctuations in temperature and the meridional or vertical wind component [*N92*; *Chang*, 2001; *Nakamura et al.*, 2002], leading to the less efficient energy conversion for eddy growth. As shown in a meridional section in Figure 7b, under the trapping, eddy amplitude rapidly decays downward and the associated heat flux was reduced by as much as 40%.

As in the South Pacific case discussed earlier, the North Pacific subarctic frontal zone remained very similar between the two types of winter regardless of the substantial changes in storm track activity (Figure 8). Only the noticeable difference is the slightly enhanced cross-frontal SST gradient for the winters with eddy-activity minimum, indicating that the anomalous surface baroclinicity was unlikely the reason for the observed changes in the activity. The axes of the upper- and lower-level storm tracks and surface westerlies were more closely located along the subarctic frontal zones in the winters of stronger eddy activity (Figure 8a). Rather, the axis of the surface westerlies nearly followed the subtropical jet axis in the winters of the suppressed storm track activity (Figure 8b).

5. DISCUSSION

The association among a storm track, polar-front jet and a subarctic frontal zone (including the APFZ) seems crucial for two-way interaction between the midlatitude atmosphere and ocean, as exemplified in an observational study by *Tanimoto et al.* [2003] and in an AGCM experiment by *Peng and Whitaker* [1999] both on the decadal variability inherent to the North Pacific [*Nakamura et al.*, 1997]. Furthermore, the whole dynamical picture of storm tracks and polar-front jets, including the localization of their core regions, can unlikely be obtained without considering their interaction with the underlying ocean, as first argued by *HV90* and recently by *NS02*, *Inatsu et al.* [2003] and *NS04*. In particular, key aspects of seasonal variations of a storm track can be interpreted reasonably well from a viewpoint of how strongly its association with the underlying subarctic frontal zone is disturbed by the seasonal intensification of a subtropical jet [*NS02*, *NS04*]. From this viewpoint, an insight can be gained into the mechanisms that cause the "midwinter activity minimum" of the North Pacific storm track [*NS02*], a puzzling feature of its seasonal cycle that cannot be explained by linear theories of baroclinic instability. The recent disappearance of the activity minimum may be interpreted as the consequence of the decadal weakening of the subtropical jet. In the absence of such a marked change in the subtropical jet, even subtle changes in the Pacific storm track activity could be observed in response to decadal SST changes in the subarctic frontal zone from the late 1960s into the 1980s [*Tanimoto et al.*, 2003]. Of course, the total baroclinicity within the troposphere must be considered in interpreting the profound seasonal march in eddy amplitude along the NH storm tracks, as discussed by *HV90*. They also emphasized the latent heat release along the storm tracks also acts to anchor them by forcing the planetary wave pattern.

It is well known that differential radiative heating acts to restore the mean baroclinicity at midlatitudes against the relaxing effect by eddy heat transport, but it provides no clear explanation why such intense surface baroclinic zones as observed are maintained. A tendency for major maritime surface baroclinic zones to be placed near midlatitude oceanic

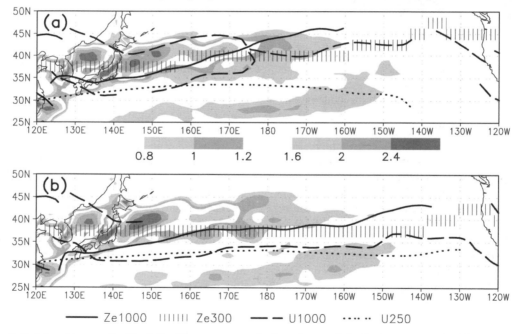

Figure 8. Relationship over the North Pacific among oceanic frontal zones (stippling), storm track axes at the 1000-hPa (solid lines) and 300-hPa (series of vertical lines) levels, and westerly jet axes at the 1000-hPa (dashed lines) and 250-hPa (dotted lines) levels. For (a) five January~February periods in 1979~95 with the weakest suppression in eddy activity and for (b) other five periods with the most distinct suppression, as in Figure 7 and *NS02*. The frontal zones are indicated as regions of intense meridional SST gradient (unit: °C/110 km), as stippled at the bottom of (a). The storm track axes are defined as local Z_e maxima. Atmospheric and SST data are based on the NCEP reanalyses and *Reynolds and Smith* [1994], respectively.

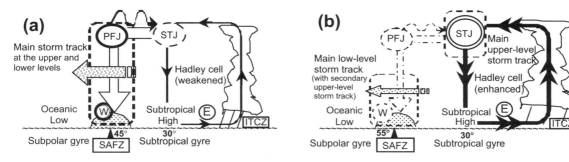

Figure 9. Schematics of different types of tropospheric general circulation over an ocean basin. (a) When a subtropical jet (STJ) is weak, the main storm track (thick dashed line) forms over a surface baroclinic zone (stippled at ~45° lat.) anchored by a subarctic frontal zone (SAFZ), as in the summertime SH, the North Atlantic or the North Pacific (in spring and fall). Wave-activity dispersion to the STJ (wavy arrow) leads to the formation of a deep polar-front jet (PFJ) above the SAFZ. Eddy downward transport (open arrow) of the mean-flow westerly momentum maintains a surface westerly jet (circled W) along the SAFZ. (b) When a STJ intensifies as in the wintertime South Pacific, the jet traps most of the upper-level eddy activity. Thus, the main branch of the upper-level storm track forms along the STJ with suppressed baroclinic eddy growth below, while the low-level storm track forms along a weak PFJ above a baroclinic zone anchored by the SAFZ.

frontal zones [*NS02, NS04*] suggests the effective restoring of the atmospheric baroclinicity, owing to the large thermal inertia of the ocean mixed layer and the differential thermal advection between to the north and south of the frontal zones by strong oceanic currents [*Kelly and Dong, 2004*]. Enhanced heat and moisture fluxes over a warm current just south of a subarctic frontal zone has been known to contribute to cyclogenesis and thus storm track formation [*HV90*]. In addition, a sharp decline of the surface heat release poleward across the frontal zone acts to restore the mean atmospheric near-surface baroclinicity, thus also contributing to the anchoring of the storm track. This anchoring, however, can be disturbed by the seasonal intensification of a subtropical jet or its interannual modulations due to a teleconnection from the tropics or an upstream continent. An important scientific issue to be clarified is how the near-surface baroclinicity is determined and maintained in the marine boundary layer.

Another important aspect of the air–sea coupling associated with a storm track is that the mean westerly momentum carried downward with upward wave-activity transfer in a storm track is organized into a surface westerly jet, which drives oceanic gyres (or the ACC) and thereby contributes to the maintenance of subarctic frontal zones. Along the ACC, the core regions of the storm track, surface westerlies and APFZ almost coincide with each other, indicative of the presence of a local feedback loop. Over each of the NH ocean basins, the frontal zone is located at the confluent region of the western boundary currents driven mainly through gyre adjustment by the surface westerlies that are strongest farther to the east (Figure 4c). A storm track acts to maintain the westerlies, especially along or slightly to the north of the subarctic frontal zones (Figure 4a). The surface westerlies along the storm track also enhance the surface evaporation, whereas precipitation associated with migratory storms largely determines the fresh water supply to the midlatitude ocean [*Lukas, 2001*]. Kinetic energy input into the ocean by the strong surface westerlies and vigorous storm activity acts to sustain the mixed layer structure. The input also becomes an important source of oceanic turbulence available for deep-layer mixing [*Nagasawa et al., 2000*].

Findings in this and related papers [*NS02, NS04*] may require some modifications to conceptual models for the zonally symmetric circulation in the wintertime troposphere, including a well-known model by *Palmén and Newton* [1969]. While resembling its original version proposed by *Rossby* [1941], it emphasizes more the concentration of westerly momentum into subtropical and polar-front jets and their respective association with the Hadley cell and a polar frontal zone. On the basis of the argument by *HV90* and our findings, a fundamental modification we would add to *Palmén's* model is the possible association among a polar-front jet, storm track, surface baroclinic zone over a subarctic frontal zone, as postulated in Figure 9a, which may add further significance to the midlatitude air–sea interaction. Unlike the polar frontal zone tilted distinctly poleward, a polar-front jet and associated baroclinic zone extend more vertically down to the surface just above the frontal zone (Figure 3). The jet is accompanied by a major storm track, and its deep structure is a manifestation of its eddy-driven nature [*Lee and Kim, 2003*].

Another point emphasized in Figure 9 is their distinct characteristics between the two types of jets, as a factor that influences the observed seasonal evolution of storm tracks. In fact, two types of schematics are presented in Figure 9 depending upon the strength of a subtropical jet, as in *Lee and Kim* [2003]. As speculated by *Palmén* [1951] and later elucidated theoretically [*Held and Hou, 1980; Lindzen and Hou, 1988*], the jet is formed through poleward transport of angular momentum by the Hadley cell, and the jet is much stronger in the winter hemisphere where the Hadley cell is stronger. Zonal asymmetries in tropical SST distribution or the presence of a tropical landmass can lead to the localization of the jet [*Inatsu et al.*, 2002]. In fact, the formation of the SH subtropical jet is related to the Asian summer monsoon. Not driven by eddies, a subtropical jet may not necessarily accompany a distinct surface baroclinic zone. Indeed, the jet axis is between the subarctic and subtropical oceanic frontal zones over the North Pacific (Figure 6). Over the SH, the jet is located above a subtropical high-pressure belt, which is unfavorable for baroclinic eddy growth. Thus, a subtropical jet is shallow and confined around its tight core at the high tropopause, unless merged with a polar-front as in the wintertime North Pacific associated with a planetary-wave trough.

Through idealized numerical experiments, *Lee and Kim* [2003] examined how storm track activity depends on the subtropical jet intensity. They found that the main storm track forms along a polar-front jet, as in Figure 9a, only when a subtropical jet is weak, consistent with the observations [*NS02, NS04*]. However, the greatest discrepancy is that a subtropical jet, as it intensifies in the model, becomes increasingly favorable for baroclinic eddy growth. As opposed to their experiments, the jet intensification in the real atmosphere is unfavorable for storm track formation. Over each of the North and South Pacific, an intensified wintertime subtropical jet traps eddies into its core, keeping them away from a surface baroclinic zone anchored by a subarctic oceanic frontal zone. The trapping thus impairs eddy growth, despite the marked baroclinicity below the jet core. Over the South Pacific, where the two jets are well separated, the trapping leads to the meridional separation of the main storm track branch between the upper and lower levels [*NS04*]. We suggest this is a typical situation of the subtropical-jet-dominant regime (Figure 9b). No such separation occurs over the North Pacific, where the two jets are merged. Still, the subtropical jet traps eddy activity, resulting in the midwinter suppression of storm track activity. This is an intermediate situation between the two prototype situations in Figure 9. The storm track activity is enhanced in fall and spring when eddies can propagate above the subarctic frontal zone. This "weak subtropical-jet regime" (Figure 9a) appears more typically over the North Atlantic and the summertime SH. In the real atmosphere, the main storm track branch exhibits an apparent preference for staying with a polar-front jet, perhaps due to the anchoring effect by an underlying oceanic frontal zone. This preference may be underestimated in the idealized experiments by *Lee and Kim*. Their experiments would have been more realistic if well-defined surface baroclinic zones as observed had been prescribed.

Of course, the schematics in Figure 9 are nothing but a working hypothesis. Further observational and modeling study is hence needed to assess how relevant they are to extracting the essence of the atmospheric general circulation observed in the extratropics. More study is also needed to assess the robustness and detailed mechanisms of the positive feedback loop, if really exists, among a polar-front jet, storm track and subarctic frontal zone, and its importance in the climate variability. Especially, the significance of the anchoring effect by oceanic frontal zones should be confirmed in experiments with an AGCM with resolution high enough to resolve the cross-frontal thermal contrasts. It is also important to study how the oceanic fronts are maintained under the forcing from overlying storm tracks.

Acknowledgments. The authors thank Prof. S.-P. Xie for stimulating discussions and encouragement and Mr. Y. Onoue for preparing Figure 4. Comments by the two anonymous referees are greatly acknowledged. The Grid Analysis and Display System (GrADS) was used for drawing the figures.

REFERENCES

Alexander, M. A., Midlatitude atmosphere–ocean interaction during El Niño, *J. Clim., 5*, 944–958, 1992.

Alexander, M. A., I. Bladé, M. Newman, J. R. Lanzante, N.-C. Lau, and J. D. Scott, The atmospheric bridge: The influence of ENSO teleconnection on air–sea interaction over the global oceans, *J. Clim., 15*, 2205–2231, 2002.

Alexander, M. A., N.-C. Lau, and J. D. Scott, Broadening the atmospheric bridge paradigm: ENSO teleconnections to the North Pacific in summer and to the tropical West Pacific–Indian Oceans over the seasonal cycle, 2004.

Bals-Elsholz, T. M., E. H. Atallah, L. F. Bosart, T. A. Wasula, M. J. Cempa, and A. R. Lupo, The wintertime Southern Hemisphere split jet: Structure, variability and evolution, *J. Clim., 14*, 4191–4215, 2001.

Barsugli, J. J., and D. S. Battisti, The basic effects of atmosphere–ocean thermal coupling on midlatitude variability, *J. Atmos. Sci., 55*, 477–493, 1998.

Berbery, E. H., and C. S. Vera, Characteristics of the Southern Hemisphere winter storm track with filtered and unfiltered data, *J. Atmos. Sci., 53*, 468–481, 1996.

Blackmon, M. L., J. M. Wallace, N.-C. Lau, and S. L. Mullen, An observational study of the Northern Hemisphere wintertime circulation, *J. Atmos. Sci., 34*, 1040–1053, 1977.

Blackmon, M. L., Y.-H. Lee, J. M. Wallace, and H.-H. Hsu, Time evolution of 500-mb height fluctuations with long, intermediate and short time scales as deduced from lag-correlation analysis, *J. Atmos. Sci., 41*, 981–991, 1984.

Bosart, L. F., Observed cyclone life cycles, in *The Life Cycles of Extratropical Cyclones*, edited by M. A. Shapiro and S. Grønås, pp. 187–213, American Meteorological Society, Boston, MA, 1999.

Cai, M., Local instability dynamics of storm tracks, *in Observation, Theory and Modeling of Atmospheric Variability,* edited by X. Zhu, X. Li, M. Cai, S. Zhou, Y. Zhu, F.-F. Jin, X. Zou and M. Zhang, pp. 3–38, World Scientific, Singapore, 2004.

Cayan, D. R., Latent and sensible heat-flux anomalies over the northern oceans – The connection to monthly atmospheric circulation, *J. Clim., 5,* 354–369, 1992a.

Cayan, D. R., Latent and sensible heat-flux anomalies over the northern oceans – Driving the sea–surface temperature, *J. Phys. Oceanogr., 22,* 859–881, 1992b.

Chang, E. K. M., Downstream development of baroclinic waves as inferred from regression analysis, *J. Atmos. Sci., 50,* 2038–2053, 1993.

Chang, E. K. M., Characteristics of wave packets in the upper troposphere. Part II: Seasonal and hemispheric variations, *J. Atmos. Sci., 56,* 1729–1747, 1999.

Chang, E. K. M., GCM and observational diagnoses of the seasonal and interannual variations of the Pacific storm track during the cool season, *J. Atmos. Sci., 58,* 1784–1800, 2001.

Chang, E. K. M., Midwinter suppression of the Pacific storm track activity as seen in aircraft observations, *J. Atmos. Sci., 60,* 1345–1358, 2003.

Chang, E. K. M., and D. B. Yu, Characteristics of wave packets in the upper troposphere. Part I: Northern Hemisphere winter, *J. Atmos. Sci., 56,* 1708–1728, 1999.

Chang, E. K. M., S. Lee, and K. L. Swanson, Storm track dynamics, *J. Clim., 15,* 2163–2183, 2002.

Charney, J. G., The dynamics of long waves in a baroclinic westerly current, *J. Meteorol., 4,* 135–163, 1947.

Chen, B., S. R. Smith, and D. H. Bromwich, Evolution of the tropospheric split jet over the South Pacific Ocean during the 1986–89 ENSO cycle, *Mon. Weather Rev., 124,* 1711–1731, 1996.

Christoph, M., U. Ulbrich, and P. Speth, Midwinter suppression of Northern Hemisphere storm track activity in the real atmosphere and in GCM experiments, *J. Atmos. Sci., 54,* 1589–1598, 1997.

Colling, A., *Ocean Circulation, 2nd Edition,* pp. 286, The Open University, Butterworth-Heinemann, Oxford, U.K., 2001.

Deser, C., and M. L. Blackmon, On the relationship between tropical and North Pacific sea–surface temperature variations, *J. Clim., 8,* 1677–1680, 1995.

Deser, C., and M. S. Timlin, Atmosphere–ocean interaction on weekly timescales in the North Atlantic and Pacific, *J. Clim., 10,* 393–408, 1997.

Dickson, R. R., and J. Namias, North American influences on the circulation and climate of the North Atlantic sector, *Mon. Weather Rev., 104,* 728–744, 1976.

Eady, E. T., Long waves and cyclone waves, *Tellus, 1,* 33–52, 1949.

Frankignoul, C., Sea surface temperature anomalies, planetary waves and air–sea feedback in the middle latitudes, *Rev. Geophys., 23,* 357–390 1985.

Frankignoul, C., and R. W. Reynolds, Testing a dynamical model for mid-latitude sea surface temperature anomalies, *J. Phys. Oceanogr., 13,* 1131–1145, 1983.

Gulev, G. K., T. Jung, and E. Ruprecht, Climatology and interannual variability in the intensity of synoptic-scale processes in the North Atlantic from the NCEP-NCAR reanalysis data, *J. Clim., 15,* 809–828 2003.

Halliwell, G. R., and D. A. Mayer, Frequency response properties of forced climatic SST anomaly variability in the North Atlantic, *J. Clim., 9,* 3575–3587, 1996.

Hanawa, K., R. Sannomiya, and Y. Tanimoto, Static relationship between anomalies of SSTs and air–sea heat fluxes in the North Pacific, *J. Meteorol. Soc. Jpn., 73,* 757–763, 1995.

Held, I. M., Planetary waves and their interaction with smaller scales, *in The Life Cycles of Extratropical Cyclones,* edited by M. A. Shapiro and S. GrØnås, pp. 101–109, American Meteorological Society, Boston, MA, 1999.

Held, I. M., and A. Y. Hou, Nonlinear axially symmetric circulations in a nearly inviscid atmosphere, *J. Atmos. Sci., 37,* 515–533, 1980.

Hoerling, M. P., and M. Ting, On the organization of extratropical transients during El Niño, *J. Clim., 7,* 745–766, 1994.

Hoerling, M. P., and A. Kumar, Atmospheric response patterns associated with tropical forcing, *J. Clim., 15,* 2184–2203, 2002.

Hoskins, B. J., and P. J. Valdes, On the existence of storm tracks, *J. Atmos. Sci., 47,* 1854–1864, 1990.

Hoskins, B. J., and K. I. Hodges, New perspectives on the Northern Hemisphere winter storm tracks, *J. Atmos. Sci., 59,* 1041–1061, 2002.

Hoskins, B. J., M. E. McIntyre, and A. W. Robertson, On the use and significance of isentropic potential vorticity maps, *Q. J. R. Meteorol. Soc., 111,* 877–946, 1985.

Inatsu, M., H. Mukougawa, and S.-P. Xie, Atmospheric response to zonal variations in mid-latitude SST: Transient and stationary eddies and their feedback, *J. Clim., 16,* 3314–3329, 2003.

Karoly, D. J., D. G. Vincent, and J. M. Schrage, General circulation, *in Meteorology of the Southern Hemisphere, Meteorol. Monogr., 49,* edited by D. J. Karoly and D. G. Vincent, pp. 47–86, American Meteorological Society, Boston, MA, 1998.

Kelly, K. A., and S. Dong, The relationship of western boundary current heat transport and storage to mid-latitude ocean–atmosphere interaction, 2004.

Klein, W. H., The frequency of cyclones and anticyclones in relation to the mean circulation, *J. Meteorol., 15,* 98–102, 1958.

Kuo, Y.-H., R. J. Reed, and S. Low-Nam, Effects of surface energy fluxes during the early development and rapid intensification stages of seven explosive cyclones in the western Atlantic, *Mon. Weather Rev., 119,* 457–476, 1991.

Kushnir, Y., W. A. Robinson, I. Bladé, N. M. J. Hall, S. Peng, and R. Sutton, Atmospheric GCM response to extratropical SST anomalies: Synthesis and evaluation, *J. Clim., 15,* 2233–2256, 2002.

Lau, N.-C., Interactions between global SST anomalies and the mid-latitude atmospheric circulation, *Bull. Am. Meteorol. Soc., 78,* 21–33, 1997.

Lau, N.-C., and E. O. Holopainen, Transient eddy forcing of the time-mean flow as identified by quasi-geostrophic tendencies, *J. Atmos. Sci., 41,* 313–328, 1984.

Lau, N.-C., and M. J. Nath, A modeling study of the relative roles of the tropical and extra-tropical SST anomalies in the variability of the global atmosphere–ocean system, *J. Clim., 7,* 1184–1207, 1994.

Lau, N.-C., and M. J. Nath, The role of the "atmospheric bridge" in linking the tropical Pacific ENSO events to extratropical SST

anomalies, *J. Clim., 9*, 2036–2057, 1996.
Lau, N.-C., and M. J. Nath, Impact of ENSO on SST variability in the North Pacific and North Atlantic: Seasonal dependence and role of extratropical air–sea coupling, *J. Clim., 14*, 2846–2866, 2001.
Lee, S., and I. M. Held, Baroclinic wave packets in models and observations, *J. Atmos. Sci., 50*, 1413–1428, 1993.
Lee, S., and H.-K. Kim, The dynamical relationship between subtropical and eddy-driven jets, *J. Atmos. Sci., 60*, 1490–1503, 2003.
Lindzen, R. S., and A. Y. Hou, Hadley circulations for zonally averaged heating centered off the equator, *J. Atmos. Sci., 45*, 2416–2427, 1998.
Lukas, R. B., Freshening of the upper pycnocline in the North Pacific subtropical gyre associated with decadal changes of rainfall, *Geophys. Res. Lett., 28*, 3485–3488, 2001.
Miller, A. J., D. R Cayan, T. P. Barnett, N. E. Graham, and J. M. Oberhuber, Interdecadal variability of the Pacific Ocean: Model response to observed heat flux and wind stress anomalies, *Clim. Dyn., 9*, 287–302, 1994.
Mohri, K., On the fields of wind and temperature over Japan and adjacent waters during winter of 1950–1951, *Tellus, 3*, 340–358, 1953.
Nagasawa, M., Y. Niwa, and T. Hibiya, Spatial and temporal distribution of the wind-induced internal wave energy available for deep water mixing in the North Pacific, *J. Geophys. Res., 105*, 13,933–13,943, 2000.
Nakamura, H., Midwinter suppression of baroclinic wave activity in the Pacific, *J. Atmos. Sci., 49*, 1629–1641, 1992.
Nakamura, H., and T. Yamagata, Recent decadal SST variability in the Northwestern Pacific and associated atmospheric anomalies, in *Beyond El Niño: Decadal and Interdecadal Climate Variability*, edited by A. Navarra, pp. 49–72, Springer-Verlag, Berlin, Germany, 1999.
Nakamura, H., and T. Sampe, Trapping of synoptic-scale disturbances into the North-Pacific subtropical jet core, *Geophys. Res. Lett., 29*, doi:1029/2002GL015335, 2002.
Nakamura, H., and A. S. Kazmin, Decadal changes in the North Pacific oceanic frontal zones as revealed in ship and satellite observations, *J. Geophys. Res., 108*, doi:10.1029/ JC19990085, 2003.
Nakamura, H., and A. Shimpo, Seasonal variations in the Southern Hemisphere storm tracks and jet streams as revealed in a reanalysis data set, *J. Clim., 17*, 1828–1842, 2004.
Nakamura, H., G. Lin, and T. Yamagata, Decadal climate variability in the North Pacific during the recent decades, *Bull. Am. Meteorol. Soc., 78*, 2215–2225, 1997a.
Nakamura, H., M. Nakamura, and J. L. Anderson, The role of high- and low-frequency dynamics in blocking formation, *Mon. Weather Rev., 125*, 2074–2093, 1997b.
Nakamura, H., T. Izumi, and T. Sampe, Interannual and decadal modulations recently observed in the Pacific storm track activity and East Asian winter monsoon, *J. Clim., 15*, 1855–1874, 2002.
Namias, J., and D. R. Cayan, Large-scale air–sea interactions and short period climate fluctuations, *Science, 214*, 869–874, 1981.
Neiman, P. J., and M. A. Shapiro, The life cycle of an extratropical marine cyclone. Part I: Frontal cyclone evolution and thermodynamic air–sea interaction, *Mon. Weather Rev., 121*, 2153–2176, 1993.
Nitta, T., and S. Yamada, Recent warming of tropical sea surface temperature and its relationship to the northern hemisphere circulation, *J. Meteorol. Soc. Jpn., 67*, 375–383, 1989.
Nonaka, M., and S.-P. Xie, Co-variations of sea surface temperature and wind over the Kuroshio and its extension: Evidence for ocean-to-atmospheric feedback, *J. Clim., 16*, 1404–1413, 2003.
Nuss, W. A., and S. I. Kamikawa, Dynamics and boundary layer processes in two Asian cyclones, *Mon. Weather Rev., 118*, 755–771, 1990.
Oberhuber, J. M., The Budget of Heat, Buoyancy and Turbulent Kinetic Energy at the Surface of the Global Ocean, *Rep. 15*, 19 pp., Max Plank Institut für Meteorologie, Hamburg, Germany, 1988.
Orlanski, I., and E. K. M. Chang, Ageostrophic geopotential fluxes in downstream and upstream development of baroclinic waves, *J. Atmos. Sci., 50*, 212–225, 1993.
Palmén, E., The role of atmospheric disturbances in the general circulation, *Q. J. R. Meteorol. Soc., 77*, 337–354, 1951.
Palmén, E., and C. W. Newton, *Atmospheric Circulation Systems: Their Structure and Physical Interpretation*, 603 pp., Academic Press, New York, N.Y., 1969.
Peng, S., and J. S. Whitaker, Mechanisms determining the atmospheric response to midlatitude SST anomalies, *J. Clim., 12*, 1393–1408, 1999.
Peng, S., and W. A. Robinson, Relationships between atmospheric internal variability and the responses to an extra-tropical SST anomaly, *J. Clim., 14*, 2943–2959, 2001.
Petterssen, S., *Weather and Forecasting, 2nd Edition, Vol. 1*, 428 pp., McGraw Hill, New York, N.Y., 1956.
Petterssen, S., and S. J. Smebye, On the development of extra-tropical cyclones, *Q. J. R. Meteorol. Soc., 97*, 457–482, 1971.
Qiu, B., Interannual variability of the Kuroshio Extension system and its impact on their wintertime SST field, *J. Phys. Oceanogr., 30*, 1486–1502, 2000.
Qiu, B., The Kuroshio extension system: Its large-scale variability and role in the midlatitude ocean–atmosphere interaction, *J. Oceanogr., 58*, 57–75, 2002.
Qiu, B., Kuroshio Extension variability and forcing of the Pacific decadal oscillations: Responses and potential feedback, *J. Phys. Oceanogr., 33*, 2465–2482, 2003.
Qiu, B., and K.A. Kelly, Upper-ocean heat balance in the Kuro-shio extension region, *J. Phys. Oceanogr., 23*, 2027–2041, 1993.
Randel, W. J., and P. A. Newman, The stratosphere in the Southern Hemisphere, in *Meteorology of the Southern Hemisphere, Meteor. Monogr., 49*, edited by D. J. Karoly and D. G. Vincent, pp. 243–282, American Meteorological Society, Boston, M.A., 1998.
Rao, V. B., A. M. C. do Carmo, and S. H. Franchito, Seasonal variations in the Southern Hemisphere storm tracks and wave propagation, *J. Atmos. Sci., 59*, 1029–1040, 2002.
Reed, R. J., G. A. Grell, and Y.-H. Kuo, The ERICA IOP-5 Storm: Part I: Analysis and simulation, *Mon. Weather Rev., 121*, 1577–1594, 1993.
Reynolds, R. W., and T. M. Smith, Improved global sea surface temperature analysis using optimum interpolation, *J. Clim., 7*, 929–948, 1994.
Rogers, J. C., Patterns of low-frequency monthly sea level pressure

variability (1899–1986) and associated wave cyclone frequencies, *J. Clim.*, *3*, 1364–1379, 1990.

Rossby, C.-G., The scientific basis of modern meteorology, in *Yearbook of Agriculture, Climate and Man*, edited by G. Hambidge, pp. 599–655, U.S. Government Printing Office, Washington, D.C., 1941.

Sanders, F., and J. R. Gyakum, Synoptic-dynamic climatology of the "bomb", *Mon. Weather Rev.*, *108*, 1589–1606, 1980.

Schneider, N., A. J. Miller and D. W. Pierce, Anatomy of North Pacific decadal variability, *J. Clim.*, *15*, 586–605, 2002.

Seager, R., Y. Kushnir, M. Visbeck, N. Naik, J. Miller, G. Krahmann, and H. Cullen, Causes of Atlantic Ocean climate variability between 1958 and 1998, *J. Clim.*, *13*, 2845–2862, 2000.

Seager, R., A. Karspeck, M. Cane, Y. Kushnir, A. Giannini, A. Kaplan, B. Kerman, and J. Miller, Predicting Pacific decadal variability, 2004.

Simmonds, I., and K. Keay, Mean Southern Hemisphere extratropical behavior in the 40-year NCEP-NCAR reanalysis *J. Clim.*, *13*, 873–885, 2000.

Simmons, A. J., J. M. Wallace, and G. W. Branstator, Barotropic wave propagation and instability, and atmospheric teleconnection patterns, *J. Atmos. Sci.*, *40*, 1363–1392, 1983.

Sinclair, M. R., An objective cyclone climatology for the Southern Hemisphere, *Mon. Weather Rev.*, *122*, 2239–2256, 1994.

Sinclair, M. R., A climatology of cyclogenesis for the Southern Hemisphere, *Mon. Weather Rev.*, *123*, 1601–1619, 1995.

Swanson, K. L., and R. T. Pierrehumbert, Nonlinear wave packet evolution on a baroclinically unstable jet, *J. Atmos. Sci.*, *51*, 384–396 1994.

Takaya, K., and H. Nakamura, A formulation of a phase-independent wave-activity flux for stationary and migratory quasigeostrophic eddies on a zonally varying basic flow, *J. Atmos. Sci.*, *58*, 608–627, 2001.

Takayabu, I., "Coupling development": An efficient mechanism of the development of extratropical cyclones, *J. Meteorol. Soc. Jpn.*, *69*, 609–628, 1991.

Tanimoto, Y., N. Iwasaka, and K. Hanawa, Relationship between sea surface temperature, the atmospheric circulation and air–sea fluxes on multiple time scales, *J. Meteorol. Soc. Jpn.*, *75*, 831–849, 1997.

Tanimoto, Y., H. Nakamura, T. Kagimoto, and S. Yamane, An active role of extratropical sea surface temperature anomalies in determining anomalous turbulent heat fluxes, *J. Geophys. Res.*, *108*, 3304, doi:10.1029/2002JC00175, 2003.

Tomita, T., B. Wang, T. Yasunari, and H. Nakamura, Global patterns of decadal-scale variability observed in sea surface temperature and lower-tropospheric circulation fields, *J. Geophys. Res.*, *106*, 26,805–26,815, 2001.

Tomita, T., S.-P. Xie, and M. Nonaka, Estimates of surface and subsurface forcing for decadal sea surface temperature variability in the mid-latitude North Pacific, *J. Meteorol. Soc. Jpn.*, *79*, 1289–1300, 2002.

Trenberth, K. E., An assessment of the impact of transient eddies on the zonal flow during a blocking episode using localized Eliassen-Palm flux diagnosis, *J. Atmos. Sci.*, *43*, 2070–2087, 1986.

Trenberth, K. E., Storm tracks in the Southern Hemisphere, *J. Atmos. Sci.*, *48*, 2159–2178, 1991.

Trenberth, K. E., W. G. Large, and J. G. Olson, The mean annual cycle in global ocean wind stress, *J. Phys. Oceanogr.*, *20*, 1742–1760, 1990.

Wallace, J. M., and D. S. Gutzler, Teleconnections in the geopotential height field during the Northern Hemisphere winter, *Mon. Weather Rev.*, *109*, 784–812, 1981.

Wallace, J. M., and Q.-R. Jiang, On the observed structure of interannual variability of the atmosphere/ocean climate system, in *Atmosphere and Oceanic Variability*, edited by H. Cattle, pp. 17–43, Royal Meteorological Society, Reading, U.K., 1987.

Wallace, J. M., G.-H. Lim, and M. L. Blackmon, Relationship between cyclone tracks, anticyclone tracks and baroclinic waveguides, *J. Atmos. Sci.*, *45*, 439–462, 1988.

Watanabe, M., and M. Kimoto, Atmosphere–ocean thermal coupling in the North Atlantic: A positive feedback, *Q. J. R. Meteorol. Soc.*, *126*, 3343–3369, 2000.

Whittaker, L. M., and L. H. Horn, Northern Hemisphere extra-tropical cyclone activity for four mid-season months, *Int. J. Climatol.*, *4*, 297–310, 1984.

Xie, S.-P., Satellite observations of cool ocean–atmosphere interaction, *Bull. Am. Meteorol. Soc.*, *85*, 195–208, 2004.

Xie, S.-P., T. Kunitani, A. Kubokawa, M. Nonaka and S. Hosoda, Interdecadal thermocline variability in the North Pacific for 1958–1997: A GCM simulation, *J. Phys. Oceanogr.*, *30*, 2798–2813, 2000.

Xie, S.-P., J. Hafner, Y. Tanimoto, W. T. Liu, H. Tokinaga, and H. Xu, Bathymetric effect on the winter sea surface temperature and climate of the Yellow and East China Seas, *Geophys. Res. Lett.*, *29*, 3261, doi:10.1029/2002GL015884, 2002.

Yasuda, I., K. Okuda, and Y. Shimizu, Distribution and modification of North Pacific intermediate water in the Kuroshio-Oyashio Interfrontal Zone, *J. Phys. Oceanogr.*, *26*, 448–465, 1996.

Yuan, X., and L. D. Talley, The subarctic frontal zone in the North Pacific: Characteristics of frontal structure from climatological data and synoptic surveys, *J. Geophys. Res.*, *101*, 16,491–16,508, 1996.

Zhang, Y., and I. M. Held, A linear stochastic model of a GCM's midlatitude storm tracks, *J. Atmos. Sci.*, *56*, 3416–3435, 1999.

H. Nakamura and T. Sampe, Department of Earth and Planetary Science, University of Tokyo, Tokyo, 113–0033, Japan. (hisashi@eps.s.u-tokyo.ac.jp; sampe@eps.s.u-tokyo.ac.jp)

A. Shimpo, Climate Prediction Division, Marine and Climate Department, Japan Meteorological Agency, Tokyo, 100–0004, Japan. (sinpo@naps.kishou.go.jp)

Y. Tanimoto, Graduate School of Environmental Earth Science, Hokkaido University, Sapporo, 060–0810, Japan. (tanimoto@ees.hokudai.ac.jp)

The Relationship of Western Boundary Current Heat Transport and Storage to Midlatitude Ocean-Atmosphere Interaction

Kathryn A. Kelly and Shenfu Dong[1]

Applied Physics Laboratory, University of Washington, Seattle, Washington

Much of the heat transported poleward by the oceans is carried in the midlatitude western boundary currents in the northern hemisphere. As these currents separate from the coastal boundaries and extend eastward into the ocean interior, they flux some of their heat to the atmosphere and store some of their heat in the recirculation gyres south of the current core; the heat content anomalies are negatively correlated with changes in the volume of an isothermal layer known as the "subtropical mode water." An analysis of upper ocean heat content observations (1955–2001) shows that there are substantial interannual variations in the amount of heat stored in the upper 400 m of the water column. About 26% of the variations in heat content in the North Atlantic and North Pacific Oceans (corresponding to the first principal component and with maxima in the western boundary current extension regions) are in phase and slightly lag the atmospheric Northern hemisphere Annular Mode (NAM or Arctic Oscillation). The simplest explanation, that changes in the westerlies cause corresponding changes in the air-sea fluxes and therefore in heat content, can be ruled out by by the sign of the correlation: strong westerlies (strong AO) are correlated with positive heat content anomalies. This conclusion is supported by previous analyses of the upper ocean heat budget, which show that the heat content anomalies are primarily caused by variations in ocean advection. The heat content anomalies, rather than being caused by changes in air-sea fluxes, instead appear to be the source of interannual variations in those fluxes. The magnitude of the flux anomalies, their association with advection and heat storage in the mode water, and the coherence between the two oceans suggest a role for ocean circulation in interannual to decadal variations in climate variability through local air-sea interaction.

1. INTRODUCTION

Strong western boundary currents in the Northern Hemisphere midlatitude oceans transport heat from the warm tropical regions to the midlatitudes, where much of the heat is fluxed to the atmosphere as the warm currents encounter cooler air masses. Some of this heat continues on into the subpolar gyre to warm the high-latitude regions and some heat is stored in a region of recirculating currents with deep wintertime mixed layers. Although the estimates vary widely, the annually averaged flux of heat from the ocean to the atmosphere over the boundary currents is at least 100 W m^{-2} [*Josey et al.*, 1999].

[1]Now at Scripps Institution of Oceanography.

Earth's Climate: The Ocean-Atmosphere Interaction
Geophysical Monograph Series 147
Copyright 2004 by the American Geophysical Union
10.1029/147GM19

In the mean the flux of heat from the ocean to the atmosphere between 30–45° N is reflected in a large drop (1.2 pW) in the amount of heat transported by the ocean, from about 1.7 pW to 0.5 pW [*Trenberth and Caron*, 2001]. In the same latitude range, the heat transported poleward by the atmosphere increases by nearly the same amount, from about 3.7 pW to 5.0 pW. Zonally integrating the net surface heat fluxes (which are concentrated in the western boundary current extension regions) between 30–45° N gives an estimated mean annual ocean-to-atmosphere flux of about 1.0 pW [*daSilva et al.*, 1994]. Given the likely large errors in these estimates, the consistency of the numbers suggests the passing of a heat-transport "baton" from the ocean to the atmosphere in the midlatitudes. With this large mean heat transfer, it seems likely that interannual variations in the heat transfer over the boundary currents would have an effect on midlatitude climate variations. Here we examine the relationship between variations in heat storage, heat fluxes, and heat advection by the boundary currents.

Observations of sea surface height (SSH) anomalies from the TOPEX/POSEIDON radar altimeter [*Fu et al.*, 1994] since 1992 suggest that there are large interannual-to-decadal variations in the structure of these current systems [*Qiu*, 2000, *Vivier et al.*, 2002; *Dong and Kelly*, 2004]. In Plate 1 the SSH anomaly has been combined with an estimate of the mean SSH [*Kelly and Gille*, 1990; *Qiu*, 2000; *Dong*, 2003] for the 10-yr altimetric record to give the absolute SSH and to illustrate the nature of the circulation changes. Large-scale changes occur in both oceans just to the south of the current core (the closely spaced SSH contours): the region of high SSH, shown in dark red, expands and contracts from year to year. These maps of SSH show similar changes in the Atlantic and in the Pacific, with an expanded region of high SSH south of the current core in 1993 and 1999, and a contracted region of high SSH in 1996. Through the geostrophic relationship (altimetric SSH anomalies correspond to height changes for a constant pressure surface), changes in the gradients of SSH represent changes in the strength of the geostrophic currents. For example, the expanded region of high SSH in 1993 and 1999 represents an increase in the geostrophic surface recirculation.

Changes in SSH also correspond to changes in heat content, because a warming water column expands, increasing SSH, whereas a cooling water column contracts, decreasing SSH. Seasonal changes in SSH (heat content) in the Gulf Stream region are caused primarily by seasonal variations in surface heating [*Kelly et al.*, 1999]. However, as discussed in Section 2, on longer time scales, changes in the heat content are more the result of ocean advection than of surface heating. In addition, changes in upper ocean heat content account for most of the observed large-scale changes in SSH. The advantage of using heat content (instead of SST) as a descriptor of ocean state and the changes in vertical thermal structure are described in Section 2.

A surprising fraction of the variations in SSH anomalies and in heat content are in phase between the North Atlantic and the North Pacific and in phase with the Northern hemisphere Annular Mode (NAM) or Arctic Oscillation (AO), as described in Section 3. We argue here that the relationship of heat content to the NAM is not simply the ocean cooling in response to stronger winds; rather, an ocean made anomalously warm by advection is increasing the flux of heat to the atmosphere. Causes of the variations in ocean advection, the effects of heat content on surface heat fluxes, and implications for ocean-atmosphere coupling are described in Section 4, followed by conclusions in Section 5.

2. BACKGROUND

The primary focus here is the analysis and interpretation of a relatively long record of upper ocean heat content and, in particular, the degree to which the anomalies in the North Atlantic and North Pacific Oceans are in phase. To understand the likely causes of the heat content anomalies and their relevance in climate studies, we summarize in this section important results from several related analyses of the midlatitude upper ocean heat content.

2.1. The Upper Ocean Heat Budget

To understand the causes of observed fluctuations in ocean heat content and SSH, parallel studies of the upper ocean heat budget have been conducted for the western North Pacific and North Atlantic [*Vivier et al.*, 2002; *Dong and Kelly*, 2004]. A simple upper ocean layer model, down to a depth of 800 meters, was forced by surface heat fluxes, winds, and currents to predict temperature for the regions shown in Plate 1. Geostrophic velocity was specified from the altimeter data using a vertical profile based on climatology and Ekman velocity was estimated from wind stress. Sea level winds were taken from the National Center for Environmental Prediction/National Center for Atmospheric Research (NCEP/NCAR) Reanalysis [*Kalnay et al.*, 1996]; air-sea fluxes were derived by using the NCEP variables in the COARE bulk flux algorithm [*Fairall et al.*, 1996]. In the upper ocean the rate of change of heat down to a fixed depth h is balanced by surface heating, the divergence of heat transport, and diffusion, as

$$\frac{\partial}{\partial t}\int_{-h}^{0} T dz = \frac{Q_{net}}{\rho c_p} - \nabla \cdot \int_{-h}^{0} \mathbf{u}T dz + \kappa \nabla^2 T, \quad (1)$$

where h is larger than the wintertime mixed layer depth. Separating the terms into their vertical and horizontal components and neglecting vertical diffusion at depth $z = -h$ gives

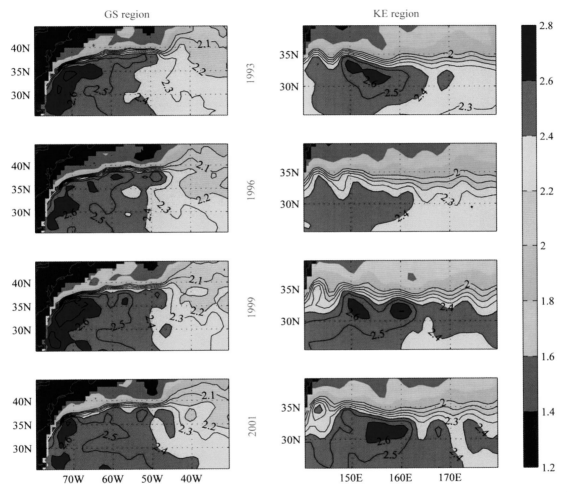

Plate 1. Sea surface height maps from the TOPEX/POSEIDON altimeter. (left) Gulf Stream region in the North Atlantic and (right) Kuroshio Extension region in the North Pacific for years 1993, 1996, 1999, and 2001. Units are meters. More positive SSH indicates more heat stored in the ocean.

$$\frac{\partial H}{\partial t} = \frac{Q_{net}}{\rho c_p} - \int_{-h}^{0} \mathbf{u_H} \cdot \nabla_H T \, dz + \kappa_H \nabla_H^2 T - w T|_{-h}, \quad (2)$$

where the vertical integral on the left-hand side in (1) is the heat content H and the second term on the right-hand side in (2) is horizontal advection. The last term on the right-hand side is the heat content change owing to vertical motion w at depth h, which is primarily the result of local convergences and divergences forced by changes in the curl of the wind stress. Two studies [Vivier et al., 2002; Dong and Kelly, 2004] showed that most of the changes in SSH can be explained by changes in the ocean heat content, primarily in the upper 400 meters. As expected, they showed that changes in both sea surface temperature (SST) and heat content are primarily forced by surface fluxes on seasonal time scales. However, the studies demonstrated that for interannual time scales the advection of heat into the region by the currents is more important in creating heat content anomalies than are air-sea fluxes (Figure 1).

Advection can be caused either by geostrophic currents, the dominant component of the boundary currents, or by ageostrophic currents, of which the Ekman component is likely the largest part. Scaling arguments [Gill and Niiler, 1973] suggest that advection by Ekman currents dominates in the interior ocean where the Ekman transport crosses isotherms. Over the separated western boundary currents, changes in the strength of the westerlies produce anomalies in the Ekman transport that cross the mean temperature gradient associated with the current, producing large contributions to advection. The geostrophic current, on the other hand, tends to flow parallel to isotherms, because it is in thermal wind balance, as given by

$$\frac{\partial u}{\partial z} = \frac{g}{f \rho_0} \frac{\partial \rho}{\partial y}, \quad (3)$$

where u is the eastward current, y is northward, and density ρ varies approximately linearly with temperature. However, a geostrophic current with a barotropic component (no verti-

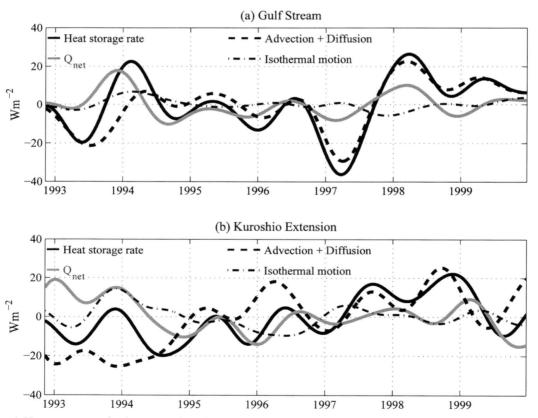

Figure 1. Heat storage rate for the western boundary current regions. (a) Gulf Stream and (b) Kuroshio Extension. The heat storage rate (bold line) is the sum of the surface heating (gray), the advection and diffusion (dashed), and the vertical motion of isotherms (dash-dot). Advection plus diffusion account for 70% of the heat storage rate variance in the Gulf Stream. The KE budget is more complicated, but advection-diffusion dominates after 1997. After *Dong and Kelly* [2004]; *Vivier et al.* [2002].

cal shear), for example, can advect temperature across isotherms. Because the boundary current velocities are so large (2 m s^{-1}), a relatively small cross-isotherm component can cause a large contribution to advection.

The North Atlantic heat budget of *Dong and Kelly* [2004] and the North Pacific estimates of *Qiu* [2000] both show that advection by the geostrophic current is substantially larger than advection by the Ekman component, consistent with previous studies using geostrophic currents derived from altimeter data [*Qiu and Kelly*, 1993; *Kelly and Qiu*, 1995]. These results differ from the heat budget analyses of *Miller et al.* [1994] and *Deser et al.* [1996], which emphasized the importance of Ekman advection. However, as noted by *Miller et al.* [1994] underestimates of geostrophic advection in the boundary currents likely occur because coarse ocean circulation models and sparse hydrographic observations underestimate the boundary current speeds. The dominance of geostrophic advection is consistent with the idea that the western boundary currents account for most of the meridional heat transport: these currents are largely (95%) geostrophic [*Johns et al.*, 1989].

Why is it important to distinguish which velocity component is causing the advection? Identification of the mechanism responsible for advection is critical to predicting the changes in ocean heat content. The Ekman current component is a local and rapid response to the wind field, whereas the geostrophic current represents the ocean's response to ocean-basin-scale forcing and may be either rapid (days for a barotropic response) or slow (years for a baroclinic response). In addition, a lag in the ocean's response to forcing, such as that associated with Rossby waves, may be the source of a coupled oscillating response [*Marshall et al.*, 2001] or the lag may allow prediction of climate variations [*Schneider and Miller*, 2001].

2.2. Heat Content as a Descriptor of Ocean State

The heat budget studies described above established that changes in ocean circulation make large contributions to the upper ocean heat content. We argue here that not only is heat content a better descriptor of the thermodynamic state of the ocean than is SST, but that it is the more relevant variable for climate studies. The heat content anomaly is an integral measure of the anomalous heat that could be fluxed to the atmosphere or that has been fluxed into the upper ocean over time periods longer than seasons.

While SST is readily available and is needed to compute an instantaneous estimate of the air-sea flux in a bulk algorithm, it is not a reliable indicator of the past or future interaction of the atmosphere and ocean. In the wintertime, when mixed layers are deep, the SST anomaly is a good indicator of anomalous heat content. However, in the summertime the SST mixed layer anomaly reflects only a small fraction of the upper ocean water column and subsurface anomalies may differ greatly in magnitude and possibly in sign. In the fall, the seasonal deepening of the mixed layer entrains the subsurface water and SST is modified by the sub-surface temperature.

The problem in using SST as an indicator of the ocean state has been quantified by *Deser et al.* [2003], who showed that SST anomalies decorrelate seasonally, whereas upper ocean heat content retains its memory beyond the seasonal time scale. Their analysis and the work of *Alexander et al.* [1995] have established that it is insufficient to characterize the ocean for climate studies in terms of its mixed layer temperature, even allowing for variations in mixed layer depth, because the entrainment of subsurface temperature anomalies into the mixed layer causes the "re-emergence" of SST anomalies. Thus, much of the ocean's heat storage capacity, and its potential contribution to interannual climate variations, is in the layer below the mixed layer.

2.3. Mode Water

We have shown that large heat content anomalies occur just south of the western boundary current extensions and we have argued that the ocean's contribution to air-sea fluxes derives from heat content anomalies, rather than from SST. For heat content anomalies to have an impact on the air-sea fluxes, they must have contact with the atmosphere at some times (that is, they must be "ventilated"); for the anomalies to have time scales longer than SST anomalies, they must be distinct from the ocean mixed layer at other times.

Such an intermittently ventilated layer of water exists just south of the western boundary currents: a thick vertically homogeneous layer known as "subtropical mode water" (STMW, or "18-degree water" in the Gulf Stream) [*McCartney and Talley*, 1982; *Suga and Hanawa*, 1995]. STMW is formed during the late winter, when the mixed layer deepens and surface cooling causes deep convection. (Some of the deepest mixed layers in the midlatitudes are found near the western boundary currents.) Mode water is in contact with the atmosphere only during the late winter; otherwise, it is shielded from air-sea fluxes by the warmer mixed layer. The large volume of the mode water allows it to resist changes in temperature from air-sea fluxes [*Warren*, 1972]. Mode water is slowly "subducted" under warmer waters and moves southwestward as part of the large-scale recirculation [*Schneider et al.*, 1999; *Joyce et al.*, 2000].

The association of cooling and mode water formation suggests that an increase in the thickness of the mode water is associated with a decrease in the upper ocean heat content. A recent study of the Gulf Stream region showed that the thickness of the 18-degree layer is indeed negatively correlated with heat content variations (Figure 2) and the correlation has been shown to be significant over a 50-year record [*Kwon*, 2003]. A similar relationship between the mode water layer thickness and fluctuations in the Kuroshio Extension has been found [*Yasuda and Hanawa*, 1997]. This correlation means that a large volume of mode water corresponds to a deficit of heat. Conversely, a positive heat content anomaly (large heat storage) corresponds to a small volume of mode water and, therefore, a stratified, warm water column.

These large heat storage reservoirs (STMW) are capable of damping seasonal and interannual variations in air-sea heat fluxes and in heat advection. The combination of the relatively large reservoir of mode water and its limited time of exposure to the atmosphere allows it to temporally integrate the effects of several winters and to resist sudden changes in temperature; hence, its long-term memory of climate variations. The mode water formed in a given winter affects SST in the subsequent fall and winter when it is entrained into the surface mixed layer as part of the seasonal cycle; the entrainment of the subsurface temperature anomalies into the mixed layer is responsible for the "re-emergence" of SST anomalies [*Alexander et al.*, 1995].

The suggestion that heat content variations are caused by anomalous advection appears to contradict the prevailing ideas about mode water being formed by air-sea fluxes. Clearly, the interaction of mode water with the overlying atmosphere, resulting in the largest heat losses, occurs in the late winter. However, if lower heat content implies more mode water (as suggested by the observed correlations), then the heat budget suggests that interannual variations in the volume of mode water are more closely related to advection than to air-sea fluxes. One resolution of these issues is that advection anomalies in the boundary current regions "precondition" the water column for mode water formation. During periods in which the boundary current has anomalously large advection, the water column is warm and stratified and normal wintertime surface heat loss will be inadequate to form mode water; conversely, during periods of weak advection, the water column is relatively cool and unstratified and the same wintertime heat loss will cause the formation of mode water.

3. INTERANNUAL VARIATIONS IN ADVECTION AND HEAT CONTENT

The studies cited in the previous section have shown that ocean heat transport makes large contributions to interannual heat storage variations. Further, much of the heat content anomaly lies below the mixed layer, associated with the subtropical mode water, and influences the mixed layer temperature through seasonal entrainment. These results, based on the heat budgets using the altimeter (Figure 1), are an intriguing indication of the role of the ocean circulation in climate; however, the altimeter record is too short to draw inferences about longer time scales. We address this issue using longer ocean temperature records, both to study the importance of advection in the heat budget and to understand the longer time scale variations in heat content.

3.1. Decadal Advection Estimates for the North Pacific

The simplicity of the vertically integrated heat budget (2) suggests that it is possible to extend the heat budget temporally by inferring heat transport variations as a residual between heat content changes and air-sea fluxes. *Kelly* [2004] used a longer record (1970–2000) of surface and subsurface temperatures to estimate a simple upper ocean heat budget in the Kuroshio Extension region in Plate 1, right panels. Tempera-

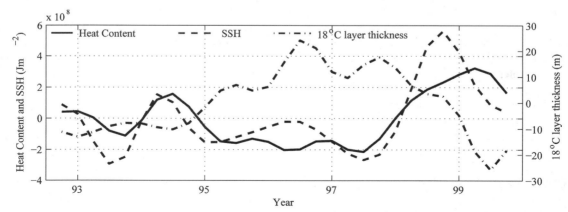

Figure 2. Heat content and mode water in the Gulf Stream region. The mean SSH (dash), heat content (solid), and the thickness of the 18° C layer (dash-dot) south of the Gulf Stream. The SSH is from the TOPEX/POSEIDON altimeter, and heat content and the layer thickness are from the GTSPP archive. After *Dong* [2004].

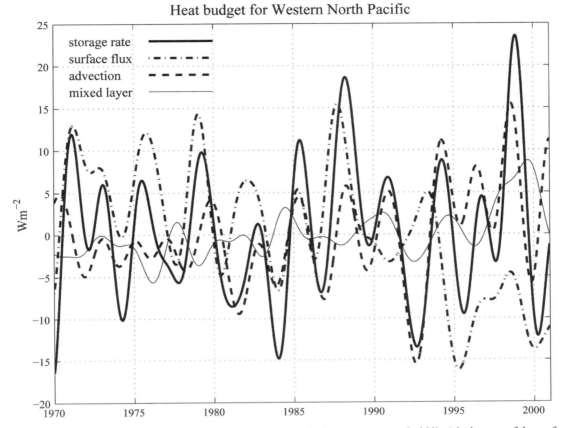

Figure 3. Heat budget of the western North Pacific: 1970–2000. The heat storage rate (bold line) is the sum of the surface heating (dash-dot), lateral (geostrophic) fluxes (dashed), and the sum of Ekman advection and a surface flux correction (thin line). Lateral fluxes are more highly correlated with the HSR than are surface fluxes. After *Kelly* [2004].

tures were taken from the Joint Environmental Data Analysis Center (JEDAC).

By forcing a one-dimensional upper ocean model with NCEP/NCAR fluxes and assimilating temperatures, lateral fluxes were inferred from the temperature adjustments required to match the model to the observations. The adjustment was made using a Kalman filter, with error covariances carefully prescribed from observations, and used the "unknown control" method of assimilation, as outlined in *Wunsch* [1996]. This analysis showed large changes in the ocean heat transport convergence with periods of about five years, consistent with the estimates from the 1990s [*Vivier et al.*, 2002] (Figure 3).

Interannual variations in heat transport from 1970–2000 are as large as the variations in air-sea fluxes, with the largest heat transport convergence occurring in the 1990s. With the exception of the 1970s, lateral fluxes make large contributions to the heat storage rate. In the mid-1970s lateral fluxes were weak and heat losses to the atmosphere were anomalously small. The correlation of the heat storage rate, left-hand side of (2), with heat transport convergence was larger than that with air-sea fluxes, consistent with the results for the shorter altimeter period. Therefore, advection appears to be more important in determining the heat content than air-sea fluxes on interannual-to-decadal time scales.

Kelly [2004] also examined the relationship between heat content and air-sea fluxes; these results will be discussed in Section 4.2.

3.2. Interannual to Decadal Variations in Heat Content

Keeping in mind the large contribution of western boundary current heat transport to heat content anomalies, we examine the available heat content data for the northern hemisphere midlatitude ocean. We compute the principal components (PCs) or empirical orthogonal functions (EOFs) of heat content for both the SSH from the altimeter from 1992–2002 and for heat content data (JEDAC) for 1955–2001. These two data sets contribute complementary information about the dominant modes of variability: the SSH time series has more uniform coverage and should yield a more accurate spatial pattern, whereas the longer record of the heat content observations should give a more accurate estimate of the contributions of each mode to total variability, as well as more robust correlations of the modes with other variables.

For the SSH the fraction of variance was 39%, 14%, and 9%, respectively, for the first three modes. A Monte Carlo test [*Preisendorfer*, 1988] on 100 sets of random data with comparable degrees of freedom spatially and temporally yielded 95% significance levels of 11%, 10%, and 9%, respectively, so that the first two modes were judged statistically significant. The first mode of SSH (Plate 2a) describes variations that are predominantly in the North Pacific, resembling that of the Pacific Decadal Oscillation (PDO) [*Hare and Mantua*, 2000]. The second mode (Plate 2b) describes variations that are in phase between the Atlantic and the Pacific, with pronounced maxima in the vicinity of the western boundary current extensions.

A principal component analysis, like any statistic, becomes a better representation of the data when derived from a longer time series; therefore the EOFs of heat content should give a more accurate indication of the importance of each mode than SSH, although with poorer spatial resolution (2° × 5°). For the JEDAC data the fraction of variance in the first three modes was 26%, 16%, and 10%, respectively (Plate 3), compared with 95% significance levels of 11%, 9%, and 9%, giving three significant modes. The first mode of heat content resembles the second mode of SSH, with the Atlantic and Pacific anomalies in phase; whereas the second mode of heat content resembles the first mode of SSH, with variations predominantly in the Pacific. The reversal of the mode order in the SSH analysis, relative to the JEDAC analysis, suggests that the Pacific mode variance was larger during the 1990s than was typical for the longer period.

The first mode of heat content (Plate 3a and c) resembles the first EOF of Pacific Ocean heat content for a similar period computed by *Giese and Carton* [1999] from fields derived by assimilating temperature observations into an ocean circulation model. The modes are similar both spatially and temporally, despite the different regions and different filtering used (Pacific Ocean down to 30° S and filtered for periods longer than 5 years). Second modes are not similar. *Giese and Carton* [1999] concluded that midlatitude decadal variations in heat content must result at least in part from meridional advection, consistent with the heat budgets described in Section 2.1.

The heat content time series (principal components) reveal similarities with climate indices: the first mode of heat content resembles the Arctic Oscillation (AO) Index [*Thompson and Wallace*, 1998], whereas the second mode of heat content resembles the PDO. The SSH modes also resemble these climate indices, with the in-phase mode (EOF 2) resembling the AO and EOF 1 resembling the PDO.

Defined originally by *Thompson and Wallace* [1998], the AO has been re-examined by *Wallace and Thompson* [2002]; in the latter article zonally coherent fluctuations in the sea level pressure (SLP) fields have been described by the name "Northern hemisphere Annular Mode" (NAM); the index of these variations is the first principal component of SLP anomalies (Figure 4). A positive NAM corresponds to anomalously strong westerlies throughout the northern hemisphere. *Wallace and Thompson* [2002] argue that the NAM is not necessarily an oscillation or a particular process, but simply describes those fluctuations that are zonally coherent. That is the sense in which we describe the corresponding heat content variations also, using the first mode of heat content.

The fact that the in-phase component of heat content variance resembles the Arctic Oscillation index (correlation of 0.49, significant at 95%, with AO leading the heat content mode by 13 months) suggests that the heat content anomalies are somehow connected with the changes in the strength of the

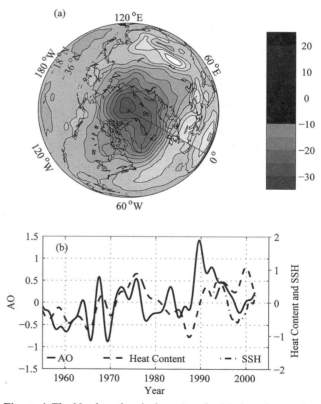

Figure 4. The Northern hemisphere Annular Mode and ocean heat content. (a) First EOF of sea level pressure (NAM). (b) Principal components of heat content (dash) and sea surface height (dash-dot). The sea level pressure (solid) time series is also known as the Arctic Oscillation Index. The correlation between the AO index and the heat content is 0.49, significant at better than 95%, with the AO leading heat content by 13 months.

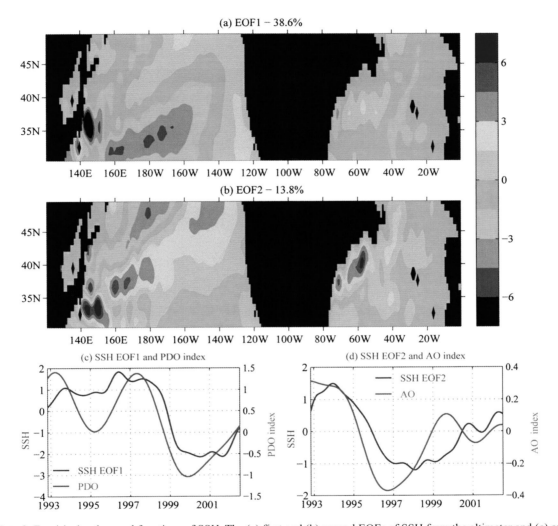

Plate 2. Empirical orthogonal functions of SSH. The (a) first and (b) second EOFs of SSH from the altimeter and (c) and (d) their respective time series. The first EOF describes SSH anomalies that are negative in the Pacific and weakly positive in the Atlantic. The second EOF describes SSH anomalies that are in phase in the two oceans and have their maxima in the western boundary currents. Indices of the Pacific Decadal Oscillation and the Arctic Oscillation are shown for comparison.

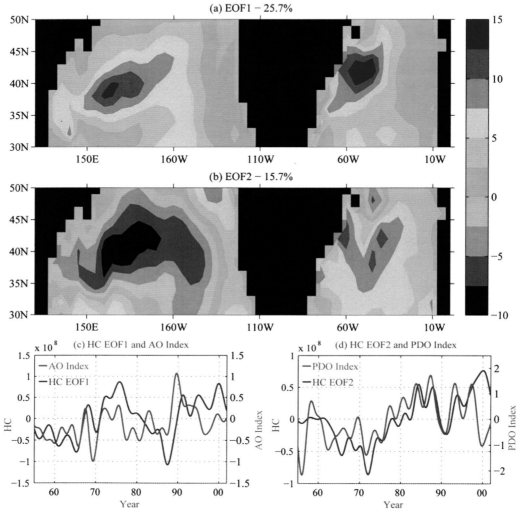

Plate 3. Empirical orthogonal functions of heat content. The (a) first and (b) second EOFs of JEDAC heat content and (c) and (d) their respective time series. The first EOF describes heat content anomalies that are in phase in the two oceans and have their maxima in the western boundary currents. The second EOF describes SSH anomalies that are negative in the Pacific and positive in the Atlantic. Indices of the Pacific Decadal Oscillation and the Arctic Oscillation are shown for comparison.

midlatitude westerly winds. Note that the SSH principal component more closely resembles the AO than does the heat content principal component (Figure 4b); we believe this problem results from poor North Atlantic heat content data, as discussed in Section 4.

Changes in the heat content corresponding to changes in the SLP fields can be seen in the observations, despite the relatively low fraction of variance (26%) described by the in-phase mode of heat content. For a period of weak AO (Plate 4a, 1985–1987), heat content anomalies are negative in both oceans; for a period of strong AO (Plate 4b, 1989–1991), heat content anomalies are positive.

The simplest connection, that changes in the westerlies cause corresponding changes in the air-sea fluxes, and therefore in heat content, can be ruled out by both the heat budget analyses and by the sign of the anomalies. One would expect strong westerlies (strong AO) to produce large heat losses from the ocean and, consequently, negative heat content anomalies. On the contrary, the heat content anomalies are positive for a strong AO. Based on the heat budget studies, we argue instead that the observed changes in heat content reflect changes in heat transport caused by changes in ocean circulation, not by air-sea fluxes.

4. CAUSES AND IMPLICATIONS OF HEAT CONTENT VARIATIONS

We have argued above that the heat content variations are being caused by advection, not air-sea fluxes, despite their correlation with the AO or NAM. If that is the case, the next question is: what is causing the changes in advection in the North Atlantic and in the North Pacific and why is a large fraction of those variations in phase between the oceans? Further, what are the climate implications of the advection-forced heat content anomalies? We address these questions in turn below.

4.1. Wind-forced Advection Changes

We now turn to the question of the causes of the changes in ocean advection. In the previous section we showed that the in-phase variations in heat content are positively correlated (stronger winds, more heat storage) with the westerlies (the AO), with the AO leading the heat content mode by 13 months. Using simple spatial averages we find similar relationships within each ocean: the heat content anomalies of the western boundary current region are significantly correlated (0.40 for the Pacific and 0.50 for the Atlantic) with variations in the (negative) wind stress curl, with curl variations leading the heat content by about one year (Figure 5b and c). The atmosphere forces ocean circulation changes through changes in the curl of the wind stress, which cause both a relatively rapid depth-independent (barotropic) and a slower, depth-varying (baroclinic) ocean response, usually characterized as a change in the Sverdrup transport [Deser et al., 1999]. The correlations found here are consistent in sign with stronger westerlies spinning up the midlatitude circulation and transporting more heat into the western boundary current regions.

To produce simultaneous increases in advection in the North Atlantic and in the North Pacific, there must be simultaneous anomalies in ocean circulation. This suggests forcing by an anomalous atmospheric circulation that is relatively uniform around the globe, such as that characterized by the NAM and discussed in Section 3.2. Although wind stress curl variations in the North Atlantic and in the North Pacific are far from uniform, when averaged over the region that forces the western boundary currents (Figure 5a), the correlation between them is 0.29, which is significant at the 95% level. For comparison, the Arctic Oscillation Index is also shown. Apparently, the ocean heat content anomalies are related to changes in the basin-scale winds, and to the extent that the winds are zonally uniform, the ocean response is coherent between the two ocean basins.

The analyses here are limited by the accuracy of the available ocean data. As in the EOF analysis (Figure 4b), the spatially averaged SSH time series (dash-dot line, Figure 5b and c) more closely resembles the wind stress curl than does the JEDAC heat content (dash line). A separate estimate of the heat content (Figure 5c, thin line) in the North Atlantic from the World Ocean Atlas [Conkright et al., 2002] more closely resembles SSH in the 1990s. The two heat content estimates were computed from different ocean data bases.

The short lag between wind stress curl forcing and the heat content variations is surprising, in light of numerous studies that suggest that the climatically important ocean response to wind forcing is the baroclinic response (e.g., Marshall et al. [2001]). The baroclinic response would lag the wind forcing by several years, longer in the wider Pacific Ocean than in the Atlantic, because Rossby waves, which propagate the adjustment to wind forcing changes, take longer to cross the Pacific. A recent study [Schneider and Miller, 2001] showed that SST anomalies in the North Pacific could be predicted using a simple Rossby wave equation and wind stress curl anomalies. Another study, also in the North Pacific, showed that SSH anomalies could be predicted using a wind-forced baroclinic ocean model [Qiu, 2002]; this model was, however, unable to describe the SSH variations just south of the Kuroshio Extension (B. Qiu, personal communication, 2003), in the subtropical mode water formation region. Gallego and Cessi [2001] formulated a coupled ocean-atmosphere model that showed the North Atlantic and North Pacific Oceans varying in phase; the ocean response was baroclinic and, unlike what is observed here, the differences in the time for a

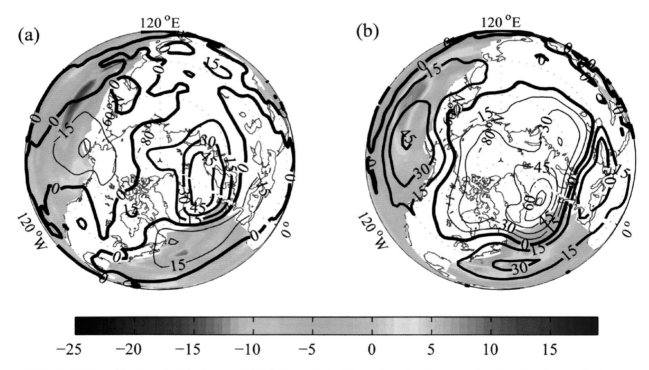

Plate 4. NAM and heat content for low and high index periods. Maps of sea level pressure (contours) and ocean heat content (color) for (a) a low index period, 1985–1987, and for (b) a high index period, 1989–1991.

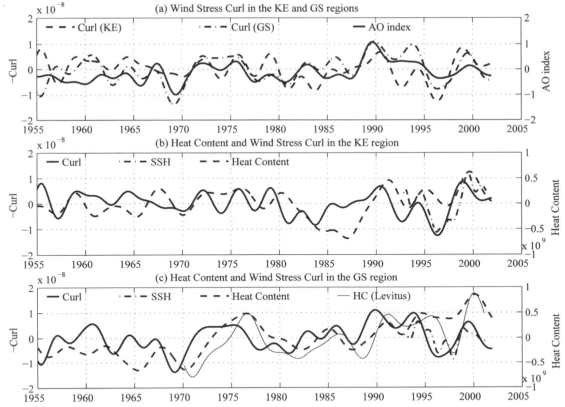

Figure 5. Wind stress curl and its relationship to heat content. (a) The negative wind stress curl in the North Atlantic [20–50° N, 10–80° W] (dash-dot), in the North Pacific [20–50° N, 140° E–130° W] (dash), and the Arctic Oscillation Index (solid). Heat content (dash), negative wind stress curl (solid), and SSH anomaly (dash-dot) in (b) the North Pacific and in (c) the North Atlantic. Heat content and SSH are domain averages over 25–45° N and 140–180° E in the Pacific and 40–80° W in the Atlantic. A second estimate for the North Atlantic heat content is shown (thin line) in (c).

Rossby wave to cross the basin in the two oceans caused the oceans to go in and out of phase.

Alternative explanations for the rapid response of heat content to wind fluctuations include advection by the more barotropic ocean circulation, a coupling between baroclinic and barotropic responses, or a nonlinear response associated with the recirculation gyre dynamics [*Dewar*, 2001]. Observations of the STMW regions, where the heat content anomalies are the largest, show that the ocean circulation there is highly barotropic [*Hogg*, 1983].

4.2. Heat Content-forced Heat Flux Anomalies

We have suggested that it is likely that winds cause changes in ocean advection, which in turn cause changes in heat content in the western boundary current regions (Plate 1). What then are the implications of these inter-annual variations in heat content for climate?

During periods of strong advection, heat accumulates near the boundary currents. Anomalously warm water below the mixed layer decreases the cooling effect of entrainment into the ocean mixed layer, causing anomalously high SSTs [*Vivier et al.*, 2002]. This, in turn, appears to cause an anomalously large heat loss to the atmosphere. Note that this scenario is entirely reversed from that of the ocean interior where heat content is relatively small and seasonal, and heat content H is determined by the net surface flux Q_{net}, as

$$\partial H/\partial t = Q_{net} \qquad (4)$$

[*Cayan*, 1992; *Bretherton and Battisti*, 2000]. In the boundary currents the predominant relationship is that of the feedback term [*Frankignoul*, 1985], as

$$H = -Q_{net}. \qquad (5)$$

During the 1990s, when the heat budgets could be computed directly using altimetric data for advection [*Vivier et al.*, 2002; *Dong and Kelly*, 2004], net surface fluxes appear to be negatively correlated with heat content as in (5).

To determine the significance of this feedback, *Kelly* [2004] examined the relationship between heat content and air-sea fluxes in the Kuroshio Extension region (Plate 1) for 1970–2000. (The relationship between heat content anomalies and advection, as examined in this study, is discussed in Section 3.1.) In this study, surface fluxes and heat content were found to be significantly negatively correlated for the region in the Pacific shown in Plate 1.

Based on this negative correlation between heat content and air-sea fluxes (with no temporal lag) and the findings of long decorrelation times for heat content [*Deser et al.*, 2003], Kelly examined the statistical skill of current heat content anomalies in predicting future anomalies of air-sea flux. Heat content anomalies show a small, but significant, skill (20% of flux variance) in predicting air-sea fluxes up to a year in advance. This skill substantially exceeds the skill of "persistence" of air-sea fluxes, which is negligible after a few months. On the other hand, consistent with the findings of *Deser* [2003], SST anomalies in the Kuroshio Extension region have no such skill in predicting air-sea fluxes. Therefore, for time scales longer than a season, heat content is a useful predictor of air-sea flux anomalies and a more robust indicator of the ocean's contribution to interannual air-sea fluxes than is SST.

To estimate the zonally coherent effect of the heat content anomalies on air-sea fluxes, we regress the NCEP/NCAR net surface fluxes onto the first (in-phase) PC of the heat content anomalies. As in the North Pacific analysis, this regression also shows that positive heat content anomalies are associated with anomalous fluxes of heat from the ocean to the atmosphere. To demonstrate the magnitude of this flux, we use the regression coefficients to compute the flux anomalies associated with the anomalous heat content for the year 2000 (Figure 6), a period of anomalously high ocean heat content. Heat fluxes from the ocean to the atmosphere in 2000 exceed 20 W m^{-2}; these anomalies are approximately 20% of the annual mean heat loss in these regions [*Josey et al.*, 1999].

These results support our assertion that heat content variations are a primary cause of air-sea flux anomalies in the western boundary current regions, rather than the result of those anomalies.

4.3. Atmosphere-Ocean Coupling

The association of larger heat losses to the atmosphere with increased ocean heat content, caused by advection, suggests that the ocean may be forcing changes in the atmosphere in the vicinity of the western boundary currents on interannual time scales. The idea that the western boundary currents force an atmospheric response has been raised by numerous authors (e.g., *Rodwell et al.* [1999]; *Nakamura et al.* [1997]; *Joyce et al.* [2000]), particularly in regard to changing the wintertime "storm tracks." The issue of the control of mid-latitude storm tracks by ocean fronts and by the jet stream is examined in *Nakamura et al.* [2004].

However, until recently, most modeling efforts failed to show a consistent or significant atmospheric response to the western boundary current variations. There are two likely causes for the difficulty in obtaining a robust atmospheric response in models: inaccurate modeling of the air-sea flux contribution and poor spatial resolution in the ocean models. These problems are primarily the result of the overwhelming computational burden of modeling the coupled atmosphere-ocean system.

Earlier modeling efforts examined the response of an atmospheric model to a fixed SST anomaly (e.g., *Kushnir and Held* [1996]; *Palmer and Sun* [1985]). These types of experiments have generally produced weak or ambiguous results, in part because fixing the SST eliminates the ocean's ability to force surface fluxes. As we have discussed here, the heat content anomaly is associated with ocean advection. For a fixed (positive) SST anomaly, the model air temperature will rise to nearly eliminate the air-sea fluxes. However, in the real ocean advection will continue to supply heat to the region, forcing heat content (and SST) even higher, until the atmosphere absorbs the heat. A more accurate way to specify the forcing by the ocean is to specify advection or the heat transport convergence. More recent studies [*Yualeva et al.*, 2001; *Sutton and Mathieu*, 2002] have specified anomalies in the ocean's heat transport convergence. *Yualeva et al.* [2001] found that heat transport convergences of up to 40 W m^{-2} produced statistically significant changes in the 500-mb geopotential height fields in the model atmosphere. *Sutton and Mathieu* [2002] found that the specified heat transport convergence produced large anomalies of latent heat flux, and that the regions of largest ocean heat loss did not have correspondingly large SST anomalies. These latter studies emphasize the importance of specifying the ocean forcing properly in climate studies; however, by using only an ocean mixed layer, these models neglected the role of subsurface heat storage to contribute to interannual flux variations.

The solution would seem to be to use coupled ocean-atmosphere models. Ideally, the coupling would solve the problem of correctly specifying the ocean's contribution to the fluxes; however, an exceedingly high spatial resolution is required to adequately resolve the western boundary currents. Ocean modelers have found [*Metzger and Hurlburt*, 2001; *Garraffo et al.*, 2002] that a spatial resolution of approximately 1/12 degree is required to obtain a realistic western boundary current, with 1–2 m s^{-1} currents within the approximately 100- km-wide boundary current.

But is this high resolution necessary to properly model the ocean's contributions to surface fluxes? One could imag-

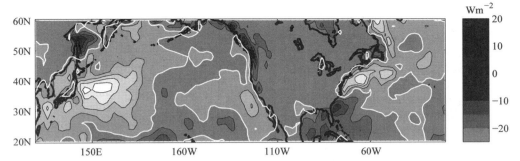

Figure 6. Relationship of net surface flux to heat content. The surface flux anomaly for the year 2000, based on a regression with the time series of the heat content in Plate 3c. Negative values indicate flux of heat from the ocean to the atmosphere. There were positive heat content anomalies in the boundary currents in 2000.

ine that a broader, slower boundary current would be sufficient for this problem, providing that the volume transports are realistic. This would likely be the case if there were no heat losses from the boundary current as it transported heat from the tropics poleward. However, a given water parcel is constantly losing heat to the atmosphere; therefore the longer it takes to transit the approximately 20° of latitude, the more heat that is lost to the atmosphere en route. This relationship between the speed of the western boundary current and the meridional heat transport was examined using a very simple model of a subtropical gyre [*Klinger*, 1996]; this analysis showed that the heat transport depends on the ratio of the time scale for the ocean temperature anomaly to be lost through air-sea interaction and the time scale to transit the western boundary. Heat transport errors (attributable to coarse ocean model resolution) of only about 20% would be sufficient to obscure the observed interannual variations in heat content in the Gulf Stream region (Figure 3) and therefore not produce the resulting interannual variations in heat fluxes over the Gulf Stream.

The question of atmosphere-ocean coupling has been addressed most often for either the Atlantic or the Pacific Oceans. An important climate issue in the Atlantic, the decadal time scale variations in the thermohaline circulation, was examined using models with varying spatial resolution [*Fanning and Weaver*, 1998]. Their analyses showed that the excessive diffusion of coarse models suppressed a decadal-scale advective-convective mechanism resulting in lower overall ocean variability, particularly at the longer time scales. *Schneider and Miller* [2001] show that the North Pacific has a predictable baroclinic response to basin-scale wind forcing, and the question of whether ocean variability associated with the PDO is a coupled atmosphere-ocean mode is being examined (N. Schneider, personal communication, 2004). The PDO-like mode in heat content shown here dominated the record of the 1990s, as evidenced by the mode order reversal in the SSH analysis in Section 3.2. That PDO mode has a larger magnitude in the central and eastern North Pacific and may not be the result of western boundary current advection.

5. CONCLUSIONS

A nearly 50-yr record of midlatitude ocean heat content variations was examined for the North Atlantic and the North Pacific Oceans. About 26% of the heat content variations are coherent between the Atlantic and the Pacific; these coherent variations have their largest amplitudes in the extension regions of the northern hemisphere western boundary currents, the Gulf Stream and the Kuroshio Extension. The coherent heat content variations are correlated with the Northern hemisphere Annular Mode (or as it is more commonly known, the Arctic Oscillation). Although it is tempting to attribute the ocean heat content variations to changes in the surface fluxes associated with changes in wind speed, the sign and phase of the correlations is inconsistent with this idea: strong westerlies are correlated with high heat content.

Previous studies of the upper ocean heat budgets near western boundary current extensions have shown that the heat content variations are caused by anomalies in ocean advection, rather than by surface fluxes. Much of the anomalous heat from advection is stored south of the boundary currents, in the same region that deep layers of uniform temperature waters are formed, known as "subtropical mode waters." Correlations between heat content and mode water show that a large volume of mode water is associated with low ocean heat content.

The source of the changes in ocean advection, which in turn cause most of the interannual changes in heat content, appears to be the midlatitude wind. The heat content anomalies lag changes in negative wind stress curl (strength of the westerlies) by about one year. This relatively short lag suggests that the ocean response may be, at least in part, barotropic, rather than baroclinic, as assumed in many simple coupled ocean-atmosphere models. A barotropic response is consistent with the known structure of the ocean currents in the mode water region.

Changes in ocean heat content have implications for the interannual-to-decadal scale climate variations through their contributions to air-sea fluxes. A negative correlation between the heat content anomalies in these regions and air-sea fluxes suggests that the ocean heat content anomalies force the fluxes on interannual to decadal time scales, rather than the other way around, as in the ocean interior. The air-sea flux anomalies associated with ocean heat content are of the order of 20% of the annual mean value; however, the variations occur in the region of the largest air-sea heat fluxes in the oceans.

The contributions of western boundary currents have likely been underestimated in climate models through the use of fixed SST or slab mixed layers, which neglect subsurface heat storage, and inadequate spatial resolution, which causes an underestimate of ocean heat advection.

Acknowledgments. The authors are grateful to Susan Hautala and LuAnne Thompson for helpful discussions and to anonymous reviewers for comments that improved the presentation of the results here. We also appreciate the efforts of S. Dickinson in data analysis and processing. Funding for KAK and SD was provided by NSF through grant OCE-9818920 with the University of Washington. The subsurface temperature data was obtained from the Joint Environmental Data Analysis Center (JEDAC) at the Scripps Institution of Oceanography. TOPEX/POSEIDON altimetry data was obtained from the Physical Oceanography Distributed Data Archive Center (PODAAC).

REFERENCES

Alexander, M. A., and C. Deser, A mechanism for the recurrence of wintertime midlatitude SST anomalies, *J. Phys. Oceanogr., 25*, 122–137, 1995.

Bretherton, C. S., and D. S. Battisti, An interpretation of the results from atmospheric general circulation models forced by the time history of the observed sea surface temperature distribution, *Geophys. Res. Lett., 27*, 767–770, 2000.

Cayan, D. R., Latent and sensible heat flux anomalies over the northern oceans: driving the sea surface temperature, *J. Phys. Oceanogr., 22*, 859–881, 1992.

Conkright, M. E., J. I. Antonov, O. Baranova, T. P. Boyer, H. E. Garcia, R. Gelfeld, D. Johnson, R. A. Locarnini, P. P. Murphy, T. D. O'Brien, I. Smolyar, and C. Stephens, *World Ocean Database 2001, Volume 1: Introduction.* Ed: Sydney Levitus, NOAA Atlas NESDIS 42, U.S. Government Printing Office, Washington, D.C., 167 pp., 2002.

daSilva, A. M., C. C. Young, and S. Levitus, *Atlas of surface Marine data 1994 Volume I. algorithms and procedures*, NOAA Atlas NESDIS 6, 51 pp., 1994.

Deser, C., M. A. Alexander, and M. S. Timlin, Evidence for a wind-driven intensification of the Kuroshio Current extension from the 1970s to the 1980s, *J. Clim., 12*, 1697–1706, 1999.

Deser, C., and M. A. Alexander, and M. S. Timlin, Upper-ocean thermal variations in the North Pacific during 1970-1991, *J. Clim., 9*, 1840–1855, 1996.

Deser, C., M. A. Alexander, and M. S. Timlin, Understanding the persistence of sea surface temperature anomalies in midlatitudes, *J. Clim., 16*, 57–72, 2003.

Dewar, W. K., On ocean dynamics in midlatitude climate, *J. Clim., 14*, 4380–4397, 2001.

Dong, S., Reconstructing the Mean Sea Surface Height in the Gulf Stream Region from TOPEX/POSEIDON Altimeter Observations and the Hydrobase Data, 2003. (http://ultrasat.apl.washington.edu/kkelly/projects/natl)

Dong, S., Interannual Variations in Upper Ocean Heat Content, Heat Transport, and Convergence in the Western North Atlantic, Ph. D. thesis, School of Oceanography, University of Washington, 2004.

Dong, S. and K. A. Kelly, The heat budget in the Gulf Stream region: the importance of heat storage and advection, *J. Phys. Oceanogr., 34*, 1214–1231, 2004.

Fairall, C. W., E. F Bradley, D. P. Rogers, J. B. Edson, and G. S. Young. Bulk parameterization of air-sea fluxes for Tropical Ocean-Global Atmosphere Coupled-Ocean Atmosphere Response Experiment, *J. Geophys. Res., 101*, 3747–3764, 1996.

Fanning, A. F., and A. J. Weaver, Thermohaline variability: the effects of horizontal resolution and diffusion, *J. Clim., 11*, 709–715, 1998.

Frankignoul, C., Sea surface temperature anomalies, planetary waves, and air-sea feedback in the middle latitudes, *Rev. Geophysics, 23*, 357–390, 1985.

Fu, L.-L., Christensen, E. J., and Yamarone, C. A., Jr., and others., TOPEX/POSEIDON mission overview in *J. Geophys. Res., 99*, 24,369–24,381, 1994.

Gallego, B., and P. Cessi, Decadal variability of two oceans and an atmosphere, *J. Clim., 14*, 2815–2832, 2001.

Garraffo, Z. D., A. J. Mariano, A. Griffa, C. Veneziani, and E. P. Chassignet, Lagrangian data in a high-resolution numerical simulation of the North Atlantic I. Comparison with in situ drifter data, *J. Marine Systems, 29*, 157–176, 2001.

Geise, B. S., and J. A. Carton, Interannual and decadal variability in the tropical and midlatitude Pacific Ocean, *J. Clim., 12*, 3402–3418, 1999.

Gill, A. E., and P. P. Niiler, The theory of seasonal variability in the ocean, *Deep-Sea Res., 20*, 141–177, 1973.

Hare, S. R., and N. J. Mantua, Empirical evidence for North Pacific regime shifts in 1977 and 1989 *Progress in Oceanography, 47*, 103–145, 2000. (http://www.jisao.washington.edu/data sets/pdo)

Hogg, N. G., A note on the deep circulation of the western North Atlantic: its nature and causes, *Deep-Sea Res., 30*, 945–961, 1983.

Johns, E., D. R. Watts, and H. T. Rossby, A test of geostrophy in the Gulf Stream, *J. Geophys. Res., 94*, 3211–3222, 1989.

Josey, S. A., E. C. Kent, and P. K. Taylor, New insights into the ocean heat budget closure problem from analysis of the SOC Air-Sea Flux Climatology, *J. Clim., 12*, 2856–2880, 1999.

Joyce, T. M., C. Deser, and M. A. Spall, Relation between decadal variability of Subtropical Mode Water and the North Atlantic Oscillation, *J. Clim., 13*, 2550–2569, 2000.

Kalnay, E. and coauthors, The NCEP/NCAR 40-Year Re-analysis Project, *Bull. Am. Meteor. Soc., 77*, 437–471, 1996.

Kelly, K. A., The relationship between oceanic heat transport and surface fluxes in the western North Pacific: 1970-2000, *J. Clim., 17*, 573–588, 2004.

Kelly, K. A., and S. T. Gille, Gulf Stream surface transport and statistics at 69 W from the GEOSAT altimeter, *J. Geophys. Res., 95*, 3149–3161, 1990.

Kelly, K. A., and B. Qiu, Heat flux estimates for the North Atlantic, Part II: The upper ocean heat budget, *J. Phys. Oceanogr., 25*, 2361–2373, 1995.

Kelly, K. A., S. Singh, and R. X. Huang, Seasonal variations of sea surface height in the Gulf Stream region, *J. Phys. Oceanogr., 29*, 313–327, 1999.

Klinger, B. A., A kinematic model of wind-driven heat trans- port, Notes and Correspondence, *J. Phys. Oceanogr., 26*, 131–135, 1996.

Kushnir, Y., and I. M. Held, 1996: Equilibrium atmospheric response to North Atlantic SST anomalies, *J. Clim., 9*, 1208–1219

Kwon, Y.-O., Observations of General Circulation and Water Mass Variability in the North Atlantic Subtropical Mode Water Region, Ph. D. thesis, School of Oceanography, University of Washington, 2003.

Marshall, J., H. Johnson, and J. Goodman, A study of the interaction of the North Atlantic Oscillation with ocean circulation, *J. Clim., 14*, 1399–1421, 2001.

McCartney, M. S., and L. D. Talley, The subpolar mode water of the North Atlantic Ocean, *J. Phys. Oceanogr., 12*, 1169–1188, 1982.

Metzger, E. J., and H. E. Hurlburt, The importance of high horizontal resolution and accurate coastline geometry in modeling South China Sea inflow, *Geophys. Res. Lett., 28*, 1059–1062, 2001.

Miller, A. J., D. R. Cayan, T. P. Barnett, N. E. Graham and J. M. Oberhuber, Interdecadal variability of the Pacific Ocean model response to observed heat flux and wind stress anomalies, *Clim. Dyn., 9*, 287–302, 1994.

Nakamura, H., G. Lin, and T. Yamagata, 1997: Decadal climate variability in the North Pacific during recent decades, *Bull Am. Meteor. Soc, 78*, 2215–2225.

Nakamura, H., T. Sampe, Y. Tanimoto, and A. Shimpo, Observed associations among storm tracks, jet streams and midlatitude oceanic fronts, this volume, 2004.

Palmer, T. N. and Z. Sun, A modeling and observational study of the relationship between sea surface temperature in the northwest Atlantic and the atmospheric general circulation. *Q. J. R. Meteorol. Soc., 111*, 947–975, 1985.

Preisendorfer, R. W., *Principal Component Analysis in Meteorology and Oceanography*, Elsevier, New York, 425 pp., 1988.

Qiu, B., Interannual variability of the Kuroshio Extension system and its impact on the wintertime SST field, *J. Phys. Oceanogr., 30*, 1486–1502, 2000.

Qiu, B., Large-scale variability in the midlatitude subtropical and subpolar North Pacific Ocean: observations and causes, *J. Phys. Oceanogr., 32*, 353–375, 2002.

Qiu, B., and K. A. Kelly, Upper-ocean heat balance in the Kuroshio Extension region, *J. Phys. Oceanogr., 23*, 2027–2041, 1993.

Rodwell, M. J., D. P. Rowell, and C. K. Folland, 1999. Oceanic forcing of the wintertime North Atlantic Oscillation and European climate, *Nature, 398*, 320–323.

Schneider, N., A. J. Miller, M. A. Alexander, and C. Deser, Subduction of decadal North Pacific temperature anomalies: observations and dynamics, *J. Phys. Oceanogr., 29*, 1056–1070, 1999.

Schneider, N., and A. J. Miller, Predicting western North Pacific Ocean climate, *J. Clim., 14*, 3997–4002, 2001.

Suga, T. and K. Hanawa, The subtropical mode water circulation in the North Pacific, *J. Phys. Oceanogr., 25*, 958–970, 1995.

Sutton, R. and P.-P. Mathieu, Response of the atmosphere-ocean mixed layer system to anomalous ocean heat flux convergence, *Q. J. R. Meteorol. Soc., 128*, 1259–1275, 2002.

Thompson, D. W. J, and J. M. Wallace, The Arctic Oscillation signature in the wintertime geopotential height and temperature fields, *Geophys. Res. Lett., 25*, 1297–1300, 1998. (http://www.jisao.washington.edu/ao)

Trenberth, K.E., and J. M Caron, Estimates of meridional atmosphere and ocean heat transports, *J. Clim., 14*, 3433–3443, 2001.

Vivier, F., K. A. Kelly, and L. Thompson, Heat budget of the Kuroshio Extension region: 1993–1999, *J. Phys. Oceanogr., 32*, 3436–3454, 2002.

Wallace, J. M. and D. W. J. Thompson, The Pacific center of action of the Northern Hemisphere annular mode: real or artifact?, *J. Clim.15*, 1987–1991, 2002.

Warren, B. A., Insensitivity of subtropical mode water characteristics to meteorological fluctuations, *Deep-sea Res., 19*, 1–19, 1972.

Wunsch, C., *The Ocean Circulation Inverse Problem*, Cambridge University Press, 441 pp., 1996.

Yasuda, T. and K. Hanawa, Decadal changes in the mode waters in the midlatitude North Pacific, *J. Phys. Oceanogr., 27*, 858–870, 1997.

Yualeva, E. N., N. Schneider, D. W. Pierce, and T. P. Barnett, Modeling of North Pacific climate variability forced by ocean heat flux anomalies, *J. Clim., 14*, 4027–4046, 2001.

S. Dong, Scripps Institution of Oceanography, University of California-San Diego, La Jolla, California 92093-0230.

K. A. Kelly, Applied Physics Laboratory, University of Washington, Box 355640, Seattle, Washington 98195-5640. (kkelly@apl.washington.edu)

Two Different Regimes of Anomalous Walker Circulation Over the Indian and Pacific Oceans Before and After the Late 1970s

Ryuichi Kawamura and Hiromitsu Aruga

Department of Earth Sciences, Toyama University, Toyama, Japan

Tomonori Matsuura and Satoshi Iizuka

National Research Institute for Earth Science and Disaster Prevention, Tsukuba, Ibaraki, Japan

Using the National Centers for Environmental Prediction/National Center for Atmospheric Research reanalysis data aided by a coupled ocean-atmosphere model, we investigated two different regimes of anomalous Walker circulation system over the Pacific and Indian Oceans before and after a climate shift, which occurred in the late 1970s. During the period before the climate shift, an upper-level velocity potential anomaly systematically moves eastward from the tropical Indian Ocean to the warm pool region of the western Pacific during the growth phase of El Niño-Southern Oscillation (ENSO). In the meantime, the activities of South Asian and Australian summer monsoon systems are directly affected by the evolution of the anomalous Walker circulation. During the period after the climate shift, in contrast, an upper-level velocity potential anomaly in the vicinity of the Philippine Sea and maritime continent is observed to expand westward into the northern Indian Ocean and South Asia during the decay phase of ENSO. This feature is identified with a major precursory signal of an anomalous South Asian summer monsoon in the preceding spring. The model captures a systematic eastward propagation similar to that observed prior to the late 1970s, but fails to reproduce the westward extension of the velocity potential anomaly observed to prevail after the late 1970s. The model results suggest that the cross-basin connection between the two oceans is a prerequisite for the turnabout of ENSO prior to the climate shift, in terms of the occurrence of westerly wind bursts.

1. INTRODUCTION

It has been recognized that significant changes in the periodicity and seasonality of El Niño-Southern Oscillation (ENSO) occurred around the late 1970s [e.g., *Wang*, 1995, *Mitchell and Wallace*, 1996, *Wang and An*, 2002]. It has also been reported that the South Asian summer monsoon-ENSO relationship through the Walker circulation obviously changed before and after the late 1970s [e.g., *Kumar et al.*, 1999, *Torrence and Webster*, 1999, *Miyakoda et al.*, 2000, *Krishnamurthy and Goswami*, 2000, *Kinter et al.*, 2002]. The variability of the western North Pacific summer monsoon [*Murakami and Matsumoto*, 1994] and the associated anomalous East Asian summer climate are also significantly different before and after that period [*Kawamura et al.*, 1998, *Wang et al.*, 2001, *Wu and Wang*, 2002].

A number of studies have examined the significant coupling of the South Asian monsoon and ENSO [e.g., *Angell*, 1981, *Shukla and Paolino*, 1983, *Rasmusson and Carpenter*, 1983, *Meehl*, 1987, *Webster and Yang*, 1992, *Ju and Slingo*, 1995, *Webster et al.*, 1998]. *Barnett* [1983, 1984] and *Yasunari* [1985, 1990, 1991] postulated that zonal propagation of anomalous Walker circulation from the Indian Ocean to the Pacific Ocean is intimately associated with ENSO, based on observational data before the early 1980s. They emphasized the active role of the Asian monsoon variability in triggering ENSO events because a weak (strong) Asian summer monsoon preceded a warm (cold) event of ENSO prevailing in boreal winter.

After the late 1970s, in contrast, a warm (cold) ENSO event tended to precede a weak (strong) Asian summer monsoon in broad-scale monsoon circulation intensity indices, such as the *Webster and Yang* [1992] index. *Yang et al.* [1996] and *Yang and Lau* [1998] pointed out that ENSO in the preceding winter and spring has an indirect impact on the summer monsoon activity through land surface hydrologic processes in the Asian continent, as well as a direct impact of ENSO. *Kawamura* [1998] showed that enhanced convective heating over the northern Indian Ocean with an ENSO signal induces anomalous upper-level anticyclonic circulation as a result of the Rossby wave response, thereby decreasing rainfall and increasing near-surface temperature in central and southwest Asia in spring prior to a strong South Asian summer monsoon. Based on observational data after the late 1970s, they emphasized the active role of ENSO in affecting the Asian summer monsoon variability.

Thus our research group postulates that the ENSO affects the variability of the Asian summer monsoon. Another claims, on the contrary, that the monsoon variability can trigger ENSO events. *Li and Zhang* [2002] noted that these contradictory results might arise from the different-period datasets used for the analyses. One possibility is that the significant changes in ENSO properties in the late 1970s might have caused changes in the monsoon-ENSO relationship. Delayed and indirect impacts of ENSO on the summer monsoon system at its decay phase have been highlighted by *Yang and Lau* [1998], *Shen et al.* [1998], *Wang et al.* [2000, 2001], and *Kawamura et al.* [2001a, b], whereas direct impacts of ENSO at its growth phase have been demonstrated by *Chen and Yen* [1994], *Ju and Slingo* [1995], *Lau and Wu* [2001], and *Kawamura et al.* [2003]. These studies suggest that indirect ENSO impacts on the monsoon system at the decay phase of ENSO are quite different from those impacts at its growth phase. Since the seasonality of the ENSO cycle changes before and after the late 1970s [*Mitchell and Wallace*, 1996], the differences in ENSO impacts between its growth and decay phases are very likely to affect the monsoon-ENSO relationship.

According to *Kawamura et al.* [2003], who examined a direct ENSO impact on the South Asian summer monsoon interannual variability at its growth phase during the period prior to the late 1970s, a combination of tropical ocean-atmosphere interactions over both the Indian and Pacific sectors is crucial for the turnabout of anomalous Walker circulation system relevant to the regular phase change of ENSO. It seems that a change in such a turnabout over the two ocean basins occurred after the late 1970s. However, few studies focus specifically on the possibility of a regime transition before and after the late 1970s in terms of the cross-basin connection between the tropical Indian and Pacific Oceans. This is surprising because the Walker circulation over the Pacific Ocean basin has been well studied in association with ENSO theories [e.g., *Schopf and Suarez*, 1988, *Weisberg and Wang*, 1997, *Jin*, 1997]. A proper understanding of its cross-basin connection might be important in clarifying why changes in ENSO properties occurred in the late 1970s and why the contradictory results were obtained on the lag relationship between the South Asian monsoon and ENSO.

The paper is organized in the following way. In section 2, we introduce the model simulation and the datasets used in this study. Section 3 demonstrates the remarkable observed features of the anomalous Walker circulation system over the tropical Indian and Pacific Oceans before and after the late 1970s. In sections 4 and 5, we examine ENSO-like phenomena simulated in a coupled ocean-atmosphere model (CGCM) and discuss the similarities and discrepancies between model results and observations in terms of the cross-basin connection between the two ocean basins. A summary is provided in section 6.

2. DATA USED AND ANALYSIS PROCEDURE

To examine the observed behavior of anomalous Walker circulation over the two ocean basins, we use the National Centers for Environmental Prediction/National Center for Atmospheric Research (NCEP/NCAR) global atmospheric reanalysis dataset [*Kalnay et al.*, 1996]. First, we apply an empirical orthogonal function (EOF) analysis to the tropical SSTs derived from the NOAA extended reconstructed sea surface temperature (ERSST) dataset [*Smith and Reynolds*, 2003] and extract dominant SST anomaly patterns during two 23-yr periods before and after the late 1970s. As is well known, the first 23-yr period 1955–1977 is the period when ENSO with a 2–3 yr periodicity dominates, whereas another 23-yr period 1979–2001 is characterized by the dominance of prolonged ENSO with a 4–5 yr periodicity [e.g., *Torrence and Webster*, 1999, *Li and Zhang*, 2002, *Kawamura et al.*, 2003]. For convenience, the former and latter periods are hereafter called period I and II, respectively. Second, a lag-regression analysis is conducted to investigate the coupling

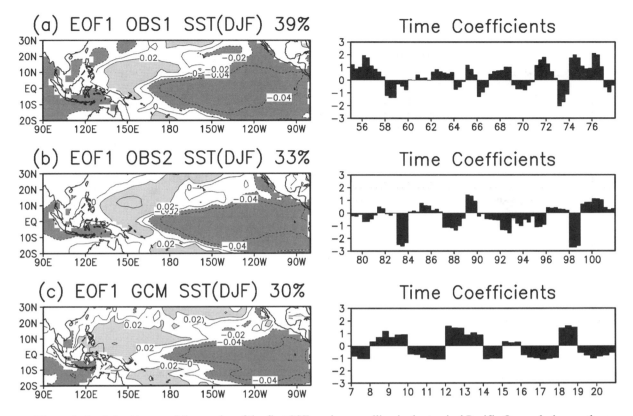

Figure 1. Spatial patterns and time series of the first SST modes prevailing in the tropical Pacific Ocean during northern winter (December–February) for period I (1955–1977) and II (1979–2001), which are obtained from the application of a simple EOF analysis to monthly SST anomaly data. Also denoted are those of model SST.

between the leading SST mode and the anomalous Walker circulation.

In this study we also analyze the model output of a CGCM developed at the National Research Institute for Earth Science and Disaster Prevention [*Iizuka et al.*, 2003]. The atmospheric component of the CGCM used in this study is the Japan Meteorological Agency global spectral model (JMA-GSM8911). The atmospheric model uses triangular truncation at wave number 42. There are 21 levels in the vertical from the surface to about 10 hPa. The model includes comprehensive physical parameterizations. A detailed description of the physical processes is provided in *Sugi et al.* [1990]. The convection scheme is the mass flux scheme proposed by *Arakawa and Schubert* [1974], which is modified for entrainment rate and for determination of mass flux at cloud base. The details of the prognostic Arakawa-Schubert scheme (PAS) are provided in *Kuma* [1996]. The ocean component of the CGCM is based on the Geophysical Fluid Dynamics Laboratory Modular Ocean Model 2.2 [*Pacanowski*, 1996]. The horizontal grid spacing is 1.125° longitude by 0.5625° latitude. There are 37 levels in vertical; the upper 400 m is divided into 25 levels. *Pacanowski and Philander's* [1981] parameterization is used in determining the coefficient of vertical mixing. The ocean model is spun up for 10 years from a static state using *Levitus* [1982] annual mean temperatures and salinities as initial conditions, and then coupled with the atmospheric model without flux corrections. The CGCM is run for 20 years, but only output for the last 15 years is analyzed when the trend in the equatorial SST is relatively small.

3. OBSERVED ENSO REGIMES

3.1. Dominant SST Patterns

Figure 1 shows the spatial patterns and time series of the first SST modes appearing in the Pacific sector during northern winter (December–February) for periods I and II, which are derived from the application of the simple EOF analysis to the monthly SST anomaly data. Note that the winter 1955 refers to the period from December 1954 through February 1955. Also exhibited is the first mode of model anomalous SST. The first mode for period I accounts for 39% of the total

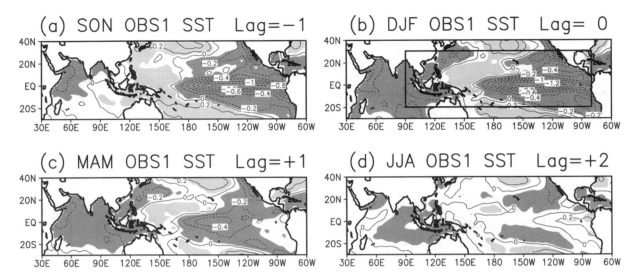

Figure 2. Lag-regression maps of SST with the time series of the first SST mode during period I (1955–1977) for September–November (lag-1), December–February (lag0), March–May (lag+1), and June–August (lag+2). Contour interval of the regression coefficients is 0.2°C. Shading indicates the regions where values satisfy the 5% level of statistical significance using a Student's *t*-test.

variance and has a 2–3 yr periodicity, whereas for period II the percentage of explained variance goes down to 33% and its periodicity shifts toward longer timescales. Their spatial patterns are very similar to each other, capturing the dominant features of ENSO. But there is a significant difference in the magnitude of SST anomaly in the Philippine Sea between the two periods. A positive SST anomaly east of the Philippines is more evident after the late 1970s than before. The CGCM is successful in simulating the basic features of ENSO-related SST anomalies. The periodicity of the first mode of the model SST field is around 3–4 years, which is comparable to observations, and its spatial pattern is also similar to that observed.

In the following subsections, we present the spatial distributions of lag-regression coefficients of SST and atmospheric variables with time series of the first SST mode in northern winter for each. Note that the lag-regression analysis is performed on a seasonal mean basis.

3.2. Regime Prior to the Late 1970s

Figure 2 shows the lag-regression maps of SST with the time series of the wintertime first mode during period I for September–November (lag-1), December–February (lag0), March–May (lag+1), and June–August (lag+2). In this figure, fall (SON) corresponds to the growth phase of the cold event of ENSO. When the cold event develops, a negative SST anomaly is indicated over the tropical Indian Ocean and persist until the subsequent spring (MAM), while in DJF a weak positive SST anomaly over the Philippine Sea becomes dissipated rapidly. A positive SST anomaly over the Arafura Sea and Timor Sea in SON is also replaced with a negative one in winter (DJF). After the mature phase of the cold event, a positive SST anomaly appears along the equator of the eastern Pacific in JJA, implying the growth phase of a warm event. Thus, MAM is the transition phase of ENSO during period I [e.g., *Rasmusson and Carpenter*, 1982].

Figure 3 displays the lag-regression maps of 200-hPa velocity potential with the time series of the first mode. In SON, a divergent anomaly expands from the eastern half of the tropical Indian Ocean to the maritime continent. South Asia and Australia are also covered by the divergent anomaly. In contrast, a convergent anomaly is dominant over the central and eastern tropical Pacific. These patterns of anomalies indicate a strong phase of the Walker circulation, which is consistent with the anomalous rainfall pattern (not shown). A positive rainfall anomaly is significant over the Indian continent and the Bay of Bengal, whereas another positive anomaly expands from the maritime continent to the eastern South Indian Ocean. These anomalies are dynamically linked to a pair of off-equatorial anomalous cyclones in the lower troposphere. According to *Kawamura et al.* [2003], such a feature of rainfall anomaly split into both hemispheres over the Indian Ocean dominates especially in late summer. It is associated with the growth phase of ENSO, indicating a direct ENSO impact on the South Asian summer monsoon system. Despite an almost unchanged convergent anomaly over the tropical Pacific from SON to DJF, the significant divergent anomaly diminishes and is confined to the vicinity of the maritime continent. In

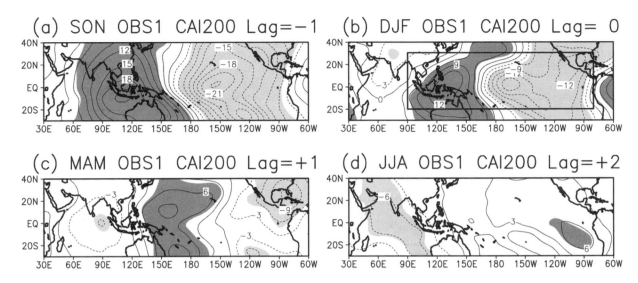

Figure 3. As in Figure 2 but for 200-hPa velocity potential during period I. Contour interval is 3×10^5 m^2 s^{-1}. Positive and negative anomalies denote divergent and convergent ones, respectively.

MAM, the divergent anomaly moves further eastward into the western tropical Pacific although weakens. In the meantime the convergent anomaly over the tropical Pacific disappears. In JJA, the Indian Ocean is covered by a new convergent anomaly, in contrast with SON regression pattern. A divergent anomaly is also seen over the eastern tropical Pacific, corresponding to the growth phase of a warm event. These features indicate the frequent occurrence of ENSO with a 2–3 yr periodicity during period I. The overall structure of this regime resembles a conceptual model of tropospheric biennial oscillation (TBO) proposed by *Meehl and Arblaster* [2002].

3.3. Regime After the Late 1970s

As mentioned earlier, prolonged ENSO cycles have occurred frequently during period II after the late 1970s. Thus, we will refer to ENSO-related circulation features during period II by the term prolonged ENSO regime for convenience. Figure 4 shows lag-regression maps of SST with the time series of the first mode when this regime dominates. Comparing it with Figure 2, there exist several distinctive differences as expected. Although, during period I, the polarity of SST anomaly in the eastern tropical Pacific reverses from MAM to JJA, it does not reverse during the prolonged ENSO regime. Looking at the tropical Indian Ocean, a negative SST anomaly is not necessarily obvious in SON, but develops in the subsequent spring and summer, while during period I it becomes less significant from MAM to JJA. Another notable difference is the presence of a remarkable positive SST anomaly in the warm pool region of the western Pacific. The persistent positive anomaly is still prominent in MAM. In contrast, there are no positive anomalies in the same season during period I. We stress that the seasonal evolution of SST anomalies over the tropical western North Pacific and Indian Oceans is significantly different between the two regimes as well as the eastern tropical Pacific.

Figure 5 displays the lag-regression maps of 200-hPa velocity potential with the time series of the first mode during period II. SON pattern is very similar to that observed during period I. In DJF, a more noticeable divergent anomaly develops over the Philippine Sea. A remarkable difference between the two regimes before and after the climate shift can be seen in the MAM panel. The overall pattern of MAM is not very significant during period I except for the western tropical Pacific. Conversely, significant areas expand in the tropics in MAM during the prolonged ENSO regime. It is found, in particular, that a divergent anomaly in the vicinity of the Philippine Sea and maritime continent extends westward into the northern Indian Ocean and South Asia in this season, which is accompanied by a positive rainfall anomaly over the Indian Ocean (not shown). This feature is also derived from an EOF analysis for outgoing longwave radiation anomaly field over the Indian Ocean in spring [*Kajikawa et al.*, 2003] and is identified with a major precursory signal of anomalous Asian summer monsoon associated with the decay phase of ENSO [e.g., *Ju and Slingo*, 1995, *Kawamura*, 1998]. *Kawamura et al.* [2001a, b] pointed out that a wind-evaporation-SST (WES) feedback operates over the tropical Indian Ocean, especially in spring, during the decay phase of prolonged ENSO and significantly affects

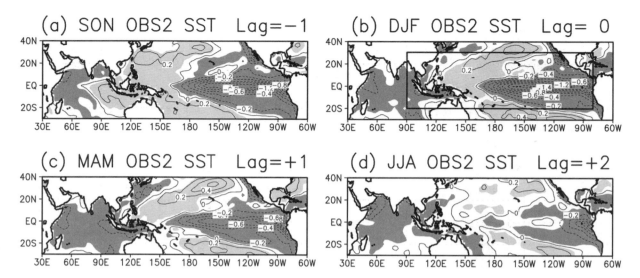

Figure 4. Lag-regression maps of SST with the time series of the first SST mode during period II (1979–2001) for September–November (lag-1), December–February (lag0), March–May (lag+1), and June–August (lag+2). Contour interval of the regression coefficients is 0.2°C. Shading indicates the regions where values satisfy the 5% level of statistical significance using a Student's t-test.

South Asian monsoon activity in the subsequent summer as a delayed and indirect impact of ENSO. They also suggested that persistent anomalous convective heating near the Philippine Sea possibly triggers a WES mode over the tropical Indian Ocean through the westward propagation of equatorially asymmetric Rossby waves. Thus, it appears that the above processes are, at least partially, responsible for the westward expansion of 200-hPa divergent anomalies into the northern Indian Ocean from DJF to MAM.

It turns out that there are significant differences between the two ENSO regimes before and after the late 1970s, in terms of the cross-basin connection between the tropical Indian and Pacific Oceans. It is thus meaningful to examine whether a CGCM simulation can capture either of the two ENSO regimes

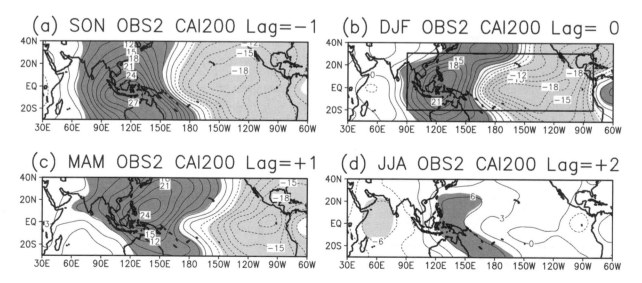

Figure 5. As in Figure 4 but for 200-hPa velocity potential during period II. Contour interval is 3×10^5 m^2 s^{-1}. Positive and negative anomalies denote divergent and convergent ones, respectively.

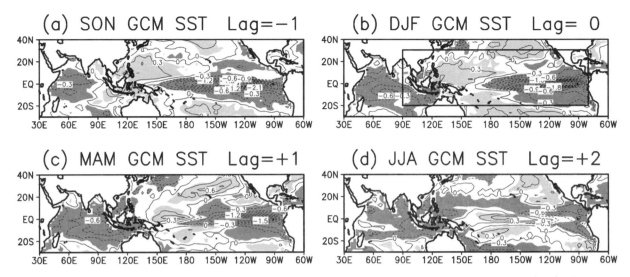

Figure 6. Lag-regression maps of SST with the time series of the first SST mode in the model simulation for September–November (lag-1), December–February (lag0), March–May (lag+1), and June–August (lag+2). Contour interval of the regression coefficients is 0.3°C. Shading indicates the regions where values satisfy the 5% level of statistical significance using a Student's *t*-test.

to clarify the mechanisms involved, which will be presented in the following section.

4. MODEL ENSO

As already shown in Figure 1, the first mode in the model SST field is similar to that observed in terms of spatial and temporal patterns. In a similar fashion, the lag-regression maps of model SST with the time series of its first mode is exhibited in Figure 6. A negative SST anomaly is prominent over the central and eastern tropical Pacific from SON to DJF, and a positive anomaly is indicated over the warm pool region of the western North Pacific. In MAM, however, the positive anomaly over the Philippine Sea disappears and another positive anomaly is established in the vicinity of the equator between 150°E–180°. This anomaly moves eastward into the

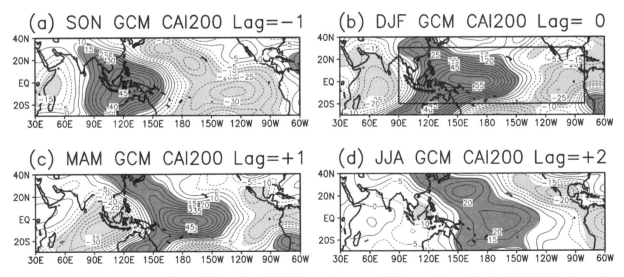

Figure 7. As in Figure 6 but for model-simulated 200-hPa velocity potential. Contour interval is 5×10^5 m^2 s^{-1}. Positive and negative anomalies denote divergent and convergent ones, respectively.

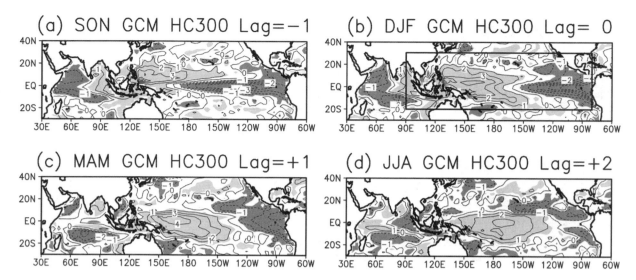

Figure 8. As in Figure 6 but for model heat content. Contour interval is 1.0×10^4 °C cm.

eastern equatorial Pacific from MAM to JJA, resulting in the retreat of the negative anomaly there. In the Indian Ocean sector, a negative anomaly is enhanced from SON to DJF and gradually weakens from MAM to JJA.

As demonstrated in Figure 7, model-simulated 200-hPa regression patterns indicate a regular eastward movement of anomalies from SON to JJA. In SON, a divergent anomaly is centered near the maritime continent, accompanied by two convergent anomalies over the eastern tropical Pacific and the western Indian Ocean. The divergent anomaly moves systematically eastward to the central equatorial Pacific from SON to JJA, corresponding well to the movement of the positive rainfall anomaly. Such an eastward-propagating feature is in good agreement with that observed during period I although its phase speed is slower. However, we cannot see any westward expansion of divergent anomalies from the warm pool region of the western North Pacific to the northern Indian Ocean which is dominant during period II. Thus, the model is not likely to simulate the indirect ENSO impact on the South Asian monsoon system at its decay phase that *Kawamura et al.* [2001a, b] pointed out.

It is plausible that an eastward-moving feature simulated in anomalous 200-hPa velocity potential field is linked to the behavior of SST anomalies in the tropics through air-sea interaction. To examine to what extent the simulated tropical SST anomalies are influenced by ocean dynamics, we present, in Figure 8, the lag-regression maps of model heat content with the time series of the first SST mode. Note that the heat content is vertically integrated from surface to a depth of 300 m. In the tropical Pacific, a positive anomaly is concentrated to the east of the Philippines in SON and it moves eastward along the equator from DJF to JJA. In DJF, a pronounced anomalous cyclone can be seen over the Philippine Sea and it looks like that the associated westerly anomaly induces equatorward and eastward movement of the positive heat content anomaly (not shown). Looking at Figure 9, showing the longitude-depth sections of subsurface temperature regression patterns along the equator, a positive subsurface temperature anomaly in the warm pool region of the western Pacific moves eastward along the equator from SON to JJA, slower than the speed of internal Kelvin Waves, which we may thus interpret as an air-sea coupled mode [e.g., *Zhang and Levitus*, 1997, *Iizuka et al.*, 2003]. A positive SST anomaly in the equatorial Pacific also moves eastward from MAM to JJA. In this simulation, the SST-thermocline feedback [*Philander et al.*, 1984] may strongly control the time evolution of SST anomalies over the equatorial Pacific.

In the tropical Indian Ocean, zonally asymmetric anomalies in the heat content field between its western and eastern parts are simulated from SON to DJF, but this structure disappears in MAM. Conversely, a reversed zonally asymmetric pattern is established in JJA. Looking at Figure 9 again, from SON to DJF a salient positive subsurface temperature anomaly is indicated in the eastern equatorial Indian Ocean and a negative anomaly is located in its western counterpart. In contrast, an alternating pattern is organized in JJA. These features may reflect the dominance of westward-propagating downwelling Rossby waves. An intriguing feature in the tropical Indian Ocean is that an apparent zonally asymmetric pattern of anomalous heat content from SON to DJF cannot be seen in the SST anomaly field. This may suggest that the role of ocean dynamics in regulating the tropical Indian Ocean SSTs is less influential than that in the tropical Pacific so far as the model ENSO-related anomalies are concerned. How-

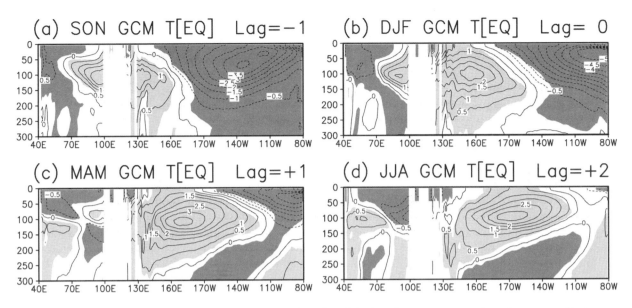

Figure 9. As in Figure 6 but for the depth-longitude section of model subsurface temperature along the equator. Contour interval is 0.5°C.

ever, we are aware, of course, that several recent studies emphasize a substantial role of the ocean dynamics in generating the tropical Indian Ocean SST anomalies [e.g., *Saji et al.,* 1999, *Webster et al.,* 1999, *Iizuka et al.,* 2000, *Rao et al.,* 2002, *Xie et al.,* 2002]. The role of surface heat flux in model SST anomalies in the Indian Ocean may become relatively important outside of the equatorial waveguide [e.g., *Yu and Rienecker,* 1999, *Lau and Nath,* 2000, 2003]. For example, *Kawamura et al.* [2003] suggested that a combination of the wind-evaporation feedback in the Indian Ocean and ocean dynamics in the tropical Pacific are prerequisites for the phase transition of the anomalous Walker circulation system over the two oceans prior to the climate shift. It is quite possible that the CGCM simulation captures such dynamic processes.

5. DISCUSSION

In the previous section, we note that there are some similarities and discrepancies between model results and observations in terms of the lag-regression analysis on a seasonal basis. For the ENSO regime prior to the late 1970s, one of the similarities is the systematic eastward movement of the anomalous rainfall and 200-hPa velocity potential from the Indian sector to the tropical Pacific sector. We expect that with this eastward movement, a low-level westerly anomaly, accompanied by enhanced rainfall, intrudes into the western tropical Pacific and an anomalous easterly wind over the tropical Pacific region retreats eastward. These changes are an important trigger of downwelling equatorial Kelvin Waves propagating eastward along the equator [e.g., *Masumoto and Yamagata,* 1991, *Moore and Kleeman,* 1999]. A pair of off-equatorial anomalous cyclones accompanies the low-level westerly anomaly over the western equatorial during the mature phase of a cold event. It looks, however, as if its origin is in the tropical Indian Ocean, as *Barnett* [1983, 1984] and *Yasunari* [1985, 1990, 1991] already pointed out. If so, this may be inconsistent with other ENSO paradigms that do not necessarily require the role of the Indian Ocean. To confirm the above process, we would like to examine how an onset of model ENSO occurs.

Figure 10 shows the behavior of an onset of model El Niño that occurred in the summer of the 18th year as a typical case. The left panel shows a time-longitude section of 5-day mean heat content and zonal wind stress anomalies along the equator, and the right panel shows the 5-day mean SST and rainfall anomalies. Positive rainfall anomalies are located over the eastern Pacific and Indian Oceans from January of to April of the 17th year, indicating the persistence of this warm event. After April, a negative SST anomaly expands into the Pacific region east of 150°E, accompanied by a negative rainfall anomaly, whereas over the Indian Ocean the positive rainfall anomaly still persists. This persistence implies the prominence of a cold event and its anomalous state continues until the end of the 17th year. In the western Indian Ocean, a negative SST anomaly is established around October of the 17th year and extends eastward by January of the 18th year. Associated with this, the positive rainfall anomaly rapidly shifts in February or March of the 18th year from the Indian Ocean to the western Pacific. With the eastward shift of

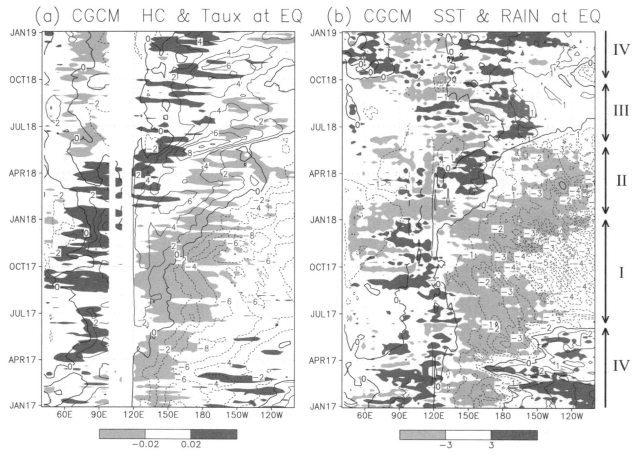

Figure 10. (a) Time-longitude section of 5-day mean model heat content and zonal wind stress anomalies along the equator. Contour interval for heat content is 2×10^4 °C cm. Heavy and light shadings denote regions of significant westerly and easterly wind stress anomalies, respectively. (b) Time-longitude section of 5-day mean model SST and rainfall anomalies. Contour interval for SST is 1°C. Heavy and light shadings denote regions of significant positive and negative rainfall anomalies, respectively.

the rainfall anomaly, a group of westerly wind stress anomalies start to intrude into the warm pool region of the western Pacific. When strong westerly bursts occur over its warm pool region in May of the 18th year, a large positive heat content anomaly propagates eastward to the central and eastern Pacific. Its phase speed is about 1.5 m s^{-1}, which is comparable to that of an internal Kelvin Wave. Corresponding to the eastward propagation of the anomalous heat content, a positive SST anomaly also shifts from the western Pacific to the central Pacific and persists after June of the 18th year. The positive rainfall anomaly also shifts eastward and is maintained over the central Pacific. These features indicate the onset phase of a warm event. The positive SST anomaly in the Pacific sector further expands, and we see two positive rainfall anomalies over the two ocean basins again from October 18th year, implying the mature phase of the warm event.

The occurrence of the westerly wind bursts is not independent of the phase transition of anomalous Walker circulation over the two ocean basins. Its phase transition provides a favorable condition for inducing the westerly wind bursts over the warm pool region of the western Pacific. We also emphasize that the cross-basin connection between the tropical Indian and Pacific Oceans is important for the turnabout of the model ENSO as an additional dynamic process, although we do not disregard the previous ENSO theories. *Kawamura et al.* [2003] demonstrated that the ENSO regime, prior to the climate shift, has four stages: i.e., onset and mature phases of a cold event, and onset and mature phases of a warm event, which will be termed phase I, II, III and IV, respectively, for convenience. If this classification is applied to Figure 10, the period from January to April corresponds to phase IV. Phase I, which is characterized by an ascending branch over the Indian Ocean and a descending branch over the central and eastern Pacific, continues until the end of the 17th year, with an anomalous east-west SST gradient between the two ocean basins. They also pointed out that the SST decrease in the Indian Ocean and the associated localization of enhanced rainfall over the warm pool region of the western Pacific result in the transition from phase I to II. The model appears to simulate such a process. At phase II, along with an ascending branch over the maritime continent and two descending branches over the Indian Ocean and the central and eastern Pacific, westerly wind bursts begin to dominate over its warm pool, eventually bringing about the onset of a model warm event (phase III). Although only a typical case is presented in this figure, similar features can also be seen in other events (figure not shown).

We note, on the other hand, that there are also discrepancies between model results and observations. If the ENSO regime prior to the late 1970s is highlighted again, one of the discrepancies is the presence of a significant model SST anomaly around the equator between 150°E–180° in MAM. In observations, there are no significant positive anomalies over that region, as seen in Figure 2c. According to *Iizuka et al.* [2003], the model used in this study has a systematic error that the mean thermocline depth in the western equatorial Pacific is shallower than that observed. Thus, model SST anomalies over the western equatorial Pacific are expected to be too sensitive to ocean dynamics, compared to observations. This systematic error might produce the difference in periodicity between the ENSO observed before the climate shift and model ENSO.

6. SUMMARY

To understand two different regimes of anomalous Walker circulation over the Pacific and Indian Oceans before and after the late 1970s, we investigated the seasonal evolutions of the two regimes in terms of cross-basin connection, using the NCEP/NCAR reanalysis aided by a coupled ocean-atmosphere model. Considering the climate shift in the late 1970s, we focused specifically on two 23-yr periods before and after the late 1970s.

Prior to the climate shift, an anomalous 200-hPa velocity potential systematically moves eastward from the tropical Indian Ocean to the warm pool region of the western Pacific during the growth phase of ENSO. In the meantime, the activities of South Asian and Australian summer monsoon systems are directly influenced by the evolution of the anomalous Walker circulation. When a cold event develops, a low-level westerly anomaly, accompanied by enhanced rainfall, intrudes into the warm pool region of the western Pacific. The model used in this study is successful in simulating the basic features of such systematic eastward movement. Westerly wind bursts over its warm pool contribute substantially to the onset of model warm event through the excitation of downwelling equatorial Kelvin waves. It is thus suggested that the cross-basin connection is a prerequisite for the turnabout of the ENSO prior to the climate shift as an additional dynamic process.

In the regime after the climate shift, on the other hand, a 200-hPa velocity potential anomaly in the vicinity of the Philippine Sea and maritime continent extends westward into the northern Indian Ocean and South Asia during the decay phase of ENSO. This westward expansion is identified with a major precursory signal of anomalous South Asian summer, which is, at least partially, a manifestation of a positive wind-evaporation-SST (WES) feedback prevailing in spring over the tropical Indian Ocean. However, the model cannot capture such westward expansion. One of the reasons may be the absence of a noticeable SST anomaly in spring over the Philippine Sea away from the equator because the excitation of a WES mode requires the persistence of equatorially asymmetric convective heating anomalies around the maritime continent.

New findings obtained in this study suggest that the cross-basin connection between the tropical Indian and Pacific Oceans is apparently different between the two regimes of anomalous Walker circulation before and after the late 1970s. This difference is very likely to bring about the contradictory results on the lag relationship between the South Asian summer monsoon and ENSO, and might also be one of the possible reasons why the changes in ENSO properties occurred around the late 1970s. An alternative hypothesis is that the Pacific climate shift causes decadal changes of ENSO properties [e.g., *Wang and An*, 2002]. We need further intensive investigation on the dynamic and thermodynamic processes of large-scale air-sea interaction in the western tropical Pacific, linked with those in the Indian Ocean and the eastern tropical Pacific.

Acknowledgments. Comments by two anonymous reviewers were extremely helpful. The authors would like to thank James Carton for a careful reading of the revised version of this paper. This research was supported by the research project "R&D of hydrological modeling and water resources system" of JST/CREST, and by the research project "Refinement of global and regional water cycle", and Grants-in-Aids (14540406) of the Japanese Ministry of Education, Sports, Culture, Science and Technology.

REFERENCES

Angell, J. K., Comparison of variations in atmospheric quantities with sea surface temperature variations in the equatorial eastern Pacific, *Mon. Weather Rev.*, *109*, 230–243, 1981.

Arakawa, A., and W. H. Schubert, Interaction of a cumulus cloud ensemble with the large-scale environment. Part I, *J. Atmos. Sci.*, *31*, 674–701, 1974.

Barnett, T. P., Interaction of the monsoon and Pacific trade wind systems at interannual time scales. Part I: The equatorial zone, *Mon. Weather Rev., 111,* 756–773, 1983.

Barnett, T. P., Interaction of the monsoon and Pacific trade wind system at interannual time scale. Part III: A partial anatomy of the Southern Oscillation, *Mon. Weather Rev., 112,* 2388–2400, 1984.

Chen, T.-C., and M.-C. Yen, Interannual variation of the Indian monsoon simulated by the NCAR community climate model: Effect of the tropical Pacific SST, *J. Clim., 7,* 1403–1415, 1994.

Iizuka, S., T. Matsuura, and T. Yamagata, The Indian Ocean SST dipole simulated in a coupled general circulation model, *Geophys. Res. Lett., 27,* 3369–3372, 2000.

Iizuka, S., K. Orito, T. Matsuura, and M. Chiba, Influence of cumulus convection schemes on the ENSO-like phenomena simulated in a CGCM, *J. Meteorol. Soc. Jpn, 81,* 805–827, 2003.

Jin, F. F., An equatorial ocean recharge paradigm for ENSO. Part I: Conceptual model, *J. Atmos. Sci., 54,* 811–829, 1997.

Ju, J., and J. M. Slingo, The Asian summer monsoon and ENSO, *Q. J. R. Meteorol. Soc., 121,* 1133–1168, 1995.

Kajikawa, Y., T. Yasunari, and R. Kawamura, The role of the local Hadley circulation over the western Pacific on the zonally asymmetric anomalies over the Indian Ocean, *J. Meteorol. Soc. Jpn, 81,* 259–276, 2003.

Kalnay, E., and Coauthors, The NCEP/NCAR 40-year reanalysis project, *Bull. Am. Meteorol. Soc., 77,* 437–471, 1996.

Kawamura, R., A possible mechanism of the Asian summer monsoon-ENSO coupling, *J. Meteorol. Soc. Jpn, 76,* 1009–1027, 1998.

Kawamura, R., M. Sugi, T. Kayahara, and N. Sato, Recent extraordinary cool and hot summers in East Asia simulated by an ensemble climate experiment, *J. Meteorol. Soc. Jpn, 76,* 597–617, 1998.

Kawamura, R., T. Matsuura, and S. Iizuka, Role of equatorially asymmetric sea surface temperature anomalies in the Indian Ocean in the Asian summer monsoon and El Niño-Southern Oscillation coupling, *J. Geophys. Res., 106,* 4681–4693, 2001a.

Kawamura, R., T. Matsuura, and S. Iizuka, Interannual atmosphere-ocean variations in the tropical western North Pacific relevant to the Asian summer monsoon-ENSO coupling, *J. Meteorol. Soc. Jpn, 79,* 883–898, 2001b.

Kawamura, R., T. Matsuura, and S. Iizuka, Equatorially symmetric impact of El Niño-Southern Oscillation on the South Asian summer monsoon system, *J. Meteorol. Soc. Jpn, 81,* 1329–1352, 2003.

Kinter, J. L., K. Miyakoda, and S. Yang, Recent change in the connection from the Asian monsoon to ENSO, *J. Clim., 15,* 1203–1215, 2002.

Krishnamurthy, V., and B. N. Goswami, Indian monsoon-ENSO relationship on interdecadal timescale, *J. Clim., 13,* 579–595, 2000.

Kuma, K., Parameterization of cumulus convection, *JMA/NPD Report, No. 42,* 93pp, 1996.

Kumar, K. K., B. Rajagopalan, and M. A. Cane, On the weakening relationship between the Indian monsoon and ENSO, *Science, 284,* 2156–2159, 1999.

Lau, K.-M., and H. T. Wu, Principal modes of rainfall-SST variability of the Asian summer monsoon: A reassessment of the monsoon-ENSO relationship, *J. Clim., 14,* 2880–2895, 2001.

Lau, N.-C., and M. J. Nath, Impacts of ENSO on the variability of the Asian-Australian monsoons as simulated in GCM experiments, *J. Clim., 13,* 4287–4309, 2000.

Lau, N.-C., and M. J. Nath, Atmosphere-ocean variations in the Indo-Pacific sector during ENSO episodes, *J. Clim., 16,* 3–20, 2003.

Levitus, S., Climatological atlas of the world ocean, *NOAA Prof. Paper 13,* 178pp., Natl. Oceanic and Atmos. Admin., Silver Spring, Md., 1982.

Li, T., and Y. Zhang, Processes that determine the quasi-biennial and lower-frequency variability of the South Asian monsoon, *J. Meteorol. Soc. Jpn, 80,* 1149–1163, 2002.

Masumoto, Y., and T. Yamagata, On the origin of a model ENSO in the western Pacific, *J. Meteorol. Soc. Jpn, 69,* 197–207, 1991.

Meehl, G. A., The annual cycle and interannual variability in the tropical Pacific and Indian Ocean regions, *Mon. Weather Rev., 115,* 27–50, 1987.

Meehl, G. A., and J. M. Arblaster, The tropospheric biennial oscillation and Asian-Australian monsoon rainfall, *J. Clim., 15,* 722–744, 2002.

Mitchell, T. P., and J. M. Wallace, ENSO seasonality: 1950–78 versus 1979–92, *J. Clim., 9,* 3149–3161, 1996.

Miyakoda, K., J. L. Kinter, and S. Yang, Analysis of the connection from the South Asian monsoon to ENSO by using precipitation and circulation indices, *COLA Technical Report, 90,* Center for Ocean-Land-Atmosphere Studies, 72pp., 2000.

Moore, A. M., and R. Kleeman, Stochastic forcing of ENSO by the intraseasonal oscillation, *J. Clim., 12,* 1199–1220, 1999.

Murakami, T., and J. Matsumoto, Summer monsoon over the Asian continent and western North Pacific, *J. Meteorol. Soc. Jpn, 72,* 719–745, 1994.

Pacanowski, R. C., Documentation user's guide and reference manual (MOM2, Version 2), *GFDL Ocean Technical Report 3.2,* 329pp, Geophys. Fluid Dyn. Lab., Princeton, N. J., 1996.

Pacanowski, R. C., and S. G. H. Philander, Parameterization of vertical mixing in numerical models of tropical oceans, *J. Phys. Oceanogr., 11,* 1443–1451, 1981.

Philander, S. G. H., T. Yamagata, and R. C. Pacanowski, Unstable air-sea interactions in the tropics, *J. Atmos. Sci., 41,* 603–613, 1984.

Rao, S. A., S. K. Behera, Y. Masumoto, and T. Yamagata, Interannual subsurface variability in the tropical Indian Ocean with a special emphasis on the Indian Ocean dipole. *Deep-Sea Res., 49B,* 1549–1572, 2002.

Rasmusson, E. M., and T. H. Carpenter, Variations in tropical sea surface temperature and surface wind fields associated with the Southern Oscillation/El Niño, *Mon. Weather Rev., 110,* 354–384, 1982.

Rasmusson, E. M., and T. H. Carpenter, The relationship between eastern equatorial Pacific sea surface temperatures and rainfall over India and Sri Lanka, *Mon. Weather Rev., 111,* 517–528, 1983.

Saji, N. H., B. N. Goswami, P. N. Vinayachandran and T. Yamagata, A dipole mode in the tropical Indian Ocean, *Nature, 401,* 360–363, 1999.

Schopf, P. S., and M. J. Suarez, Vacillations in a coupled ocean-atmosphere model, *J. Atmos. Sci., 45,* 549–566, 1988.

Shen, X., M. Kimoto, and A. Sumi, Role of land surface processes associated with interannual variability of broad-scale Asian summer monsoon as simulated by the CCSR/NIES AGCM, *J. Meteorol. Soc. Jpn, 76,* 217–236, 1998.

Shukla, J., and D. A. Paolino, The Southern Oscillation and long-range forecasting of the summer monsoon rainfall over India, *Mon. Weather Rev., 111,* 1830–1837, 1983.

Smith, T. M., and R. W. Reynolds, Extended reconstruction of global sea surface temperatures based on COADS data (1854–1997), *J. Clim., 16,* 1495–1510, 2003.

Sugi, M., K. Kuma, K. Tada, K. Tamiya, N. Hasegawa, T. Iwasaki, S. Yamada, and T. Kitade, Description and performance of the JMA operational global spectral model (JMA-GSM88), *Geophys. Mag., 43,* 105–130, 1990.

Torrence, C., and P. J. Webster, Interdecadal changes in the ENSO-monsoon system, *J. Clim., 12,* 2679–2690, 1999.

Wang, B., Interdecadal changes in El Niño onset in the last four decades, *J. Clim., 8,* 267–285, 1995.

Wang, B., R. Wu, and X. Fu, Pacific-East Asian teleconnection: How does ENSO affect East Asian climate?, *J. Clim., 13,* 1517–1536, 2000.

Wang, B., R. Wu, K.-M. Lau, Interannual variability of the Asian summer monsoon: Contrasts between the Indian and the western North Pacific-East Asian monsoons, *J. Clim., 14,* 4073–4090, 2001.

Wang, B., and S. I. An, A mechanism for decadal changes of ENSO behavior: roles of background wind changes, *Clim. Dyn., 18,* 475–486, 2002.

Weisberg, R., and C. Wang, A western Pacific oscillator paradigm for the El Niño-Southern Oscillation, *Geophys. Res. Lett., 24,* 779–782, 1997.

Webster, P. J., and S. Yang, Monsoon and ENSO: selectively interactive systems, *Q. J. R. Meteorol. Soc., 118,* 877–926, 1992.

Webster, P. J., V. O. Magana, T. N., Palmer, J. Shukla, R. A. Tomas, M. Yanai, and T. Yasunari, Monsoons: Processes, predictability, and prospects for prediction, *J. Geophys. Res., 103,* 14451–14510, 1998.

Webster, P. J., A. M. Moore, J. P. Loschnigg, and R. R. Leben, Coupled ocean-atmospheric dynamics in the Indian Ocean during 1997–1998, *Nature, 401,* 356–360, 1999.

Wu, R., and B. Wang, A Contrast of the East Asian Summer Monsoon and ENSO Relationship between 1962–1977 and 1978–1993, *J. Clim., 15,* 3266–3279, 2002.

Xie, S.-P., Structure and mechanisms of South Indian Ocean climate variability, *J. Clim., 15,* 864–878, 2002.

Yang, S., K.-M. Lau, and M. Sankar-Rao, Precursory signals associated with the interannual variability of the Asian summer monsoon, *J. Clim., 9,* 949–964, 1996.

Yang, S., and K.-M. Lau, Influences of sea surface temperature and ground wetness on Asian summer monsoon, *J. Clim., 11,* 3230–3246, 1998.

Yasunari, T., Zonally propagating modes of the global east-west circulation associated with the Southern Oscillation, *J. Meteorol. Soc. Jpn, 63,* 1013–1029, 1985.

Yasunari, T., Impact of Indian monsoon on the coupled atmosphere ocean system in the tropical Pacific, *Meteorol. Atmos. Phys., 44,* 29–41, 1990.

Yasunari, T., The monsoon year – A new concept of the climate year in the tropics, *Bull. Amer. Meteorol. Soc., 72,* 1331–1338, 1991.

Yu, L., and M. M. Rienecker, Mechanisms for the Indian Ocean warming during the 1997–1998 El Niño, *Geophys. Res. Lett., 26,* 735–738, 1999.

Zhang, R.-H., and S. Levitus, Interannual variability of the coupled tropical Pacific Ocean-Atmosphere system associated with the El Niño-Southern Oscillation, *J. Clim, 10,* 1312–1330, 1997.

H. Aruga and R. Kawamura, Department of Earth Sciences, Toyama University, 3190 Gofuku, Toyama 930-8555, Japan. (kawamura@sci.toyama-u.ac.jp)

S. Iizuka and T. Matsuura, National Research Institute for Earth Science and Disaster Prevention, 3-1 Tennodai, Tsukuba, Ibaraki 305-0006, Japan. (iizuka@bosai.go.jp; matsuura@bosai.go.jp)

Tropical Tropospheric Temperature and Precipitation Response to Sea Surface Temperature Forcing

Hui Su and J. David Neelin

Department of Atmospheric Sciences, and Institute of Geophysics and Planetary Physics, University of California, Los Angeles, Los Angeles, California

Joyce E. Meyerson

Department of Atmospheric Sciences, University of California, Los Angeles, Los Angeles, California

During an El Niño event, there are substantial tropospheric temperature anomalies across the tropics associated with sea surface temperature (SST) warming in the central and eastern Pacific. The typical spatial scale for teleconnection response of tropospheric temperature tends to be large. On the other hand, the precipitation response exhibits strong compensation between positive response over warm SST anomalies and a complex negative response remotely. The tropical spatial averages of tropospheric temperature and precipitation thus yield an interesting contrast in behavior. Anomalies of tropical averaged precipitation for 3-month averages appear quite scattered in relation to tropical SST anomalies, while the tropical mean tropospheric temperature obeys an approximately linear relationship to SST. This different behavior of tropical mean precipitation and tropospheric temperature in relation to SST is examined in detail using observational data, GCM simulations and idealized experiments with the quasi-equilibrium tropical circulation model (QTCM). Theoretical understanding is provided through a simple analytical model, which suggests that the integral constraint on tropical average precipitation is dominated by dry static energy transport into or out of the tropics. Convection acts to keep tropospheric temperature in quasi-equilibrium (QE) with boundary layer moist static energy, which is in turn held toward SST by surface fluxes. To maintain QE, the tropical average convective heating (i.e., precipitation) anomalies react to oppose any processes that tend to cool the tropical troposphere. Thus, while tropical average tropospheric temperature is closely related to SST, unrelated heating or cooling anomalies such as those due to the tropical-midlatitude transports can create large scatter in tropical average precipitation anomalies.

1. INTRODUCTION

1.1. Background

On interannual time scales, variability of tropical tropospheric-averaged temperature is dominated by El Niño/Southern Oscillation [Horel and Wallace, 1981; Pan and Oort, 1983; Sun and Oort, 1995]. When sea surface temperature (SST) is warmer than normal in the eastern and central Pacific, warm tropospheric temperature anomalies are observed across the entire tropical band [Yulaeva and Wallace, 1994; Wallace et al., 1998]. At the same time, substantial precipitation anomalies occur within the tropics. Unlike wide-spread tropospheric temperature warming, precipitation anomalies exhibit strong spatial variations. The spatial scale for teleconnection of tropospheric temperature (and associated wind vectors) appears larger than that for precipitation [Wallace et al., 1998; Su et al., 2001]. In regions of warm SST anomalies, the amount of precipitation is increased. Away from the directly heated regions, precipitation is reduced, which is primarily a remote response to the warmest SST anomalies [Su et al., 2001]. When the tropical average is considered, the near-cancellation of positive and negative precipitation anomalies yields scattered tropical-mean precipitation anomalies $\langle P'\rangle$ in relation to tropical SST anomalies $\langle T_s'\rangle$. However, the tropical mean tropospheric temperature anomalies $\langle \widehat{T}'\rangle$ are approximately linearly related to SST anomalies.

The approximately linear relationship of $\langle \widehat{T}'\rangle$ to $\langle T_s'\rangle$ is examined in detail in *Su et al.* [2003, hereafter SNM]. The scatter of $\langle P'\rangle$ with respect to $\langle T_s'\rangle$ is presented in Su and Neelin [2003, hereafter SN03]. Here, we combine the results from *SNM* and *SN03* to provide an overview of tropical tropospheric temperature and precipitation response to SST forcing on interannual time scales. The different characteristics of tropical tropospheric temperature and precipitation in relation to SST are compared using a variety of observational data and model results. Subsequently, with a simple analytical model we aim to examine the physical processes involved quantitatively, and thus gain insight into the dynamics governing the behavior of tropospheric temperature and precipitation response to interannual SST forcing. For present purposes, ENSO SST anomalies are discussed as a forcing to the atmosphere, and model simulations with specified SST are used. We focus on the simultaneous relationship of atmospheric variables to SST at 3-month averages, noting the caveat that for some phenomena ocean-atmosphere coupling needs to be considered.

The structure of this article is as follows: In Section 1.2, the $\langle \widehat{T}'\rangle$ and $\langle T_s'\rangle$ relationship is shown. The $\langle P'\rangle$ and $\langle T_s'\rangle$ relation is then presented in Section 1.3. Section 2 provides additional cases for the relationship of $\langle \widehat{T}'\rangle$, $\langle P'\rangle$ and $\langle T_s'\rangle$ using numerical model results. An analytical model is introduced in Section 3 to unravel the approximately linear $\langle \widehat{T}'\rangle$ but scattered $\langle P'\rangle$ with respect to $\langle T_s'\rangle$. In Section 4, the dynamics for the $\langle \widehat{T}'\rangle$, $\langle P'\rangle$ and $\langle T_s'\rangle$ relationship are described. Finally, conclusion and discussion are given in Section 5.

1.2. The Approximately Linear Relationship of $\langle \widehat{T}'\rangle$ and $\langle T_s'\rangle$

The linear relationship between tropical mean tropospheric temperature and SST anomalies has been documented extensively [e.g., Newell and Weare, 1976; Angell, 1981; Horel and Wallace, 1981; Pan and Oort, 1983; Soden, 2000; Kumar and Hoerling, 2003], although many have used 200 hPa geopotential height or temperature as a proxy for tropospheric averaged temperature. Sobel et al. [2002] showed that interannual anomalies of tropical tropospheric temperature are correlated not only with SST anomalies averaged over the precipitating regions, but also with SST anomalies averaged over the entire tropics. In Figure 1a, the tropical averaged (25°S–25°N) tropospheric (850–200 hPa) temperature anomalies are plotted against observed tropical SST anomalies [Reynolds and Smith, 1994]. Three temperature datasets are used, one of which is the NCEP/NCAR (National Center for Environmental Prediction (NCEP)/National Center for Atmospheric Research (NCAR)) reanalysis data from 1982 to 1998 [Kalnay et al., 1996]. Another dataset consists of the satellite measurement (1982-1993) from the microwave sounding unit (MSU) associated with the vertical weighting functions of Channel 2-3 [Spencer and Christy, 1992]. The third is from a simulation with the quasi-equilibrium tropical circulation model [QTCM, Neelin and Zeng, 2000] driven by observed SST anomalies from 1982-1998. A similar scatterplot using tropical averaged temperature at 200 hPa is shown in Soden [2000] Figure 5c. It is clear that there is an approximate linearity between $\langle \widehat{T}'\rangle$ and $\langle T_s'\rangle$ for the NCEP/NCAR reanalysis and MSU estimate. The QTCM simulation with observed SST also shows a prominent linear relationship. The slopes of the linear fits to each dataset are surprisingly close, all around 1.4 C C^{-1}. The model results have less scatter than the observed datasets due to reduced internal variability.

1.3. The Scatter of $\langle P'\rangle$ in Relation to $\langle T_s'\rangle$

Tropical mean precipitation anomalies, on the other hand, appear rather scattered with SST anomalies, as shown in Figure 1b for four datasets. Besides the NCEP/NCAR reanalysis and the QTCM simulation driven by observed SST from 1982-1998, two combined satellite and rain gauge measurements of precipitation are used. One is from the Global Precipitation Climatology Project (GPCP) [Huffman et al., 1997] and the other is from the the Climate Prediction Center (CPC) Merged Analysis of Precipitation (CMAP) [Xie and Arkin, 1997]. These two satellite products use similar satellite infrared, microwave emission and in-situ rain-gauge measurements, but with different algorithms. The MSU precipitation [Spencer, 1993] used in Soden [2000]

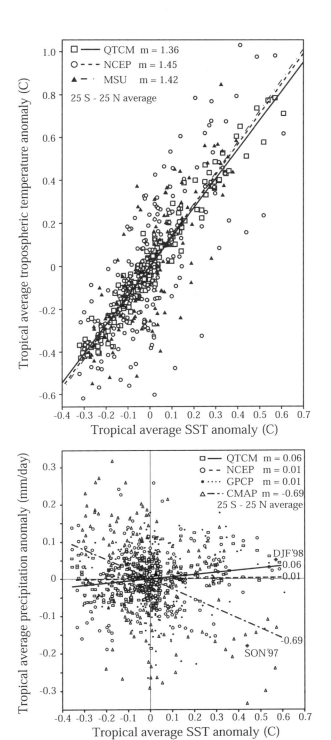

Figure 1. Tropical averaged (25°S–25°N) (a) tropospheric temperature and (b) precipitation anomalies versus tropical averaged SST anomalies for various datasets (see text for details, Figure 1a modified after SNM Figure 1) The solid lines are the linear fits to the datasets (the NCEP and GPCP precipitation have the same linear fits). The slopes of the linear fits are listed by the values of m.

covers oceanic regions only, and thus is not displayed in Figure 1b. The monthly precipitation anomalies are smoothed by a 3-month running mean, a typical filter for examining interannual anomalies. It is clear that tropical mean precipitation anomalies appear scattered with respect to tropical SST anomalies for all datasets. There is no simple relation between $\langle P' \rangle$ and $\langle T_s' \rangle$. Linear fits to the datasets show a wide range of slopes, with even negative slope for the CMAP precipitation. Defining P_i' and P_{Li}' as actual and linearly fitted precipitation anomalies at month i, respectively, the root-mean-square errors of the linear fits

$$\left(\frac{1}{N}\sum_{i=1}^{N}(P_i' - P_{Li}')^2\right)^{1/2}$$

are comparable to the standard deviations of the datasets themselves, suggesting the linear fits are not representative of the relationship between $\langle P' \rangle$ and $\langle T_s' \rangle$. Similar scatter of $\langle P' \rangle$ versus $\langle T_s' \rangle$ is also found for a recent satellite product, the Tropical Rainfall Measuring Mission (TRMM) precipitation, although the temporal coverage of the TRMM data is much shorter (available since January 1998). The linear fit to the TRMM precipitation also gives a negative slope of -0.2 mm day^{-1}C^{-1}(SN03). The linear correlations between $\langle P' \rangle$ and $\langle T_s' \rangle$ are 0.01 for the NCEP/NCAR reanalysis and the GPCP data, 0.21 for the QTCM result and -0.38 for the CMAP.

For a given tropical mean SST anomaly, the tropical mean precipitation anomaly could be of either sign. For example, the $\langle T_s' \rangle$ is about 0.4°C for September to November 1997, and 0.6°C for December 1997 to February 1998. The corresponding $\langle P' \rangle$ from the GPCP data is -0.2 mm day^{-1} for SON and 0.05 mm day^{-1} for DJF. Furthermore, the opposite signs of $\langle P' \rangle$ are associated with rather similar spatial distributions of precipitation anomalies as in Figure 2. By viewing the horizontal maps alone, one cannot predict the sign of the tropical mean precipitation anomaly because of the near-cancellation of positive anomalies against negative anomalies.

The approximately linear relation between $\langle \widehat{T}' \rangle$ and $\langle T_s' \rangle$ and the poor correlation between $\langle P' \rangle$ and $\langle T_s' \rangle$ on the interannual time scales seem contradictory to global warming scenario simulations where tropical mean precipitation increases as tropical mean SST increases [Mitchell et al., 1987; Dai et al., 2001]. It also seems contradictory to a traditional view that increased SST enhances convective activity, and thus increases the precipitation rate and intensifies the hydrological cycle in the tropics [Holton, 1992]. For local precipitation anomalies near warm SST anomalies, the traditional theory holds [Ropelewski and Halpert, 1987; Kiladis and Diaz, 1989]. Thus it is of interest to understand why it fails in the tropical means on interannual time scales.

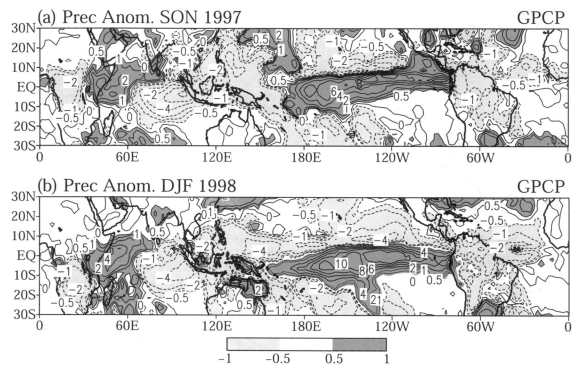

Figure 2. Spatial patterns of precipitation anomalies from the GPCP data for (a) September to November 1997 and (b) for December 1997–January 1998. After SN03 (their Figures 2b and 3b).

Soden [2000] pointed out that the performance of existing climate models in simulating the interannual variations of tropical mean tropospheric temperature is better than the performance in simulating the tropical mean precipitation variations. He attributed this discrepancy to two possible sources. One is the error of satellite measurements of tropical mean precipitation. The other is associated with problems of various physical parameterization schemes employed in climate models.

We concur with Soden that both observational data and numerical models might require improvement. However, the drastically different behaviors of $\langle \widehat{T}' \rangle$ versus $\langle T_s' \rangle$ and $\langle P' \rangle$ versus $\langle T_s' \rangle$ (Figure 1), and the difference in the performance of climate models in simulating the interannual variability of $\langle \widehat{T}' \rangle$ and $\langle P' \rangle$ suggest that the dynamics governing the temperature and precipitation response to SST forcing may be fundamentally different. While SNM showed that tropical mean tropospheric temperature has a rather simple relationship with SST, *Su and Neelin* [2002] found the mechanisms for anomalous subsidence forced by ENSO warm SST anomalies involve a complex set of pathways. These involve feedbacks dependent on local climatology, and tend to vary among regions and events, although many share a unifying theme, namely, they are initiated by the temperature response. Thus,

it is not surprising that GCMs are able to capture tropospheric temperature response but have more difficulty reproducing precipitation response.

2. ADDITIONAL CASES OF THE RELATIONSHIP OF $\langle \widehat{T}' \rangle$, $\langle P' \rangle$ AND $\langle T_s' \rangle$

The precipitation data shown in Figure 1b exhibit a wide range of linear correlation with $\langle T_s' \rangle$. The differences in the $\langle P' \rangle$ and $\langle T_s' \rangle$ relation for the observational data could potentially be due to errors in the satellite retrieval algorithms. Because observational data lack constraints in the moisture budget, the tropical mean precipitation values may be inconsistent with mass conservation. Hence, we extend our analysis to atmospheric model results that have a consistent moisture budget. First, we examine ensemble simulations from the National Aeronautics and Space Administration (NASA) Seasonal-to-Interannual Prediction Project (NSIPP) atmospheric GCM [*Bacmeister et al.*, 2000; *Pegion et al.*, 2000].

2.1. NSIPP Experiments

Five NSIPP GCM simulations with different initial conditions are used. They are all driven by observed SST from

1930 to present. Only the results from 1982 to 1998 are shown for comparison to other datasets. Figure 3a shows the scatterplot of $\langle P'\rangle$ versus $\langle T_s'\rangle$ for all five runs where both $\langle P'\rangle$ and $\langle T_s'\rangle$ are smoothed by a 3-month running mean. It is evident that the tropical mean precipitation anomalies have a large degree of variability and poor correlation with tropical average SST anomalies for all runs. The standard deviations of $\langle P'\rangle$ for individual runs are around 0.05 mm day^{-1}C^{-1}. A linear fit to all values of $\langle P'\rangle$ is shown by the dashed line, with a positive slope of 0.07 mm day^{-1}C^{-1}. The slopes of linear fits to individual runs range from 0.05 to 0.08 mm day^{-1}C^{-1}. However, the r.m.s. errors of the linear fits are very close to the standard deviations of $\langle P'\rangle$. Thus, a linear fit to $\langle T_s'\rangle$ explains little of the $\langle P'\rangle$ behavior as the linear correlation of $\langle P'\rangle$ to $\langle T_s'\rangle$ is only 0.01.

In contrast, the tropical averaged tropospheric temperature anomalies are strikingly linear with respect to the tropical mean SST forcing. Figure 3b shows the scatterplot of $\langle \widehat{T}'\rangle$ versus $\langle T_s'\rangle$ for the five AGCM experiments. The approximate linearity is prominent, with a relatively large departure from linearity occuring at large warm SST anomalies. The slope of the linear fit to all data is about 1.76 C C^{-1}, which is slightly higher than values obtained from observational estimates of $\langle \widehat{T}'\rangle$ and , approximately 1.4 C C^{-1} (Figure 1a). The correlation of $\langle \widehat{T}'\rangle$ to $\langle T_s'\rangle$ is 0.91 in Figure 3b. The corresponding correlations of $\langle \widehat{T}'\rangle$ to $\langle T_s'\rangle$ from the MSU and the NCEP/NCAR reanalysis are 0.80 and 0.77, respectively.

2. QTCM Regional SST Anomaly Experiments

Besides analyzing the GCM results, we conduct idealized experiments with the intermediate climate model–QTCM [*Neelin and Zeng*, 2000; *Zeng et al.*, 2000]. In this set of experiments, positive SST anomalies are specified over particular regions of the Pacific, while the rest of oceanic model domain uses climatological SST. The SST anomalies are of various shapes, areas, amplitudes and locations, although all are based on observed SST anomalies during JFM 1998 El Niño in some way. About 40 experiments are conducted, and each of them is an ensemble of 3–10 members with slightly different initial conditions. The spatial averages of tropospheric (850–200 hPa) temperature over the tropical band (25°S–25°N) from the QTCM simulations are displayed as a function of SST forcing for the 40 experiments shown in Figure 4a. The side panels show the simulated tropospheric temperature anomalies for different SST forcings, indicated by the heavy outlines. The panel on the right margin shows the distribution of positive SST anomalies observed during JFM 1998, with outlines indicating areas for regional SST forcings used in the four side panels. The ordinate of Figure 4a is the *spatial integral* of the SST anomaly. It can be seen that there is a remarkable degree of linearity between the trop-

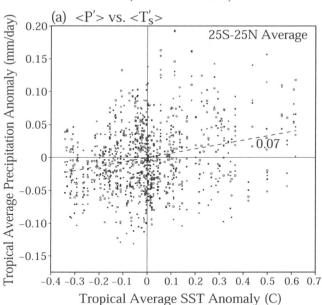

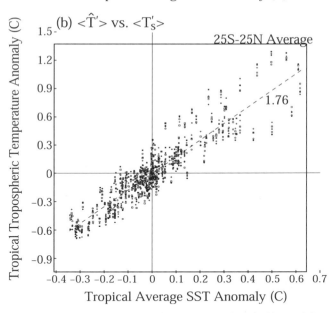

Figure 3. Scatterplot of tropical mean (25°S–25°N) (a) precipitation anomalies (in mm day^{-1}) and (b) tropospheric temperature (850–200 mb) anomalies (in C) as a function of tropical average SST anomalies (in C) from 1982 to 1998 for the five NSIPP AGCM ensemble simulations. Corresponding least-square linear fits are shown by dashed lines, with slopes marked (units of mm day^{-1}C^{-1} and unitless, respectively). From SN03 (their Figure 6).

ical mean tropospheric temperature and SST anomalies, despite the large range of regional size and spatial patterns sampled.

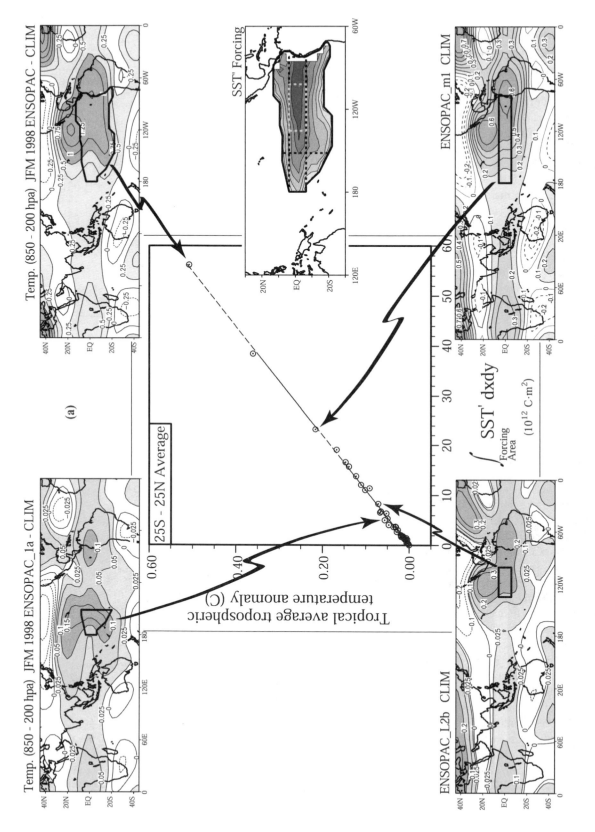

Figure 4. (a) Tropical averaged (25°S–25°N) tropospheric temperature anomalies versus the spatial integral of SST anomaly forcing for a number of experiments with subregions of the 1998 JFM El Niño SST anomaly. The side-panels show examples of the simulated tropospheric temperature anomalies, with heavy outlines indicated the regional SST forcings used. (b) Tropical averaged (25°S–25°N) precipitation anomalies versus the spatial integral of SST anomaly forcing for the same experiments as in (a). The side-panels show examples of the spatial distribution of precipitation anomalies for runs in Figure 4a side panels. Modified after SNM (their Figures 6b, 8b, 9b and 10, plus one case not shown in SNM).

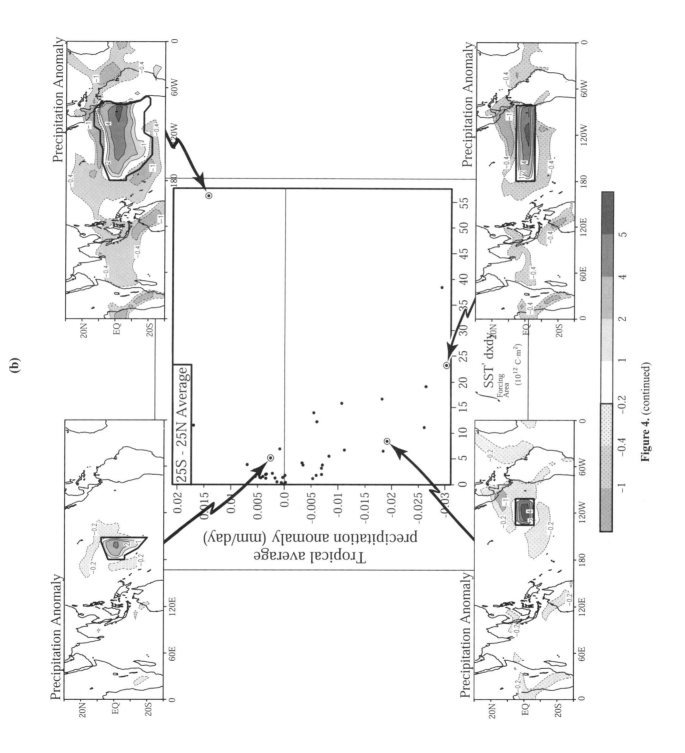

Figure 4. (continued)

On the other hand, tropical mean precipitation anomalies bear no simple relation to SST forcing, nor to the tropospheric temperature anomaly, as shown in Figure 4b. The intra-ensemble standard deviations for $\langle P'\rangle$ are 0.002 to 0.01 mm day^{-1}, depending on the ensemble size. Examples of the spatial distribution of precipitation anomalies are shown in side panels, corresponding to those in Figure 4a. While the spatial patterns are qualitatively similar, the tropical average is set by the sum of positive anomalies local to the SST forcing, against weaker negative anomalies that may occur over a larger area. The near-cancellation can produce either sign in the tropical average.

In terms of spatial distribution of anomalies, the horizontal maps of tropospheric anomalies tend to be similar regardless of the configuration of SST forcing (Figure 4a). The spatial pattern generally resembles the traditional wave response to a localized heat source, although with large longitudinal extent compared to heavily damped simple models [*Gill*, 1980]. In subregional SST anomaly runs (e.g., the upper and lower left panels in Figure 4a), although positive SST anomalies are highly localized, the tropospheric temperature anomalies display broad warming in the whole tropical band. The scope of the warming in both zonal and meridional directions is similar to that for the run with the entire tropical Pacific positive SST anomaly (the upper right panel), indicating the effectiveness of wave dynamics in transporting the heating anomalies. On the other hand, the direct, positive response of precipitation to a positive SST anomaly is much more local and tends to be surrounded by negative anomalies. Although there exist negative precipitation anomalies in far-field, the amplitude is very weak. The tropical averaged precipitation is thus a sum of large positive values in a localized region and weak negative values over a broad area, leaving room for nonlinearity to complicate the response.

Because the scatter in tropical mean precipitation anomalies is common in all datasets and model simulations, it appears that this is not an artifact of observational error or imperfect numerical models. We conjecture that it is an inherent feature of the moist convective response to tropical SST forcing and other factors. In the next section, a simple model based on the equations of the QTCM [*Neelin and Zeng*, 2000] is used to illustrate the dynamics governing the tropical mean precipitation and temperature variations.

3. ANALYTICAL CONSIDERATIONS FOR THE RELATIONSHIP OF $\langle \widehat{T}'\rangle$, $\langle P'\rangle$ AND $\langle T_s'\rangle$

3.1. Derivations

Following *SN03*, $\langle \widehat{T}'\rangle$ and $\langle P'\rangle$ solutions can be derived for a steady-state atmospheric response to SST and mid-latitude forcing. We write the column-averaged temperature and moisture perturbation equations as:

$$D_T' = \widehat{Q}_c' + F_{rad}' + H' \quad (1)$$

$$-D_q' = \widehat{Q}_q' + E', \quad (2)$$

where D_T and $-D_q$ are the horizontal divergences of the vertically-integrated dry static energy and moisture transports by the dynamics. The signs are chosen because the two tend to cancel on the tropical average. Using $\langle\ \rangle$ to denote the averages over the whole tropical band, we define $F_T' \equiv \langle D_T'\rangle$ and $F_q' \equiv \langle D_q'\rangle$, where F_T' and F_q' are the anomalous dry static energy and moisture fluxes across the boundaries (25°S–25°N) between the tropics and mid-latitudes, respectively. Positive values of F_T' imply an export of energy out of the tropics, while for F_q', export of moisture out of the tropics corresponds to negative values. The anomalous moist static energy transport from the tropics, given by $F_T' - F_q'$, is usually less than the individual terms. The atmospheric column radiative heating rate is denoted as F_{rad}. The surface sensible and latent heat fluxes are H and E. The column-averaged convective heating and moisture sink are $\widehat{Q}_c$ and $\widehat{Q}_q$, respectively, and they satisfy

$$-\widehat{Q}_q = \widehat{Q}_c = P \quad (3)$$

where ($\widehat{\ }$) denotes vertical averaging over the troposphere and P is the precipitation rate. The ($'$) indicates perturbations relative to climatological means. All quantities are presented in energy units.

Combining (1) and (2), we obtain the moist static energy perturbation equation

$$F_{rad}' + H' + E' = D_T' - D_q'. \quad (4)$$

Similar to *SNM*, the flux balance can be approximated as linear functions of atmospheric temperature, moisture and SST. For simplicity, we neglect sensible heat flux anomalies because they are relatively small compared to latent heat flux anomalies. The coefficient for radiative flux anomalies due to atmospheric moisture changes can be combined with that for temperature changes because the tropical mean moisture anomalies can be approximately linearly fitted to tropospheric temperature anomalies. So we use only one proportionality parameter ϵ_T here. The evaporation anomalies are parameterized using the conventional bulk-aerodynamic formula. The evaporation anomaly due to changes in wind speed is not easily linearized, so it is denoted as $\tilde{E}$. All other nonlinear effects in the fluxes can be incorporated into $\tilde{E}$. The cloud-radiative forcing amounts to roughly 10% of the surface heat flux forcing and is omitted here. Thus, we have

$$-\epsilon_T \widehat{T}' + \epsilon_{T_s} T_s' + \epsilon_H (\gamma T_s' - q_a') + \tilde{E} = D_T' - D_q'. \quad (5)$$

In (5), $\widehat{T}'$ represents the tropospheric temperature anomalies and T_s' and q_a are sea surface temperature and near-surface air moisture (in units of K). The constants ϵ_T and ϵ_{TS} are proportionality coefficients for the atmospheric radiative heating rate dependence on temperature and SST anomalies, with $\epsilon_T \approx 6$ W m^{-2}K^{-1} and $\epsilon_{T_s} \approx 6$ W m^{-2}K^{-1}. We use $\epsilon_H = \rho_a C_H V_s$, where ρ_a is surface air density and C_H is the drag coefficient. The surface wind speed is denoted as V_s. For a tropical mean wind speed of 5 m s^{-1}, the value of ϵ_H is about 5 W m^{-2}K^{-1}. The surface saturation moisture q_{sat} is a function of SST, with the dependence of $\gamma = \left(\frac{dq_{sat}}{dT}\right)_{T_s}$. Because the value of γ is nearly constant in the normal range of observed SST variations, we use $\gamma \approx 3$ K K^{-1}, corresponding to an SST of 300 K.

When the tropical average of equation (5) is considered, the tropical mean moisture change can be related to the tropical mean tropospheric temperature variations due to the constraint on large-scale circulation by deep convection. In other words, convection vigorously adjusts tropospheric temperature to a value set by boundary layer moist static energy, which is largely determined by surface air moisture. Outside the region of deep convection, tropospheric temperature is not strongly tied to boundary layer moisture. However, the fraction of non-precipitating regions in the tropics is not large, so we try the approximation $\langle q_a' \rangle \approx \gamma n \langle \widehat{T}' \rangle + \xi$ and fit it against model output. The perturbation term ξ indicates the contribution to tropical average moisture change not directly related to tropospheric temperature change, such as that over the non-precipitating regions. Its effect can be incorporated into $\tilde{E}$ in equation (5), so it is omitted hereafter. The parameter n is a scale factor, considering the boundary layer sub-saturation and the ratio of surface air temperature to the tropospheric average temperature. The value of γn is 1.73 for the NSIPP model results, based on the linear regression of $\langle q_a' \rangle$ to $\langle \widehat{T}' \rangle$ for the period of 1982 to 1998.

3.2. Analytical Relationship of $\langle \widehat{T}' \rangle$, $\langle P' \rangle$ and $\langle T_s' \rangle$

Taking the tropical average of equation (5) and rearranging it, SN03 obtained the relationship between tropical average tropospheric temperature and SST anomalies

$$\langle \widehat{T}' \rangle = [(\epsilon_{T_s} + \epsilon_H \gamma) \langle T_s' \rangle - F_T' + F_q' + \langle \tilde{E} \rangle] \\ \times (\epsilon_T + \epsilon_H \gamma n)^{-1}. \quad (6)$$

Substituting equation (6) into (1) or (2), the $\langle P' \rangle$ and $\langle T_s' \rangle$ relation can thus be expressed as

$$\langle P' \rangle = [\epsilon_H \gamma (\epsilon_T - n \epsilon_{T_s}) \langle T_s' \rangle + (\epsilon_H \gamma n) F_T' + \epsilon_T F_q' \\ + \epsilon_T \langle \tilde{E} \rangle](\epsilon_T + \epsilon_H \gamma n)^{-1}. \quad (7)$$

Comparing equations (6) and (7), we notice that both $\langle \widehat{T}' \rangle$ and $\langle P' \rangle$ have an approximately linear relation to $\langle T_s' \rangle$, with super-imposition of nonlinear terms such as transport anomalies and contributions to evaporation anomalies by variations of wind speed. Because these tend not to be simply related to SST, they produce scatter in the relationship to $\langle T_s' \rangle$. The proportionality constant of $\langle \widehat{T}' \rangle$ to $\langle T_s' \rangle$ is approximately 1.4 C C^{-1}, which is close to what is shown in Figure 1a. The dependence of tropical mean precipitation on $\langle T_s' \rangle$ results from competing effects of column radiative cooling and surface emissive warming. The current choice of parameters yields a rate of 0.09 mm day^{-1}C^{-1}. However, it is possible to have a negative slope of $\langle P' \rangle$ versus $\langle T_s' \rangle$ if the value of n varies. For example, n is generally higher when a larger area of non-precipitation regions is involved. This could result in a negative tropical mean precipitation anomaly for a given positive SST anomaly.

Most importantly, the transport anomaly terms in equation (7) play a greater role in producing scatter in $\langle P' \rangle$ compared to the $\langle T_s' \rangle$ term than occurs in equation (6) for $\langle \widehat{T}' \rangle$. Contributing to this, (i) F_T' and F_q' tend to cancel in equation (6), and (ii) the $\langle T_s' \rangle$ term in equation (7) is multiplied by a small time scale, $(\epsilon_T - \eta \epsilon_T)$.

3.3. The Simplest Case

Let us consider a simple case in which only evaporation is taken into account as the dominant driving force for the tropical atmospheric response to $\langle T_s' \rangle$, and sensible heat and radiative flux anomalies are neglected. In this case, setting ϵ_{T_s} and ϵ_T to zero in equations (6) and (7) yields

$$\langle \widehat{T}' \rangle \approx \frac{\langle T_s' \rangle}{n} - (F_T' - F_q' - \langle \tilde{E} \rangle)(\epsilon_H \gamma n)^{-1} \quad (8)$$

$$\langle P' \rangle \approx F_T'. \quad (9)$$

Equivalently, equation (9) can be derived directly from (1), neglecting F_{rad}' and H', and using $F_T' \equiv \langle D_T' \rangle$. This simply states that convective heating balances dry static energy transport. Here, the tropical mean precipitation anomalies are dominated by the midlatitude-tropical dry static energy transport anomalies and are not necessarily related to SST changes, while the tropospheric temperature anomalies still approximately linearly follow the SST anomalies. The scale factor n^{-1} gives the slope of $\langle\widehat{T}'\rangle$ to $\langle T_s'\rangle$, approximately 1.73, close to the slope of the linear fit for the NSIPP ensemble simulations.

Given equation (9), the moisture equation (2) then implies that evaporation must balance the precipitation minus moisture transport, i.e., $\langle E'\rangle = F_T' - F_q'$. We note that this implies $\langle E'\rangle$ should be poorly related to $\langle T_s'\rangle$ on these time scales as well [Robertson et al., 2004, personal communication].

3.4. Testing of the Dominant Balance for $\langle P'\rangle$

According to equation (9), the dry static energy transport anomaly between the tropics and mid-latitudes is a dominant factor in determining tropical mean precipitation variability. Since heat and moisture budgets tend not to be well closed in data such as the NCEP/NCAR reanalysis [Trenberth and Guillemot, 1998; Su and Neelin, 2002], SN03 computed the dry static energy and moisture transport anomalies for one of the NSIPP ensemble experiments. This tests the extent to which equation (9) gives the dominant balance for interannual $\langle P'\rangle$ variations in this model. Figure 5 shows the scatterplot of $\langle P'\rangle$ against F_T', both in units of W m^{-2}. The tropical mean precipitation anomalies follow the dry static energy transport anomalies, with a correlation coefficient of 0.8. The linear regression gives a slope of 0.74, somewhat less than the slope of 1 predicted by the simplest case equation (9). The more general case, equation (7), gives a slope of $(1+\epsilon_T(\epsilon_H \gamma n)^{-1})^{-1} \approx 0.6$ for the parameters given above. The slight scatter in Figure 5 about the F_T' regression line would be due to other terms in equation (7). This confirms that variations of tropics-midlatitude transports can indeed play an important role in the variability of tropical mean precipitation anomalies. This holds for 3-month averages, as are shown here. At much longer time scales, the variations explained by random fluctuations of the transports would be smaller. SN03 also verified that F_T' is not closely related to $\langle T_s'\rangle$, as the correlation coefficient between the two is only –0.1. Our claim that $F_T' - F_q'$ has a smaller effect in producing scatter in $\langle\widehat{T}'\rangle$ than F_T' does in $\langle P'\rangle$ was also verified in SN03 by examining the standard deviations associated with each term.

4. DYNAMICS BEHIND THE $\langle\widehat{T}'\rangle$, $\langle P'\rangle$ AND $\langle T_s'\rangle$ RELATIONS

Considering the analytical explorations in Section 3.1, one notices that convective heating anomalies ($\widehat{Q}_c'$) do not appear explicitly in the moist static energy equation (5). Does this mean that convection is not important for the tropospheric temperature response to SST? The answer is that convection is essential but the convective heating is a by-product. SN03 used a simple convective adjustment scheme and associated fast time scale τ_c to illustrate that the amount of convective heating is not relevant to the tropospheric temperature anomalies because of the small value of τ_c. In this case, tropospheric temperature is held close to a convective quasi-equilibrium (QE) profile whose variations depend primarily on boundary layer moist static energy. Departures from QE are only on the order of τ_c, regardless of the value of the heating. Convection itself is an important player in communicating between boundary layer forcing and deep tropospheric temperature response, but the amount of convective heating is subject to the balance with various cooling terms in the temperature equation. On the tropical average, the dominant cooling term is dry static energy transport anomaly because sensible heat and radiative fluxes are associated with relatively small damping rates. The dry static energy transport, which has a large contribution from mid-latitude transients and correlates poorly with tropical SST anomalies, is thus able to create large scatter in the tropical mean precipitation anomalies with respect to SST anomalies.

Figure 6 illustrates a schematic for the dynamical processes involved in the tropical tropospheric response to SST forc-

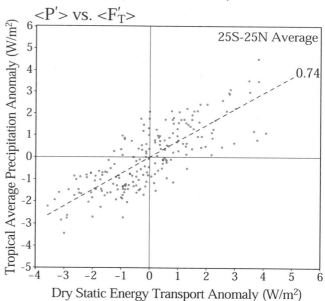

Figure 5. Scatterplot of tropical mean (25°S–25°N) precipitation anomalies (in W m^{-2}) as a function of the anomalies in the export of dry static energy from the tropics by atmospheric dynamical transports across 25°S and 25°N (in W m^{-2}). The results are from 1982 to 1998 for one of the NSIPP AGCM simulations. Corresponding least-square linear fit is shown by the dashed line, with its slope marked. From SN03 (their Figure 7).

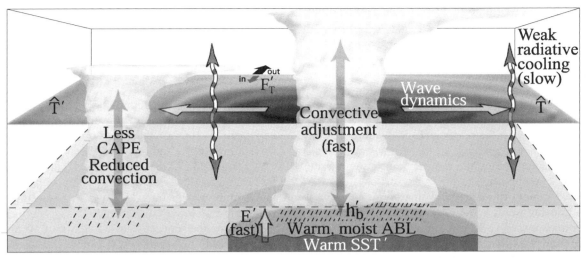

Figure 6. Schematic of the dynamic processes involved in the tropical tropospheric temperature and precipitation response to SST forcing. After SN03 (their Figure 8).

ing. Different arrows are used to indicate these processes, with associated time scales marked in parentheses. The short open arrow within the boundary layer indicates evaporation anomalies (E_s'). During an El Niño, warm SST anomalies increase evaporation, and to a lesser extent, sensible heat fluxes and radiative emission into the atmosphere. Consequently, boundary layer moist static energy h_b' is increased. The time scale associated with the boundary layer adjustment is relatively fast, as represented by the parameter $(\epsilon_H \gamma)^{-1}$. Positive h_b' would tend to yield a larger convective available potential energy (CAPE) for the atmospheric column. Stronger convection occurs over the warm water, indicated by the long solid arrow. It constrains the tropospheric temperature to a value in equilibrium with boundary layer moist static energy. The convective adjustment time scale τ_c is less than a day [Bretherton et al., 2004]. Because convection establishes the equilibrium on a fast time scale, the amount of convective heating does not explicitly determine the tropospheric temperature anomaly. Across the tropics, the tropospheric temperature warming is spread horizontally by wave dynamics, as indicated in the upper plane by two horizontal arrows pointing outwards from the origin of the warming and by gray shading representing the propagating warm anomalies. The corresponding time scale results from a combination of tropical moist and dry Kelvin or Rossby waves propagating across the domain, with phase speeds on the order of 10–50 m s^{-1}. For the tropical band, this yields a time scale of 1–2 weeks to 1–2 months. The wave dynamics does not show up explicitly in our analysis for tropical averages, but is important to the large-scale features of the temperature anomalies. Hence, the tropospheric temperature anomalies are wide-spread and its tropical average is approximately linear with tropical mean SST anomalies. Nonlinearity due to the mid-latitudes transports and the dependence of fluxes on the $\langle T_s'\rangle$ and $\langle \hat{T}'\rangle$ are weak because of the predominance of approximate linear boundary layer flux adjustment. The curly dashed arrows in Figure 6 indicate radiative cooling associated with tropospheric temperature anomalies. It is a relatively slow damping process with a characteristic time scale of 15 days, proportional to ϵ_T^{-1}.

Associated with the warming of tropospheric temperature across the tropics, the CAPE tends to decrease in regions away from the warm SST anomaly, unless atmospheric boundary layer (ABL) moist static energy is able to compensate. Reduction of precipitation thus tends to occur, as shown in Figure 6 by the shorter solid arrow with less dense rainfall. The mechanisms for the relative subsidence involve complicated pathways [Su and Neelin, 2002]. Anomalous moisture or temperature advection, evaporation anomalies due to changes in surface wind speed or air-sea moisture differences and changes in gross moist stability due to moisture variations, etc, can all come into play to balance the adiabatic warming associated with the relative descent. On the tropical average, the dry static energy transport anomalies, F_T', appear to be an important constraint for precipitation anomalies. Suppose there is net dry static energy export from the tropics to the midlatitude, as indicated by the dark slanted arrow in Figure 6. It produces a cooling tendency in the dry static energy equation. Thus the tropical mean convective heating anomalies have to be positive to compensate the cooling effect. This fits in the "normal" picture that tropical mean precipitation anomalies increase as tropical SST warms during an El Niño. On the other hand, dry static energy transport anomalies between the tropics and mid-latitude are presumably substantially due to midlatitude transient eddy activity.

The F_T' can be negative (import into the tropics) even during warm SST years. In this case, the warming effect caused by dry static energy import would tend to reduce CAPE and suppress convection over climatologically convecting regions. This could result in negative precipitation anomalies on the tropical average, which seems counter-intuitive.

The dry static energy transport anomalies F_T' can have contributions from atmospheric internal variability, and contributions associated with the response to SST anomalies. In the latter case, the precipitation scatter persists even in an ensemble average over many atmospheric model runs, as seen in Figure 4b. The important factor in the scatter is that the tropics to midlatitude dry static energy transport is not closely related to tropical average SST anomalies. Nonlinearity in the transport response to SST can thus play a role.

Because the dry static energy transport anomalies bear no simple relation to tropical SST anomalies, the tropical mean precipitation anomalies are scattered in relation to SST. For the tropospheric temperature anomalies, the magnitude of moist static energy transport anomalies is much smaller than that of the forcing from the boundary layer, thus the scatter of $\langle \widehat{T}' \rangle$ is small. For tropical mean precipitation anomalies, the dry static energy transport anomalies are competing against relatively small damping rates associated with radiation, and thus transport anomalies are able to create large scatter in $\langle P' \rangle$ in relation to $\langle T_s' \rangle$.

5. CONCLUSION AND DISCUSSION

On interannual time scales, there are substantial tropospheric temperature and precipitation anomalies associated with the variations of tropical SST. The spatial pattern of tropospheric temperature response tends to be smooth and wide-spread because of the effectiveness of wave dynamics in spreading the temperature anomalies. Because the spatial extent of the temperature response is large even when the SST forcing is highly localized, the tropical mean tropospheric temperature anomaly $\langle \widehat{T}' \rangle$ is highly relevant as a measure of the dynamical response, in addition to being of interest in global warming related studies. It bears an approximately linear relationship to tropical mean SST anomalies. On the other hand, tropical precipitation response to ENSO SST tends to be highly non-uniform in spatial pattern. Positive precipitation anomalies are local to warm SST anomalies. Strong negative precipitation anomalies tend to occur within the convective zones and relatively near the positive SST anomalies, while weaker negative precipitation anomalies occur in the far-field. The near-cancellation of positive and negative precipitation anomalies produces a scattered tropical mean precipitation anomalies in relation to tropical mean SST anomalies. For this reason, the tropical mean precipitation anomaly $\langle P' \rangle$ is a poor measure of tropical hydrological cycle on interannual time scales.

The dynamics governing the different relationships between $\langle \widehat{T}' \rangle$ and $\langle T_s' \rangle$ and between $\langle P' \rangle$ and $\langle T_s' \rangle$ are examined using a simple analytical model. This suggests that the scatter of $\langle P' \rangle$ against $\langle T_s' \rangle$ is associated with dry static energy transport anomalies between the tropics and mid-latitudes, changes in evaporation due to wind speed variations and other nonlinear effects. In contrast, the effects of anomalous midlatitude-tropical transports and nonlinearities are secondary on the $\langle \widehat{T}' \rangle$ relation to $\langle T_s' \rangle$ because of the strong linkage between SST, boundary layer moist static energy, and tropospheric temperature through boundary layer flux adjustment and tropospheric convective adjustment.

Following the convective quasi-equilibrium point of view [Arakawa and Schubert, 1974; Emanuel et al., 1994], the linear relationship between the tropical averaged tropospheric temperature and SST regardless of the value of tropical mean precipitation is not surprising. Convection establishes a link between tropospheric temperature and ABL moist static energy, which in turn is constrained toward SST by surface fluxes. The convective heating anomaly itself is simply a by-product that can be positive or negative depending on other terms in the temperature equation, such as random variations in midlatitude-tropical transports. When convection communicates boundary layer anomalies upward to constrain tropospheric temperature, the convective heating anomaly must react to oppose any process that would tend to cool or warm the troposphere away from balance with SST. We note that the relationship found in this study holds on interannual time scales. On shorter time scales and smaller spatial scales, QE constrain is less strictly satisfied [Brown and Bretherton, 1997; Sobel et al., 2004]. For cases at longer time scales and global spatial scales, such as global warming or paleoclimate applications, it is possible that the dominant cooling effects (for example, the radiative effects) have a rather simple relationship to tropospheric temperature, which in turn would produce a simple precipitation and SST relationship. For the interannual variations examined here, the dominant anomalous cooling process on the tropical average is the tropics-to-midlatitude dry static energy transport anomalies, which have little relationship to SST. Hence there is large scatter in tropical mean precipitation anomalies in relation to SST due to the midlatitude dry static energy transport variations.

The approximately linear relationship of $\langle \widehat{T}' \rangle$ with $\langle T_s' \rangle$ and scattered $\langle P' \rangle$ with $\langle T_s' \rangle$ are results of the combined effects of quasi-equilibrium convective adjustment processes and midlatitude-tropical transports. The contrasting behavior of tropical mean tropospheric temperature and precipitation in relation to SST is a nice illustration of the usefulness of convective QE approaches to understanding the interaction of tropical convection with large-scale dynamics.

Acknowledgments. This work was supported under National Science Foundation Grant ATM-0082529, National Oceanographic and Atmospheric Administration Grant NA16-GP2003 and National Aeronautics and Space Administration Grant NA-GS-9358. The authors thank I. Held, A. H. Sobel, C. S. Bretherton, and M. P. Hoerling for discussions.

REFERENCES

Angell, J. K., Comparison of variations in atmospheric quantities with sea surface temperature variations in the equatorial eastern Pacific, *Mon. Weather Rev.*, *109*, 230–243, 1981.

Arakawa, A. and W. H. Schubert, Interaction of a cumulus cloud ensemble with the large-scale environment, Part I, *J. Atmos. Sci.*, *31*, 674–701, 1974.

Bacmeister, J., P. J. Pegion, S. D. Schubert, and M. J. Suarez, Atlas of Seasonal Means Simulated by the NSIPP 1 Atmospheric GCM, *NASA/TM-2000-104505*, volume 17, 2000.

Bretherton, C. S., M. E. Peters, and L. E. Back, Relationships between water vapor path and precipitation over the tropical oceans, *J. Clim.*,*17*, 1517–1528, 2004.

Dai, A., T. M. L. Wigley, B. A. Boville, J. T. Kiehl, and L. E. Buja, Climates of the twentieth and twenty-first centuries simulated by the NCAR climate system model, *J. Clim.*, *14*, 485–519, 2001.

Emanuel, K. A., J. D. Neelin and C. S. Bretherton, On large-scale circulations in convecting atmospheres, *Q. J. R. Meteorol. Soc.*, *120*, 1111–1143, 1994.

Gill, A. E., Some simple solutions for heat induced tropical circulation, *Q. J. R. Meteorol. Soc.*, *106*, 447–462, 1980.

Holton, J. R., *An introduction to dynamic meteorology* (third edition), Academic Press, San Diego, Calif., 511 pp, 1992.

Horel, J. D., and J. M. Wallace, Planetary-scale atmospheric phenomena associated with the Southern Oscillation, *Mon. Weather Rev.* *109*, 813–829, 1981.

Huffman, G. J. and co-authors, The Global Precipitation Climatology Project (GPCP) combined data set, *Bull. Am. Meteorol. Soc.*, *78*, 5–20, 1997.

Kalnay, E., et al., The NCEP/NCAR 40-year reanalysis project, *Bull. Am. Meteorol. Soc.*, *77*, 437–471, 1996.

Kiladis, G. N., and H. F. Diaz, Global climatic anomalies associated with extremes in the Southern Oscillation, *J. Clim.*, *2*, 1069–1090, 1989.

Kumar, A., and M. P. Hoerling, The nature and causes for the delayed atmospheric response to El Niño, *J. Clim.*, *16*, 1391–1403, 2003.

Mitchell, J. F. B., C. A. Wilson, and W. M. Cunnington, On CO_2 climate sensitivity and model dependence of results, *Q. J. R. Meteorol. Soc.*, *113*, 293–332, 1987.

Neelin, J. D., and N. Zeng, A quasi-equilibrium tropical circulation model—formulation, *J. Atmos. Sci.*, *57*, 1741–1766, 2000.

Newell, R. E., and B. C. Weare, Ocean temperatures and large scale atmospheric variations, *Nature*, *262*, 40–41, 1976.

Pan, Y. H., and A. H. Oort, Global climate variations connected with sea surface temperature anomalies in the eastern equatorial Pacific Ocean for the 1958-73 period, *Mon. Weather Rev.* *111*, 1244–1258, 1983.

Pegion, P. J., S. D. Schubert, and M. J. Suarez, An Assessment of the Predictability of Northern Winter Seasonal Means with the NSIPP 1 AGCM, *NASA/TM-2000-104505*, volume 18, 2000.

Reynolds, R. W., and T. M. Smith, Improved global sea surface temperature analyses using optimum interpolation, *J. Clim.*, *7*, 929–948, 1994.

Ropelewski, C. F. and M. S. Halpert, Global and regional scale precipitation associated with El Niño/Southern Oscillation, *Mon. Weather Rev.*, *115*, 1606–1626, 1987.

Spencer, R. W., Global oceanic precipitation from the MSU during 1979-91 and comparisons to other climatologies, *J. Clim.*, *6*, 1301–1326, 1993.

Spencer, R. W., and J. R. Christy, Precision and Radiosonde Validation of Satellite Grid-point Temperature Anomalies. Part II: A Tropospheric Retrieval and Trends during 1979–90, *J. Clim.*, *5*, 858–866, 1992.

Sobel, A. H., I. M. Held, and C. S. Bretherton, The ENSO signal in tropical tropospheric temperature, *J. Clim.*, *15*, 2702–2706, 2002.

Sobel, A. H., S. E. Yuter, C. S. Bretherton, and G. N. Kiladis, Large-scale meteorology and deep convection during TRMM KWAJEX, *Mon. Weather Rev.*, *132*, 422–444, 2004.

Soden, B. J., The sensitivity of the tropical hydrological cycle to ENSO, *J. Clim.*, *13*, 538–549, 2000.

Su, H., J. D. Neelin, and C. Chou, Tropical teleconnection and local response to SST anomalies during the 1997-1998 El Niño, *J. Geophys. Res.*, *106*, 20,025–20,043, 2001.

Su, H., and J. D. Neelin, Teleconnection mechanisms for tropical Pacific descent anomalies during El Niño, *J. Atmos. Sci.*, *59*, 2682-2700, 2002.

Su, H., J. D. Neelin, and J. E. Meyerson, Sensitivity of tropical tropospheric temperature to sea surface temperature forcing, *J. Clim.*, *8*, 1283-1301, 2003.

Su, H., and J. D. Neelin, The scatter in tropical average precipitation anomalies, *J. Clim.*, *16*, 3966–3977, 2003.

Sun, D. Z., and A. H. Oort, Humidity-temperature relationships in the tropical troposphere, *J. Clim.*, *8*, 1974–1987, 1995.

Trenberth, K. E., and C. J. Guillemot, Evaluation of the atmospheric moisture and hydrological cycle in the NCEP/NCAR reanalysis, *Clim. Dyn.*, *14*, 213–231, 1998.

Wallace, J. M., E. M. Rasmusson, T. P. Mitchell, V. E. Kousky, E. S. Sarachik and H. von Storch, On the structure and evolution of ENSO-related climate variability in the tropical Pacific: Lessons from TOGA, *J. Geophys. Res.*, *103*, 14241–14260, 1998.

Xie, P., and P. A. Arkin, Global precipitation: A 17-year monthly analysis based on gauge observations, satellite estimates and numerical model outputs, *Bull. Am. Meteorol. Soc.*, *78*, 2539–2558, 1997.

Yulaeva, E., and J. M. Wallace, The signature of ENSO in global temperature and precipitation fields derived from the microwave sounding unit, *J. Clim., 7*, 1719-1736, 1994.

Zeng, N., J. D. Neelin, and C. Chou, A quasi-equilibrium tropical circulation model—implementation and simulation, *J. Atmos. Sci., 57*, 1767–1796, 2000.

―――――――――

J. E. Meyerson, J. D. Neelin, and H. Su, Department of Atmospheric Sciences, University of California, Los Angeles, Los Angeles, California 90095-1565. (hobo@atmos.ucla.edu; neelin@atmos.ucla.edu; hui@atmos.ucla.edu)

Convection, Cloud-Radiative Feedbacks and Thermodynamic Ocean Coupling in Simple Models of the Walker Circulation

Adam H. Sobel[1], Christopher S. Bretherton[2], Hezi Gildor[3], and Matthew E. Peters[2]

The authors consider a set of simple models for the divergent component of the tropical atmospheric circulation, and the associated precipitation field. A number of strong simplifying assumptions are made, leaving deep convection and radiation (both simply parameterized) as the key processes in the models. The first case considered is that of fixed sea surface temperature (SST), with an SST gradient imposed across the domain, in the limit of zero convective time scale or strict quasi-equilibrium (SQE). Steady solutions are found for sufficiently weak cloud-radiative feedback. As the parameter controlling the strength of the cloud-radiative feedback is increased, the region of nonzero precipitation shrinks in size and the precipitation grows stronger there. For cloud-radiative feedback stronger than a particular value, steady solutions cannot be found, and numerically obtained time-dependent solutions blow up. In the next case considered, the slab ocean lower boundary condition is used. In this case the behavior is dramatically different. In the steady solutions, the strength of the radiative feedback now has little effect on the size of the precipitating region. The finite convective time scale changes the stability criterion, and when the steady solutions become unstable, rather than blowing up, well-behaved radiative-convective oscillations set in. The mechanism for these oscillations is essentially local radiative-convective or surface flux convective instability, so they can be studied at a single point. Their frequencies are intraseasonal, and some inferences can be drawn from them about how ocean coupling affects the Madden-Julian oscillation. The simplicity of the models allows a relatively complete understanding of their behavior. The sensitivity of the solutions to the key parameters, especially the convective time scale and the cloud radiative feedback parameter, are discussed in some detail.

1. INTRODUCTION

We describe a set of simple models for aspects of tropical climate dynamics. We focus on feedbacks between deep convection, radiative transfer, the quasi-steady, divergent component of the atmospheric circulation, and the energy budget of the ocean mixed layer, all parameterized as simply as possible. Ocean dynamics and planetary rotation are neglected.

In many ways the model solutions are inadequate representations of the real system. Nonetheless, we believe that

[1]Columbia University, New York, New York.
[2]University of Washington, Seattle, Washington.
[3]Weizmann Institute of Science, Rehovot, Israel.

Earth's Climate: The Ocean-Atmosphere Interaction
Geophysical Monograph Series 147
Copyright 2004 by the American Geophysical Union
10.1029/147GM22

they have enough in them that is right that they can help us understand the roles of the various physical processes and some important aspects of the feedbacks between them.

2. UNDERLYING MODEL FORMULATION

2.1. Equations

We use a set of models which differ in detail but share an underlying formulation. Our starting point is the quasi-equilibrium tropical circulation model (QTCM) introduced by Neelin and Zeng [2000] and Zeng et al. [2000]. This model represents the vertical structure of the atmospheric circulation by a set of fixed basis functions, two (one barotropic and one baroclinic) for velocity, and one each for temperature and moisture.

If we exclude the barotropic mode, as we will here, the system is essentially of shallow-water type, though coupled to a moisture equation. It thus has much in common with other reduced models of the tropical atmosphere, but is derived from a different set of assumptions, primarily those behind quasi-equilibrium theory for deep convection as embodied by the Betts-Miller scheme [Betts, 1986] and related theory as reviewed by Emanuel et al. [1994] and Neelin [1997]. The reader is referred to Neelin and Zeng [2000] for more on the model formulation. Examples of climate simulations by the full model (nonlinear, two modes for velocity etc.) can be found in subsequent papers by Neelin's group at UCLA [e.g., Zeng et al., 2000; Su and Neelin, 2002; Su et al., this volume].

In addition to the simplifications assumed in deriving the QTCM itself, we make a number of additional simplifications to obtain the models discussed here:

1. The Weak Temperature Gradient (WTG) approximation (e.g., Sobel and Bretherton, 2000; Sobel et al., 2001; and references therein): we neglect both the tendency and horizontal advection of atmospheric temperature, as both are small terms (compared to diabatic heating and vertical advection of potential temperature) in the tropical free troposphere.

2. As mentioned above, we omit the barotropic mode; the model flow is purely baroclinic.

3. We neglect rotation, and assume slab-symmetry in the meridional direction, leaving horizontal variability only in the longitudinal direction (vertical structure is already determined by the QTCM basis functions). These are models only of the purely divergent, or Walker circulation.

4. We simplify the parameterizations of radiation and surface fluxes, as described below.

With all these assumptions, the atmospheric temperature and moisture equations are:

$$M_s \frac{\partial u}{\partial x} = P - R - \frac{1}{L}\int_0^L (P-R)\,dx, \quad (1)$$

$$\hat{b}\frac{\partial q}{\partial t} + A_q u \frac{\partial q}{\partial x} - M_q \frac{\partial u}{\partial x} = E - P. \quad (2)$$

The temperature and humidity are represented as deviations from fixed reference profiles, T_{ref}, q_{ref}[1]:

$$\tilde{T}(x,p,t) = T_{ref}(p) + a(p)T(t), \quad (3)$$

$$\tilde{q}(x,p,t) = q_{ref}(p) + b(p)q(x,t). \quad (4)$$

T is the temperature, and q is the specific humidity times the latent heat of vaporization of water (L_v, assumed constant) and divided by the heat capacity for air (c_p, also constant), so that q, like T, has units of degrees. x, p, t are longitudinal distance, pressure, and time; $a(p), b(p)$ are the nondimensional basis functions, and tildes denote the total physical fields. We write T as a function of t only because of WTG; in general T can be a function of position also. R is the radiative cooling. P and E are the precipitation and surface evaporation, each multiplied by L_v / c_p. Besides WTG, equation (1) has had the domain average subtracted from it; the integral on the RHS is necessary to insure that the divergence $\partial u/\partial x$ integrates to zero, as necessary to satisfy boundary conditions (either periodic or walls at $x = 0, L$). The domain-average heating goes to change the horizontally uniform temperature, satisfying the domain-averaged heat budget:

$$\hat{a}\frac{\partial T}{\partial t} = \frac{1}{L}\int_0^L (P-R)\,dx, \quad (5)$$

where $x = 0, L$ are the domain boundaries.[2]

M_s and M_q are the dry static stability and gross moisture stratification, which in the earlier versions of the QTCM depended on T and q as:

---

[1] Different profiles were used in different papers reviewed here. Bretherton and Sobel [2002] and Peters and Bretherton [2004] defined T_{ref} and q_{ref} to be solutions in radiative-convective equilibrium, whereas Sobel [2003] and Sobel and Gildor [2003] simply used default profiles from the QTCM. The two choices differ only by constants times $a(p), b(p)$, and have no effect on the total solutions.

[2] Equation (3) in Sobel [2003] should have been identical to equation (5) here, but the former erroneously omitted the factor $1/L$ on RHS. The factor L was eventually set to 1, but this had not yet been done at the point where Sobel's equation (3) was presented.

$$M_q = M_{qr} + M_{qp}q, \qquad (6)$$

$$M_S = M_{Sr} + M_{Sp}T, \qquad (7)$$

with M_{qr}, M_{qp}, M_{Sr}, M_{Sp} constants. Here, for reasons to be discussed below [and see *Yu et al.* 1998], the formula for M_S has been modified to

$$M_S = M_{Sr} + M_{Sp}\max(T, q). \qquad (8)$$

The difference of the dry static stability and the gross moisture stratification is the gross moist stability,

$$M = M_S - M_q, \qquad (9)$$

an important parameter in analyses of the moist static energy budget [*Neelin and Held*, 1987; *Yu et al.*, 1998]. The order-unity constants $\hat{b}$ and A_q (the latter being negative) are derived from the Galerkin projection of the original primitive equations on the QTCM basis functions.

The horizontal velocity also has a basis function, $V(p)$, so

$$\tilde{u}(x, p, t) = V(p)u(x, t),$$

which changes sign once with pressure in the troposphere, so that it is positive in the upper troposphere and negative in the lower.

We do not show a momentum equation, because under our assumptions (particularly WTG, no rotation, and slab-symmetry) one is not necessary. The flow is purely divergent, and the divergence $\partial u/\partial x$ is determined by the heating through equation (1), which states that the heating is balanced by horizontal divergence of dry static energy. u can then be found by integration in x, given boundary conditions. If desired, the momentum equation can be used to infer the pressure perturbations necessary to drive this flow. Hydrostatically, these must be associated with small horizontal temperature perturbations neglected under WTG, e.g., in equation (1). The validity of WTG can be assessed by checking that these diagnosed temperature perturbations are much smaller than the horizontal variations of surface temperature which drive the atmospheric circulation.

2.2. Parameterizations

As in the standard QTCM, the precipitation, or equivalently convective heating, is parameterized by the simplified Betts-Miller scheme:

$$P = \mathcal{H}(q - T)\frac{(q - T)}{\tau_c}, \qquad (10)$$

where $\mathcal{H}$ is the Heaviside function and τ_c is a specified convective time scale.

Under WTG, $q - T$ can be regarded as the excess of a column-averaged relative humidity above a convective threshold at which $q = T$. *Bretherton et al.* [2004] showed that over the tropical oceans, monthly-mean satellite-observed P is an exponentially increasing function of relative humidity. Their results suggest that $\tau_c = 16$ hours was the best match to the above observations. Were one mainly interested in transient variation of convection on daily time scales, rather than (as here) long-term quasi-steady mean behavior, their results would suggest a slightly shorter time scale. Taking this large value for τ_c allows q to be somewhat larger in heavily precipitating regions than in lightly precipitating regions, allowing complexities due to horizontal moisture advection and, possibly, also due to q-induced horizontal variations of gross moist stability (see Section 3.2.4).

A theoretically appealing, though less realistic limit is $\tau_c \to 0$, also known as hard convective adjustment or strict quasi-equilibrium [SQE; *Emanuel et al.*, 1994], in which $q \leq T$ everywhere, with the equality holding in convective regions (where $P > 0$). In this case, we compute precipitation by first computing the divergence from the moist static energy equation, which is obtained by adding the temperature and moisture equations together:

$$M\frac{\partial u}{\partial x} = E - R, \qquad (11)$$

where we have neglected horizontal moisture advection. This is consistent with SQE, under which $q = T$ in convective regions, while T has already been assumed horizontally uniform throughout the domain. Direct substitution in equation (1) then yields

$$P = \frac{M_S}{M}E - R(\frac{M_S}{M} - 1). \qquad (12)$$

We parameterize evaporation by a bulk formula with fixed surface wind speed and exchange coefficient:

$$E = \frac{q_S^* - \tilde{q}(p_S)}{\tau_E} \qquad (13)$$

where q_S^* is the saturation specific humidity at the sea surface temperature, $\tilde{q}(p_S)$ is computed from equation (4) with p_S the surface pressure, and τ_E is an exchange time scale, kept constant, whose value is on the order of 5–15 d. The neglect of surface wind speed variations in particular is a strong limitation of these models, but one which can be relaxed.

The atmospheric radiative cooling is parameterized by

$$R = R^{clr} - rP = \frac{T - T_e}{\tau_R} - rP. \qquad (14)$$

The first term, R^{clr}, represents clear-sky conditions by relaxation to an equilibrium temperature T_e on a time scale τ_R, whose value is on the order of a month. The second term models the greenhouse effect of high clouds as proportional to the precipitation with a coefficient r. This cloud-radiative feedback on the atmosphere is assumed to be all in the longwave band. Shortwave variations are assumed to be significant only at the surface (i.e. constant shortwave absorption in the atmosphere).

The surface forcing, which is defined here as the heating of the ocean mixed layer due to the sum of ocean heat transport and total surface longwave radiative energy flux, is modeled by

$$S = S^{clr} - rP, \qquad (15)$$

where S^{clr} is the clear-sky value and the term rP models the shortwave cloud-radiative feedback (strictly the total surface cloud-radiative forcing, but we assume that to be dominated by the shortwave component); ocean heat transport and surface longwave are assumed constant. Use of the same coefficient r in equations (14) and (15) builds in the assumption that longwave and shortwave effects of high clouds cancel at the top of the atmosphere, as is approximately observed [Ramanathan et al., 1989; Harrison et al., 1990; Kiehl, 1994; Hartmann et al., 2001]. Since variations in atmospheric shortwave absorption are neglected, the shortwave modulation represented by equation (15) only enters the surface energy budget and is therefore irrelevant if a fixed-SST lower boundary condition is used.

2.3. Lower Boundary Condition

We use two different lower boundary conditions: fixed SST, and a dynamically passive ocean mixed layer. In the latter case, the thermodynamic equation for the mixed layer is

$$C \frac{dT_S}{dt} = S - E, \qquad (16)$$

where T_S is the SST and C a bulk heat capacity proportional to the mixed layer depth; C is dimensionless if S and E are expressed in degrees per day and T_S in degrees.

In steady state, equation (16) tells us that the net surface forcing (ocean heat transport plus surface radiative forcing) must balance evaporation:

$$S = E. \qquad (17)$$

Because of the use of the same r in both the longwave and shortwave cloud feedbacks, equation (11) can be combined with equations (17), (14), and (15) to yield

$$M \frac{\partial u}{\partial x} = S^{clr} - R^{clr}, \qquad (18)$$

which can be combined with equation (1) to yield, instead of equation (12),

$$P = (1+r)^{-1} \left[\frac{M_S}{M} S^{clr} - \left(\frac{M_S}{M} - 1 \right) R^{clr} \right]. \qquad (19)$$

In other words, the precipitation is controlled by the *clear-sky* values of the radiative cooling and surface forcing. The cloud-radiative feedback drops out because of the cancellation of the shortwave and longwave components at the top of the atmosphere.

3. SPECIFIC MODELS AND RESULTS

3.1. Specified, Time-Independent, Spatially Varying SST; SQE Model

The first case we consider is that studied by Bretherton and Sobel [2002], in which the sea surface temperature is specified, time-independent, and has a sinusoidal profile in longitude, with specified amplitude Δ and mean value T_{S_0}:

$$T_S = T_{S_0} + \Delta \cos(\pi x/L),$$

here the domain extends from $x = -L$ to L. In the convective scheme, SQE is assumed. The aim here is to vary the magnitude of the SST gradient and the value of r, and to see how these modify the extent of the convective region and the strength of the circulation.

For sufficiently small Δ, convection occurs everywhere, and the equations become linear and analytically tractable. Precipitation and upward motion are stronger over the higher SST region, but increasing the cloud-radiative feedback increases the precipitation contrast across the domain and strengthens the circulation (see Section 4, and Figure 3, from Bretherton and Sobel [2002]).

For Δ larger than about $0.15\ K$, the precipitation becomes zero in part of the domain. This introduces two nonlinearities: one due to the nonnegativity of precipitation, and one due to horizontal moisture advection, as the humidity field in the nonprecipitating region is not constrained by SQE (as it is in the precipitating region) so that horizontal gradients develop in response to the SST gradient. Bretherton and Sobel [2002] obtained steady solutions directly, by treating the precipitating and nonprecipitating regions separately and matching them at a boundary whose position is determined, together with the tropospheric temperature, as part of the solution.

Figure 1 shows the solutions for $\Delta SST = 2K$, for $r = 0$ and $r = 0.2$. As in the linear case, the cloud-radiative feedback increases the contrast between the regions of low and high SST, now shrinking the precipitating region and intensifying the precipitation and vertical motion there. Solutions with the horizontal moisture advection term omitted are also shown, and we can see that this term also plays an important role in setting the horizontal extent of the precipitating region. Since the low-level flow is convergent into that region, it flows from the region of lower humidity, and thus dry air advection suppresses convection near the boundary and pushes the boundary back to higher SST. This effect is not as strong as that of the cloud-radiative feedback for $r = 0.2$.

As r is increased, the convective region shrinks and intensifies further until eventually, for large enough r, no steady solution can be obtained. This indicates the existence of a radiative-convective instability, which can be understood by simple analysis of the eqations. Some manipulation of equations (1) and (11) by using equation (14) yields:

$$M_{eff} \frac{\partial u}{\partial x} = E - R^{clr}, \qquad (20)$$

where we have defined the effective gross moist stability [Su and Neelin, 2002; Bretherton and Sobel, 2002].

$$M_{eff} \equiv \frac{M - rM_q}{1 + r}. \qquad (21)$$

As with the definition of the standard gross moist stability,

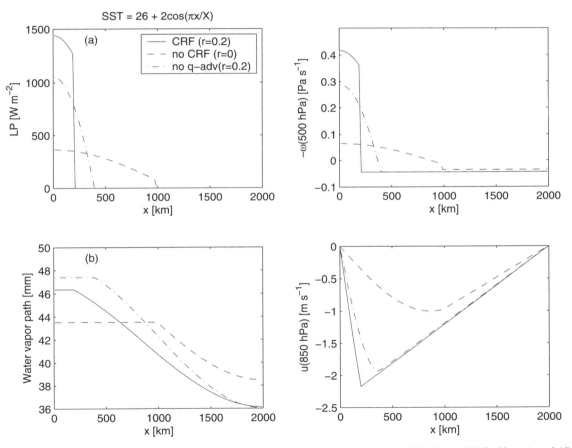

Figure 1. (a) Precipitation, (b) water vapor path, (c) pressure vertical velocity (ω) at 500 hPa, multiplied by -1, and (d) zonal velocity at 850 hPa, for $\Delta SST = 2$ K. In the latter two panels, the full two-dimensional fields are obtained by multiplying by the relevant vertical structure functions [see Section 2.1 and Neelin and Zeng (2000), and note these variables were displayed before multiplication by basis functions in the corresponding figure in Bretherton and Sobel (2002)]. Solutions are shown with and without cloud-radiative forcing (CRF). A solution with CRF but without horizontal moisture advection is also shown. Figure adapted from Bretherton and Sobel (2002).

M_{eff} tells us the strength of the mass circulation through a convective region given a rate of moist static energy supply through the vertical boundaries, but where the latter supply comes from the surface fluxes minus total tropospheric radiative cooling in the case of M, now it is defined as the surface fluxes minus only the clear-sky component of the radiative cooling, with the cloud-radiative feedback treated as part of the response. For negative M_{eff}, the system can sustain upward motion ($\partial u/\partial x > 0$) even if the clear-sky component of the radiative cooling removes more moist static energy than the surface fluxes add, because the cloud-radiative feedback can more than compensate for the difference. This means that convection can occur spontaneously even where the boundary conditions do not favor it, in the sense that descent and a lack of convection would also be an acceptable steady solution. This is an indication of instability. $M_{eff} = 0$ corresponds to the bifurcation point at which the instability sets in, past which no steady solution can be obtained.

The very strong response of this model to the cloud-radiative feedback turns out to be misleading in an important way when we consider the coupled atmosphere–ocean system. In the fixed-SST model described above, the cloud-radiative feedback acts as a net energy gain to the system, whose magnitude is determined as part of the solution. Observations show, however, that the net effect of high tropical clouds (the only ones we can presume to model as related directly to precipitation) is approximately zero at the top of the atmosphere (TOA), because the shortwave albedo effect approximately balances the longwave greenhouse effect. The dominant effect of the shortwave albedo of high clouds is to reduce the incident shortwave flux at the surface, cooling the ocean. This important effect is neglected in the fixed-SST model, but is considered below.

The momentum equation (not used above) can be used to infer the small spatially-varying temperature perturbation which is neglected under WTG. In these models, the amplitude of this perturbation grows as the domain size does, so that WTG is a better approximation for small domains (exactly how small depends on how strong friction is). This result comes about because of the single vertical mode. With a viscous boundary layer in which WTG is relaxed, coupled to an inviscid free troposphere, WTG in fact holds better for larger spatial domains than for smaller ones [Shaevitz and Sobel, 2004].

3.2. Model With a Dynamically Passive Ocean Mixed Layer

3.2.1. Results. Sobel [2003] and Peters and Bretherton [2004] considered the system including an ocean mixed layer as represented by equation (16). The model is then forced not by a given SST distribution, but instead by a given x-dependent surface forcing, $S^{clr}(x)$. Other differences from the study of Bretherton and Sobel [2002] include generalization from SQE to a finite τ_c, and a linear, rather than sinusoidal forcing[3]:

$$S^{clr} = S_0 + \Delta S x. \quad (22)$$

For sufficiently small forcing gradient, this system also has a linear regime in which precipitation occurs throughout the entire domain. We focus here on the nonlinear regime, in which both Sobel [2003] and Peters and Bretherton [2004] obtained solutions by numerical integration of the time-dependent equations, as opposed to the steady algorithm used by Bretherton and Sobel [2002].

Figure 2, taken from Sobel [2003], shows steady solutions for SST, humidity, evaporation, precipitation, and radiative cooling, obtained for $r = 0.2$, $\tau_c = 0.2$d, $S_0 = 130$W m^{-2}, and $\Delta S = 100$W m^{-2}. We see a qualitatively similar structure in q as in the fixed-SST model, with a relatively large gradient in the nonprecipitating region and a much smaller gradient in the precipitating region. The gradient is not quite zero in the latter due to the finiteness of τ_c, which allows q to differ slightly from T, and to track the SST to some extent. The SST is strongly controlled by the shortwave cloud feedback. As the surface forcing creates stronger precipitation at larger x, the shortwave albedo effect intensifies and prevents the SST from warming in response to the surface forcing increase. This negative feedback, a local version of the "thermostat" [Ramanathan et al., 1989; Waliser 1996], turns the precipitating region into a "warm pool" of nearly homogenized SST.

3.2.2. Analytic solution for SQE limit. Peters and Bretherton [2004] obtained analytical solutions for the coupled model, in the limit $\tau_c \to 0$ (SQE), and with a linearization of the Clausius-Clapeyron equation so that

$$qs(T) = qs(T_{S_0}) + \gamma T',$$

with T_{S_0} the SST in radiative-convective equilibrium and T' the departure from T_{S_0}. With these assumptions, the atmospheric temperature is equal to the value it would have at radiative-convective equilibrium, with the surface forcing, S, set to its mean value [S_0, in Peters and Bretherton's notation; again in Sobel's notation S_0 is the minimum value of the forcing, see equation (22)] over the domain. The horizontal distribution of P is obtained from equation (19), by noting that R^{clr} is uniform and equal to R_0^{clr}. By also invoking domain-integrated energy balance, we deduce that the fraction of the domain in which $P > 0$ is

---

[3] Sobel [2003] used the domain $x = [0,1]$, so that $S = S_0$ at $x = 0$, while Peters and Bretherton [2004] defined the domain symmetrically, so that the domain average of S is equal to S_0.

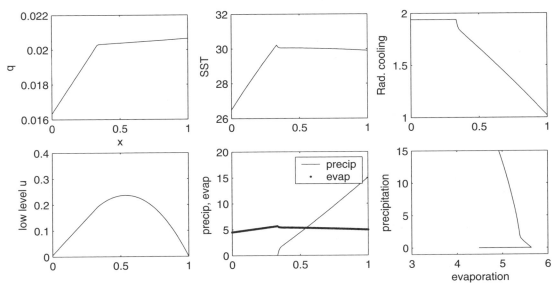

Figure 2. Steady solutions of the model coupled to an ocean mixed layer, with the cloud-radiative feedback parameter $r = 0.2$ and convective time scale $\tau_c = 1$ d; see text for further details. From left to right and top to bottom, the first five panels show surface specific humidity (kg kg^{-1}), SST (°C), radiative cooling rate (K d^{-1}), lower level wind velocity [arbitrary units; note upper level wind was shown in Sobel (2003)], and precipitation and evaporation (mm d^{-1}), vs. the horizontal coordinate x, while the last (bottom right) panel shows precipitation on the y axis and evaporation on the x axis. Figure adapted from Sobel (2003).

$$f_c = \max\left(2 \frac{R_0^{clr} M_0}{\Delta S M_{S_0}}, 1\right),$$

where ΔS is the variation in S across the domain. In SQE, $q = T$ in the convecting region, so that M_q and thus M are determined once T is. The subscript "0" refers to the RCE value. The warm pool SST is

$$T_S = \frac{\Delta S \tau_E}{\gamma} \frac{M_{eff}}{M_0} x,$$

while the cold pool SST is

$$T_S = \frac{\tau_E}{\gamma}\left[rE_0 + \Delta Sx + \frac{b_S}{\tau_E} \frac{M_0}{M_{qp}}(1 - x/f_c L)\right], \quad (23)$$

where M_{qp} is the proportionality constant relating M_q to q, and b_S is the surface value of $b(p)$. The precipitation is

$$P = \max\left(P_0 + \frac{\Delta S(x)}{1+r} \frac{M_{S_0}}{M_0}, 0\right) \quad (24)$$

An interesting feature of this solution is that the convective area fraction f_c is independent of the cloud-radiative feedback parameter r, in stark distinction to the fixed-SST case. This is because the precipitation is now controlled by the clear-sky surface forcing, as expressed by equation (19). Numerical results (not shown) show that even when r is made large enough that instability sets in and large oscillations occur in the convective region (see below), the spatial structure of the time-mean precipitation field, including the value of f_c, changes extremely little from the steady-state values, as long as the simulation is carried out long enough for the statistics to become stationary. Cloud-radiation feedback flattens the SST gradient in the convective region instead of reducing the spatial extent of the convective region.

3.2.3. Dependence on r. The precipitation, quite unlike the fixed-SST case, is actually slightly weakened by the cloud radiative feedbacks, as can be seen clearly in the analytical expression equation (24) obtained in the SQE limit. The cloud-radiative feedbacks act to reduce S (which in steady state is equal to E) in convective regions at the same time that they decrease R by the same amount, so that the net forcing of the divergent circulation $E - R$ [see equation (11)] and therefore the circulation strength is unchanged. In steady state $E = S$, and then using equations (14) and (15) we have

$$E - R = S^{clr} - R^{clr}.$$

However, the decrease in E results in a decrease in precipitation, since the precipitation must equal evaporation plus moisture convergence, and moisture convergence is the product of the circulation strength and the gross moisture stratification M_q. M_q is a function of q only, and hence is fixed in convective regions in the SQE limit since then $q = T$. The precipitation itself can also be found explicitly in terms of the clear-sky forcings, e.g. by first obtaining P as shown in equation (25) below, and then substituting $E = S = S_0 - rP$.

3.2.4. Dependence on τ_c.

Much of the discussion in this paper, including the analysis immediately above, concerns the SQE limit $\tau_c \to 0$. As previously noted, observations have shown suitable values of τ_c to be approximately 16 hours. This is smaller than any other time scale in the problem, but not by so much that deviations from SQE can be assumed unimportant a priori. It is reasonable to ask how the solutions may change as τ_c is increased towards realistic values.

As τ_c is increased, q becomes increasingly horizontally non-uniform, in proportion to $\tau_c P$, in the convecting region. Whether these variations in q have other major effects on the simulation depends on how the gross moist stability depends on q. If equation (7) is used for M_S, then increases in q, by increasing M_q while M_S does not change, will lead to decreases in M. Consideration of equation (11) shows that this will lead to increases in upward motion, and hence precipitation, in the convective region. If τ_c is made large enough, M can actually become negative, leading to a strong instability which will cause the solution to blow up.

However, we now believe that this behavior may be unrealistic. Yu et al. [1998] showed that increasing low-level moisture (here q) need not always decrease M. Yu et al. argued that there is another effect, due to variations in the height attained by convection. The higher the convective updrafts go, the larger M_S should become, since air then outflows with higher values of moist static energy. Yu et al. related the height of convection to the degree of instability in the column, which is mostly dependent on near-surface moisture. These considerations have led to the formulation of equation (8). With this formulation (which was not used by Sobel [2003]; it was not needed since he used the small value $\tau_c = 0.2$ d), M becomes nearly independent of τ_c in the convective region, and as a result the precipitation and other variables exhibit only very small changes as τ_c is increased up to values on the order of 1 d; the changes are sufficiently small that they are not shown in Figure 2. Thus, in this regime SQE is a good approximation for the purposes of understanding many properties of the steady solutions. On the other hand, when we consider the stability of those solutions to time-dependent perturbations, the picture changes considerably and the model behavior becomes, even within this range $\tau_c \leq 1$ d, quite sensitive to variations in τ_c, as will be discussed in Section 3.2.6.

3.2.5. Coexistence of evaporation minimum and precipitation maximum.

In presenting the results from the coupled model, Sobel [2003] emphasized the coexistence of a local minimum in evaporation and maximum in precipitation at $x = 1$. Such a feature is observed, broadly speaking, in the western Pacific warm pool, where climatological evaporation is small although precipitation is large. The moist static energy budget in this situation requires some explanation. If radiative cooling and M were both constant, equation (12) tells us that features in P would have to mimic those in E, albeit with larger amplitude. Leaving aside the possibility of significant variations in M for the moment, we can see that variations in R cannot change this feature of the solutions, if we parameterize R as a decreasing function of P, as in equation (14), and neglect horizontal moisture advection (consistent with the SQE limit $\tau_c \to 0$). In that case we can rewrite equation (12) as

$$P = \frac{\frac{M_S}{M_{eff}}(E - R^{clr}) + R^{clr}}{1 + r}. \qquad (25)$$

As long as $M_{eff} > 0$, equation (25) means that if $M_{eff} > 0$ in a region where $P > 0$, gradients in P and E will have the same sign, $\partial P/\partial E > 0$, so one cannot have a minimum where the other has a maximum.

The coexistent evaporation minimum and precipitation maximum therefore require $M_{eff} < 0$. We found in the fixed-SST model that this condition led to inability to find a solution (strictly, a steady solution, but in fact not even a well-behaved time-dependent solution is obtainable in this case). This was true because the fixed-SST model assumed SQE. If instead τ_c is finite, the stability criterion for the system is changed, as discussed in more detail in Section 3.3 and in Sobel and Gildor [2003]. The system is stabilized, and stable steady solutions can be obtained for M_{eff} slightly negative. Horizontal moisture advection in general becomes nonzero for finite τ_c, rendering equation (25) not strictly correct. However, for reasonably small τ_c, this is a small effect; the primary stabilization results from a negative feedback between q and E, and negative M_{eff} is still required for a stable steady solution with a coexistent evaporation minimum and precipitation maximum.

Obtaining the coexistent evaporation minimum and precipitation maximum in the coupled system, as in Figure 2, is fairly easy, as long as τ_c is finite, r is large enough to render $M_{eff} < 0$, but not too large (in which case the steady solution becomes unstable to time-dependent disturbances, discussed below). For reasonable parameters there is a sizeable range of r values in which these conditions hold. $S^{clr} - R^{clr}$ is maximum at $x = 1$, which leads to the precipitation maximum there; at the same time S, and thus in steady state also E, is minimized there by the cloud-radiative feedbacks. These same feedbacks flatten the SST in the convective region.

Obtaining the coexistent evaporation minimum and precipitation maximum in the fixed-SST system with finite τ_c is possible, but difficult, because the negative feedbacks that constrain E and the SST in the coupled system are not present. The SST has to be prescribed to a profile very close to one obtained as a solution (with the desired coexistent E minimum and P maximum) from the coupled model. In the absence of the coupled feedbacks, slight deviations will eliminate either the E minimum or the P maximum.

The physical argument this leads to is that both the evaporation minimum and precipitation maximum result from the maximum in the clear-sky surface forcing in the warm pool (which results from there being a longitudinal minimum in the divergence of ocean heat transport and an equatorial maximum of annually-averaged clear-sky insolation), cloud-radiative feedbacks and the surface energy budget (i.e., thermal ocean coupling). The surface wind speed is viewed as irrelevant because it can be viewed as determining E only for fixed SST. Once the surface energy budget must balance, we have $S = E$ in steady state, so E is constrained independently of the wind speed.

While this argument is an improvement on any based on a fixed SST, there are still (at least) two weaknesses in it. One is that there may be spatial variations in M, which if large enough could allow P to differ in qualitative structure from $E - R$. We do not address this further here, but note that the latest estimates do not suggest large variations in M [Yu et al., 1998].

The more serious problem may be our neglect of potential atmospheric feedbacks on ocean dynamics. In our model, ocean dynamics are encapsulated by specifying $S^{clr}(x)$, which can include a contribution due to ocean heat flux divergence. $S^{clr}(x)$ is treated as an external control which does not interact with the predicted atmospheric circulation [Seager et al., 1988; Gent, 1991]. On the contrary, Seager et al. [2003] show, in an ocean general circulation model (GCM) coupled to an atmospheric mixed layer model, that interactive ocean heat transport plays an important role in producing the evaporation minimum in the equatorial warm pool.

3.2.6. Stability and time-dependence. For the fixed-SST, SQE model, as mentioned above $M_{eff} = 0$ delineates the critical value of r at which the steady solution is marginally stable. Once τ_c becomes finite, whether the system is coupled or not, the stability criterion is modified and stable solutions can persist for slightly negative M_{eff}. However, qualitatively the situation is the same as in SQE, in that for sufficiently large r, the solution becomes unstable to time-dependent disturbances. For the fixed-SST case, numerical integrations show that these time-dependent disturbances are not well behaved, but that the precipitation blows up in amplitude as the precipitating region shrinks to the grid scale. For the coupled case, well-behaved oscillations develop, first at $x = 1$, and then for larger r, increasing their domain of influence until soon the entire precipitating region exhibits time-dependent behavior. These oscillatory solutions were mentioned but not explicitly presented by Sobel [2003]. Here we show a Hovmoeller plot of the precipitation field from such a solution, in Figure 3. The solution shown in Figure 3 thus uses a model and parameter values identical to those used by Sobel [2003] to produce Figure 2, except:

• $\tau_c = 0.7$ d rather than 2 d, to be consistent with Bretherton et al. [2004] and the mixed layer depth is 20 m (Sobel used 1 m, without discussing the choice, since only steady solutions were shown)
• r is raised to 0.25 to render the steady solution unstable and generate time-dependent behavior, and
• equation (8) is used, for reasons discussed in Section 2.1.

The figure shows eastward-propagating oscillations in rainfall in the eastern part of the domain. The period is on the order of 40 d. We view the propagation dynamics of these disturbances as unworthy of study, since the system as formulated contains none of the standard equatorial wave modes (Kelvin, Rossby etc.), which if present would presumably modify the propagation dynamics qualitatively as well as quantitatively. The fact that the propagation is eastward is not

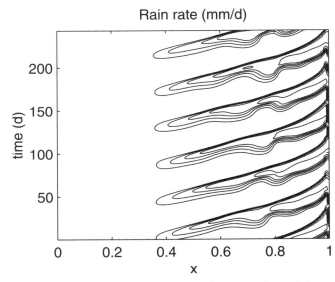

Figure 3. The precipitation rate (mm d^{-1}; contour interval 5 mm d^{-1}), as a function of longitude (x) and time, for a time-dependent solution of the 1D model coupled to an ocean mixed layer. The model and parameters are as in Sobel (2003), except that the mixed layer depth is 20 m, convective time scale $\tau_c = 0.7$ d, and cloud-radiative feedback parameter $r = 0.25$, and the formulation (8) rather than (7) is used to compute the dry static stability.

deemed to be significant, and in fact disturbances sometimes propagate westward for other parameter values, particularly right near $x = 1$. The phase speed of the disturbances is particularly irrelevant since its dimensional value scales linearly with the domain size, for the same reason that u does—namely, that the domain size drops out of the dynamical equations of this model due to WTG!

What is of most interest is the dynamics of the instability which generates the disturbances and selects the time scale. Since horizontal advection has only a weak effect on the disturbances due to the weak moisture gradient in the convective region, it seems reasonable to simplify the problem still further, to a single horizontal location, in order to study them.

3.3. Single-Column, Time-Dependent Model

Sobel and Gildor [2003] studied the equations from the preceding section with horizontal moisture advection omitted, in which case the dynamics of a single horizontal location can be treated independently of others and we have a zero-dimensional coupled model:

$$M_S \delta = P - R, \quad (26)$$

$$\hat{b}\frac{\partial q}{\partial t} - M_q \delta = E - P, \quad (27)$$

$$\frac{\partial T_S}{\partial t} = S - E. \quad (28)$$

Sobel and Gildor viewed this as a model for variations of SST on an intraseasonal time scale, in an idealized context where the horizontal propagation associated with the Madden-Julian Oscillation (MJO) is ignored. The best observational context for thinking about intraseasonal SST variability without directly considering the MJO is provided by Waliser [1996], who looked at the dynamics of "hot spots", or locations where the SST becomes greater than 29.5°C for at least a month.

In this model, the parameter r in equation (15) may be viewed as representing not only shortwave cloud-radiative feedbacks associated with precipitation, but also the feedback of increased surface fluxes associated with the development of 'cold pools' and convectively enhanced gustiness where there is more precipitation. Under the assumption that the total cloud-radiative feedback vanishes at TOA, it acts just like a surface flux, simply moving energy between ocean and atmosphere. Additionally, variations in surface turbulent fluxes (mainly evaporation) and radiative fluxes are found to be nearly in phase on intraseasonal time scales. Because of these two facts, both feedbacks can be empirically modeled by the single parameterization of equation (15).

3.3.1. Linear results. Figure 4 shows growth rate and frequency as a function of parameters for the system obtained by linearizing equations (26)–(28). The linear equations, and their closed-form solutions from which these plots were generated, can be found in Sobel and Gildor [2003]. Sensitivity is shown to the two most important parameters, r and τ_c, and also to the mixed layer depth. The blacked-out regions are those in which the linear frequency is zero, that is, the linear calculation predicts pure, non-oscillatory growth or decay. The growth rate is positive only for large r and small τ_c. There is only a fairly narrow region around the marginal stability curve, in $r - \tau_c$ space, in which the linear frequency is nonzero. In that range, the frequency is intraseasonal to subannual (0.1 corresponds to $10 \times 2\pi$ or around 60 d). In the SQE limit $\tau_c \to 0$, the range of r supporting oscillations shrinks to zero width, and the marginal stability criterion for growing (but now non-oscillatory) solutions reduces to $M_{eff} = 0$. This again emphasizes that while SQE is a useful guide to the finite-τ_c behavior when $M_{eff} > 0$ and steady-state solutions are obtained, it is not so useful when M_{eff} becomes negative.

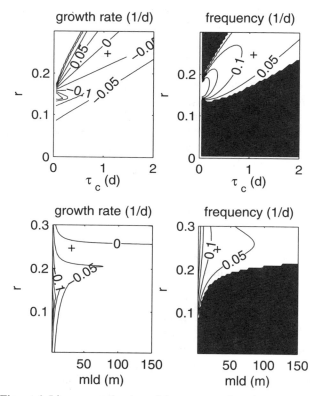

Figure 4. Linear growth rate and frequency as functions of cloud-radiative feedback parameter r and convective time scale (top) and mixed-layer depth (bottom). In the frequency plots, regions where the frequency is zero are filled. Plus symbols indicate the default parameter values given in Table 1 of Sobel and Gildor (2003), from which this figure is adapted.

3.3.2. Nonlinear results. Figure 5 shows a nonlinear simulation in the regime in which the system is linearly unstable. The precipitation P, evaporation E, radiative cooling R, surface shortwave flux S, SST, and surface humidity q_s, are shown as functions of time. Note that the evaporation is that computed directly from equation (13); it does not explicitly include the "extra" evaporation variations which we might take to be included with the cloud-radiative feedback.

We see vaguely square-wave type variations in precipitation, R and S. Detailed features, such as the particular shape of the precipitation pulses and their duration relative to the duration of nonprecipitating periods, are dependent on parameters. The SST has more sawtooth-like variations. The oscillation overall has characteristics of a recharge-discharge oscillation. The "discharge" is the flux of moist static energy from ocean to atmosphere (by radiative processes and surface fluxes), and horizontal export of that moist static energy in the atmosphere associated with the precipitating phase, and the "recharge" is the buildup of energy in the ocean associated with the suppressed phase, when the atmosphere is descending and the SST is increasing.

3.3.3. Dependence on mixed layer depth. An interesting prediction of this model is that the growth rate for the instability increases (weakly) with mixed layer depth. This is contrary to what we might expect, since with larger mixed layer depth, the SST can vary less and we might expect this to lead to less overall variability, but viewing the variability as a recharge-discharge oscillation helps us to understand this, since the mixed layer is the capacitor for the system; without storage capacity there can be no recharge-discharge.

Watterson [2002] indeed found, in an atmospheric GCM coupled to a constant-depth ocean mixed layer, that intraseasonal variability tended to decrease with mixed layer depth. This contradicts the results above, if we view the growth rate as a proxy for the variance we expect in a nonlinear simulation. To explain this, Sobel and Gildor [2003] forced the linear model, in a weakly stable regime, with an imposed atmospheric oscillation in the heating $(P - R)$, of intraseasonal frequency, to represent an atmospheric MJO which would exist in the absence of coupling. The model response showed a peak amplitude near 10–20 m (the value of mixed layer depth which leads to a linear frequency near the forcing value) with decay at larger depths, as found by Watterson, but also decay at smaller depths, not investigated by Watterson. More recent simulations with an atmospheric general circulation model coupled to a constant-depth ocean mixed layer, bear out the simple model prediction [Maloney and Sobel, 2004]. These simulations show a dramatic decrease in the magnitude of precipitation variability as mixed layer depth is reduced below about 10 meters.

4. CONCLUSIONS

We have described a set of simple models for aspects of the tropical climate. These models represent the divergent tropical circulation as a product of feedbacks between SST, surface fluxes, deep convection, and radiation, with the humidity field being the primary prognostic atmospheric variable modulating these feedbacks. The basic model ingredients are the Neelin-Zeng QTCM equations, the weak temperature gradient approximation, the neglect of rotation and the assumption of slab-symmetry in latitude so that only the along-equatorial circulation is modeled. The strict quasi-equilibrium (SQE) approximation may also be used, either in the models themselves or in analytical work aimed at understanding the results when the convective time scale, τ_c, is finite. SQE is a good approximation in the sense that many features of the steady solutions are insensitive to increases in the convective time scale from zero up to at least realistic values on the order of ~1 d. The stability of the steady solutions to time dependent perturbations, on the other hand, is quite sensitive to variations in τ_c in this range, so that SQE may under some circumstances be quite inaccurate for understanding the stability and possible time-dependent behavior of finite τ_c solutions.

When SST is fixed, the cloud-radiative feedback is a net source of energy to the system, and strongly shrinks the convective region in steady state. For strong enough feedback, the convective region shrinks to a single point and no steady solution can be found. In fact, the system becomes badly behaved, and numerical simulations suggest that even time-dependent simulations blow up after a finite time.

When coupling to an ocean mixed layer is added, the cloud-radiative feedback has no effect on the size of the convective region, the latter being controlled by the clear-sky surface energy budget. The model in this configuration is able to robustly produce a coexistent evaporation minimum and precipitation maximum, as are observed, but perhaps for reasons which are only partly correct. More recent work suggests that feedbacks between ocean heat transport and the atmospheric circulation, which are neglected in our model, plays a large role in creating this feature.

When the cloud-radiative feedback parameter is increased past a threshold value, the coupled model becomes unstable to free oscillations of intraseasonal to subannual period. The strength of the instability is most sensitive to the cloud-radiative feedback parameter (which can also be thought of as representing the surface flux feedback) and the convective time scale. The onset of instability is well captured by a linear analysis, but the oscillations in the full system tend to be nonlinear, with a recharge-discharge character and periods of zero precipitation alternating with rainy periods. Although the growth rate of the instabilities increases (slowly) with mixed

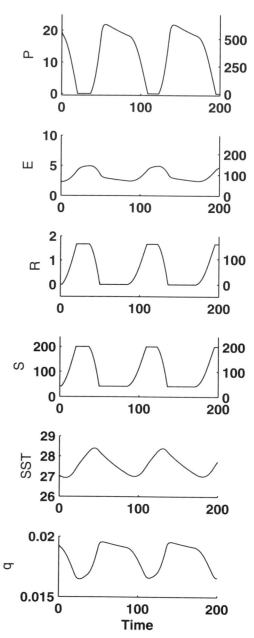

Figure 5. A nonlinear solution to the Sobel-Gildor model with $r = 0.3$ and $\tau_c = 1$ d. From top to bottom: Precipitation (left axis mm d^{-1}, right axis W m^{-2}), evaporation (left axis mm d^{-1}, right axis W m^{-2}), radiative cooling (left axis K d^{-1}, right axis W m^{-2}), net flux of short wave and long wave at the ocean surface (both axes (W m^{-2}), sea surface temperature (°C) and specific humidity (kg kg^{-1}). Taken from Sobel and Gildor (2004).

layer depth, in a forced calculation in a weakly stable parameter regime the amplitude of the response has a peak near 10–20 m, decreasing for both larger and smaller mixed layer depths. This dependence of intraseasonal variability on mixed layer depth has also been found in atmospheric GCMs coupled to slab ocean mixed layer models.

Acknowledgments. We thank Isaac Held, David Neelin and Richard Seager for discussions. AHS acknowledges support from NSF grant DMS-01-39830 and a fellowship from the David and Lucile Packard foundation. CB and MP acknowledge support from NSF grant DMS-0139794. Any opinions, findings, conclusions or recommendations expressed in this paper are those of the authors and do not necessarily reflect the views of the sponsors. Figures 1, 2, 4, and 5 were adapted from earlier publications in journals of the American Meteorological Society, which holds copyright to these figures.

REFERENCES

Betts, A. K., A new convective adjustment scheme.1. observational and theoretical basis, Q. J. R. Meteorol. Soc., 112, 677–691, 1986.

Bretherton, C. S., and A. H. Sobel, A simple model of a convectively-coupled Walker circulation using the weak temperature gradient approximation, J. Clim., 15, 2907–2920, 2002.

Bretherton, C. S., M. E. Peters, and L. Back, Relationships between water vapor path and precipitation over the tropical oceans, J. Clim., 17, 1517–1528, 2004.

Emanuel, K. A., J. D. Neelin, and C. S. Bretherton, On large-scale circulations in convecting atmospheres, Q. J. R. Meteorol. Soc., 120, 1111–1143, 1994.

Gent, P. R., The heat budget of the TOGA COARE domain in an ocean model, J. Geophys. Res., 96, 3323–3330, 1991.

Harrison, E. F., P. Minnis, B. R. Barkstrom, V. Ramanathan, R. D. Cess, and G. G. Gibson, Seasonal-variation of cloud radiative forcing derived from the earth radiation budget experiment, J. Geophys. Res., 95, 18,687–18,703, 1990.

Hartmann, D. L., L. A. Moy, and Q. Fu, Tropical convection and the energy balance at the top of the atmosphere, J. Clim., 14, 4495–4511, 2001.

Kiehl, J., On the observed near cancellation between longwave and shortwave cloud forcing in tropical regions, J. Clim., 7, 559–565, 1994.

Maloney, E. D., and A. H. Sobel, Surface fluxes and ocean coupling in the tropical intraseasonal oscillation, J. Clim., in press, 2004.

Neelin, J., Implications of convective quasi-equilibrium for the large-scale flow, in The physics and parameterization of moist atmospheric convection, 413–446, Kluwer Academic Publisher, 1997.

Neelin, J. D., and I. M. Held, Modeling tropical convergence based on the moist static energy budget, Mon. Weather Rev., 115, 3–12, 1987.

Neelin, J. D., and N. Zeng, A quasi-equilibrium tropical circulation model: Formulation, J. Atmos. Sci., 57, 1741–1766, 2000.

Peters, M. E. and C. S. Bretherton, An idealized Walker circulation coupled to an ocean mixed layer. J. Clim., in preparation, 2004.

Ramanathan, V., R. D. Cess, E. F. Harrison, P. Minnis, B. R. Barkstrom, E. Ahmad, and D. Hartmann, Cloud-radiative forcing and climate: results from the earth radiation budget experiment, Science, 243, 57–63, 1989.

Seager, R., S. E. Zebiak, and M. A. Cane, A model of the tropical Pacific sea–surface temperature climatology, J. Geophys. Res., 93, 1265–1280, 1988.

Seager, R., R. Murtugudde, A. Clement, and C. Herweijer, Why is there an evaporation minimum at the equator?, J. Clim., 16, 3793–3802, 2003.

Shaevitz, D. A., and A. H. Sobel, Implementing the weak temperature gradient approximation with full vertical structure, Mon. Weather Rev., 132, 662–669, 2004.

Sobel, A. H., On the coexistence of an evaporation minimum and precipitation maximum over the warm pool, J. Clim., 16, 1003–1009, 2003.

Sobel, A. H., and C. S. Bretherton, Modeling tropical precipitation in a single column, J. Clim., 13, 4378–4392, 2000.

Sobel, A., and H. Gildor, A simple time-dependent model of SST hot spots, J. Clim., 16, 3978–3992, 2003.

Sobel, A. H., J. Nilsson, and L. M. Polvani, The weak temperature gradient approximation and balanced tropical moisture waves, J. Atmos. Sci., 58, 3650–3665, 2001.

Su, H., and J. D. Neelin, Teleconnection mechanisms for tropical Pacific descent anomalies during El Nino, J. Atmos. Sci., 59, 2694–2712, 2002.

Su, H., J. D. Neelin, and J. E. Meyerson, Tropical tropospheric temperature and precipitation response to sea surface temperature forcing, this volume.

Waliser, D. E., Formation and limiting mechanisms for very high sea surface temperature: Linking the dynamics and the thermodynamics, J. Clim., 9, 161–188, 1996.

Watterson, I. G., The sensitivity of subannual and intraseasonal tropical variability to model ocean mixed layer depth, J. Geophys. Res., 107, 4020, doi:10.1029/2001JD000,671, 2002.

Yu, J.-Y., J. D. Neelin, and C. Chou, Estimating the gross moist stability of the tropical atmosphere, J. Atmos. Sci., 55, 1354–1372, 1998.

Zeng, N., J. D. Neelin, and C. Chou, A quasi-equilibrium tropical circulation model: Implementation and simulation, J. Atmos. Sci., 57, 1767–1796, 2000.

C. S. Bretherton, Department of Atmospheric Sciences, University of Washington, Seattle, Washington. (breth@atmos.washington.edu)

H. Gildor, Weizmann Institute of Science, Rehovot, Israel. (hezi.gildor@Weizmann.ac.il)

M. E. Peters, Department of Applied Mathematics, University of Washington, Seattle, Washington. (peters@atmos.washington.edu)

A. H. Sobel, Department of Applied Physics and Applied Mathematics and Department of Earth and Environmental Sciences, Columbia University, 500 W. 120th St., Rm. 217, New York, New York 10027. (ahs129@columbia.edu)